Human Orthopaedic Biomechanics

Fundamentals, Devices and Applications

Human Orthopaedic Biomechanics

Fundamentals, Devices and Applications

Edited by

Bernardo Innocenti

Department of Bio Electro and Mechanical Systems (BEAMS),
Universitè Libre de Bruxelles, Brussels, Belgium

Fabio Galbusera

IRCCS Istituto Ortopedico Galeazzi, Milan, Italy

ACADEMIC PRESS

An imprint of Elsevier

Academic Press is an imprint of Elsevier
125 London Wall, London EC2Y 5AS, United Kingdom
525 B Street, Suite 1650, San Diego, CA 92101, United States
50 Hampshire Street, 5th Floor, Cambridge, MA 02139, United States
The Boulevard, Langford Lane, Kidlington, Oxford OX5 1GB, United Kingdom

Notices

Knowledge and best practice in this field are constantly changing. As new research and experience broaden our
understanding, changes in research methods, professional practices, or medical treatment may become necessary.

Practitioners and researchers must always rely on their own experience and knowledge in evaluating and using any
information, methods, compounds, or experiments described herein. In using such information or methods they
should be mindful of their own safety and the safety of others, including parties for whom they have a professional
responsibility.

To the fullest extent of the law, neither the Publisher nor the authors, contributors, or editors, assume any liability
for any injury and/or damage to persons or property as a matter of products liability, negligence or otherwise, or
from any use or operation of any methods, products, instructions, or ideas contained in the material herein.

British Library Cataloguing-in-Publication Data
A catalogue record for this book is available from the British Library

Library of Congress Cataloging-in-Publication Data
A catalog record for this book is available from the Library of Congress

ISBN: 978-0-12-824481-4

For Information on all Academic Press publications
visit our website at https://www.elsevier.com/books-and-journals

Publisher: Mara Connor
Acquisitions Editor: Carrie Bolger
Editorial Project Manager: John Leonard
Production Project Manager: Anitha Sivaraj
Cover Designer: Miles Hitchen

Typeset by MPS Limited, Chennai, India

Contents

PART 2 HUMAN JOINTS BIOMECHANICS

PART 3 BIOMECHANICS AND DESIGN OF ORTHOPAEDIC DEVICES

CHAPTER 18 Biomaterials and biocompatibility 341

Ludovica Cacopardo

CHAPTER 19 Hip prosthesis: biomechanics and design 361

Edoardo Bori, Fabio Galbusera and Bernardo Innocenti

CHAPTER 20 Knee prosthesis: biomechanics and design 377

Bernardo Innocenti

CHAPTER 23 Devices for traumatology: biomechanics and design 459

Pankaj Pankaj

CHAPTER 24 Regeneration and repair of ligaments and tendons 485

Rocco Aicale, Nicola Maffulli and Francesco Oliva

CHAPTER 27 Computer-assisted orthopedic surgery **533**
Nicola Francesco Lopomo

PART 4 APPLICATIONS IN ORTHOPAEDIC BIOMECHANICS

CHAPTER 28 Experimental orthopedic biomechanics.............................. **557**
Luigi La Barbera, Tomaso Villa, Bernardo Innocenti and Fabio Galbusera

List of contributors

Rocco Aicale
Department of Trauma and Orthopaedic Surgery, Surgery and Dentistry, University of Salerno School of Medicine, Salerno, Italy; Department of Musculoskeletal Disorders, Faculty of Medicine and Surgery, University of Salerno, Baronissi, Salerno, Italy

Alessandra Aldieri
Department of Mechanical and Aerospace Engineering, Politecnico di Torino, Turin, Italy; Polito^{BIO}Med Lab, Politecnico di Torino, Turin, Italy

Michael Skipper Andersen
Department of Materials and Production, Aalborg University, Aalborg, Denmark

Federica Armaroli
Department of Bio Electro and Mechanical Systems (BEAMS), École Polytechnique de Bruxelles, Université Libre de Bruxelles, Brussels, Belgium

Alberto Audenino
Department of Mechanical and Aerospace Engineering, Politecnico di Torino, Turin, Italy; Polito^{BIO}Med Lab, Politecnico di Torino, Turin, Italy

Edoardo Bori
Department of Bio Electro and Mechanical Systems (BEAMS), École Polytechnique de Bruxelles, Université Libre de Bruxelles, Brussels, Belgium

Alexander Cleveland Breen
Centre for Biomechanics Research, AECC University College, Bournemouth, United Kingdom

Ludovica Cacopardo
Research Center 'E. Piaggio', University of Pisa, Pisa, Italy

Emmannuel J. Camus
IMPPACT - Lille Sud Clinic, Lesquin, France; Functional Anatomy Laboratory, Université Libre de Bruxelles, Brussels, Belgium; IMPPACT - Val de Sambre Clinic, Maubeuge, France; IMPPACT - Clinic of Cambresis, Cambrai, France

Vee San Cheong
Insigneo Institute for in Silico Medicine, University of Sheffield, Sheffield, United Kingdom; Department of Automatic Control and Systems Engineering, University of Sheffield, Sheffield, United Kingdom

Paolo Dalla Pria
Waldemar Link GmbH & Co. KG, Hamburg, Germany

Enrico Dall'Ara
Insigneo Institute for in Silico Medicine, University of Sheffield, Sheffield, United Kingdom; Department of Oncology and Metabolism, University of Sheffield, Sheffield, United Kingdom

Lutz Dürselen
Institute of Orthopaedic Research and Biomechanics, Ulm University, Ulm, Germany

Marwan El-Rich
Department of Mechanical Engineering, Khalifa University, Abu Dhabi, United Arab Emirates

Maren Freutel
Institute of Orthopaedic Research and Biomechanics, Ulm University, Ulm, Germany

Fabio Galbusera
IRCCS Istituto Ortopedico Galeazzi, Milan, Italy

Markus O. Heller
Department of Mechanical Engineering, University of Southampton, Southampton, United Kingdom

Yana Hoepfner
Institute of Orthopaedic Research and Biomechanics, Ulm University, Ulm, Germany

Bernardo Innocenti
Department of Bio Electro and Mechanical Systems (BEAMS), École Polytechnique de Bruxelles, Université Libre de Bruxelles, Brussels, Belgium

René Jonas
Institute of Orthopaedic Research and Biomechanics, Ulm University, Ulm, Germany

Luigi La Barbera
Laboratory of Biological Structure Mechanics, Department of Chemistry, Materials and Chemical Engineering "Giulio Natta", Politecnico di Milano, Milan, Italy

Luc Labey
Department of Mechanical Engineering, KU Leuven, Geel, Belgium

Nicola Francesco Lopomo
Department of Information Engineering, University of Brescia, Brescia, Italy

Nicola Maffulli
Department of Trauma and Orthopaedic Surgery, Surgery and Dentistry, University of Salerno School of Medicine, Salerno, Italy; Department of Musculoskeletal Disorders, Faculty of Medicine and Surgery, University of Salerno, Baronissi, Salerno, Italy; School of Pharmacy and Bioengineering, Keele University, Stoke-on-Trent, United Kingdom; Centre for Sports and Exercise Medicine, Queen Mary University of London, London, United Kingdom

Fabian Moungondo
Department of Orthopedics and Traumatology, Faculty of Medicine, Erasme University Hospital, Université Libre de Bruxelles, Brussels, Belgium

Cornelia Neidlinger-Wilke
Institute of Orthopaedic Research and Biomechanics, Ulm University, Ulm, Germany

Francesco Oliva
Department of Trauma and Orthopaedic Surgery, Surgery and Dentistry, University of Salerno School of Medicine, Salerno, Italy; Department of Musculoskeletal Disorders, Faculty of Medicine and Surgery, University of Salerno, Baronissi, Salerno, Italy

Luc Van Overstraeten
Department of Orthopedics and Traumatology, Faculty of Medicine, Erasme University Hospital, Université Libre de Bruxelles, Brussels, Belgium

Pankaj Pankaj
Computational Biomechanics, School of Engineering, Institute for Bioengineering, The University of Edinburgh, United Kingdom

Silvia Pianigiani
Adler Ortho, Milano, Italy

Giovanni Putame
Department of Mechanical and Aerospace Engineering, Politecnico di Torino, Turin, Italy; Polito[BIO]Med Lab, Politecnico di Torino, Turin, Italy

Dan Robbins
School of Allied Health, Anglia Ruskin University, Chelmsford, United Kingdom

Lennart Scheys
KU Leuven, Department of Development and Regeneration, Institute for Orthopaedic Research and Training (IORT), Leuven, Belgium

Benedikt Schlager
Institute of Orthopaedic Research and Biomechanics, Ulm University, Ulm, Germany

Andreas Martin Seitz
Institute of Orthopaedic Research and Biomechanics, Ulm University, Ulm, Germany

Victor Sholukha
Laboratory of Anatomy, Biomechanics and Organogenesis, Faculty of Medicine, Université Libre de Bruxelles, Brussels, Belgium; Department of Applied Mathematics, Peter the Great St. Petersburg Polytechnic University, St. Petersburg, Russia

Graciosa Quelhas Teixeira
Institute of Orthopaedic Research and Biomechanics, Ulm University, Ulm, Germany

Mara Terzini
Department of Mechanical and Aerospace Engineering, Politecnico di Torino, Turin, Italy; Polito[BIO]Med Lab, Politecnico di Torino, Turin, Italy

Serge Van Sint Jan
Laboratory of Anatomy, Biomechanics and Organogenesis, Faculty of Medicine, Université Libre de Bruxelles, Brussels, Belgium

Tomaso Villa
IRCCS Istituto Ortopedico Galeazzi, Milan, Italy; Laboratory of Biological Structure Mechanics, Department of Chemistry, Materials and Chemical Engineering "Giulio Natta", Politecnico di Milano, Milan, Italy

Daniela Warnecke
Institute of Orthopaedic Research and Biomechanics, Ulm University, Ulm, Germany

Hans-Joachim Wilke
Institute of Orthopaedic Research and Biomechanics, Ulm University, Ulm, Germany

Sara Zacchetti
IRCCS Orthopedic Institute Galeazzi, Milano, Italy

Preface

Orthopedic surgery is an important field for both scientific research and medical technology. Undeniably, the orthopedic industry constitutes one of the most relevant job opportunities for many biomedical engineers after their graduation. Several universities offer courses in orthopedic biomechanics, which cover the joints and anatomical districts that are the subject of orthopedic surgery, as well as the biomechanical aspects of the most common orthopedic implants, such as joint replacements, spinal implants, and devices used in traumatology. However, the study of the implants themselves must also include other important aspects such as mechanical design, biocompatibility, biomechanical function, safety, risk of failure, and related complications (as loosening, fracture, wear, etc.), which require a deep knowledge of mechanics as well as the anatomy, physiology, and pathology of the anatomical structures of interest.

For this reason, we have decided to write this book as a textbook for engineering students pursuing courses in orthopedic biomechanics. To this end, we intend to cover, with a high level of detail typical of undergraduate courses, all the important topics that are necessary to acquire in-depth knowledge in the field. Therefore the book is divided into four main parts:

The first part (Chapters 1−11) is devoted to the *Orthopedic Biomechanics Theory*, and it describes all theoretical aspects of orthopedic biomechanics, covering mechanics, stress analysis, constitutive laws for the various musculoskeletal tissues, and mechanobiology.

The second part (Chapters 12−17) is related to the *Human Joint Biomechanics*. In particular, it covers the biomechanics of the most important joints and anatomical structures of interest for orthopedic applications, such as hip, knee, ankle, spine, shoulder, elbow, and wrist.

The third part (Chapters 18−27) is devoted to the *Biomechanics and Design of Orthopedic Devices*. Following the same structure as the second part, it illustrates the main concepts related to the design of the main implants used in orthopedic and traumatology together with all aspects concerning biomaterials and biocompatibility, regulatory and quality, computer-assisted orthopedic surgery, and clinical trials.

The fourth and final part (Chapters 28−34) concludes the book by describing *Applications in Orthopedic Biomechanics*. The state-of-the-art methods used in orthopedic biomechanics and in orthopedic implant design are described in detail, covering in vitro experimental methods, in silico numerical modeling, and in vivo fluoroscopic and gait analysis.

As mentioned, the primary audience for this book is engineering students engaged in courses on orthopedic biomechanics. In addition, all bioengineers, researchers, and manufacturers of medical devices involved in orthopaedic biomechanics might be interested in using this book as a reference or handbook. Another potential audience is orthopedic surgeons involved in clinical research and are willing to enhance their knowledge on the biomechanical aspects.

Acknowledgments

Editing a book with such a large scope as *Human Orthopaedic Biomechanics: Fundamentals, Devices and Applications* has been an extremely challenging but also rewarding experience. None of this would have been possible without the valuable contribution of the Authors of the chapters. Forty-three scientists from several institutions around the world shared their expertise in their respective research fields: we are immensely grateful to all of them. We would also like to thank the publishing team at Elsevier, who supported us through numerous technical difficulties that we faced during the editing of the material.

We thank our colleagues at the IRCCS Istituto Ortopedico Galeazzi, Université Libre de Bruxelles, and Ulm University, who supported us throughout this challenging endeavor, always available and happy to help.

We would also like to thank all our PhDs, assistants, and students for their questions, comments, and, especially, for their interest in biomechanics: honestly, this book is especially for you!

Finally, a special thanks to all those who are part of our daily journey: our families, parents, kids, and partners; in particular, a strong mention must go to Serena, together with Morgana and Edoardo, and Marta for always being the persons to turn to in this 2-year journey.

Fabio Galbusera[1] and Bernardo Innocenti[2]
[1]Laboratory of Biological Structures Mechanics,
IRCCS Istituto Ortopedico Galeazzi, Milan, Italy
[2]Department of Bio Electro and Mechanical Systems (BEAMS),
Universitè Libre de Bruxelles, Brussels, Belgium

ORTHOPAEDIC BIOMECHANICS THEORY

INTRODUCTION: FROM MECHANICS TO BIOMECHANICS

Fabio Galbusera[1] and Bernardo Innocenti[2]

[1]*IRCCS Istituto Ortopedico Galeazzi, Milan, Italy* [2]*Department of Bio Electro and Mechanical Systems (BEAMS), École Polytechnique de Bruxelles, Université Libre de Bruxelles, Brussels, Belgium*

It is impossible to talk and discuss about biomechanics without talking about mechanics.

Mechanics is a branch of physics that is concerned with the description of motion and deformation of the bodies and how forces induce motions and deformation. Historically, mechanics was born based on observation and experience, and the first mechanical objects were born based on practice. Among all the physical sciences, mechanics is the oldest, dating back to the time of the ancient Greeks. The first writings in which mechanical problems were dealt with were made by Aristotle (Fig. 1.1), which defined the first principle of statics and dynamics, that there can be no movement without a motive force, which remained in effect until the times of Galileo. Subsequently, the credit for the first theoretical research work on the barycenter, theory of the equilibrium of the lever, and the law of hydrostatics goes to Archimedes.

Applied mechanics (also called engineering mechanics) is the science of applying the principles of mechanics. In other words, applied mechanics links the theory to the industrial application. In particular, the concepts of kinematics are used in the field of functional design, which is the study of the shape to be given to different structures of a mechanical system to allow a certain trajectory and a certain level of performance. The notions of statics make it possible to determine, from the different resistant forces and moments, the equilibrium motor forces and moments (generally unknown), that is, no forces or moments of inertia arise in the system. The kinetostatics analysis enables, gaining the knowledge of the velocity of one or more members of the structure, to graphically tracing the unknown forces that produce the motion. Lastly, the dynamics allows us to study the forces acting on a mechanical system and ultimately develop a system of equations of motion that permits us to obtain the motion of the system (inverse problem of dynamics), or from gaining the knowledge of the acceleration of the system to know the forces acting on it (direct problem of dynamics), thus providing analytical methods (coupled with graphical methods), to determinate the moments of inertia and the main solutions for their total or partial balance. Leonardo da Vinci (Fig. 1.2), for his machines, and Galileo and Newton for defining the equations of motion are regarded as the fathers of applied mechanics. On the basis of Newtonian principles, any mechanical problem can be mathematically calculated, but its actual resolution requires the use of appropriate mathematical tools: the concepts and methods of infinitesimal analysis by Newton and their application by G.W. Leibniz; mechanics then ends up from experimental to become rational.

Human Orthopaedic Biomechanics. DOI: https://doi.org/10.1016/B978-0-12-824481-4.00031-7

FIGURE 1.1

Aristotle (384−322 BCE).

Biomechanics (from Ancient Greek: βίος "life" and μηχανική "mechanics") is the application of mechanical principles to living organisms, such as humans, animals, and also plants. Dealing with human motion and performance, it is now widely recognized that biomechanics plays an important role in this context; biomechanics has a very long history as Aristotle (Fig. 1.1) in his book "*De Motu Animalium*" (On the Movement of Animals) defined animals' bodies as mechanical systems and described the actions of the muscles and subjected them to geometric analysis for the first time. Hippocrates (often referred to as the "Father of Medicine") used the force of gravity to relieve the pressure on the intervertebral discs in patients and to reduce the onset and effects of back pain. To this end, he used a sort of ladder to which the patient was tied. He also designed a bed for vertebral traction, which he called "*Scamnum*."

Leonardo da Vinci is regarded as the first bioengineer, and the father of this field; his Vitruvian Man drawing (Fig. 1.3) is regarded as an icon or a symbolic representation of biomechanics. He was able to dissect human corps at the Hospital of Santa Maria Nuova in Florence and reported the results of these studies in his anatomical drawings. He was able to study the mechanical functions of the skeleton and the muscular forces that are applied to it. Leonardo's dissections and

FIGURE 1.2

Leonardo da Vinci (1452−1519 CE).

documentation of muscles, nerves, and vessels helped describe the physiology and mechanics of movement. Galileo in "*De Animalium Motibus*" studied the biomechanics of jumping in humans, the analysis of the steps of horses and insects, and the study of human flotation. He also studied the behavior of biomaterials such as bone. After Galileo, Alfonso Borelli published (posthumously) the book "*De Motu Animalium*," describing different complex human activities such as walking, running, and analyzing statically the forces in the joints and in the muscle during such activities, providing the basics of the musculoskeletal modeling.

Eadweard Muybridge (Fig. 1.4) in 1878 first performed dynamic analysis using a cinematographic technique. He used the technique of chronophotography to study the movement of animals and people. Christian Wilhelm Braune and Otto Fisher in 1889 conducted research involving the position of the center of gravity in the human body and its various segments. By first determining the planes of the "gravitational centers" of the longitudinal, sagittal, and frontal axes of a frozen human cadaver in a given position, and then dissecting the cadaver with a saw, they were able to establish the center of gravity of the body and its component parts. Moreover, they performed

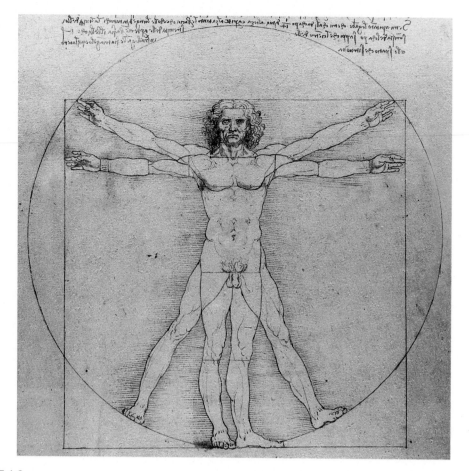

FIGURE 1.3

Vitruvian Man drawing (1485 CE), the symbol of biomechanics.

anatomical studies of the human gait, published in the book "*Der Gang des Menschen*" (The Walk of Men, 1895). Braune's study of the biomechanics of gait covered free walking and walking with a load and the underlying methodology paved the way for modern gait-analysis.

Since then, the field of biomechanics has blossomed, and, today, several subfields are categorized as orthopedic biomechanics, mainly focused on studying human joint performance and relative structures, cardiovascular biomechanics, the relationship between the mechanics of the cardiovascular system and biological function under healthy and diseased conditions, forensic biomechanics related to the study, and the effect of injury and accident and ergonomics, focused on the design of products and systems to allow a proper interaction among humans and other elements of the environment.

FIGURE 1.4

Eadweard Muybridge (1830–1904 CE).

Recently, in the field of orthopedic biomechanics, more sophisticated equipment and analyses have become available to perform advanced experiments enabling a better understanding of joint kinematics and tissue function during walking, running, and other daily activities. Thanks to the fruitful interaction with the computer scientists, advanced mathematical modeling and improved engineering design of orthopedic implants have also taken great strides. Additionally, biomechanical engineers in collaboration with orthopedic surgeons have applied biomechanical principles to study clinically relevant problems, improving patient treatments and outcomes.

However, despite all advances made in this field, there are still many questions that remain to be answered and a great deal of knowledge yet to be gained. Looking into the future, it is imperative to move forward in research and education through interaction with other players of this multidisciplinary field to make our contributions available for the future generations.

MECHANICAL PROPERTIES OF BIOLOGICAL TISSUES

2

Bernardo Innocenti

Department of Bio Electro and Mechanical Systems (BEAMS), École Polytechnique de Bruxelles, Université Libre de Bruxelles, Brussels, Belgium

INTRODUCTION: MATERIAL PROPERTIES AND STRUCTURAL PROPERTIES

Each object shows a specific mechanical response to a certain applied load. The behavior and the magnitude of such response are quantified by its mechanical properties, which are related to the material and to the shape of the object itself; indeed, a tube made of rubber has a different mechanical response (e.g., elasticity or deformation) compared with an identical tube made of steel, and, similarly, there are differences in the mechanical behavior between a hollow and a solid tube made of the same material.

In general, talking about the mechanical properties of a structure, it is possible to identify the following two main subdivisions:

1. *Material properties,* that are independent of the shape of the object, i.e. they do not depend on the amount of material; therefore, they are specific to the material of the structure. Such properties are usually expressed in terms of the stress−strain relationship of the material and related parameters. Examples of material properties include elastic modulus, yield point, ultimate strength, and so on.
2. *Structural properties,* that depend both on the shape and on the material of the structure. Examples of structural properties include bending stiffness, torsional stiffness, axial stiffness, and so on. Such properties, which represent the relation between a certain force and the relative deformation, are usually structure/object-specific, as different hip prosthesis designs, even if made of the same material, could have different torsional stiffness and are also anatomically different and patient-specific, for example, the tibia of a person has a different bending stiffness compared than the scapula, or to the tibia of another person.

In the case of tissues, time-dependent parameters (e.g., viscoelasticity) and parameters such as aging, pathology, gender, and lifestyle could influence their material and structural properties.

Human Orthopaedic Biomechanics. DOI: https://doi.org/10.1016/B978-0-12-824481-4.00034-2

MATERIAL PROPERTIES: GENERAL CONCEPT
FORCE−DISPLACEMENT CURVE AND STIFFNESS OF A MATERIAL

Given a certain material, it is always possible to characterize the material's properties by performing some experimental tests. The mechanical behavior of a material under a tensile force is determined following a mono-axial tension test consisting of fixing a specimen of a certain material to a specific machine and then applying a certain force and registering the specimen displacement (under a force-controlled test) or by applying a certain displacement and recording the force required to guarantee such displacement (under a displacement-controlled test). At the end of the test, we are able to plot a force−displacement curve as shown in Fig. 2.1 (limited to the elastic region). Such a graph is, at first approximation, linear and the slope of such a curve is able to quantify the stiffness of the material as:

$$\text{Stiffness} = \frac{\text{Force}}{\text{Displacement}} \tag{2.1}$$

As clearly illustrated in Fig. 2.1, depending on the slope, two materials can be compared to determine which is more rigid and which is more flexible.

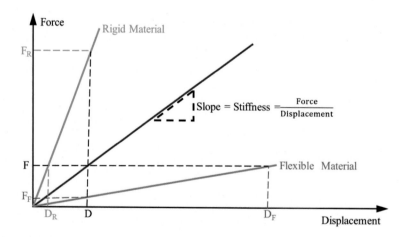

FIGURE 2.1

Force−displacement curve for a certain material (*red line*) limited to the elastic region. The force−displacement curves for a more rigid and a more flexible material are also reported, respectively, as green and orange curves. The same force F will induce a small displacement (D_R) in the rigid material and a big displacement (D_F) in the flexible material; however, the same displacement D will require a higher force in the rigid material (F_R) and a lower force in the flexible material (F_F).

STRESS–STRAIN CURVE AND ELASTIC MODULUS

Practically, it is more efficient, instead of the force-displacement curve, to plot the stress-strain curve. Using the stress (σ), which is defined as the force applied on a certain area of a material, and the strain (ε), which is defined as the ratio between the change in length (ΔL) of the specimen and its initial length (L_0), we can convert the force–displacement curve into a stress–strain curve (Fig. 2.2):

$$\sigma = \frac{\text{Force}}{\text{Area}}; \varepsilon = \frac{\Delta L}{L_0} \tag{2.2}$$

In this graph, the slope of the initial linear tract (elastic region) is called Young's modulus, which depends only on the material and not on the shape of the specimen used for the test.

Table 2.1 reports some values of the elastic modulus (in GPa) for some common materials used in orthopedic biomechanics. In the biological tissues, these values could change according to the anatomical structure, age, and pathology of the specimen/donor.

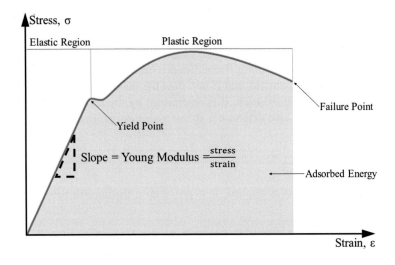

FIGURE 2.2

Stress–strain curve for a generic material.

Table 2.1 Elastic modulus (in GPa) for some common materials used in orthopedic biomechanics.	
Material	**Elastic modulus (GPa)**
Stainless steel	200
Titanium	100
Cortical bone	7–21
Cancellous bone	0.7–4.9
Bone cement (PMMA)	2.5–3.5
UHMWPE	1.4–4.2

In a mono-axial test (both in tension and in compression), the linear trend between stress and strain is usually limited to small deformations; increasing the force applied, we obtain a curve that is not linear, but it shows a certain plateau as illustrated in Fig. 2.2. In detail, we can identify two regions:

- *Elastic region (linear tract)*. It is characterized by a fully reversible state, in which if we unload the material, the deformation goes back to its initial unloaded value. In other words, the material has no memory; it remains as it was before the application of the force.
- *Plastic region*. In this region, the material starts to deform permanently, meaning that when it is unloaded, it does not go back to the initial state, but we have a residual deformation, which means the material has a memory of the force applied.

In comparison with the force–displacement curve, the benefit of the stress–strain curve lies in the fact that it provides more information. Together with the elastic modulus of the material, it provides information on the following parameters:

- *Yield point*: It is the point when the material goes from elastic to plastic.
- *Failure/fracture point*: It is the point of fracture of the specimen.
- *Brittle/ductile behavior of the material*: It is the distance between the yield point and the failure point (span of the plastic region). If the distance between these two points is large, then the material is considered ductile, and if not, then the material is brittle (e.g., ceramic).
- *Energy absorbed during the deformation*: It is determined by the area under the curve until failure. The amount of energy also determines the tough/weak nature of the material.

NORMAL AND SHEAR STRESS

Based on the directions of the forces applied, it is possible to distinguish between different stress conditions (Fig. 2.3):

1. *Tensile stress*: If we pull at the ends of a specimen, we exert a tensile stress. When a tensile force is applied, the specimen length is increased, while the section is reduced.

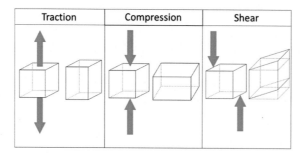

FIGURE 2.3

Example of tensile, compressive, and shear stress and relative specimen deformation.

2. *Compressive stress*: If we push at the ends of a specimen, we exert a compressive stress. In this case, the specimen length is reduced, while the section is increased.
3. *Shear stress*: If we push/pull on opposite sides of an object, we exert a shear stress. Because the forces are no more aligned, the deformation of the specimen is completely different from that under compressive or tensile stress.

Tensile and compressive stress (and strains) are also called "normal" as the resulting stress (strain) is perpendicular to the face of an element, while a shear stress (strain) is parallel to it and hence called tangential stress.

MATERIAL ISOTROPY AND ANISOTROPY

Apart from the normal and shear stress, the mechanical response of a material could also change based on the direction of application of the force. It is therefore possible to identify different behaviors as:

- *Isotropic*: The mechanical responses are not dependent on the direction of loading; therefore, the material properties remain constant in all orientations.
- *Anisotropic*: The mechanical responses (and therefore the mechanical properties) of the material change according to the direction of application of the force.

Metals and polymers are examples of isotropic materials. Biological materials, such as bone and soft tissues, are usually anisotropic.

STRESS TENSOR AND HOOKE'S LAW

In general, a certain load condition induces a stress and a strain distribution in each point of the specimen. However, a general stress state of a point inside a solid needs nine components to be completely specified (see Fig. 2.4) since each of the three components of the stress must be defined not only along the three normal directions x_1, x_2, and x_3, described, respectively, by the stress components σ_{11}, σ_{22}, and σ_{33}, but also by the two tangential components in the planes orthogonal to them. (For instance, for the plane orthogonal to x_1, the two tangential stresses are described by σ_{12} and σ_{13}.) Therefore, we could represent the different components of the stress using the so-called stress tensor (or Cauchy stress tensor), a 3×3 matrix representing the normal stress along the diagonal and the tangential stress (Fig. 2.4).

Even if the stress tensor contains nine values, due to its symmetry, it is possible to express the stress as a six-dimensional vector using the following double-index or single-index (Voigt) notations:

$$[\sigma] = [\sigma_{11}\sigma_{22}\sigma_{33}\sigma_{23}\sigma_{13}\sigma_{12}]^T = [\sigma_1\sigma_2\sigma_3\sigma_4\sigma_5\sigma_6]^T \qquad (2.3)$$

In a generic anisotropic material, each single stress component σ_i (with $i = 1.6$) can cause normal and shear strains, and, therefore, under the hypothesis of elasticity of the material, the relation

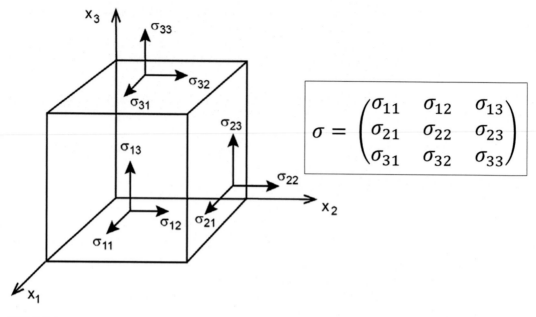

FIGURE 2.4

Stress components in three dimensions and relative stress tensor. The first index of the stress component specifies the direction in which the stress component acts, while the second index identifies the orientation of the surface on which it is acting. Stresses are positive if they act on positive planes in the positive direction or on negative planes in the negative direction; else, they are negative.

between stress and strain could be represented by the anisotropic form of Hooke's law in matrix notation as follows:

$$
\begin{bmatrix} \sigma_{11} \\ \sigma_{22} \\ \sigma_{33} \\ \sigma_{23} \\ \sigma_{31} \\ \sigma_{12} \end{bmatrix} = \begin{bmatrix} C_{1111} & C_{1122} & C_{1133} & C_{1123} & C_{1131} & C_{1112} \\ C_{2211} & C_{2222} & C_{2233} & C_{2223} & C_{2231} & C_{2212} \\ C_{3311} & C_{3322} & C_{3333} & C_{3323} & C_{3331} & C_{3312} \\ C_{2311} & C_{2322} & C_{2333} & C_{2323} & C_{2331} & C_{2312} \\ C_{3111} & C_{3122} & C_{3133} & C_{3123} & C_{3131} & C_{3112} \\ C_{1211} & C_{1222} & C_{1233} & C_{1223} & C_{1231} & C_{1212} \end{bmatrix} \begin{bmatrix} \varepsilon_{11} \\ \varepsilon_{22} \\ \varepsilon_{33} \\ \varepsilon_{23} \\ \varepsilon_{31} \\ \varepsilon_{12} \end{bmatrix}
$$ (2.4)

or in a compact form as $[\sigma] = [C][\varepsilon]$ or using the Voight notation as

$$
\begin{bmatrix} \sigma_1 \\ \sigma_2 \\ \sigma_3 \\ \sigma_4 \\ \sigma_5 \\ \sigma_6 \end{bmatrix} = \begin{bmatrix} C_{11} & C_{12} & C_{13} & C_{14} & C_{15} & C_{16} \\ C_{21} & C_{22} & C_{23} & C_{24} & C_{25} & C_{26} \\ C_{31} & C_{32} & C_{33} & C_{34} & C_{35} & C_{36} \\ C_{41} & C_{42} & C_{43} & C_{44} & C_{45} & C_{46} \\ C_{51} & C_{52} & C_{53} & C_{54} & C_{55} & C_{56} \\ C_{61} & C_{62} & C_{63} & C_{64} & C_{65} & C_{66} \end{bmatrix} \begin{bmatrix} \varepsilon_1 \\ \varepsilon_2 \\ \varepsilon_3 \\ \varepsilon_4 \\ \varepsilon_5 \\ \varepsilon_6 \end{bmatrix}
$$ (2.5)

where $[\varepsilon] = [\epsilon_1 \quad \epsilon_2 \quad \epsilon_3 \quad \epsilon_4 \quad \epsilon_5 \quad \epsilon_6]^T = [\varepsilon_{11} \quad \varepsilon_{22} \quad \varepsilon_{33} \quad \varepsilon_{23} \quad \varepsilon_{13} \quad \varepsilon_{12}]^T$.

The above equations can be inverted so that the strains are explicitly expressed in terms of the stresses:

$$[\varepsilon] = [S][\sigma],$$

where C and S are, respectively, called elastic stiffness tensor and elastic compliance tensor.

Even if the tensors consist of 36 components, due to their symmetry, only 21 components in each tensor are independent. Looking at the structure of the stiffness tensor, it is important to highlight how it could be considered subdivided into four submatrixes:

$$[C] = \begin{bmatrix} [N] & [A] \\ [B] & [S] \end{bmatrix},$$

where:

- $[N]$ is a 3×3 square matrix containing the elements C_{ij} (for $i, j = 1, 2, 3$), and it links the normal strains to the normal stresses;
- $[S]$ is a 3×3 square matrix containing the elements C_{ij} (for $i, j = 4, 5, 6$), expressing the relation between the shear strains and shear stresses;
- $[A]$ and $[B]$ are two 3×3 square matrixes containing the elements C_{ij} (where $i = 1, 2, 3; j = 4, 5, 6$ for $[A]$ and $i = 4, 5, 6; j = 1, 2, 3$ for $[B]$), expressing, respectively, the relations between the shear strains and the normal stress $[A]$ and vice versa $[B]$.

ORTHOTROPIC, TRANSVERSALLY ISOTROPIC, AND ISOTROPIC MATERIAL MODELS

As in a generic anisotropic elastic material the stiffness and the compliance tensors are associated with 21 linearly independent coefficients; sometimes it becomes necessary to simplify the material models to be able to express each tensor with a lower number of constants. In biomechanics, especially for commonly used orthopedic devices for bones, the following material models are frequently used:

- orthotropic material model;
- transverse isotropic material model;
- isotropic material model.

The selection of one of these material models (whichever is applicable) will reduce the number of coefficients to, respectively, nine, five, and two independent constants.

ORTHOTROPIC MATERIAL

Different materials present orthotropic material behaviour, meaning that they have material properties at a particular point, which differs along three mutually orthogonal axes. Therefore, their properties depend on the direction in which they are measured; moreover, orthotropic materials have three planes/axes of symmetry. As the common coordinate system used with these materials is the cylindrical-polar, this type of material model is also called polar orthotropy.

Using the approximation for the material models that there is no interaction between the normal stress and the shear strain, we can simplify the generic stiffness tensor as the submatrixes [A] and [B] become null and the matrix S becomes a diagonal matrix. Therefore, the stiffness matrix finally consists of 12 nonzero components and nine independent components as reported below:

$$
\begin{bmatrix} \sigma_1 \\ \sigma_2 \\ \sigma_3 \\ \sigma_4 \\ \sigma_5 \\ \sigma_6 \end{bmatrix} = \begin{bmatrix} \dfrac{1 - v_{23}v_{32}}{E_2 E_3 \Delta} & \dfrac{v_{21} + v_{31}v_{23}}{E_2 E_3 \Delta} & \dfrac{v_{31} + v_{21}v_{32}}{E_2 E_3 \Delta} & 0 & 0 & 0 \\[2mm] \dfrac{v_{12} + v_{13}v_{32}}{E_3 E_1 \Delta} & \dfrac{1 - v_{31}v_{13}}{E_3 E_1 \Delta} & \dfrac{v_{32} + v_{31}v_{12}}{E_3 E_1 \Delta} & 0 & 0 & 0 \\[2mm] \dfrac{v_{13} + v_{12}v_{23}}{E_1 E_2 \Delta} & \dfrac{v_{23} + v_{13}v_{21}}{E_1 E_2 \Delta} & \dfrac{1 - v_{12}v_{21}}{E_1 E_2 \Delta} & 0 & 0 & 0 \\[2mm] 0 & 0 & 0 & 2G_{23} & 0 & 0 \\ 0 & 0 & 0 & 0 & 2G_{31} & 0 \\ 0 & 0 & 0 & 0 & 0 & 2G_{12} \end{bmatrix} \begin{bmatrix} \varepsilon_1 \\ \varepsilon_2 \\ \varepsilon_3 \\ \varepsilon_4 \\ \varepsilon_5 \\ \varepsilon_6 \end{bmatrix} \qquad (2.6)
$$

where $\Delta = \dfrac{1 - v_{12}v_{21} - v_{23}v_{32} - v_{31}v_{13} - 2v_{12}v_{23}v_{31}}{E_1 E_2 E_{31}}$

or

$$
\begin{bmatrix} \varepsilon_1 \\ \varepsilon_2 \\ \varepsilon_3 \\ \varepsilon_4 \\ \varepsilon_5 \\ \varepsilon_6 \end{bmatrix} = \begin{bmatrix} \dfrac{1}{E_1} & -\dfrac{v_{21}}{E_2} & -\dfrac{v_{31}}{E_3} & 0 & 0 & 0 \\[2mm] -\dfrac{v_{12}}{E_1} & \dfrac{1}{E_2} & -\dfrac{v_{32}}{E_3} & 0 & 0 & 0 \\[2mm] -\dfrac{v_{13}}{E_1} & -\dfrac{v_{23}}{E_2} & \dfrac{1}{E_3} & 0 & 0 & 0 \\[2mm] 0 & 0 & 0 & \dfrac{1}{2G_{23}} & 0 & 0 \\[2mm] 0 & 0 & 0 & 0 & \dfrac{1}{2G_{31}} & 0 \\[2mm] 0 & 0 & 0 & 0 & 0 & \dfrac{1}{2G_{12}} \end{bmatrix} \begin{bmatrix} \sigma_1 \\ \sigma_2 \\ \sigma_3 \\ \sigma_4 \\ \sigma_5 \\ \sigma_6 \end{bmatrix} \qquad (2.7)
$$

Therefore, the stiffness tensor could be fully determined as a function of three Young's moduli (E_i) obtained by following a uniaxial strain along each axis, three shear moduli ($G_{ij} = G_{ji}$) obtained by three planar shear tests, and six Poisson's ratios, v_{ij}, obtained by measuring the strain in the j-direction when loaded in the i-direction. It is important to note that the six Poisson's ratios are not linearly independent as the following equations are valid due to the symmetry of the tensor:

$$
\frac{v_{23}}{E_2} = \frac{v_{32}}{E_3}; \frac{v_{31}}{E_3} = \frac{v_{13}}{E_1}; \frac{v_{12}}{E_1} = \frac{v_{21}}{E_2} \qquad (2.8)
$$

TRANSVERSALLY ISOTROPIC MATERIAL

A special case of orthotropic materials is when the material has the same properties in one plane (e.g., the x_1–x_2 plane, named "transverse plane" or "plane of isotropy") and different properties in the direction normal to this plane (e.g., the x_3-axis). Such materials are called transversally isotropic, and they are described by five independent elastic constants, instead of nine for a fully orthotropic. For example, human long bones (e.g., femur and tibia) have material properties that could be approximated with a transversally isotropic model using the anatomical axis as the normal main direction (with greater differences between the axial and transverse directions than between radial and circumferential). With the approximation of being transversally isotropic and with the x_3-axis as the normal direction, we have ("a" = axial and "t" = transverse)

- $E_1 = E_2 = E_t$;
- $E_3 = E_a$;
- $v_{12} = v_{21} = v_t$;
- $G_{12} = G_t$;
- $G_{23} = G_{31} = G_a$.

Therefore, the stiffness and the tensor tensors could be rewritten as a function of the five elastic constants as follows:

$$
\begin{bmatrix} \sigma_1 \\ \sigma_2 \\ \sigma_3 \\ \sigma_4 \\ \sigma_5 \\ \sigma_6 \end{bmatrix} =
\begin{bmatrix}
\dfrac{1 - v_{ta}v_{at}}{E_t E_a \Delta} & \dfrac{v_t + v_{at}v_{ta}}{E_t E_a \Delta} & \dfrac{v_{at} + v_t v_{at}}{E_t E_a \Delta} & 0 & 0 & 0 \\
\dfrac{v_t + v_{ta}v_{at}}{E_a E_t \Delta} & \dfrac{1 - v_{at}v_{ta}}{E_a E_t \Delta} & \dfrac{v_{at} + v_{at}v_t}{E_a E_t \Delta} & 0 & 0 & 0 \\
\dfrac{v_{ta} + v_t v_{ta}}{E_t^2 \Delta} & \dfrac{v_{ta} + v_{ta}v_t}{E_t^2 \Delta} & \dfrac{1 - v_t^2}{E_t^2 \Delta} & 0 & 0 & 0 \\
0 & 0 & 0 & 2G_a & 0 & 0 \\
0 & 0 & 0 & 0 & 2G_a & 0 \\
0 & 0 & 0 & 0 & 0 & 2G_t
\end{bmatrix}
\begin{bmatrix} \varepsilon_1 \\ \varepsilon_2 \\ \varepsilon_3 \\ \varepsilon_4 \\ \varepsilon_5 \\ \varepsilon_6 \end{bmatrix}
\tag{2.9}
$$

where $\Delta = \dfrac{(1 + v_t)(1 - v_t - 2v_{at}v_{ta})}{E_t^2 E_a}$
   or

$$
\begin{bmatrix} \varepsilon_1 \\ \varepsilon_2 \\ \varepsilon_3 \\ \varepsilon_4 \\ \varepsilon_5 \\ \varepsilon_6 \end{bmatrix} =
\begin{bmatrix}
\dfrac{1}{E_t} & -\dfrac{v_t}{E_t} & -\dfrac{v_{at}}{E_a} & 0 & 0 & 0 \\
-\dfrac{v_t}{E_t} & \dfrac{1}{E_t} & -\dfrac{v_{at}}{E_a} & 0 & 0 & 0 \\
-\dfrac{v_{ta}}{E_t} & -\dfrac{v_{ta}}{E_t} & \dfrac{1}{E_a} & 0 & 0 & 0 \\
0 & 0 & 0 & \dfrac{1}{2G_a} & 0 & 0 \\
0 & 0 & 0 & 0 & \dfrac{1}{2G_a} & 0 \\
0 & 0 & 0 & 0 & 0 & \dfrac{1}{2G_t}
\end{bmatrix}
\begin{bmatrix} \sigma_1 \\ \sigma_2 \\ \sigma_3 \\ \sigma_4 \\ \sigma_5 \\ \sigma_6 \end{bmatrix}
\tag{2.10}
$$

It is important to note that the three Poisson's ratios are not linearly independent as the following equations are valid due to the symmetry of the tensor: $v_{at}/E_a = v_{ta}/E_t$, and the tangential shear modulus could be estimated as $G_t = E_t/2(1 + v_t)$.

ISOTROPIC MATERIAL

Most metallic alloys could be considered isotropic; their response to a certain load is independent of the direction of the force, meaning that it has identical material properties in all directions at every given point. Such materials have only two independent variables (i.e., elastic constants) in their stiffness and compliance matrices, as opposed to the 21 elastic constants in the general anisotropic case. With the approximation of being isotropic material, we have the following:

- E_1, E_2, and E_3 are equal and are usually called elastic modulus or Young's modulus E of the material.
- ν_{12}, ν_{23}, and ν_{31} are equal and are usually called Poisson's ratio of the material ν.

With these hypotheses, the stiffness and the compliance tensors could be rewritten as follows:

$$\begin{bmatrix} \sigma_1 \\ \sigma_2 \\ \sigma_3 \\ \sigma_4 \\ \sigma_5 \\ \sigma_6 \end{bmatrix} = \frac{E}{(1+\nu)(1-2\nu)} \begin{bmatrix} 1-\nu & \nu & \nu & 0 & 0 & 0 \\ \nu & 1-\nu & \nu & 0 & 0 & 0 \\ \nu & \nu & 1-\nu & 0 & 0 & 0 \\ 0 & 0 & 0 & 1-2\nu & 0 & 0 \\ 0 & 0 & 0 & 0 & 1-2\nu & 0 \\ 0 & 0 & 0 & 0 & 0 & 1-2\nu \end{bmatrix} \begin{bmatrix} \varepsilon_1 \\ \varepsilon_2 \\ \varepsilon_3 \\ \varepsilon_4 \\ \varepsilon_5 \\ \varepsilon_6 \end{bmatrix} \tag{2.11}$$

and

$$\begin{bmatrix} \varepsilon_1 \\ \varepsilon_2 \\ \varepsilon_3 \\ \varepsilon_4 \\ \varepsilon_5 \\ \varepsilon_6 \end{bmatrix} = \frac{1}{E} \begin{bmatrix} 1 & -\nu & -\nu & 0 & 0 & 0 \\ -\nu & 1 & -\nu & 0 & 0 & 0 \\ -\nu & -\nu & 1 & 0 & 0 & 0 \\ 0 & 0 & 0 & 1+\nu & 0 & 0 \\ 0 & 0 & 0 & 0 & 1+\nu & 0 \\ 0 & 0 & 0 & 0 & 0 & 1+\nu \end{bmatrix} \begin{bmatrix} \sigma_1 \\ \sigma_2 \\ \sigma_3 \\ \sigma_4 \\ \sigma_5 \\ \sigma_6 \end{bmatrix} \tag{2.12}$$

The two elastic constants are usually named Young's modulus (E) and Poisson's ratio (ν) and they are defined as:

$$E = \frac{\sigma_{11}}{\varepsilon_{11}}; \nu = -\frac{\sigma_{22}}{\varepsilon_{11}} = -\frac{\sigma_{33}}{\varepsilon_{11}} \tag{2.13}$$

Then, defining the shear modulus, G, for an isotropic material by:

$$G = \frac{\sigma_{12}}{2\varepsilon_{12}} \tag{2.14}$$

we get:

$$G = \frac{E}{2(1+\nu)} \tag{2.15}$$

For an isotropic material, the bulk modulus 'K', which is defined as the ratio between the mean normal pressure and the volume change, could be easily determined as a function of E and ν as follows:

$$K = \frac{\frac{1}{3}\sum_k \sigma_{kk}}{\sum_k \varepsilon_{kk}} = \frac{E}{3(1-2\nu)} \tag{2.16}$$

In general, for an isotropic material, it is always possible to express two of the four material constants (E, ν, G, and K) as a function of the other two. Fig. 2.5 reports such relations.

	G	K	E	v
G, E	---	$\dfrac{GE}{3(3G-E)}$	---	$\dfrac{E-2G}{2G}$
G, v	---	$\dfrac{2G(1+v)}{3(1-2v)}$	$2G(1+v)$	---
G, K	---	---	$\dfrac{9KG}{3K+G}$	$\dfrac{1}{2}\left[\dfrac{3K-2G}{3K+G}\right]$
E, v	$\dfrac{E}{2(1+v)}$	$\dfrac{E}{3(1-2v)}$	---	---
E, K	$\dfrac{3EK}{9K-E}$	---	---	$\dfrac{1}{2}\left[\dfrac{3K-E}{3K}\right]$
v, K	$\dfrac{3K(1-2v)}{2(1+v)}$	---	$3K(1-2v)$	---

FIGURE 2.5

Relation between material constants under the hypothesis of linear elastic isotropy.

It is possible to demonstrate that for a linear elastic isotropic material, the Poisson's ratio always lies between 0 and 0.5 ($0 < v < 0.5$); however, this is not true for anisotropic or orthotropic material, where the value can be higher than 0.5; moreover, some materials known as "auxetic materials" present an even negative Poisson's ratio. Such materials when subjected to a positive strain in a longitudinal axis induce a positive transverse strain in the material (i.e., it would increase the cross-sectional area) and vice versa.

HYPERELASTIC MATERIAL

An elastic material is a linear material, which means that the stress varies linearly with strain. This material is very common and is frequently used for different material models, such as metal, ceramic, bone, and some plastic material, as long as the deformation is very small.

Another class of materials are the hyperelastic materials, which are normally used for rubber-like material models in which the elastic deformation can be extremely large (Fig. 2.6).

To be able to correlate stress and strain in an hyperelastic material, it is common to introduce the strain energy density function. Using such an approach, it is possible to model the relationship even when the strain is between 100% and 700%, depending on the exact hyperelastic model that is used.

Commonly used hyperelastic material models are as follows:

- Mooney—Rivlin (phenomenological model);
- Ogden (phenomenological model);
- Neo—Hooke (mechanistic model);
- Arruda—Boyce (mechanistic model);
- Gent (hybrid model).

Fig. 2.7 reports several stress—strain curves for various hyperelastic material models. To build a complete hyperelastic material model, it is necessary to first select a constitutive model, and then

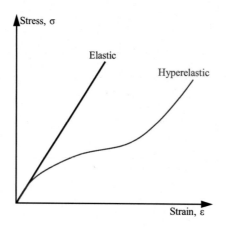

FIGURE 2.6

Stress−strain general curve for an elastic (*blue curve*) and an hyperelastic material (*red curve*).

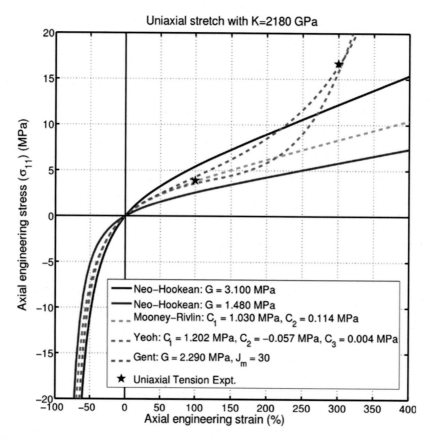

FIGURE 2.7

Stress−strain curves for various hyperelastic material models.

From (n.d.). Available at: https://commons.wikimedia.org/wiki/File:Hyperelastic.svg.

find the material parameters by calibrating the material parameters to experimental data. Since one or more experimental stress−strain curves are used for the calibration, the mathematical procedure of determining the material parameters involves solving an overconstrained set of equations. Hence, it is typically not possible to develop a model perfect fitting the experimental data.

VISCOELASTICITY AND VISCOELASTIC MODELS

From a mechanical point of view, comparing bone's and soft tissues' mechanical properties, it can be noted that the former fail at 5% deformation, and before failure, they could be considered as linear elastic materials up to yield point, while the latter usually fail at high strain values (hyperelastic materials), and before failure, they behave nonlinearly (nonlinear materials). Moreover, their load−deformation curve is highly time-dependent (viscoelastic), and they are incompressible and anisotropic. Therefore, viscoelasticity theory needs to be introduced for the analysis of soft tissues.

While in an elastic material the stress is linearly proportional to the strain $\sigma = \sigma(\varepsilon)$ and it is fully reversible (i.e., when the stress is removed, the material will return to its original shape, implying that the energy used to deform the material will be fully recovered in the inverse process to allow the object to return to its original shape), in viscoelastic materials, the applied stress is proportional to the strain and also to the time rate of change of the strain: $\sigma = \sigma(\dot{\varepsilon},\varepsilon)$. Unlike elastic materials, the viscous behavior is irreversible; therefore, it will induce a permanent (nonrecoverable) deformation.

As a matter of fact, all the biological materials have viscoelastic properties; however, for some of them (e.g., bone), the change in elasticity induced by a different loading rate for the majority of physiological activities, with low strain rates ranging from 0.01% to 1.0% strain per second, could be neglected. Nevertheless, for a more accurate investigation, in particular under dynamic conditions, or for specific materials (such as soft tissues), it is necessary to model the materials with viscoelastic properties.

In general, there are different approaches that could be used to model viscoelasticity. The most common are the Maxwell model, the Kelvin−Voight model, and the standard linear solid model; however, more complex models could be adopted if necessary. It is important to note that viscoelasticity should be considered only if the viscoelastic behavior is necessary for the analysis; otherwise, its effect could be neglected and a linear elastic analysis could be performed.

Considering the mono-dimensional viscoelasticity for simplicity, it is possible to model the viscoelastic behavior of a material considering two main structures (Fig. 2.8):

- a Hookean elastic spring, responsible for the linear elastic behavior, in which $\sigma = E\varepsilon$, with $E =$ Young's modulus (MPa);
- a Newtonian viscous dashpot element, responsible for the linear viscous behavior, in which $\sigma = \eta\dot{\varepsilon}$, with η being the coefficient of viscosity (Pa•s).

It is worth noting that not all the fluids are Newtonian; in reality, Newtonian fluids are the simplest mathematical models of fluids that account for viscosity, considering a linear relation between the stress and the strain rates. While no real fluid fits the definition perfectly, many common liquids and gases, such as water and air, can be assumed to be Newtonian for practical calculations under

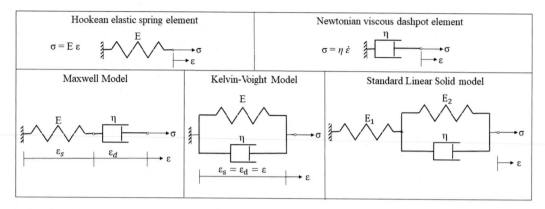

FIGURE 2.8

Viscoelastic structures and main viscoelastic models.

ordinary conditions. However, there are many non-Newtonian fluids that significantly deviate from this behavior, with change in viscosity under force.

Springs and dashpots constitute the unit elements that are used, connected in various forms, to build empirical viscoelastic models. Springs are used to account for the elastic solid behavior and dashpots are used to describe the viscous fluid behavior. It is assumed that a constantly applied force (stress) produces a constant deformation (strain) in a spring and a constant rate of deformation (strain rate) in a dashpot. The deformation in a spring is completely recoverable upon the release of applied forces, whereas the deformation that the dashpot undergoes is permanent. For simplicity, the constants of proportionality between stress and strain/strain rates will be considered constant during the deformation.

MAXWELL AND KELVIN–VOIGHT MODELS

The simplest forms of empirical viscoelastic models are obtained by connecting a spring and a dashpot together in series (Maxwell model) and in parallel (Kelvin–Voight model) (Fig. 2.8).

In the Maxwell model, as the two elements are in series:

- the stress σ applied to the entire system is applied equally on the spring and the dashpot ($\sigma = \sigma_s = \sigma_d$);
- the total strain ε is the sum of the strain in the spring and of the strain in the dashpot ($\varepsilon = \varepsilon_s + \varepsilon_d$).

Therefore, it is possible to derive the constitutive relation of the Maxwell model, expressed as a linear first-order differential equation, as a function of the two structural constituent parameters (E and η):

$$\sigma + \frac{\eta}{E}\dot{\sigma} = \eta\dot{\varepsilon} \tag{2.17}$$

In the Kelvin–Voight model, as the two elements are in parallel:

- the stress σ applied to the entire system will produce the stresses σ_s and σ_d in the spring and in the dashpot element, respectively ($\sigma = \sigma_s + \sigma_d$);

- the total strain ε of the system will be equal to the strains ε_s and ε_d in the spring and in the dashpot, respectively ($\varepsilon = \varepsilon_s = \varepsilon_d$).

Therefore, it is possible to derive the constitutive relation of the Kelvin–Voight model, which relates stress to strain and to strain rate and is a first-order, linear ordinary differential equation given by

$$\sigma = E\varepsilon + \eta\dot{\varepsilon} \tag{2.18}$$

It is important to note that in both models, the spring, used to represent the elastic solid behavior, can deform up to a certain limit given by the value of σ, while the dashpot, used to represent the fluid behavior, is assumed to deform continuously (flow) as long as there is a force acting on it. Therefore, we conclude the following:

- In the case of a Maxwell model, a force applied will cause both the spring and the dashpot to deform. The deformation of the spring will be finite. The dashpot will keep deforming as long as the force is maintained. Therefore, the overall behavior of the Maxwell model is more like a fluid than a solid, and, in fact, the Maxwell model is known to be the viscoelastic fluid model.
- In the case of a Kelvin–Voight model, the deformation of a dashpot connected in parallel to a spring is restricted by the response of the spring to the applied loads. The dashpot in the Kelvin–Voight model cannot undergo continuous deformations. Therefore, the Kelvin–Voight model represents the viscoelastic solid behavior.

Considering the effects of the two models under creep and relaxation tests, by the effects of a material under a constant stress (creep) or a constant strain (relaxation), we note the following:

- In the Maxwell model, if the material is put under a constant strain, the stresses gradually relax. When a material is put under a constant stress, the elastic stress (due to the spring) happens instantaneously and relaxes immediately upon the release of the stress (fully reversible); however, the viscous component induces a strain, which increases with time as long as the stress is applied and it is not reversible. Therefore, the Maxwell model correctly predicts relaxation but not creep.
- In the Kelvin–Voight model, under the application of a constant stress, the material deforms at a decreasing rate, asymptotically approaching the steady-state strain. When the stress is released, the material gradually relaxes to its undeformed state. (The spring is in parallel with the dashpot making the system fully reversible.) Therefore, this model is quite realistic in modeling creep in materials; however, the Kelvin–Voight model is not accurate in modeling the relaxation of the material as a constant strain in the dashpot will never induce full recovery due to the irreversible deformation.

STANDARD LINEAR SOLID MODEL

Unfortunately, the Maxwell model does not reproduce accurately the creep test of a material, while the Kelvin–Voigt model does not describe stress relaxation properly. To be able to consider both behaviors with the same model, it is necessary to introduce the standard linear solid (SLS), which is the simplest model that describes both the creep and stress relaxation behaviors of a viscoelastic material properly. Such a model is made of three constitutional elements (two springs and one dashpot) and could be considered by a spring in series with a Kelvin–Voight model (see Fig. 2.8).

In the SLS model, the following are considered:

- Spring 1 and the Kelvin—Voight model are in series; therefore, the overall stress σ is also the stress in spring 1 ($\sigma = \sigma_1 = \sigma_{KV}$).
- The total strain ε of the system is the sum of the strain in spring 1 and the strain in the Kelvin—Voight model ($\varepsilon = \varepsilon_1 + \varepsilon_{KV}$).
- In the Kelvin—Voight model, the Kelvin—Voight model's constitutive equation is valid.

Therefore, it is possible to derive the SLS model constitutive relation as

$$(E_1 + E_2)\sigma + \eta\dot{\sigma} = E_1 E_2 \varepsilon + E_1 \eta \dot{\varepsilon} \tag{2.19}$$

The SLS model is a three-parameter (E_1, E_2, and η) model and is used to describe the viscoelastic behavior of a number of biological materials such as the ligament, tendon and muscles (Hill's model), cartilage, and the white blood cell membrane.

Looking at the creep and relaxation behavior, the SLS model, having two springs, is able to respect both behaviors that limit the use of the Maxwell and Kelvin—Voight models.

It is also important to note that the model obtained with a spring and a Maxwell model in parallel could also be considered as an alternative SLS model; however, the constitutive equation is different from the one reported above.

ORTHOPEDIC BIOMECHANICS: STRESS ANALYSIS

Marwan El-Rich

Department of Mechanical Engineering, Khalifa University, Abu Dhabi, United Arab Emirates

STATICS REVIEW

Mechanics is a branch of physical science that deals with the state of rest and motion of bodies when subjected to loading. Rigid body mechanics includes statics, which studies bodies being at rest or moving with a constant velocity, and dynamics, which accounts for the accelerated motion of bodies. Study of mechanics of materials facilitates the analysis and design of various machines and load-bearing structures, including the determination of stress and strain.

A rigid body is said to be in equilibrium if and only if all forces and moments acting on the free-body diagram (FBD) of this rigid body are balanced, that is, the resultant (net) force equals zero and the resultant (net) couple moment at any arbitrary point O on or off the body equals zero:

$$\sum \vec{F} = \vec{0}, \sum \vec{M}_{/O} = \vec{0} \tag{3.1}$$

These are the equilibrium equations, and $\vec{F}$ and $\vec{M}$ are the net vector force and net vector moment about point O of all external forces acting on the rigid body, respectively. The vector form allows for the determination of the magnitude and sense of these vectors, and it can be expressed in terms of the vectors components written with respect to a coordinate system like the Cartesian orthogonal system $x - y - z$.

An example of an FBD of the upper body (assumed as a rigid body), which allows for the determination of the spinal forces and moments at a spinal lumbar level in the sagittal plane under the upper body weight and a handheld weight using the statics equilibrium equations, is shown in Fig. 3.1.

If the number of unknown forces and moments equals the number of equilibrium equations, the problem is called statically determinate. However, if the number of unknown forces and moments exceeds the number of equilibrium equations, the problem is called statically indeterminate and, in this case, additional equations such as relations between displacements, named compatibility of displacement equations, are needed to solve the unknowns of the problem.

STRESS AND STRAIN CONCEPT

When a body is loaded, stresses are generated in its constituent materials. The distribution of these stresses, their magnitudes, and orientations throughout the body depend not only on the loading

Human Orthopaedic Biomechanics. DOI: https://doi.org/10.1016/B978-0-12-824481-4.00008-1

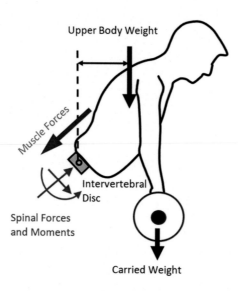

FIGURE 3.1

A free-body diagram of the upper body to determine spinal forces and moments at a lumbar level.

scenario but also on the geometry of the body and the properties of its constituent materials. The magnitude and orientation of these stresses can be determined analytically using the internal forces and moments developed in the constituent materials. These internal loads are calculated at a specified location on the body by virtually cutting the body and satisfying the equilibrium conditions at that location.

ONE-DIMENSIONAL SIMPLE STRESSES AND STRAINS

AXIAL STRESS DUE TO AXIAL LOADING

When the internal force is perpendicular to the surface of the cut through the body, it creates axial (also called normal) stress on that surface. In the example of pullout test of the pedicle screw shown in Fig. 3.2 (Lee et al., 2019), the head of the screw is subjected to a tensile force, which creates internal axial elementary forces perpendicular to the cross-section of the screw head, which, in turn, creates an axial stress of magnitude $\sigma = \frac{F}{A}$,

where F is the resultant of all elementary forces distributed over the entire area A of the cross-section and σ represents the average value of the stress over the cross-section rather than the stress at a specific point of the cross-section. In practice, it is assumed that the distribution of normal stress due to axial force is uniform (i.e., has the same magnitude everywhere along the cross-section) away from the region of load application. When the SI metric unit system is used, the force F is expressed in newton (N), the cross-sectional area A in square meters (m^2), and the stress σ in newton per square meters (N/m^2) or pascal (Pa).

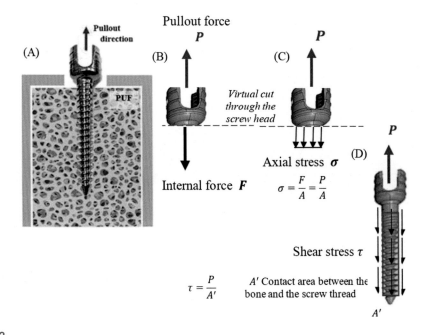

FIGURE 3.2

(A) Schematic diagram of pullout test, (B) free-body diagram of the screw head, (C) axial stress on the screw head, and (D) shear stress on the screw thread.

If the internal force is pointing out of the surface, it creates a tensile stress, while if it is pointing into the surface, it creates a compressive stress.

If the stress is calculated using the cross-sectional area A_0 measured before any deformation has taken place, the stress is called the engineering stress. In this case, the change in the cross-sectional area due to the applied load is neglected. However, the true stress accounts for this change and uses the actual cross-sectional area, which varies with applied load.

P also creates shear stress of average magnitude equals $\tau = \frac{F}{A}$ on the contact area between the bone and the screw thread along the insertion length.

In the lumbar intervertebral disc example shown in Fig. 3.3 (El-Rich, Arnoux, Wagnac, Brunet, & Aubin, 2009), the pressure in the nucleus pulposus increases with the compressive load, which creates tensile stresses in annular fibers and compressive stresses at almost all locations and directions of the annulus ground substance (Shirazi-Adl, Shrivastava, & Ahmed, 1984).

SAMPLE PROBLEM

A sheep leg bone AB is subjected to a tensile force of magnitude 1000N as shown in Fig. 3.4. Determine the average normal stress developed at the cross-section at C. Assume that the cross-section of the bone at C be annular and its outer diameter is 30 mm and inner diameter 25 mm.

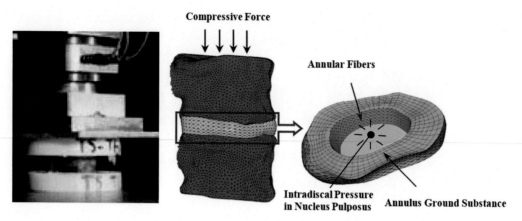

FIGURE 3.3

Lumbar disc under compressive force.

From El-Rich, M., Arnoux, P.-J., Wagnac, E., Brunet, C., & Aubin, C.-E. (2009). Finite element investigation of the loading rate effect on the spinal. Journal of Biomechanics, 42(9), 1252–1262. https://doi.org/10.1016/j.jbiomech.2009.03.036

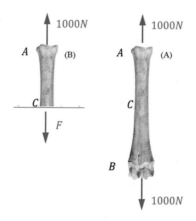

FIGURE 3.4

Sample problem about tensile loading.

Solution

The leg bone is in equilibrium as shown by its FBD in Fig. 3.4A. By cutting virtually the bone at C (where the stress is to be calculated) and drawing the FBD of part AC of the bone (Fig. 3.4B), the internal force F can be determined using the force equilibrium condition. In this example, it is obvious that the force F is a tension force and its magnitude equals 1000N.

The average normal stress at section C would be

$$\sigma = \frac{F}{A} = \frac{1000(\text{N})}{\frac{\pi}{4} \times (30^2 - 25^2) \times 10^{-6}(\text{m}^2)} = 4.63 \times 10^6 \frac{\text{N}}{\text{m}^2} = 4.63\text{MPa}, \text{ which is a tensile stress.}$$

STRESSES ON AN OBLIQUE SECTION UNDER AXIAL LOADING

If an oblique cut is made on a body that is subjected to axial force P (Fig. 3.5), the normal and tangential components of the internal force F on the oblique surface create normal and shear stresses, respectively. Denoting by α the angle formed by the oblique section with the normal plane, and by A the area of a section perpendicular to the internal force F, the cross-sectional area of the body cut by the oblique plane is $\frac{A}{\cos\alpha}$ and the average normal stress σ and the average shear stress τ on the oblique section are

$\sigma = \frac{F}{A}\cos^2\alpha$, $\tau = \frac{F}{A}\sin\alpha\cos\alpha$, respectively.

The normal and shear stresses σ and τ vary with the angle α. σ is maximum when $\alpha = 0$, whereas τ reaches its maximum when $\alpha = 45°$. This maximum is $\tau_{max} = \frac{F}{2A}$.

In the example of the FBD of the upper body shown in Fig. 3.1 (Naserkhaki, Jaremko, & El-Rich, 2016), the lumbar intervertebral disk is subjected to normal and shear stresses due to its orientation with respect to the body gravitational force. The magnitude and direction of these stresses vary along the lumbar spine levels. The spinal force and moment profile shown in Fig. 3.6 illustrates variation in compressive and shear forces along the lumbar spine under follower load simulating gravitational and muscle forces combined with forward flexion moment, which produces compressive and shear stresses of different magnitudes and in different directions along the spine.

The magnitude of the normal and shear stresses at any given point in a body depends on the direction of the surface on which the stress acts. Therefore, to determine the complete state of stress at a given point in a body, it is necessary to determine the stresses acting on all possible planes.

Unlike the normal stress, the distribution of the shear stress cannot be assumed to be uniform. The actual magnitude of τ varies from zero at the surface of the body to a maximum value, which may exceed the average value. More details will be provided in the shear stress section.

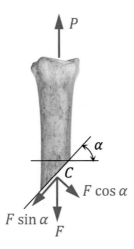

FIGURE 3.5

The components of the internal axial force F taken perpendicular and parallel to an oblique cut making an angle α with the plan normal to the axial force.

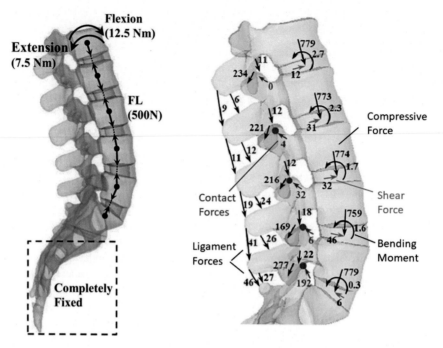

FIGURE 3.6

Spinal force (N) and moment (N m) profile along the lumbar spine under 500 N follower load simulating gravitational and muscle forces and 12.5 N m forward flexion (Naserkhaki et al., 2016). The *arrows* and *solid circles* illustrate the actual direction of forces and location of contact, respectively.

From Naserkhaki, S., Jaremko, J. L., & El-Rich, M. (2016). Effects of inter-individual lumbar spine geometry variation on load-sharing: Geometrically personalized Finite Element study. Journal of Biomechanics, 49(13), 2909–2917. https://doi.org/10.1016/j.jbiomech.2016.06.032

NORMAL AND SHEAR STRAIN

Deformations are changes in the form or dimensions of a nonrigid body caused by external loads.

Longitudinal deformations Δl refer to the lengthening or shortening of the body in one direction:

$$\Delta l = (\text{deformed length } l - \text{original length } l_0), \begin{Bmatrix} \Delta l > 0 \rightarrow \text{lengthening of the body} \\ \Delta l < 0 \rightarrow \text{shortening of the body} \end{Bmatrix}. \qquad (3.2)$$

Normal strain, a dimensionless number denoted by ε, is calculated by dividing the longitudinal deformation by the length:

$$\varepsilon = \frac{\Delta l}{l} \qquad (3.3)$$

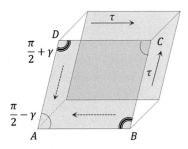

FIGURE 3.7

Deformation of a cube due to shear stress.

Similar to the normal stress, if the normal strain is calculated using the original length l_0 measured before any deformation has taken place, it is called engineering strain, while the true strain accounts for the variation in length due to the applied load and uses the actual length of the body:

$$\epsilon_{true} = \int_{l_0}^{l} \frac{dl}{l} \tag{3.4}$$

Angular deformation γ refers to a change of angle between the faces of a cube taken from the body. If only shear stresses τ are applied to the faces of the cube, as shown in Fig. 3.7, the latter deforms into rhomboid (oblique parallelepiped). Two of the angles formed by the four faces under stress are decreased from $\frac{\pi}{2}$ to $\frac{\pi}{2} - \gamma$, while the other two are increased from $\frac{\pi}{2}$ to $\frac{\pi}{2} + \gamma$. The small angle γ (expressed in radians) defines the shear strain. When the deformation of the element involves a decrease in the angle, the shear strain is positive (corners A and C), while an increase in the angle produces a negative shear strain (corners B and D).

NORMAL STRESS DUE TO PURE BENDING (SIMPLE BEAM THEORY)

If a prismatic beam with a vertical plane of symmetry and a rectangular cross-section is subjected to a pair of opposite bending moments of magnitude M at its ends, and if the cross-section of the beam is symmetric with respect to the plane containing the moments, the beam is said to be in pure bending and will deflect into a circular arc in the same plane and remains symmetric with respect to that plane. Every cross-section will be subjected to a similar stress and strain due to symmetry. If the deflection is assumed to be small as compared to the beam length, and if a Cartesian orthogonal system $x - y - z$ with origin taken at the centroid of one of the cross-sections is chosen, as shown in Fig. 3.8, the deflection of the centroidal surface, that is, the surface with the y-axis as normal and located at $y = 0$ in the undeformed configuration, defines the deflection of the beam. This centroidal surface is called *neutral surface* as its fibers have zero stress and zero strain. Let us now consider two neighboring cross-sections AB and CD that are perpendicular to the plane $y = 0$ when the beam is undeformed. After loading, the two cross-sections AB and CD are deformed into $A'B'$ and $C'D'$, which remain perpendicular to the centroidal surface, which is deformed into a

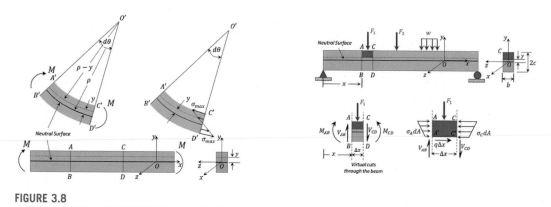

FIGURE 3.8

Prismatic beam subjected to pure bending (left). Beam bending test (right).

thick arc. ρ denotes the radius of curvature of the neutral surface. The centroidal arc between the cross-sections $A'B'$ and $C'D'$ has length $\rho d\theta$, where $d\theta$ is the relative angle formed by the cross-sections $A'B'$ and $C'D'$. A surface at distance y above the centroidal surface will have a length $(\rho - y)d\theta$ and the change in length is $y d\theta$.

The normal strain will be $\varepsilon = \frac{(\rho - y)d\theta - \rho d\theta}{\rho d\theta} = \frac{-y}{\rho}$. The minus sign indicates that the surface has shortened as it is in compression. If the beam material is homogeneous and the resulting stress remains below the proportional and elastic limits, Hooke's law for uniaxial stress applies, and the normal (bending) stress will be $\sigma = E\varepsilon = -E\frac{y}{\rho}$

As it is not easy to determine the radius of curvature, the bending stress can be calculated in terms of the bending moment and the moment of inertia, also called the second moment of area of the cross-section about the bending axis (z-axis in the example shown) using the following relation $\sigma = -\frac{My}{I}$. The stress is compressive ($\sigma < 0$) above the neutral surface when the moment M is positive and tensile ($\sigma > 0$) when M is negative. The stress has a linear distribution, and its maximum magnitudes occur at the free top and bottom surfaces of the beam, located at the largest distance from the neutral surface.

SHEAR STRESS DUE TO BENDING

If a prismatic beam with a vertical plane of symmetry and a rectangular cross-sectional area is subjected to concentrated forces and/or distributed loads, the cross-section will be subjected to shear forces V and bending moments M. The cross-sections AB and CD located at distances x and $x + \Delta x$, respectively, from the left end of the simply supported beam show the positive directions of these internal loads. The moments M_{AB} and M_{CD} are replaced by their equivalent forces, as shown in the FBD of the element $ACC'A'$. The magnitudes of these forces are equal to the magnitudes of the normal stresses produced by the moments multiplied by the cross-sectional area dA of the infinitesimal element $ACC'A'$.

The force $q\Delta x$ represents the horizontal shear force at the lower side of the element, which is located at a distance y from the neutral surface. This shear force results from the unbalanced horizontal forces produced by the moments. The horizontal shear force per unit length, q, also named the shear flow, is defined as $q = \frac{VQ}{I}$, where V is the shear force produced by the increment moment $M_{CD} - M_{AB}$, Q is the first moment with respect to the neutral surface of the portion of the cross-section located either below or above the location at which q is calculated, and I is the moment of inertia of the entire cross-section about the neutral axis (z-axis in the example shown).

The average shear stress on the face $A'C'$, which is parallel to the neutral surface, is $\tau_{\text{average}} = \frac{VQ}{Ib}$, where b is the width of the face $A'C'$. The shear stresses τ_{yx} and τ_{xy} exerted on the horizontal face $A'C'$ and the transverse face CC', respectively, are equal. Distribution of the shear stress τ_{xy} in a transverse section of a prismatic beam that has a rectangular cross-section of height $2c$ and width b can be calculated as $\tau_{yx} = \frac{3}{2} \frac{V}{A} \left(1 - \frac{y^2}{c^2} \right)$, where $A = 2bc$ is the cross-sectional area of the beam and c is the y distance from the neutral axis of the upper and lower surfaces of the beam. The maximum shear stress is $\tau_{\max} = \frac{3}{2} \frac{V}{A}$ and occurs at the neutral surface ($y = 0$). This relationship demonstrates that the maximum value of the shear stress in a beam of rectangular cross-section is 50% greater than the value $\frac{V}{A}$ obtained by incorrectly assuming a uniform stress distribution across the entire cross-section (Beer, Johnston, DeWolf, & Mazurek, 2014). $\tau = 0$ at the top and bottom of the cross-section ($y = mc$).

In the four-point bending test used to compare the performance of three different screw—plate and cable—plate systems for the fixation of periprosthetic femoral fractures near the tip of a total hip arthroplasty (Lever, Zdero, Nousiainen, Waddell, & Schemitsch, 2010) (Fig. 3.9), the region near to the shaft fracture (between the downward arrows) is subjected to a pure bending moment, which produces normal stress only. However, the regions located between the support reaction forces (upward arrows) and the applied forces (downward arrows) are subjected to both normal and shear stresses. Also, in the axial load scenarios for both 20-degree abduction and 20-degree forward flexion, the neck of the hip joint implant is subjected to normal and shear stresses. The normal stress is produced by the axial force simulating the upper body weight and the bending moment caused by the eccentricity of this force as its line of action is not parallel to the implant femoral stem and does not pass through the centroid of its cross-section. The shear stress is produced by the shear force, which represents the projection of the body weight force on the cross-section of the implant femoral stem. These loading conditions are illustrated through the proximal femur loading conditions under upper body weight (Solórzano, Ojeda, & Diaz Lantada, 2020). Denoting by d the perpendicular distance from the centroid of the femur shaft cross-section to the line of action of the body weight force, the normal stress would be $\sigma = \frac{W/2}{A} - \frac{My}{I}$, where W is the body weight force, A is the area of the cross-section, I its centroidal moment of inertia, and y is measured from the centroidal axis of the cross-section. The moment M is equal to $\frac{W}{2} d$. The shear stress is due to bending, and it can be calculated using the approach described previously.

SHEAR STRAIN DUE TO TORSION

In this section, shear stress in circular and noncircular members subjected to torsion (twist) will be analyzed. Torsion results from a moment, called torque, applied around the member axis.

Circular solid shaft

Let us consider a solid circular shaft of length L and radius c subjected to torsional moment (torque) T. Due to the axisymmetry conditions, any given cross-section of the circular shaft will remain

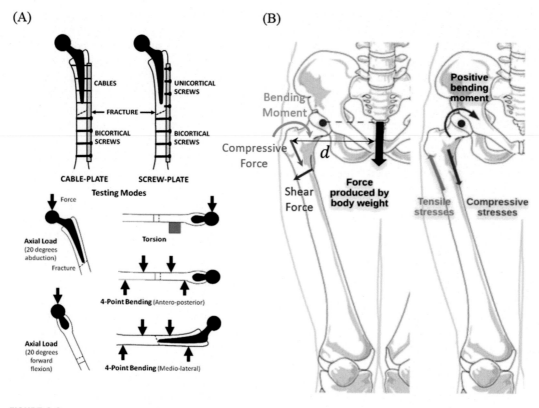

FIGURE 3.9

(A) Mechanical tests used to biomechanically examine three different screw—plate and cable—plate systems for fixation of periprosthetic femoral fractures near the tip of a total hip arthroplasty. (B) Proximal femur loading under upper body weight (Solórzano et al., 2020).

From Lever, J. P., Zdero, R., Nousiainen, M. T., Waddell, J. P., & Schemitsch, E. H. (2010). The biomechanical analysis of three plating fixation systems for. Journal of Orthopaedic Surgery and Research, 5, 45. https://doi.org/10.1186/1749-799X-5-45A.

plane and undistorted. The resulting deformations will be uniform throughout the entire length of the shaft. Shear strain γ at a given point of the shaft is proportional to the angle of torsion ϕ and the distance r from the axis of the shaft to that point. The shear strain varies linearly with the distance from the axis of the shaft and equals $\gamma = \frac{r\phi}{L}$ with a maximum value on the surface of the shaft where r is maximum and equals c:

$$\gamma_{max} = \frac{c\phi}{L}.$$

(3.5)

Shear stress due to torsion

The shear forces dF acting on elements dA of the cross-section and located at distances r from the shaft axis are equivalent to the internal torque $T = \int r dF$. These forces create shear stress on elements dA ($\tau dA = dF$). If the torque produces elastic shear stress, that is, stress not exceeding the proportional and elastic limits, Hook's law applies to calculate shear stress $\tau = G\gamma$, where G is the modulus of rigidity or shear modulus of the material. The shear stress in the shaft varies linearly with radius r and can be calculated as a function of the maximum shear stress, which occurs on the surface of the shaft where $r = c$:

$$\tau = \frac{r}{c}\tau_{max}. \tag{3.6}$$

τ_{max} can be calculated in terms of the internal torque T as $\tau_{max} = \frac{Tc}{J}$, where J is the polar moment of inertia of the cross-section with respect to its center O. The shear stress at any distance r from the axis of the shaft is

$$\tau = \frac{Tr}{J}. \tag{3.7}$$

In the example shown in Fig. 3.10, the internal torque is constant along the entire length of the shaft and is equal to the applied torque T due to the equilibrium requirements. It is important to mention that if the shaft is subjected to various torques as it will be shown in the following example, the above equation is valid in the portion of the shaft where the internal torque is constant.

If an element with faces making arbitrary angles with the shaft axis is considered, then the internal torque will create normal stress in addition to shear stress. If the faces of the element are at

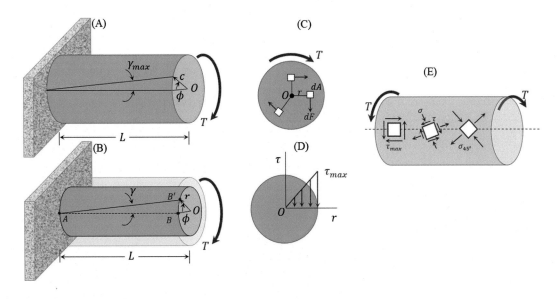

FIGURE 3.10

(A) The twist angle ϕ and maximum shear strain γ_{max} on a circular solid shaft fixed at one end and subjected to a torque T at the other end. (B) Deformed portion of the shaft. (C) Shear forces dF, equivalent to the internal torque T. (D) Distribution of the shear stress in the shaft. (E) Elements on the surface of the shaft with faces parallel, normal, or at 45 degrees to its axis.

Axial Rotation

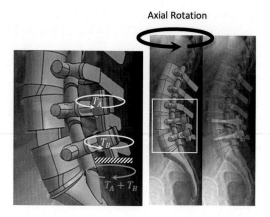

FIGURE 3.11

A lumbar spine model simulating a wide laminectomy with L4-L5 PEEK cages and posterior fusion. Under axial rotation of the spine, the rod and the bilateral pedicle screw fixation are subjected to torques T_A and T_B, while the distal end of the rod is assumed fixed completely.

From Nikkhoo, M., Khoz, Z., Cheng, C.-H., Niu, C.-C., El-Rich, M., & Khalaf, K. (2020). Development of a novel geometrically-parametric patient-specific finite. Journal of Biomechanics, 102, 109722. https://doi.org/10.1016/j.jbiomech.2020.109722A.

$45°$ or $45° \pm 90°$ to the axis of the shaft, then normal stress with magnitude $\sigma_{45°} = \tau_{\max} = \pm \frac{Tc}{J}$ is produced.

In the SI metric unit system, T will be in N.m, J in m⁴, and τ in N/m² or Pa.

In the instrumented lumbar spine shown in Fig. 3.11 (Nikkhoo et al., 2020), let us assume that when the spine is subjected to axial rotation, the spinal rod will be subjected to torques T_A and T_B at pedicle screw fixations A and B, respectively. If the distal end of the rod is supposed to be completely fixed, the maximum shear stress in the rod segment between the two fixations A and B will be $\tau_{\max} = \frac{T_A c}{J}$, where c and J are the radius and the polar moment of inertia of the rod with respect to its center, respectively.

The maximum shear stress in the rod segment between the fixation B and the distal end of the rod will be $\tau_{\max} = \frac{(T_A + T_B)c}{J}$.

Noncircular thin-walled hollow shaft

Consider a hollow shaft with a noncircular thin cross-section with constant thickness t subjected to torque T (Fig. 3.12). The shear stress can be written as $\tau = \frac{q}{t}$, where q is the shear flow in the shaft wall and is assumed constant. The internal torque will be $T = 2q\mathscr{A}$, where $\mathscr{A}$ is the area bounded by the centerline of the wall cross-section. This relation is obtained by equating T to the moments of the elementary shear forces exerted on the wall section about an arbitrary point O located inside the hollow shaft. By substituting $q = \frac{T}{2\mathscr{A}}$, the shear stress at any given point of the wall can be written as

$$\tau = \frac{T}{2t\mathscr{A}}.$$

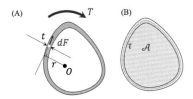

FIGURE 3.12

(A) *dF* is a shear force applied at one element on the shaft cross-section and resulting from the internal torque. (B) Shear flow area.

REFERENCES

Beer, F. P., Johnston, E. R., DeWolf, J. T., & Mazurek, D. F. (2014). *Mechanics of materials* (7th Edition). New York: McGraw-Hill.

El-Rich, M., Arnoux, P. J., Wagnac, E., Brunet, C., & Aubin, C. E. (2009). Finite element investigation of the loading rate effect on the spinal load-sharing changes under impact conditions. *Journal of Biomechanics*, *42*, 1522-1262.

Lee, E. S., Goh, T. S., Heo, J.-Y., Kim, Y.-J., Lee, S.-E., Kim, Y. H., & Lee, C.-S. (2019). Experimental evaluation of screw pullout force and adjacent bone damage. *NATO Advanced Science Institutes Series E Applied Sciences*, *9*(3), 586. Available from https://doi.org/10.3390/app9030586.

Lever, J. P., Zdero, R., Nousiainen, M. T., Waddell, J. P., & Schemitsch, E. H. (2010). The biomechanical analysis of three plating fixation systems for periprosthetic femoral fracture near the tip of a total hip arthroplasty. *Journal of Orthopaedic Surgery and Research*, *5*, 45. Available from https://doi.org/10.1186/1749-799X-5-45.

Naserkhaki, S., Jaremko, J. L., & El-Rich, M. (2016). Effects of inter-individual lumbar spine geometry variation on load-sharing: Geometrically personalized Finite Element study. *Journal of Biomechanics*, *49*(13), 2909−2917. Available from https://doi.org/10.1016/j.jbiomech.2016.06.032.

Nikkhoo, M., Khoz, Z., Cheng, C.-H., Niu, C.-C., El-Rich, M., & Khalaf, K. (2020). Development of a novel geometrically-parametric patient-specific finite. *Journal of Biomechanics*, *102*, 109722. Available from https://doi.org/10.1016/j.jbiomech.2020.109722.

Shirazi-Adl, S. A., Shrivastava, S. C., & Ahmed, A. M. (1984). Stress analysis of the lumbar disc-body unit in compression. A three dimensional non linear finite element study. *Spine (Philadelphia, PA: 1986)*, *9*(2), 120−134. Available from https://doi.org/10.1097/00007632-198403000-00003.

Solórzano, W., Ojeda, C., & Diaz Lantada, A. (2020). Biomechanical study of proximal femur for designing stems for total hip. *NATO Advanced Science Institutes Series E Applied Sciences*, *10*(12), 4208. Available from https://doi.org/10.3390/app10124208.

ORTHOPEDIC BIOMECHANICS: MULTIBODY ANALYSIS

Giovanni Putame[1,2], Alessandra Aldieri[1,2], Alberto Audenino[1,2] and Mara Terzini[1,2]

[1]Department of Mechanical and Aerospace Engineering, Politecnico di Torino, Turin, Italy
[2]Polito[BIO]Med Lab, Politecnico di Torino, Turin, Italy

INTRODUCTION

To date, computational methods have been widely used in biomechanical studies, with finite element analysis (FEA) and multibody (MB) analysis representing the two mostly used computational approaches, to address musculoskeletal research topics. Based on concepts related to continuum mechanics, FEA mainly aims at predicting stress and strain distributions within hard and soft biological tissues. However, due to its high computational costs, FEA is generally applied to small-scale systems under static loading conditions. On the other hand, based on the dynamic equations of motion, MB analyses allow us to investigate the kinematical and dynamical behavior of systems, consisting of multiple rigid or flexible bodies interconnected by joints. In contrast to FEA, traditional MB analyses are not able to provide information about stress or strain distributions, but they offer the possibility to deal with small- as well as large-scale systems, considerably reducing computational costs. In this framework, the musculoskeletal apparatus can be partially or totally modeled as a complex MB system, and different valuable quantities, such as forces related to ligaments, tendons, muscles, and articular contacts, can be rapidly obtained as simulation outcomes, together with translations and rotations of the bone segments. In light of the above-mentioned advantages, MB simulations are particularly suitable for clinical applications, especially for those related to the orthopedic field. This explains the growing popularity of the methodology in biomechanics, starting from the first applications in 1906, when it was employed to study human locomotion (Schiehlen, 1997). On top of that, the implementation of new specifically built software, such as Anybody (AnyBody Technology, Aalborg, Denmark) and OpenSim (open-source software, available from https://opensim.stanford.edu/), has made MB modeling for biomechanics more accessible, further increasing its diffusion. In addition, general-purpose MB software, such as ADAMS View (MSC Software Corporation, Santa Ana, CA), can also be employed in MB modeling of biomechanical structures, with the advantage of guaranteeing a full control on the created model at the expense of a greater initial effort in developing each biomechanics-specific modeling element.

In the literature, many different applications of the MB approach in biomechanics can be found, adopted to develop and assess the performance of prosthetic devices (Grosu, Cherelle, Verheul, Vanderborght, & Lefeber, 2014; Müller, Zakaria, van der Merwe, & D'Angelo, 2016; Pascoletti,

Human Orthopaedic Biomechanics. DOI: https://doi.org/10.1016/B978-0-12-824481-4.00014-7

Cianetti, Putame, Terzini, & Zanetti, 2018); to predict articular contact forces in human and animal joints (Chen et al., 2014; Fregly et al., 2012; Harrison, Whitton, Kawcak, Stover, & Pandy, 2010; Hu, Chen, Xin, Zhang, & Jin, 2018); to investigate the effect of ligament deficiency on joint kinematics and dynamics (Guess & Razu, 2017; Guess, Stylianou, & Kia, 2014; Kang et al., 2017; Rahman, Sharifi Renani, Cil, & Stylianou, 2018); to evaluate athletes' performance during sport activities (Mahadas, Mahadas, & Hung, 2019); and to understand typical and impaired neuromuscular control (Seth et al., 2018), among others.

Focusing on arthroplasty, MB models of anatomical structures aim at supporting surgeons throughout the pre-, intra-, and postoperative phases providing additional information useful to improve surgical outcomes. Reported studies referring to the preoperative phase have the primary purpose of predicting consequences of surgical interventions, analyzing how the surgical variables will affect (1) the postoperative kinematics, (2) soft tissue functionality (e.g., ligament tension/strain, muscle strengths, cartilage compressions), or (3) intraarticular contact forces. For instance, by means of a patient-specific model, Viceconti, Ascani, and Mazzà (2019) performed a preoperative prediction of the knee ligament elongation following a total knee arthroplasty (TKA), where the elongation was analyzed as a function of the position and orientation of the knee implant during level walking. In Geier et al. (2017), instead, a systematic analysis using a lower limb model integrated into a hardware-in-the-loop simulation was carried out to study how implant position, design, and impaired muscle function could impact total hip implant dislocation. Other parametric studies confirmed the efficiency of the MB approach in predicting the range of motion of a prosthetic joint in relation to the positioning of its components (Morra, Rosca, Greenwald, & Greenwald, 2008; Pianigiani, Chevalier, Labey, Pascale, & Innocenti, 2012; Putame et al., 2019). Moreover, the analysis of postoperative data successfully helps correlate surgical choices with relative outcomes, thus improving the actual surgical techniques. In this context, MB simulations allow us to estimate significant quantities that are usually difficult to obtain experimentally, such as the ligament elongation patterns (Hosseini Nasab et al., 2019), tibiofemoral contact forces (Navacchia et al., 2016), and accurate joint kinematics (Twiggs et al., 2018). Over the past few years, the combination of patient-specific virtual models with surgical navigation systems has made it possible to accomplish very accurate spatial measurements during the intraoperative phase (e.g., relative positions and orientations between body segments and implant components). Nevertheless, obtaining intraarticular contact pressure and soft tissue tensioning still remains a challenge or at times even impossible to measure, unless expensive and/or highly invasive sensing devices are used. Intraoperative computational analyses could potentially provide such desired information as long as the following two main requirements are met: (1) compatibility of simulation time with the fast timing of surgical procedures; and (2) integration of patient-specific anatomical geometries as well as of tissue mechanical properties into the model. Despite the increasingly available computational power and the continuous advances made in research, only few studies have tried to bring biomechanical simulations directly to the operating theater to date. Among these studies, Armand et al. (2018) presented a simulation-based guidance system for periacetabular osteotomy, which aims, by using a discrete-element analysis, at providing joint contact pressures during the surgical intervention.

In summary, further research is necessary to make computational models even more reliable, thereby increasing their adoption by surgeons. Besides, desirable outcomes could result from the combination of increasingly sophisticated sensing devices with more functionally accurate MB models.

MODELING STRATEGIES

Generally, computational MB analyses allow us to study systems consisting of multiple rigid bodies interconnected by joints. In this context, the musculoskeletal apparatus can actually be modeled as a complex MB system made of rigid bodies interconnected through joints (Fig. 4.1).

Commonly, biomechanical models involve idealized joints (e.g., spherical, revolute, universal) in order to represent the main degrees of freedom (DOFs) of the considered anatomical joint. For instance, the elbow arthrodial joint is often idealized as a simple 1 DOF revolute joint, and the hip as a 3 DOF spherical joint. However, to better investigate intraarticular contact forces and the real kinematics of anatomical or prosthetic joints, a detailed representation of the joint is mandatory. Therefore, it is necessary to take into account the constraining forces due to biological soft tissues, the geometry of the articulating surfaces, and their material properties. In this section, some fundamental concepts and modeling strategies are presented for the creation of musculoskeletal multibody models including soft anatomical structures (i.e., ligaments and menisci), contacts, and muscle actions.

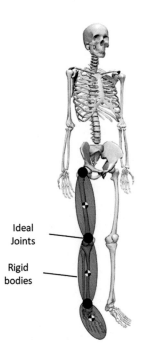

Ideal
Joints

Rigid
bodies

FIGURE 4.1

Exemplificative picture illustrating the skeletal apparatus with the lower limb represented as an open kinematic chain by means of rigid bodies (*red ellipsoids*) connected by generic ideal joints (*black dots*).

LIGAMENTS

Over the past few decades, several research groups have investigated the biomechanical behavior of ligaments in terms of stress and strain during specific articular movements. In this context, different models have been proposed to mathematically describe the force–strain relationship of ligaments under different loading conditions. In general, ligaments are modeled by using line (1D), planar (2D), or solid (3D) elements (Galbusera et al., 2014). Although 2D and 3D elements can potentially add more information compared to the 1D elements, such as local strains, the latter present some distinct advantages in the form of a straightforward numerical implementation and high computational efficiency. One of the most used 1D ligament models was proposed by Blankevoort, Huiskes, and de Lange (1991). This model represents each ligament bundle as a single tension-only spring element connecting the origin and insertion points. In particular, the force–strain relationship of each spring is described by the following nonlinear piecewise function:

$$f = \begin{cases} -k(\varepsilon - \varepsilon_L), \varepsilon > 2\varepsilon_L \\ -0.25k\dfrac{\varepsilon^2}{\varepsilon_L}, 0 \le \varepsilon \le 2\varepsilon_L \\ 0, \varepsilon < 0 \end{cases} \tag{4.1}$$

where ε is the ligament strain, ε_L is a reference value for the strain assumed to be 0.03, and k is the stiffness parameter, expressed as force per unit strain, of each different ligament bundle. The ligament strain ε is defined as:

$$\varepsilon = \frac{(l - l_0)}{l_0} \tag{4.2}$$

where l is the actual ligament length and l_0 is the zero-load length, also called slack length, which is the maximum linear distance between the ligament attachment points above which the ligament gets taut. The force–strain curve obtained for a ligament modeled through the function proposed by Blankevoort et al. (1991) is depicted in Fig. 4.2. In detail, two regions can be identified: the toe region for lower strains ($\varepsilon < 2\varepsilon_L$) and the linear region for higher strains ($\varepsilon > 2\varepsilon_L$), where the

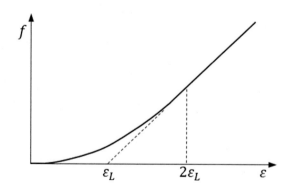

FIGURE 4.2

Force–strain curve obtained for a ligament modeled through the function proposed by Blankevoort et al. (1991). The strain value equal to $2\varepsilon_L$ defines the limit between the toe region ($\varepsilon < 2\varepsilon_L$) and the linear region ($\varepsilon > 2\varepsilon_L$).

ligament force is determined by the stiffness parameter k (i.e., the slope of the straight line). In order to consider the different recruitment patterns of the fiber bundles within the same ligament during the joint movement, each bundle is treated as a single parallel spring: The total stiffness value is thus equally divided among the springs.

Following the described constitutive model, the viscoelastic behavior of the ligaments is neglected. Nevertheless, in order to avoid high-frequency vibrations during the MB simulation, a parallel damper with a damping coefficient of 0.5 N s/mm can be added to each spring element (Guess, Thiagarajan, Kia, & Mishra, 2010). An important aspect deserving attention is the determination of the zero-load length l_0 for each considered ligament bundle: this parameter is indeed closely related to the reference strain ε_R, that is, the strain of the ligament at a given joint reference position. The following expression relates the zero-load length l_0 with the reference strain ε_R:

$$l_0 = \frac{l_R}{\varepsilon_R + 1} \tag{4.3}$$

where l_R is the ligament length at the reference joint position. Unfortunately, reference strains are difficult to measure experimentally. Therefore, they are often adapted from previous experimental studies (Blankevoort et al., 1991; Trent, Walker, & Wolf, 1976; Wismans, Veldpaus, Janssen, Huson, & Struben, 1980). Starting from values reported in the literature, two different strategies may be adopted to refine the zero-load lengths. The first strategy consists of assessing the joint behavior by simulating a laxity test. In this case, the joint is subjected to specific loading conditions (e.g., the clinical drawer test) and the related displacements are measured and compared with experimental data. Then, the zero-load lengths are refined until numerical results match the experimentally reported data. The second strategy is derived from the study of Guess and Razu (2017), where a systematic strategy, specifically developed for the knee joint, is proposed based on two assumptions: during the passive movement of the joint, throughout its range of motion, (1) every ligament bundle is stretched and (2) the force exerted by each bundle should be below 50 N.

MENISCI

In order to investigate the behavior of an anatomical joint, it is important to include all the significant constraining elements. From a modeling point of view, the implementation of an anatomical joint model is more challenging than the implementation of a prosthetic one because of the presence of biological structures, which are sacrificed during an arthroplasty. For instance, focusing on the knee joint, the anterior or both anterior and posterior cruciate ligaments as well as the menisci and part of the articular capsule might be resected depending on the knee prosthesis design. In the anatomical knee, besides ligaments, the menisci contribute to improving the joint stability too, as far as the femoral sliding on the tibial plateau is concerned. In the MB context, three different approaches were proposed in the literature to model menisci. The first, introduced by Li, Gil, Kanamori, and Woo (1999), only emulates the constraining forces due to the menisci. Therefore, details regarding the meniscal geometry as well as femoro-meniscal and tibio-meniscal contact forces are neglected. Each meniscus (medial and lateral) is represented by two perpendicular springs acting on the tibial plateau in the anteroposterior and mediolateral directions, respectively. Moreover, each spring is constrained, at one end, to the tibial plateau, and, at the other end, to the

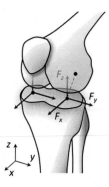

FIGURE 4.3

Illustration of the knee showing the implementation of the menisci through bushing elements connecting the tibial plateau with the femoral condyles. Red arrows represent the constraining translational forces generated by the bushings.

femoral condyle. Such an approach can be easily implemented by using a simplified formulation of the bushing element,[1] considering the only effect of two translational forces as represented below:

$$\begin{bmatrix} F_X \\ F_Y \\ F_Z \end{bmatrix} = \begin{bmatrix} K_{11} & 0 & 0 \\ 0 & K_{22} & 0 \\ 0 & 0 & 0 \end{bmatrix} \cdot \begin{bmatrix} x \\ y \\ z \end{bmatrix} \tag{4.4}$$

where F_X and F_Y are the meniscal constraining forces in the anteroposterior and mediolateral directions, respectively. The force F_Z, normal to the tibial plateau, is null. The K_{ij} coefficients represent the stiffness values. Finally, X, Y, and Z represent the translations along the respective perpendicular directions (Fig. 4.3).

A second approach takes into account the geometry of each meniscus as a single rigid body connected to the tibial plateau (Hu et al., 2018). Therefore, two additional meniscus—cartilage contacts pairs are included into the model and the respective forces can be obtained. The last approach, presented by Guess et al. (2010), implies the discretization of each meniscus into elements connected through bushing elements defined by a 6×6 stiffness matrix. Intuitively, since this approach makes the model more complex to be solved, it should be adopted only when the aim of the study is the investigation of the meniscal behavior.

CONTACTS

The selection of a proper contact representation for the simulation of biomechanical MB systems is crucial not only to obtain of correct results but also to reduce the required computational time. Poorly defined contact conditions may indeed negatively affect the whole simulation due to the

---

[1]The bushing element allows us to generate a six-component force (three translational forces and three torques) acting between two bodies depending on their relative linear and angular displacements and velocities. Thus, in general, the bushing element is defined by a 6×6 stiffness matrix and a 6×6 damping matrix.

generation of high-frequency vibrations. Here, some of the most common formulations used to include contacts into MB models are presented (Flores, 2011; Flores, Flores, & Lankarani, 2016; Sherman, Seth, & Delp, 2011). Briefly, the first is described by an interpenetration formulation derived from Hertz's contact theory, the second is based on the elastic foundation theory, and the third is based on the coefficient of restitution. In addition to the treated contact formulations, the contribution of the friction force during contacts can be also considered, for instance, using a Coulomb's model.

Hertz's law–based formulation

Hertz's law–based formulation is derived from that proposed in Hunt and Crossley (1975). The contact force F_C is calculated based on both the penetration depth δ between the undeformed contacting bodies and penetration velocity as follows:

$$F_C = K\delta^e + C(\delta, \delta_{max}, C_{max})\dot{\delta} \tag{4.5}$$

where K is the contact stiffness constant, δ is the penetration depth (Fig. 4.4), e is the nonlinear power exponent, $\dot{\delta}$ is the penetration velocity, and C is a sigmoid damping function, which depends on the penetration depth and is defined by a maximum penetration constant δ_{max} at which the damping function reaches its maximum value C_{max}. On the right side of the equation, the first term is the elastic force component, while the second term represents the energy loss during the contact. In order to avoid discontinuities at the initial instant of contact, in contrast to Hunt and Crossley (1975), the second term is multiplied by a sigmoid function, which is proportional to the penetration. The single vector force F_C is applied at the centroid of the interpenetration volume. Although this formulation is numerically efficient, its goodness depends on the defined stiffness and damping parameters whose determination is often difficult due to the lack of data.

Elastic foundation theory–based formulation

An alternative approach for the prediction of the contact force is derived from the elastic foundation theory. Based on this theory, one of the two contacting bodies is considered as partially deformable, whereas the other one is considered to be rigid. Therefore, a layer of independent

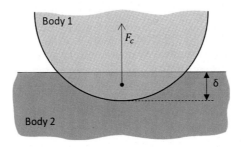

FIGURE 4.4

Schematic representation of Hertz's law–based contact between two rigid bodies. The distance δ is the penetration depth.

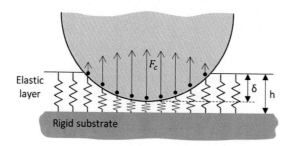

FIGURE 4.5

Schematic representation of the elastic foundation theory—based contact, where h is the layer thickness and δ is the spring deformation.

FIGURE 4.6

Discretization process applied to a polyethylene insert of a total knee prosthesis. From left to right: upper view of the insert geometry; discretized insert geometry; and contact pressure map resulted from a multibody simulation of a squat, respectively.

elastic elements is implemented on the contact surface of the deformable body. In particular, the pressure p acting on each elastic element can be expressed by the following formula:

$$p = \frac{(1 - \nu)E\delta}{(1 + \nu)(1 - 2\nu)h} \tag{4.6}$$

where E is Young's modulus of the deformable layer, ν is Poisson's ratio of the layer, h is the layer thickness, and δ represents the spring deformation, that is, the interpenetration between the contacting bodies (Fig. 4.5).

In detail, for each elastic element, a contact is defined using the elastic term of the Eq. (4.5) ($K\,\delta^e$), where the nonlinear power exponent in considered to be unitary ($e = 1$); the stiffness constant K can then be determined from Eq. (4.6). Specifically, the ratio p/δ is multiplied by the cross-section area A of the elastic element (i.e., discretization area) as shown below:

$$K = \frac{p}{\delta}A = \frac{(1 - \nu)EA}{(1 + \nu)(1 - 2\nu)h} \tag{4.7}$$

The major advantage of this approach consists of the possibility to derive the contact stiffness as well as to directly obtain the contact pressure distribution within the MB framework. Furthermore, it allows an easier identification of the location of the contact area and its extension (Fig. 4.6).

Coefficient of restitution method

An additional approach to computing the contact force can be obtained from the combination of a penalty parameter and a specific coefficient of restitution. On one hand, the penalty parameter is used to enforce a unilateral constraint: a large penalty value ensures that the penetration of one geometry into the other and vice versa is small. On the other hand, the coefficient of restitution controls the energy dissipation at the contact. In particular, the coefficient of restitution can range between 0 and 1, which would result in a perfectly anelastic or elastic contact, respectively. This contact method is suitable for cases when the coefficient of restitution is known a priori or when the modeled contact can be approximated as perfectly anelastic or elastic (Bersini, Sansone, & Frigo, 2016).

MUSCLES

In musculoskeletal modeling, muscles are considered as actuators, which, under the application of a traction force, move the articulated bony segments. Muscles are often represented using vector forces connecting, directly or through via-points, origin and insertion sites. The implementation of a muscle-driven simulation of movement requires the determination of the amount of force generated by each muscle involved in the specific motion task. To estimate the muscle forces within an MB model, a possible strategy consists of the integration of a closed-loop control algorithm. In the following, the control algorithm is based on a proportional–integral–derivative (PID) controller (Fitzpatrick, Baldwin, Clary, Maletsky, & Rullkoetter, 2012; Guess et al., 2014; Rahman et al., 2018; Stylianou, Guess, & Kia, 2013). The PID control algorithm represents a straightforward tool, widely used in different engineering applications, which allows us to iteratively adjust a state variable of a dynamic system in order to obtain a desired state of the system itself. To simply explain how the PID controller can be applied for muscle force determination, an example is used in the following. It estimates the quadriceps muscle forces needed to simulate a weight-bearing squat movement. Specifically, the considered lower limb model includes the quadriceps muscle group (Fig. 4.7A), which consists of four distinct muscles and which represents the main extensor muscle of the knee.

In the example, the femoral head actual position q_m and its velocity $\dot{q}_m$, both referring to the vertical direction Y (Fig. 4.7B), are measured at each time-step of the model simulation. Iteratively, the measured values are compared with the respective desired values q_d and $\dot{q}_d$, defined a priori by smooth time-dependent functions (Fig. 4.8), which prescribe the desired position and velocity of the femoral head during the squat movement. In the PID controller framework, the measured variables represent the inputs, while the desired values represent the so-called setpoints (Fig. 4.9).

Since the setpoints change over time, the PID controller continuously calculates the errors q_e and $\dot{q}_e$ as the difference between the measured and setpoint values: hence, it applies corrections aimed at minimizing the errors over time by adjusting its output, which represents the generated muscle force. The force derived by the PID formulation is reported below:

$$q_{out}(t) = K_p q_e(t) + K_i \int_0^t q_e(\tau)d\tau + K_d \dot{q}_e(t) \tag{4.8}$$

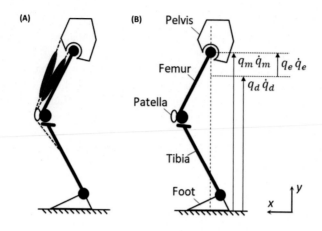

FIGURE 4.7

Explanatory 2D diagram of a lower limb on the sagittal plane. (A) Model configuration presenting the quadriceps muscle group (only two muscles are depicted); and (B) measured coordinates (q_m, $\dot{q}_m$), desired coordinates (q_d, $\dot{q}_d$), and relative errors (q_e, $\dot{q}_e$).

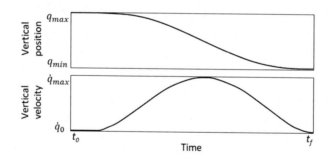

FIGURE 4.8

Explanatory charts showing the smooth time-dependent functions used to prescribe the desired vertical position (top) and the consequent velocity (bottom) of the femoral hip. The vertical position is defined between a maximum (q_{max}) and a minimum (q_{min}) limit, while the velocity ranges between zero ($\dot{q}_0$) and a maximum value ($\dot{q}_{max}$) reached at the middle of the motion duration (from t_0 up to t_f).

where $q_{out}(t)$ is the PID output, $q_e(t)$ is the position error, $\dot{q}_e(t)$ is the velocity error, and K_p, K_i, and K_d are the proportional, integral, and derivative gains, respectively. In addition, the muscle force needs to be conditioned so that only a traction force within a physiological range can be generated. In particular, an upper limit can be estimated as the physiologic cross-sectional area (PCSA) of the muscle multiplied by a maximum isometric muscle stress equal to 1 MPa (An, Kaufman, & Chao, 1989). A lower limit can be also defined, for example, equal to 1% of the maximum force value (Shelburne & Pandy, 2002), as a preload to compensate for slackening of tendons when these are included into the model. In addition, in order to take the different muscles

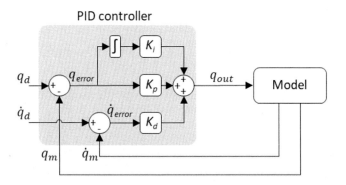

FIGURE 4.9

Closed-loop control algorithm used to estimate the muscle force.

contribution into account, the controller output force might be multiplied by the specific PCSA of each muscle normalized by the mean PCSA. In order to achieve the best performance of the controller, the K gains need to be tuned. For a limb model where tendons are not modeled and few main muscles are considered, the Ziegler–Nichols tuning method, also called continuous cycling method, can be adopted to refine the K gains (Ziegler & Nichols, 1995). This method requires to set the integral and derivative gains to zero. Subsequently, the proportional gain is increased from zero until it reaches the ultimate gain K_U, at which the controller shows stable oscillations of its output. The ultimate gain K_U and the measured oscillation period P_U are then used to set the K_p, K_i, and K_d gains by multiplying by predefined constants, that is, $K_p = 0.2 \cdot K_U$, $K_i = 0.2 \cdot K_U$ $(2/P_U)$ and $K_d = 0.2 \cdot K_U$ $(P_U/3)$. When dealing with more complex models, where, for instance, tendons and multiple muscle forces may be considered, a trial-and-error method could be carried out by first increasing K_p until a position error lower than an arbitrary acceptable tolerance is obtained. Afterwards, the K_d and K_i gains might be defined as fractions of K_p. It should be noted that, although being computationally efficient and able to produce good outcomes in replicating joint kinematics, the described muscle modeling strategy implies some simplifications. For instance, unlike a Hill-type muscle model, physiological parameters that affect force generation, such as contraction velocity and actual muscle length, are not explicitly taken into account. Moreover, the approach is not able to account for the redundancy of muscles recruited to accomplish a motion task. In light of this, such a redundancy problem might indeed be solved adopting optimization algorithms able to minimize the force generated by the cocontraction of the antagonist muscles. Nevertheless, in general, the PID controller can represent a versatile tool in the context of MB simulations. It has the advantage of allowing the selection of the variables considered as inputs and outputs of the algorithm according to the type of analysis to be carried out. For instance, the measured state variable could be the muscle lengthening during a specific movement, while the controlled variable could be the generated muscle force to obtain the desired lengthening; alternatively, the measured state variable might be the trajectory of a segment in space and the controlled variable a set of three orthogonal forces applied to the segment in order to make it follow the desired trajectory.

CASE STUDIES

MULTIBODY MODEL FOR LIGAMENT BALANCING IN TOTAL KNEE ARTHROPLASTY

Background

Chronic degenerative pathologies of the knee, such as osteoarthritis, arthrosis, and osteoporosis, are among the main causes for disability, morbidity, and pain in the modern society (Zheng & Nolte, 2015). When the nonsurgical treatments are not sufficient, the implant of joint prosthesis after bone resection is usually recommended. TKA is the most common surgical procedure for the treatment of these pathologies, and it involves the replacement of the damaged articular surfaces of the knee with metal and plastic components. A successful surgical outcome depends on a number of technical factors such as adequate bone resection, correct alignment of the implant components, and, above all, adequate soft tissue balancing. In particular, ligament balancing represents a pivotal factor in postoperative patient satisfaction as well as in implant longevity (Smith, Vignos, Lenhart, Kaiser, & Thelen, 2016). Traditionally, in order to achieve an appropriate balance, surgeons can adjust implant alignment or perform soft-tissue releases. Nonetheless, these methods commonly rely on the surgeon's experience and his subjective evaluations. In light of this, the presented study aimed at the implementation of an MB model of a prothesized knee able to predict the traction forces exerted by the collateral ligaments throughout a passive knee flexion and in the presence of different implant configurations.

Methods

The multibody model of a prothesized left knee (Fig. 4.10) was created by putting together the geometries of the femur and tibia bones (Sawbones Europe AB, Malmoe, Sweden) with a total

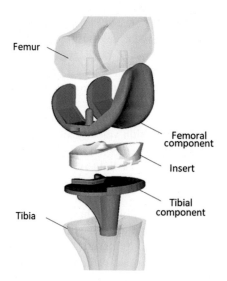

FIGURE 4.10

Exploded view showing the prothesized knee model geometry.

knee prosthesis (Gruppo BioImpianti s.r.l., Milan, Italy). The prosthesis insert consisted of polyethylene (UHMWPE) and presented a posterior stabilized design characterized by a symmetrical mediolateral shape. The insert geometry was split along its sagittal plane obtaining two symmetrical rigid bodies, in order that medial and lateral tibiofemoral contact forces could be distinguished. Average densities were assigned to the femur (1000 kg/m³), the tibia (2740 kg/m³), the implant metal components (4850 kg/m³), and the polyethylene insert (960 kg/m³).

Deformable contacts between the femoral component and each of the two insert parts were defined by means of the Hertz's law−based formulation (Eq. 4.5), considering a friction force between the articular surfaces (Stylianou et al., 2013) as well. Table 4.1 lists the adopted contacts parameters.

The medial and lateral (LCL) collateral ligaments (MCLs and LCLs, respectively) were considered in the model. Each ligament was split into three different bundles, therefore resulting in six bundles in total. This split allowed us to take into account the actual ligament structure in bundles with their different constraining contributions. Ligament bundles were modeled as nonlinear springs (Morra et al., 2008; Pianigiani et al., 2012; Putame, Pascoletti, Franceschini, Dichio, & Terzini, 2019) whose force−strain relationship is described by Eq. (4.1). The adopted stiffness parameter K for each ligament bundle is shown in Table 4.2 (Marra et al., 2015).

The zero-load length of each ligament bundle was tuned by iteratively simulating a passive knee flexion and simultaneously adjusting the ligament zero-load lengths following the strategy suggested by Guess and Razu (2017), which is based on two hypotheses concerning the passive movement of the knee throughout its range of motion: (1) every ligament bundle is stretched and (2) the force exerted by each bundle should be below 50 N. Moreover, the bundle attachment points were adjusted to be in agreement with the bundle recruitment patterns reported in literature (Blankevoort et al., 1991).

Table 4.1 Contact parameters and their values (see Eq. 4.5).

Contact parameter	Contact parameter value
Contact stiffness (K)	72,800 N/mm
Damping coefficient (C)	966 N s/mm
Maximum penetration constant (δ_{max})	0.01 mm
Exponent (e)	2.5
Static friction coefficient	0.03
Dynamic friction coefficient	0.01

Table 4.2 Ligament stiffness parameters.

Ligament bundle	K (N)
aMCL	2750
iMCL	2750
pMCL	2750
aLCL	2000
sLCL	2000
pLCL	2000

a, *Anterior;* i, *intermediate;* p, *posterior;* s, *superior.*

Once the zero-load ligament lengths were optimized, the resulting model was taken as the balanced reference configuration useful for assessing the influence of the insert thickness and tibial tray slope variation on the ligament forces. The insert thickness was therefore varied in the range ± 2 mm in steps of 1 mm, whereas the varus/valgus and posterior/anterior slopes of the tibial component were varied in the range ± 4 degrees in steps of 2 degrees. Inside the model, the insert distance from the upper surface of the tibial tray was increased and decreased to mimic a higher and lower insert thickness, respectively. Similarly, the tibial component was tilted with respect to a point centered on the tibial tray. A passive flexion simulation was performed in order to study the model behavior.

Results and discussion

The initial ligament refinement allowed us to achieve a balanced knee model, which was then considered as the baseline to investigate the impact of the different implant variations in terms of the generated ligament forces and tibiofemoral contact forces. As far as the implant positioning is concerned, Fig. 4.11 points out that valgus and varus slope changes mainly caused an increase in the lateral and medial ligament forces, respectively. In particular, higher valgus slopes mostly affected the lateral ligament, which resulted notably tauter than the medial one. In contrast, varus slopes induced the medial ligament tautening and, secondarily, a partial lateral loosening.

Conversely, changes in the slope of the tibial component on the sagittal plane did not significantly affect ligament force trends with respect to the reference condition (Fig. 4.12). It is visible from the sLCL graph that the highest difference from the reference condition corresponded to barely 1.5 times the related reference value. However, higher anterior slopes led, in accordance with the literature (Chen et al., 2015; Kebbach et al., 2019), to more considerable joint forces and vice versa.

As far as the insert thickness is concerned, both MCL and LCL were equally affected by an increase/decrease in the bundle force following insert thickening/thinning (Fig. 4.13). What turned

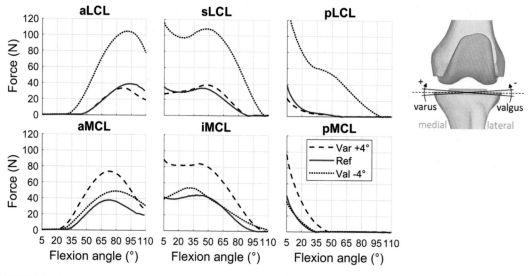

FIGURE 4.11

Impact of the varus/valgus tilt on the ligament forces.

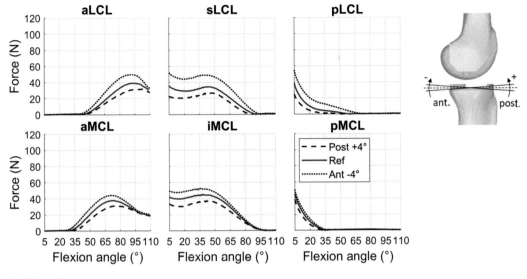

FIGURE 4.12

Impact of the anterior/posterior tilt on the ligament forces.

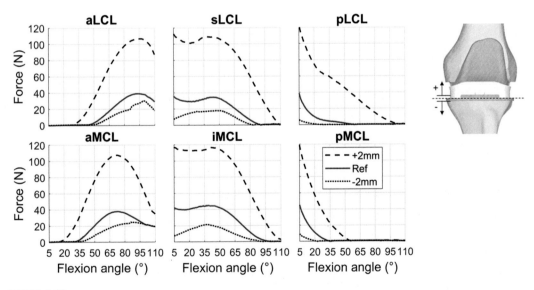

FIGURE 4.13

Impact of the insert thickness on the ligament forces.

out to be surprising was the more significant force changes consequent to the insert thickening compared to the insert thinning. This was evident in the case of the +2 mm insert for the pLCL, where a 3.3-fold increase with respect to the reference values is visible at 5 degrees of flexion.

Despite presenting some limitations, such as the oversimplification of the soft tissue structures surrounding the knee joint and the lack of the quadriceps passive resistance, the model developed in this study has proven to be able to rapidly predict significant quantitative information in agreement with experimental studies. In addition, the performed analysis could highlight the high sensitivity of the ligament balancing to changes of the insert thickness within 2 mm as well as of tibial tray slope within 4 degrees.

Conclusions

Thanks to its computational efficiency (i.e., short simulation time), the model presented here, employed in combination with a further improved software interface and the already available sensing devices, might be the first step toward the implementation of a valuable tool aimed at assisting the surgeon in both preoperative and intraoperative evaluations of ligament balancing during TKA.

EXPERIMENTAL KINEMATIC DATA FOR HUMAN ELBOW STABILITY ESTIMATION

Background

The elbow ligamentous and bony structures play an important role in the biomechanical stability of the joint. However, the individual contributions of the different structures to joint stability are not completely understood yet (Terzini et al., 2019). Computational modeling of articular joints, based on experimental data, represents a powerful tool for predicting the joint behavior, along with offering the possibility of simulating different pathological conditions overcoming the difficulties faced in performing experimental studies with cadaveric specimens (e.g., specimen unavailability). In this case study, insights are given regarding the use of kinematic experimental data within MB models. Here, the effects of ligamentous and/or bony deficiency on elbow stability have been investigated by combining an elbow joint MB model with ulna-to-humerus relative motions, quantified using a motion capture—based methodology involving clusters of markers (Panero et al., 2019).

Methods

Humerus, ulna, and radius geometries, forming a medium-size physiological human right arm, were preassembled in the extended position, and a density of 1600 kg/m^3 assigned to the three osseous components (Rahman, Cil, & Stylianou, 2016). In the extension—flexion range required for daily activities (20—120 degrees), elbow stability is guaranteed by the MCL, while a stability for lower and higher degrees of daily activity is provided by interlocking of bones. Therefore, both ligament forces and contacts between bodies need to be implemented for stability assessment. Contact forces describing the interaction between the bones of the upper limb were defined using Hertz's law——based formulation, and due to the unavailability of the articular cartilage geometry, the presence of this deformable body was fictitiously reproduced exploiting a compliant contact between the osseous components, which greatly reduced computational costs. The parameters required for the

Table 4.3 Humerus–ulna, humerus–radius, and ulna–radius contact parameters and their values (see Eq. 4.5).

Contact parameter	Contact parameter value
Contact stiffness (K)	8000 N/mm
Damping coefficient (C)	400 N s/mm
Maximum penetration constant (δ_{max})	0.001 mm
Exponent (e)	2

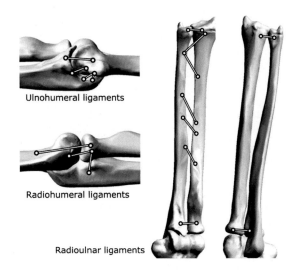

Ulnohumeral ligaments

Radiohumeral ligaments

Radioulnar ligaments

FIGURE 4.14

Ligament origins and insertions on humerus, radius, and ulna.

contact formulation, derived from literature (Fisk & Wayne, 2009; Spratley & Wayne, 2011), are listed in Table 4.3.

The MCL complex is composed of the anterior, posterior, and transverse bundles (AB, PB, and TB, respectively), with the latter not involved in maintaining the stability of the elbow (Morrey & An, 1985). Both AB and PB were included in the model using Eq. (4.1), together with the LCL, the radial collateral ligament, the interosseous membrane, and the distal radioulnar ligaments (Fig. 4.14). Localizations and stiffnesses were derived from anatomical and biomechanical data in literature (de Haan et al., 2011; Noda et al., 2009; Schuind et al., 1991; Terzini et al., 2019), adapting origin and insertion points thorough iterative procedures aimed at reaching reported ligament activation ranges and force trends (Regan, Korinek, Morrey, & An, 1991).

The designed elbow MB model was used to evaluate joint instability under physiologic and injured conditions through the application of a common dislocation maneuver. In detail, cadaveric

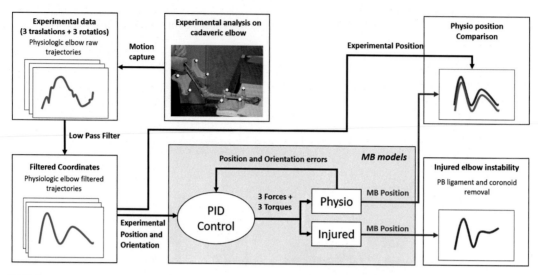

FIGURE 4.15

Workflow of the study: starting from experimental motion data (Panero et al., 2019), trajectories have been imposed as target variables within the PID-controlled derived loads applied by the surgeon. Eventually, loads have been applied to an injured elbow model to assess instability.

bone trajectories at different flexion degrees (30, 60, and 90 degrees) were acquired in an experimental session with surgeons. Filtered trajectories were thus used as target variables for a PID-controlled motion application in the MB environment, obtaining as output variables the forces and the torques applied by the surgeon on the cadaveric elbow. The obtained loads were eventually applied to an injured MB elbow model, in order to derive PB and coronoid process contributions to elbow stability (Fig. 4.15).

Results and discussion

While the unquestioned importance of the AB as a primary stabilizer of the elbow to valgus stress was investigated by several authors (Hotchkiss & Weiland, 1987), with gold standard reconstruction techniques addressing only the AB, the PB is always sacrificed in many common surgical procedures, with its role in elbow stability not clearly defined. Similarly, the coronoid process is often injured in dislocated elbows, but it is not reconstructed in many surgical procedures depending on the fracture level. In light of this, the benefits of studying elbow kinematics by MB modeling are numerous, since it allows us to investigate joint stability in different configurations of ligament injuries, bony resections, and types of motion.

A representative result of the maneuverability simulation is shown in Fig. 4.16, compared to experimental trajectories. Here, the dislocation maneuver has been performed at 30 degrees of flexion, and trajectories represent the relative movements with respect to the 30-degree starting point. Physiologic elbow trajectories are in good agreement with experimental ones, while injured trajectories highlight an increased instability, characterized by an increased range of motion (ROM) of

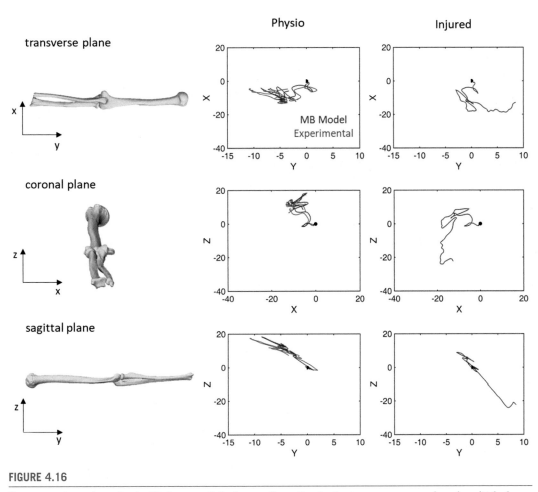

FIGURE 4.16

Representative trajectories in 30 degrees of flexion configuration in the transverse, coronal, and sagittal planes. Both superimposed experimental/numerical physio results and injured model trajectories are shown.

the joint when the same loads are applied. ROMs derived from each configuration (shown in Fig. 4.17) allow us to compare the physiologic elbow stability at different flexion degrees from the injured one. Here, both the PB ligament and coronoid process were removed, causing an instability increment, with a preponderant effect at lower degrees.

Conclusions

The presented case study confirms the great synergy between experimental kinematic data and MB orthopedic model implementation. Thanks to reference point trajectories acquired through motion capture on physiological joints, loads required to obtain movements can be derived and exploited

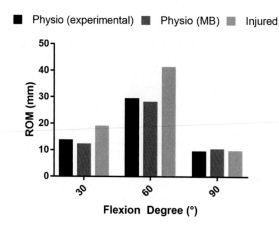

FIGURE 4.17

ROMs computed as the movement amplitude due to load application under the experimental physiological condition in the physiological and injured elbow models. Results are reported for the three flexion degrees analyzed.

to evaluate the ROM of pathological joints. Here, as a representative example, the effect of a ligamentous and bony deficiency has been evaluated, highlighting the relative contribution of ligamentous/bony structures to elbow stability at lower flexion degrees.

DESIGN OF EXPERIMENT FOR PROSTHETIC HIP RANGE OF MOTION ESTIMATION

Background

With an overall incidence of 2%−3% in primary surgery patients, dislocation is a persistent problem in total hip arthroplasty (Morrey, 1997). The occurrence of this event is dictated by several factors, including (1) acetabular cup geometry and orientation, (2) stem neck diameter, and (3) stem anteversion. Clinically, only the effects of coupled factors on ROM can be observed, while factorial in vitro and in silico experiments allow us to isolate each factor contribution. This type of analysis finds its natural collocation in the MB modeling context, where factorial experiments can be implemented, ensuring, in addition, a very low computational cost (Pascoletti et al., 2018). Within this framework, loads and displacements are applied to predict the resulting joint kinematics and dynamics. In addition, the possibility of detecting contacts and/or interferences between bodies allows the computation of the resulting ROM. Commonly, the criteria adopted for evaluation include the ROM from impingement to the onset of subluxation, together with the resisting moment increase during the dislocation event. The latter is rarely considered because it requires the simulation of both bone-prosthesis geometries and forces acting on the hip joint, such as articular and muscle forces. The proposed MB model was built aiming at determining, through a multivariate analysis, the influence of acetabular cup inclination and anteversion on the ROM.

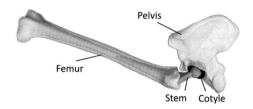

FIGURE 4.18

Prosthetic hip model.

Methods

Geometries of the femur and the pelvis were obtained from CT scans through a segmentation process, and a prosthetic hip was virtually implanted with an initial anteversion of 20 degrees (Fig. 4.18).

The prosthetic components were completely constrained to the respective bones (i.e., the stem to the femur and the cotyle to the acetabulum). The pelvis was left free to translate along three orthogonal directions, while the femur, positioned at 90-degree flexion, was subjected to internal/external rotations in order to simulate a typical dislocating movement (e.g., the sit-to-stand/stand-to-sit activity). Contact functions were defined between the prosthetic components (the stem and the cotyle), between bones (the femur and the pelvis), and between the bones and the prosthetic components. The implemented functions followed a simplified Hertz's law—based formulation, which neglects the damping effects:

$$F_C = K\delta^e \tag{4.9}$$

where K is the contact stiffness constant, δ is the penetration depth, and e is the nonlinear power exponent. Stiffness and nonlinear power exponents were determined through experimental tests on head—cotyle assemblies (Zanetti, Bignardi, & Audenino, 2012), obtaining e equal to 1.2 and K equal to 8800 N/mm. The impingement detection was implemented through sensors able to monitor the contacts between bodies and, thus, to trigger the simulation end when the contact occurred. To exploit the potential of MB methodology, a factorial design was implemented, able to assess 225 prosthetic positions obtained through the combination of 15 acetabular anteversions (−15 to 55 degrees in steps of 5 degrees) and 15 acetabular inclinations (0−70 degrees in steps of 5 degrees) (Fig. 4.19), considering a 28 mm head size as a benchmark. The maximum allowed internal/external rotations resulting from each configuration were automatically collected into the MB software framework.

Results and discussion

The analysis of the whole set of 225 different configurations required few minutes on a standard PC (i7-6500U CPU and 8 GB RAM), confirming the low computational costs of the MB methodology. The computed maximum allowed internal and external rotations versus acetabular anteversion/inclination are reported in Figs. 4.20 and 4.21, respectively. Each dot on the surface represents the given anteversion/inclination combination: a total of 225 dots are therefore shown.

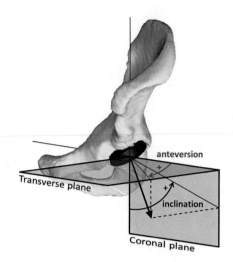

FIGURE 4.19

Sign convention and planes on which anteversion and inclination angles lie.

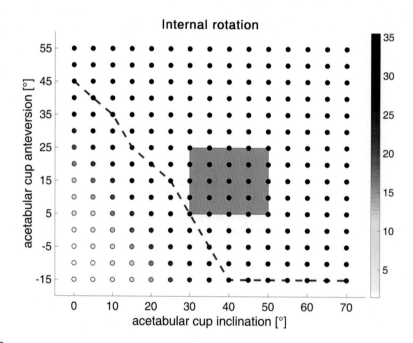

FIGURE 4.20

Maximum allowed internal rotation (in degrees) at 90-degree flexion vs acetabular cup anteversion and acetabular cup inclinations. The red line highlights the transition from prosthetic impingement to bone impingement. The shaded box corresponds to Lewinnek's "safety area" (Lewinnek, Lewis, Tarr, Compere, & Zimmerman, 1978).

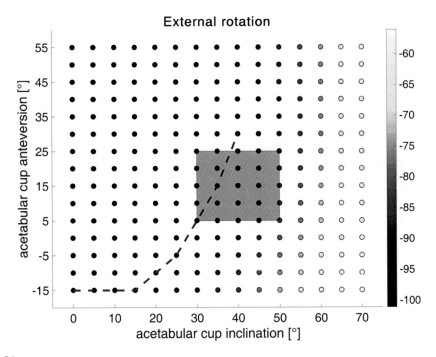

FIGURE 4.21

Maximum allowed external rotation (in degrees) at 90-degree flexion versus acetabular cup anteversion and acetabular cup inclination. The red line highlights the transition from prosthetic impingement to bone impingement. The shaded box corresponds to Lewinnek's "safety area" (Lewinnek et al., 1978).

As is visible, some anteversion/inclination couples produced interference even in the absence of femoral rotation, and, as a consequence, all the respective dots have the same value. The red line marks the transition from prosthetic impingement to bone impingement. The explanation for this behavior is that as soon as the bone impingement occurs (at +35 degrees for the internal rotation and at −101 degrees for the external rotation), the acetabular cup implant position becomes irrelevant. In the internal rotation movement, the acetabular anteversion range where prosthetic impingement occurs is wider for lower inclination angles. As an example, considering 0 degree of inclination, the prosthetic impingement is triggered at 45 degrees of anteversion, while at 30 degrees of inclination, internal rotation ROM decreases considerably. In the external rotation movement, 0−40 degrees of inclination produces the bone impingement condition, and referring to this range, the acetabular anteversion "transition angle" increases as the inclination grows (red line in Fig. 4.21). As an example, the acetabular anteversion "transition angle" is equal to 5 degrees in the case of 30 degrees of inclination, while it reaches 30 degrees for a 40-degree inclination. In general, it is advisable to keep the acetabular cup inclination below 50 degrees, in order to guarantee an higher external rotation ROM, but it should also be more than 15 degrees, in order to ensure a sufficient internal rotation ROM. Once this range is set, an anteversion equal to 25 degrees allows the optimization of the ROM. However, the absolute ranges for acetabular inclination/anteversion might be arbitrary (Yoshimine, 2006), since these two variables are strongly related. Low inclinations need to be coupled to high anteversions in order to widen the internal rotation ROM. To compare this study results with "golden rules" typically applied in the orthopedic practice, Figs. 4.20 and 4.21 also report a Lewinnek's "safety area" (Lewinnek et al., 1978), which is delimited by an inclination ranging from 30 to 50 degrees, and an anteversion ranging from 5 to 25 degrees. According to the results of this study, the "safety area" guarantees a peak internal rotation of 35 degrees (Fig. 4.20), and at least −82 degrees of external rotation (which is 18% less than the maximum external rotation obtainable with a 30-degree acetabular inclination, as visible in Fig. 4.21).

Conclusions

The MB framework presented here confirms the great potential of the methodology in the study of patient-specific prosthetic component positioning and design. The proposed model could be easily extended in order to consider a number of other parameters (such as femoral head size, the acetabular cup depth, femoral neck lengths, neck−shaft angle) but still maintaining computational costs at the minimum.

IMPACT OF THE MODULAR HIP IMPLANT DESIGN ON REACTION FORCES AT THE NECK−STEM JOINT DURING WALKING

Background

Nowadays, total hip replacement surgery represents one of the most common orthopedic operations. Driven by the developments in the surgical technique, hip prosthesis has become more and more sophisticated, including the introduction of modular designs, which allow patient-specific customization thanks to the possibility of combining implant components of different sizes, thus facilitating the application of different neck offsets, head diameters, and stem lengths. However, in spite of the intrinsic flexibility of the modular implants, they involve the presence of additional couplings, for example, between neck and stem, by means of taper junctions, which increase the risk of wear phenomena such as the trunnionosis, which accounts for up to 3% of the total revision surgeries (Cantrell, Samuel, Sultan, Acuña, & Kamath, 2019). The onset of trunnionosis depends on a variety of concomitant factors, including implant design and loading conditions. The relationship between such factors and trunnionosis is still not well known (Baker, French, Brian, Thomas, & Davis, 2021; Mistry et al., 2016). Therefore, in this case study, an MB model was created to investigate, by means of a parametric analysis, the effect of two geometric parameters, the head diameter and neck offset, on the reaction loads generated at the neck−stem joint of a hip modular implant.

Methods

The MB model of a lower left limb was created based on standard geometries of pelvis, femur, and tibia (Sawbones, Malmoe, Sweden), also including a schematic geometry of a modular hip implant consisting of four main components: polyethylene acetabular cup, ceramic head, metal stem, and metal neck connecting the head to the stem (Fig. 4.22A). The acetabular cup and the pelvis were

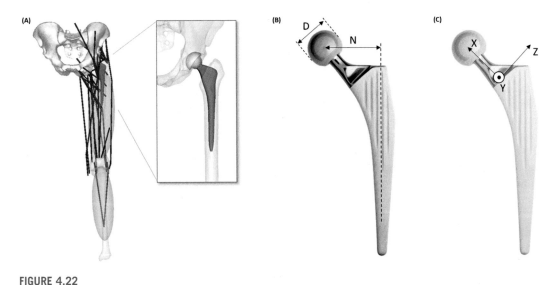

FIGURE 4.22

Multibody model of the lower limb. (A) Model showing the muscular actuators (*red lines*) and the integrated hip implant geometries (*magnification box*); (B) representative hip implant showing the studied design parameters, that is, the head diameter (*D*) and the neck offset (*N*); and (C) representative hip implant showing the reference system defined for the measurement of the reaction forces and torques at the neck–stem connection.

considered as a single body, whereas the head, neck, and stem were connected to each other by fixed constraints. A contact pair was defined between the acetabular cup and the head by means of Hertz's law–based formulation. Bony geometries were scaled to fit patient-specific data reported in the literature (Bergmann et al., 2001; Heller et al., 2001; Schwachmeyer et al., 2013). In addition, ellipsoidal bodies were added to the femur and tibia segments in order to obtain coherent mass and inertial properties. Mass values were computed as percentage of body weight for a subject of 85 kg. Moreover, a revolute joint was defined between femur and tibia, to which the patella was rigidly attached. Muscles were modeled using 33 linear actuators, each of which was controlled by a PID algorithm. Specifically, an inverse kinematic analysis was carried out driving the model by means of a set of motion agents connected to the limb segments through bushing elements and whose trajectories, computed relative to the pelvis, were determined based on experimental data of a walking cycle (Bergmann et al., 2001; Heller et al., 2001; Schwachmeyer et al., 2013). During this phase, the lengthening of the muscles was recorded. Subsequently, a forward dynamic analysis was performed by deactivating the motion agents and applying reaction forces at the ankle while the pelvis is fixed in space. During this phase, the PID algorithm takes the previously recorded length variations of each muscle as setpoints and controls the traction force generated by the actuators in order to minimize the difference between the recorded muscle length and the actual measured length. A parametric analysis was performed by modifying the offset distance (20, 24, and 40 mm) and the head diameter (22, 28, and 44 mm) of the hip implant (Fig. 4.22B). For each simulation trial, the reaction forces and torques generated at the joint between the neck and stem components were measured relative to a predefined reference system (Fig. 4.22C).

Results and discussion

With reference to the initial configuration of the model (neck offset = 24 mm, head diameter 28 mm), results revealed a maximum magnitude of the measured reaction force at the neck−stem joint equal to 4.6 times the body weight (BW). In detail, looking separately at the three components of the reaction force (Fig. 4.23), the highest one resulted along the X direction, which coincides with the neck axis, followed by the Z component, and, lastly, by the Y component. In summary, the amplitude of the force components varied inversely with respect to the offset length. In particular, the X component of the force assumed values from 3.5 up to 4 BW for offset lengths of 40 and 20 mm, respectively. This inverse relationship between the reaction force amplitude and neck offset length can be explained by a corresponding variation of the muscle level arms. A higher offset implies a longer muscle lever arm: hence, lower muscle forces will generate the torque necessary

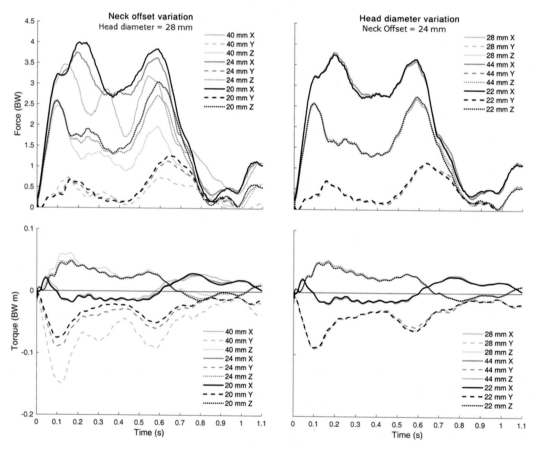

FIGURE 4.23

Reaction forces and torques measured at the neck−stem joint by varying the neck offset (head diameter = 28 mm) and head diameter (neck offset = 24 mm) of the hip implant.

to accomplish the prescribed walking motion, and vice versa. Regarding the computed reaction torques, the highest absolute value of 0.15 BW/m was measured around the Y axis, which is initially parallel to the adduction/abduction axis of the hip, with a simulated offset length equal to 40 mm. This value decreases to about 0.08 BW/m when the offset length is decreased to 20 mm. Torque components around the X and Z axes resulted in less than 0.07 BW/m, with the Z component reaching slightly higher values than the X one, especially for an offset length of 40 mm. In light of the previous considerations, a longer neck offset might reduce all the reaction forces at the neck—stem joint, although at the expense of a higher reaction torque around the adduction/abduction joint axis. Conversely, analyses that involved the variation of the head diameter showed that this variation did not importantly affect either the reaction forces or torques. Such behavior derives from a less pronounced effect of the diameter variation on the muscle spatial configuration. In this perspective, it should be noticed that the relative position of the center of the implant head with respect to the pelvis remained unaltered in spite of the head diameter variations. Therefore, the overall muscle configuration was not affected by these changes.

Conclusions

The presented parametric MB model has proven to be a powerful tool for the biomechanical investigation of the reaction loads generated at the component-coupling interface of a modular hip implant. The model, characterized by a high versatility, can be exploited to perform numerically efficient design-specific as well as patient-specific comparative studies involving a variety of implant geometries or motions, along with offering the possibility of producing realistic boundary conditions for detailed finite-element analyses.

REFERENCES

An, K. N., Kaufman, K. R., & Chao, E. Y. S. (1989). Physiological considerations of muscle force through the elbow joint. *Journal of Biomechanics, 22*(11–12), 1249–1256. Available from https://doi.org/10.1016/0021-9290(89)90227-3.

Armand, M., Grupp, R., Murphy, R., Hegman, R., Armiger, R., & Taylor, R. (2018). Biomechanical guidance system for periacetabular osteotomy. *Advances in Experimental Medicine and Biology, 1093*, 169–179.

Baker, E., French, C., Brian, P., Thomas, J., & Davis, C. M. (2021). Impending trunnion failure: An uncommon radiographic presentation of total hip arthroplasty failure. *Arthroplasty Today, 7*, 230–234. Available from https://doi.org/10.1016/j.artd.2020.12.015.

Bergmann, G., Deuretzbacher, G., Heller, M., Graichen, F., Rohlmann, A., Strauss, J., & Duda, G. N. (2001). Hip contact forces and gait patterns from routine activities. *Journal of Biomechanics, 34*(7), 859–871. Available from https://doi.org/10.1016/S0021-9290(01)00040-9.

Bersini, S., Sansone, V., & Frigo, C. A. (2016). A dynamic multibody model of the physiological knee to predict internal loads during movement in gravitational field. *Computer Methods in Biomechanics and Biomedical Engineering, 19*(5), 571–579. Available from https://doi.org/10.1080/10255842.2015.1051972.

Blankevoort, L., Huiskes, R., & de Lange, A. (1991). Recruitment of knee joint ligaments. *Journal of Biomechanical Engineering, 113*(1), 94. Available from https://doi.org/10.1115/1.2894090.

Cantrell, W. A., Samuel, L. T., Sultan, A. A., Acuña, A. J., & Kamath, A. F. (2019). Operative times have remained stable for total hip arthroplasty for >15 years systematic review of 630,675 procedures. *JBJS Open Access, 4*(4), e0047. Available from https://doi.org/10.2106/JBJS.OA.19.00047.

Chen, Z., Wang, L., Liu, Y., He, J., Lian, Q., Li, D., & Jin, Z. (2015). Effect of component mal-rotation on knee loading in total knee arthroplasty using multi-body dynamics modeling under a simulated walking gait. *Journal of Orthopaedic Research, 33*(9), 1287−1296. Available from https://doi.org/10.1002/jor.22908.

Chen, Z., Zhang, X., Ardestani, M. M., Wang, L., Liu, Y., Lian, Q., . . . Jin, Z. (2014). Prediction of in vivo joint mechanics of an artificial knee implant using rigid multi-body dynamics with elastic contacts. *Proceedings of the Institution of Mechanical Engineers, Part H: Journal of Engineering in Medicine, 228* (6), 564−575. Available from https://doi.org/10.1177/0954411914537476.

de Haan, J., Schep, N. W. L., Eygendaal, D., Kleinrensink, G.-J., Tuinebreijer, W. E., & den Hartog, D. (2011). Stability of the elbow joint: Relevant anatomy and clinical implications of in vitro biomechanical studies. *The Open Orthopaedics Journal, 5*, 168−176. Available from https://doi.org/10.2174/1874325001105010168.

Fisk, J. P., & Wayne, J. S. (2009). Development and validation of a computational musculoskeletal model of the elbow and forearm. *Annals of Biomedical Engineering, 37*(4), 803−812. Available from https://doi.org/10.1007/s10439-009-9637-x.

Fitzpatrick, C. K., Baldwin, M. A., Clary, C. W., Maletsky, L. P., & Rullkoetter, P. J. (2012). Evaluating knee replacement mechanics during ADL with PID-controlled dynamic finite element analysis. *Computer Methods in Biomechanics and Biomedical Engineering, 17*(4), 360−369. Available from https://doi.org/10.1080/10255842.2012.684242.

Flores, P. (2011). Compliant contact force approach for forward dynamic modeling and analysis of biomechanical systems. *Procedia IUTAM, 2*, 58−67. Available from https://doi.org/10.1016/j.piutam.2011.04.006.

Flores, P., Flores, P., & Lankarani, H. M. (2016). *Contact force models for multibody dynamics (Issue April). Solid Mechanics and Its Applications* (Vol. 226, pp. 1−13). Springer.

Fregly, B. J., Besier, T. F., Lloyd, D. G., Delp, S. L., Banks, S. A., Pandy, M. G., & D'Lima, D. D. (2012). Grand challenge competition to predict in vivo knee loads. *Journal of Orthopaedic Research, 30*(4), 503−513. Available from https://doi.org/10.1002/jor.22023.

Galbusera, F., Freutel, M., Dürselen, L., D'Aiuto, M., Croce, D., Villa, T., . . . Innocenti, B. (2014). Material Models and properties in the finite element analysis of knee ligaments: A literature review. *Frontiers in Bioengineering and Biotechnology, 2*, 1−11. Available from https://doi.org/10.3389/fbioe.2014.00054.

Geier, A., Kluess, D., Grawe, R., Herrmann, S., D'Lima, D., Woernle, C., & Bader, R. (2017). Dynamical analysis of dislocation-associated factors in total hip replacements by hardware-in-the-loop simulation. *Journal of Orthopaedic Research, 35*(11), 2557−2566. Available from https://doi.org/10.1002/jor.23549.

Grosu, S., Cherelle, P., Verheul, C., Vanderborght, B., & Lefeber, D. (2014). Case study on human walking during wearing a powered prosthetic device: Effectiveness of the system "human-Robot. *Advances in Mechanical Engineering, 6*, 2014. Available from https://doi.org/10.1155/2014/365265.

Guess, T. M., & Razu, S. (2017). Loading of the medial meniscus in the ACL deficient knee: A multibody computational study. *Medical Engineering & Physics, 41*(3), 26−34. Available from https://doi.org/10.1016/j.medengphy.2016.12.006.

Guess, T. M., Stylianou, A. P., & Kia, M. (2014). Concurrent prediction of muscle and tibiofemoral contact forces during treadmill gait. *Journal of Biomechanical Engineering, 136*(2), 021032. Available from https://doi.org/10.1115/1.4026359.

Guess, T. M., Thiagarajan, G., Kia, M., & Mishra, M. (2010). A subject specific multibody model of the knee with menisci. *Medical Engineering and Physics, 32*(5), 505−515. Available from https://doi.org/10.1016/j.medengphy.2010.02.020.

Harrison, S. M., Whitton, R. C., Kawcak, C. E., Stover, S. M., & Pandy, M. G. (2010). Relationship between muscle forces, joint loading and utilization of elastic strain energy in equine locomotion. *Journal of Experimental Biology, 213*(23), 3998−4009. Available from https://doi.org/10.1242/jeb.044545.

Heller, M. O., Bergmann, G., Deuretzbacher, G., Claes, L., Haas, N. P., & Duda, G. N. (2001). Influence of femoral anteversion on proximal femoral loading: Measurement and simulation in four patients. *Clinical Biomechanics, 16*(8), 644−649. Available from https://doi.org/10.1016/S0268-0033(01)00053-5.

Hosseini Nasab, S. H., Smith, C. R., Schütz, P., Postolka, B., List, R., & Taylor, W. R. (2019). Elongation patterns of the collateral ligaments after total knee arthroplasty are dominated by the knee flexion angle. *Frontiers in Bioengineering and Biotechnology*, 7, 1−12. Available from https://doi.org/10.3389/fbioe.2019.00323.

Hotchkiss, R. N., & Weiland, A. J. (1987). Valgus stability of the elbow. *Journal of Orthopaedic Research: Official Publication of the Orthopaedic Research Society*, 5(3), 372−377. Available from https://doi.org/10.1002/jor.1100050309.

Hu, J., Chen, Z., Xin, H., Zhang, Q., & Jin, Z. (2018). Musculoskeletal multibody dynamics simulation of the contact mechanics and kinematics of a natural knee joint during a walking cycle. *Proceedings of the Institution of Mechanical Engineers, Part H: Journal of Engineering in Medicine*, 232(5), 508−519. Available from https://doi.org/10.1177/0954411918767695.

Hunt, K. H., & Crossley, F. R. E. (1975). Coefficient of restitution interpreted as damping in vibroimpact. *Journal of Applied Mechanics*, 42(2), 440−445. Available from https://doi.org/10.1115/1.3423596.

Kang, K.-T., Koh, Y.-G., Jung, M., Nam, J.-H., Son, J., Lee, Y. H., ... Kim, S.-H. (2017). The effects of posterior cruciate ligament deficiency on posterolateral corner structures under gait- and squat-loading conditions. *Bone & Joint Research*, 6(1), 31−42. Available from https://doi.org/10.1302/2046-3758.61.bjr-2016-0184.r1.

Kebbach, M., Grawe, R., Geier, A., Winter, E., Bergschmidt, P., Kluess, D., ... Bader, R. (2019). Effect of surgical parameters on the biomechanical behaviour of bicondylar total knee endoprostheses − A robot-assisted test method based on a musculoskeletal model. *Scientific Reports*, 9(1), 14504. Available from https://doi.org/10.1038/s41598-019-50399-3.

Lewinnek, G. E., Lewis, J. L., Tarr, R., Compere, C. L., & Zimmerman, J. R. (1978). Dislocations after total hip-replacement arthroplasties. *The Journal of Bone and Joint Surgery. American Volume*, 60(2), 217−220.

Li, G., Gil, J., Kanamori, A., & Woo, S. L.-Y. (1999). A validated three-dimensional computational model of a human knee joint. *Journal of Biomechanical Engineering*, 121(6), 657. Available from https://doi.org/10.1115/1.2800871.

Mahadas, S., Mahadas, K., & Hung, G. K. (2019). Biomechanics of the golf swing using OpenSim. *Computers in Biology and Medicine*, 105, 39−45. Available from https://doi.org/10.1016/j.compbiomed.2018.12.002.

Marra, M. A., Vanheule, V., Fluit, R., Koopman, B. H. F. J. M., Rasmussen, J., Verdonschot, N., & Andersen, M. S. (2015). A subject-specific musculoskeletal modeling framework to predict in vivo mechanics of total knee arthroplasty. *Journal of Biomechanical Engineering*, 137(2), 020904. Available from https://doi.org/10.1115/1.4029258.

Mistry, J. B., Chughtai, M., Elmallah, R. K., Diedrich, A., Le, S., Thomas, M., & Mont, M. A. (2016). Trunnionosis in total hip arthroplasty: A review. *Journal of Orthopaedics and Traumatology*, 17, 1−6. Available from https://doi.org/10.1007/s10195-016-0391-1.

Morra, E. A., Rosca, M., Greenwald, J. F., & Greenwald, A. S. (2008). The influence of contemporary knee design on high flexion: A kinematic comparison with the normal knee. *The Journal of Bone and Joint Surgery-American*, 90(Suppl. 4), 195−201. Available from https://doi.org/10.2106/JBJS.H.00817.

Morrey, B. F. (1997). Difficult complications after hip joint replacement. Dislocation. *Clinical Orthopaedics and Related Research*, 344, 179−187.

Morrey, B. F., & An, K. N. (1985). Functional anatomy of the ligaments of the elbow. *Clinical Orthopaedics and Related Research*, 201, 84−90.

Müller, J. H., Zakaria, T., van der Merwe, W., & D'Angelo, F. (2016). Computational modelling of mobile bearing TKA anterior−posterior dislocation. *Computer Methods in Biomechanics and Biomedical Engineering*, 19(5), 549−562. Available from https://doi.org/10.1080/10255842.2015.1045499.

Navacchia, A., Rullkoetter, P. J., Schütz, P., List, R. B., Fitzpatrick, C. K., & Shelburne, K. B. (2016). Subject-specific modeling of muscle force and knee contact in total knee arthroplasty. *Journal of Orthopaedic Research*, 34(9), 1576−1587. Available from https://doi.org/10.1002/jor.23171.

Noda, K., Goto, A., Murase, T., Sugamoto, K., Yoshikawa, H., & Moritomo, H. (2009). Interosseous membrane of the forearm: An anatomical study of ligament attachment locations. *The Journal of Hand Surgery*, 34(3), 415−422. Available from https://doi.org/10.1016/j.jhsa.2008.10.025.

Panero, E., Gastaldi, L., Terzini, M., Bignardi, C., Sard, A., & Pastorelli, S. (2019). Biomechanical role and motion contribution of ligaments and bony constraints in the elbow stability: A preliminary study. *Bioengineering*, *6*(3), 68. Available from https://doi.org/10.3390/bioengineering6030068.

Pascoletti, G., Cianetti, F., Putame, G., Terzini, M., & Zanetti, E. M. (2018). Numerical simulation of an intra-medullary elastic nail: Expansion phase and load-bearing behavior. *Frontiers in Bioengineering and Biotechnology*, *6*, 1–11. Available from https://doi.org/10.3389/fbioe.2018.00174.

Pianigiani, S., Chevalier, Y., Labey, L., Pascale, V., & Innocenti, B. (2012). Tibio-femoral kinematics in different total knee arthroplasty designs during a loaded squat: A numerical sensitivity study. *Journal of Biomechanics*, *45*(13), 2315–2323. Available from https://doi.org/10.1016/j.jbiomech.2012.06.014.

Putame, G., Pascoletti, G., Franceschini, G., Dichio, G., & Terzini, M. (2019). *Prosthetic hip ROM from multibody software simulation. Annual international conference of the IEEE engineering in medicine and biology society, 2019* (pp. 5386–5389). IEEE.

Putame, G., Terzini, M., Bignardi, C., Beale, B., Hulse, D., Zanetti, E., & Audenino, A. (2019). Surgical treatments for canine anterior cruciate ligament rupture: Assessing functional recovery through multibody comparative analysis. *Frontiers in Bioengineering and Biotechnology*, *7*, 1–11. Available from https://doi.org/10.3389/fbioe.2019.00180.

Rahman, M., Cil, A., & Stylianou, A. P. (2016). Prediction of elbow joint contact mechanics in the multibody framework. *Medical Engineering & Physics*, *38*(3), 257–266. Available from https://doi.org/10.1016/j.medengphy.2015.12.012.

Rahman, M., Sharifi Renani, M., Cil, A., & Stylianou, A. (2018). Musculoskeletal model development of the elbow joint with an experimental evaluation. *Bioengineering*, *5*(2), 31. Available from https://doi.org/10.3390/bioengineering5020031.

Regan, W. D., Korinek, S. L., Morrey, B. F., & An, K. N. (1991). Biomechanical study of ligaments around the elbow joint. *Clinical Orthopaedics and Related Research*, *271*, 170–179.

Schiehlen, W. (1997). Multibody system dynamics: Roots and perspectives. *Multibody System Dynamics*, *1*(2), 149–188. Available from https://doi.org/10.1023/A:1009745432698.

Schuind, F., An, K. N., Berglund, L., Rey, R., Cooney, W. P., 3rd, Linscheid, R. L., & Chao, E. Y. (1991). The distal radioulnar ligaments: A biomechanical study. *The Journal of Hand Surgery*, *16*(6), 1106–1114. Available from https://doi.org/10.1016/s0363-5023(10)80075-9.

Schwachmeyer, V., Damm, P., Bender, A., Dymke, J., Graichen, F., & Bergmann, G. (2013). In vivo hip joint loading during post-operative physiotherapeutic exercises. *PLoS One*, *8*(10), e77807. Available from https://doi.org/10.1371/journal.pone.0077807.

Seth, A., Hicks, J. L., Uchida, T. K., Habib, A., Dembia, C. L., Dunne, J. J., ... Delp, S. L. (2018). OpenSim: Simulating musculoskeletal dynamics and neuromuscular control to study human and animal movement. *PLoS Computational Biology*, *14*(7), e1006223. Available from https://doi.org/10.1371/journal.pcbi.1006223.

Shelburne, K. B., & Pandy, M. G. (2002). A dynamic model of the knee and lower limb for simulating rising movements. *Computer Methods in Biomechanics and Biomedical Engineering*, *5*(2), 149–159. Available from https://doi.org/10.1080/10255840290010265.

Sherman, M. A., Seth, A., & Delp, S. L. (2011). Simbody: Multibody dynamics for biomedical research. *Procedia IUTAM*, *2*, 241–261. Available from https://doi.org/10.1016/j.piutam.2011.04.023.

Smith, C. R., Vignos, M. F., Lenhart, R. L., Kaiser, J., & Thelen, D. G. (2016). The influence of component alignment and ligament properties on tibiofemoral contact forces in total knee replacement. *Journal of Biomechanical Engineering*, *138*(2), 021017. Available from https://doi.org/10.1115/1.4032464.

Spratley, E. M., & Wayne, J. S. (2011). Computational model of the human elbow and forearm: Application to complex varus instability. *Annals of Biomedical Engineering*, *39*(3), 1084–1091. Available from https://doi.org/10.1007/s10439-010-0224-y.

Stylianou, A. P., Guess, T. M., & Kia, M. (2013). Multibody muscle driven model of an instrumented prosthetic knee during squat and toe rise motions. *Journal of Biomechanical Engineering*, *135*(4), 041008. Available from https://doi.org/10.1115/1.4023982.

Terzini, M., Zanetti, E. M., Audenino, A. L., Putame, G., Gastaldi, L., Pastorelli, S., ... Bignardi, C. (2019). Multibody modelling of ligamentous and bony stabilizers in the human elbow. *Muscle Ligaments and Tendons Journal*, *07*(04), 493. Available from https://doi.org/10.32098/mltj.04.2017.03.

Trent, P. S., Walker, P. S., & Wolf, B. (1976). Ligament length patterns, strength, and rotational axes of the knee joint. *Clinical Orthopaedics and Related Research*, *117*, 263−270. Available from https://doi.org/10.1097/00003086-197606000-00034.

Twiggs, J. G., Wakelin, E. A., Roe, J. P., Dickison, D. M., Fritsch, B. A., Miles, B. P., & Ruys, A. J. (2018). Patient-specific simulated dynamics after total knee arthroplasty correlate with patient-reported outcomes. *The Journal of Arthroplasty*, *33*(9), 2843−2850. Available from https://doi.org/10.1016/j.arth.2018.04.035.

Viceconti, M., Ascani, D., & Mazzà, C. (2019). Pre-operative prediction of soft tissue balancing in knee arthroplasty part 1: Effect of surgical parameters during level walking. *Journal of Orthopaedic Research*, *37*(7), 1537−1545. Available from https://doi.org/10.1002/jor.24289.

Wismans, J., Veldpaus, F., Janssen, J., Huson, A., & Struben, P. (1980). A three-dimensional mathematical model of the knee-joint. *Journal of Biomechanics*, *13*(8), 677−685. Available from https://doi.org/10.1016/0021-9290(80)90354-1.

Yoshimine, F. (2006). The safe-zones for combined cup and neck anteversions that fulfill the essential range of motion and their optimum combination in total hip replacements. *Journal of Biomechanics*, *39*(7), 1315−1323. Available from https://doi.org/10.1016/j.jbiomech.2005.03.008.

Zanetti, E. M., Bignardi, C., & Audenino, A. L. (2012). Human pelvis loading rig for static and dynamic stress analysis. *Acta of Bioengineering and Biomechanics*, *14*(2), 61−66. Available from https://doi.org/10.5277/abb120208.

Zheng, G., & Nolte, L. P. (2015). Computer-assisted orthopedic surgery: Current state and future perspective. *Frontiers in Surgery*, *2*, 66. Available from https://doi.org/10.3389/fsurg.2015.00066.

Ziegler, J. G., & Nichols, N. B. (1995). Optimum settings for automatic controllers. *InTech*, *42*(6), 94−100. Available from https://doi.org/10.1115/1.2899060.

FUNDAMENTALS OF MECHANOBIOLOGY

5

Graciosa Quelhas Teixeira, Yana Hoepfner and Cornelia Neidlinger-Wilke
Institute of Orthopaedic Research and Biomechanics, Ulm University, Ulm, Germany

BIOMECHANICAL SIGNALING

Mechanical forces are involved in the regulation of numerous structures and signaling molecules, including stretch-activated ion channels, caveolae, integrins, cadherins, growth factor receptors, myosin motors, cytoskeletal filaments, and nuclei or extracellular matrix (ECM) proteins (Ingber, 2006). They have been shown to contribute to changes in gene expression, adhesion, migration, and cell fate, with great impact on a balanced tissue homeostasis (Bajpai, Li, & Chen, 2021). Mechanobiology focuses on investigating the interactions between mechanical stimuli (including elasticity of the cell matrix or applied forces) and the biological response to the stimuli (from the cell to organismal level) via mechanotransduction (i.e., various mechanisms by which cells convert a mechanical stimulus into a biological activity) (Ingber, 2006; Lim, Bershadsky, & Sheetz, 2010). Cells can sense their mechanical/physical environment, being this phase called mechanoreception. After signal reception, cells perform signal transmission and the mechanical stimulus is converted into a biological reaction (Fig. 5.1). These activities are generally designated as mechanotransduction and support the basic functions of human tissues and organ systems (Bajpai et al., 2021; Ingber, 2006).

In bone cells, for instance, several cellular structures (Fig. 5.2) have been identified to contribute to sensing mechanical stimuli such as matrix strain, fluid flow–induced shear stress. Integrins connect bone cells with their ECM; therefore, the matrix–integrin–cytoskeleton pathway facilitates a fast transmission of extracellular stimuli through the cell membrane and intracellular matrix to the nucleus. The glycocalyx and primary cilium allow communication with integrins and the cytoskeleton. Membrane-bound proteins (e.g., connexins) allow exchange of molecules between adjacent cells, whereas fluid flow–sensitive ion channels (piezo, transient receptor potential vanilloid 4, etc.) allow ion exchange, including Ca^{2+} (Wittkowske, Reilly, Lacroix, & Perrault, 2016). Stretch-activated ion channels, primary cilia, and integrins are the major components of mechanotransduction (Fearing, Hernandez, Setton, & Chahine, 2018; Gilbert & Blain, 2018).

With aging, several changes occur in the musculoskeletal system, including, for instance, loss of bone mass, articular cartilage and intervertebral disc (IVD) degeneration, which are associated with structural changes of the tissues. An increase in tissue stiffness and loss of elasticity together with decreased muscle strength may contribute to pain and loss of mobility (Bajpai et al., 2021; Roberts et al., 2016). These degenerative changes of the musculoskeletal system influence the

Human Orthopaedic Biomechanics. DOI: https://doi.org/10.1016/B978-0-12-824481-4.00022-6

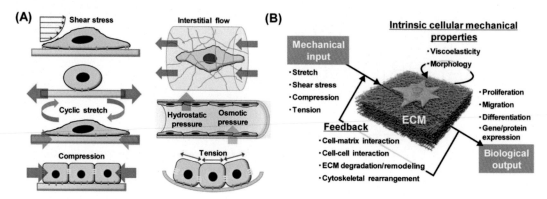

FIGURE 5.1

(A) Mechanical forces sensed by cells in the musculoskeletal system. (B) Mechanotransduction of a mechanical input into a biological output is a dynamic process involving feedback mechanisms.

From Uto, K., Tsui, J. H., DeForest, C. A., & Kim, D. H. (2017). Dynamically tunable cell culture platforms for tissue engineering and mechanobiology. Progress in Polymer Science, 65, 53–82. https://doi.org/10.1016/j.progpolymsci.2016.09.004; Polacheck, W. J., Li, R., Uzel, S. G. M., & Kamm, R. D. (2013). Microfluidic platforms for mechanobiology. Lab on a Chip, 13(12), 2252–2267. https://doi.org/10.1039/c3lc41393dA.

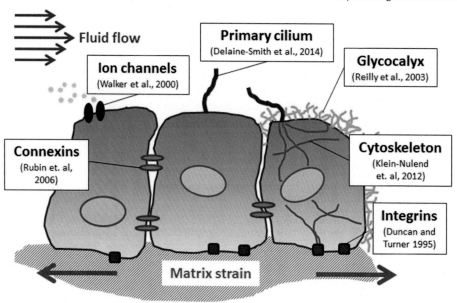

FIGURE 5.2

In bone mechanotransduction, cellular structures such as connexins, ion channels, primary cilium, glycocalyx, cytoskeleton and integrings are responsible for sensing mechanical stimuli including matrix strain and fluid flow.

From Wittkowske, C., Reilly, G. C., Lacroix, D., & Perrault, C. M. (2016). In vitro bone cell models: impact of fluid shear stress on bone formation. Frontiers in Bioengineering and Biotechnology, 4, 87. https://doi.org/10.3389/fbioe.2016.00087.

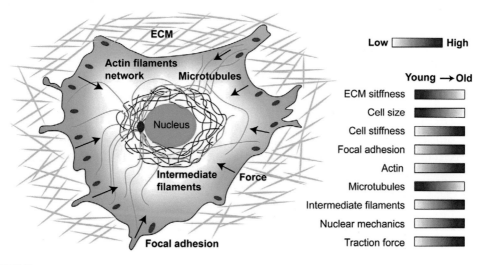

FIGURE 5.3

Cellular alterations with aging and degeneration. Several changes occur at cellular level with aging and degeneration that contribute to a dysfunctional metabolism and disease. Extracellular matrix (ECM) stiffness, cell size, microtubules and focal adhesion decrease with aging and degeneration, whereas cell stiffness, focal adhesion, actin content, presence of intermediate filaments, nuclear mechanics and traction force increase.

From Bajpai, A., Li, R., & Chen, W. (2021). The cellular mechanobiology of aging: From biology to mechanics. Annals of the New York Academy of Sciences, 1491, 3–24. https://doi.org/10.1111/nyas.14529.

mechanobiological environment and force the cells to adapt to altered extracellular biomechanical signals through mechanotransduction. Generally, aging and degeneration processes influence several cellular components of tissue-specific cells, thereby affecting the kinetics of the intracellular and intercellular forces and contributing to age-related dysregulation and disease (Fig. 5.3) (Bajpai et al., 2021). These changes often result in an imbalance of matrix turnover with proceeding tissue degradation and weakening of the mechanical properties of the musculoskeletal tissues: pathological loss of bone mass can contribute to osteoporosis and an increased risk of fractures, reduction of cartilage thickness in synovial joints may develop into osteoarthritis, and disc degeneration may lead to a decreasing structural integrity of the IVDs, loss of disc height, or tissue failure with disc herniation (Roberts et al., 2016).

Therefore, changes in cell cytoskeleton structures, mechanosensitive signaling, and the extracellular microenvironment affect cell shape and motility, as well as their metabolism (balance between catabolism and anabolism). Tissue functions are altered with aging (and early degeneration onsets) (Li, Duance, & Blain, 2008; Lim et al., 2010). The cellular response can lead to remodeling or, in cases of high mechanical demand and stress, can contribute to degenerative changes and tissue failure (Lim et al., 2010; Setton & Chen, 2006).

The knowledge of mechanobiological effects on a tissue is a determinant of its physiology, related pathologies, and regenerative potential for tissue engineering applications to halt and/or reverse degeneration (Lim et al., 2010; Setton & Chen, 2006). Successful tissue engineering is determined by three crucial components: (1) suitable cell selection, (2) suitable 3D scaffold for the cell culture, and (3) the presence of adequate signals, including biomechanical cues, to recreate the native tissue, which is fundamental for controlling cell functions toward the development of a functionalized cell–scaffold construct. In different research fields, including bone (Haffner-Luntzer, Liedert, & Ignatius, 2020), articular cartilage (Massari et al., 2019), and IVD (Molladavoodi, McMorran, & Gregory, 2020), further work on understanding the mechanisms that regulate the effect of mechanical stimuli at the cellular level is needed (Bajpai et al., 2021).

Bone, articular cartilage, and IVD will be addressed in more details in this chapter. Cells and matrix components responsible for different biomechanical functions are discussed in what follows.

MECHANICAL STIMULATION AND STUDY MODELS
BONE

The functions of the skeleton include providing load support, providing mechanical stability, protecting inner organs, and shaping the body. Bones enable complex movements, and are reservoirs of nutrients, minerals, and lipids, and long bones play an important role in hematopoiesis and in the regulation of the mineral balance (Clarke, 2008). Bone structure, morphology, and biomechanics are detailed in Chapter 7. Bone tissue is in constant adaptation to different loading situations as described in Wolff's Law (Clarke, 2008; Wolff, 1892), and adequate bone resorption and formation relies on the quality and frequency of the mechanical stimulus as investigated by Frost (1987). In bone tissue, osteocytes are the most abundant and mechanoresponsive cells. They play a key role in mechanical loading adaption and maintenance of bone mass by regulating both osteoblast (responsible for synthesizing new bone matrix) and osteoclast (responsible for bone resorption) activities (Burger & Klein-Nulen, 1999; Clarke, 2008). The interstitial bone space is filled with interstitial fluid, which moves upon loading, and osteocytes can sense, among other stimuli, the fluid flow–induced shear stress caused by tissue loading through the different sensing mechanosensing structures and receptors summarized in Fig. 5.2 (Clarke, 2008; Wittkowske et al., 2016). Bone tissue remodeling occurs in the so-called "basic molecular unit" (Fig. 5.4), and it is mainly driven by osteoblasts and osteoclasts, although there are also other mechanoresponsive cells (such as bone-lining cells or mesenchymal stem cells) (Qin & Hu, 2014). Gravity, mechanical stimuli, and matrix (micro)damage activate bone remodeling. During the activation phase of the remodeling process, osteoblasts increase the production of receptor activator of nuclear factor-κB ligand (RANKL), which induces osteoclast activation through binding to RANK receptors in osteoclast precursors, promoting their proliferation and differentiation into mature osteoclasts. Production of osteoprotegerin (OPG), which competitively can bind to RANKL, is decreased (Haffner-Luntzer et al., 2020; Wittkowske et al., 2016). Osteoclasts then bind to the bone matrix, become polarized, secrete H^+ ions and ECM-degrading enzymes such as metalloproteinases (MMPs) and cathepsin K, and actively resorb the degraded matrix during the bone-resorption phase (Clarke, 2008; Haffner-Luntzer et al., 2020). A reversal period then follows; osteoclasts undergo apoptosis, while osteoblasts are recruited, which begin to differentiate into mature cells. Transition signals stop bone

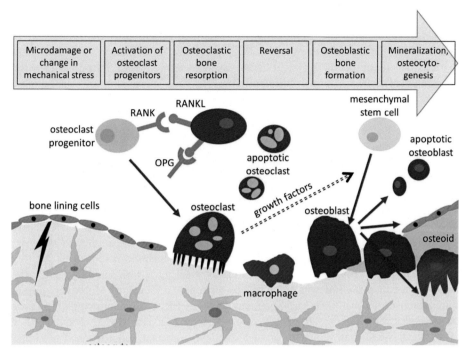

FIGURE 5.4

The "basic multicellular unit." Cells and molecules involved in different bone-remodeling phases.

From Wittkowske, C., Reilly, G. C., Lacroix, D., & Perrault, C. M. (2016). In vitro bone cell models: impact of fluid shear stress on bone formation. Frontiers in Bioengineering and Biotechnology, 4, 87. https://doi.org/10.3389/fbioe.2016.00087.

resorption and stimulate bone formation. In the formation phase, bone formation occurs: osteoblasts secrete collagen type I (COL1), as well as other proteins, including osteocalcin and alkaline phosphatase. In the termination phase, bone formation ceases, and the tissue is mineralized and returns to a resting state. Osteoblasts enter apoptosis or become bone-lining cells (Haffner-Luntzer et al., 2020; Wittkowske et al., 2016).

Degradation of the osteocyte networks may contribute to loss of bone mass, delayed mechanosensitive calcium signaling, and impaired bone mechanosensitivity, leading to an imbalance in osteoclast/osteoblast activity and tissue homeostasis (Bajpai et al., 2021; Wittkowske et al., 2016). Disruption of the balanced bone regulation triggers several diseases, including osteoporosis (increased bone turnover and loss of bone mass) or osteopetrosis (increased bone mass). Several signaling pathways—nitric oxide (NO), prostaglandin E_2 (PGE_2), Wnt, kinases (mitogen-activated protein kinase [MAPK], extracellular signal-regulated kinase 1/2, focal adhesion kinase), and Ca^{2+} signaling—are likely to be involved in osteoblast and osteoclast activities, depending on the mechanical stimulus (Qin & Hu, 2014; Wittkowske et al., 2016). Some studies are summarized in the following.

Loading and fluid flow

Bone tissue deformation (even to a small extent) can be directly sensed by the cells. Bone location, as well as the magnitude and frequency of loading, affects the tissue matrix. Bones are exposed to different strains depending on the physical activity. About 0.03%−0.15% strain with a frequency of 1−3 Hz is characteristic of normal daily activities, whereas sports activities, including running and jumping, can lead to bone strains of about 0.2%−0.35% in a range of locations, and are described to have the greatest impact on loading-induced bone remodeling (Burr et al., 1996; Fritton, McLeod, & Rubin, 2000; Lynch & Fischbach, 2014). Interestingly, failure is expected to occur at 1%−3% strain (Reilly & Burstein, 1975). However, intracellular responses have been observed at strains of around 10%, indicating that tissue-level strain must be amplified at the cellular level, with cells sensing microstructural strains near lacunae and microcracks within the bone (Lynch & Fischbach, 2014).

Tibial osteocytes have been shown to respond to both strain (600 cycles, 1 Hz) and estrogen by upregulating the expression of stimulating transient estrogen receptor (ER)-α translocation to the nucleus and transient changes in its gene expression. Reduction in ER-α number/activity is associated with lower estrogen concentration, which decreases the anabolic response to strain on bone cells (Zaman et al., 2006). In ovariectomized/postmenopausal mice with low bone mass, functional ER and Wnt/β-catenin signaling was impaired by estrogen deficiency. Moreover, physiological ulna loading increases cortical bone formation and activates ER or β-catenin signaling in the ovariectomized mice (Liedert et al., 2020). Production of cyclooxygenase (COX)-2 and PGE$_2$ induced by fluid-flow shear stress was shown to protect osteocytes against apoptosis and to contribute to bone formation through the cAMP/protein kinase A and β-catenin signaling pathways (Kitase et al., 2010). Moreover, Piezo1, one of the two mechanically activated ion channels of the Piezo family, has been shown to have a mechanosensory function essential for bone formation. In osteocyte cultures, Piezo1 was shown to be required for changes in gene expression induced by fluid shear stress (15 dynes/cm^2 oscillatory fluid shear stress, 1 Hz) (Li et al., 2008; Lim et al., 2010). Recently, mice with targeted deletion of Piezo1 or Piezo2 in either osteoblasts, osteoclasts, or chondrocytes developed severe osteoporosis with numerous spontaneous fractures, particularly the mice with Piezo1 deletion, corroborating its important role in bone remodeling (Hendrickx et al., 2021).

Vibration

Whole-body low-magnitude high-frequency vibration (LMHFV) has been recently described as a promising bone anabolic therapy. Its influence on bone cells and tissue regeneration has been recently reviewed (Steppe, Liedert, Ignatius, & Haffner-Luntzer, 2020). For instance, human cancellous bone (osteoblast-like) cells have been shown to upregulate ALP expression after LMHFV (30−60 Hz, 1.0−10 m/s^2) (Rosenberg, Levy, & Francis, 2002). Gao et al. (2017) investigated the effect of LMHFV (45 Hz, 0.3 g) in primary human osteoblasts. The results demonstrated that the vibration upregulated the expression of ALP, osteocalcin, runt-related transcription factor 2 (RUNX2), bone morphogenic protein (BMP)-2, and osteoprotegerin, and downregulated sclerostin. Osteoblastic proliferation and ECM mineralization were also increased with vibration, being the Wnt signaling shown to be involved in cytoskeletal remodeling (Gao et al., 2017). Cytoskeletal remodeling, increased cell metabolic activity, and proliferation were also observed in vibrated (45 Hz, 0.3 g) MC3T3-E1 cells and primary C57BL/6 mouse osteoblasts, compared with nonvibrated. Using a mouse model of estrogen deficiency/osteoporosis, ER-α signaling was shown to be involved in the vibration-induced effects on fracture healing (Haffner-Luntzer, Kovtun, et al., 2018; Haffner-Luntzer, Lackner, Liedert, Fischer, & Ignatius, 2018). Interestingly, Sakamoto et al. (2019) investigated the effect of vibration on osteoclastogenesis. Vibration (48.3 Hz, 0.5 g) applied to preosteoclast cell line MLO-Y4 upregulated RANKL expression via nuclear factor-κB (NF-κB) signaling (Sakamoto et al., 2019).

ARTICULAR CARTILAGE

A comprehensive summary of cartilage structure, morphology, and biomechanics can be found in Chapter 10. Briefly, cartilage is an elastic tissue that has high tear resistance and compressive elasticity. In adults, a subchondral plate separates the articular cartilage from the bone. Nutrients are transported to the condrocytes mainly by diffusion from the synovial fluid. The cartilage ECM, in which chondrocytes are starkly distributed, consists mainly of water, collagen, representing collagen type II (COL2) comprising 90%−95% of the ECM collagen, and proteoglycans, such as aggrecan (ACAN), with other noncollagenous proteins and glycoproteins present in lesser amounts, and almost no blood vessels and nerves or nociceptors. It can attract sodium ions and bind water due to the negatively charged molecular components of proteoglycans with high affinity for water molecules. Cartilage tissues present zonal variations in structure and composition, what is not addressed in this chapter. Chondrocytes occupy about 2% of the total volume of articular cartilage, and each chondrocyte is responsible for the ECM turnover in its immediate vicinity. They respond to stimuli, including mechanical loads, hydrostatic pressures, piezoelectric forces, and molecular factors. Proinflammatory cytokines such as tumor necrosis factor (TNF)-α and interleukins (IL-1 and IL-7) have been shown to play a role in both catabolism and anabolism (Huang et al., 2018; Sophia Fox, Bedi, & Rodeo, 2009).

Articular cartilage can be damaged by injury or wear and tear due to aging or abnormal mechanical stress. Moreover, diseases such as arthritis or osteoarthritis are also characterized by loss of articular cartilage and sclerosis of subchondral bone. This leads to loss of joint function, limitation of movement, and pain. At the molecular level, initial catabolic stimuli increase matrix synthesis and chondrocyte proliferative response. Chondrocytes form clusters, hypertrophic differentiation occurs, and the expression of hypertrophic markers such as RUNX2 and collagen type X is upregulated. Furthermore, proinflammatory molecules stimulate chondrocytes to produce MMPs, including collagenases MMP-1 and MMP-13, and aggrecanases such as disintegrin and metalloproteinase with thrombospondin motifs (ADAMTS)-4 and -5. Furthermore, calcium crystal deposition, osteophytes, and protrusions may contribute to pathologic cartilage degradation. Ultimately, chondrocyte apoptosis and loss of cartilage may occur (Leong, Hardin, Cobelli, & Sun, 2011; Sandell & Aigner, 2001).

Several in vitro and in vivo investigations have shown the effects of mechanical loading on the mechanotransduction of chondrocytes isolated from healthy or osteoarthritic tissues, as well as induction of an osteoarthritic environment by medium supplementation with physiological osteoarthritis concentrations of IL-1α or IL-1β. In chondrocytes, the effects of the stimuli are subsequently mediated via a series of intracellular signaling pathways, including the cytoskeletal elements, Wnt molecules, and miRNAs, that influence both cell metabolism and ECM synthesis (Fig. 5.5). All these have been reviewed in detail by several authors (Gilbert & Blain, 2018; Grodzinsky,

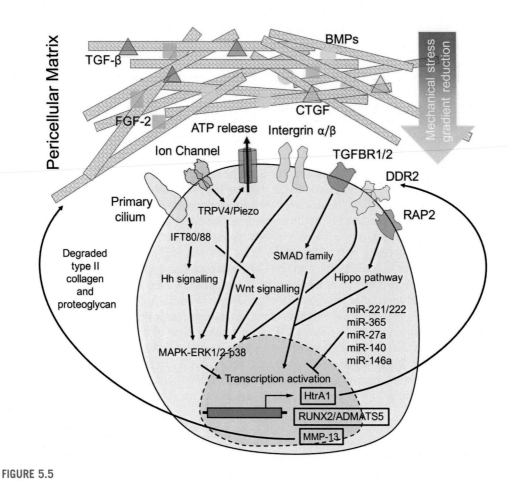

FIGURE 5.5

Molecular and intracellular pathways in regulating mechanotransduction signaling in articular chondrocytes.

From Zhao, Z., Li, Y., Wang, M., Zhao, S., Zhao, Z., & Fang, J. (2020). Mechanotransduction pathways in the regulation of cartilage chondrocyte homoeostasis. Journal of Cellular and Molecular Medicine, 24(10), 5408–5419. https://doi.org/10.1111/jcmm.15204.

Levenston, Jin, & Frank, 2000; Leong et al., 2011; Zhao et al., 2020). Chondrocyte mechanical stimulation has been associated with the activation of pathways such as MAPK, NF-κB, or Janus kinase/signal transducers and activators of transcription, and reported to regulate the expression of proinflammatory cytokines such as IL-1β, IL-6, and TNF-α, thereby identifying a mechanism by which mechanical load elicits an inflammatory effect in cartilage degeneration (Leong et al., 2011; Mariani, Pulsatelli, & Facchini, 2014). Some example studies are described in the following.

Static or dynamic compression

Application of dynamic (10% or 15%, 1 Hz) or static compression (15%) results in cytoskeletal actin remodeling in bovine chondrocytes from healthy articular cartilage (Knight, Toyoda, Lee, &

Bader, 2006). In mouse chondrocyte lines subjected to compression (15%, 1 Hz), primary cilia were shown to be involved in promoting proteoglycan synthesis, through compression-induced Ca^{2+} signaling mediated by ATP release (Wann et al., 2012). The effect of mechanical stimulation on signal transduction pathways activated by IL-1β has also been investigated in chondrocyte cultures. Chowdhury, Bader, and Lee (2001, 2003) have demonstrated that physiological dynamic compression (0%−15% strain, 1 Hz) can counteract IL-1β−induced COX-2 and inducible nitric oxide synthase (iNOS) activity, inhibiting PGE_2 and NO synthesis by human chondrocytes, possibly through a p38 MAPK−dependent pathway (Chowdhury et al., 2001, 2003). Moreover, their studies also suggest that the mechanotransduction may occur via integrin signaling (Chowdhury, Appleby, Salter, Bader, & Lee, 2006). Bovine knee cartilage explants have been subjected to 0.2 or 0.5 MPa (0.5 Hz) compression individually and in combination with IL-1α. While IL-1α stimulation alone induced collagen and proteoglycan degradation, this effect was inhibited when the IL-1α stimulation was combined with 0.5 MPa dynamic loading (Torzilli, Bhargava, Park, & Chen, 2010). In contrast, pig articular cartilage explants subjected to compression (0.1 MPa, 0.5 Hz) increased the production of NO and PGE_2 as a result of mechanical compression, suggesting that mechanical load may also induce an inflammatory phenotype in cartilage (Fermor et al., 2002).

Cyclic tensile strain

Cyclic tensile strain (CTS, 20%, 0.05 Hz) has also been shown to downregulate IL-1β−dependent NO production and to increase proteoglycan synthesis by rabbit chondrocytes (Gassner et al., 1999). In another study, rat articular chondrocytes exposed to CTS (3%, 0.25 Hz) downregulated the expression of IL-1β−induced iNOS, COX-2, MMP-9, and MMP-13, together with higher ACAN synthesis, when compared with chondrocytes stimulated only with IL-1β. In Fig. 5.6, results for COX-2 and ACAN production are depicted, also demonstrating that different periods of strain and rest influence the factors analyzed (Madhavan et al., 2006). Similar observations were later associated with the possibility of CTS acting at multiple sites within the NF-B signaling cascade to inhibit IL-1β−induced proinflammatory gene expression (Dossumbekova et al., 2007). Human osteoarthritic chondrocytes subjected to CTS (7%, 0.5 Hz) increased expression of the α2 and α5 integrin subunits, as well as an increased expression of vimentin, together with intracellular reconfiguration of the enzyme protein kinase C, implicated in regulating integrin affinity for ECM ligands (Lahiji, Polotsky, Hungerford, & Frondoza, 2004). The expression of different miRNAs has also been investigated in osteoarthritis. For instance, Yang et al. (2016) have stimulated (10%, 1 Hz) chondrocytes obtained from macroscopically normal regions of osteoarthritic cartilage and found that miR-365 was up-regulated by cyclic loading and IL-1β. Overexpression of miR-365 upregulated the expression of MMP-13 and collagen type X, whereas inhibition of miR-365 downregulated them. Moreover, this study demonstrated that histone deacetylase 4 (HDAC4), which mediates mechanical stress and inflammation in osteoarthritis, is a direct target of miR-365 (Yang et al., 2016). In vivo, application of continuous passive motion by flexion of the joint ranging between 40 and 110 degrees (0.022 Hz) to the knees of rabbits with induced arthritis inhibited the expression of catabolic mediators such as IL-1β, COX-2, and MMP-1, while inducing the expression of the antiinflammatory cytokine, IL-10, by articular cartilage chondrocytes, when compared to immobilization of the joint (Ferretti et al., 2006). One hour of daily passive joint motion (65 and 115 degrees joint flexion, 0.033 Hz) applied to the knees of osteoarthritic rats has also been shown

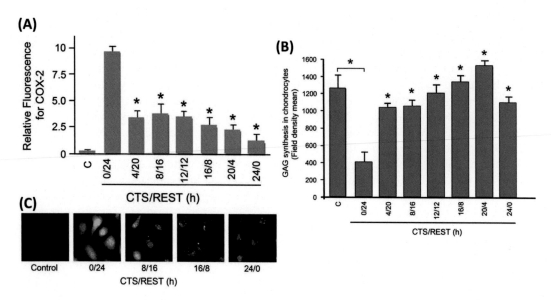

FIGURE 5.6

Effect of cyclic tensile strain (CTS) on (A) IL-1β—dependent cyclooxygenase (COX)-2 induction and (B) glycosaminoglycan (GAG) synthesis in (C) articular chondrocytes. Cells were exposed to different periods of CTS and at rest, in the constant presence of IL-1β.

Modified from Madhavan, S., Anghelina, M., Rath-Deschner, B., Wypasek, E., John, A., Deschner, J., ... Agarwal, S. (2006). Biomechanical signals exert sustained attenuation of proinflammatory gene induction in articular chondrocytes. Osteoarthritis and Cartilage, 14(10), 1023–1032. https://doi.org/10.1016/j.joca.2006.03.016.

to inhibit upregulation of MMP-3 and ADAMTS-5 by chondrocytes and to prevent changes in proteoglycan loss, in contrast to immobilization (Leong et al., 2010).

These data suggest that loading of arthritic joints may be beneficial to preserving cartilage integrity by counteracting cytokine-induced proinflammatory and catabolic effects. However, as observed for bone, overload can induce by itself catabolic mechanisms (Leong et al., 2011).

INTERVERTEBRAL DISC

In humans, the IVD enables stability, absorption, and dispersion of loads, while allowing spine's multiaxial motions, such as flexion-extension, rotation, and lateral bending. This can generate fluid flow with high magnitudes of osmotic pressures, as well as compressive, tensile, and shear stresses and strains within the IVD matrix (Fig. 5.7) (Setton & Chen, 2006). The IVD is mainly composed of water, proteoglycans, and collagen, with their relative proportions varying between its different regions, which respond differently to mechanical loading (Adams & Muir, 1976; Setton & Chen, 2006). Several studies have suggested that the IVD cells can experience fluid flow, pressure, electrokinetic changes such as streaming potentials, and deformation in tension, compression, or shear,

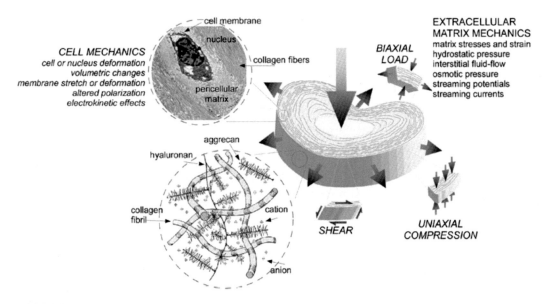

FIGURE 5.7

Illustration of mechanical loading applied to the disc (axial compressive loading represented by the *large arrow*), which can induce shearing and tensile stresses and dynamic compression. The proteoglycan–collagen rich matrix (illustrated on the bottom left) will promote the distribution of loads. This contributes to changes in hydrostatic pressure and pressure gradients, which drive interstitial fluid flow. Mechanically induced changes are transduced, activating different responses (enumerated on the top left).

From Setton, L. A., & Chen, J. (2006). Mechanobiology of the intervertebral disc and relevance to disc degeneration. The Journal of Bone and Joint Surgery, 88*(Suppl. 2), 52–57. https://doi.org/10.2106/00004623-200604002-00011.*

and depending on the IVD cell type, cells respond differently to both magnitude and frequency of the stimuli (Molladavoodi et al., 2020; Neidlinger-Wilke et al., 2014; Setton & Chen, 2006).

In adults, the cartilaginous endplate (CEP) consists of relatively thin layers of hyaline cartilage (Roberts, Menage, & Urban, 1989), being mainly exposed to compression loads (Ochia, Tencer, & Ching, 2003). It is on one side connected to the vertebral endplate, and, on the other side, to the annulus fibrosus (AF) and nucleus pulposus (NP) (Roberts et al., 1989). The CEP cells resemble chondrocytes, and its matrix consists largely of proteoglycans and collagen fibers, namely COL2, with a water content lower than that of the NP and AF (Raj, 2008; Roberts et al., 1989). The CEP has been considered the main route of diffusion for oxygen, nutrients, and residues, which occurs predominantly through passive diffusion (Holm, Maroudas, Urban, Selstam, & Nachemson, 1981; Ogata & Whiteside, 1981). The AF is formed by a concentric lamellar structure of regularly arranged COL1 fibers and about 60%–80% of water. The lamellae are interconnected by a network of elastin and fibrillin, which contribute to providing tissue mechanical support and elasticity (Adams & Muir, 1976; Setton & Chen, 2006). The AF provides lateral NP confinement and resistance to tensile and compressive stresses during physiological loading, with changes in the loading environment from more tension in the outer AF to more compression toward the NP (Eyre, 1979).

The NP is a viscoelastic material rich in ACAN entrapped in a randomly arranged COL2 fiber network, and with elastin fibers radially distributed. Of note, ACAN is responsible for mediating the osmotic pressure within the NP and the resistance to compressive loads by forming large aggregates of sulfate-rich chondroitin and keratan glycosaminoglycan chains, which are linked to a central hyaluronan protein core. Due to the high negative charge density, this mesh retains water (70%–90% of the NP wet weight, depending on the age) and limits diffusion (Adams & Roughley, 2006; Raj, 2008). NP cells, chondrocyte-like cells in adulthood, are exposed to hydrostatic pressure due to the matrix water-binding capacity and due to tensile and shear stresses during compression. Moreover, these cells can withstand the harsh IVD microenvironment with low oxygen and nutrition (Pattappa et al., 2012; Raj, 2008).

Degeneration can be triggered by various risk factors, among which mechanical stress plays an important role in favoring catabolism (Fig. 5.8). This process is characterized by cell senescence and apoptosis, as well as an inflammatory/immune response, including the production of increased

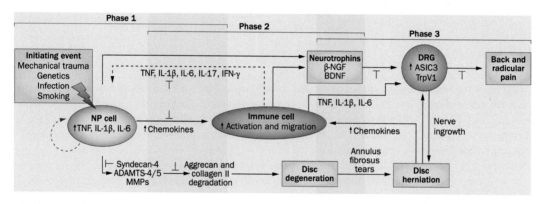

FIGURE 5.8

Summary of the different molecules produced by the different intervertebral disc or immune cells during the different phases of disc degeneration. After an initial event (often associated with mechanical trauma or stress), IVD cells upregulate the expression of inflammatory cytokines (including tumor necrosis factor [TNF], interleukin [IL]-1β and IL6) and chemokines. This induces an upregulation of catabolic molecules by disc cells (for instance, syndecan-4, a disintegrin and metalloproteinase with thrombospondin motifs [ADAMTS]-4/5 and metalloproteinases [MMPs]) and degradation of matrix components (particularly aggrecan and collagen type II). Continuous structural breakdown of matrix molecules of the NP and AF results in mechanical instability, annular tears, and, in many instances, disc herniation. In a second phase, the recruitment and infiltration of leukocytes into the disc will be activated, further amplifying the inflammatory response, and promoting neovascularization and the appearance of nociceptive nerve fibers that arise from the dorsal root ganglion (DRG). In a third phase, neurogenic factors (mainly nerve growth factor [NGF] and brain derived neurotrophic factor [BDNF]) produced by disc cells and leukocytes induce pain by the expression of pain-associated cation channels (e.g., acid-sensing ion channel [ASIC]3 and transient receptor potential cation channel subfamily V member 1 [TrpV1]). Possible therapeutic intervention sites are indicated by *red inhibitory lines.*

amounts of nitric oxide (NO) and inflammatory mediators such as PGE_2 or interleukins (IL-1, -6, and -8), among others, and immune cell recruitment. There is an increase in the production of matrix-degrading enzymes such as MMP-1, -3, and 13, ADAMTS-4 and -5, and a shift from COL2 to COL1 production by NP cells (Molinos et al., 2015; Teixeira, Barbosa, & Goncalves, 2018; Vergroesen et al., 2015). The loss of proteoglycans and water-retention capacity within the NP contributes to a pressure reduction, which increases compressive stress applied to the AF. The cells' ability to sense mechanical stimuli is affected by the organization and amount of cytoskeletal elements, particularly actin, tubulin, and vimentin. Changes in these proteins may contribute to the imbalance in ECM production, leading to a decrease in IVD height, AF bulging, and tears (Li, Jia, Duance, & Blain, 2011; Lim et al., 2010; Vergroesen et al., 2015). CEP calcification, by limiting the supply of nutrients to the IVD, also contributes to disc degeneration (Huang et al., 2018; Sophia Fox et al., 2009), and increased growth of blood vessels and nerve fibers into the IVD contributes to pain (Risbud & Shapiro, 2014).

Further details on IVD composition, biomechanics, and degeneration can be found in Chapter 12. The IVD is one of the most important tissues involved in the transmission of forces, as suggested by different authors (Molladavoodi et al., 2020; Neidlinger-Wilke et al., 2014; Setton & Chen, 2006). Several studies are summarized in the following text.

Static or dynamic compression

Static compression (0.1 MPa, 25% compressive strain) on alginate scaffolds seeded with pig AF and NP cells has been shown to upregulate the expression of ECM proteins ACAN, biglycan, COL1A1, COL2A1, decorin and lumican, and cytoskeletal protein vimentin by AF cells, after 30 hours of stimulus, in comparison to uncompressed control samples, whereas NP cells did not show any changes after stimulation (Chen, Yan, & Setton, 2004). In contrast, high dynamic load (1.3 MPa, 1 Hz) was shown to induce a catabolic effect through the upregulation of MMP-3 and MMP-13 gene expressions by rat NP and AF cells in IVD organ culture (Kurakawa et al., 2015). An in vitro study by Korecki, MacLean, and Iatridis (2008) showed that the IVD responds to dynamic compression in a magnitude-dependant manner. Bovine IVDs in organ culture stimulated with low (0.2−1 MPa, 1 Hz) vs high (0.2−2.5 MPa, 1 Hz) dynamic load displayed altered gene expression responses affected by disc region and dynamic compression magnitude, particularly for COL1 and MMP-3 (Korecki et al., 2008). Le Maitre et al. (2009) have shown that mechanosensing in human NP from nondegenerate IVDs subjected to dynamic compressive loading (0.35−0.95 MPa, 1 Hz) occurs via RGD-integrins, possibly via the $\alpha5\beta1$ integrin, while cells from degenerated discs showed a different signaling pathway, which did not seem to involve RGD-integrins (Le Maitre et al., 2009). The contribution of RGD-integrin signaling in the mechanoresponse of AF cells derived from nondegenerated IVDs, but not degenerate IVDs, was also confirmed by others (Fearing et al., 2018; Gilbert & Blain, 2018), particularly the involvement of the $\alpha5\beta1$ subtype (Kurakawa et al., 2015). Moreover, compression of agarose-encapsulated human NP cells (0.004 MPa, 1.0 Hz) for 1 hour showed that, at a pH representative of degenerated IVDs (pH 6.5), ACAN downregulation was RGD-integrin dependent, while MMP-3 upregulation was RGD-independent (Hodson, Patel, Richardson, Hoyland, & Gilbert, 2018). Recently, the role of integrin $\alpha5\beta1$ in the homeostasis of notochordal cells using rat tail functional spinal units cultured under dynamic compressive loading (1.3 MPa, 1 Hz) has also been investigated (Kanda et al., 2021).

In vivo studies in rats have shown a frequency-dependent response. The increase in frequency of dynamic compression has been shown to increase the disc height loss and a shift particularly in NP cells from expressing anabolic markers to increasing the expression of catabolic enzymes, which might be due to the increase in fluid pressurization, which accompanies an increase in loading frequency. Low-frequency dynamic compression (1 MPa, 0.01 Hz) applied to rat tail motion segments in vivo led to an upregulation of anabolic genes such as ACAN, COL1, and COL2, but also MMP-13 and ADAMTS-4 by NP cells. However, high-frequency compression (1 MPa, 1 Hz) only upregulated the proteases MMP-3, MMP-13, and ADAMTS-4. Interestingly, AF cells upregulated the expression of catabolic markers after exposure to both frequencies. This study indicates that while NP cells might be affected by both load magnitude and frequency, AF cells might be more strongly affected by load magnitude (MacLean, Lee, Alini, & Iatridis, 2004). In the same year, Walsh and Lotz (2004) also investigated the effects of dynamic compression in a mouse model in which a single IVD in the tail of each mouse was placed under dynamic compression. In this study, discs loaded with 1.3 MPa and 0.001−0.1 Hz increased proteoglycan content, but also cell apoptosis, in comparison to sham animals (Walsh & Lotz, 2004). Rabbits treated with axial compression (2.4 MPa) of the lumbar spine motion segment L4/L5 using an external fixator for up to 56 days showed an initial upregulation of COL1, COL2, biglycan, and decorin followed by a later downregulation to control levels. Fibromodulin, fibronectin, MMP-3, tissue inhibitor of metalloproteinase (TIMP)-1, and BMP-2 were continuously upregulated and the IVDs displayed signs of degeneration after 56 days of treatment, suggesting an important role of these molecules in the degeneration process (Omlor et al., 2006).

Static and dynamic compression studies have been largely applied to IVD cells, discs in organ culture, and in animal experiments. Among several relevant publications, the ones described in this chapter demonstrate that IVD cells respond to mechanical stimuli and that the response might differ depending on the magnitude of the stimulus, if the cells are isolated form healthy or generated IVDs, and if they are AF or NP cells. Compression at low magnitude increases the disc metabolic rate and enhances anabolic remodeling (Chen et al., 2004; Korecki et al., 2008). However, high magnitudes for prolonged periods can promote a shift to catabolism (Kurakawa et al., 2015).

Cyclic tensile strain

Rat AF cells exposed to an inflammatory stimulus has been shown to increase catabolic gene expression (MMP-3, MMP-13), which could be decreased by approximately 50% after exposure to both inflammatory stimulus and CTS (6% strain, 0.05 Hz) (Sowa & Agarwal, 2008). In line with these findings, stimulation of human AF cells seeded in collagen gels with cyclic strain alone (up to 8% strain, 1 Hz) has been shown to contribute to upregulation of ACAN and COL2A1 expression, as well as downregulation of MMP-3 (Neidlinger-Wilke et al., 2005). Later, CTS at 8% and 1 Hz of AF cells has also been shown to induce MAPK activation, with upregulation of proinflammatory genes including COX-2, IL-6, and IL-8 (Pratsinis et al., 2016). Li et al. (2011) have analyzed the organization and expression of cytoskeletal elements in bovine AF and NP cells after CTS stimulation (10% strain, 1 Hz) and observed F-actin reorganization in AF and NP cells subjected to tensile strain, with increased β-tubulin expression by AF cells, whereas NP cells were less responsive to strain (Li et al., 2011). In contrast, high mechanical strain (20%, 0.5 Hz) was demonstrated to induce apoptosis in rat AF cells with partial mediation by endoplasmic reticulum stress through NO production (Zhang, Zhao, Jiang, & Dai, 2011). Moreover, stretching of human AF and

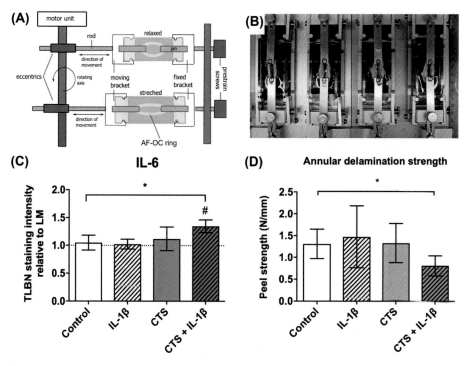

FIGURE 5.9

(A) Schematic representation of the (B) custom-made electromechanical device for the application of cyclic tensile strain (CTS) to deformable silicone dishes. (C) Interleukin (IL)-6 staining intensity quantification in the translamellar bridging network (TLBN) relative to lamella (LM) tissue ($n = 5$). (D) Annular delamination strength as a function of displacement rate (N/mm) ($n = 6$). (*) indicates a significant effect of treatment and (#) indicates a significant effect of region ($P < .05$).

Modified from Saggese, T., Teixeira, G. Q., Wade, K., Moll, L., Ignatius, A., Wilke, H. J., ... Neidlinger-Wilke, C. (2019). Georg Schmorl Prize of the German Spine Society (DWG) 2018: Combined inflammatory and mechanical stress weakens the annulus fibrosus: Evidences from a loaded bovine AF organ culture. European Spine Journal, 28*(5), 922–933. https://doi.org/10.1007/ s00586-019-05901-wA.*

NP cells at 20% (0.001 Hz) was also shown to upregulate the expression of inflammatory receptors and cytokines, namely Toll-like receptor 2 (TLR2), TLR4, TNF-α, and nerve growth factor (NGF), and to promote the secretion of growth-related oncogene, IL-6, IL-8, IL-15, monocyte chemoattractant protein (MCP)-1, MCP-3, monokine induced by interferon-γ, transforming growth factor (TGF)-β1, TNF-α, and NGF, which are associated with disc degeneration and pain (Gawri et al., 2014). Recently, Saggese et al. (2019) have shown that CTS (9% strain, 1 Hz; Fig. 5.9A and B) applied to bovine AF rings, in combination with a proinflammatory environment, may contribute to an increase in the production of proinflammatory molecules such as prostaglandin E2 and IL-6 (Fig. 5.9C), which ultimately may have contributed to a decrease in the tissue's delamination strength (Fig. 5.9D) (Saggese et al., 2019).

CTS (10% strain, 0.5 Hz) has also been applied to primary human CEP chondrocytes (Feng et al., 2018; Xiao et al., 2018; Zheng et al., 2019). This degenerative stimulation was shown to alter the expression of different miRNAs, being several predicted to target the MAPK and Wnt signaling pathways (Feng et al., 2018). Interestingly, CTS was shown to upregulate the expression of miR-455-5p by CEP cells via activation of the TGF-β/SMAD signaling pathway. miR-455-5p was specifically bound to RUNX2, which is known to play a role in CEP degeneration. Overall, this study showed that the TGF-β/SMAD signaling pathway can inhibit the intermittent cyclic tension—induced CEP degeneration through the regulation of the miR-455-5p/RUNX2 axis (Xiao et al., 2018). In another study, miR-365 was also identified as mechanosensitive, regulating human chondrocyte degeneration by directly targeting HDAC4. Its dysregulation was shown to induce the activation of the Wnt/β—catenin signaling pathway. Moreover, downregulation of miR-365 after intermittent cyclic tension stimulation led to COL2A and ACAN downregulation by CEP chondrocytes (Zheng et al., 2019). Taking this into consideration, CTS is another mechanical factor that contributes to changes in cellular mechanotransduction, with impact on AF and CEP cells, and on the regulation inflammation-related pathways.

Hydrostatic pressure

NP cells experience high hydrostatic pressure in nondegenerated discs during both rest and loading. The magnitude of the local hydrostatic pressure strongly depends on the discs water-binding capacity. With aging and degeneration, hydrostatic pressure decreases due to an incremental degradation of disc matrix proteoglycans. In an early study by Handa et al. (1997) using human IVD explants, the physiologic level of hydrostatic pressure (3 atm, atm = atmospheres) was shown to stimulate proteoglycan synthesis and upregulate TIMP-1, whereas abnormal pressure (30 atm) inhibited proteoglycan synthesis and upregulated MMP-3 (Handa et al., 1997). To better understand the complex effect of environmental stimuli on a cellular level and its influence on the gene expression of matrix proteins and matrix-degrading enzymes, several investigations have been performed by Neidlinger-Wilke and colleagues using a cyclic hydrostatic pressure model (Fig. 5.10A) (Neidlinger-Wilke et al., 2005, 2006, 2012). Intermittent hydrostatic pressure (0.25 MPa, 0.1 Hz) tended to upregulate ACAN and COL1A1 expression by human NP cells cultured in a collagen gel, and to downregulate MMP-2 and MMP-3 expression (Neidlinger-Wilke et al., 2005). A study comparing low hydrostatic pressure (0.25 MPa) and high hydrostatic pressure (2.5 MPa) confirmed downregulated expression of ACAN and COL1A1 (Fig. 5.10B) and upregulated expression of MMP1, MMP3 and MMP13 under high hydrostatic pressure conditions when compared to low hydrostatic pressure (Neidlinger-Wilke et al., 2006). In a following study with different hydrostatic pressure conditions (2.5 MPa, 0.1 Hz), and variation of different environmental conditions (e.g., glucose, pH, oxygen), hydrostatic pressure was shown to have a little direct effect on bovine NP cell gene expression but appeared to counteract matrix degradation by reducing or inverting some of the adverse effects of other stimuli (Neidlinger-Wilke et al., 2012). Le Maitre, Frain, Fotheringham, Freemont, and Hoyland (2008) have also investigated the response of human AF and NP cells isolated from degenerate and nondegenerate IVDs and cultured in alginate substrates, following application of dynamic hydrostatic pressure (0.8−1.7 MPa, 0.5 Hz). The authors showed that cell viability was unaffected by the loading regime; however, NP cells from degenerate IVDs responded differently from nondegenerated, suggesting an altered mechanotransduction pathway. On one hand, in nondegenerate NP cells, hydrostatic pressure was shown to upregulate C-FOS and

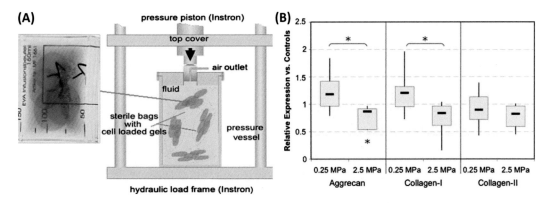

FIGURE 5.10

(A) Scheme of the stimulation device for the application of high hydrostatic pressure (2.5 MPa) and photo of a sterile bag with cell-seeded collagen gels. (B) Relative changes of mRNA expression of stimulated NP−seeded collagen constructs (0.25 and 2.5 MPa) normalized to the housekeeping gene (GAPDH) and to the respective unstimulated controls ($n = 14$ for 0.25 MPa, $n = 9$ for 2.5 MPa). (*) indicates significant pressure-induced differences ($P < .007$) and (*) indicates significant differences between the two loading groups ($P < .0062$).

Modified from Neidlinger-Wilke, C., Würtz, K., Urban, J. P. G., Börm, W., Arand, M., Ignatius, A., ... Claes, L. E. (2006). Regulation of gene expression in intervertebral disc cells by low and high hydrostatic pressure. European Spine Journal, 15(3), S372–S378. https://doi.org/10.1007/s00586-006-0112-1A.

ACAN gene expression, but not MMP-3. On the other hand, hydrostatic pressure did not seem to have any effect on the gene expression of degenerated cells (Le Maitre et al., 2008).

In general, different studies were in agreement that changes in hydrostatic pressure associated with IVD degeneration may play a role in further regulating disc matrix turn toward catabolism.

Microgravity

Understanding the effect of prolonged bed rest and microgravity in IVD degeneration has gained interest over the past decade. Prolonged microgravity exposure has been described to be associated with low back pain and higher risk of postflight disc herniation (Bailey et al., 2018). Different models have been established to simulate microgravity experimentally. For instance, microgravity conditions can be simulated by rotating samples in motion patterns that equally distribute the gravity vector spatially and temporally, averaging gravity to zero (Franco-Obregón et al., 2018; Jin et al., 2013). An example device is depicted in Fig. 5.11A. In vitro, a decrease in the proliferative capacity of IVD cells, induced senescence (Fig. 5.11B), and downregulation of transient receptor potential cation channel, subfamily C, member 6 (TRPC6) (Fig. 5.11C), by primary human IVD cells cultured under simulated microgravity has been observed. Interestingly, TRPC6 is a member of Ca^{2+}-conductive channel in the TRPC family and is involved in fundamental cellular functions such as regulation of Ca^{2+} signaling cascade by reactive oxygen species (Franco-Obregón et al., 2018). In rat disc organ cultures, microgravity was also shown to downregulate glycosaminoglycan content, and increase MMP-3 production and cell apoptosis (Jin et al., 2013). In rat (Li, Han, et al.,

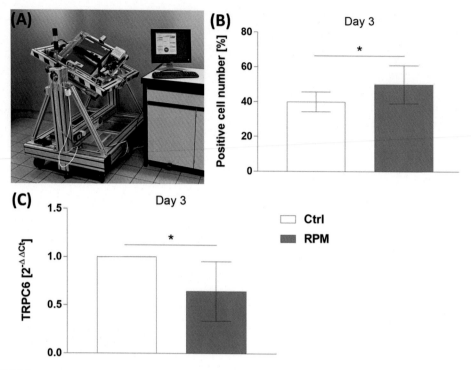

FIGURE 5.11

(A) Microgravity simulated in a random positioning machine (RPM). (B) Effects of simulated microgravity on cell senescence. Percentage of SA-β-galactosidase-positive human intervertebral disc (IVD) cells. (C) Relative gene expression of TRPC6 under microgravity. Results are shown as 2—$\Delta\Delta Ct$ values relative to the housekeeping gene and to the untreated control. Human IVD cells were either kept at 1 g (Ctrl) or exposed to simulated microgravity (RPM) for 3 days ($n = 4$). *$P < .05$.

Modified from Franco-Obregón, A., Cambria, E., Greutert, H., Wernas, T., Hitzl, W., Egli, M., ... Wuertz-Kozak, K. (2018). TRPC6 in simulated microgravity of intervertebral disc cells. European Spine Journal, 27(10), 2621–2630. https://doi.org/10.1007/s00586-018-5688-8A.

2019) and rabbit (Wu et al., 2017, 2019) animal experiments, microgravity/weightlessness has been simulated using tail-suspension models, in which the animals are hung with no weight bearing on the legs. Yasuoka et al. (2007) observed ACAN downregulation, MMP-3 up-regulation, and a significant decrease in the glycosaminoglycan content of rat NP and AF in the tail-suspended group in comparison to unloaded animals (Yasuoka et al., 2007). Furthermore, Li, Cao, et al. (2019) and Li, Han, et al. (2019) found abnormal expression levels of tumor suppressor proteins p53 and p16 and increased expression of inflammatory factors in simulated microgravity, which correlated with the degree of lumbar disc degeneration of rats (Li, Cao, et al., 2019). It has also been observed that there is an increase in rabbit IVD cell apoptosis, an increase in MMP-1, MMP-3, and TIMP-1 production under microgravity conditions (Wu et al., 2017), and a decrease in the proteoglycan content of lumbar IVDs with time (Wu et al., 2019).

Cells are constantly experiencing multiple loads. Overall, different studies summarized have shown that abnormal or excessive loading can trigger cells in transducing a mechanical signal into a catabolic response, contributing to aging/degeneration (Bajpai et al., 2021). Studies with focus on a particular stimulus are fundamental to better understanding the pathomechanism of load-induced mechanisms, and to identifying molecules and pathways that could be targeted with specific therapeutic approaches.

REFERENCES

Adams, M. A., & Roughley, P. J. (2006). What is intervertebral disc degeneration, and what causes it? *Spine (Philadelphia, PA: 1986), 31*(18), 2151−2161. Available from https://doi.org/10.1097/01.brs.0000231761.73859.2c.

Adams, P., & Muir, H. (1976). Qualitative changes with age of proteoglycans of human lumbar discs. *Annals of the Rheumatic Diseases, 35*(4), 289−296. Available from https://doi.org/10.1136/ard.35.4.289.

Bailey, J. F., Miller, S. L., Khieu, K., O'Neill, C. W., Healey, R. M., Coughlin, D. G., … Lotz, J. C. (2018). From the international space station to the clinic: How prolonged unloading may disrupt lumbar spine stability. *Spine Journal, 18*(1), 7−14. Available from https://doi.org/10.1016/j.spinee.2017.08.261.

Bajpai, A., Li, R., & Chen, W. (2021). The cellular mechanobiology of aging: From biology to mechanics. *Annals of the New York Academy of Sciences, 1491, 3−24*. Available from https://doi.org/10.1111/nyas.14529.

Burger, E. H., & Klein-Nulen, J. (1999). Responses of bone cells to biomechanical forces in vitro. *Advances in Dental Research, 13*, 93−98. Available from https://doi.org/10.1177/08959374990130012201.

Burr, D. B., Milgrom, C., Fyhrie, D., Forwood, M., Nyska, M., Finestone, A., … Simkin, A. (1996). In vivo measurement of human tibial strains during vigorous activity. *Bone, 18*(5), 405−410. Available from https://doi.org/10.1016/8756-3282(96)00028-2.

Chen, J., Yan, W., & Setton, L. A. (2004). Static compression induces zonal-specific changes in gene expression for extracellular matrix and cytoskeletal proteins in intervertebral disc cells in vitro. *Matrix Biology, 22*(7), 573−583. Available from https://doi.org/10.1016/j.matbio.2003.11.008.

Chowdhury, T. T., Appleby, R. N., Salter, D. M., Bader, D. A., & Lee, D. A. (2006). Integrin-mediated mechanotransduction in IL-1β stimulated chondrocytes. *Biomechanics and Modeling in Mechanobiology, 5* (2−3), 192−201. Available from https://doi.org/10.1007/s10237-006-0032-3.

Chowdhury, T. T., Bader, D. L., & Lee, D. A. (2001). Dynamic compression inhibits the synthesis of nitric oxide and PGE2 by IL-1β-stimulated chondrocytes cultured in agarose constructs. *Biochemical and Biophysical Research Communications, 285*(5), 1168−1174. Available from https://doi.org/10.1006/bbrc.2001.5311.

Chowdhury, T. T., Bader, D. L., & Lee, D. A. (2003). Dynamic compression counteracts IL-1β-induced release of nitric oxide and PGE2 by superficial zone chondrocytes cultured in agarose constructs. *Osteoarthritis and Cartilage, 11*(9), 688−696. Available from https://doi.org/10.1016/S1063-4584(03)00149-3.

Clarke, B. (2008). Normal bone anatomy and physiology. *Clinical Journal of the American Society of Nephrology, 3*, S131−S139. Available from https://doi.org/10.2215/CJN.04151206.

Dossumbekova, A., Anghelina, M., Madhavan, S., He, L., Quan, N., Knobloch, T., & Agarwal, S. (2007). Biomechanical signals inhibit IKK activity to attenuate NF-κB transcription activity in inflamed chondrocytes. *Arthritis and Rheumatism, 56*(10), 3284−3296. Available from https://doi.org/10.1002/art.22933.

Eyre, D. R. (1979). Biochemistry of the intervertebral disc. *International Review of Connective Tissue Research, 8*, 227−291. Available from https://doi.org/10.1016/b978-0-12-363708-6.50012-6.

Fearing, B. V., Hernandez, P. A., Setton, L. A., & Chahine, N. O. (2018). Mechanotransduction and cell biomechanics of the intervertebral disc. *JOR Spine*, *1*, e1026. Available from https://doi.org/10.1002/jsp2.1026.

Feng, C., Liu, M., Fan, X., Yang, M., Liu, H., & Zhou, Y. (2018). Intermittent cyclic mechanical tension altered the microRNA expression profile of human cartilage endplate chondrocytes. *Molecular Medicine Reports*, *17*(4), 5238−5246. Available from https://doi.org/10.3892/mmr.2018.8517.

Fermor, B., Weinberg, J. B., Pisetsky, D. S., Misukonis, M. A., Fink, C., & Guilak, F. (2002). Induction of cyclooxygenase-2 by mechanical stress through a nitric oxide-regulated pathway. *Osteoarthritis and Cartilage*, *10*(10), 792−798. Available from https://doi.org/10.1053/joca.2002.0832.

Ferretti, M., Gassner, R., Wang, Z., Perera, P., Deschner, J., Sowa, G., . . . Agarwal, S. (2006). Biomechanical signals suppress proinflammatory responses in cartilage: Early events in experimental antigen-induced arthritis. *Journal of Immunology*, *177*(12), 8757−8766. Available from https://doi.org/10.4049/jimmunol.177.12.8757.

Franco-Obregón, A., Cambria, E., Greutert, H., Wernas, T., Hitzl, W., Egli, M., . . . Wuertz-Kozak, K. (2018). TRPC6 in simulated microgravity of intervertebral disc cells. *European Spine Journal*, *27*(10), 2621−2630. Available from https://doi.org/10.1007/s00586-018-5688-8.

Fritton, S. P., McLeod, K. J., & Rubin, C. T. (2000). Quantifying the strain history of bone: Spatial uniformity and self-similarity of low-magnitude strains. *Journal of Biomechanics*, *33*(3), 317−325. Available from https://doi.org/10.1016/S0021-9290(99)00210-9.

Frost, H. M. (1987). Bone "mass" and the "mechanostat": A proposal. *The Anatomical Record*, *219*(1), 1−9. Available from https://doi.org/10.1002/ar.1092190104.

Gao, H., Zhai, M., Wang, P., Zhang, X., Cai, J., Chen, X., . . . Jing, D. (2017). Low-level mechanical vibration enhances osteoblastogenesis via a canonical Wnt signaling-associated mechanism. *Molecular Medicine Reports*, *16*(1), 317−324. Available from https://doi.org/10.3892/mmr.2017.6608.

Gassner, R., Buckley, M. J., Georgescu, H., Studer, R., Stefanovich-Racic, M., Piesco, N. P., . . . Agarwal, S. (1999). Cyclic tensile stress exerts antiinflammatory actions on chondrocytes by inhibiting inducible nitric oxide synthase. *Journal of Immunology*, *163*(4), 2187−2192.

Gawri, R., Rosenzweig, D. H., Krock, E., Ouellet, J. A., Stone, L. S., Quinn, T. M., & Haglund, L. (2014). High mechanical strain of primary intervertebral disc cells promotes secretion of inflammatory factors associated with disc degeneration and pain. *Arthritis Research and Therapy*, *16*(1), R21. Available from https://doi.org/10.1186/ar4449.

Gilbert, S. J., & Blain, E. J. (2018). Chapter 4 − Cartilage mechanobiology: How chondrocytes respond to mechanical load. In S. Verbruggen (Ed.), *Mechanobiology in health and disease* (1st ed., pp. 99−126). Academic Press. Available from https://doi.org/10.1016/B978-0-12-812952-4.00004-0.

Grodzinsky, A. J., Levenston, M. E., Jin, M., & Frank, E. H. (2000). Cartilage tissue remodeling in response to mechanical forces. *Annual Review of Biomedical Engineering*, *2*, 691−713. Available from https://doi.org/10.1146/annurev.bioeng.2.1.691.

Haffner-Luntzer, M., Kovtun, A., Lackner, I., Mödinger, Y., Hacker, S., Liedert, A., . . . Ignatius, A. (2018). Estrogen receptor α-(ERα), but not ERβ-signaling, is crucially involved in mechanostimulation of bone fracture healing by whole-body vibration. *Bone*, *110*, 11−20. Available from https://doi.org/10.1016/j.bone.2018.01.017.

Haffner-Luntzer, M., Lackner, I., Liedert, A., Fischer, V., & Ignatius, A. (2018). Effects of low-magnitude high-frequency vibration on osteoblasts are dependent on estrogen receptor α signaling and cytoskeletal remodeling. *Biochemical and Biophysical Research Communications*, *503*(4), 2678−2684. Available from https://doi.org/10.1016/j.bbrc.2018.08.023.

Haffner-Luntzer, M., Liedert, A., & Ignatius, A. (2020). Mechanobiology of bone remodeling and fracture healing in the aged organism. *Innovative Surgical Sciences*, *1*(2), 57−63. Available from https://doi.org/10.1515/iss-2016-0021.

Handa, T., Ishihara, H., Ohshima, H., Osada, R., Tsuji, H., & Obata, K. (1997). Effects of hydrostatic pressure on matrix synthesis and matrix metalloproteinase production in the human lumbar intervertebral disc. *Spine (Philadelphia, PA: 1986), 22*(10), 1085−1091. Available from https://doi.org/10.1097/00007632-199705150-00006.

Hendrickx, G., Fischer, V., Liedert, A., von Kroge, S., Haffner-Luntzer, M., Brylka, L., ... Schinke, T. (2021). Piezo1 inactivation in chondrocytes impairs trabecular bone formation. *Journal of Bone and Mineral Research, 36, 369−384*. Available from https://doi.org/10.1002/jbmr.4198.

Hodson, N. W., Patel, S., Richardson, S. M., Hoyland, J. A., & Gilbert, H. T. J. (2018). Degenerate intervertebral disc-like pH induces a catabolic mechanoresponse in human nucleus pulposus cells. *JOR Spine, 1*(1), e1004.

Holm, S., Maroudas, A., Urban, J. P. G., Selstam, G., & Nachemson, A. (1981). Nutrition of the intervertebral disc: Solute transport and metabolism. *Connective Tissue Research, 8*(2), 101−119. Available from https://doi.org/10.3109/03008208109152130.

Huang, K., Li, Q., Li, Y., Yao, Z., Luo, D., Rao, P., & Xiao, J. (2018). Cartilage tissue regeneration: The roles of cells, stimulating factors and scaffolds. *Current Stem Cell Research and Therapy, 13*(7), 547−567. Available from https://doi.org/10.2174/1574888X12666170608080722.

Ingber, D. E. (2006). Cellular mechanotransduction: Putting all the pieces together again. *FASEB Journal, 20*(7), 811−827. Available from https://doi.org/10.1096/fj.05-5424rev.

Jin, L., Feng, G., Reames, D. L., Shimer, A. L., Shen, F. H., & Li, X. (2013). The effects of simulated microgravity on intervertebral disc degeneration. *Spine Journal, 13*(3), 235−242. Available from https://doi.org/10.1016/j.spinee.2012.01.022.

Kanda, Y., Yurube, T., Morita, Y., Takeoka, Y., Kurakawa, T., Tsujimoto, R., ... Kakutani, K. (2021). Delayed notochordal cell disappearance through integrin α5β1 mechanotransduction during ex-vivo dynamic loading-induced intervertebral disc degeneration. *Journal of Orthopaedic Research, 39(9), 1933−1944*. Available from https://doi.org/10.1002/jor.24883.

Kitase, Y., Barragan, L., Qing, H., Kondoh, S., Jiang, J. X., Johnson, M. L., & Bonewald, L. F. (2010). Mechanical induction of PGE2 in osteocytes blocks glucocorticoid-induced apoptosis through both the β-catenin and PKA pathways. *Journal of Bone and Mineral Research, 25*(12), 2657−2668. Available from https://doi.org/10.1002/jbmr.168.

Knight, M. M., Toyoda, T., Lee, D. A., & Bader, D. L. (2006). Mechanical compression and hydrostatic pressure induce reversible changes in actin cytoskeletal organisation in chondrocytes in agarose. *Journal of Biomechanics, 39*(8), 1547−1551. Available from https://doi.org/10.1016/j.jbiomech.2005.04.006.

Korecki, C. L., MacLean, J. J., & Iatridis, J. C. (2008). Dynamic compression effects on intervertebral disc mechanics and biology. *Spine (Philadelphia, PA: 1986), 33*(13), 1403−1409. Available from https://doi.org/10.1097/BRS.0b013e318175cae7.

Kurakawa, T., Kakutani, K., Morita, Y., Kato, Y., Yurube, T., Hirata, H., ... Nishida, K. (2015). Functional impact of integrin α5β1 on the homeostasis of intervertebral discs: A study of mechanotransduction pathways using a novel dynamic loading organ culture system. *Spine Journal, 15*(3), 417−426. Available from https://doi.org/10.1016/j.spinee.2014.12.143.

Lahiji, K., Polotsky, A., Hungerford, D. S., & Frondoza, C. G. (2004). Cyclic strain stimulates proliferative capacity, α2 and α5 integrin, gene marker expression by human articular chondrocytes propagated on flexible silicone membranes. *In Vitro Cellular and Developmental Biology - Animal, 40*(5−6), 138−142. Available from https://doi.org/10.1290/1543-706X(2004)40 < 138:CSSPCA > 2.0.CO;2.

Le Maitre, C. L., Frain, J., Fotheringham, A. P., Freemont, A. J., & Hoyland, J. A. (2008). Human cells derived from degenerate intervertebral discs respond differently to those derived from non-degenerate intervertebral discs following application of dynamic hydrostatic pressure. *Biorheology, 45*(5), 563−575. Available from https://doi.org/10.3233/BIR-2008-0498.

Le Maitre, C. L., Frain, J., Millward-Sadler, J., Fotheringham, A. P., Freemont, A. J., & Hoyland, J. A. (2009). Altered integrin mechanotransduction in human nucleus pulposus cells derived from degenerated discs. *Arthritis and Rheumatism*, *60*(2), 460−469. Available from https://doi.org/10.1002/art.24248.

Leong, D. J., Gu, X. I., Li, Y., Lee, J. Y., Laudier, D. M., Majeska, R. J., . . . Sun, H. B. (2010). Matrix metalloproteinase-3 in articular cartilage is upregulated by joint immobilization and suppressed by passive joint motion. *Matrix Biology*, *29*(5), 420−426. Available from https://doi.org/10.1016/j.matbio.2010.02.004.

Leong, D. J., Hardin, J. A., Cobelli, N. J., & Sun, H. B. (2011). Mechanotransduction and cartilage integrity. *Annals of the New York Academy of Sciences*, *1240*(1), 32−37. Available from https://doi.org/10.1111/j.1749-6632.2011.06301.x.

Li, S., Duance, V. C., & Blain, E. J. (2008). Zonal variations in cytoskeletal element organization, mRNA and protein expression in the intervertebral disc. *Journal of Anatomy*, *213*(6), 725−732. Available from https://doi.org/10.1111/j.1469-7580.2008.00998.x.

Li, S., Jia, X., Duance, V. C., & Blain, E. J. (2011). The effects of cyclic tensile strain on the organisation and expression of cytoskeletal elements in bovine intervertebral disc cells: An in vitro study. *European Cells and Materials*, *21*, 508−522. Available from https://doi.org/10.22203/eCM.v021a38.

Li, X., Han, L., Nookaew, I., Mannen, E., Silva, M. J., Almeida, M., & Xiong, J. (2019). Stimulation of piezo1 by mechanical signals promotes bone anabolism. *ELife*, *8, e4963*. Available from https://doi.org/10.7554/eLife.49631.

Li, Y., Cao, L., Li, J., Sun, Z., Liu, C., Liang, H., . . . Tian, J. (2019). Influence of microgravity-induced intervertebral disc degeneration of rats on expression levels of p53/p16 and proinflammatory factors. *Experimental and Therapeutic Medicine*, *17*(2), 1367−1373. Available from https://doi.org/10.3892/etm.2018.7085.

Liedert, A., Nemitz, C., Haffner-Luntzer, M., Schick, F., Jakob, F., & Ignatius, A. (2020). Effects of estrogen receptor and wnt signaling activation on mechanically induced bone formation in a mouse model of postmenopausal bone loss. *International Journal of Molecular Sciences*, *21*(21), 1−12. Available from https://doi.org/10.3390/ijms21218301.

Lim, C. T., Bershadsky, A., & Sheetz, M. P. (2010). Mechanobiology. *Journal of the Royal Society Interface*, *7*(3), S291−S293. Available from https://doi.org/10.1098/rsif.2010.0150.focus.

Lynch, M. E., & Fischbach, C. (2014). Biomechanical forces in the skeleton and their relevance to bone metastasis: Biology and engineering considerations. *Advanced Drug Delivery Reviews*, *79*, 119−134. Available from https://doi.org/10.1016/j.addr.2014.08.009.

MacLean, J. J., Lee, C. R., Alini, M., & Iatridis, J. C. (2004). Anabolic and catabolic mRNA levels of the intervertebral disc vary with the magnitude and frequency of in vivo dynamic compression. *Journal of Orthopaedic Research*, *22*(6), 1193−1200. Available from https://doi.org/10.1016/j.orthres.2004.04.004.

Madhavan, S., Anghelina, M., Rath-Deschner, B., Wypasek, E., John, A., Deschner, J., . . . Agarwal, S. (2006). Biomechanical signals exert sustained attenuation of proinflammatory gene induction in articular chondrocytes. *Osteoarthritis and Cartilage*, *14*(10), 1023−1032. Available from https://doi.org/10.1016/j.joca.2006.03.016.

Mariani, E., Pulsatelli, L., & Facchini, A. (2014). Signaling pathways in cartilage repair. *International Journal of Molecular Sciences*, *15*(5), 8667−8698. Available from https://doi.org/10.3390/ijms15058667.

Massari, L., Benazzo, F., Falez, F., Perugia, D., Pietrogrande, L., Setti, S., . . . Cadossi, R. (2019). Biophysical stimulation of bone and cartilage: State of the art and future perspectives. *International Orthopaedics*, *43* (3), 539−551. Available from https://doi.org/10.1007/s00264-018-4274-3.

Molinos, M., Almeida, C. R., Caldeira, J., Cunha, C., Gonçalves, R. M., & Barbosa, M. A. (2015). Inflammation in intervertebral disc degeneration and regeneration. *Journal of the Royal Society, Interface*, *12*(108), 429. Available from https://doi.org/10.1098/rsif.2015.0429.

Molladavoodi, S., McMorran, J., & Gregory, D. (2020). Mechanobiology of annulus fibrosus and nucleus pulposus cells in intervertebral discs. *Cell and Tissue Research*, *379*(3), 429−444. Available from https://doi.org/10.1007/s00441-019-03136-1.

Neidlinger-Wilke, C., Galbusera, F., Pratsinis, H., Mavrogonatou, E., Mietsch, A., Kletsas, D., & Wilke, H. J. (2014). Mechanical loading of the intervertebral disc: From the macroscopic to the cellular level. *European Spine Journal*, *23*(3), S333−S343. Available from https://doi.org/10.1007/s00586-013-2855-9.

Neidlinger-Wilke, C., Mietsch, A., Rinkler, C., Wilke, H. J., Ignatius, A., & Urban, J. (2012). Interactions of environmental conditions and mechanical loads have influence on matrix turnover by nucleus pulposus cells. *Journal of Orthopaedic Research*, *30*(1), 112−121. Available from https://doi.org/10.1002/jor.21481.

Neidlinger-Wilke, C., Würtz, K., Liedert, A., Schmidt, C., Börm, W., Ignatius, A., ... Claes, L. (2005). A three-dimensional collagen matrix as a suitable culture system for the comparison of cyclic strain and hydrostatic pressure effects on intervertebral disc cells. *Journal of Neurosurgery Spine*, *2*(4), 457−465. Available from https://doi.org/10.3171/spi.2005.2.4.0457.

Neidlinger-Wilke, C., Würtz, K., Urban, J. P. G., Börm, W., Arand, M., Ignatius, A., ... Claes, L. E. (2006). Regulation of gene expression in intervertebral disc cells by low and high hydrostatic pressure. *European Spine Journal*, *15*(3), S372−S378. Available from https://doi.org/10.1007/s00586-006-0112-1.

Ochia, R. S., Tencer, A. F., & Ching, R. P. (2003). Effect of loading rate on endplate and vertebral body strength in human lumbar vertebrae. *Journal of Biomechanics*, *36*(12), 1875−1881. Available from https://doi.org/10.1016/S0021-9290(03)00211-2.

Ogata, K., & Whiteside, L. A. (1981). 1980 Volvo award winner in basic science. Nutritional pathways of the intervertebral disc. An experimental study using hydrogen washout technique. *Spine (Philadelphia, PA: 1986)*, *6*(3), 211−216.

Omlor, G. W., Lorenz, H., Engelleiter, K., Richter, W., Carstens, C., Kroeber, M. W., & Guehring, T. (2006). Changes in gene expression and protein distribution at different stages of mechanically induced disc degeneration − An in vivo study on the New Zealand white rabbit. *Journal of Orthopaedic Research*, *24*(3), 385−392. Available from https://doi.org/10.1002/jor.20055.

Pattappa, G., Li, Z., Peroglio, M., Wismer, N., Alini, M., & Grad, S. (2012). Diversity of intervertebral disc cells: Phenotype and function. *Journal of Anatomy*, *221*(6), 480−496. Available from https://doi.org/10.1111/j.1469-7580.2012.01521.x.

Pratsinis, H., Papadopoulou, A., Neidlinger-Wilke, C., Brayda-Bruno, M., Wilke, H. J., & Kletsas, D. (2016). Cyclic tensile stress of human annulus fibrosus cells induces MAPK activation: Involvement in proinflammatory gene expression. *Osteoarthritis and Cartilage*, *24*(4), 679−687. Available from https://doi.org/10.1016/j.joca.2015.11.022.

Qin, Y. X., & Hu, M. (2014). Mechanotransduction in musculoskeletal tissue regeneration: Effects of fluid flow, loading, and cellular-molecular pathways. *BioMed Research International*, *2014*. Available from https://doi.org/10.1155/2014/863421.

Raj, P. P. (2008). Intervertebral disc: Anatomy-physiology-pathophysiology-treatment. *Pain Practice*, *8*(1), 18−44. Available from https://doi.org/10.1111/j.1533-2500.2007.00171.x.

Reilly, D. T., & Burstein, A. H. (1975). The elastic and ultimate properties of compact bone tissue. *Journal of Biomechanics*, *8*, 393−405. Available from https://doi.org/10.1016/0021-9290(75)90075-5.

Risbud, M. V., & Shapiro, I. M. (2014). Role of cytokines in intervertebral disc degeneration: Pain and disc content. *Nature Reviews Rheumatology*, *10*(1), 44−56. Available from https://doi.org/10.1038/nrrheum.2013.160.

Roberts, S., Colombier, P., Sowman, A., Mennan, C., Rölfing, J. H. D., Guicheux, J., & Edwards, J. R. (2016). Ageing in the musculoskeletal system: Cellular function and dysfunction throughout life. *Acta Orthopaedica*, *87*, 15−25. Available from https://doi.org/10.1080/17453674.2016.1244750.

Roberts, S., Menage, J., & Urban, J. P. (1989). Biochemical and structural properties of the cartilage end-plate and its relation to the intervertebral disc. *Spine (Philadelphia, PA: 1986)*, *14*(2), 166−174. Available from https://doi.org/10.1097/00007632-198902000-00005.

Rosenberg, N., Levy, M., & Francis, M. (2002). Experimental model for stimulation of cultured human osteoblast-like cells by high frequency vibration. *Cytotechnology*, *39*(3), 125−130. Available from https://doi.org/10.1023/A:1023925230651.

Saggese, T., Teixeira, G. Q., Wade, K., Moll, L., Ignatius, A., Wilke, H. J., ... Neidlinger-Wilke, C. (2019). Georg Schmorl Prize of the German Spine Society (DWG) 2018: Combined inflammatory and mechanical stress weakens the annulus fibrosus: Evidences from a loaded bovine AF organ culture. *European Spine Journal*, *28*(5), 922−933. Available from https://doi.org/10.1007/s00586-019-05901-w.

Sakamoto, M., Fukunaga, T., Sasaki, K., Seiryu, M., Yoshizawa, M., Takeshita, N., & Takano-Yamamoto, T. (2019). Vibration enhances osteoclastogenesis by inducing RANKL expression via NF-κB signaling in osteocytes. *Bone*, *123*, 56−66. Available from https://doi.org/10.1016/j.bone.2019.03.024.

Sandell, L. J., & Aigner, T. (2001). Articular cartilage and changes in Arthritis: Cell biology of osteoarthritis. *Arthritis Research & Therapy*, *3*, 107. Available from https://doi.org/10.1186/ar148.

Setton, L. A., & Chen, J. (2006). Mechanobiology of the intervertebral disc and relevance to disc degeneration. *The Journal of Bone and Joint Surgery. American*, *88*(Suppl. 2), 52−57. Available from https://doi.org/10.2106/00004623-200604002-00011.

Sophia Fox, A. J., Bedi, A., & Rodeo, S. A. (2009). The basic science of articular cartilage: Structure, composition, and function. *Sports Health*, *1*(6), 461−468. Available from https://doi.org/10.1177/1941738109350438.

Sowa, G., & Agarwal, S. (2008). Cyclic tensile stress exerts a protective effect on intervertebral disc cells. *American Journal of Physical Medicine and Rehabilitation*, *87*(7), 537−544. Available from https://doi.org/10.1097/PHM.0b013e31816197ee.

Steppe, L., Liedert, A., Ignatius, A., & Haffner-Luntzer, M. (2020). Influence of low-magnitude high-frequency vibration on bone cells and bone regeneration. *Frontiers in Bioengineering and Biotechnology*, *8*. Available from https://doi.org/10.3389/fbioe.2020.595139.

Teixeira, G., Barbosa, M., & Goncalves, R. (2018). Immunomodulation in degenerated intervertebral disc. In R. Goncalves, & M. Barbosa (Eds.), *Gene and cell delivery for intervertebral disc degeneration* (1st ed., p. 48). CRC Press. Available from https://doi.org/10.1201/9781351030182.

Torzilli, P., Bhargava, M., Park, S., & Chen, C. (2010). Mechanical load inhibits IL-1 induced matrix degradation in articular cartilage. *Osteoarthritis and Cartilage*, *18*(1), 97−105. Available from https://doi.org/10.1016/j.joca.2009.07.012.

Vergroesen, P. P. A., Kingma, I., Emanuel, K. S., Hoogendoorn, R. J. W., Welting, T. J., van Royen, B. J., ... Smit, T. H. (2015). Mechanics and biology in intervertebral disc degeneration: A vicious circle. *Osteoarthritis and Cartilage*, *23*(7), 1057−1070. Available from https://doi.org/10.1016/j.joca.2015.03.028.

Walsh, A. J. L., & Lotz, J. C. (2004). Biological response of the intervertebral disc to dynamic loading. *Journal of Biomechanics*, *37*(3), 329−337. Available from https://doi.org/10.1016/S0021-9290(03)00290-2.

Wann, A. K. T., Zuo, N., Haycraft, C. J., Jensen, C. G., Poole, C. A., McGlashan, S. R., & Knight, M. M. (2012). Primary cilia mediate mechanotransduction through control of ATP-induced Ca2 + signaling in compressed chondrocytes. *FASEB Journal*, *26*(4), 1663−1671. Available from https://doi.org/10.1096/fj.11-193649.

Wittkowske, C., Reilly, G. C., Lacroix, D., & Perrault, C. M. (2016). In vitro bone cell models: Impact of fluid shear stress on bone formation. *Frontiers in Bioengineering and Biotechnology*, *4*, 87. Available from https://doi.org/10.3389/fbioe.2016.00087.

Wolff, J. (1892). Das Gesetz der transformation der Knochen. Berlin, Hirschwald.

Wu, D., Zheng, C., Wu, J., Huang, R., Chen, X., Zhang, T., & Zhang, L. (2017). Molecular biological effects of weightlessness and hypergravity on intervertebral disc degeneration. *Aerospace Medicine and Human Performance*, *88*(12), 1123−1128. Available from https://doi.org/10.3357/AMHP.4872.2017.

Wu, D., Zhou, X., Zheng, C., He, Y., Yu, L., Qiu, G., . . . Liu, Y. (2019). The effects of simulated + Gz and microgravity on intervertebral disc degeneration in rabbits. *Scientific Reports*, *9*(1), 16608. Available from https://doi.org/10.1038/s41598-019-53246-7.

Xiao, L., Xu, S., Xu, Y., Liu, C., Yang, B., Wang, J., & Xu, H. (2018). TGF-β/SMAD signaling inhibits intermittent cyclic mechanical tension-induced degeneration of endplate chondrocytes by regulating the miR-455-5p/RUNX2 axis. *Journal of Cellular Biochemistry*, *119*(12), 10415−10425. Available from https://doi.org/10.1002/jcb.27391.

Yang, X., Guan, Y., Tian, S., Wang, Y., Sun, K., & Chen, Q. (2016). Mechanical and IL-1β responsive miR-365 contributes to osteoarthritis development by targeting histone deacetylase 4. *International Journal of Molecular Sciences*, *17*(4), 436. Available from https://doi.org/10.3390/ijms17040436.

Yasuoka, H., Asazuma, T., Nakanishi, K., Yoshihara, Y., Sugihara, A., Tomiya, M., . . . Nemoto, K. (2007). Effects of reloading after simulated microgravity on proteoglycan metabolism in the nucleus pulposus and anulus fibrosus of the lumbar intervertebral disc: An experimental study using a rat tail suspension model. *Spine (Philadelphia, PA: 1986)*, *32*(25), E734−E740. Available from https://doi.org/10.1097/BRS.0b013e31815b7e51.

Zaman, G., Jessop, H. L., Muzylak, M., De Souza, R. L., Pitsillides, A. A., Price, J. S., & Lanyon, L. L. (2006). Osteocytes use estrogen receptor α to respond to strain but their ERα content is regulated by estrogen. *Journal of Bone and Mineral Research*, *21*(8), 1297−1306. Available from https://doi.org/10.1359/jbmr.060504.

Zhang, Y. H., Zhao, C. Q., Jiang, L. S., & Dai, L. Y. (2011). Cyclic stretch-induced apoptosis in rat annulus fibrosus cells is mediated in part by endoplasmic reticulum stress through nitric oxide production. *European Spine Journal*, *20*(8), 1233−1243. Available from https://doi.org/10.1007/s00586-011-1718-5.

Zhao, Z., Li, Y., Wang, M., Zhao, S., Zhao, Z., & Fang, J. (2020). Mechanotransduction pathways in the regulation of cartilage chondrocyte homoeostasis. *Journal of Cellular and Molecular Medicine*, *24*(10), 5408−5419. Available from https://doi.org/10.1111/jcmm.15204.

Zheng, Q., Li, X. X., Xiao, L., Shao, S., Jiang, H., Zhang, X. L., . . . Xu, H. G. (2019). MicroRNA-365 functions as a mechanosensitive microRNA to inhibit end plate chondrocyte degeneration by targeting histone deacetylase 4. *Bone*, *128*, 115052. Available from https://doi.org/10.1016/j.bone.2019.115052.

BONE BIOMECHANICS

6

Enrico Dall'Ara[1,2] and Vee San Cheong[1,3]

[1]*Insigneo Institute for in Silico Medicine, University of Sheffield, Sheffield, United Kingdom* [2]*Department of Oncology and Metabolism, University of Sheffield, Sheffield, United Kingdom* [3]*Department of Automatic Control and Systems Engineering, University of Sheffield, Sheffield, United Kingdom*

BONE PHYSIOLOGY

Bone contitutes a connective tissue of the muscular—skeletal system together with muscles, ligaments, tendons, and cartilage. Its main functions are to support the body transmit forces generated by the muscles to enable movement, to protect the organs, and to store minerals (e.g., calcium and phosphorus). Bone is a complex heterogeneous, anisotropic, nonlinear, and hierarchical material, with mechanical properties driven by its geometry, density, microstructure, and material properties.

Bone can be considered a composite material consisting of 65% minerals (mainly impure hydroxyapatite [$Ca_{10}(PO_4)_6(OH)_2$]) and the remaining 35% water and organic matrix (90% collagen Type I and 10% noncollagenous proteins; Cowin, 2001). Nevertheless, variations in the proportions of these constituents can be found in different types of bone across different anatomical sites and species; for example, the rostrum of the whale *Mesoplodon densirostris* is mainly composed of minerals, whereas the bones of some fish are composed mainly of collagen and water (Macesic & Summers, 2012; Zioupos, Currey, & Casinos, 2000).

The components of bone are arranged together to form complex structures at each dimensional scale. This hierarchical arrangement (Fig. 6.1) provides optimal mechanical properties for bone. In particular, bone is stiff and hard enough to resist external loads yet ductile enough to resist against fracture propagation (Turner, 2006). The stiffness and hardness are driven mainly by the mineral phase, while the ductility is mainly driven by the organic phase.

Collagen fibrils are intercalated with mineral crystals, which are embedded in the organic matrix. Packages of fibrils form collagen fibers, which form bone lamellae. The lamellae, $3-7\,\mu m$ in thickness, are arranged in different microstructures and form trabecular and cortical bone. In trabecular bone (also referred to as spongeous bone or cancellous bone), lamellae are mainly arranged parallel to each other and form an organized matrix of interconnected trabeculae. The mean thickness of the trabeculae can vary between 100 and $640\,\mu m$ (Cowin, 2001). In cortical bone (also referred to as compact bone or cortex), the lamellae form cylindrical substructures called osteons, where circular rings of lamellae surround a longitudinal (Haversian canals) or transverse (Volkmann canals) vascular channel. The external diameter of the Haversian and Volkmann systems (blood vessel plus the concentric lamellae around it) is approximately $100-300\,\mu m$, and they

Human Orthopaedic Biomechanics. DOI: https://doi.org/10.1016/B978-0-12-824481-4.00007-X

97

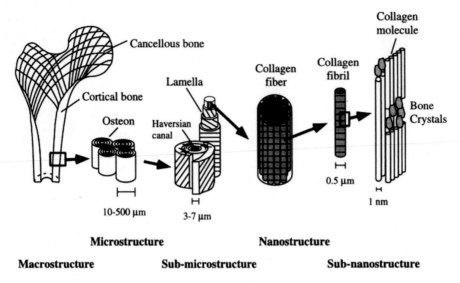

FIGURE 6.1

Hierarchical structure of bone.

Reprinted with permission from Rho, J. Y., Kuhn-Spearing, L., & Zioupos, P. (1998). Mechanical properties and the hierarchical structure of bone. Medical Engineering and Physics, 20*(2), 92–102. https://doi.org/10.1016/S1350-4533(98)00007-1.*

can be up to a few millimeters long. These systems are embedded in the interstitial tissue, which is the remnant of old osteons (Cowin, 2001). The two main macrostructures (cortical or trabeculae bone) are distributed in the bone to optimize its properties: Cortical bone constitutes the diaphysis of long bones (e.g., femur) and the external shell of short bones (e.g., tarsus, carpus), flat bones (e.g., pelvis), and irregular bones (e.g., vertebrae). Trabecular bone is located in the epiphysis of the long bones and in the central core of flat, short, and irregular bones (Fig. 6.2).

Furthermore, thanks to the activity of the cells embedded in it, bone is capable of repairing microcracks and microdamage induced by local overloading. More details about bone cells and their functions are provided in "Bone Cells and (Re)Modeling" section. It is important to mention that within the bone extracellular matrix, a dense network of pores is present at the different dimensional scales (Peyrin, Dong, Pacureanu, & Langer, 2014). These pores are important for cell communication and for hosting soft tissues (blood vessels, bone marrow, nerves) fundamental for cell survival and maintenance of the tissue. These pores affect the biomechanical properties of bone and should be factored in when designing new orthopedic implants or pharmaceutical treatments.

Long bones have a large internal longitudinal cylindrical hole along the longitudinal axis of the bone, called the medullary canal. This large portion of the bone is filled with bone marrow, blood vessels, and nerves. In the cortical bone, around the medullary canal, there are osteonal canals that host blood vessels and nerves. In trabecular bone, the space between the trabeculae are also filled with bone marrow. The spacing between the trabeculae depends on the anatomical site, age, and disease state. Within the extracellular matrix, there is an extensive network of micropores, termed the lacuno-canalicular network (LCN). The main function of this network of pores is to connect the

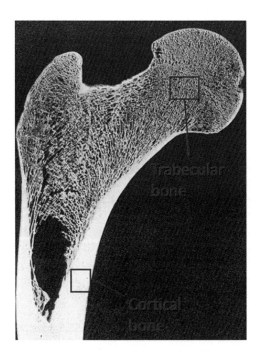

FIGURE 6.2

Bone microstructure. Frontal section of the proximal human femur highlighting regions of cortical and trabecular bone.

Adapted from Grey, H. (1918). Anatomy of the human body. Lea and Febiger. (public domain).

osteocytes—cells embedded in the bone extracellular matrix ("Bone Cells and (Re)Modeling" section). Osteocyte lacunae are ellipsoidal in shape, a few micrometers in size, and are spread quite homogenously across the mineralized portion of bone (Peyrin et al., 2014). The lacunae are connected with a dense network of canaliculi, a few hundreds of nanometers in diameter and several micrometers in length. The LCN is formed by several thousand lacunae per mm^3 and hundreds of thousands of canaliculi per mm^3. Its structure can be evaluated in 3D with confocal microscopy (Van Tol et al., 2020) or synchrotron X-ray tomography (Peyrin et al., 2014).

BONE CELLS AND (RE)MODELING

Bone tissue comprises four main types of bone cells: osteoblasts, osteocytes, osteoclasts, and bone lining cells. Osteoblasts, osteoclasts, and bone lining cells are located on bone surfaces, while osteocytes are found within the lacunae—empty spaces of the bone matrix.

Osteoblasts are derived from mesenchymal stem cells and are found on the surfaces of new bone. They synthesize and secrete the organic constituents of the bone matrix, termed the osteoids, and are predominately composed of Type I collagen (85%−95%). Thereafter, they control calcium and mineral deposition. As the collagen mineralizes, some osteoblasts may get trapped, and differentiate to become osteocytes.

Osteocytes are mature, postproliferative bone cells at the final stage of the osteoblast lineage, and bone contains 10 times more osteocytes than osteoblasts (Manolagas, 2000). Osteocytes are stellate shaped, with dendrites that project from the cell body. The cell body sits in the lacuna, while the dendrites extend into canaliculi that form the LCN. Osteocytes are joined to one another via gap junctions that facilitate exchange of nutrients, waste, and cell-to-cell communication and coordination. Osteocytes are also the primary mechanoreceptors in bone, responsible for bone homeostasis and integrity (Bonewald, 2011). Research has shown that osteocytes produce signaling proteins such as sclerostin and insulin-like growth factor 1 (IGF-1) in response to mechanical loads (Qin, Liu, Cao, & Xiao, 2020).

Bone lining cells are postproliferative osteoblasts that are flat in shape, and they line the external surfaces of bone (e.g., periosteum and endosteum) that are in a quiescent state. They are connected to the osteocyte network and are thought to regulate calcium homeostasis, hematopoiesis, and bone remodeling, by communicating with osteocytes through gap junctions (Everts et al., 2002; Matsuo & Otaki, 2012; Miller, De Saint-Georges, Bowman, & Jee, 1989). Bone lining cells are also responsible for the characteristics of the bone surface, as they remove nonmineralized collagen fibrils, and deposit a smooth layer of collagen over the surface (Everts et al., 2002; Matsuo & Otaki, 2012).

Osteoclasts are large, multinucleated cells involved in bone resorption. They are derived from hematopoietic stem cells found in the bone marrow and are formed by the fusion of multinucleate osteoclasts or monocytes, thus having 2−12 nuclei per cell. They secrete acid and proteases, which dissolve the bone mineral matrix and collagen, releasing the minerals back to the circulatory system (Blair, 1998). Osteoclasts also produce factors that inhibit osteoblast differentiation and activity.

BONE FORMATION AND REMODELING

Bone formation (ossification) occurs either by endochondral ossification or by intramembranous ossification. Endochondral ossification is the process where the existing cartilage is replaced by bone, and it occurs mainly at the physis and epiphysis of long bones, the cuboidal bones of the carpus and tarsus, and the base of the skull and vertebrae. In intramembranous ossification, mesenchymal stem cells proliferate and differentiate directly into osteoblasts at the ossification center to form bone. It is responsible for the formation of flat bones such as the clavicle, the skull, and the mandible.

Bone homeostasis: Bone is a dynamic tissue, and under physiological conditions, the amount of bone resorption is matched by the amount of bone formation, maintaining a balanced, homeostatic amount of bone. Multiple factors are involved in the "coupling" of bone resorption and formation for bone homeostasis, including the transforming growth factor β (TGF-β), which is secreted during bone resorption, to initiate local bone formation, and bone morphogenetic proteins (BMPs), which are deposited during bone formation to regulate osteoclast differentiation. On top of maintaining the structural integrity, bone homeostasis is also responsible for ionic composition in blood and interstitial fluids in the body. In particular, calcium is responsible for neuromuscular excitability, contraction coupling in muscles, cytoskeleton function, formation of bones, and so on. Bone is the largest reservoir of calcium in the body, and three hormonal loops exist to regulate calcium and phosphorous ion concentration. Parathyroid hormone (PTH) tends to increase the blood calcium level, while calcitonin decreases the calcium level. Vitamin D enhances the absorption of calcium and phosphate, increasing the concentration of these ions in the blood.

Bone remodeling is the replacement of old bone tissue by new bone tissue, to repair microdamaged bone, maintain bone integrity, and to regulate calcium homeostasis and hematopoiesis. Bone remodeling is affected by the local mechanical stimulus, and has been found to be reduced adjacent to metallic orthopedic implants due to stress shielding, which may cause the implant to fail eventually (Cheong, Fromme, Mumith, Coathup, & Blunn, 2018). Imbalances in bone remodeling can result in bone disorders ("Ageing and Bone Diseases" section). Bone turnover is mediated by osteocytes through paracrine cell signaling, which acts on osteoblasts and osteoclasts. This coupling of the action of small packets of osteoclasts and osteoblasts is referred to as a bone remodeling unit or basic multicellular unit (BMU). Bone remodeling in humans has six phases (Fig. 6.3): (1) resting (or quiescence); (2) activation, where osteoclasts are recruited and bone lining cells are retracted; (3) resorption, by osteoclasts, which create depressions called Howship's lacunae in trabecular bone or cutting cones in cortical bone; (4) reversal, where mesenchymal stem cells migrate to the resorption pit and differentiate into osteoblasts; (5) formation and mineralization of bone matrix, by osteoblasts, which initially lay down ostcoid, which is mineralized with calcium and phosphate; and (6) return to resting. The activation phase can be triggered by microfractures, mechanical loading, or mineral deficiency due to pregnancy or diet.

Bone modeling refers to the process where bone undergoes changes in morphology due to the displacement of bone surfaces. This was first described by Frost and colleagues, who observed that

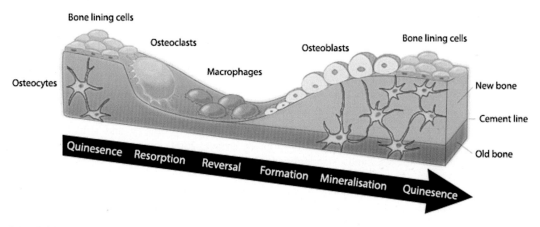

FIGURE 6.3

Bone remodeling cycle. The bone remodeling cycle begins when old bone is resorbed by activated osteoclasts. Macrophages prepare the resorbed surface for stem cells, which migrates and differentiates to form osteoblasts in the reversal phase. As bone forms and mineralizes, some osteoblasts differentiate to form osteocytes, which completes the bone remodeling cycle.

some of the cement lines in healthy adult bones were smooth, suggesting that bone formation was not preceded by bone resorption like in bone remodeling (Kobayashi et al., 2003). Hence, unlike bone remodeling, where the actions of osteoblasts and osteoclasts are coupled spatiotemporally, in modeling, osteoblasts and osteoclasts act independently of each other. Bone modeling is most active during periods of active bone turnover, but it can also occur after skeletal maturity as a response to mechanical load (Paiva & Granjeiro, 2017). For example, the higher bone mass and cortical area in the dominant arm of the tennis player have been shown to be caused by periosteal expansion from bone modeling—based bone formation (Kontulainen, Sievänen, Kannus, Pasanen, & Vuori, 2002). Disruptions in bone modeling can cause dysmorphias or skeletal dysplasias ("Ageing and Bone Diseases" section).

BONE MECHANICAL PROPERTIES
BONE DENSITY AND STRUCTURE

Even though cortical and trabecular bones have similar composition and, therefore, show a similar microscopic mechanical behavior, the differences in density and microstructures lead to different macroscopic mechanical properties. Cortical bone is stiffer and less ductile than trabecular bone, as measured in mechanical tests. The differences in mechanical properties are mainly due to differences in bone volume fraction (BV/TV) and architecture. In fact, the porosities of these microstructures play a dominant role in their mechanical properties. In this chapter, we focus on the relationship between mechanical and morphometric properties of trabecular bone, but most concepts can be extended to cortical bone.

Macroscopically, trabecular bone is highly heterogeneous and anisotropic, with properties optimized through its adaptation to external physiological loading. The two main determinants of trabecular bone mechanical properties are its density (Ohman et al., 2007), and the architecture and orientation of its trabeculae (Matsuura, Eckstein, Lochmüller, & Zysset, 2008). The quantification of the 3D morphology of human trabecular bone can be done ex vivo with micro-computed tomography (micro-CT, voxel size of approximately $5-20\,\mu m$) or in vivo with high-resolution peripheral computed tomography (HR-pQCT, voxel size of approximately $40-80\,\mu m$) for peripheral anatomical sites (typically distal radius and tibia). The trabecular bone geometry and structure can be reconstructed in 3D (Boyd & Müller, 2006; Varga & Zysset, 2009) by segmenting the acquired micro-CT/HR-pQCT images, and standardized morphological properties can be evaluated (Fig. 6.4).

Typical morphometric measurements are bone volume (BV), total volume (TV), bone volume fraction (BV/TV), trabecular thickness (Tb.Th), trabecular number (Tb.N), and trabecular spacing (Tb.Sp) (Hildebrand, Laib, Müller, Dequeker, & Rüegsegger, 1999). The morphological anisotropy of trabecular bone can be estimated with the trabecular orientation (also called fabric). The most common method to measure fabric in 3D is the mean intercept length (MIL), adapted from histomorphometric analysis (Harrigan & Mann, 1984; Whitehouse, 1974). Briefly, the method measures the mean interceptions of a ray field superimposed on the image with the bone in different

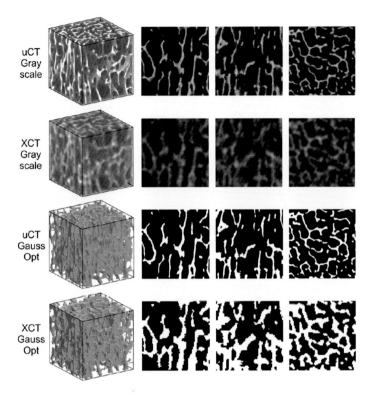

FIGURE 6.4

Trabecular bone structure. Three-dimensional structure of trabecular bone acquired with registered microCT (uCT) and HR-pQCT (XCT) images before (top) and after (bottom) image segmentation.

Modified with permission from Varga, P., & Zysset, P.K. (2009). Assessment of volume fraction and fabric in the distal radius using HR-pQCT. Bone, 45(5), 909–917. https://doi.org/10.1016/j.bone.2009.07.001.

directions. The MIL values can be represented as a quadratic form of a second-order tensor (fabric tensor M) (Harrigan & Mann, 1984):

$$M = \sum_{i=1}^{3} m_i M_i = \sum_{i=1}^{3} m_i (m_i \otimes m_i) \qquad (6.1)$$

The normalized eigenvectors in the fabric tensor provide the orientation of the principal directions (m_i), which represent the main orientation of the trabeculae. The degree of anisotropy (DA) is defined as the ratio between the maximum and minimum eigenvalues m_i.

BONE ELASTICITY AND ANISOTROPY

The elastic behavior of cortical and trabecular bone is linear between strain and stress at the macroscopic level at low strains (Fig. 6.5), with similar elastic constants for tensile and compressive loads (Rohl, Larsen, Linde, Odgaard, & Jorgensen, 1991).

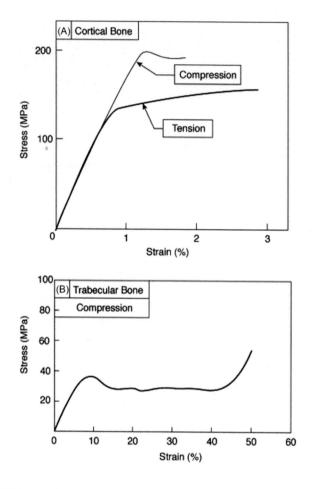

FIGURE 6.5

Mechanical properties of cortical and trabecular bone. Typical stress—strain curves for cortical (A) and trabecular (B) bone specimens tested in compression or tension.

Reproduced with permission from Mercer, C., He, M. Y., Wang, R., & Evans, A. G. (2006). Mechanisms governing the inelastic deformation of cortical bone and application to trabecular bone. Acta Biomaterialia, 2(1), 59–68. https://doi.org/10.1016/j.actbio.2005.08.004.

The stress (S) and strain (E) second-order tensors are related to each other by Hooke's law of elasticity:

$$S = \mathbb{S}E \tag{6.2}$$

$$E = \mathbb{E}S \tag{6.3}$$

where $\mathbb{S}$ and $\mathbb{E}$ are symmetric, positive, and definite fourth-order stiffness and compliance tensors, respectively. For a generic anisotropic material, 21 independent material constants are required to defined these tensors. Bone can be considered an orthotropic material (i.e., with three planes of elastic symmetry), and only nine constants remain independent. Expanding Eq. (6.3) gives:

$$
\begin{bmatrix} E_{11} \\ E_{22} \\ E_{33} \\ \sqrt{2}E_{23} \\ \sqrt{2}E_{31} \\ \sqrt{2}E_{12} \end{bmatrix} =
\begin{bmatrix}
\dfrac{1}{\varepsilon_1} & \dfrac{-v_{21}}{\varepsilon_2} & \dfrac{-v_{31}}{\varepsilon_3} & 0 & 0 & 0 \\
\dfrac{-v_{12}}{\varepsilon_1} & \dfrac{1}{\varepsilon_2} & \dfrac{-v_{32}}{\varepsilon_3} & 0 & 0 & 0 \\
\dfrac{-v_{13}}{\varepsilon_1} & \dfrac{-v_{23}}{\varepsilon_2} & \dfrac{1}{\varepsilon_3} & 0 & 0 & 0 \\
0 & 0 & 0 & \dfrac{1}{2\mu_{23}} & 0 & 0 \\
0 & 0 & 0 & 0 & \dfrac{1}{2\mu_{31}} & 0 \\
0 & 0 & 0 & 0 & 0 & \dfrac{1}{2\mu_{12}}
\end{bmatrix}
\begin{bmatrix} S_{11} \\ S_{22} \\ S_{33} \\ \sqrt{2}S_{23} \\ \sqrt{2}S_{31} \\ \sqrt{2}S_{12} \end{bmatrix}
\tag{6.4}
$$

where the engineering constants ε_i, v_{ij}, and μ_{ij} are Young's moduli, Poisson's ratios, and the shear moduli, respectively, and the indices $i = 1, 2, 3$ and $j = 1, 2, 3$ represent the three principal axes of symmetry. Both stiffness and compliance tensors are symmetric; therefore:

$$
\frac{v_{ij}}{\varepsilon_i} = \frac{v_{ji}}{\varepsilon_j}
\tag{6.5}
$$

From the mechanical testing performed on biopsies of trabecular bone previously scanned with micro-CT to evaluate the fabric M and BV/TV (in the equations, represented as ρ), it was shown that the elastic modulus is related to the density and fabric through power laws (Zysset & Curnier, 1995; Zysset, 2003). Therefore, the compliance tensor can be described as:

$$
\mathbb{E}(\rho, \boldsymbol{M}) = \frac{1}{\rho^k}
\begin{bmatrix}
\dfrac{1}{\varepsilon_0 m_1{}^{2l}} & \dfrac{-v_0}{\varepsilon_0 m_1{}^l m_2{}^l} & \dfrac{-v_0}{\varepsilon_0 m_1{}^l m_3{}^l} & 0 & 0 & 0 \\
\dfrac{-v_0}{\varepsilon_0 m_1{}^l m_2{}^l} & \dfrac{1}{\varepsilon_0 m_2{}^{2l}} & \dfrac{-v_0}{\varepsilon_0 m_2{}^l m_3{}^l} & 0 & 0 & 0 \\
\dfrac{-v_0}{\varepsilon_0 m_1{}^l m_3{}^l} & \dfrac{-v_0}{\varepsilon_0 m_2{}^l m_3{}^l} & \dfrac{1}{\varepsilon_0 m_3{}^{2l}} & 0 & 0 & 0 \\
0 & 0 & 0 & \dfrac{1}{2\mu_0 m_2{}^l m_3{}^l} & 0 & 0 \\
0 & 0 & 0 & 0 & \dfrac{1}{2\mu_0 m_3{}^l m_1{}^l} & 0 \\
0 & 0 & 0 & 0 & 0 & \dfrac{1}{2\mu_0 m_1{}^l m_2{}^l}
\end{bmatrix}
\tag{6.6}
$$

where the material properties of a pore-less material ($\rho = \text{BV/TV} = 1$), which can be extrapolated from experimental tests, are reported as ε_0, v_0, μ_0, k, and l (Rincón-Kohli & Zysset, 2009).

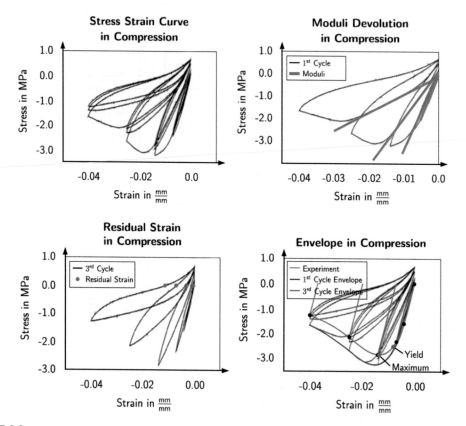

FIGURE 6.6

Trabecular bone damage test. Typical stress—strain curves from cyclic compressive tests of cylindrical trabecular bone specimens. The effect of damage and plasticity in the specimen can be observed as a reduction of elastic modulus and increase in residual strain.

Reproduced with permission from Wolfram, U., Wilke, H. J., & Zysset, P. K. (2011). Damage accumulation in vertebral trabecular bone depends on loading mode and direction. Journal of Biomechanics, 44(6), 1164—1169. https://doi.org/10.1016/j. jbiomech.2011.01.018.

BONE POSTELASTIC BEHAVIOR

When loaded above the elastic regime, bone exhibits a nonlinear behavior between stress and strain until its failure (Fig. 6.6). This is due to the presence of microcracks and/or diffused microdamages. Mechanical testing on trabecular bone biopsies have shown that bone failure load (or failure stress) is related to the BV/TV through a power law with an exponent close to 2 (Wolfram, Wilke, & Zysset, 2011). Bone failure stress depends on the trabecular morphology (Bayraktar, Gupta, Kwon, Papadopoulos, & Keaveny, 2004), is asymmetric for tension and compression (Rincón-Kohli & Zysset, 2009), and is similar for different anatomical sites (Turner, 1989).

Several studies in the literature have modeled the apparent yield and damage behavior of trabecular bone. Some have modeled bone using a perfectly plastic material where the elastic modulus is reduced when bone is loaded beyond yield (Costa et al., 2019), or using a Drucker−Prager yield surface together with a fixed yield strain (Bessho et al., 2007). More complex yield surfaces can be used to consider also the anisotropy of trabecular bone: Tsai-Wu (Wolfram et al., 2012), piecewise Hill (Garcia, Zysset, Charlebois, & Curnier, 2009), or modified superellipsoidal (Bayraktar et al., 2004). An elegant way of generalizing the constitutive laws for anisotropic quadratic yield criteria has been presented by Schwiedrzik and Zysset (2013).

In these cases, the constitutive laws become more complex. As an example, we report here the equations for postyield behavior simulating only isotropic damage for trabecular bone, without modeling the plasticity, which is adequate for studying the deformation of the bone under monotonic loading. The relationship between the stress and strain becomes:

$$S = (1 - D)\mathbb{S}E \tag{6.7}$$

where D is a scalar variable that represents the damage accumulation, which reduces the local stiffness of the material. D can lie between 1 (total damage) and 0 (no damage). The damage starts after the material is loaded beyond a damage surface, which can be described as a function of the density (ρ or BV/TV) and fabric (M). For example, with a generalized Hill criterion, which is asymmetric for tension and compression, the damage can be expressed as:

$$Y^D = \begin{cases} \sqrt{S:\mathbb{F}^+S} - r^D(D) \leq 0 \text{ if } m(S) \geq 0 \\ \sqrt{S:\mathbb{F}^-S} - r^D(D) \leq 0 \text{ if } m(S) < 0 \end{cases} \tag{6.8}$$

where $m(S)$ is the plane separating tension and compression in the stress domain and $r^D(D)$ is the radius of the damage criterion. $\mathbb{F}^+$ and $\mathbb{F}^-$ are two fourth-order tensors defined for tension and compression, respectively:

$$\mathbb{F}^\pm(\rho, M) = \frac{1}{\sigma_0^\pm{}^2 \rho^{2p}} \begin{bmatrix} \dfrac{1}{m_1^{4q}} & \dfrac{-\chi_0^\pm}{m_1^{2q}m_2^{2q}} & \dfrac{-\chi_0^\pm}{m_1^{2q}m_3^{2q}} & & & \\ \dfrac{-\chi_0^\pm}{m_1^{2q}m_2^{2q}} & \dfrac{1}{m_2^{4q}} & \dfrac{-\chi_0^\pm}{m_2^{2q}m_3^{2q}} & & \begin{matrix} 0 & 0 & 0 \\ 0 & 0 & 0 \\ 0 & 0 & 0 \end{matrix} & \\ \dfrac{-\chi_0^\pm}{m_1^{2q}m_3^{2q}} & \dfrac{-\chi_0^\pm}{m_1^{2q}m_3^{2q}} & \dfrac{1}{m_3^{4q}} & & & \\ & & & \dfrac{\sigma_0^\pm{}^2}{2\tau_0 m_2^{2q}m_3^{2q}} & 0 & 0 \\ \begin{matrix} 0 & 0 & 0 \\ 0 & 0 & 0 \\ 0 & 0 & 0 \end{matrix} & & 0 & \dfrac{\sigma_0^\pm{}^2}{2\tau_0 m_1^{2q}m_3^{2q}} & 0 \\ & & 0 & 0 & \dfrac{\sigma_0^\pm{}^2}{2\tau_0 m_1^{2q}m_2^{2q}} \end{bmatrix}$$

$$\tag{6.9}$$

The following coefficients extrapolated for pore-less bone ($\rho = $ BV/TV $= 1$) are used to define the tensors: uniaxial tensile and compressive strength, σ_0^+ and σ_0^-, respectively; multiaxial coupling terms in tension and compression, χ_0^+ and χ_0^-, respectively; and ultimate shear stress τ_0.

These coefficients, including also the exponents p and q, can be estimated from multiaxial tests performed on trabecular bone specimens with different density and fabric, as described in Rincón-Kohli and Zysset (2009).

Different types of damage evolution can be modeled. A typical isotropic damage can be modeled with an exponential law:

$$r^D(D) = R\left(1 + \chi_D\left(1 - e^{-\nu D}\right)\right) \tag{6.10}$$

where the damage hardening coefficients χ_D and ν can be estimated from uniaxial tests on trabecular bone specimens (Garcia et al., 2009). R represents the radius of the damage criterion. Examples of coefficients to be used for the elastic and damage laws can be found in Dall'Ara, Luisier, Schmidt, Kainberger, et al. (2013), Dall'Ara, Luisier, Schmidt, Pretterklieber, et al. (2013), Rincón-Kohli and Zysset (2009), Schwiedrzik and Zysset (2013), and Wolfram et al. (2012).

BONE TIME-DEPENDENT PROPERTIES

Bone's mechanical properties depend on the loading rate (i.e., bone is a viscoelastic material), with increased stiffness for higher loading rates. In order to study this phenomenon, more complex constitutive laws can be defined to account for the viscoelastic properties of bone in the elastic and postyield regimes. While the description of these complex constitutive laws is not the main aim of this chapter, the interested reader is encouraged to read the specialized literature where laws based on compression, relaxation, and creep tests have been laid out for trabecular bone (Manda, Wallace, Xie, Levrero-Florencio, & Pankaj, 2017; Xie et al., 2020).

ASSESSMENT OF BONE BIOMECHANICAL PROPERTIES AT DIFFERENT DIMENSIONAL LEVELS

EX VIVO ASSESSMENT OF BONE MECHANICAL PROPERTIES

Bone is a hierarchical material, and, therefore, its mechanical properties should be evaluated at different dimensional scales. In the following paragraphs, we will present typical ways of evaluating ex vivo bone mechanical properties at the organ (whole bone), tissue (bone biopsies), and bone structural unit (BSU; single trabecula or packages of lamellae) levels. The description is not meant as a comprehensive review of all possible methods for evaluating bone mechanical properties but to provide an overview of approaches that the reader is encouraged to know by reading the specialized literature.

ORGAN LEVEL

Mechanical testing on whole bones (e.g., the femur, one vertebra, or a spine unit) from cadaveric studies can be performed to study the bone mechanical properties such as apparent stiffness and strength (Fig. 6.7). The sample preparation (e.g., keeping or removing the soft tissues), the fixation in the machine (e.g., fully constrained vs freeing some degrees of freedom; loading direction; distribution of the load on regions of the bone), the specimen conditions during the test (e.g., dried or

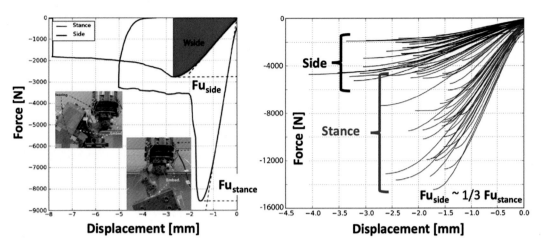

FIGURE 6.7

Mechanical tests on human femurs. Left: Load—displacement curves acquired for a pair of femurs from the same donor: one tested simulating the position of the femur in a fall on the side (side) and the other tested simulating a one-legged stance (stance). The ultimate force (F_u) can be computed from the curves under the two conditions. The area under the curve in red represents the work to failure for the side configuration. Right: load—displacement curves from 36 pairs of femurs tested under the two loading conditions.

Adapted from Dall'Ara, E., Luisier, B., Schmidt, R., Pretterklieber, M., Kainberger, F., Zysset, P., & Pahr, D. (2013). DXA predictions of human femoral mechanical properties depend on the load configuration. Medical Engineering and Physics, 35(11), 1564–1572. *https://doi.org/10.1016/j.medengphy.2013.04.008.*

hydrated), and the loading parameters (e.g., loading rate, monotonic vs cyclic loads) play an important role in the measurement of the mechanical properties. In standard mechanical tests, usually the applied load and the resulting displacement of a portion of the organ is measured with internal sensors of the machine (e.g., load cells, linear variable differential transformer (LVDTs)) or with external sensors applied in specific regions of the bone. Load—displacement curves can be obtained from the test, and apparent stiffness is measured as the slope of the linear portion of the curve. The failure load is calculated as the maximum force in the first peak of the load—displacement curve. Particular attention should be given to the effect of the loading direction on the properties of the bones, which usually exhibit a strong anisotropic mechanical behavior. For example, when testing pairs of human femurs from the same donor, simulating the position during a physiological standing condition (stance) or a fall onto the greater trochanter (side), the failure load of the femur in the "stance" condition is approximately three times larger than the failure load measured in the "side" condition (Fig. 6.7) (Dall'Ara, Luisier, Schmidt, Kainberger, et al., 2013; Rincón-Kohli & Zysset, 2009; Schwiedrzik & Zysset, 2013; Wolfram et al., 2012).

Recent development of full-field methods like digital image correlation (DIC) or digital volume correlation (DVC) in combination with mechanical testing within imaging devices enables the measurement of the full-field deformation in the external surface of the bone (Palanca, Tozzi, & Cristofolini, 2016) or the internal strain induced by time-lapsed mechanical testing (Grassi & Isaksson, 2015). For example, these approaches have shown the failure mechanism in the human proximal femur (Martelli, Giorgi, Dall'Ara, & Perilli, 2021) and the heterogeneous local deformation that occurs in a large portion of the bone due to osteoarthritis (Ryan, Oliviero, Costa, Mark Wilkinson, & Dall'ara, 2020) or injections of bone cement (Danesi, Tozzi, & Cristofolini, 2016).

TISSUE/BIOPSY LEVEL

To determine the relationship between the local bone mineral density (BMD) or BV/TV and the mechanical properties of the bone, mechanical testing on small portions of trabecular or cortical bone can be performed (Fig. 6.8). Cylindrical or cubic specimens are usually extracted from the bones of interest using a combination of band saws and core drills with diamond coated tools, to minimize the damage induced by heating during sample extraction. Micro-CT scans can be performed before testing to acquire the geometry of the specimens and evaluate its densitometric properties. The specimens can be mounted in the machine by optimizing the alignment with brass or polymethylmethacrylate (PMMA) endcaps (Ohman et al., 2007). External sensors such as extensometers can be used to measure the local deformation of the bone, reducing the effect of the compliance due to the setup. The specimen has to be kept hydrated before and during the test to avoid overestimating the mechanical properties due to dehydration. The applied load is usually measured with a load-cell, and monotonic or cyclic tests can be performed on displacement or load control. The load−displacement curves acquired during the mechanical tests can be normalized into stress−strain curves by using the geometrical properties of the specimens.

From the stress−strain curves, important mechanical parameters such as the yield and ultimate stress, the yield and ultimate strain, the elastic modulus, and the energy to failure can be calculated. Several studies have correlated these properties to the densitometric and microstructural properties of the bone in order to define the constitutive laws mentioned above, or to evaluate the effect of the specimen orientation, hydration conditions, loading rate, and diseases on the tissue mechanical

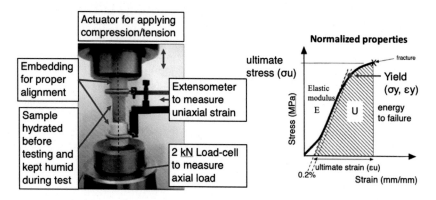

FIGURE 6.8

Mechanical test on a trabecular bone specimen. Left: typical setup to test trabecular bone specimen. Embedding caps are added to align the specimen in the testing machine and an extensometer is used to evaluate the apparent strain in the specimen. Right: typical stress−strain curve obtained from the normalized displacement and force measured during the experiment to calculate the mechanical properties.

properties. Examples are reported in the *Bone Mechanics Handbook* edited by Cowin (2001). Advanced time-lapsed mechanical testing and DVC measurements have been used to study the failure mechanism of trabecular bone specimens at the tissue level (Roberts, Perilli, & Reynolds, 2014).

BONE STRUCTURAL UNIT/LAMELLAR LEVEL

Trabecular bone is composed of trabeculae, while cortical bone is composed of osteons and interstitial bone. These structures, consisting of packages of bone lamellae properly oriented in space, are also called BSUs.

Three-point bending tests of single trabeculae (Jungmann et al., 2011) (Fig. 6.9) or single osteons (Ascenzi, Baschieri, & Benvenuti, 1990) have been performed to evaluate their mechanical properties. These tests are associated with experimental challenges in aligning and fixing the tiny portions of bone (a couple millimeters in size) within the testing devices, which are usually connected to a microscope to visualize the specimen under load. Nevertheless, they have provided important insights on the local mechanical properties of the BSU depending on the local arrangement of the lamellae. Young's moduli ranging from 1.1 to 11.4 GPa have been reported in the literature for single trabeculae (Wu, Isaksson, Ferguson, & Persson, 2018).

To measure the local hardness and stiffness of bone tissues, micro- and nanoindentation (Fig. 6.9) have been used extensively over the past few decades. These techniques extend previous generations of microhardness testers (Dall'Ara, Ohman, Baleani, & Viceconti, 2007), and consist of performing an indentation test with a black diamond microindenter of known shape on a flat and

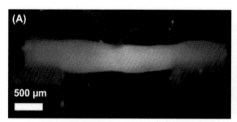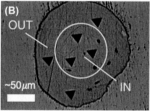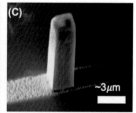

FIGURE 6.9

Mechanical tests at the BSU level. *BSU*, Bone structural unit. Typical mechanical tests performed at the BSU and lamellar levels. (A) Three points bending on a single trabecula; (B) microindentation test on axial sections of a trabecula; and (C) micropillar of bone within lamellae after mechanical testing.

Reproduced with permission from (A) Jungmann, R., Szabo, M. E., Schitter, G., Yue-Sing Tang, R., Vashishth, D., Hansma, P. K., & Thurner, P. J. (2011). Local strain and damage mapping in single trabeculae during three-point bending tests. Journal of the Mechanical Behavior of Biomedical Materials, 4(4), 523–534. https://doi.org/10.1016/j.jmbbm.2010.12.009; (B) Wolfram, U., Wilke, H. J., & Zysset, P. K. (2010). Rehydration of vertebral trabecular bone: Influences on its anisotropy, its stiffness and the indentation work with a view to age, gender and vertebral level. Bone, 46(2), 348–354. https://doi.org/10.1016/j. bone.2009.09.035; (C) Luczynski, K. W., Steiger-Thirsfeld, A., Bernardi, J., Eberhardsteiner, J., and Hellmich, C. (2015). Extracellular bone matrix exhibits hardening elastoplasticity and more than double cortical strength: Evidence from homogeneous compression of non-tapered single micron-sized pillars welded to a rigid substrate. Journal of the Mechanical Behavior of Biomedical Materials, 52, 51–62. https://doi.org/10.1016/j.jmbbm.2015.03.001.

polished bone surface (Zysset, 2009). Recording the penetration depth and the applied load during the test, micro- and nanoindentation have been used to evaluate the effect of hydration (Wolfram et al., 2010), microdamage (Dall'Ara et al., 2012), anisotropy, and diseases such as osteoporosis (Kim, Cole, Boskey, Baker, & Van Der Meulen, 2014), osteoarthritis (Li, Dai, Jiang, & Qiu, 2012), and osteogenesis imperfecta (Imbert, Aurégan, Pernelle, & Hoc, 2014) on the bone mechanical properties.

Recently, improvement in bone microspecimen preparation with dual focus ion beam (FIB) has allowed researchers to produce and mechanically test cylindrical or prismatic micropillars of a few micrometers in cross-section using scanning electron microscopes (Luczynski et al., 2015; Schwiedrzik & Zysset, 2013) (Fig. 6.9). In these studies, the local failure behavior of portions of single lamellae has been measured and their elastoplastic behavior identified (Zysset, Schwiedrzik, & Wolfram, 2016).

IN VIVO ASSESSMENT OF BONE MECHANICAL PROPERTIES

All the above-mentioned studies have contributed to developing approaches to evaluating the mechanical properties of bone at different dimensional levels ex vivo. Nevertheless, approaches to evaluating the bone mechanical properties in vivo in patients have also been developed. From laboratory experiments, we have learnt that the mechanical properties of the bone depend on its size, geometry, density, microarchitecture, and material properties. Quantitative computed tomography (QCT) images calibrated with densitometric phantoms can be used to acquire the three-dimensional geometry (image resolution of approximately 1 mm) and distribution of BMD of a bone in vivo. Based on the QCT images, finite element (FE) models (Fig. 6.10), after proper validation against experiments performed in the laboratory on cadaveric studies, can be used to create computer models of the bone of a subject at a certain time point and evaluate its mechanical properties under

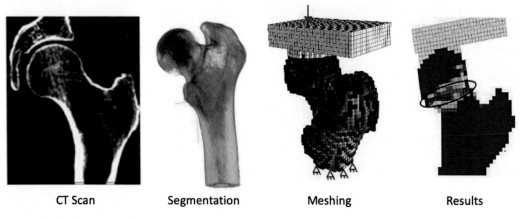

CT Scan   Segmentation   Meshing   Results

FIGURE 6.10

Finite element model of the human femur. Pipeline to create a CT-based subject-specific finite element model of the human proximal femur.

certain loading conditions (Zysset & Curnier, 1995; Zysset, 2003). For example, subject-specific FE models have been used extensively to study in vivo the femoral strength (Keaveny et al., 2012), the risk of osteoporotic femoral fractures (Qasim et al., 2016), and the biomechanical properties of vertebrae with bone metastases (Costa et al., 2019).

To account for the influence of bone microarchitecture on the bone mechanical properties in vivo, FE models based on HR-pQCT of the distal radius (Fig. 6.11) and tibia have been developed. HR-pQCT images of small portions of bone can be acquired in vivo with a low radiation dose (image resolution of approximately $50-100\,\mu m$), enabling the assessment of trabecular microarchitecture and cortical thickness in the scanned region. These images can be converted into FE models to study, for example, the mechanical properties of the distal radius (Varga, Dall'Ara, Pahr, Pretterklieber, & Zysset, 2011).

With these techniques, the effect of geometry, density, and microarchitecture on bone biomechanical properties can be evaluated. However, the material properties of the bone (i.e., the stiffness of the local extracellular matrix) have to be assumed in the models or estimated from ex vivo studies. To evaluate the material properties of the bone in vivo, a particular indentation technique has been developed: OsteoProbe (Hansma et al., 2009). This portable indenter can perform percutaneous indentations approximately 50 m in size in different locations of the anterior side of the tibia in patients, where the skin is thinner. While the approach is similar to standard depth-sensing microindentation, the OsteoProbe estimates the indentation distance increase (IDI) from cyclic indentation tests under the same load. The IDI is calculated to estimate the local resistance to fracture by comparing the relative penetration depth measured in the last cycles with the penetration depth measured in the cycle, which is affected by a thin layer of soft tissues between the bone and the indenter. The OsteoProbe has been used in vivo to evaluate the material properties of patients who have sustained fragility fractures (Diez-Perez et al., 2010).

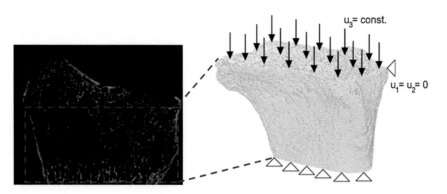

FIGURE 6.11

Finite element model of the distal radius. HR-pQCT-based finite element model of the human distal radius for the assessment of the mechanical properties of the bone in vivo.

Reproduced with permission from Varga, P., Dall'Ara, E., Pahr, D. H., Pretterklieber, M., & Zysset, P. K. (2011). Validation of an HR-pQCT-based homogenized finite element approach using mechanical testing of ultra-distal radius sections. Biomechanics and Modeling in Mechanobiology, 10(4), 431–444. https://doi.org/10.1007/s10237-010-0245-3.

AGEING AND BONE DISEASES

The effect of ageing is a reduction in the strength and stiffness of bone. Studies have shown that the elderly are 10 times more at risk of suffering from a fracture than a younger adult with the same BMD (Kanis, 2002). This is primarily due to a reduction in bone ductility with age, demineralization, and a loss of calcium from bone, caused by a reduction in osteoblast activity (Razi, Birkhold, Zaslansky, et al., 2015). This presents as a weakening of the trabecular bone with age (Nagaraja, Lin, & Guldberg, 2007). Moreover, the remodeling response to mechanical loading reduces with skeletal maturity, and while bones of young mice have a clear range of strains where bone formation and resorption occur, this specificity is lost in the bones of aged mice (Razi, Birkhold, Weinkamer, et al., 2015). Ageing also causes a reduction in the production of extracellular matrix proteins (especially collagen fibers), making bones more brittle, more susceptible to fracture, and more prone to osteoporosis. Clinically, the assessment of the bone health is typically done using dual-energy X-ray absorptiometry (DEXA). The results are returned in the form of a T-score, which compares the BMD score of the patient with that of healthy sex-matched 25 year olds, in units of standard deviations (SDs). The lower is the T-Score, the higher is the risk of suffering from a fracture.

Osteoporosis is the most prevalent metabolic bone disease. Affecting one in three postmenopausal women, and one in five men, osteoporosis increases bone fragility and causes hundred thousands of fractures annually, and is associated with morbidity and increased mortality (Ström et al., 2011). Osteoporosis is characterized by the loss of bone mass, the deterioration of bone microarchitecture (see Fig. 6.12 for a mouse model of osteoporosis), and brittle bones, as

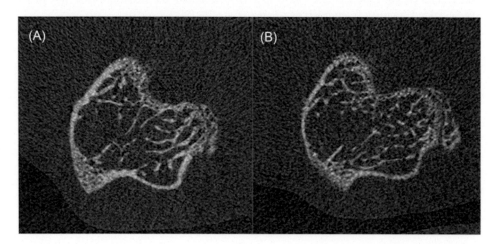

FIGURE 6.12

Mouse model of osteoporosis. Micro-CT scans of a mouse tibia at 18 weeks of age. (A) Healthy bone tissue. (B) Microarchitectural deterioration of bone tissue in a mouse model of osteoporosis (ovariectomy performed at Week 14 of age). Lower trabecular number and lower trabecular thickness can be observed in the model of osteoporosis. More details in Roberts et al. (2019).

a result of increased resorption and some concomitant slowing down of bone formation (Cheong, Roberts, Kadirkamanathan, & Dall'Ara, 2020). According to the World Health Organization (WHO), a patient is osteopenic when their BMD is 1−2.5 SDs below that of a healthy young adult, and osteoporotic when their BMD is more than 2.5 SDs below the healthy young adult population. Osteoporosis can be classified into type I primary, type II primary, or type II secondary. Type I primary osteoporosis, or postmenopausal osteoporosis, is caused by estrogen deficiency, which disrupts the feedback mechanism between OPG and receptor activator of nuclear factor kappa-B ligand (RANKL), reducing osteoclast apoptosis, leading to an increase in bone resorption (Nakamura et al., 2007). Type II primary osteoporosis (senile osteoporosis) is caused by deficiencies in calcium and/or vitamin D due to malabsorption, low dietary intake, or inadequate exposure to sunlight. Type II secondary osteoporosis is caused due to certain medical conditions (such as hyperparathyroidism, diabetes) and adverse effects to medications, which lead to imbalances between bone production and bone resorption (Feng & McDonald, 2011).

Rickets and osteomalacia are disorders in which bones fail to calcify, caused typically by a deficiency of vitamin D. Although osteoids are still produced by the osteoblasts, the failure to ossify causes the bone to be soft and rubbery. Rickets in children cause bowed legs and deformities in the pelvis, rib cage, and skull. In adults, osteomalacia is characterized by pain and tenderness in the bone, and bone fracture from minor trauma.

Osteoarthritis (OA) is a degenerative joint disease affecting 303 million people worldwide in 2017 (GBD, 2017), and can be divided into two types: primary OA and secondary OA. Primary OA is caused by the degeneration of articular cartilage, causing bones to be in direct contact. Secondary OA occurs when the cartilage becomes damaged by another disease or medical condition. Factors that increase the risk of developing secondary OA include abnormal and/or excessive stress on the joints, injury or surgery to the joint, pathological kinematics, rheumatoid arthritis, and gout, among others. Patients typically present with joint pain, stiffness, and decreased range of motion. The resulting friction between the bones worsens the condition, and in later stages of OA, the pain may be continuous.

Skeletal dysplasias, also known as osteochondrodysplasias, is a heterogeneous group of genetic disorders characterized by abnormalities of cartilage and bone growth, resulting in an abnormal shape and disproportionate size of the long bones, spine, and head. These disorders are broadly classified into three types: osteodysplasias, chondrodysplasias, and dystoses. One of the most commonly known types of osteodysplasias is *osteogenesis imperfecta* (brittle-bone disease; OI), which affects 1 in 15,000 births (Forlino, Cabral, Barnes, & Marini, 2011). OI is the result of a genetic mutation that disrupts the production or structure of collagen molecules, resulting in an abnormal or inadequate amount of type I collagen to be produced. Patients with OI have increased bone turnover from the higher number of osteoblasts and osteoclasts present (Rauch, Travers, Parfitt, & Glorieux, 2000). However, the mineral apposition rate is low, and the matrix produced is little but hypermineralized, leading to low cortical and trabecular bone volume and bone fragility in patients (Glorieux et al., 2002). In more severe cases of OI, the disruption in collagen production weakens connective tissues, leading to severe bone abnormalities and problems with normal bone development. Bone material properties are dramatically affected by OI (Bishop, 2016).

REFERENCES

Ascenzi, A., Baschieri, P., & Benvenuti, A. (1990). The bending properties of single osteons. *Journal of Biomechanics*, *23*(8), 763−771. Available from https://doi.org/10.1016/0021-9290(90)90023-V.

Bayraktar, H. H., Gupta, A., Kwon, R. Y., Papadopoulos, P., & Keaveny, T. M. (2004). The modified super-ellipsoid yield criterion for human trabecular bone. *Journal of Biomechanical Engineering*, *126*(6), 677−684. Available from https://doi.org/10.1115/1.1763177.

Bessho, M., Ohnishi, I., Matsuyama, J., Matsumoto, T., Imai, K., & Nakamura, K. (2007). Prediction of strength and strain of the proximal femur by a CT-based finite element method. *Journal of Biomechanics*, *40*(8), 1745−1753. Available from https://doi.org/10.1016/j.jbiomech.2006.08.003.

Bishop, N. (2016). Bone material properties in osteogenesis imperfecta. *Journal of Bone and Mineral Research*, *31*(4), 699−708. Available from https://doi.org/10.1002/jbmr.2835.

Blair, H. C. (1998). How the osteoclast degrades bone. *Bioessays: News and Reviews in Molecular, Cellular and Developmental Biology*, *20*(10), 837−846, https://doi.org/10.1002/(SICI)1521-1878(199810) 20:10 < 837::AID-BIES9 > 3.0.CO;2-D.

Bonewald, L. F. (2011). The amazing osteocyte. *Journal of Bone and Mineral Research*, *26*(2), 229−238. Available from https://doi.org/10.1002/jbmr.320.

Boyd, S. K., & Müller, R. (2006). Smooth surface meshing for automated finite element model generation from 3D image data. *Journal of Biomechanics*, *39*(7), 1287−1295. Available from https://doi.org/10.1016/j.jbiomech.2005.03.006.

Cheong, V. S., Fromme, P., Mumith, A., Coathup, M. J., & Blunn, G. W. (2018). Novel adaptive finite element algorithms to predict bone ingrowth in additive manufactured porous implants. *Journal of the Mechanical Behavior of Biomedical Materials*, *87*, 230−239. Available from https://doi.org/10.1016/j.jmbbm.2018.07.019.

Cheong, V. S., Roberts, B. C., Kadirkamanathan, V., & Dall'Ara, E. (2020). Bone remodelling in the mouse tibia is spatio-temporally modulated by oestrogen deficiency and external mechanical loading: A combined in vivo/in silico study. *Acta Biomaterialia*, *116*, 302−317. Available from https://doi.org/10.1016/j.actbio.2020.09.011.

Costa, M. C., Eltes, P., Lazary, A., Varga, P. P., Viceconti, M., & Dall'Ara, E. (2019). Biomechanical assessment of vertebrae with lytic metastases with subject-specific finite element models. *Journal of the Mechanical Behavior of Biomedical Materials*, *98*, 268−290. Available from https://doi.org/10.1016/j.jmbbm.2019.06.027.

Cowin, S. C. (Ed.), (2001). *Bone mechanics handbook* (2nd ed.). CRC Press. Available from https://doi.org/10.1201/b14263.

Dall'Ara, E., Ohman, C., Baleani, M., & Viceconti. (2007). The effect of tissue condition and applied load on Vickers hardness of human trabecular bone. *Journal of Biomechanics*, *40*(14), 3267−3270.

Dall'Ara, E., Schmidt, R., & Zysset, P. (2012). Microindentation can discriminate between damaged and intact human bone tissue. *Bone*, *50*(4), 925−929. Available from https://doi.org/10.1016/j.bone.2012.01.002.

Dall'Ara, E., Luisier, B., Schmidt, R., Kainberger, F., Zysset, P., & Pahr, D. (2013). A nonlinear QCT-based finite element model validation study for the human femur tested in two configurations in vitro. *Bone*, *52*(1), 27−38.

Dall'Ara, E., Luisier, B., Schmidt, R., Pretterklieber, M., Kainberger, F., Zysset, P., & Pahr, D. (2013). DXA predictions of human femoral mechanical properties depend on the load configuration. *Medical Engineering & Physics*, *35*(11), 1564−1572.

Danesi, V., Tozzi, G., & Cristofolini, L. (2016). Application of digital volume correlation to study the efficacy of prophylactic vertebral augmentation. *Clinical Biomechanics*, *39*, 14−24. Available from https://doi.org/10.1016/j.clinbiomech.2016.07.010.

Diez-Perez, A., Güerri, R., Nogues, X., Cáceres, E., Peñ, M. J., Mellibovsky, L., ... Hansma, P. K. (2010). Microindentation for in vivo measurement of bone tissue mechanical properties in humans. *Journal of Bone and Mineral Research*, 25(8), 1877–1885. Available from https://doi.org/10.1002/jbmr.73.

Everts, V., Delaissié, J. M., Korper, W., Jansen, D. C., Tigchelaar-Gutter, W., Saftig, P., & Beertsen, W. (2002). The bone lining cell: Its role in cleaning Howship's lacunae and initiating bone formation. *Journal of Bone and Mineral Research*, 17(1), 77–90. Available from https://doi.org/10.1359/jbmr.2002.17.1.77.

Feng, X., & McDonald, J. M. (2011). Disorders of bone remodeling. *Annual Review of Pathology: Mechanisms of Disease*, 6, 121–145. Available from https://doi.org/10.1146/annurev-pathol-011110-130203.

Forlino, A., Cabral, W. A., Barnes, A. M., & Marini, J. C. (2011). New perspectives on osteogenesis imperfecta. *Nature Reviews Endocrinology*, 7(9), 540–557. Available from https://doi.org/10.1038/nrendo.2011.81.

Garcia, D., Zysset, P. K., Charlebois, M., & Curnier, A. (2009). A three-dimensional elastic plastic damage constitutive law for bone tissue. *Biomechanics and Modeling in Mechanobiology*, 8, 149–165.

GBD. (2017). Disease and injury incidence and prevalence collaborators. Global, regional, and national incidence, prevalence, and years lived with disability for 354 diseases and injuries for 195 countries and territories, 1990–2017: A systematic analysis for the Global Burden of Disease Study 2017. *Lancet, 2018* (392), 1789–1858.

Glorieux, F. H., Ward, L. M., Rauch, F., Lalic, L., Roughley, P. J., & Travers, R. (2002). Osteogenesis imperfecta type VI. A form of brittle bone disease with a mineralization defect. *Journal of Bone and Mineral Research*, 17(1), 30–38. Available from https://doi.org/10.1359/jbmr.2002.17.1.30.

Grassi, L., & Isaksson, H. (2015). Extracting accurate strain measurements in bone mechanics: A critical review of current methods. *Journal of the Mechanical Behavior of Biomedical Materials*, 50, 43–54. Available from https://doi.org/10.1016/j.jmbbm.2015.06.006.

Hansma, P., Yu, H., Schultz, D., Rodriguez, A., Yurtsev, E. A., Orr, J., ... Lotz, J. (2009). The tissue diagnostic instrument. *The Review of Scientific Instruments*, 80(5), 054303.

Harrigan, T. P., & Mann, R. W. (1984). Characterization of microstructural anisotropy in orthotropic materials using a second rank tensor. *Journal of Materials Science*, 19(3), 761–767. Available from https://doi.org/10.1007/BF00540446.

Hildebrand, T., Laib, A., Müller, R., Dequeker, J., & Rüegsegger, P. (1999). Direct three-dimensional morphometric analysis of human cancellous bone: Microstructural data from spine, femur, iliac crest, and calcaneus. *Journal of Bone and Mineral Research*, 14(7), 1167–1174. Available from https://doi.org/10.1359/jbmr.1999.14.7.1167.

Imbert, L., Aurégan, J. C., Pernelle, K., & Hoc, T. (2014). Mechanical and mineral properties of osteogenesis imperfecta human bones at the tissue level. *Bone*, 65, 18–24. Available from https://doi.org/10.1016/j.bone.2014.04.030.

Jungmann, R., Szabo, M. E., Schitter, G., Yue-Sing Tang, R., Vashishth, D., Hansma, P. K., & Thurner, P. J. (2011). Local strain and damage mapping in single trabeculae during three-point bending tests. *Journal of the Mechanical Behavior of Biomedical Materials*, 4(4), 523–534. Available from https://doi.org/10.1016/j.jmbbm.2010.12.009.

Kanis, J. A. (2002). Diagnosis of osteoporosis and assessment of fracture risk. *Lancet*, 359, 1929–1936.

Keaveny, T. M., McClung, M. R., Wan, X., Kopperdahl, D. L., Mitlak, B. H., & Krohn, K. (2012). Femoral strength in osteoporotic women treated with teriparatide or alendronate. *Bone*, 50(1), 165–170. Available from https://doi.org/10.1016/j.bone.2011.10.002.

Kim, G., Cole, J. H., Boskey, A. L., Baker, S. P., & Van Der Meulen, M. C. H. (2014). Reduced tissue-level stiffness and mineralization in osteoporotic cancellous bone. *Calcified Tissue International*, 95(2), 125–131. Available from https://doi.org/10.1007/s00223-014-9873-4.

Kobayashi, S., Takahashi, H. E., Ito, A., Saito, N., Nawata, M., Horiuchi, H., ... Takaoka, K. (2003). Trabecular minimodeling in human iliac bone. *Bone, 32*(2), 163−169. Available from https://doi.org/10.1016/S8756-3282(02)00947-X.

Kontulainen, S., Sievänen, H., Kannus, P., Pasanen, M., & Vuori, I. (2002). Effect of long-term impact-loading on mass, size, and estimated strength of humerus and radius of female racquet-sports players: A peripheral quantitative computed tomography study between young and old starters and controls. *Journal of Bone and Mineral Research, 17*(12), 2281−2289. Available from https://doi.org/10.1359/jbmr.2002.17.12.2281.

Li, Z. C., Dai, L. Y., Jiang, L. S., & Qiu, S. (2012). Difference in subchondral cancellous bone between post-menopausal women with hip osteoarthritis and osteoporotic fracture: Implication for fatigue microdamage, bone microarchitecture, and biomechanical properties. *Arthritis and Rheumatism, 64*(12), 3955−3962. Available from https://doi.org/10.1002/art.34670.

Luczynski, K. W., Steiger-Thirsfeld, A., Bernardi, J., Eberhardsteiner, J., & Hellmich, C. (2015). Extracellular bone matrix exhibits hardening elastoplasticity and more than double cortical strength: Evidence from homogeneous compression of non-tapered single micron-sized pillars welded to a rigid substrate. *Journal of the Mechanical Behavior of Biomedical Materials, 52*, 51−62. Available from https://doi.org/10.1016/j.jmbbm.2015.03.001.

Macesic, L. J., & Summers, A. P. (2012). Flexural stiffness and composition of the batoid propterygium as predictors of punting ability. *Journal of Experimental Biology, 215*(12), 2003−2012. Available from https://doi.org/10.1242/jeb.061598.

Manda, K., Wallace, R. J., Xie, S., Levrero-Florencio, F., & Pankaj, P. (2017). Nonlinear viscoelastic characterization of bovine trabecular bone. *Environmental Economics and Policy Studies, 16*(1), 173−189. Available from https://doi.org/10.1007/s10237-016-0809-y.

Manolagas, S. C. (2000). Birth and death of bone cells: Basic regulatory mechanisms and implications for the pathogenesis and treatment of osteoporosis. *Endocrine Reviews, 21*(2), 115−137. Available from https://doi.org/10.1210/er.21.2.115.

Martelli, S., Giorgi, M., Dall'Ara, E., & Perilli, E. (2021). Damage tolerance and toughness of elderly human femora. *Acta Biomaterialia, 123*, 167−177.

Matsuo, K., & Otaki, N. (2012). Bone cell interactions through Eph/ephrin: Bone modeling, remodeling and associated diseases. *Cell Adhesion and Migration, 6*(2), 148−156. Available from https://doi.org/10.4161/cam.20888.

Matsuura, M., Eckstein, F., Lochmüller, E. M., & Zysset, P. K. (2008). The role of fabric in the quasi-static compressive mechanical properties of human trabecular bone from various anatomical locations. *Biomechanics and Modeling in Mechanobiology, 7*(1), 27−42. Available from https://doi.org/10.1007/s10237-006-0073-7.

Mercer, C., He, M. Y., Wang, R., & Evans, A. G. (2006). Mechanisms governing the inelastic deformation of cortical bone and application to trabecular bone. *Acta Biomaterialia, 2*(1), 59−68.

Miller, S. C., De Saint-Georges, L., Bowman, B. M., & Jee, W. S. S. (1989). Bone lining cells: Structure and function. *Scanning Microscopy, 3*(3), 953−961.

Nagaraja, S., Lin, A. S. P., & Guldberg, R. E. (2007). Age-related changes in trabecular bone microdamage initiation. *Bone, 40*(4), 973−980. Available from https://doi.org/10.1016/j.bone.2006.10.028.

Nakamura, T., Imai, Y., Matsumoto, T., Sato, S., Takeuchi, K., Igarashi, K., ... Kato, S. (2007). Estrogen prevents bone loss via estrogen receptor α and induction of fas ligand in osteoclasts. *Cell, 130*(5), 811−823. Available from https://doi.org/10.1016/j.cell.2007.07.025.

Ohman, C., Baleani, M., Perilli, E., Dall'Ara, E., Tassani, S., Baruffaldi, F., & Viceconti, M. (2007). Mechanical testing of cancellous bone from the femoral head: Experimental errors due to off-axis measurements. *Journal of Biomechanics, 40*, 2426−2433.

Paiva, K. B. S., & Granjeiro, J. M. (2017). *Matrix metalloproteinases in bone resorption, remodeling, and repair,* . *Progress in molecular biology and translational science* (Vol. 148, pp. 203−303). Elsevier B.V. Available from https://doi.org/10.1016/bs.pmbts.2017.05.001.

Palanca, M., Tozzi, G., & Cristofolini, L. (2016). The use of digital image correlation in the biomechanical area: A review. *International Biomechanics, 3*(1), 1−21. Available from https://doi.org/10.1080/23335432.2015.1117395.

Peyrin, F., Dong, P., Pacureanu, A., & Langer, M. (2014). Micro- and nano-CT for the study of bone ultra-structure. *Current Osteoporosis Reports, 12*(4), 465−474. Available from https://doi.org/10.1007/s11914-014-0233-0.

Qasim, M., Farinella, G., Zhang, J., Li, X., Yang, L., Eastell, R., & Viceconti, M. (2016). Patient-specific finite element estimated femur strength as a predictor of the risk of hip fracture: The effect of methodological determinants. *Osteoporosis International, 27*(9), 2815−2822. Available from https://doi.org/10.1007/s00198-016-3597-4.

Qin, L., Liu, W., Cao, H., & Xiao, G. (2020). Molecular mechanosensors in osteocytes. *Bone Research, 8*(1), 23. Available from https://doi.org/10.1038/s41413-020-0099-y.

Rauch, F., Travers, R., Parfitt, A. M., & Glorieux, F. H. (2000). Static and dynamic bone histomorphometry in children with osteogenesis imperfecta. *Bone, 26*(6), 581−589. Available from https://doi.org/10.1016/S8756-3282(00)00269-6.

Razi, H., Birkhold, A. I., Weinkamer, R., Duda, G. N., Willie, B. M., & Checa, S. (2015). Aging leads to a dysregulation in mechanically driven bone formation and resorption. *Journal of Bone and Mineral Research, 30*(10), 1864−1873. Available from https://doi.org/10.1002/jbmr.2528.

Razi, H., Birkhold, A. I., Zaslansky, P., Weinkamer, R., Duda, G. N., Willie, B. M., & Checa, S. (2015). Skeletal maturity leads to a reduction in the strain magnitudes induced within the bone: A murine tibia study. *Acta Biomaterialia, 13*, 301−310. Available from https://doi.org/10.1016/j.actbio.2014.11.021.

Rho, J. Y., Kuhn-Spearing, L., & Zioupos, P. (1998). Mechanical properties and the hierarchical structure of bone. *Medical Engineering & Physics, 20*, 92−102.

Rincón-Kohli, L., & Zysset, P. K. (2009). Multi-axial mechanical properties of human trabecular bone. *Biomechanics and Modeling in Mechanobiology, 8*(3), 195−208. Available from https://doi.org/10.1007/s10237-008-0128-z.

Roberts, B. C., Giorgi, M., Oliviero, S., Wang, N., Boudiffa, M., & Dall'Ara, E. (2019). The longitudinal effects of ovariectomy on the morphometric, densitometric and mechanical properties in the murine tibia: A comparison between two mouse strains. *Bone, 127*, 260−270.

Roberts, B. C., Perilli, E., & Reynolds, K. J. (2014). Application of the digital volume correlation technique for the measurement of displacement and strain fields in bone: A literature review. *Journal of Biomechanics, 47*(5), 923−934. Available from https://doi.org/10.1016/j.jbiomech.2014.01.001.

Rohl, L., Larsen, E., Linde, F., Odgaard, A., & Jorgensen, J. (1991). Tensile and compressive properties of cancellous bone. *Journal of Biomechanics, 24*, 1143−1149.

Ryan, M. K., Oliviero, S., Costa, M. C., Mark Wilkinson, J., & Dall'ara, E. (2020). Heterogeneous strain distribution in the subchondral bone of human osteoarthritic femoral heads, measured with digital volume correlation. *Materials, 13*(20), 1−18. Available from https://doi.org/10.3390/ma13204619.

Schwiedrzik, J. J., & Zysset, P. K. (2013). An anisotropic elastic-viscoplastic damage model for bone tissue. *Biomechanics and Modeling in Mechanobiology, 12*(2), 201−213. Available from https://doi.org/10.1007/s10237-012-0392-9.

Ström, O., Borgström, F., Kanis, J. A., Compston, J., Cooper, C., McCloskey, E. V., & Jönsson, B. (2011). Osteoporosis: Burden, health care provision and opportunities in the EU. *Archives of Osteoporosis, 6*(1−2), 59−155. Available from https://doi.org/10.1007/s11657-011-0060-1.

Turner, C. H. (1989). Yield behavior of bovine cancellous bone. *Journal of Biomechanical Engineering, 111* (3), 256–260. Available from https://doi.org/10.1115/1.3168375.

Turner, C. H. (2006). Bone strength: Current concepts. *Annals of the New York Academy of Sciences, 1068*(1), 429–446. Available from https://doi.org/10.1196/annals.1346.039, Blackwell Publishing Inc.

Van Tol, A. F., Schemenz, V., Wagermaier, W., Roschger, A., Razi, H., Vitienes, I., ... Weinkamer, R. (2020). The mechanoresponse of bone is closely related to the osteocyte lacunocanalicular network architecture. *Proceedings of the National Academy of Sciences of the United States of America, 117*(51), 32251–32259. Available from https://doi.org/10.1073/pnas.2011504117.

Varga, P., & Zysset, P. K. (2009). Assessment of volume fraction and fabric in the distal radius using HR-pQCT. *Bone, 45*(5), 909–917. Available from https://doi.org/10.1016/j.bone.2009.07.001.

Varga, P., Dall'Ara, E., Pahr, D. H., Pretterklieber, M., & Zysset, P. K. (2011). Validation of an HR-pQCT-based homogenized finite element approach using mechanical testing of ultra-distal radius sections. *Biomechanics and Modeling in Mechanobiology, 10*(4), 431–444. Available from https://doi.org/10.1007/s10237-010-0245-3.

Whitehouse, W. J. (1974). The quantitative morphology of anisotropic trabecular bone. *Journal of Microscopy, 101*(2), 153–168. Available from https://doi.org/10.1111/j.1365-2818.1974.tb03878.x.

Wolfram, U., Gross, T., Pahr, D. H., Schwiedrzik, J., Wilke, H. J., & Zysset, P. K. (2012). Fabric-based Tsai-Wu yield criteria for vertebral trabecular bone in stress and strain space. *Journal of the Mechanical Behavior of Biomedical Materials, 15,* 218–228. Available from https://doi.org/10.1016/j.jmbbm.2012.07.005.

Wolfram, U., Wilke, H. J., & Zysset, P. K. (2010). Rehydration of vertebral trabecular bone: Influences on its anisotropy, its stiffness and the indentation work with a view to age, gender and vertebral level. *Bone, 46* (2), 348–354. Available from https://doi.org/10.1016/j.bone.2009.09.035.

Wolfram, U., Wilke, H. J., & Zysset, P. K. (2011). Damage accumulation in vertebral trabecular bone depends on loading mode and direction. *Journal of Biomechanics, 44*(6), 1164–1169. Available from https://doi.org/10.1016/j.jbiomech.2011.01.018.

Wu, D., Isaksson, P., Ferguson, S. J., & Persson, C. (2018). Young's modulus of trabecular bone at the tissue level: A review. *Acta Biomaterialia, 78,* 1–12. Available from https://doi.org/10.1016/j.actbio.2018.08.001.

Xie, S., Wallace, R. J., & Pankaj, P. (2020). Time-dependent behaviour of demineralised trabecular bone - Experimental investigation and development of a constitutive model. *Annals of Biomedical Engineering, 45*(5), 229–248.

Zioupos, P., Currey, J. D., & Casinos, A. (2000). Exploring the effects of hypermineralisation in bone tissue by using an extreme biological example. *Connective Tissue Research, 41*(3), 229–248. Available from https://doi.org/10.3109/03008200009005292.

Zysset, P. K. (2003). A review of morphology-elasticity relationships in human trabecular bone: Theories and experiments. *Journal of Biomechanics, 36*(10), 1469–1485. Available from https://doi.org/10.1016/S0021-9290(03)00128-3.

Zysset, P. K. (2009). Indentation of bone tissue: A short review. *Osteoporosis International, 20*(6), 1049–1055. Available from https://doi.org/10.1007/s00198-009-0854-9.

Zysset, P. K., & Curnier, A. (1995). An alternative model for anisotropic elasticity based on fabric tensors. *Mechanics of Materials, 21*(4), 243–250. Available from https://doi.org/10.1016/0167-6636(95)00018-6.

Zysset, P. K., Schwiedrzik, J., & Wolfram, U. (2016). European society of biomechanics S.M. Perren Award 2016: A statistical damage model for bone tissue based on distinct compressive and tensile cracks. *Journal of Biomechanics, 49*(15), 3616–3625. Available from https://doi.org/10.1016/j.jbiomech.2016.09.045.

MUSCLE BIOMECHANICS

7

Dan Robbins

School of Allied Health, Anglia Ruskin University, Chelmsford, United Kingdom

INTRODUCTION

Depending on age, gender, and training status, muscles make between 60% and 85% of body mass. There are three broad categories of muscle: skeletal, cardiac, and smooth. Skeletal muscles control movement and posture. Cardiac muscle controls the functioning of the heart and smooth muscles control the so called "hollow organs," that is, those found in the gastrointestinal tract, in addition to being a component in blood vessel walls. There are similarities between the three muscle types. This chapter, will focus on skeletal muscles.

Skeletal muscles consist of thousands of small fibers grouped and aligned, optimizing the performance of its functions. The vast majority of muscles attach proximally and distally to bones via tendons; there are some exceptions to the rule, for example, muscles that attach to the eyeball have no distal bony attachment. However, an overwhelming majority of muscles in the body are attached to bones. The gross structure of muscles is divided into groups or bundles defined by the soft tissues that enclose them (see Fig. 7.1).

TERMINOLOGY

To understand which sections of muscle structure the names refer to, it is important to know about the naming conventions. In terms of muscle structure, the words can be broken down into three parts. The first part of the words is referred to as "**prefix**," the middle section of the word is referred to as a "**root**," and the end part is referred to as a "**suffix**." Table 7.1 displays the word sections and their meanings.

Note: The table displays the root as my(o). The (o) indicates that the most common connecting vowel for this root is "o." However, when my(o) is used in words describing muscle structure, the root is used without a connecting vowel, though the suffix has a preceding connecting vowel "i," for example, epimysium. While it is very useful to learn medical terminology to understand, or interpret, the complex words that are used in medical science, it is outside the scope of this chapter to cover medical terminology in detail. There are a variety of online resources and/or textbooks that can be referred to learn medical terminology, and the reader is encouraged to do so.

Human Orthopaedic Biomechanics. DOI: https://doi.org/10.1016/B978-0-12-824481-4.00009-3

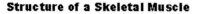

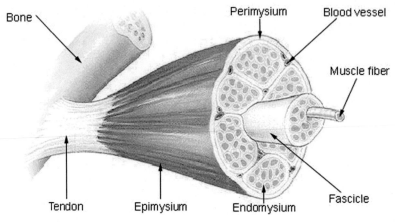

FIGURE 7.1

Muscle structure.

Table 7.1 Medical terminology/word components.

Prefix	Meaning	Root	Meaning	Suffix	Meaning
Epi	Upon/outer	my(o)	Muscle	um	Referring to
Peri	Around	my(o)	Muscle	um	Referring to
Endo	Within/inside	my(o)	Muscle	um	Referring to

ANATOMY

The outermost layer of muscle tissue is the **epimysium**, which is also known as the **epimysial fascia**, a component of the **deep fascia** that surrounds entire muscles. The epimysial fascia is different from the deep **aponeurotic fascia**, which is connected to tendons (Gatt & Zito, 2018). While the deep fasciae have a similar structure, when creating mathematical models of muscle structure and force, the epimysial and aponeurotic fasciae must be considered separately. The perimysium encircles groups of muscle fibers, collectively known as a **muscle fascicle**. Finally, the **endomysium**, which is a thin sheath of connective tissue, surrounds the muscle fibers. The smallest component shown in Fig. 7.1 is a muscle fiber, which is the contractile element of muscles and can be classified as an individual muscle cell (Herzog & Epstein, 1998).

A muscle fiber is composed of two types of **myofibrils**, **actin** and **myosin**, which are highly organized in structure and give muscle a striped, or "striated," appearance, which can be viewed using a standard light microscope. Actin filaments are globular proteins that form double helical strands, whereas myosin filaments consist of two components: a globular head that interacts with the actin filaments and a tail that interacts with other myosin molecules to form the combined myosin filaments (Jones, Round, & de Haan, 2004). The size and location of the myofibril striations are used to develop a mapping process where regions of a myofibril are divided into "bands" (see Fig. 7.2).

A region within a myofibril can be measured from one "z-disc," sometimes referred to as a "z-line," to next and is referred to as a "**sarcomere**." The core components of the sarcomere are the actin and myosin filaments, though it should be noted that the myosin filaments are fixed to the z-discs at each end by **titin**, a giant protein that acts as a molecular spring to fix the myosin in place (Granzier & Labeit, 2004; Tskhovrebova & Trinick, 2003).

There are two "bands": the "I-band" and "A-band." I-band represents the section where the z-disc is located. Typically, the I-band will therefore appear lighter and contribute to the striated appearance of muscle fibers. However, due to the contractile nature of muscle fibers, the size of the I-bands changes (Goodman & Zimmer, 2007). During rest and/or at times when muscles fiber lengths are increased, the I-band contains only the z-disc and the actin filament. However, during times of maximal shortening of muscle fiber lengths, the myosin filaments is also found within the I-band region. The majority of the "A-band" represents a section that contains both the actin and

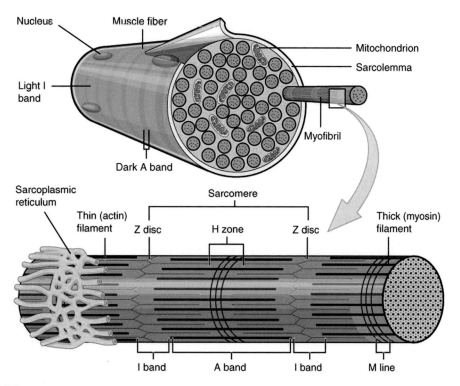

FIGURE 7.2

Muscle structure with striated myofibril.

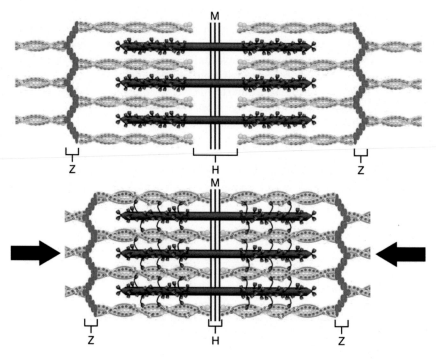

FIGURE 7.3

Sarcomere structure during contraction.

myosin filaments, with the middle of the A-band being a region referred to as the "H-zone," which normally contains only myosin filaments (Tortora & Derrickson, 2009). During maximally shortened muscle fiber states, the H-zone disappears due to the actin filaments moving closer to each other, though the length of the A-band remains the same (Goodman & Zimmer, 2007).

Considering representations such as that shown in Fig. 7.3, it is easy to consider filaments working in two dimensions, interaction with filaments above and below, to create contractions. However, muscles are three-dimensional structures. Each myosin filament has five actin fibers creating a pentagon shape around them. Each actin fiber interacts with three myosin filaments creating a three-dimensional structure with a distance of 42 nm between each myosin filament (Herzog & Epstein, 1998).

SLIDING FILAMENT THEORY

The molecular process of muscle contraction is defined by the **sliding filament theory**. A detailed overview of the sliding filament theory is outside the scope of this chapter. In brief, the sliding filament theory explains how the actin and myosin filaments interact to create a muscle contraction. In a resting state, actin filaments are surrounded by a regulatory protein called *tropomyosin*, which follows an alpha helical groove of the actin filament and is bounded by a second regulatory protein *troponin*

(Tortora & Derrickson, 2009). Tropomyosin does not make direct contact with actin, but maintains a distance of 3.9 − 4.0 nm directly above the surface (Geeves, 2012). When calcium released from the sarcoplasmic reticulum it binds with troponin C and causes a structural change in tropomyosin altering the alignment between tropomyosin and actin. The reconfigured alignment of tropomyosin uncovers the myosin binding sites, allowing crossbridges to be formed when the globular heads of the myosin filaments engage with myosin binding sites on the actin filaments (Leiber, 2002). Once the myosin heads are bound to the actin filaments, energy from adenosine triphosphate creates movement known as a *power stroke* by the rotation of the globular head of myosin. The power stroke changes the relative positions of actin and myosin and subsequently creates a muscle contraction (see Fig. 7.4). The reader is encouraged to refer to a textbook (or chapter) on muscle physiology to gain a better understanding of the underlying molecular mechanisms of muscle contraction.

BIOMECHANICS

From a mechanical point of view, there are two primary factors that influence how much force a muscle fiber can generate: the size of the muscle fiber and the orientation of the muscle fiber. Human sarcomeres are typically between 2.5 and 4.5 μm in length (Zatsiorsky & Prilutsky, 2012), though it should be noted that the overall muscle length affects the length of the sarcomere when taking measurements (Lichtwark, Farris, Chen, Hodges, & Delp, 2018; Son, Indresano, Sheppard, Ward, & Lieber, 2018). The diameter of the fibers will depend on the fiber type and training status, as exercise obviously influences muscle size and structure. The orientations of muscle fibers within the human body are diverse (see Fig. 7.5).

DASHPOT DIAGRAMS

Muscle fibers that align end to end with each other are referred to as being in **series**, while fibers that align side by side are referred to as **parallel** fibers. The orientation of fibers influences the overall function of the muscle (Richards, 2008). The same terminology is used when referring to the relative alignment of the noncontractile tissues, for example, tendons and epimysium. Muscles with fibers arranged in series generally contracts more, that is, the distance between the proximal and distal tendons is reduced more than in muscles with fibers arranged in a parallel orientation. Muscles with fibers arranged in a parallel orientation generally generate greater force. The alignment of the structure of muscles and tendons can be modeled and represented visually using **dashpot diagrams** (Herzog & Epstein, 1998; Nordin & Frankel, 2001) (see Fig. 7.6).

In dashpot diagrams, the **series elastic element** refers to the proximal or distal tendons and the deep aponeurotic fascia. Typically, only one series elastic element is displayed as the diagram is representative of one sarcomere. Within the body, sarcomeres connect with each other; therefore, to model a whole muscle, many dashpot diagrams would be required (each connected to one another). The **contractile element** of the dashpot diagram represents the muscle fibers, which provide the contraction and active resistance as required. The **parallel elastic element** of dashpot diagrams represents the epimysial fascia, which surrounds muscle tissue. Note that while the overall orientation of muscle fibers might be classed as parallel, the structural components still include series components (Robertson, Caldwell, Hamill, Kamen, & Whittlesey, 2014). When generating models

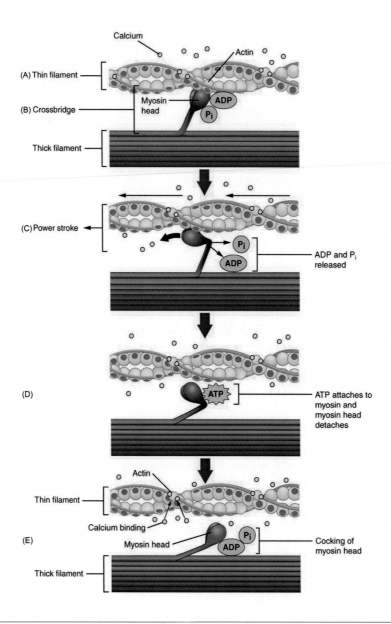

FIGURE 7.4

Interaction between actin and myosin filaments during muscle fiber contractions. (A) Displays the actin filament. (B) Displays the connection form between myosin and actin filaments at the crossbridge. (C) Displays the power cycle creating tension. (D) Displays the point at which crossbridge detaches and ends the tension producing capabilities of the crossbridge. (E) Displays the realignment of the myosin head in preparation for a new power cycle.

Muscle Types

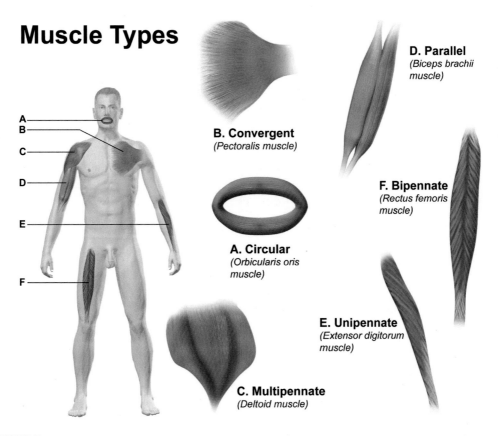

D. Parallel
(Biceps brachii muscle)

B. Convergent
(Pectoralis muscle)

F. Bipennate
(Rectus femoris muscle)

A. Circular
(Orbicularis oris muscle)

E. Unipennate
(Extensor digitorum muscle)

C. Multipennate
(Deltoid muscle)

FIGURE 7.5

Examples of muscle fibers with different fiber orientations.

of pennate fiber orientations, the pennation angle should be included—particularly if the force produced and its resultant effects are calculated. The contribution of contractile and noncontractile (series and parallel) elements to overall **tension** can be represented graphically via **length—tension curves** (Bartel, Davy, & Keaveny, 2006) (see Fig. 7.7).

LENGTH—TENSION CURVES

Length—tension curves show the relation between the length of one **sarcomere** and the tension that can be generated. The length—tension relationship of a whole musculotendinous unit is different and should not be confused with the relationship presented in Fig. 7.7. The **active tension** in the length—tension curve is generated by the contractile element of the sarcomere, that is, the interaction between the myofibrils. The peak tension is produced when sarcomeres are at their resting length, as this provides the optimum alignment between the actin and myosin filaments. When the length of

Basic Hill Model (a)

Supplemented Hill Model (b)

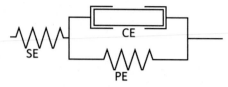

Supplemented Hill Model (c)

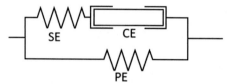

Pennate Muscle Model (d)

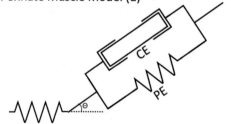

FIGURE 7.6

Examples of dashpot diagrams: Panel (A) denotes a parallel muscle with components in series; panels (B) and (C) denote a parallel muscle with components both in series and parallel; and panel (D) denotes a pennate muscle with components both in series and in parallel.

sarcomeres shortens, the actin filaments are pulled along the myosin filaments, which, in turn, pulls the z-lines closer to the myosin filaments. At extremely shortened lengths, titin is compressed between the myosin and the z-lines, and the actin filaments overlap each other centrally. A consequence of extreme shortening is that there is no further potential for myofibrils to contract, and little to no force can be generated. During sarcomere lengthening, actin moves in the opposite direction along the myosin. At extreme lengths, the myosin binding sites on the actin filaments move out of the range of the globular heads of the myosin filaments. As fewer myosin heads can bind with the actin filaments, lower force is generated from myofibril contractions (McMahon, 1984). **Passive tension** in the length–tension curve is generated by the parallel and series elements and therefore does not contribute to the generation of tension with shortened lengths (Nordin & Frankel, 2001). According to Fig. 7.7, below resting lengths, the passive tension line is absent. Technically, the

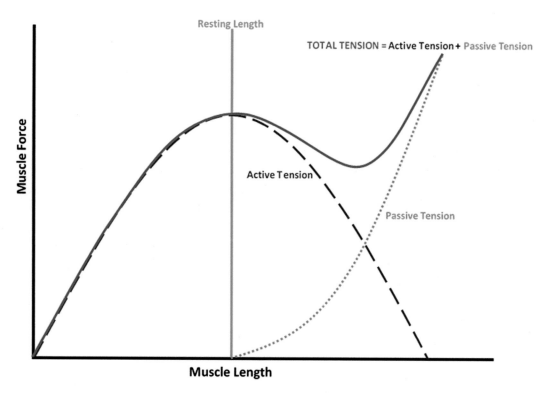

FIGURE 7.7

Graphical representation of the length–tension curve of one sarcomere.

passive line should be at zero force for lengths shorter than resting length, though, typically, the line is absent in this region of length–tension curves. As sarcomere lengths increase, the passive tissues reach their full length and start to provide resistance to further increases in length. Note that the ratio of the contribution of the passive tissues to the total tension increases until the maximum lengths are reached. At maximal lengths, tension, or resistance to increases in length, is virtually all passive.

CONTRACTION TYPES

While the length–tension curve is a useful tool to conceptualize the relationship between active and passive components of musculotendinous units, there are a variety of factors that can influence the relationship between the force a muscle generates and a muscle length. Muscular contractions are divided into three categories:

1. **Concentric**
   a. During contractions, the distance between the proximal and distal ends of the muscle reduces.
2. **Isometric**
   a. During contractions, the distance between the proximal and distal ends of the muscle does not change.

3. Eccentric

a. During contractions, the distance between the proximal and distal ends of the muscle increases.

The three muscle contraction types are sometimes described in terms of velocity, which is a vector and therefore has both a magnitude and a direction. A concentric contraction results in a positive velocity, an isometric contraction has zero velocity, and an eccentric contraction results in a negative velocity (see Fig. 7.8).

The greatest forces recorded in muscles result from eccentric contractions, followed by isometric contractions and finally concentric contractions. **Power** can be calculated by multiplying the velocity of a muscle contraction by the force generated; note that **peak power** occurs at one-third positive velocity and one-third concentric power. Intuitively, the greatest powers would be linked

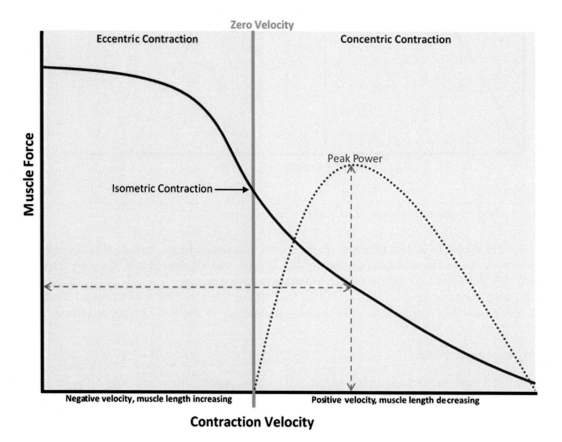

FIGURE 7.8

Force–velocity–power curve. Note that peak power is achieved at one-third positive velocity and one-third concentric force.

to the greatest forces, though it is important to remember that the highest forces occur in eccentric contractions which have a negative velocity; therefore, the resulting power is also negative.

FIBER LEVEL MODELING

An additional factor in force production is the alignment of fibers, both whether the alignment is a series of parallel orientation and the pennation angle. In pennate muscles, the force, fiber length, and pennation angle all change during all contraction types (see Fig. 7.9).

At the whole muscle level, there is no change in the muscle length during an isometric contraction, though at the sarcomere level a reduction in muscle length and increase in pennation angle can be observed (see row A in Fig. 7.9). Note that the forces presented in the concentric force (row B in Fig. 7.9) are equivalent to the active tension line displayed in the length–tension curve in Fig. 7.7 though as the contraction starts from resting length only half of the active tension line is displayed. Both concentric and isometric contractions result in an increase in pennation angle and force, while the fiber length decreases. Eccentric contractions result in increase in fiber length and force, while pennation angle decreases. For illustrative purposes, Fig. 7.9 is generated on the basis that both concentric and eccentric contractions have continuous changes in length. In reality, when testing whole muscles contractions will either reach an end of range and change to an isometric contraction at that position or the contracting muscle will change to an antagonistic muscle and the movement will reverse. It is important to note that the relationships presented in Fig. 7.9 are generic representations and that due to the biomechanical and physiological differences in muscle fibers it is not possible to represent all muscles in a simple figure. For the purposes of this chapter, the biomechanical difference in fiber type is limited to that displayed in Fig. 7.10.

Fast-twitch fibers respond quickly and explosively but are not resistant to fatigue. Slow-twitch fibers do not respond as quickly and generate less force but are much more resistance to fatigue. Intermediate fibers, as the name suggests, perform somewhere in between fast- and slow-twitch fibers. Intermediate fibers are most easily influenced by lifestyle; therefore, people who engage in regular long runs will have intermediate fibers, which are more similar to slow fibers, and people who engage in fast, explosive style of activities will have intermediate fibers, which are more similar to fast-twitch fibers. There is a range of terminology used to represent muscle fiber types, Table 7.2 outlines the most common naming conventions.

In pennate muscles, the contraction force is not equal to the force generated by the individual fiber due to the difference in alignment. The difference in force can be calculated using trigonometry, provided the muscle fiber length and pennation angle can be measured, which is typically achieved using ultrasound imaging techniques. During force analysis, muscle fiber length is considered the length of the hypotenuse of the force triangle, the adjacent side of the triangle is typically aligned with superficial or deep aponeurosis, and is therefore the adjacent side of the force triangle. The force along the adjacent line is effectively the tensile force which is transmitted between the proximal and distal tendons. The opposite side of the force triangle is a stress force which is transmitted in to hydrostatic force (Sporrong & Styf, 1999). The length of the opposite side of the force triangle is the distance between the superficial and deep fasciae (see Fig. 7.11).

While trigonometry offers a simple and straightforward method to analyze forces, it should be noted that there is no joint in the human body that moves in a straight line. In order for a

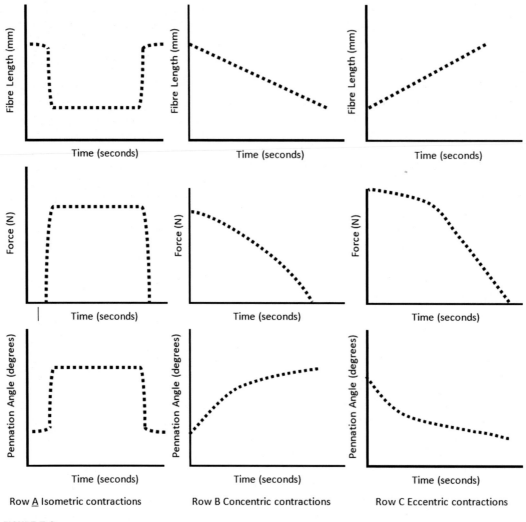

FIGURE 7.9

Contraction parameters. The relationship between force, length, and pennation angle in isometric, concentric, and eccentric muscle contractions.

portion of the human body to move in a straight line, multiple joints are required to work in coordination. Muscle contractions are transmitted to bone via tendons close to articulations (joints), the tensile forces generated results in a circular movement with a distal bone moving towards a proximal bone. When considering forces such as those from human joints, which move on a circular path (curvilinear motion), the ultimate force produced is calculated by multiplying the linear force applied (in this case the muscle force) and the moment arm or the displacement between the line of force and the point of rotation (in this case the articulation).

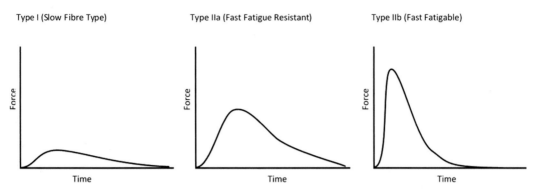

FIGURE 7.10

Fiber type contraction characteristics. Differences in rate of force development, magnitude, and duration of force generated across fiber types.

Table 7.2 Muscle fiber naming conventions.					
Slow	**Slow twitch**	**Type I**	**Type I**	**Slow oxidative**	**Slow**
Intermediate	Fast twitch A	Type IIa	Type IIA	Fast oxidative glycolytic	Fast, fatigue resistant
Fast	Fast twitch B	Type IIb	Type IIX	Fast twitch glycolytic	Fast, fatigable

FIGURE 7.11

Constituent forces for muscle fibers. Calculating constituent forces for muscle fibers. Note that the muscle fiber lengths represent the hypotenuse of the force triangle. Angles Θa and Θb will the same, and therefore either can be measured and used for calculations.

The resulting product of **force** multiplied by **moment arm is torque**, which is measured in newton meters (N m). As multiple muscles typically contribute to movement at any given joint, the relative contribution can be calculated using the physiological cross-sectional areas of muscles, combined with the moment arm and angle at which each muscle attaches to the distal bone (Watkins, 2013).

ELECTROMYOGRAPHY

Due to the technical challenges of directly measuring muscle force, one tool that is frequently used to assess muscle activation is **electromyography**, typically referred to by the abbreviation **EMG**. EMG measures electrical activity generated by the nervous system and propagated by muscles fibers to initiate and control muscle contractions (Criswell, 2011). EMG signals are measured in both the time and frequency domains (Robertson et al., 2014), providing parameters such as onset (start of contraction), offset (end of contraction), and magnitude in the time domain, plus frequency spectrums and mean or median frequencies in the frequency domain. Advanced techniques can combine domains to complete time−frequency analyses such as the short-term Fourier transform, the Hilbert−Huang transform, Wavelets, and Empirical Mode Decomposition. Time−frequency analysis has the advantage of containing more information than analyses that separate the time and frequency domains, though subsequently it can be more difficult to interpret. Both frequency and time−frequency analyses are popular in applications investigating muscle fatigue. There is a growing interest in machine learning and deep learning techniques for analysis of EMG signals, particularly in clinical applications.

EMG signals can be acquired using a variety of different sensor types, broadly grouped into either **surface EMG** or **intramuscular EMG**. Intramuscular EMG is acquired either using fine wires inserted into muscles using small needles or needle electrodes that are inserted directly into muscles. Both fine wire and needle electrodes record the electrical activity of muscles in very specific locations and are much less likely to record electrical activity of muscles located closely to the muscle being examined, an issue referred to as "cross-talk." Due to the highly specific nature of the signal acquired, intramuscular EMG is the tool of choice for most clinical examinations by medical doctors specializing in neurophysiology. Surface EMG sensors include monopolar, that is, single site, bipolar, i.e., two sites very close together with a differential signal acquired, and arrays which have multiple sensors. The number of sensors in an array is only limited by the space on the surface of the skin and the number of inputs the recording amplifier can receive. Due to the noninvasive nature of the signal acquisition, surface EMG is far more common than intramuscular EMG in both research and sporting applications.

Advances in the design of amplifiers, sensors, signal processing, and software have resulted in a renewed interest in surface EMG decomposition, a process that decomposes the EMG signal into individual motor units to allow a more detailed comparison of the motor unit contributions to muscle contractions. For a detailed review of advances in EMG decomposition, please review by De Luca, Chang, Roy, Kline, & Hamid Nawab (2015). Incorporating EMG signals into computer simulations can generate EMG-driven forward-dynamic estimations of muscle force and free software such as OpenSim, making such simulations much more accessible (Seth et al., 2018). There is also increasing interest in high-density EMG which is frequently acquired via EMG array electrodes.

REFERENCES

Bartel, D., Davy, D., & Keaveny, T. (2006). *Orthopaedic biomechanics: Mechanics and design in musculo-skeletal systems*. Pearson.

Criswell, E. (2011). *Cram's introduction to surface electromyography* (2nd ed.). Jones and Bartlett Learning.

De Luca, C. J., Chang, S. S., Roy, S. H., Kline, J. C., & Hamid Nawab, S. (2015). Decomposition of surface EMG signals from cyclic dynamic contractions. *Journal of Neurophysiology*, *113*(6), 1941–1951. Available from https://doi.org/10.1152/jn.00555.2014.

Gatt, A., & Zito, P. M. (2018). *Anatomy, skin, fascias*. Treasure Island (FL): StatPearls Publishing.

Geeves, M. A. (2012). Thin filament regulation. *Comprehensive Biophysics*, *4*, 251–267. Available from https://doi.org/10.1016/B978-0-12-374920-8.00416-1.

Goodman, S. R., & Zimmer, W. E. (2007). *Cytoskeleton. Medical cell biology* (3rd ed., pp. 59–100). Elsevier Inc. Available from https://doi.org/10.1016/B978-0-12-370458-0.50008-6.

Granzier, H. L., & Labeit, S. (2004). The giant protein titin: A major player in myocardial mechanics, signaling, and disease. *Circulation Research*, *94*(3), 284–295. Available from https://doi.org/10.1161/01. RES.0000117769.88862.F8.

Herzog, W., & Epstein, M. (1998). *Theoretical models of skeletal muscle, biological and mathematical considerations. Biological and mathematical considerations*. John Wiley and Sons, Inc.

Jones, D., Round, J., & de Haan, A. (2004). *Skeletal muscle from molecules to movement: A textbook of muscle physiology for sport, exercise, physiotherapy and medicine* (pp. 1–202). Elsevier Inc. Available from https://doi.org/10.1016/B978-0-443-07427-1.X5001-8.

Leiber, R. (2002). *Skeletal muscle structure, function, and plasticity* (pp. 374–386). Lippincott Williams & Wilkins.

Lichtwark, G. A., Farris, D. J., Chen, X., Hodges, P. W., & Delp, S. L. (2018). Microendoscopy reveals positive correlation in multiscale length changes and variable sarcomere lengths across different regions of human muscle. *Journal of Applied Physiology*, *125*(6), 1812–1820. Available from https://doi.org/ 10.1152/japplphysiol.00480.2018.

McMahon, T. (1984). *Muscles, reflexes, and locomotion*. Princeton University Press.

Nordin, M., & Frankel, V. (2001). *Basic biomechanics of the musculoskeletal system*. Lippincott Williams & Wilkins.

Richards, J. (2008). *Biomechanics in clinic and research*. Churchill Livingstone Elsevier.

Robertson, D., Caldwell., Hamill, J., Kamen, G., & Whittlesey, S. N. (2014). *Research methods in biomechanics* (2nd ed.). Human Kinetics.

Seth, A., Hicks, J., Uchida, T., Habib, A., Dembia, C., Dunne, J., et al. (2018). Simulating musculoskeletal dynamics and neuromuscular control to study human and animal movement. *PLoS Computational Biology*, *14*(7), e1006223.

Son, J., Indresano, A., Sheppard, K., Ward, S. R., & Lieber, R. L. (2018). Intraoperative and biomechanical studies of human vastus lateralis and vastus medialis sarcomere length operating range. *Journal of Biomechanics*, *67*, 91–97. Available from https://doi.org/10.1016/j.jbiomech.2017.11.038.

Sporrong, H., & Styf, J. (1999). Effects of isokinetic muscle activity on pressure in the supraspinatus muscle and shoulder torque. *Journal of Orthopaedic Research*, *17*(4), 546–553. Available from https://doi.org/ 10.1002/jor.1100170413.

Tortora, G., & Derrickson, B. (2009). (12th ed.). *Principles of anatomy and physiology (Maintenance and continuity of the human body)*, (vol. 2). John Wiley and Sons, Inc.

Tskhovrebova, L., & Trinick, J. (2003). Titin: Properties and family relationships. *Nature Reviews Molecular Cell Biology*, *4*(9), 679–689. Available from https://doi.org/10.1038/nrm1198.

Watkins, J. (2013). *Structure and function of the musculoskeletal system*. Human Kinetics.

Zatsiorsky, V., & Prilutsky, B. (2012). *Biomechanics of skeletal muscle*. Human Kinetics.

FURTHER READING

Enoka, R. (2002). *Neuromechanics of human movement* (3rd ed.). Human Kinetics.

LIGAMENT AND TENDON BIOMECHANICS

Fabio Galbusera[1] and Bernardo Innocenti[2]

[1]*IRCCS Istituto Ortopedico Galeazzi, Milan, Italy* [2]*Department of Bio Electro and Mechanical Systems (BEAMS), École Polytechnique de Bruxelles, Université Libre de Bruxelles, Brussels, Belgium*

ANATOMY, STRUCTURE, AND FUNCTION

Ligaments and tendons are fibrous connective tissues that connect the structures of the musculo-skeletal systems, enabling proper force transmission and motion of the human joints. Ligaments link two distinct bones; for example, the cruciate ligaments of the knee joint connect the intercondylar area of the tibia with the intercondylar notch of the femur. Tendons provide the link between muscles and bone, thus transmitting the force generated by the muscles themselves; an example is the Achilles tendon, which connects the plantaris, the soleus, and the gastrocnemius muscle with the calcaneus (heel) bone. It should be noted that other anatomical structures denominated ligaments exist in the human body, namely the periodontal ligaments, the peritoneal ligaments, as well as several remnants of tubular structures from the fetal period, but they are distinct from the true musculoskeletal ligaments and are not described in this chapter.

Unlike muscles, which are active tissues, ligaments and tendons are passive tissues, that id, not capable of contracting and generating mechanical loads, sharing many similarities between them in terms of anatomy, composition, and structure. They have a string-like or membrane-like shape with either a round or flattened cross-section depending on their location in the body and their functions; for example, extensor tendons are flat in shape, while flexors are more round or oval (Benjamin, Kaiser, & Milz, 2008; Buschmann & Burgisser, 2017). Both ligaments and tendons are composed of packed, oriented collagen fibers embedded in a water-based ground substance. While water constitutes 60%−80% of the wet weight, 65%−85% of the dry weight consists of collagen, almost exclusively type I; proteoglycans (1%−5% in tendons, 1%−3% in ligaments), and elastin (1%−3% in tendons, 1%−15% in ligaments) are also present (Buckwalter, Einhorn, & Simon, 2000; Buschmann & Burgisser, 2017; Kannus, 2000). Being in series with the muscle, in tendons, the collagen fibrils are aligned with the direction of the muscle, while in ligaments, the fibrils have a more complex orientation; for example, in the anterior cruciate ligament, the fascicles are spirally wound about each other.

Ligaments and tendons have a specific hierarchical structure (Fig. 8.1), which has the precursor of the collagen I molecule, or procollagen, as its elementary unit. Three procollagen molecules join to form tropocollagen, a triple helix that forms the basis of a connective tissue found in fibrous tissues such as ligaments, tendons, cartilage, intervertebral discs, and blood vessels.

Human Orthopaedic Biomechanics. DOI: https://doi.org/10.1016/B978-0-12-824481-4.00016-0

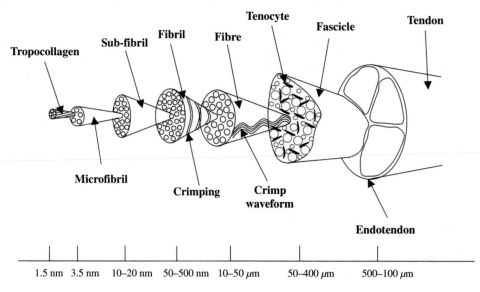

FIGURE 8.1

The hierarchical structure of tendons, from the triple-helix tropocollagen molecule to the fascicle and, finally, the whole tendon. The architecture of ligaments shares many similarities with that of tendons.

From Screen, H. R. C., Bader, D. L., Lee, D. A., & Shelton, J. C. (2004). Local strain measurement within tendon. Strain, 40(4), 157–163. https://onlinelibrary.wiley.com/doi/pdf/10.1111/j.1475-1305.2004.00164.x?casa_token = WGI2JUpm96MAAAAA: ZYzvDx7qv1Ab9ZgFaqE4fumdvHesXkNpcDBb320acjAzglU78My_y1b2Zb-gNfOg9LMEOk1PdWpfiLg.

A bundle of five tropocollagen molecules forms a collagen microfibril, which, in turn, join to form subfibrils, fibrils, and, finally, fibers, also named primary bundles (Fig. 8.1). Fibrils are organized in the fibers in a crimped pattern, which disappears when the tissue is subjected to elongation. The successive hierarchical level is the fascicle, or secondary bundle, which also includes cells and extracellular matrix; a set of fascicles then builds up the whole tendon or ligament. In tendons, fascicles are separated by endotenons, thin connective layers with high cell density, which facilitate the interfiber sliding. Other layers encompassing fascicles, namely epitenon and paratenon, exist in some tendons.

BIOMECHANICAL PROPERTIES

Coherently with their string-like nature, most ligaments and tendons can transmit forces mostly along their long (main) axis, while their response to bending and torsion is less significant and commonly neglected. Uniaxial testing of cadaveric specimens by means of standard material testing systems is therefore the most common and preferred method for the investigation of the biomechanics of these tissues and has indeed been used in a large number of studies. However, it should be noted that some ligaments have more complex, membrane-like or capsule-like shapes with a spatially variable orientation of the collagen fibers, which may require biaxial testing for an appropriate biomechanical characterization.

The uniaxial load–displacement curves of ligaments and tendons are similar and exhibit a marked difference between the tension and compression responses (Fig. 8.2), while the collagen fibers are effective in opposing resistance to tension due to their orientation and the hierarchical organization, and they do not provide a significant response to compression.

In tension, the load–displacement curve shows three distinct regions. In the toe region, the tissue has a low stiffness, which can be attributed to the crimped fibers, which progressively straighten during elongation; when the crimps disappear, the fibers start to offer significant resistance to tension, resulting in higher stiffness, i.e., in the linear region. The threshold between the toe and the linear regions typically occurs at 1.5%–3% strain (Martin et al., 2015). The relevance and usefulness of the toe region have not been fully understood from a biomechanical point of view; ligaments and tendons are normally pretensioned in vivo, and their response therefore reflects that of the linear region (Gillis et al., 1995).

The linear region covers the functional physiological loading and strain range of tendons and ligaments in vivo. In this region, the collagen molecules are aligned along the direction of loading and are tensioned. The slope of the linear region describes the elastic modulus of the tissue, typically in the range between 0.2 and 2 GPa (Martin et al., 2015) with significant variability depending on the biomechanical function and location of the tissue (Fig. 8.3). Under the action of higher loads, collagen fibers start to undergo damage and failure, resulting in yielding and failure of the specimen; the ultimate load and stress describe the condition under which the complete failure of the tissue sample occurs (Fig. 8.3).

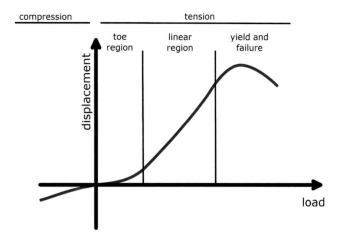

FIGURE 8.2

A typical uniaxial load–displacement curve of ligaments and tendons, depicting the toe region, the linear region, and the region where yield and failure occur.

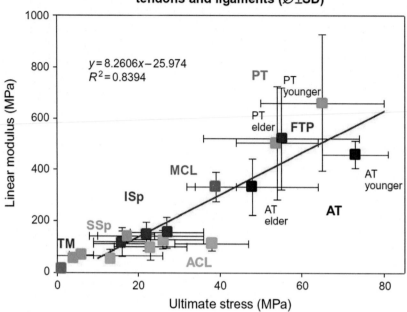

FIGURE 8.3

Mechanical properties (linear modulus and ultimate stress) of human tendons and ligaments measured experimentally in tension on cadaver specimens. *ACL*, Anterior cruciate ligament; *AT*, Achilles tendon; *FDP*, flexor digitorum profundus; *ISp*, infraspinatus; *MCL*, medial collateral ligament; *PT*, patellar tendon; *SSp*, supraspinatus; *TM*, teres minor.

From Buschmann, J., & Meier, G. B. (2017). Biomechanical properties of tendons and ligaments in humans and animals. In Biomechanics of tendons and ligaments (pp. 31−61). Amsterdam: Elsevier.

Ligaments and tendons have similar mechanical behavior; however, due to their different composition, their mechanical properties are different (Fig. 8.4). Due to the higher presence of collagen (up to 85%) and lower elastin, tendons are, in general, stiffer and stronger; the toe region is longer for ligaments since its collagen fibers are less aligned. In general, they can be stretched up to 5%− 7% without any damage, whereas failure strains are typically around 12%−15% (Martin et al., 2015) and depend on the elastin content.

The stiffness of ligaments and tendons is highly variable; differences in shape and mechanical behavior can be found, for example, even between the medial and lateral collateral ligaments of the same knee (Wilson, Deakin, Payne, Picard, & Wearing, 2012). The differences depend on the composition and the architecture of the tissue; ligaments and tendons with high rates of elastin, such as the yellow ligament (ligamentum flavum) or the ligamentum nuchae of the spine, have low stiffness and are able to sustain high strains without any damage; their main function is to store elastic energy which is then exploited to restore a convenient position, for example, to

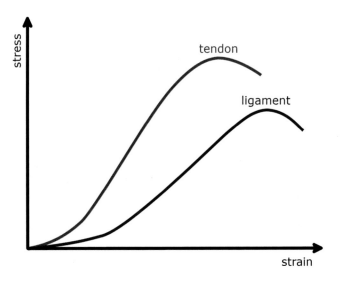

FIGURE 8.4

The distinct stress—strain curves of tendons and ligaments, highlighting the higher stiffness and strength of the former.

regain the standing posture after the forward flexion of the trunk. On the other side, ligaments and tendons which are subjected to sustained high axial loads have higher stiffness and may show signs of calcification.

Ligaments and tendons exhibit a clearly time-dependent response, that is, they are viscoelastic. The values of the material properties measured experimentally (elastic modulus, failure strain, etc.) therefore depend on the loading rate, which should be appropriately considered when planning an experimental campaign. Higher strain rates normally induce an increase in failure load and stretch (Martin et al., 2015). Ligaments and tendons also show the creep and stress relaxation behaviors typical of viscoelastic materials; samples subjected to a constant load will elongate over time, whereas if subjected to a constant strain, the load required to maintain the deformation will decrease together with the internal stresses. Viscoelastic effects will decrease over time until an equilibrium condition is reached.

Ligaments and tendons show another biomechanical behavior typical of viscoelastic materials, the hysteresis (Fig. 8.5). When subjected to a loading—unloading cycle, the stress—strain curves describing the loading and unloading phases do not exactly correspond as it would be expected in a purely elastic material; the area between them describes the energy dissipated by the material during the cycle. Studies showed that the hysteresis depends on the biomechanical function of the tissue, with tendons involved in gait dissipating less energy than others (Martin et al., 2015). When several loading—unloading cycles are performed, the curves corresponding to the various cycles are not fully overlapping but show differences especially in the first cycle (Fig. 8.5); stability, that is, stress—strain curves not changing any more in a significant way, is typically achieved after 15—20 loading—unloading cycles.

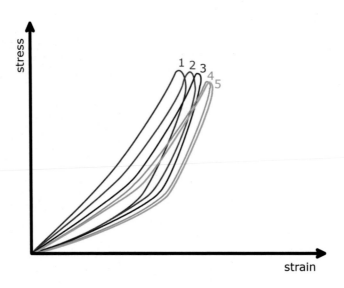

FIGURE 8.5

The hysteresis shown by ligaments and tendons under cyclic testing conditions (five cycles of repeated loading and unloading).

EXPERIMENTAL MEASUREMENT OF THE BIOMECHANICAL PROPERTIES

As mentioned in the previous paragraph, the measurement of the material properties of ligaments and tendons is routinely performed by subjecting cadaver specimens to tensile testing in a material testing system. This procedure involves several technical challenges such as achieving gripping of the sample, maintaining proper hydration throughout the duration of the test, and estimating the cross-sectional area of the specimen and its length at rest, which are necessary to convert the load—displacement curve (directly measured with load cells and displacement sensors) to a stress—-strain curve.

The fixation of the specimen in the clamps tends to be problematic due to the slippery nature of the surfaces to be gripped. Several solutions have been proposed and are employed with different degrees of success, depending on the size and characteristics of the specimens: small samples can be effectively gripped by using adhesives; screws, sutures, and cryoclamps, that is, cooled grips which freeze the tissue in contact with them (Riemersa & Schamhardt, 1982), are used for larger specimens. If feasible, an excellent solution is the preservation of the enthesis, that is, the insertion of the specimen into bone, which provides a solid grip even with standard clamps.

The viscoelastic properties of ligaments and tendons can in large part be attributed to their high water content, which is imbibed and exuded during tissue deformation and therefore determines the response based on the permeability of the tissue (Thornton, Shrive, & Frank, 2001). It is therefore important that the physiological hydration of the specimen is preserved throughout the duration of the test, which is achieved by keeping it immersed in a fluid bath, by spraying it with fluid or by wrapping with a humid cloth. Saline solutions (physiological, Ringer's solution) at body temperature should be used instead of plain water, in order to avoid hypo- or hyperhydration due to osmotic effects.

The measurement of the cross-sectional area and that of the length at rest of ligaments and tendons is made challenging by the viscoelastic nature of the tissue and its capability of exuding water, which makes the tissue volume variable depending on the testing conditions. A common method for the estimation of the cross-sectional area is fitting the sample into a vacant lot with known width, pressing it with a plunge, and measuring the height of the space necessary to contain the sample; if rigorously performed, this measurement method is accurate and repeatable (Martin et al., 2015). More sophisticated methods using laser systems have been presented (Pokhai, Oliver, & Gordon, 2009), and while they are expensive they provide the possibility of performing measurement both at rest and under loaded conditions. The resting length is conventionally measured in a relatively simple way, by applying a small load sufficient to strengthen the specimen and measuring the length in that condition. Alternatively, after performing a standard tensile test, the length at rest of the specimen may be estimated by performing a regression on the linear region of the load−displacement curve; the resting length is then the one in correspondence with zero load (Martin et al., 2015).

In the case of ligaments with complex anatomy which significantly differ from the most common string-like shape, other methods may be preferable to simple tensile testing. For example, Heuer and colleagues introduced the concept of "stepwise reduction" to assess the biomechanical relevance of spinal ligaments (Heuer, Schmidt, Klezl, Claes, & Wilke, 2007), which have membrane-like shapes and complex interconnections with bone and other soft tissues such as the intervertebral disc. In stepwise reduction, the ligaments of a functional spinal unit are resected one by one, and the specimen is tested after each iteration. The difference in the biomechanical response between two consecutive iterations can be attributed to the ligament which has been resected. Stepwise reduction allows testing of each structure in its own mechanical environment, and avoids difficulties associated with gripping and applying loads in the physiological directions; on the other side, it does not permit a direct mechanical characterization of the tissue, that is, the calculation of a stress−strain curve, which can still be achieved indirectly by coupling this technique with numerical analysis (Schmidt et al., 2007).

IN VIVO ASSESSMENT OF THE BIOMECHANICAL PROPERTIES

In vivo and in vitro behaviors of a biological tissue could be quite different (Zhang, Adam, Hosseini Nasab, Taylor, & Smith, 2020). In tendons, the toe region is usually not observable in vivo due to the preload associated with muscle activity; the strength decreases substantially with age (e.g., the strength of an anterior cruciate ligament is around 2.200 N at 20−35 years but it drops to 650 N at 60−80 years); and the tissues (including bone) remodel according to the change in loading environment. For these reasons, assessing the biomechanical properties of tissues in vivo, that is, in their own biomechanical and physiological environment, assumes a capital importance.

Forces and strains have been measured in vivo by means of force transducers and strain gauges, in both animals and humans. Several structures including the Achilles tendon and the cruciate ligaments of the knee (Howe et al., 1990) have been investigated with such techniques with good success. It should be noted that the implantation of sensors and transducers, despite it can be performed with minimally invasive techniques, always requires a surgical incision and is associated with a risk of perioperative and postoperative complications such as infection, especially if wired data transmission is used. Another challenge associated with implantable sensors and transducers is the need for calibration, that is, the calculation of the quantity of interest (force, strain) from the physical output of the transducer. While calibration is easily performed in in vivo tests by subjecting the specimen to standardized loads, this procedure cannot be used in vivo; calibration can then be conducted indirectly exploiting motion analysis equipment (Martin et al., 2015).

Ligaments and tendons can also be investigated in vivo noninvasively, by means of imaging techniques. Ultrasonography is routinely used in the clinical practice for the diagnosis of several disorders of connective soft tissues such as tendinitis, tendosynovitis, tears, and bursitis (Rasmussen, 2000); its high resolution and the capability of dynamically capturing motion and deformation makes it a valuable tool for the estimation of the ligament and tendon strains in vivo. Ultrasonography also allows estimating the stiffness of the material under investigation; however, similarly to all medical imaging techniques, it does not offer the possibility of measuring forces and stresses. Several structures, mostly tendons, have been investigated by means of ultrasound imaging, most notably the patellar tendon (Pearson, Burgess, & Onambele, 2007) and the Achilles tendon (Maganaris & Paul, 2002; Suydam & Buchanan, 2014). Results revealed material properties and deformations during daily activities compatible with those measured in vivo with tensile testing (Maganaris & Paul, 2002).

ENTHESES AND APONEUROSES

Ligaments and tendons enter bones at the so-called entheses, specialized structures with a transitional composition and architecture with respect to that of soft and bone tissues. Entheses are relatively diverse depending on their location and biomechanical environment and are classified into two subtypes (Benjamin et al., 2006) (Fig. 8.6). Fibrocartilaginous entheses include a transitional layer of fibrocartilage, mineralized on the side facing bone, and are typical of tendons and ligaments inserting into bone at large angles, that is, far from parallel to the bone surface. If the ligament or tendon inserts into bone at acute angles, the enthesis does not possess the fibrocartilaginous layer and is then labeled as fibrous; in this case, the collagen fibers are directly attached to the periosteum and extend into the underlying bone tissue. Insertion sites can show characteristics of both types, depending on the region. Entheses are mechanically strong, and do not commonly undergo traumatic injury even if they are subjected to high loads in the physiological environment (Butler, Sheh, Stouffer, Samaranayake, & Levy, 1990); ligaments and tendons are indeed more prone to failure in sites outside from the bone insertions (Gao, Räsänen, Persliden, & Messner, 1996). Avulsion fractures, that is, fractures of the underlying bone which can be attributed to loads applied by ligaments and tendons, are also more common than failures of the entheses.

Muscles are attached to tendons at the aponeurosis or myotendinous junction, a flat structure mostly consisting of collagen that provides a large insertion area to the muscle fibers due to its

(A)

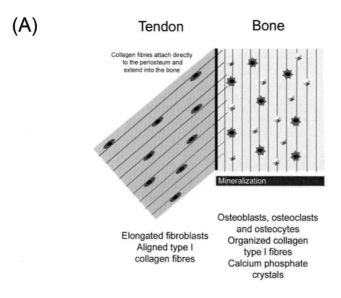

(B)

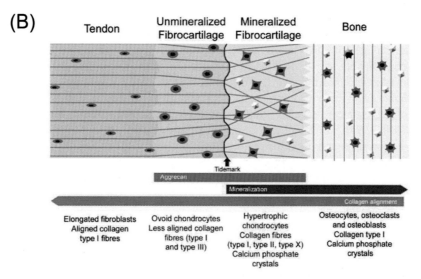

FIGURE 8.6

The two types of entheses: the fibrous (A) and the fibrocartilaginous (B) entheses. In the fibrous type, the collagen fibers directly attach on the periosteum and extend into bone. In fibrocartilaginous enthesis, there is a transitional zone consisting of a layer of fibrocartilage, mineralized on the bone side; a tidemark clearly indicates the extent of the mineralized region.

From Paxton, J. Z., Baar, K., & Grover, L. M. (2012). Current progress in enthesis repair: Strategies for interfacial tissue. Orthopedic & Muscular System, 1, 3. https://doi.org/10.4172/2161-0533.S1-003.

folded architecture. In contrast with most tendons, which are commonly studied under uniaxial loading conditions, the myotendinous junction should not be considered as a uniaxial structure since it is subjected to high transverse strains that play a role in determining its longitudinal compliance (Azizi & Roberts, 2009); this characteristic is believed to be involved in the regulation of the muscle function. Besides, it was shown that the mechanical properties of the aponeurosis, including the failure strength, increase when the muscle is active with respect to those recorded in passive tests (Tidball, Salem, & Zernicke, 1993). In comparison with the enthesis, the aponeurosis shows lower stiffness and larger strains under physiological loading (Maganaris & Paul, 2002).

MUSCULOSKELETAL MATURATION, AGING, AND EXERCISE

The biomechanical properties of ligaments and tendons are affected by musculoskeletal maturation and aging. Tensile testing on animal models at different stages of maturity revealed that stiffness and failure loads tend to increase with age (Woo, Peterson, Ohland, Sites, & Danto, 1990). These changes are associated with three distinct phenomena (Martin et al., 2015). First, an increase in collagen and a consequent decrease of water collagen content have been observed, which have direct implications on stiffness and failure strength. Second, collagen molecules tend to cross-link, further contributing to enhancing the stiffness. Finally, the crimping of the collagen fibers decreases with aging, determining a more linear, straightly oriented organization; the latter factor determines a decrease of the extent of the toe region. The three changes are interrelated and sustain each other; for example, cross-linking may be favored by straightening of the fibers, and a straighter orientation induces higher rates of cross-linking. The mechanical properties of ligaments and tendons do not change significantly any more after musculoskeletal maturity is achieved, whereas a deterioration is observed in elderly subjects (Narici, Maffulli, & Maganaris, 2008; Neumann, Ekström, Keller, Perry, & Hansson, 1994). These mechanical changes have an effect on the mobility of the joints as well as on the balance and mobility of the whole body, which tend to decrease with age (Onambele, Narici, & Maganaris, 2006).

Exercise has also a determinant effect on the mechanical properties of connective soft tissues, especially on those of tendons that have been documented in detail (Svensson, Heinemeier, Couppé, Kjaer, & Magnusson, 2016). Indeed, the mechanical properties and cross-sectional area of tendons increase when subjected to exercise programs, indicating that the tendon cells respond to mechanical stimuli by increasing their production of proteins of the extracellular matrix and growth factors. Studies conducted on humans showed that mechanical loading also increases collagen synthesis (Langberg, Skovgaard, Petersen, Bulow, & Kjaer, 1999), demonstrating that exercise affects the mechanical properties of the tissues in addition to their size.

ANIMAL MODELS

As preclinical studies often use animal models to investigate novel surgical techniques as well as tissue engineered constructs, there is an inevitable need for animal models of human ligaments and tendons, appropriate in terms of biomechanics (Hast, Zuskov, & Soslowsky, 2014). Soft tissues of both small and large mammals commonly used as animal models such as rats, rabbits, sheep, horses, pigs, and primates share many anatomical and biomechanical characteristics with human ones, and were indeed shown to be suitable for basic research or preclinical testing of novel repair and replacement strategies (Buschmann & Meier, 2017).

Murine tendons and ligaments are not commonly used in biomechanical research due to their small size and low material properties, in a lower order of magnitude with respect to human samples; however, they are commonly employed in basic research to investigate biological and mechanobiological aspects such as the healing properties of the tissue. Ligaments and tendons of rats are significantly larger, stiffer, and have higher strengths than murine ones; these properties open the possibility of using rats for testing tissue repair strategies such as cell-seeded grafts (Pietschmann et al., 2013). Among small mammals, rabbits show the properties closest to human samples and their use as animal models for tendon and ligament repair techniques is therefore widespread. Besides, decellularized rabbit tendons have been proposed to be used as allografts in human patients (Tischer et al., 2007).

Large mammals such as sheep, pigs, dogs, horses, and monkeys show larger dimensional compatibility with humans with respect to rodents and rabbits, but the material properties may show relevant differences due to the different behaviors and locomotion strategies. The use of large animal models tends to be limited by high costs and challenging aspects related to housing and care; nevertheless, some models have proved to be successful, such as rhesus monkeys for the in vivo investigation of defects and repair techniques for the anterior cruciate ligament of the knee (Stone et al., 2007), which show mechanical properties very similar to that of human specimens (Noyes & Grood, 1976). Besides, ligaments and tendons of horses have been widely investigated in the frame of veterinary medicine, that is, not considering the animals as models of the human being but rather as patients. Indeed, horses tend to show relatively high rates of tendonitis and ligament injuries, especially when subjected to high athletic demands such as in competitions (Singer, Barnes, Saxby, & Murray, 2008).

REFERENCES

Azizi, E., & Roberts, T. J. (2009). Biaxial strain and variable stiffness in aponeuroses. *The Journal of Physiology*, *587*(Pt 17), 4309−4318. Available from https://doi.org/10.1113/jphysiol.2009.173690.

Benjamin, M., Kaiser, E., & Milz, S. (2008). Structure-function relationships in tendons: A review. *Journal of Anatomy*, *212*(3), 211−228. Available from https://onlinelibrary.wiley.com/doi/pdf/10.1111/j.1469-7580.2008.00864.x.

Benjamin, M., Toumi, H., Ralphs, J. R., Bydder, G., Best, T. M., & Milz, S. (2006). Where tendons and ligaments meet bone: attachment sites ('entheses'). *Journal of Anatomy*, *208*(4), 471−490. Available from https://doi.org/10.1111/j.1469-7580.2006.00540.x.

Buckwalter, J. A., Einhorn, T. A., & Simon, S. R. (2000). *Biology and biomechanics of the musculoskeletal system. Orthopedic basic science* (pp. 548−555). Rosemont: American Academy of Orthopedic.

Buschmann, J., & Burgisser, G. M. (2017). *Structure and function of tendon and ligament tissues. Biomechanics of tendons and ligaments (pp* (pp. 3−30). Cambridge: Woodhead.

Buschmann, J., & Meier, G. B. (2017). *Biomechanical properties of tendons and ligaments in humans and animals. Biomechanics of tendons and ligaments* (pp. 31−61). Amsterdam: Elsevier.

Butler, D. L., Sheh, M. Y., Stouffer, D. C., Samaranayake, V. A., & Levy, M. S. (1990). Surface strain variation in human patellar tendon and knee cruciate. *Journal of Biomechanical Engineering*, *112*(1), 38−45. Available from https://doi.org/10.1115/1.2891124.

Gao, J., Räsänen, T., Persliden, J., & Messner, K. (1996). The morphology of ligament insertions after failure at low strain. *Journal of Anatomy*, *189*(Pt 1), 127−133.

Gillis, C., Sharkey, N., Stover, S. M., Pool, R. R., Meagher, D. M., & Willits, N. (1995). Effect of maturation and aging on material and ultrasonographic properties. *American Journal of Veterinary Research*, *56*(10), 1345−1350.

Hast, M. W., Zuskov, A., & Soslowsky, L. J. (2014). The role of animal models in tendon research. *Bone & Joint Research*, *3*(6), 193−202. Available from https://doi.org/10.1302/2046-3758.36.2000281.

Heuer, F., Schmidt, H., Klezl, Z., Claes, L., & Wilke, H.-J. (2007). Stepwise reduction of functional spinal structures increase range of motion and change lordosis angle. *Journal of Biomechanics*, *40*(2), 271−280. Available from https://doi.org/10.1016/j.jbiomech.2006.01.007.

Howe, J. G., Wertheimer, C., Johnson, R. J., Nichols, C. E., Pope, M. H., & Beynnon, B. (1990). Arthroscopic strain gauge measurement of the normal anterior cruciate. *Arthroscopy*, *6*(3), 198−204. Available from https://doi.org/10.1016/0749-8063(90)90075-o.

Kannus, P. (2000). Structure of the tendon connective tissue. *Scandinavian Journal of Medicine & Science in Sports*, *10*(6), 312−320. Available from https://doi.org/10.1034/j.1600-0838.2000.010006312.x.

Langberg, H., Skovgaard, D., Petersen, L. J., Bulow, J., & Kjaer, M. (1999). Type I collagen synthesis and degradation in peritendinous tissue after. *The Journal of Physiology*, *521*(Pt 1), 299−306. Available from https://doi.org/10.1111/j.1469-7793.1999.00299.x.

Maganaris, C. N., & Paul, J. P. (2002). Tensile properties of the in vivo human gastrocnemius tendon. *Journal of Biomechanics*, *35*(12), 1639−1646. Available from https://doi.org/10.1016/s0021-9290(02)00240-3.

Martin, R. B., Burr, D. B., Sharkey, N. A., Fyhrie, D. P., Martin, R. B., Burr, D. B., ... Fyhrie, D. P. (2015). *Mechanical properties of ligament and tendon. Skeletal tissue mechanics* (pp. 175−225). New York: Springer. Available from https://doi.org/10.1007/978-1-4939-3002-9_4.

Narici, M. V., Maffulli, N., & Maganaris, C. N. (2008). Ageing of human muscles and tendons. *Disability and Rehabilitation*, *30*(20−22), 1548−1554. Available from https://doi.org/10.1080/09638280701831058.

Neumann, P., Ekström, L. A., Keller, T. S., Perry, L., & Hansson, T. H. (1994). Aging, vertebral density, and disc degeneration alter the tensile. *Journal of Orthopaedic Research*, *12*(1), 103−112. Available from https://doi.org/10.1002/jor.1100120113.

Noyes, F. R., & Grood, E. S. (1976). The strength of the anterior cruciate ligament in humans and Rhesus monkeys. *The Journal of Bone and Joint Surgery. American Volume*, *58*(8), 1074−1082. Available from https://www.academia.edu/download/47739036/The_strength_of_the_anterior_cruciate_li20160802-19348-1flqfo.pdf.

Onambele, G. L., Narici, M. V., & Maganaris, C. N. (2006). Calf muscle-tendon properties and postural balance in old age. *Journal of Applied Physiology (Bethesda, MD: 1985)*, *100*(6), 2048−2056. Available from https://doi.org/10.1152/japplphysiol.01442.2005.

Pearson, S. J., Burgess, K., & Onambele, G. N. L. (2007). Creep and the in vivo assessment of human patellar tendon mechanical. *Clinical Biomechanics*, *22*(6), 712−717. Available from https://doi.org/10.1016/j.clinbiomech.2007.02.006.

Pietschmann, M. F., Frankewycz, B., Schmitz, P., Docheva, D., Sievers, B., Jansson, V., ... Müller, P. E. (2013). Comparison of tenocytes and mesenchymal stem cells seeded on biodegradable. *Journal of Materials Science. Materials in Medicine*, *24*(1), 211−220. Available from https://doi.org/10.1007/s10856-012-4791-3.

Pokhai, G. G., Oliver, M. L., & Gordon, K. D. (2009). A new laser reflectance system capable of measuring changing. *Journal of Biomechanical Engineering*, *131*(9), 094504. Available from https://doi.org/10.1115/1.3194753.

Rasmussen, O. S. (2000). Sonography of tendons. *Scandinavian Journal of Medicine & Science in Sports, 10* (6), 360−364. Available from https://doi.org/10.1034/j.1600-0838.2000.010006360.x.

Riemersa, D. J., & Schamhardt, H. C. (1982). The cryo-jaw, a clamp designed for in vitro rheology studies of horse. *Journal of Biomechanics, 15*(8), 619−620. Available from https://doi.org/10.1016/0021-9290(82)90073-2.

Schmidt, H., Heuer, F., Drumm, J., Klezl, Z., Claes, L., & Wilke, H.-J. (2007). Application of a calibration method provides more realistic results for a a finite element model of a lumbar spinal segment. *Clinical Biomechanics, 22*(4), 377−384. Available from https://doi.org/10.1016/j.clinbiomech.2006.11.008.

Singer, E. R., Barnes, J., Saxby, F., & Murray, J. K. (2008). Injuries in the event horse: training vs competition. *Veterinary Journal (London, England: 1997), 175*(1), 76−81. Available from https://doi.org/10.1016/j.tvjl.2006.11.009.

Stone, K. R., Walgenbach, A. W., Turek, T. J., Somers, D. L., Wicomb, W., & Galili, U. (2007). Anterior cruciate ligament reconstruction with a porcine xenograft: A serologic, histologic, and biomechanical study in primates. *Arthroscopy, 23*(4), 411−419. Available from https://doi.org/10.1016/j.arthro.2006.12.024.

Suydam, S. M., & Buchanan, T. S. (2014). Is echogenicity a viable metric for evaluating tendon properties in vivo? *Journal of Biomechanics, 47*(8), 1806−1809. Available from https://doi.org/10.1016/j.jbiomech.2014.03.030.

Svensson, R. B., Heinemeier, K. M., Couppé, C., Kjaer, M., & Magnusson, S. P. (2016). Effect of aging and exercise on the tendon. *Journal of Applied Physiology (Bethesda, MD: 1985), 121*(6), 1237−1246. Available from https://doi.org/10.1152/japplphysiol.00328.2016.

Thornton, G. M., Shrive, N. G., & Frank, C. B. (2001). Altering ligament water content affects ligament prestress and creep. *Journal of Orthopaedic Research, 19*(5), 845−851. Available from https://doi.org/10.1016/S0736-0266(01)00005-5.

Tidball, J. G., Salem, G., & Zernicke, R. (1993). Site and mechanical conditions for failure of skeletal muscle. *Journal of Applied Physiology (Bethesda, MD: 1985), 74*(3), 1280−1286. Available from https://doi.org/10.1152/jappl.1993.74.3.1280.

Tischer, T., Vogt, S., Aryee, S., Steinhauser, E., Adamczyk, C., Milz, S., . . . Imhoff, A. B. (2007). Tissue engineering of the anterior cruciate ligament: A new method using. *Archives of Orthopaedic and Trauma Surgery, 127*(9), 735−741. Available from https://doi.org/10.1007/s00402-007-0320-0.

Wilson, W. T., Deakin, A. H., Payne, A. P., Picard, F., & Wearing, S. C. (2012). Comparative analysis of the structural properties of the collateral. *The Journal of Orthopaedic and Sports Physical Therapy, 42*(4), 345−351. Available from https://doi.org/10.2519/jospt.2012.3919.

Woo, S. L., Peterson, R. H., Ohland, K. J., Sites, T. J., & Danto, M. I. (1990). The effects of strain rate on the properties of the medial collateral. *Journal of Orthopaedic Research, 8*(5), 712−721. Available from https://doi.org/10.1002/jor.1100080513.

Zhang, Q., Adam, N. C., Hosseini Nasab, S. H., Taylor, W. R., & Smith, C. R. (2020). Techniques for in vivo measurement of ligament and tendon strain: A review. *Annals of Biomedical Engineering, 49(1), 7−28*. Available from https://doi.org/10.1007/s10439-020-02635-5.

CARTILAGE BIOMECHANICS

Andreas Martin Seitz, Daniela Warnecke and Lutz Dürselen
Institute of Orthopaedic Research and Biomechanics, Ulm University, Ulm, Germany

INTRODUCTION

A typical synovial joint is covered with hyaline articular cartilage (AC) at the bony ends, allowing the transmission of high loads with low friction. Healthy AC is white, and its surface is smooth and shiny. In addition to low-friction load transmission, the AC plays a very important mechanical role in our musculoskeletal system by increasing the contact area of the bones, thereby reducing stress. Unfortunately, the AC can degenerate prematurely either idiopathically or due to acquired trauma or congenital joint misalignment, resulting in osteoarthritis (OA). In the final OA stadium, the degenerated AC requires the implantation of a joint replacement. Therefore the evident relationship between OA and the mechanical stresses in a diarthrotic joint covered with hyaline cartilage has been drawing the attention of researchers for decades toward finding different strategies to heal, regenerate, or preventively protect the AC. To evaluate and develop such methods, it is important to understand the basic biomechanical functioning of the AC, which is mainly attributed to the biphasic composition of this tissue. Basically, cartilage is not innervated and lacks blood supply. Moreover, AC contains chondrocytes, which form an extracellular matrix (ECM) with collagen and proteoglycans as its main components. In addition to this solid matrix, AC possesses a significant amount of water as a fluid component, thus explaining the biphasic structure. While the collagen is mainly responsible for the mechanical strength of the solid matrix, the charged proteoglycans are able to store water by osmosis, which generates the time-dependent behavior leading to creep and relaxation phenomena. Degenerative processes can lead to structural changes or changes in the composition of the ECM. These changes then in turn lead to changes in the biomechanical behavior of AC. Therefore the biomechanical characterization of AC is an important and useful tool for assessing its structural integrity and can be used in many ways in research. For example, fundamental questions about its functionality can be answered as well as changes in the structure–function relationship induced by degenerative processes. Furthermore, new approaches towards regeneration or tissue replacement can be investigated and advanced on the basis of the results of biomechanical investigations. Based on this motivation, a wide variety of biomechanical testing methods have been developed in recent years. This chapter provides an overview of the structural composition of the AC as well as standardized methods for assessing its time-dependent biomechanical properties also including frictional behavior.

Human Orthopaedic Biomechanics. DOI: https://doi.org/10.1016/B978-0-12-824481-4.00029-9

STRUCTURAL COMPOSITION

Compared to other tissues, AC exhibits a number of structural peculiarities. In order to fulfill its weight-bearing and load-distribution function, it does not possess vulnerable structures such as nerves, blood vessels, or lymphatic vessels (Buckwalter & Mankin, 1997; Dijkgraaf, de Bont, Boering, & Liem, 1995). AC consists of a dense ECM with a sparse distribution of cells called chondrocytes (Fig. 9.1). The chondrocytes are highly specialized and play a crucial role in the development, maintenance, and repair of the ECM (Sophia Fox, Bedi, & Rodeo, 2009). They vary in shape, number, and size depending on their localization in the AC. Each chondrocyte creates its own microenvironment and is also entrapped within it, thus preventing migration to other areas

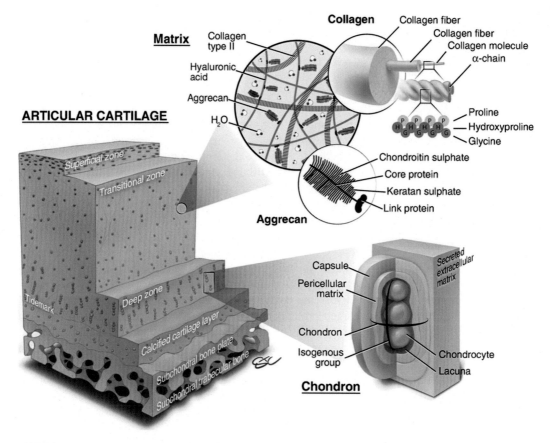

FIGURE 9.1

Structure of the osteochondral unit and the unit's individual components.

From Baumann, C. A., Hinckel, B. B., Bozynski, C. C., Farr, J., Yanke, A. B., & Cole, B. J. (2019). Articular cartilage: Structure and restoration. In Joint preservation of the knee: A clinical casebook *(pp. 3–24). Springer International Publishing. https://doi.org/10.1007/978-3-030-01491-9_1.*

(Poole, Flint, & Beaumont, 1987). The ECM is composed primarily of water [65%−80% (Mow, Wang, & Hung, 1999)], collagen (60% of the dry weight), and proteoglycans (10%−15% of the dry weight), with other noncollagenous proteins and glycoproteins present in lesser amounts (Buckwalter & Mankin, 1997). Collagen type II is the most dominant collagen in AC, representing 95%, while type I, IV, V, VI, IX, and XI are also present to a minor proportion and predominantly help to form and stabilize the type II collagen network. The proteoglycans are heavily glycosylated protein monomers and are, besides the collagen, the second largest representative of macromolecules in the ECM. Aggrecan is one of the most important proteoglycans, while being responsible for the normal functioning of the AC. Moreover, Aggrecan interacts with hyaluronan (HA), occupies the interfibrillar space of the ECM, and provides AC with its osmotic properties (Buckwalter, Mow, & Ratcliffe, 1994). The other proteoglycans are interacting with the collagen while also playing a role in fibrillogenesis and interfibril interaction. Together, the ECM components with a major contribution from the proteoglycans help retain water in the ECM, which is critical to maintaining its unique biphasic biomechanical properties. The biochemical composition of the matrix varies both interindividually and intraindividually, depending on age, joint, cartilage depth, and cartilage region (Dijkgraaf et al., 1995). However, chemical composition, molecular organization, and biomechanical properties may also change during differentiation, physical training, and mineralization (Modis, Botos, Kiviranta, Lukacsko, & Helminen, 1996).

Depending on the cartilage localization, zone, and matrix region, differences exist in the structure, organization, and mechanical properties of the matrix. Cell morphology, cell density, metabolic activity, and probably cell function also vary depending on cartilage depth (Buckwalter & Mankin, 1997; Quinn, Hunziker, & Hauselmann, 2005). Both, the biochemical composition and physical properties of the ECM, such as collagen fiber orientation, size, and amount of proteoglycan composition, and aggregation properties, also vary depending on the particular cartilage zone (Chen, Falcovitz, Schneiderman, Maroudas, & Sah, 2001; Hunziker, Quinn, & Hauselmann, 2002).

ZONES

Because of these structural differences, AC can be divided into four zones: superficial zone, transitional zone, deep zone, and the calcified cartilage layer (Fig. 9.1) (Buckwalter & Mankin, 1997; Wong & Carter, 2003). Because the individual cartilage zones respond differently to mechanical loading, it is clear that the zonal arrangement of cartilage components is also important for the function of cartilage.

The superficial zone (10%−20% of the total thickness) shows a tangential arrangement of densely packed collagen types II and IV fibers, providing tensile strength and giving resistance to shear forces that occur during joint motion. The surface layer is of great importance for nutrition as well as for the wear and lubrication potential of the AC. The cell density at the superficial zone is higher, while the matrix content is relatively low with high collagen and low proteoglycan concentration (Buckwalter & Mankin, 1997) compared to the deeper AC zones. Moreover, the content of fibronectin and water is highest in this cartilage zone. Because of its low proteoglycan content (Muehleman et al., 2004; Orford & Gardner, 1985), this cartilage zone is not nearly as resistant to pressure as the deeper proteoglycan-rich zones (Orford & Gardner, 1985), but it nevertheless provides an important contribution with respect to the behavior of cartilage to compressive forces. Indeed, removal of this zone increases the permeability of the tissue and the stress on the macromolecular scaffold during compression, thereby altering the mechanical behavior of AC (Guilak, Ratcliffe, Lane, Rosenwasser, & Mow, 1994). As the name suggests, the transitional zone is a transition area from the superficial to the deep cartilage zone in terms of its morphology and matrix composition. Its volume is many times that of the superficial zone (Buckwalter & Mankin, 1997). Because the matrix content, including the proteoglycan content, is higher than in the superficial zone, the collagen fibers, which become thicker with increasing cartilage depth, are more widely spaced. The fiber arrangement also appears randomly oriented without a preferred direction. In addition, the water content in this zone is also lower than in the superficial zone (Buckwalter & Mankin, 1997). Both cell size and cell concentration are greater in the transition zone than in the deep cartilage zone. The oval and hemispherical chondrocytes, respectively, are usually arranged in pairs or groups with no definite orientation (Errington, Fricker, Wood, Hall, & White, 1997; Poole et al., 1987; Weiss, Rosenberg, & Helfet, 1968). The thickest collagen fibers, the highest concentration of proteoglycans, and the lowest content of water are found in the deep transitional zone (Weiss et al., 1968). Like the chondrocyte columns, the collagen fibers are perpendicular to and parallel to the articular surface (Poole et al., 1987). Compared with the other cartilage zones, the cells of the transitional zone are the largest, round to spheroidal, and normally arrange themselves in groups of six to nine cells in a columnar fashion, perpendicular to the articular surface at low volume density (Buckwalter & Mankin, 1997; Errington et al., 1997; Poole et al., 1987; Quinn et al., 2005). The main function of the deep zone is to anchor the noncalcified in the calcified cartilage (Poole et al., 1987). However, this zone also contributes to the tension of the collagen fiber network due to its high proteoglycan content and the resulting swelling pressure. From a functional point of view, the transition zone is the first to withstand compressive forces, while the deep zone is even more important for the resistance to compressive forces due to the orthogonal alignment of the collagen fibers in the deep zone. As a boundary between the calcified and noncalcified cartilage lies a histologically distinct line, named the tidemark (Bullough & Jagannath, 1983). The AC is interlocked with the rough surface of the underlying subchondral bone by a thin layer of calcified cartilage. Nevertheless, cartilage and bone remain separated. The calcified cartilage zone represents the transition between cartilage and bone (Vanwanseele, Lucchinetti, & Stussi, 2002). The amount of calcified cartilage is seen as a parameter for its adaptation to long-term stresses. The chondrocytes of the calcified cartilage zone are relatively small and embedded in the surface of the calcification front, and each cell is surrounded by some calcified tissue. The calcification of AC and subsequent endochondral ossification seem to be important in determining the shape of the joint and thus the load distribution. The exact feedback mechanism controlling the formation and progression of the calcification front is still unknown but seems to be related to the load on the articular surface. In any case, changes in this mechanism have an effect on the joint shape and thus on the load distribution (Bullough & Jagannath, 1983).

NUTRITION

Since AC is exposed to considerable stress and deformation, it has no vulnerable structures such as lymphatic and blood vessels. Because of the avascularity of cartilage, there are only two possible pathways for its nutrition: either from the articular surface via the synovial fluid or from the underlying subchondral bone (Mow, Ratcliffe, & Poole, 1992). The relative importance of these two pathways has long been a matter of controversy. Meanwhile, the synovial fluid nutrition pathway has been widely substantiated, as healthy AC cannot survive without it (Mauck, Hung, & Ateshian, 2003). However, the nutrition of mature cartilage through the subchondral bone as another nutritional pathway is controversial. The mobility of nutrients, cell products, enzymes, cytokines, and growth factors with the interstitial fluid is essential not only for chondrocyte viability but also for the structural integrity of cartilage tissue (Torzilli, Adams, & Mis, 1987). Both the nutrition of AC and the continuous removal of its waste products are mainly achieved not only by molecular transport via diffusion between cartilage matrix and synovial fluid (Dijkgraaf et al., 1995; Mauck et al., 2003), but also by mechanical, compression-induced convection of the interstitial fluid (Torzilli, Arduino, Gregory, & Bansal, 1997).

ELECTROMECHANICAL EFFECTS

The electromechanical effects include the increasing repulsive force of the proteoglycans during cartilage deformation due to the closer compression of their side chains (Fig. 9.2). The increased negative charge density leads to increased resistance to fluid flow until equilibrium between load and swelling pressure is reached (Buschmann & Grodzinsky, 1995; Buschmann, Gluzband, Grodzinsky, Kimura, & Hunziker, 1992; Szafranski et al., 2004). This process determines not only the mechanical response of cartilage tissue to deformation but also its lubricating properties and capacity for the exchange of nutrients and metabolic waste products (Torzilli, Dethmers, Rose, & Schryuer, 1983). The effects of the fixed negative charges of the proteoglycans of cartilage were also investigated by Sun et al. (Sun, Guo, Likhitpanichkul, Lai, & Mow, 2004), who concluded that a charged tissue always withstands a greater load than an uncharged tissue with comparable intrinsic elastic moduli while the load capacity is based on intrinsic matrix stiffness, and hydraulic and osmotic pressure.

BIOMECHANICS

AC is a thin layer of connective tissue that exhibits time-dependent viscoelastic properties. As mentioned above, AC consists of tissue fluid (mainly water) and a solid porous-permeable matrix (Urban, 1994). The high fluid content of cartilage results from the hydrophilic nature of its proteoglycans, which is also responsible for the swelling behavior of cartilage due to high internal osmotic pressure (Torzilli et al., 1983). Thus the biphasic nature of cartilage, including the interactions of its constituents' water, electrolytes, and matrix, is of great importance for its mechanical properties and, consequently, for its load-bearing capacity (Mow et al., 1992; Urban, 1994). It reduces the pressure on the matrix when the cartilage is loaded, thus protecting it. Since the water concentration at the surface of AC is up to 85%, only 15% of the loading pressure acts on the solid matrix. The physiological pressure on the cartilage surfaces is usually not more than 3−10 MPa and the compressive pressure on the proteoglycan-collagen matrix is thus not more than 0.45−1.5 MPa (Ateshian, Lai, Zhu, & Mow, 1994). Experiments also show that both the modulation of intra- and intermolecular repulsive forces between the proteoglycans and the disruption of electrostatic interactions between collagen and proteoglycans through changes in the counterion environment can affect the compression, tension, and shear properties of the tissue. Further, the movement of

(A)

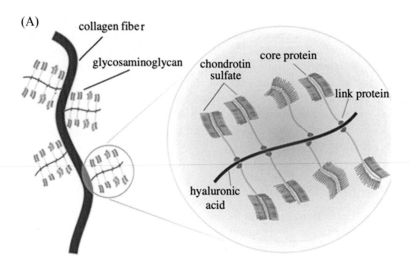

(B)

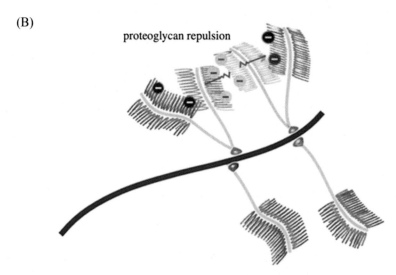

FIGURE 9.2

(A) Schematic of the cartilage network. It is composed of cross-linked type II collagen with proteoglycan aggrecan molecules bound. These proteoglycan aggregates consist of chondroitin-sulfate- and keratin-sulfate-rich regions (forming glycosaminoglycan molecules) linked to hyaluronic acid. (B) Proteoglycan repulsion phenomena owing to negative fixed charges attached in glycosaminoglycans.

From Manzano, S., Manzano, R., Doblare, M., & Doweidar, M. H. (2015). Altered swelling and ion fluxes in articular cartilage as a biomarker in osteoarthritis and joint immobilization: A computational analysis. Journal of The Royal Society Interface, *12(102), 20141090. https://doi.org/10.1098/rsif.2014.1090.*

interstitial fluid and the rate of ion transport are of paramount importance for cartilage function (Mow et al., 1992). A decrease in matrix stiffness results in the formation of a thicker fluid film due to greater deformation of the cartilage surface. Cartilage with a stiffness of 0.5 MPa therefore forms a thicker fluid film with lower peak pressure than cartilage with a stiffness of 2.5 MPa (Hou, Mow, Lai, & Holmes, 1992). In general, there are different ways to determine the (bio)mechanical properties of healthy and degenerated AC.

VISCOELASTICITY

Viscoelasticity is defined as the time-dependent response of a material subjected to a constant load or deformation (Cohen, Foster, & Mow, 1998). As long as the loading time is significantly shorter than the recovery time, viscoelastic materials behave as elastic materials (Torzilli & Mow, 1976). AC is a viscoelastic material consisting of two separate phases: the solid phase (ECM) and the fluid phase (interstitial water) (Vanwanseele et al., 2002). The change in pressure that occurs during compression of the AC results in a flow of interstitial fluid through the solid porous-permeable ECM. This flow generates significant frictional resistance, which is mainly responsible for the viscoelastic property of cartilage during compression (Hlavacek, 1993a; Mow et al., 1992; Vanwanseele et al., 2002). Flow-independent viscoelastic mechanisms as well as the viscoelastic properties of the matrix itself influence the deformation of AC (Mow, Holmes, & Lai, 1984; Vanwanseele et al., 2002; Zhu, Mow, Koob, & Eyre, 1993). In the case where the duration of loading is longer than the recovery time or the compression of the AC is constant, fluid flow is substantially increased (Cohen et al., 1998; Torzilli & Mow, 1976) and AC responds as a viscoelastic material with its characteristic creep deformation and stress relaxation behavior.

UNCONFINED COMPRESSION AND TENSILE TESTING

The easiest way to determine the material properties of AC is unconfined compression or tensile testing of either geometrically undefined or standardized specimens. From such experiments, parameters can be derived such as stiffness (K in N/mm) from the force-elongation diagram, and the elastic or Young's modulus (E in N/mm^2 or MPa) from the stress—strain diagram. For the latter, geometrically standardized specimens are required. This means that both details about the dimensions including the cross-sectional area and the anticipated material behavior like the Poisson's ratio, which defines the negative relationship between transverse strain and axial compression needs to be known. Although both tensile and unconfined compression testing lead to the same result parameters, the results are different, depending on heterogeneity and anisotropy of the tissue (Federico & Herzog, 2008). For example, the cartilage tensile modulus has been reported to be one to two orders of magnitude greater than the corresponding compressive modulus (Elliott, Narmoneva, & Setton, 2002). When testing AC under compressive loads, cylindrical or cubic specimens are widely used, which can be easily harvested with punches or hollow drills. On the other hand, tensile tests typically use rectangular strips or dumbbell-shaped samples, which are more difficult to produce. The compressive testing method is advantageous because of its relatively simple test configuration, which involves compressing the specimen to a previously specified load, stress, deformation, or strain using a flat-ended rod (Fig. 9.3). Due to the varying Poisson's ratio of different soft tissues, the cross-section of the cartilage specimen changes differently during compressive

[$\nu = 0.2$; (Kiviranta et al., 2006)] or tensile [$\nu > 0.5$; (Elliott et al., 2002)] loads. Tensile testing is more complicated than unconfined compression testing because it involves clamping soft tissue, which is a common pitfall in soft tissue biomechanics. Usually, a typical tensile testing system includes two clamps to ensure sufficient sample fixation during tensile loading (Fig. 9.3) (Oinas et al., 2018).

Both test setups are not intended to simulate physiological or in-vivo loading conditions, including the integrity of the AC sample, loading characteristics, or environmental conditions. Special care must be taken during the tests to keep the specimen moist as drying can result in a significant alteration in the obtained properties. The aforementioned anisotropic and inhomogeneous material conditions together with the biphasic material constitution result in a typical AC strain-rate-dependent response during both tensile and unconfined compression testing. For biphasic materials like AC, an increased strain rate is directly associated with an increase in both stiffness and/or Young's modulus (Fig. 9.4) (Langelier & Buschmann, 2003; Oloyede, Flachsmann, & Broom, 1992). Additionally, when applying alternating loading and unloading, cartilage samples exhibit so-called hysteresis loops (Fig. 9.4) (Park, Nicoll, Mauck, & Ateshian, 2008).

DYNAMIC COMPRESSION, CREEP AND STRESS RELAXATION TESTING

However, the biomechanical behavior of AC is best understood when the tissue is viewed as a biphasic medium while focusing on the elaboration of its time-dependent properties. The easiest way to determine these time-dependent properties is to repeat the above-mentioned hysteresis loop for several times. In doing so, the loading and unloading curve will converge, and the more repetitions are made. When using a time-resolving illustration of these stress and strain curves, it leads to a characteristic phase shift of the AC response. This phase shift also occurs during sinusoidal or

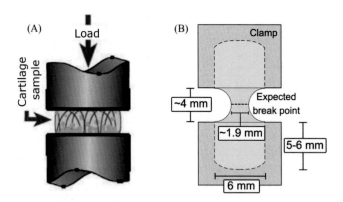

FIGURE 9.3

(A) A schematic of an unconfined compression measurement system. Reprinted with permission from (Kiviranta et al., 2006). (B) Positioning of a dumbbell-shaped specimen between two clamps for tensile testing.

From Oinas, J., Ronkainen, A. P., Rieppo, L., Finnila, M. A. J., Iivarinen, J. T., van Weeren, P. R., Helminen, H. J., Brama, P. A. J., Korhonen, R. K., & Saarakkala, S. (2018). Composition, structure and tensile biomechanical properties of equine articular cartilage during growth and maturation. Scientific Reports, 8(1), 11357. https://doi.org/10.1038/s41598-018-29655-5.

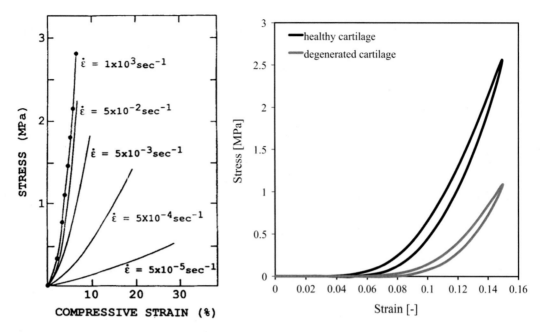

FIGURE 9.4

(Left) Stress—strain curves obtained from a cartilage specimen over the strain rate. Reprinted with permission from (Oloyede et al., 1992). (Right) Stress—strain curves of healthy and degenerated cartilage samples stimulated at 1 Hz and showing a typical hysteresis, characteristic of viscoelastic tissues.

From Abdel-Sayed, P., Moghadam, M. N., Salomir, R., Tchernin, D., & Pioletti, D. P. (2014). Intrinsic viscoelasticity increases temperature in knee cartilage under physiological loading. Journal of the Mechanical Behavior of Biomedical Materials, 30, 123–130. https://doi.org/10.1016/j.jmbbm.2013.10.025.

alternating loading schemes. The dynamic mechanical analysis (DMA) is a method that enables the characterization of AC viscoelastic properties under dynamic loading and can be combined with a frequency sweep (Lawless et al., 2017; Mountcastle et al., 2019; Temple, Cederlund, Lawless, Aspden, & Espino, 2016). Usually, during DMA, the viscoelastic behavior of the AC samples is determined in terms of its storage modulus, E', and a loss modulus, E''. The storage modulus represents the elastic part of the response (where energy is stored and used for elastic recoil of the specimen when stress is removed) and the loss modulus represents the viscous response (where energy is dissipated and the material flows). E' and E'' are related to the dynamic (complex) modulus, E^*, and phase angle, δ, of a viscoelastic material according to (Hukins, Leahy, & Mathias, 1999):

$$|E^*| = \sqrt{E'^2 + E''^2}$$

$$\delta = \tan^{-1}\left(\frac{E''}{E'}\right)$$

A phase angle of 0 degree would indicate that a material is purely elastic, while a purely viscous material would have a phase angle of 90 degrees. Isolated cartilage samples indicate under load at frequencies across a physiological range a frequency-dependent behavior of the storage and loss moduli (Fig. 9.5) (Fulcher et al., 2009).

In addition to the frequency analysis of AC, where a dynamic material-testing machine is required, other more simple tests can be carried out to determine the time-dependent properties of AC (Mow et al., 1992).

In relaxation testing (Fig. 9.6), a constant displacement is applied to the tissue, and the force needed to maintain the displacement is measured. The deformation results in increased stress and leads to fluid displacement, after which relaxation occurs. Due to the relaxation behavior of the AC tissue tension ceases and results in increased pressure distribution over a larger area.

In creep testing, on the other hand, the cartilage deforms under a constant load, but not as quickly as it does in a single-phase elastic material like a spring. The displacement of the cartilage is a function of time since the fluid cannot escape from the matrix instantaneously. The initial, relatively rapid displacement corresponds to a large fluid flow out of the cartilage. Then this fluid flow decreases as the rate of displacement slows until it reaches a constant value. The displacement is stable at equilibrium, and fluid flow has ceased. In general, it takes several thousand seconds to reach the equilibrium displacement.

Different mathematical models have been developed to simulate the cartilage during its time-dependent response to load or displacement stimuli (Frank & Grodzinsky, 1987a,b; Hayes & Mockros, 1971; Mow, Kuei, Lai, & Armstrong, 1980; Selyutina, Argatov, & Mishuris, 2015). By

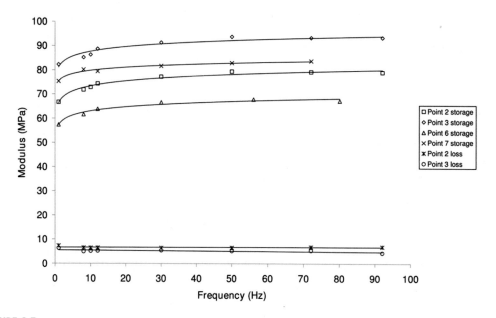

FIGURE 9.5

Storage and loss moduli against frequency, for frequencies of up to 92 Hz.

From Fulcher, G. R., Hukins, D. W., & Shepherd, D. E. (2009). Viscoelastic properties of bovine articular cartilage attached to subchondral bone at high frequencies. BMC Musculoskeletal Disorders, *10, 61. https://doi.org/10.1186/1471-2474-10-61.*

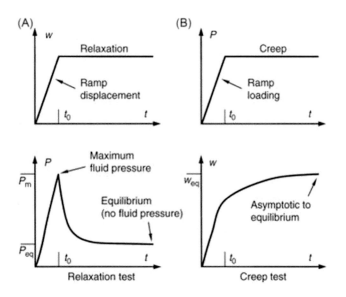

FIGURE 9.6

(A) Stress−relaxation test: A displacement is applied on the tissue at a constant rate until a desired level of compression is reached. This displacement results in a force increase followed by stress relaxation until an equilibrium is reached. (B) Creep test: A step force is suddenly applied resulting in an increase in deformation (i.e., creep). Usually, the sample diminishes until an asymptotic deformation is reached which is also called equilibrium state.

fitting these mathematical models to the measured displacement or force, two AC material properties are commonly determined: the aggregate modulus (H_a) and permeability (k). The aggregate modulus is a measurement of the stiffness of the tissue at equilibrium when all fluid flow has ceased. Depending on the mathematical model used, this modulus is also called the equilibrium modulus (E_{eq}). The higher the modulus at the equilibrium state is, the less the tissue reacts to the stimulus. Commonly H_a is in the range of 0.5−0.9 MPa (Athanasiou, Rosenwasser, Buckwalter, Malinin, & Mow, 1991; Mansour & Mow, 1976). Combining the aggregate modulus and representative values of Poisson's ratio (<0.4), the Young's modulus of cartilage is in the range of 0.45−0.80 MPa.

The second parameter that is determined is the (hydraulic) permeability, which is a measure for the interstitial fluid flow through the porous-permeable matrix. It is inversely proportional to the friction exerted by the fluid on the matrix (Cohen et al., 1998). Although the porosity of cartilage tissue is very high, about 80%, its permeability is relatively low because of the small pore size (Hlavacek, 1993b; Mow et al., 1992; Vanwanseele et al., 2002), which means that large frictional forces are exerted on the matrix even at very low flow rates. This frictional resistance, together with the flow of interstitial fluid, appears to be the main mechanism for load support in the joints (Cohen et al., 1998). Under compression, the permeability of the AC decreases. In this process, the permeability rate decreases at a slower rate under high load than under lower load (Mansour & Mow, 1976). This mechanical feedback strengthens and protects the cartilage by limiting the flow rate (Cohen et al., 1998). As the load and consequently the pressure are lower at the beginning, the permeability of the cartilage is higher and a large amount of fluid can leak into the joint. An increase in load and pressure subsequently leads to a decrease in cartilage permeability. Fluid cannot easily flow back from the joint space into the cartilage and the lubricating film between the articulating joint surfaces is maintained. When the load is removed, the cartilage tissue returns to its initial state with the increased permeability facilitating the re-swelling of the cartilage back to its hydration state (Mansour & Mow, 1976).

The testing methods can be performed using different test setups. Indentation testing is performed pushing a flat, spherical, or hemispherical indenter into the cartilaginous surface. The respective strain or stress decrease is measured over time. The advantage of such testing setups is that no time-consuming preparation of standardized, round-shaped, or cylindrical specimens is required. This is because the specimen can be mounted as a whole to the respective testing setup (Fig. 9.7). In order to be able to determine the instantaneous modulus (IM) (in MPa), which equals the Young's modulus in other test setups, it is necessary to determine the cartilage thickness in a second experiment by means of needle penetration (Jurvelin, Rasanen, Kolmonen, & Lyyra, 1995). After the thickness is determined, the IM can be determined by solving the respective mathematical model (Hayes & Mockros, 1971). Another very common test configuration is a confined compression test (Fig. 9.7). A coplanar, flat-shaped specimen is required to perform this test. The cartilage

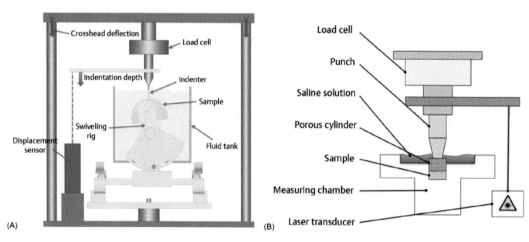

FIGURE 9.7

(A) Setup for indentation testing: An indenter is adapted to a load cell applying a defined load perpendicular to the cartilage surface, which is inserted in an adjustable measuring chamber, filled with saline solution. The indentation depth is measured using an accurate displacement transducer. (B) Schematic drawing of a test setup for a compression test under confined conditions: The geometrically defined sample is radially constrained allowing fluid flow only in the axial direction. The load is applied via a punch and a porous cylinder, which is connected to a load cell. The strain is accurately measured using a high-precision laser displacement transducer.

From Dürselen, L., & Seitz, A. M. (2015). 12 Biomechanics of cartilage. In Experimental research methods in orthopedics and *trauma. Georg Thieme Verlag. https://doi.org/10.1055/b-0035-122012.*

sample is placed in a radially confining impervious chamber between one or two permeable porous cylinders. Saline solution is commonly used as a fluid isotone to prevent the sample from dehydration during testing. The AC sample is loaded either with constant strain or stress and the resultant target value is measured over time. Since the compression chamber itself is impervious, the fluid flow through the AC is only possible in the vertical direction.

However, it should be kept in mind that each test method and underlying constitutive mathematical formulation might lead to different results for the time-dependent, viscoelastic parameters. This is due to the fact that the used mathematical models are simplified models that do not necessarily match physiology. Therefore care must be taken when comparing own results with already existing data in the literature.

BIOMECHANICAL MAPPING OF THE JOINT SURFACES IN HUMAN KNEES

Recent technological advancements led to the introduction of methods that allow us to measure multiple measuring points along joint surfaces (Pordzik et al., 2020; Seidenstuecker et al., 2019; Sim, Chevrier, Garon, Quenneville, Lavigne, et al., 2016; Sim, Chevrier, Garon, Quenneville, Yaroshinsky, et al., 2014). Despite several authors have investigated the biomechanical properties of isolated meniscal and tibial and femoral cartilage specimens utilizing indentation mapping, a combined complementary characterization of the articulating partners within the knee joint remains lacking. A very recent study enlightened this important issue (Seitz et al., 2021).

An exact understanding of the interplay between the articulating tissues of the knee joint in relation to the OA-related degeneration process is of considerable interest. Therefore the aim of the present study was to characterize the biomechanical properties of mildly and severely degenerated human knee joints, including their menisci and tibial and femoral AC surfaces. Spatial biomechanical mapping of the articulating knee joint surfaces of 12 mildly and 12 severely degenerated knee joints was assessed using a multiaxial mechanical testing machine. To do so, indentation stress relaxation tests were combined with thickness measurements at the lateral and medial menisci and the AC of the tibial plateau and femoral condyles to calculate the IM (Fig. 9.8), the modulus after a relaxation time of 20 s (Et20), the maximum applied load (Pmax), and the relaxation percentage over the maximum stress ($\Delta\sigma$relax). With progressing joint degeneration, we found an increase in the lateral and the medial meniscal instantaneous moduli ($P < .02$), relaxation moduli ($P < .01$), and maximum applied forces ($P < .01$), while for the underlying tibial AC, the IM ($P = .01$) and maximum applied force ($P < .01$) decreased only at the medial compartment. Degeneration had no influence on the relaxation percentage of the soft tissues. While the water content of the menisci did not change with progressing degeneration, the severely degenerated tibial AC contained more water ($P < .04$) compared to the mildly degenerated tibial cartilage. The results of this study indicate that degeneration-related (bio-) mechanical changes seem likely to be first detectable in the menisci before the articular knee joint cartilage is affected. Should these findings be further reinforced by structural and imaging analyses, the treatment and diagnostic paradigms of OA might be modified, focusing on the early detection of meniscal degeneration and its respective treatment, with the final aim to delay OA onset.

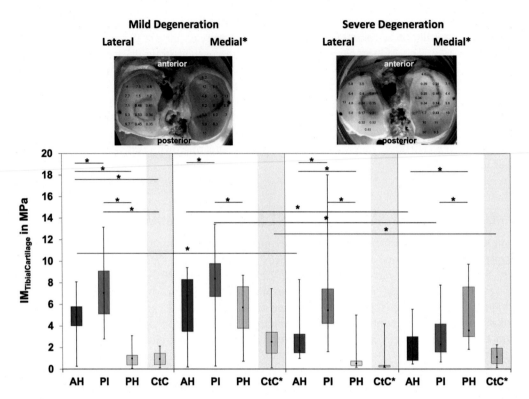

FIGURE 9.8

Representative biomechanical mappings of the instantaneous modulus (IM) measurements of the mildly and severely degenerated tibial plateau articular cartilage (AC), with all values given in megapascal. Middle row: Box plots (minimum, maximum, median, and 25th and 75th percentiles) of the lateral and medial IM TibialCartilage values in megapascal. Subdivided anatomical regions are as follows: AH, anterior horn; PI, pars intermedia; PH, posterior horn; CtC, cartilage-to-cartilage contact area. Nonparametric statistical analyses: $n = 12$; *$P = .05$. For reasons of readability, we marked significant differences between the mild and severe degeneration of the medial and lateral sides above the representative biomechanical mappings and also between the CtC and other anatomical compartments only at the legend of the category axis (e.g., CtC*).

From Seitz, A. M., Osthaus, F., Schwer, J., Warnecke, D., Faschingbauer, M., Sgroi, M., Ignatius, A., & Dürselen, L. (2021). Osteoarthritis-related degeneration alters the biomechanical properties of human menisci before the articular cartilage. Frontiers in Bioengineering and Biotechnology, 9(296), 659989. https://doi.org/10.3389/fbioe.2021.659989.

CARTILAGE FRICTION

The synovial joint is subjected to a variety of different loading conditions with a wide range of different velocities, all accompanied by remarkably low frictional forces. This reduction in friction occurs through lubrication of the AC and serves to protect the cartilage from wear and mechanical damage (Torzilli & Mow, 1976). Healthy joints have a very low coefficient of friction (0.002−0.02) and increased to 0.2−0.4 under constant load for several hours (Krishnan, Mariner,

& Ateshian, 2005). Joint lubrication is provided on the one hand by a 0.1–0.4-μm-thick layer of highly concentrated phospholipids located on the outermost cartilage surface and on the other hand by the synovial fluid (Fig. 9.9) (Hlavacek, 1995; Mow et al., 1992; Vanwanseele et al., 2002). The synovial fluid forms a boundary layer on the AC. It consists of molecules with a high molecular weight, such as hyaluronic acid, proteins, and a glycoprotein, respectively (Charnley, 1960; Dijkgraaf et al., 1995) forming a hyaluronic acid–protein–macromolecule complex. This complex forms a macromolecular network that gives the synovial fluid its viscoelastic property while exhibiting non-Newtonian behavior under shear loads. The glycoprotein lubricin, for example, is partly responsible for the lubrication properties of the synovial fluid (Jay, Tantravahi, Britt, Barrach, & Cha, 2006; Sarma, Powell, & LaBerge, 2006). It is mainly composed of protein and carbohydrates, but also a small number of phospholipids probably enabling its ability to hydrophobically coat the surfaces of AC by aggregation of the individual lubricin polymers (Jay et al., 2006).

The squeeze-film formed during cartilage deformation is the direct result of the pressure gradient and the indirect result of matrix compaction. The movement of the lubricating film can be described by Darcy's law:

$$Q = k\Delta p/\Delta h$$

where Q is the volume flow rate, Δp is the pressure drop, Δh is the cartilage sample thickness, and k is the permeability. Darcy's law is independent of matrix deformation and only the fluid pressure gradient is important. Thus it depends on the surface porosity and matrix permeability (Torzilli & Mow, 1976).

Regarding joint lubrication, there are several theories that function in different combinations depending on the types of loading involved (Fig. 9.10). Back in 1960, Charnley postulated boundary lubrication as the responsible mechanism for lubrication of AC (Charnley, 1960). According to his theory, the coefficient of friction is independent of the viscosity of the lubricant while only the quality of the lubricant is crucial. Further, the multiphospholipid surface layer is also actively involved. According to Charnley (1960) and Dijkgraaf et al. (1995), boundary lubrication occurs primarily at low loads, whereas Sarma et al. (2006) postulated that it occurs under heavy loads. On the other hand, the "weeping" theory suggests that the loading of cartilage results in an increase in internal hydrostatic pressure. When this pressure exceeds the osmotic pressure, interstitial fluid from the compressed cartilage region is exuded over the cartilage surface into the joint space. The exuded interstitial fluid creates a lubricating film, which is maintained by a self-pressure mechanism called weeping lubrication. Outside the loaded region, fluid resorption occurs. This mechanism primarily occurs during severe cartilage loading (Dijkgraaf et al., 1995). Another theory for the lubrication of the cartilage surface is the so-called "boosted lubrication." It is based on the assumption that the soluble components of the synovial fluid are trapped between the articulating surfaces and are slowly impelled into the underlying cartilage. Due to the small pore size of cartilage, this filtration process, together with the roughness of the articular surface, creates a condensed lubricating film that is particularly rich in hyaluronic acid-protein complexes. This film cannot permeate into the cartilage and consequently remains on the articular surface (Walker, Dowson, Longfield, & Wright, 1968). This hypothesis is based on the assumption that as the thickness of the fluid film decreases, the viscous resistance to lateral flow due to movement of the lubricating film becomes greater than the resistance to flow into the cartilage (Mow et al., 1984). This theory is based on the assumption that the articular surface is irregular and not smooth. However, in

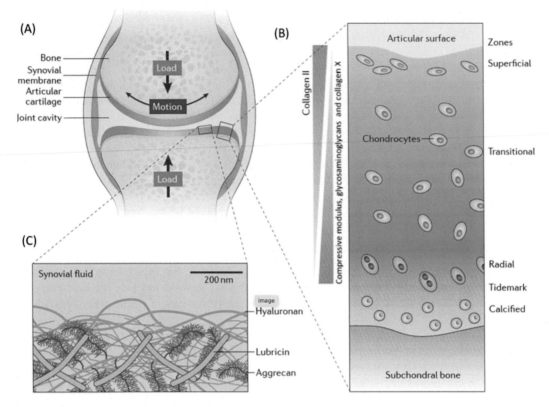

FIGURE 9.9

Structure and constituents of articular cartilage. (A) A synovial joint enclosed within the synovial membrane which is lined with specialized cells (synoviocytes) and contains the synovial fluid within the joint cavity. (B) The articular cartilage layer is organized into four main zones located above the subchondral bone. Chondrocytes (cartilage cells) occupy less than 5% (volume fraction) of cartilage tissue. The matrix composition of collagen II, collagen X, and glycosaminoglycans (GAGs) of the tissue as well as its modulus has a depth dependence as indicated. (C) Structure of the most superficial layer of articular cartilage. Linear hyaluronic acid (HA, green), bottle-brush like aggrecan (violet), and lubricin (light blue) are major macromolecules involved in cartilage lubrication, with lubricin reported both at the surface of the cartilage and at the outer superficial zone, while phospholipids are also present on the cartilage surface but are not shown here.

From Lin, W., and Klein, J. (2021). Recent progress in cartilage lubrication. Advanced Materials, *33(18), e2005513. https://doi.org/10.1002/adma.202005513.*

experiments on AC samples, Kirk, Wilson, & Stachowiak (1993) were able to demonstrate that the cartilage surface is extremely smooth in an unloaded, hydrated state. At the end of loading, the pressure in the synovial fluid is removed and only the pressure at the interface between synovial fluid and cartilage remains. The fluid ideally flows back out of the cartilage into the joint space. This pumping effect can also be observed during pulsating continuous loading. A latter theory

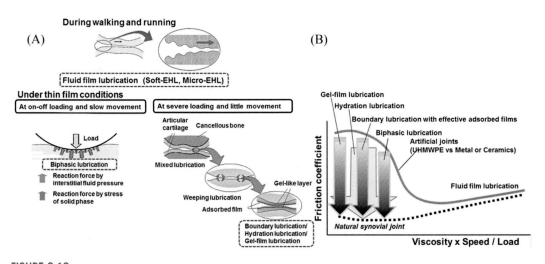

FIGURE 9.10

(A) Schema for adaptive multimode lubrication mechanism. (B) Stribeck curve for adaptive multimode lubrication.

From Murakamia, T., Yarimitsu, S., Sakai, N., Nakashima, K., Yamaguchi, T., & Sawae, Y. (2017). Importance of adaptive multimode lubrication mechanism in natural synovial joints. Tribology International, 113, 306–315. https://doi.org/10.1016/j. triboint.2016.12.052.

concerning joint lubrication is as follows: during movement between the loaded joint surfaces, interstitial fluid is squeezed out of one sliding partner. Just below and behind the other sliding partner, the fluid is reabsorbed by the cartilage as a result of the elastic recovery of the tissue. In this way, cartilage can form its own lubricating film by exudation and imbibition of interstitial fluid. Hou et al. described this lubrication mode during joint loading as a squeeze-film lubrication mechanism (Hou et al., 1992). The lubricating film thickness is reduced from a fairly thick film (100 μm) to a film thickness of only a few microns within 1 second. After that, the reciprocal approach of the joint surfaces occurs much more slowly. This process also serves to transport nutrients and waste products. In conclusion, due to the fact that there are so many theoretical lubrication modes, it has been speculated that not only one particular mode is predominant, it is rather assumed that there is a combination of several modes acting to provide smooth friction of AC surfaces (Katta, Jin, Ingham, & Fisher, 2008).

Tribological studies demonstrated that friction in synovial joints is multifactorial and depends not only on the load but also on the sliding velocity, the load duration, and the used lubricant (Forster & Fisher, 1996; Gleghorn & Bonassar, 2008; Warnecke, Schild et al., 2017). To determine the frictional properties of cartilage and meniscus, specimens are extracted and examined in isolation from the natural joint environment in translating (pin-on-plate) or rotating (pin-on-disk) test devices (Forster & Fisher, 1996; Gleghorn & Bonassar, 2008). Coulomb's friction law is used to calculate the coefficient of friction (μ) by relating the frictional force (F_R) induced from motion to the applied load (F_N):

$$F_R = \mu * F_N \mu = \frac{F_R}{F_N}$$

However, the tests based on pin-on devices are carried out with highly simplified test conditions. The so-called pendulum test rigs are suitable for investigating the friction of entire knee joints (Akelman et al., 2013; Crisco, Blume, Teeple, Fleming, & Jay, 2007; Elmorsy et al., 2014; Kawano et al., 2003). In addition to an intact joint environment, they allow physiological force transmission in the joint and thus also to its structures. This is accomplished by initially deflecting the joint from the horizontal to determine the total friction by recording the resulting pendulum oscillation. From the decreasing oscillation amplitude in the flexion/extension plane, the friction coefficient can be determined as a measure of the intraarticular friction according to Stanton (1923), or alternatively according to Crisco et al. (2007).

Friction analysis under simulated physiological loading and motion conditions

Due to the complex nature of cartilage tribology, it is not easy to measure the friction coefficient of AC, especially under physiological loading situations as a quite high number of testing parameters must be taken into consideration. However, in a recent publication, the tribology of cartilage was investigated under physiological conditions (Forster & Fisher, 1996; Gleghorn & Bonassar, 2008; Warnecke, Schild, et al., 2017):

Within the knee joint, the (axial) loading and motion conditions vary considerably during activities of daily life (Taylor, Heller, Bergmann, & Duda, 2006). During one walking—gait cycle, which can be divided into a stance phase (60%) and a swing phase (40%), the tibiofemoral contact forces rise up to two to three times body weight during the stance phase, while during the low-loaded swing phase, the velocities are the highest (Neu, Komvopoulos, & Reddi, 2009). Nevertheless, most previous friction studies investigated the friction properties of AC and meniscus under constant axial loading conditions and sliding velocities and therefore neglected these variations. Out of these, several tribological theories were postulated based on the three lubrication modes to describe the extremely low but complex friction properties of both cartilaginous tissues: boundary, mixed, and fluid lubrication (Hou et al., 1992; Jahn, Seror, & Klein, 2016; Neu et al., 2009; Walker et al., 1968). Consequently, the question arises of how these dynamic or physiological conditions affect the friction within the knee joint. To answer this question, a new dynamic friction-testing device was developed in a pin-on-plate configuration. Its centerpiece was a dynamic material-testing machine (ElectroForce 5500, BOSE/TA Instruments, United States), which was equipped with a linear motor (VT-75, PImiCos, Germany) mounted on a customized aluminum frame (Fig. 9.11). Using the testing device, it was possible to investigate the friction properties under defined but more physiological testing conditions, as the dynamic normal forces, which are typically acting in the knee joint during gait, were applied to cylindrical samples (pin) of the silk fibroin scaffold, meniscus, or AC. Synchronously, a flat cartilage sample (plate) slid against it with varying velocities derived from the stance and swing phase of a human gait cycle.

This was the first study, synchronously varying axial load as well as sliding velocity, comprising three test scenarios (Fig. 9.12).

Both native forms of cartilaginous tissues showed higher friction coefficients during the low-loaded swing phase than during the stance phase when testing under the most physiological testing conditions (FT-III, meniscus: μ (stance phase) $\cong 0.015$, μ (swing phase) $\cong 0.03$, and cartilage: μ (stance phase) $\cong 0.02$, μ (swing phase) $\cong 0.03$; Figure 8b). This is in line with the literature, where

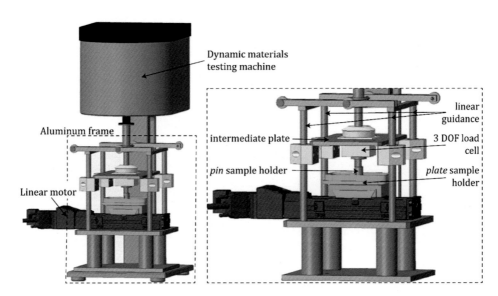

FIGURE 9.11

The dynamic friction testing device was developed to investigate the friction properties under defined testing conditions occurring during gait. It basically consists of a dynamic material testing machine (ElectroForce 5500, BOSE/TA Instruments, United States), which was equipped with a linear motor mounted on a customized aluminum frame (detailed view).

From Warnecke, D., Messemer, M., de Roy, L., Stein, S., Gentilini, C., & Walker, R. (2019). Articular cartilage and meniscus reveal higher friction in swing phase than in stance phase under dynamic gait conditions. Scientific Reports, 9(1), 5785. https://doi.org/ 10.1038/s41598-019-42254-2.

Krishnan, Kopacz, & Ateshian (2006) showed an increased friction coefficient when testing AC against glass during low-loaded phases. A reason for that might be the simultaneously detectable negative values of the fluid load support, leading to the assumption that suction might occur when the load changes to a lower level (Krishnan et al., 2005). Next to the soaking effect and the resultant rise in the friction, it is also known that a consistent fluid film is formed during the swing phase that additionally can be maintained throughout the entire gait cycle (Neu et al., 2009). This fluid film in connection with the different loads and velocities during a gait cycle leads to the assumption that the already postulated lubrication modes, especially hydro- and elastohydrodynamic lubrication synergistically contribute to the remarkably low friction properties of the physiologically articulating surfaces in the knee joint (Neu et al., 2009; Thier & Tonak, 2018). Although friction testing under static testing conditions and defined lubrication regimes might be important, the current study showed that testing under physiological testing conditions revealed higher friction coefficients, especially during the low-loaded swing phase. Consequently, this leads to the assumption that static friction tests can underestimate friction coefficients rather than reflecting the complex in-vivo behavior, which especially might be important for potential replacement materials.

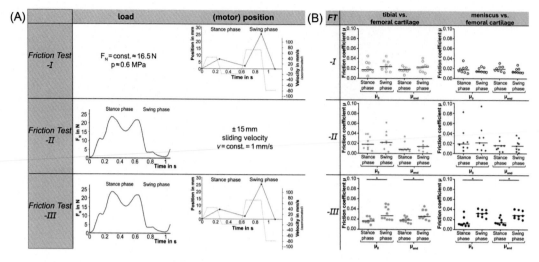

FIGURE 9.12

(A) Overview of the three friction test scenarios and the resulting combinations of axial load (F_N in N) and loading velocities in mm/s. (B) Plots of the resulting friction coefficient (μ) between tibial and femoral articular cartilage, and between meniscus and femoral articular cartilage, determined under the three different test scenarios.

From Warnecke, D., Messemer, M., de Roy, L., Stein, S., Gentilini, C., & Walker, R. (2019). Articular cartilage and meniscus reveal higher friction in swing phase than in stance phase under dynamic gait conditions. Scientific Reports, *9(1), 5785. https://doi.org/ 10.1038/s41598-019-42254-2.*

REFERENCES

Akelman, M. R., Teeple, E., Machan, J. T., Crisco, J. J., Jay, G. D., & Fleming, B. C. (2013). Pendulum mass affects the measurement of articular friction coefficient. *Journal of Biomechanics, 46*(3), 615−618. Available from https://doi.org/10.1016/j.jbiomech.2012.09.034.

Ateshian, G. A., Lai, W. M., Zhu, W. B., & Mow, V. C. (1994). An asymptotic solution for the contact of two biphasic cartilage layers. *Journal of Biomechanics, 27*(11), 1347−1360. Available from https://doi.org/ 10.1016/0021-9290(94)90044-2.

Athanasiou, K. A., Rosenwasser, M. P., Buckwalter, J. A., Malinin, T. I., & Mow, V. C. (1991). Interspecies comparisons of in situ intrinsic mechanical properties of distal femoral cartilage. *Journal of Orthopaedic Research: Official Publication of the Orthopaedic Research Society, 9*(3), 330−340. Available from https://doi.org/10.1002/jor.1100090304.

Buckwalter, J. A., & Mankin, H. J. (1997). Articular cartilage 1. Tissue design and chondrocyte-matrix interactions. *Journal of Bone and Joint Surgery-American Volume, 79*(4), 600−611. Available from https://doi.org/10.2106/00004623-199704000-00021.

Buckwalter, J. A., Mow, V. C., & Ratcliffe, A. (1994). Restoration of injured or degenerated articular cartilage. *The Journal of the American Academy of Orthopaedic Surgeons, 2*(4), 192−201. Available from https://doi.org/10.5435/00124635-199407000-00002.

Bullough, P. G., & Jagannath, A. (1983). The morphology of the calcification front in articular cartilage. Its significance in joint function. *The Journal of Bone and Joint Surgery. British Volume*, *65*(1), 72−78. Available from https://doi.org/10.1302/0301-620X.65B1.6337169.

Buschmann, M. D., Gluzband, Y. A., Grodzinsky, A. J., Kimura, J. H., & Hunziker, E. B. (1992). Chondrocytes in agarose culture synthesize a mechanically functional extracellular matrix. *Journal of Orthopaedic Research: Official Publication of the Orthopaedic Research Society*, *10*(6), 745−758. Available from https://doi.org/10.1002/jor.1100100602.

Buschmann, M. D., & Grodzinsky, A. J. (1995). A molecular model of proteoglycan-associated electrostatic forces in cartilage mechanics. *Journal of Biomechanical Engineering*, *117*(2), 179−192. Available from https://doi.org/10.1115/1.2796000.

Charnley, J. (1960). The lubrication of animal joints in relation to surgical reconstruction by arthroplasty. *Annals of the Rheumatic Diseases*, *19*(1), 10−19. Available from https://doi.org/10.1136/ard.19.1.10.

Chen, S. S., Falcovitz, Y. H., Schneiderman, R., Maroudas, A., & Sah, R. L. (2001). Depth-dependent compressive properties of normal aged human femoral head articular cartilage: Relationship to fixed charge density. *Osteoarthritis and Cartilage / OARS, Osteoarthritis Research Society*, *9*(6), 561−569. Available from https://doi.org/10.1053/joca.2001.0424.

Cohen, N. P., Foster, R. J., & Mow, V. C. (1998). Composition and dynamics of articular cartilage: Structure, function, and maintaining healthy state. *The Journal of Orthopaedic and Sports Physical Therapy*, *28*(4), 203−215. Available from https://doi.org/10.2519/jospt.1998.28.4.203.

Crisco, J. J., Blume, J., Teeple, E., Fleming, B. C., & Jay, G. D. (2007). Assuming exponential decay by incorporating viscous damping improves the prediction of the coefficient of friction in pendulum tests of whole articular joints. *Proceedings of the Institution of Mechanical Engineers, Part H*, *221*(3), 325−333. Available from https://doi.org/10.1243/09544119JEIM248.

Dijkgraaf, L. C., de Bont, L. G., Boering, G., & Liem, R. S. (1995). Normal cartilage structure, biochemistry, and metabolism: A review of the literature. *Journal of Oral and Maxillofacial Surgery: Official Journal of the American Association of Oral and Maxillofacial Surgeons*, *53*(8), 924−929. Available from https://doi.org/10.1016/0278-2391(95)90283-x.

Elliott, D. M., Narmoneva, D. A., & Setton, L. A. (2002). Direct measurement of the Poisson's ratio of human patella cartilage in tension. *Journal of Biomechanical Engineering*, *124*(2), 223−228. Available from https://doi.org/10.1115/1.1449905.

Elmorsy, S., Funakoshi, T., Sasazawa, F., Todoh, M., Tadano, S., & Iwasaki, N. (2014). Chondroprotective effects of high-molecular-weight cross-linked hyaluronic acid in a rabbit knee osteoarthritis model. *Osteoarthritis and Cartilage / OARS, Osteoarthritis Research Society*, *22*(1), 121−127. Available from https://doi.org/10.1016/j.joca.2013.10.005.

Errington, R. J., Fricker, M. D., Wood, J. L., Hall, A. C., & White, N. S. (1997). Four-dimensional imaging of living chondrocytes in cartilage using confocal microscopy: A pragmatic approach. *American Journal of Physiology-Cell Physiology*, *272*(3), C1040−C1051. Available from https://doi.org/10.1152/ajpcell.1997.272.3.C1040.

Federico, S., & Herzog, W. (2008). On the anisotropy and inhomogeneity of permeability in articular cartilage. *Biomechanics and Modeling in Mechanobiology*, *7*(5), 367−378. Available from https://doi.org/10.1007/s10237-007-0091-0.

Forster, H., & Fisher, J. (1996). The influence of loading time and lubricant on the friction of articular cartilage. *Proceedings of the Institution of Mechanical Engineers, Part H*, *210*(2), 109−119. Available from https://doi.org/10.1243/PIME_PROC_1996_210_399_02.

Frank, E. H., & Grodzinsky, A. J. (1987a). Cartilage electromechanics−I. Electrokinetic transduction and the effects of electrolyte pH and ionic strength. *Journal of Biomechanics*, *20*(6), 615−627. Available from https://doi.org/10.1016/0021-9290(87)90282-x.

Frank, E. H., & Grodzinsky, A. J. (1987b). Cartilage electromechanics–II. A continuum model of cartilage electrokinetics and correlation with experiments. *Journal of Biomechanics, 20*(6), 629–639. Available from https://doi.org/10.1016/0021-9290(87)90283-1.

Fulcher, G. R., Hukins, D. W., & Shepherd, D. E. (2009). Viscoelastic properties of bovine articular cartilage attached to subchondral bone at high frequencies. *BMC Musculoskeletal Disorders, 10*, 61. Available from https://doi.org/10.1186/1471-2474-10-61.

Gleghorn, J. P., & Bonassar, L. J. (2008). Lubrication mode analysis of articular cartilage using Stribeck surfaces. *Journal of Biomechanics, 41*(9), 1910–1918. Available from https://doi.org/10.1016/j.jbiomech.2008.03.043.

Guilak, F., Ratcliffe, A., Lane, N., Rosenwasser, M. P., & Mow, V. C. (1994). Mechanical and biochemical changes in the superficial zone of articular cartilage in canine experimental osteoarthritis. *Journal of Orthopaedic Research: Official Publication of the Orthopaedic Research Society, 12*(4), 474–484. Available from https://doi.org/10.1002/jor.1100120404.

Hayes, W. C., & Mockros, L. F. (1971). Viscoelastic properties of human articular cartilage. *Journal of Applied Physiology (Bethesda, Md.: 1985), 31*(4), 562–568. Available from https://doi.org/10.1152/jappl.1971.31.4.562.

Hlavacek, M. (1993a). The role of synovial fluid filtration by cartilage in lubrication of synovial joints–I. Mixture model of synovial fluid. *Journal of Biomechanics, 26*(10), 1145–1150. Available from https://doi.org/10.1016/0021-9290(93)90062-j.

Hlavacek, M. (1993b). The role of synovial fluid filtration by cartilage in lubrication of synovial joints–II. Squeeze-film lubrication: Homogeneous filtration. *Journal of Biomechanics, 26*(10), 1151–1160. Available from https://doi.org/10.1016/0021-9290(93)90063-k.

Hlavacek, M. (1995). The role of synovial fluid filtration by cartilage in lubrication of synovial joints–IV. Squeeze-film lubrication: The central film thickness for normal and inflammatory synovial fluids for axial symmetry under high loading conditions. *Journal of Biomechanics, 28*(10), 1199–1205. Available from https://doi.org/10.1016/0021-9290(94)00178-7.

Hou, J. S., Mow, V. C., Lai, W. M., & Holmes, M. H. (1992). An analysis of the squeeze-film lubrication mechanism for articular cartilage. *Journal of Biomechanics, 25*(3), 247–259. Available from https://doi.org/10.1016/0021-9290(92)90024-u.

Hukins, D. W. L., Leahy, J. C., & Mathias, K. J. (1999). Biomaterials: Defining the mechanical properties of natural tissues and selection of replacement materials. *Journal of Materials Chemistry, 9*(3), 629–636. Available from https://doi.org/10.1039/a807411i.

Hunziker, E. B., Quinn, T. M., & Hauselmann, H. J. (2002). Quantitative structural organization of normal adult human articular cartilage. *Osteoarthritis and Cartilage / OARS, Osteoarthritis Research Society, 10* (7), 564–572. Available from https://doi.org/10.1053/joca.2002.0814.

Jahn, S., Seror, J., & Klein, J. (2016). Lubrication of articular cartilage. *Annual Review of Biomedical Engineering, 18*, 235–258. Available from https://doi.org/10.1146/annurev-bioeng-081514-123305.

Jay, G. D., Tantravahi, U., Britt, D. E., Barrach, H. J., & Cha, C. J. (2006). Homology of lubricin and superficial zone protein (SZP): Products of megakaryocyte stimulating factor (MSF) gene expression by human synovial fibroblasts and articular chondrocytes localized to chromosome 1q25. *Journal of Orthopaedic Research: Official Publication of the Orthopaedic Research Society, 19*(4), 677–687. Available from https://doi.org/10.1016/S0736-0266(00)00040-1.

Jurvelin, J. S., Rasanen, T., Kolmonen, P., & Lyyra, T. (1995). Comparison of optical, needle probe and ultrasonic techniques for the measurement of articular cartilage thickness. *Journal of Biomechanics, 28*(2), 231–235. Available from https://doi.org/10.1016/0021-9290(94)00060-h.

Katta, J., Jin, Z., Ingham, E., & Fisher, J. (2008). Biotribology of articular cartilage – A review of the recent advances. *Medical Engineering & Physics, 30*(10), 1349–1363. Available from https://doi.org/10.1016/j.medengphy.2008.09.004.

Kawano, T., Miura, H., Mawatari, T., Moro-Oka, T., Nakanishi, Y., Higaki, H., & Iwamoto, Y. (2003). Mechanical effects of the intraarticular administration of high molecular weight hyaluronic acid plus phospholipid on synovial joint lubrication and prevention of articular cartilage degeneration in experimental osteoarthritis. *Arthritis and Rheumatism*, *48*(7), 1923−1929. Available from https://doi.org/10.1002/art.11172.

Kirk, T. B., Wilson, A. S., & Stachowiak, G. W. (1993). The effects of dehydration on the surface-morphology of articular-cartilage. *Journal of Orthopaedic Rheumatology*, *6*(2−3), 75−80.

Kiviranta, P., Rieppo, J., Korhonen, R. K., Julkunen, P., Toyras, J., & Jurvelin, J. S. (2006). Collagen network primarily controls Poisson's ratio of bovine articular cartilage in compression. *Journal of Orthopaedic Research: Official Publication of the Orthopaedic Research Society*, *24*(4), 690−699. Available from https://doi.org/10.1002/jor.20107.

Krishnan, R., Kopacz, M., & Ateshian, G. A. (2006). Experimental verification of the role of interstitial fluid pressurization in cartilage lubrication. *Journal of Orthopaedic Research: Official Publication of the Orthopaedic Research Society*, *22*(3), 565−570. Available from https://doi.org/10.1016/j.orthres.2003.07.002.

Krishnan, R., Mariner, E. N., & Ateshian, G. A. (2005). Effect of dynamic loading on the frictional response of bovine articular cartilage. *Journal of Biomechanics*, *38*(8), 1665−1673. Available from https://doi.org/10.1016/j.jbiomech.2004.07.025.

Langelier, E., & Buschmann, M. D. (2003). Increasing strain and strain rate strengthen transient stiffness but weaken the response to subsequent compression for articular cartilage in unconfined compression. *Journal of Biomechanics*, *36*(6), 853−859. Available from https://doi.org/10.1016/s0021-9290(03)00006-x.

Lawless, B. M., Sadeghi, H., Temple, D. K., Dhaliwal, H., Espino, D. M., & Hukins, D. W. L. (2017). Viscoelasticity of articular cartilage: Analysing the effect of induced stress and the restraint of bone in a dynamic environment. *Journal of the Mechanical Behavior of Biomedical Materials*, *75*, 293−301. Available from https://doi.org/10.1016/j.jmbbm.2017.07.040.

Lin, W., & Klein, J. (2021). Recent progress in cartilage lubrication. *Advanced Materials*, *33*(18), e2005513. Available from https://doi.org/10.1002/adma.202005513.

Mansour, J. M., & Mow, V. C. (1976). The permeability of articular cartilage under compressive strain and at high pressures. *The Journal of Bone and Joint Surgery. American Volume*, *58*(4), 509−516. Available from https://www.ncbi.nlm.nih.gov/pubmed/1270471.

Mauck, R. L., Hung, C. T., & Ateshian, G. A. (2003). Modeling of neutral solute transport in a dynamically loaded porous permeable gel: Implications for articular cartilage biosynthesis and tissue engineering. *Journal of Biomechanical Engineering*, *125*(5), 602−614. Available from https://doi.org/10.1115/1.1611512.

Modis, L., Botos, A., Kiviranta, I., Lukacsko, L., & Helminen, H. J. (1996). Differences in submicroscopic structure of the extracellular matrix of canine femoral and tibial condylar articular cartilages as revealed by polarization microscopical analysis. *Acta Biologica Hungarica*, *47*(1−4), 341−353. Available from http:// < Go to ISI >://WOS:000171777000027.

Mountcastle, S. E., Allen, P., Mellors, B. O. L., Lawless, B. M., Cooke, M. E., Lavecchia, C. E., & Cox, S. C. (2019). Dynamic viscoelastic characterisation of human osteochondral tissue: understanding the effect of the cartilage-bone interface. *BMC Musculoskeletal Disorders*, *20*(1), 575. Available from https://doi.org/10.1186/s12891-019-2959-4.

Mow, V. C., Holmes, M. H., & Lai, W. M. (1984). Fluid transport and mechanical properties of articular cartilage: A review. *Journal of Biomechanics*, *17*(5), 377−394. Available from https://doi.org/10.1016/0021-9290(84)90031-9.

Mow, V. C., Kuei, S. C., Lai, W. M., & Armstrong, C. G. (1980). Biphasic creep and stress relaxation of articular cartilage in compression? Theory and experiments. *Journal of Biomechanical Engineering*, *102*(1), 73−84. Available from https://doi.org/10.1115/1.3138202.

Mow, V. C., Ratcliffe, A., & Poole, A. R. (1992). Cartilage and diarthrodial joints as paradigms for hierarchical materials and structures. *Biomaterials*, *13*(2), 67−97. Available from https://doi.org/10.1016/0142-9612(92)90001-5.

Mow, V. C., Wang, C. C., & Hung, C. T. (1999). The extracellular matrix, interstitial fluid and ions as a mechanical signal transducer in articular cartilage. *Osteoarthritis and Cartilage / OARS, Osteoarthritis Research Society, 7*(1), 41−58. Available from https://doi.org/10.1053/joca.1998.0161.

Muehleman, C., Majumdar, S., Issever, A. S., Arfelli, F., Menk, R. H., Rigon, L., & Mollenhauer, J. (2004). X-ray detection of structural orientation in human articular cartilage. *Osteoarthritis and Cartilage / OARS, Osteoarthritis Research Society, 12*(2), 97−105. Available from https://doi.org/10.1016/j.joca.2003.10.001.

Neu, C. P., Komvopoulos, K., & Reddi, A. H. (2009). The interface of functional biotribology and regenerative medicine in synovial joints. *Tissue Engineering. Part B, Reviews, 14*(3), 235−247. Available from https://doi.org/10.1089/ten.teb.2008.0047.

Oinas, J., Ronkainen, A. P., Rieppo, L., Finnila, M. A. J., Iivarinen, J. T., van Weeren, P. R., & Saarakkala, S. (2018). Composition, structure and tensile biomechanical properties of equine articular cartilage during growth and maturation. *Scientific Reports, 8*(1), 11357. Available from https://doi.org/10.1038/s41598-018-29655-5.

Oloyede, A., Flachsmann, R., & Broom, N. D. (1992). The dramatic influence of loading velocity on the compressive response of articular cartilage. *Connective Tissue Research, 27*(4), 211−224. Available from https://doi.org/10.3109/03008209209006997.

Orford, C. R., & Gardner, D. L. (1985). Ultrastructural histochemistry of the surface lamina of normal articular cartilage. *The Histochemical Journal, 17*(2), 223−233. Available from https://doi.org/10.1007/BF01003221.

Park, S., Nicoll, S. B., Mauck, R. L., & Ateshian, G. A. (2008). Cartilage mechanical response under dynamic compression at physiological stress levels following collagenase digestion. *Annals of Biomedical Engineering, 36*(3), 425−434. Available from https://doi.org/10.1007/s10439-007-9431-6.

Poole, C. A., Flint, M. H., & Beaumont, B. W. (1987). Chondrons in cartilage: Ultrastructural analysis of the pericellular microenvironment in adult human articular cartilages. *Journal of Orthopaedic Research: Official Publication of the Orthopaedic Research Society, 5*(4), 509−522. Available from https://doi.org/10.1002/jor.1100050406.

Pordzik, J., Bernstein, A., Watrinet, J., Mayr, H. O., Latorre, S. H., Schmal, H., & Seidenstuecker, M. (2020). Correlation of biomechanical alterations under gonarthritis between overlying menisci and articular cartilage. *Applied Sciences-Basel, 10*(23), 8673. Available from https://doi.org/10.3390/app10238673.

Quinn, T. M., Hunziker, E. B., & Hauselmann, H. J. (2005). Variation of cell and matrix morphologies in articular cartilage among locations in the adult human knee. *Osteoarthritis and Cartilage / OARS, Osteoarthritis Research Society, 13*(8), 672−678. Available from https://doi.org/10.1016/j.joca.2005.04.011.

Sarma, A. V., Powell, G. L., & LaBerge, M. (2006). Phospholipid composition of articular cartilage boundary lubricant. *Journal of Orthopaedic Research: Official Publication of the Orthopaedic Research Society, 19*(4), 671−676. Available from https://doi.org/10.1016/S0736-0266(00)00064-4.

Seidenstuecker, M., Watrinet, J., Bernstein, A., Suedkamp, N. P., Latorre, S. H., Maks, A., & Mayr, H. O. (2019). Viscoelasticity and histology of the human cartilage in healthy and degenerated conditions of the knee. *Journal of Orthopaedic Surgery and Research, 14*(1), 256. Available from https://doi.org/10.1186/s13018-019-1308-5.

Seitz, A. M., Osthaus, F., Schwer, J., Warnecke, D., Faschingbauer, M., Sgroi, M., & Dürselen, L. (2021). Osteoarthritis-related degeneration alters the biomechanical properties of human menisci before the articular cartilage. *Frontiers in Bioengineering and Biotechnology, 9*(296), 659989. Available from https://doi.org/10.3389/fbioe.2021.659989.

Selyutina, N. S., Argatov, I. I., & Mishuris, G. S. (2015). On application of Fung's quasi-linear viscoelastic model to modeling of impact experiment for articular cartilage. *Mechanics Research Communications, 67*, 24−30. Available from https://doi.org/10.1016/j.mechrescom.2015.04.003.

Sim, S., Chevrier, A., Garon, M., Quenneville, E., Lavigne, P., Yaroshinsky, A., & Buschmann, M. D. (2016). Electromechanical probe and automated indentation maps are sensitive techniques in assessing early degenerated human articular cartilage. *Journal of Orthopaedic Research: Official Publication of the Orthopaedic Research Society, 35*(4), 858−867. Available from https://doi.org/10.1002/jor.23330.

Sim, S., Chevrier, A., Garon, M., Quenneville, E., Yaroshinsky, A., Hoemann, C. D., & Buschmann, M. D. (2014). Non-destructive electromechanical assessment (Arthro-BST) of human articular cartilage correlates with histological scores and biomechanical properties. *Osteoarthritis and Cartilage / OARS, Osteoarthritis Research Society*, *22*(11), 1926−1935. Available from https://doi.org/10.1016/j.joca.2014.08.008.

Sophia Fox, A. J., Bedi, A., & Rodeo, S. A. (2009). The basic science of articular cartilage: Structure, composition, and function. *Sports Health*, *1*(6), 461−468. Available from https://doi.org/10.1177/1941738109350438.

Stanton, T. E. (1923). Boundary lubrication in engineering practice. *The Engineer*, *29*, 678−680. Available from https://ci.nii.ac.jp/naid/10029608887/en/.

Sun, D. D., Guo, X. E., Likhitpanichkul, M., Lai, W. M., & Mow, V. C. (2004). The influence of the fixed negative charges on mechanical and electrical behaviors of articular cartilage under unconfined compression. *Journal of Biomechanical Engineering*, *126*(1), 6−16. Available from https://doi.org/10.1115/1.1644562.

Szafranski, J. D., Grodzinsky, A. J., Burger, E., Gaschen, V., Hung, H. H., & Hunziker, E. B. (2004). Chondrocyte mechanotransduction: Effects of compression on deformation of intracellular organelles and relevance to cellular biosynthesis. *Osteoarthritis and Cartilage / OARS, Osteoarthritis Research Society*, *12*(12), 937−946. Available from https://doi.org/10.1016/j.joca.2004.08.004.

Taylor, W. R., Heller, M. O., Bergmann, G., & Duda, G. N. (2006). Tibio-femoral loading during human gait and stair climbing. *Journal of Orthopaedic Research: Official Publication of the Orthopaedic Research Society*, *22*(3), 625−632. Available from https://doi.org/10.1016/j.orthres.2003.09.003.

Temple, D. K., Cederlund, A. A., Lawless, B. M., Aspden, R. M., & Espino, D. M. (2016). Viscoelastic properties of human and bovine articular cartilage: A comparison of frequency-dependent trends. *BMC Musculoskeletal Disorders*, *17*(1), 419. Available from https://doi.org/10.1186/s12891-016-1279-1.

Thier, S., & Tonak, M. (2018). Influence of synovial fluid on lubrication of articular cartilage in vitro − A review. *Zeitschrift fur Orthopadie und Unfallchirurgie*, *156*(2), 205−213. Available from https://doi.org/10.1055/s-0043-117959.

Torzilli, P. A., Adams, T. C., & Mis, R. J. (1987). Transient solute diffusion in articular cartilage. *Journal of Biomechanics*, *20*(2), 203−214. Available from https://doi.org/10.1016/0021-9290(87)90311-3.

Torzilli, P. A., Arduino, J. M., Gregory, J. D., & Bansal, M. (1997). Effect of proteoglycan removal on solute mobility in articular cartilage. *Journal of Biomechanics*, *30*(9), 895−902. Available from https://doi.org/10.1016/s0021-9290(97)00059-6.

Torzilli, P. A., Dethmers, D. A., Rose, D. E., & Schryuer, H. F. (1983). Movement of interstitial water through loaded articular-cartilage. *Journal of Biomechanics*, *16*(3), 169−179. Available from https://doi.org/10.1016/0021-9290(83)90124-0.

Torzilli, P. A., & Mow, V. C. (1976). On the fundamental fluid transport mechanisms through normal and pathological articular cartilage during function − II. The analysis, solution and conclusions. *Journal of Biomechanics*, *9*(9), 587−606. Available from https://doi.org/10.1016/0021-9290(76)90100-7.

Urban, J. P. (1994). The chondrocyte: A cell under pressure. *British Journal of Rheumatology*, *33*(10), 901−908. Available from https://doi.org/10.1093/rheumatology/33.10.901.

Vanwanseele, B., Lucchinetti, E., & Stussi, E. (2002). The effects of immobilization on the characteristics of articular cartilage: Current concepts and future directions. *Osteoarthritis and Cartilage / OARS, Osteoarthritis Research Society*, *10*(5), 408−419. Available from https://doi.org/10.1053/joca.2002.0529.

Walker, P. S., Dowson, D., Longfield, M. D., & Wright, V. (1968). "Boosted lubrication" in synovial joints by fluid entrapment and enrichment. *Annals of the Rheumatic Diseases*, *27*(6), 512−520. Available from https://doi.org/10.1136/ard.27.6.512.

Warnecke, D., Messemer, M., de Roy, L., Stein, S., Gentilini, C., Walker, R., & Durselen, L. (2019). Articular cartilage and meniscus reveal higher friction in swing phase than in stance phase under dynamic gait conditions. *Scientific Reports*, *9*(1), 5785. Available from https://doi.org/10.1038/s41598-019-42254-2.

Warnecke, D., Schild, N. B., Klose, S., Joos, H., Brenner, R. E., Kessler, O., & Durselen, L. (2017). Friction properties of a new silk fibroin scaffold for meniscal replacement. *Tribology International*, *109*, 586−592. Available from https://doi.org/10.1016/j.triboint.2017.01.038.

Weiss, C., Rosenberg, L., & Helfet, A. J. (1968). An ultrastructural study of normal young adult human articular cartilage. *The Journal of Bone and Joint Surgery. American Volume*, *50*(4), 663−674. Available from https://doi.org/10.2106/00004623-196850040-00002.

Wong, M., & Carter, D. R. (2003). Articular cartilage functional histomorphology and mechanobiology: A research perspective. *Bone*, *33*(1), 1−13. Available from https://doi.org/10.1016/s8756-3282(03)00083-8.

Zhu, W., Mow, V. C., Koob, T. J., & Eyre, D. R. (1993). Viscoelastic shear properties of articular cartilage and the effects of glycosidase treatments. *Journal of Orthopaedic Research: Official Publication of the Orthopaedic Research Society*, *11*(6), 771−781. Available from https://doi.org/10.1002/jor.1100110602.

MENISCUS BIOMECHANICS

10

Andreas Martin Seitz, Maren Freutel and Lutz Dürselen

Institute of Orthopaedic Research and Biomechanics, Ulm University, Ulm, Germany

INTRODUCTION

Traumatic situations, premature degeneration, or pathophysiologic aging are the main causes for meniscal lesions. In case of trauma, different mechanisms can lead to different meniscus tear patterns. Meniscus injuries are commonly associated with pain, joint locking, or hematoma. In case a surgical treatment is required, meniscal tears are commonly treated arthroscopically by means of sutures, partial or total meniscectomy. Arthroscopic intervention involving the meniscus is one of the most common surgeries, while partial meniscectomy accounts by far the largest proportion of these procedures. However, clinical mid- and long-term studies as well as biomechanical investigations revealed that such meniscectomies lead to an increase in tibiofemoral contact pressure causing premature degeneration of the articular tibial and femoral surfaces. Therefore, surgeons are encouraged to resect as less meniscal tissue as possible or, alternatively, reconstruct or replace the damaged tissue if possible.

ANATOMY

The lateral and medial menisci are located between the articulating surfaces of the tibia and femur (Fig. 10.1).

The meniscus body can be subdivided into five parts: anterior root, anterior horn, pars intermedia, posterior horn, and posterior root (Fig. 10.2).

Both menisci are crescent shaped and indicate a wedge-shaped cross-section. Besides the capsular connection to the tibial plateau via the meniscotibial connection, the medial meniscus has an additional rigid connection with the medial collateral ligament. Therefore, the medial meniscus is less mobile than the lateral meniscus during the flexion—extension movement of the knee joint. The shape of the menisci together with their microstructural composition defines their biomechanical properties and thus their function inside the knee joint.

The fibrocartilaginous meniscus is usually characterized by its two phases (biphasic), and consists of:

- fluid phase, interstitial fluid (60%−70%);
- solid phase, porous solid matrix mainly consisting of collagen type I (15%−25%) and proteoglycans (1%−2%).

Human Orthopaedic Biomechanics. DOI: https://doi.org/10.1016/B978-0-12-824481-4.00012-3

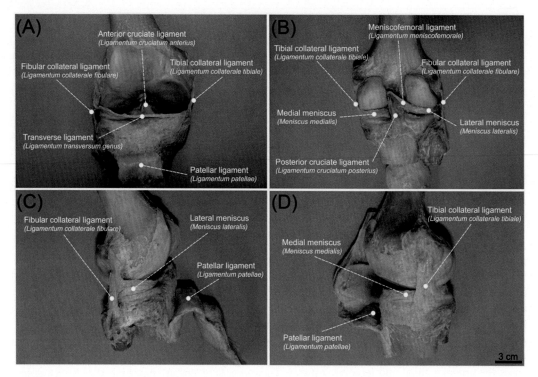

FIGURE 10.1

Anatomical knee joint visualization. Ventral (A), dorsal (B), lateral (C), and medial (D) views are depicted. Scale = 3 cm.

From Meyer, J. J., Obmann, M. M., Giessler, M., Schuldis, D., Bruckner, A. K., Strohm, P. C., Sandeck, F., & Spittau, B. (2017). Interprofessional approach for teaching functional knee joint anatomy. Annals of Anatomy, 210, *155–159. https://doi.org/10.1016/j.aanat.2016.10.011.*

Microstructurally, scanning electron microscopy images reveal a three-layered structure (Petersen & Tillmann, 1998). The 10-m thick surface consists of a friction-optimized, tightly knitted layer with randomly distributed collagen fibrils. This is followed on the femoral and tibial sides by a lamellar layer in which the collagen fibers are crossed. This layer is approximately 150 μm thick. The major portion of the inner meniscus consists of circumferential collagen fiber bundles, which are sporadically bound by radial tie fibers (Fig. 10.3).

Besides the microstructural composition, the vascularization of the menisci is essential for their regenerative potential after an injury. While the menisci of neonates are completely vascularized, adult menisci can be divided into three blood supply zones in their cross-section (Fig. 10.4): the red-red zone is located at the area close to the capsule and indicates vascularization by small radial branches of the superior and inferior genus medial and lateral arteries (Arnoczky & Warren, 1982). It is thus well supplied with blood and, as a result, has good

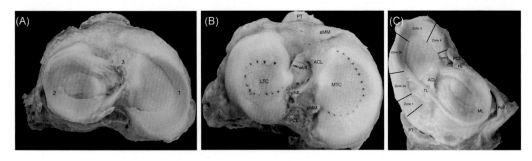

FIGURE 10.2

(A) Anatomical dissection of proximal tibial articular surface (plan view, femur removed). 1, Medial meniscus; 2, lateral meniscus; 3, tibial attachment of anterior cruciate ligament; and 4, tibial attachment of posterior cruciate ligament. (B) The medial meniscus covers up to 50%−60% of the articular surface of medial tibial condyle. *ACL*, anterior cruciate ligament; *PCL*, posterior cruciate ligament; *MTC*, medial tibial condyle; *LTC*, lateral tibial condyle; *aMM*, anterior root of medial meniscus; *pMM*, posterior root of medial meniscus; *aML*, anterior root of lateral meniscus; *pML*, posterior root of lateral meniscus. (C) Anatomical dissection showing five anatomical zones within medial meniscus. ACL, anterior cruciate ligament; *tl*, transverse ligament (anterior intermeniscal ligament); *PT*, patellar tendon; *PCL*, posterior cruciate ligament; *ML*, lateral meniscus; *PoT*, Popliteus tendon; hl, Humphry ligament (anterior menisco-femoral ligament).

From Smigielski, R., Becker, R., Zdanowicz, U., & Ciszek, B. (2015). Medial meniscus anatomy-from basic science to treatment. Knee Surgery, Sports Traumatology, Arthroscopy: Official Journal of the ESSKA, 23(1), 8–14. https://doi.org/10.1007/s00167-014-3476-5.

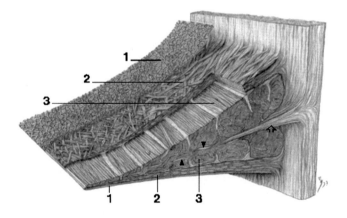

FIGURE 10.3

Synoptic drawing. Scanning electron microscopy reveals three distinct layers in the meniscus cross section. (1) The superficial network: the tibial and femoral sides of the meniscus surface are covered by a meshwork of thin fibrils. (2) Lamellar layer: beneath the superficial network there is a layer of lamellae of collagen fibrils on the tibial and femoral surface. In the area of the external circumference of the anterior and posterior segments the bundles of collagen fibrils are arranged in a radial direction. In all other parts, the collagen fibril bundles intersect at various angles. (3) Central main layer: the main portion of the meniscus collagen fibrils is located in the central region between the femoral and the tibial surface layers. Everywhere in the central main layer of the meniscus the bundles of collagen fibrils are orientated in a circular manner. In the region of the internal circumference, a few radial collagen fibrils are interwoven with the circular fibril bundles (*arrowheads*). In the external circumference, loose connective tissue from the joint capsule penetrates radially between the circular fibril bundles (*arrow*).

From Petersen, W., & Tillmann, B. (1998). Collagenous fibril texture of the human knee joint menisci. Anatomy and Embryology (Berlin), 197(4), 317–324. https://doi.org/10.1007/s004290050141.

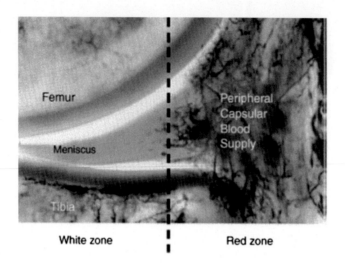

FIGURE 10.4

Blood supply meniscus indicating that radial branches penetrating the peripheral third of the lateral meniscus.

From Karia, M., Ghaly, Y., Al-Hadithy, N., Mordecai, S., & Gupte, C. (2019). Current concepts in the techniques, indications and outcomes of meniscal repairs. European Journal of Orthopaedic Surgery & Traumatology, *29(3), 509–520. https://doi.org/10.1007/s00590-018-2317-5.*

regenerative potential. The middle third (red-white zone) is a zone of moderate blood supply, accompanied by rather low healing potential. The inner third of the meniscus (white zone) has no blood supply, is nourished by diffusion, and has poor healing potential after an injury (Petersen & Tillmann, 1999).

At the anterior and posterior horns, a clear transition to the fiber-rich meniscus root attachments can be identified (Fig. 10.5). The geometry of these ligamentous structures shows clear differences to the meniscal body, whereas the tibial insertions can be determined very well on the basis of anatomical landmarks (Johannsen et al., 2012).

The anteromedial root attachment is flat fan-shaped and inserts at the area intercondylaris, approximately 7 mm anterior to the insertion of the anterior cruciate ligament (ACL). The posterior root attachment of the medial meniscus inserts in the posterior region of the intercondylar fossa, anterior to the posterior cruciate ligament insertion. The insertions of the medial meniscus are more widely spanned than those of the lateral meniscus, making the latter much more mobile. The anterolateral root attachment inserts anterior to the lateral tubercle of the area intercondylaris and lateral to the ACL. Posteriorly, the flat and broad root attachment of the lateral meniscus inserts at the area from the medial to the lateral eminentia intercondylaris.

Like the meniscus, the root attachments also possess a biphasic structure (Abraham et al., 2011; Hauch et al., 2010), with the interstitial fluid accounting for about two-thirds of the total weight. The solid matrix consists mainly of type I collagen fibers (80%, Villegas & Donahue, 2010), which merge into the meniscus horns (Freutel et al., 2015).

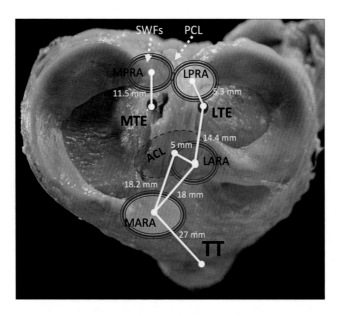

FIGURE 10.5

Medial and lateral tibial plateau. Useful landmarks in relation to meniscal roots are shown. *MPRA*, medial posterior root attachment; *LPRA*, lateral posterior root attachment; *LARA*, lateral anterior root attachment; *MARA*, medial anterior root attachment; *ACL*, anterior cruciate ligament; *TT*, tibial tubercle; *MTE*, medial tibial eminence; *LTE*, lateral tibial eminence; *SWF*, shiny white fibers of the posterior medial meniscus root; *PCL*, posterior cruciate ligament. The distances between meniscal root attachments and landmarks, such as MTE and LTE, have been established by cadaver studies and are important for anatomical meniscal root repair. Nonanatomical meniscal root repair is often equivalent to meniscectomy in terms of progression to osteoarthritis. Therefore, a precise knowledge of meniscal root anatomy is necessary to reduce the menisci to anatomical position. Sometimes, this requires an extensive release with arthroscopic biters or comparable tools if the menisci have scarred into an extruded position against the capsule. The SWFs and PCL are represented by *dashed lines*, indicating their attachments are actually inferior to the level of the tibial plateau.

From LaPrade, R. F., Floyd, E. R., Carlson, G. B., Moatshe, G., Chahla, J., & Monson, J. K. (2021). Meniscal root tears: Solving the silent epidemic. Journal of Arthroscopic Surgery and Sports Medicine, 2, 47–57. https://doi.org/10.25259/jassm_55_2020.

FUNCTION

The main function of the wedge-shaped menisci is to increase the force-transmitting contact area between the femur and tibia, resulting in a reduction of tibiofemoral contact pressure. Due to their geometry, structure, and bony anchorage, they reduce peak stress during load transmission and thus protect the articular cartilage from overload-induced injuries or premature degeneration.

Furthermore, the menisci compensate the incongruence between the articular surfaces of the distal femur and proximal tibia. Considering that the medial femoral condyle articulates over a concave tibia and the lateral over a flat or slightly convex tibia leads to a better congruency of the medial compartment compared to the lateral compartment. Their mobility, flexibility, and shape enable the menisci to perform this task in any flexion (up to 160 degrees) and rotation position of the tibia relative to the femur. The distal part of the condyles shows a large radius of curvature and contacts the entire meniscus when the knee joint is extended. Under flexion the lesser radii of the femoral condyles result in a decreased contact area which moves posteriorly under further knee flexion until the posterior horns are reached. Thus, the menisci are displaced outwards, which is associated with a significant displacement of the anterior horns.

Due to their wedge-shaped cross-section, the menisci tend to extrude out of the knee joint space under axial loads. Since they are rigidly fixed to the bone via their anterior and posterior attachments, tensile stresses develop in the circumferential direction of the menisci (McDermott et al., 2008), preventing the meniscus from this extrusion (Fig. 10.6). As a result, the menisci are able to absorb pressure and transmit up to 81% of the axial joint forces (Pena et al., 2005; Walker & Erkman, 1975).

Based on the irregular movements of the lateral and medial femoral condyles during knee flexion and extension, the lateral meniscus is required to perform larger movements than the medial meniscus. In order to fulfill this task, the lateral meniscus indicates no rigid connection to the lateral collateral ligament like the medial meniscus does and, as already explained before, the anterior and posterior attachments of the lateral meniscus are closer together compared to those of the medial meniscus. Greater displacements were determined for the anterior horns of the lateral and medial meniscus compared to the posterior horns. During flexion of the knee joint from 0 to 90 degrees under axial loading, a mean posterior displacement of 7.1 mm in the anterior horn and 3.9 mm in the posterior horn of the medial meniscus was observed. The anterior horn of the lateral

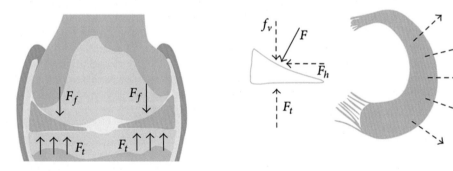

FIGURE 10.6

Schematic diagram of the meniscus force-bearing mechanism. Meniscal configuration adapts well to the corresponding shape of the femoral condyles and the tibial plateau in the knee joint. The axial load force (F) perpendicular to the meniscus surface and horizontal force (f_r) are created by compressing the femur (F_f). F rebounds due to the tibial upgrade force (F_t), whereas f_r leads to meniscal extrusion radially, which is countered by the pulling force from the anterior and posterior insertional ligaments. Consequently, tensile hoop stress is created along the circumferential directions during axial compression.

From Guo, W., Liu, S., Zhu, Y., Yu, C., Lu, S., Yuan, M., Gao, Y., Huang, J., Yuan, Z., Peng, J., Wang, A., Wang, Y., Chen, J., Zhang, L., Sui, X., Xu, W., & Guo, Q. (2015). Advances and prospects in tissue-engineered meniscal scaffolds for meniscus regeneration. Stem Cells International, 2015, 517520. https://doi.org/10.1155/2015/517520.

meniscus shifted posteriorly by 9.5 mm and the posterior horn by 5.6 mm (Thompson et al., 1991; Vedi et al., 1999). The rigid connection of the medial meniscus to its adjacent medial collateral ligament and the joint capsule does not allow the posterior horn of the medial meniscus to posteriorly displace during deep knee flexion. As shown in a study by Seedhom & Hargreaves (2016) in such deep flexion positions, the contact between the femur and the tibia is solely carried between the medial femoral condyle and the posterior horn of the medial meniscus, which might explain why the medial meniscus indicated a higher risk for tears at the posterior horn of the medial meniscus compared to any other meniscus location.

In addition to the most important function of redistribution of the contact force across the tibiofemoral articulation, the menisci contribute to secondary joint stabilization. This function has been identified during experimental and clinical studies where injuries of the ACL were compared to ACL ruptures associated with a concurrent meniscus injury. Similarly, the menisci act as secondary restraints for tibial rotation. Furthermore, the knee joint menisci contribute to the joint lubrication (Danzig et al., 1987; McDermott et al., 2008; Renstrom & Johnson, 1990), nutrient distribution (Danzig et al., 1987; McDermott et al., 2008; Renstrom & Johnson, 1990), and proprioception (Day et al., 1985; Gray, 1999; Seitz, Schall et al., 2021) of the knee joint. Due to its viscoelastic nature, the meniscus also potentially plays a role as a shock absorber, which remains controversial as a review (Andrews et al., 2011) questions the conclusions of previous work.

BIOMECHANICAL PROPERTIES

As for all (biphasic) materials of the human body, the microstructure of the meniscus also defines its mechanical properties and thus the mechanical behavior of the tissue. Numerous studies have been published on the characterization of the biomechanical properties of the healthy meniscus. The findings from such tests are essential for a better understanding of their function, but serve also as a rationale for the development of new meniscus replacement devices.

TENSILE MATERIAL PROPERTIES

The anisotropy of the meniscus requires testing in all spatial directions. Therefore, tensile testing in the main collagen fiber direction (circumferentially) and perpendicular to it (radially) is highly recommended for complete determination of the tensile material properties (Fig. 10.7). The Young's modulus in the circumferential direction is about one order of magnitude higher and ranges from 85 MPa (Lechner et al., 2000) and 175 MPa (Fithian et al., 1990) compared with those in the radial direction, which exhibits up to 11 MPa (Tissakht & Ahmed, 1995). This difference in strength clearly reflects the predominance of circumferential fibers and might explain the higher frequency of circumferential tears compared with radial tears.

The following factors further contribute to the anisotropic tensile material properties: localization (anterior horn, pars intermedia, posterior horn); lateral or medial meniscus; and the layer (surface, lamellar or internal), where the tensile sample was harvested from.

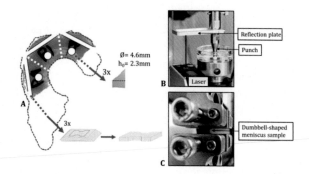

FIGURE 10.7

(A) Meniscus preparation for complementary mechanical testing. (B) For confined compression relaxation tests according to Seitz et al. (Seitz et al., 2021), the cylindrical meniscus sample was mounted in a radial confining measuring chamber filled with PBS. A porous Al2O3 cylinder was placed on its uncut tibial side and a material testing machine (BXC-EZ001.A50000, ZwickRoell GmbH & Co. KG, Germany) equipped with a stainless-steel punch applied the respective strain level/axial deformation, which was registered using a laser (OptoNCDT 2200, Micro Epsilon, Germany). (C) For tensile testing, the dumbbell-shaped meniscus sample (dashed in red, C) was fixed within fixation clamps of the materials testing machine (ElectroForce 5500, BOSE/TA Instruments Inc., USA) and tested until it failed.

From Warnecke, D., Balko, J., Haas, J., Bieger, R., Leucht, F., Wolf, N., Schild, N. B., Stein, S. E. C., Seitz, A. M., Ignatius, A., Reichel, H., Mizaikoff, B., & Durselen, L. (2020). Degeneration alters the biomechanical properties and structural composition of lateral human menisci. Osteoarthritis Cartilage, 28(11), 1482–1491. https://doi.org/10.1016/j.joca.2020.07.004.

COMPRESSIVE MATERIAL PROPERTIES

Fewer studies investigated the compressive properties of the meniscus tissue. Compression tests can be conducted under creep (force is held constant while the deformation changes) or relaxation (deformation constant while the force changes) conditions to identify the viscoelastic (time-dependent) properties of the meniscus (Figs 10.7 and 10.8). The target parameter of such tests is permeability, which measures the ability of the fluid to flow through the porous matrix and lies in the range of $1.9 - 2.1 \times 10^{-15}$ m^4/N s (Joshi et al., 1995; Martin Seitz et al., 2013; Sweigart et al., 2004). The permeability can be related to the proteoglycan content of the meniscus. An increased proteoglycan content leads to an increased water-binding capacity and thus to a lower permeability of the tissue. Furthermore, the outcome of compression testing also relates to the so-called aggregate modulus (H_A) and the equilibrium modulus (E_{eq}) can be determined after reaching the equilibrium state, which has been describe to be in the range of 0.11 MPa in the meniscus (Martin Seitz et al., 2013; Sweigart et al., 2004). Both values represent a measure for the stiffness of the solid matrix. Likewise, the compression properties vary depending on the medial and lateral meniscus and the location in the meniscus [anterior horn, pars intermedia, posterior horn (Martin Seitz et al., 2013; Sweigart et al., 2004)]. In conclusion, the meniscus appears to be 1000 times stiffer in tension than in compression. This enables the menisci to deform more easily during the axial load

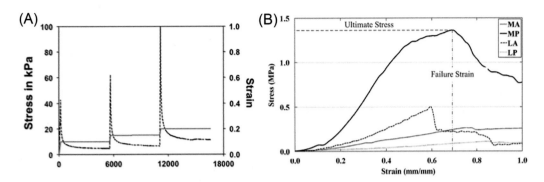

FIGURE 10.8

(A) Experimental stress in kPa (black, dotted) and belonging strain curves (green) of a representative meniscus sample at three consecutive strain levels ($\varepsilon = 0.1, 0.15, 0.2$) during confined compression stress–relaxation testing. (B) Representative tensile stress–strain curves in the transverse direction for human meniscal attachments using machine displacement to obtain ultimate stress and failure strain. Elastic moduli were computed from optical data (not shown) that was comparable to machine displacement. On average, the medial and lateral posterior attachments exhibited the greatest ultimate stress and failure strain, respectively.

From Abraham, A. C., Moyer, J. T., Villegas, D. F., Odegard, G. M., & Haut Donahue, T. L. (2011). Hyperelastic properties of human meniscal attachments. Journal of Biomechanics, 44(3), 413–418. https://doi.org/10.1016/j.jbiomech.2010.10.001.

application, thus adapting to the adjacent femoral condyles and tibial plateau. Furthermore, it suggests that the meniscus is built to resist rather circumferential tensile stress than compressive stress.

ROOT ATTACHMENT PROPERTIES

Tensile tests on the meniscal root attachments indicated a higher Young's modulus for the anterior attachments (169 MPa) compared to the posterior ones (91 MPa) (Day et al., 1985; Gray, 1999; Seitz, Schall et al., 2021). The stiffness of the ligamentous structures ranges from 102 N/mm (Seitz, Kasisari, Claes, Ignatius, & Durselen, 2012) to 216 N/mm (Hauch et al., 2010), which is in the same range as those of the ACL [129−182 N/mm (Noyes & Grood, 1976)]. The failure load lies between 400 and 650 N, which is an indication of their importance for load transformation and transmission (Hauch et al., 2010; Seitz, Kasisari et al., 2012). However, forces in the root attachments of only 25 N have been determined in human knee joint specimens under quasi-static movement and loading conditions, which seems very low compared to these high failure loads (Seitz, Kasisari et al., 2012). It can therefore be assumed that there must be loading situations that place much greater stress on the ligaments. More recent investigations have shown that these forces increase up to 10-fold under dynamic movements with physiological muscle force simulations (Day, Mackenzie, Shim, & Leung, 1985; Gray, 1999; Seitz et al., 2021). This is a clear indicator of the need for physiological testing conditions when aiming to determine the loading profiles of the meniscus root attachments.

INJURY IMPACT ON MENISCUS PERFORMANCE

Acute trauma or degenerative processes can lead to meniscal lesions of different forms.

Based on the classification system of the International Society of Arthroscopy, Knee Surgery and Orthopaedic Sports Medicine (ISAKOS), they are classified as follows: longitudinal (or bucket handle) tear, horizontal tear, radial tear, vertical or horizontal leaf tear (Fig. 10.9). Moreover, the root attachments might also be involved in traumatic situations and are differentiated into five different tear groups (LaPrade et al., 2015): partial root tear, complete radial root tear, complete root tear with associated bucket handle tear, oblique tear into root attachment, root avulsion fracture. In terms of biomechanical function, it has been shown that a complete tear radial root tear is equivalent to total meniscectomy. The contact area decreases by 20% leading to an increase of 32% of

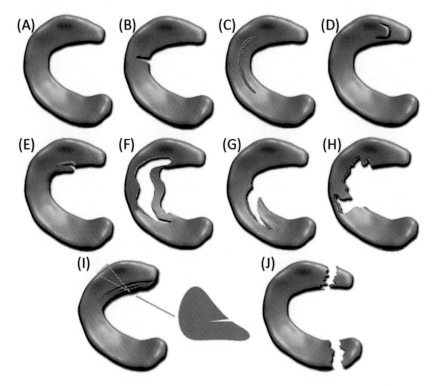

FIGURE 10.9

Illustration of normal meniscus (A), and common types of meniscus tears: radial tear (B), longitudinal tear (C), horizontal flap (D), vertical flap (E), bucket- handle tear (F), oblique/parrot- beak lesion (G), complex degenerative (H), horizontal tear (I), and root tears (J).

From Cengiz, I. F., Pereira, H., Espregueira-Mendes, J., Oliveira, J. M., & Reis, R. L. (2017). Treatments of meniscus lesions of the knee: Current concepts and future perspectives. Regenerative Engineering and Translational Medicine, 3(1), 32–50. https://doi.org/ 10.1007/s40883-017-0025-z.

maximum contact pressure in the affected compartment (Marzo & Gurske-DePerio, 2009). Thus, it can be concluded that a complete radial root tear increases the contact pressure in the knee joint by up to 32% which is comparable to total meniscectomy. As a result, the load transfer function of the meniscus is no longer guaranteed, as meniscus extrusion out of the knee joint gap might occur when axially loaded. A tear involving the attachment of the posterior horn is defined as a meniscus ramp lesion, which further involves the disruption of the meniscotibial ligament and/or the peripheral attachment of the posterior horn.

The gold standard for the treatment of meniscal tears is still arthroscopic partial meniscectomy. Biomechanical studies have shown that this leads to reduced functionality of the meniscus and negative long-term effects such as early osteoarthritis (Higuchi et al., 2000; Petty & Lubowitz, 2011; Rangger et al., 1995). For both radial meniscus tears and after partial meniscectomy, the meniscus functionality is limited by the disrupted circumferential collagen fibers, which are no longer able to carry the load (Hauch et al., 2010; Seitz, Kasisari et al., 2012).

PARTIAL MENISCECTOMY

In vitro studies revealed that partial meniscectomy is mainly associated with dramatic disturbance of meniscotibial contact mechanics. The intact menisci carry about 80% of the total knee load (Pena et al., 2005) and cover on average 71%−82% of the contact area between femur and tibia. The more meniscal tissue is resected, the more dramatic the impact on the contact pressure becomes. Lee et al. (2006) showed that a combination of an axial load of 2.2 times bodyweight, as indicated during daily activities like during gait, in combination with a 50% resection at the medial posterior horn leads to a 21% decreased contact area, whereas the contact pressure increased by 23%. Using an ovine model indicated a superior outcome in terms of tibiofemoral contact mechanics when only a single leaf is resected in horizontal cleavage tears compared to a double leaf resection (Haemer et al., 2007). There is clinical evidence indicating a worse outcome for lateral partial meniscectomy compared with medial partial meniscectomy, which might be explained by the increased mobility of the lateral meniscus compared to the medial (Chatain et al., 2003; Simpson et al., 1986). In general, it can be concluded that the more meniscus tissue is removed, the higher the tibiofemoral contact and accordingly the higher the risk for premature osteoarthritis (Ihn et al., 1993).

TOTAL MENISCECTOMY

In the long term, clinical follow-up studies indicate generally a much worse outcome for totally meniscectomized compared to partially meniscectomized knee joints (Andersson-Molina et al., 2002; Englund et al., 2001; Hede et al., 1992). A complete resection of the posterior horn of the medial meniscus leads to an increase of the maximum contact pressure in the medial compartment by up to 68%, and a 50% decrease in the contact area, compared with the intact state (Fig. 10.10) (Seitz, Lubomierski et al., 2012). Thus, it is similar to total meniscectomy.

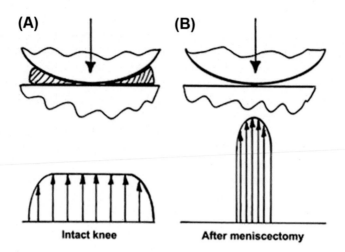

FIGURE 10.10

Tibiofemoral pressure distribution before (A) and after (B) meniscectomy, indicating an increased peak load after meniscus removal.

From McDermott, I. D., Masouros, S. D., & Amis, A. A. (2008). Biomechanics of the menisci of the knee. Current Orthopaedics, *22 (3), 193–201. https://doi.org/10.1016/j.cuor.2008.04.005.*

CHANGES IN MENISCUS BIOMECHANICS IN OSTEOARTHRITIS

The periarticular soft tissues and the menisci of the knee joint are also affected by degenerative processes during osteoarthritis. Macroscopically, calcification, surface fraying, microinjuries, and tissue tears up to loss of surface integrity can be observed (Pauli et al., 2011; Sun et al., 2010; Warnecke et al., 2020). Microscopically, Desrochers et al. (2012) observed a degradation of the collagen network, while other authors reported only a decreasing collagen content (Herwig et al., 1984; Pauli et al., 2011; Sun et al., 2010). Changes in proteoglycan content remain controversial indicating either an increased (Kwok et al., 2014; Pauli et al., 2011; Sun et al., 2010; Warnecke et al., 2020) or decreased glycosaminoglycan content (Desrochers et al., 2012; Fischenich et al., 2015; Herwig et al., 1984; Ishimaru et al., 2014) with progressing degeneration. However, the water content in degenerated menisci seems to increase as also indicated in articular cartilage (Herwig et al., 1984; Pauli et al., 2011; Warnecke et al., 2020). The time-dependent compressive properties of the meniscus indicate that the equilibrium modulus decreases with progressive degeneration, i.e., the material becomes softer (Fischenich et al., 2015; Katsuragawa et al., 2010; Sandmann et al., 2013). According to a study by Katsuragawa et al. (2010), hydraulic permeability increases with progressing degeneration only in the medial meniscus. Only Fischenich et al. (2015) have published studies on the tensile properties of osteoarthritic menisci and did not detect relevant changes of the Young's modulus during degeneration.

RESTORING THE MENISCUS

Both the awareness of resecting as little meniscus tissue as possible during meniscectomy proce-dures and the development of innovative suturing techniques and special implants to preserve the injured meniscus whenever possible changed the treatment paradigm during the last decades. However, only some types of meniscal tears are suitable for repair. These include, in particular, tears in the peripheral vascularized "red" zone. The avascular inner third of the meniscus is not vas-cularized and has low cellularity, explaining its poor regenerative potential in this zone.

SUTURES

It is essential to adapt the tear surfaces in order to allow the meniscus to heal. Thus, the sutures (Fig. 10.11) or anchors (Fig. 10.12) used for this purpose must maintain this required contact until the tissue has healed without causing the tear to gap during knee joint movement and loading. An important biomechanical parameter is therefore the pullout force of meniscal sutures and anchors. A number of papers exist on this subject, comparing several procedures (Brucker et al., 2010; Durselen et al., 2003; Kurzweil & Friedman, 2002). Depending on suture thickness and technique, it was found that sutures typically exhibit pullout forces between 60 and 120 N. Generally, thicker or vertical sutures showed a higher load capacity than sutures with smaller diameters and horizontal sutures.

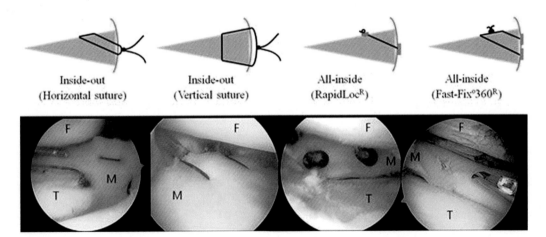

| Inside-out | Inside-out | All-inside | All-inside |
| (Horizontal suture) | (Vertical suture) | (RapidLoc[R]) | (Fast-Fix°360[R]) |

FIGURE 10.11

Illustrations and arthroscopy images showing the various suture techniques that are frequently applied (F = femur, T = tibia, M = meniscus).

From Lee, H., Lee, S. Y., Na, Y. G., Kim, S. K., Yi, J. H., Lim, J. K., & Lee, S. M. (2016). Surgical techniques and radiological findings of meniscus allograft transplantation. European Journal of Radiology, *85(8), 1351–1365. https://doi.org/10.1016/j. ejrad.2016.05.006.*

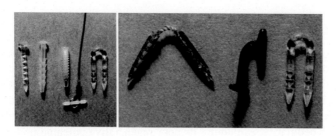

FIGURE 10.12

Meniscus repair devices: from left to right, the Bionx Meniscus Arrow, the Linvatec BioStinger, the Innovasive Clearfix screw, the Smith & Nephew T-Fix, and the Biomet staple, the Surgical Dynamics SDsorb meniscal staple, the Mitek meniscal repair system, and the Biomet staple.

From Barber, F. A., & Herbert, M. A. (2000). Meniscal repair devices. Arthroscopy, *16(6), 613–618. https://doi.org/10.1053/ jars.2000.4819.*

Anchor implants in the form of absorbable screws, arrows, and other forms mostly showed lower maximum holding forces between 20 and 60 N. The stiffness of fixation also varies and is, for example, about 10 N/mm for the FAST-FIX suture anchor and about 5 N/mm for other implants (Brucker et al., 2010). However, the lower pullout forces of the implants compared with the suture techniques do not necessarily have to be a disadvantage, since it has been shown that, for example, the use of three meniscus screws (pullout force 75 N) can sufficiently prevent gapping, even when the knee joint is under weight-bearing (Durselen et al., 2007). Nevertheless, it should be noted that some of these implants are designed with implant heads that are in contact with the surface of the meniscus and with the femoral condyle. During joint loading, this can result in articular cartilage damage due to abrasion. Implants that are embedded in the meniscal tissue can also cause the anchor to migrate out of the meniscus and thus interact with the articular cartilage finally leading to the same abrasion effect.

Suture techniques are in principle preferable to meniscal implants. However, since meniscal suturing is surgically quite demanding, particularly at the difficult-to-access posterior horn region, combinations of suture and anchor implants have been applied in which small absorbable suture anchors are placed behind the meniscus base using an arthroscopic approach and fixed to the meniscal surface by means of pre-tied knots. The FAST-Fix suture anchor can withstand pullout forces of approximately 80 N. Most in vitro studies measured pullout forces immediately after suture or implantation. However, dynamic loading, which occur in the knee joint, can reduce the pullout forces of sutures and anchors over time (Durselen et al., 2003). This should be taken into account when choosing the fixation technique.

MENISCUS REPLACEMENT

Unfortunately, tears often occur in the nonvascularized zone of the meniscus, which are challenging to suture. In such cases, partial or total removal of the meniscus may be indicated, with the previously described adverse effects on force transmission in the knee joint. Alternatively, the replacement of meniscus tissue by implants is an option (Fig. 10.13). A number of research approaches with different strategies have now emerged for this purpose. On the one hand, there are regenerative methods using partially or fully resorbable porous scaffolds being inserted into the meniscal defect as placeholders. In this case, the goal is to initiate remodeling processes to regenerate meniscal tissue. So far, only two regenerative implants are in clinical use (Vrancken et al., 2013): the Collagen Meniscal Implant (CMI), which is made of highly purified bovine collagen type I, and the Actifit implant, which consists of a composite of fast-degrading polycaprolactone and very slowly degrading urethane components. Clinically, good results with pain relief are described in part for both implants (Grassi et al., 2014; Hutchinson et al., 2014). However, there are only few prospective studies with control groups that scientifically demonstrate an added value of the implant compared to a partial meniscectomy (Zaffagnini et al., 2011). When the implant is inserted, the biomechanical properties of both materials are initially inadequate (Sandmann et al., 2013). To date, it has not been proven whether a tissue with biomechanical material properties similar to those of the meniscus is created after degradation or remodeling of the material.

An exciting approach is the use of decellularized human menisci as scaffolds for tissue engineering of the meniscus. Sandmann et al. showed that both stiffness and residual force of the decellularized scaffold under cyclic loading did not differ from those of native menisci (Sandmann et al., 2013). However, it remains to be proven that even this material supports the formation of a biomechanically sufficient biological tissue in vivo. In addition to regenerative approaches, there are also ideas to permanently replace meniscal tissue with a nonresorbable implant. Permanent implants are currently more likely to achieve adequate mechanical properties than regenerative methods. Polycarbonate (McKeon et al., 2020; Shemesh et al., 2020; Zur et al., 2011), silk fibroin (Stein, Hose et al., 2019; Stein, von Luebken et al., 2019), and other polymers (Vrancken et al., 2013) are used to partially or completely replace the meniscus. However, these implants have not yet found their way into the clinic or have been clinically tested only in small series (McKeon et al., 2020).

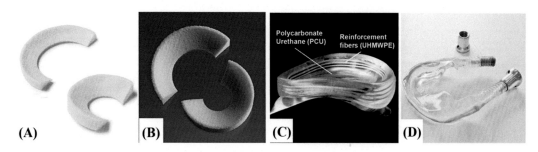

(A) **(B)** **(C)** **(D)**

FIGURE 10.13

Different synthetic total meniscus replacements: (A) medial (left) and lateral (right) collagen meniscus implant (CMI). (B) Medial (top) and lateral (bottom) Actifit implant. (C) Disk-shaped, free-floating NUsurface total meniscal prosthesis for replacement of the human medial meniscus. (D) Anatomically shaped version of the NUsurface total meniscal prosthesis, including fixation bolts at both horns.

From Vrancken, A. C., Buma, P., & van Tienen, T. G. (2013). Synthetic meniscus replacement: A review. International Orthopaedics, 37(2), 291–299. https://doi.org/10.1007/s00264-012-1682-7.

In contrast, meniscal grafts from human donors (allograft) are clinically used (Fig. 10.14). While several long-term studies report quite positive results (Kazi et al., 2015; Kempshall et al., 2015; Samitier et al., 2015), the fixation technique is of great importance for allograft integration. As described above, the native meniscal root attachments accommodate the circumferential stresses in the meniscus, thereby preventing it from extruding. Of course, this also applies to the transplanted meniscus. Thus, fixation of an allograft with transtibial sutures leads to increased extrusion and thus necessarily to diminished biomechanical function of the meniscus compared with more rigid fixation with bone blocks (Abat et al., 2012). Furthermore, due to an idealized size fitting between the donor meniscus and the recipient knee, the use of meniscal allografts also represents a particular regulatory and logistical challenge. In order to provide a biomechanically sufficient performance in the recipient knee, both above-mentioned requirements (fixation, sizing) should be fulfilled (Seitz & Durselen, 2019).

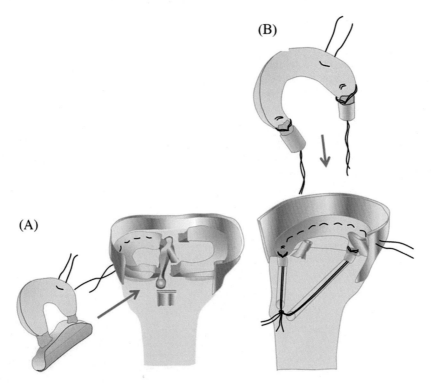

FIGURE 10.14

Illustration of lateral meniscus allograft transplantation using a bone block. (B) Illustration of medial meniscus allograft transplantation. Compared to the lateral meniscus, two separated bone plugs are used to fix the allograft meniscus to the host proximal tibia.

From Lee, H., Lee, S. Y., Na, Y. G., Kim, S. K., Yi, J. H., Lim, J. K., & Lee, S. M. (2016). Surgical techniques and radiological findings of meniscus allograft transplantation. European Journal of Radiology, 85(8), 1351–1365. *https://doi.org/10.1016/j. ejrad.2016.05.006.*

REFERENCES

Abat, F., Gelber, P. E., Erquicia, J. I., Pelfort, X., Gonzalez-Lucena, G., & Monllau, J. C. (2012). Suture-only fixation technique leads to a higher degree of extrusion than bony fixation in meniscal allograft transplantation. *The American Journal of Sports Medicine*, *40*(7), 1591–1596. Available from https://doi.org/10.1177/0363546512446674.

Abraham, A. C., Moyer, J. T., Villegas, D. F., Odegard, G. M., & Haut Donahue, T. L. (2011). Hyperelastic properties of human meniscal attachments. *Journal of Biomechanics*, *44*(3), 413–418. Available from https://doi.org/10.1016/j.jbiomech.2010.10.001.

Andersson-Molina, H., Karlsson, H., & Rockborn, P. (2002). Arthroscopic partial and total meniscectomy: A long-term follow-up study with matched controls. *Arthroscopy*, *18*(2), 183–189. Available from https://doi.org/10.1053/jars.2002.30435.

Andrews, S., Shrive, N., & Ronsky, J. (2011). The shocking truth about meniscus. *Journal of Biomechanics*, *44*(16), 2737–2740. Available from https://doi.org/10.1016/j.jbiomech.2011.08.026.

Arnoczky, S. P., & Warren, R. F. (1982). Microvasculature of the human meniscus. *The American Journal of Sports Medicine*, *10*(2), 90–95. Available from https://doi.org/10.1177/036354658201000205.

Brucker, P. U., Favre, P., Puskas, G. J., von Campe, A., Meyer, D. C., & Koch, P. P. (2010). Tensile and shear loading stability of all-inside meniscal repairs: An in vitro biomechanical evaluation. *The American Journal of Sports Medicine*, *38*(9), 1838–1844. Available from https://doi.org/10.1177/0363546510368131.

Chatain, F., Adeleine, P., Chambat, P., Neyret, P., & Societe Francaise, d'Arthroscopie. (2003). A comparative study of medial vs lateral arthroscopic partial meniscectomy on stable knees: 10-year minimum follow-up. *Arthroscopy*, *19*(8), 842–849. Available from https://doi.org/10.1016/s0749-8063(03)00735-7.

Danzig, L. A., Hargens, A. R., Gershuni, D. H., Skyhar, M. J., Sfakianos, P. N., & Akeson, W. H. (1987). Increased transsynovial transport with continuous passive motion. *Journal of Orthopaedic Research*, *5*(3), 409–413. Available from https://doi.org/10.1002/jor.1100050313.

Day, B., Mackenzie, W. G., Shim, S. S., & Leung, G. (1985). The vascular and nerve supply of the human meniscus. *Arthroscopy*, *1*(1), 58–62. Available from https://doi.org/10.1016/s0749-8063(85)80080-3.

Desrochers, J., Amrein, M. W., & Matyas, J. R. (2012). Viscoelasticity of the articular cartilage surface in early osteoarthritis. *Osteoarthritis and Cartilage / OARS, Osteoarthritis Research Society*, *20*(5), 413–421. Available from https://doi.org/10.1016/j.joca.2012.01.011.

Durselen, L., Hebisch, A., Wagner, D., Claes, L. E., & Bauer, G. (2007). Meniscal screw fixation provides sufficient stability to prevent tears from gapping. *Clinical Biomechanics (Bristol, Avon)*, *22*(1), 93–99. Available from https://doi.org/10.1016/j.clinbiomech.2006.07.010.

Durselen, L., Schneider, J., Galler, M., Claes, L. E., & Bauer, G. (2003). Cyclic joint loading can affect the initial stability of meniscal fixation implants. *Clinical Biomechanics*, *18*(1), 44–49. Available from https://doi.org/10.1016/s0268-0033(02)00139-0.

Englund, M., Roos, E. M., Roos, H. P., & Lohmander, L. S. (2001). Patient-relevant outcomes fourteen years after meniscectomy: Influence of type of meniscal tear and size of resection. *Rheumatology (Oxford)*, *40*(6), 631–639. Available from https://doi.org/10.1093/rheumatology/40.6.631.

Fischenich, K. M., Lewis, J., Kindsfater, K. A., Bailey, T. S., & Haut Donahue, T. L. (2015). Effects of degeneration on the compressive and tensile properties of human meniscus. *Journal of Biomechanics*, *48*(8), 1407–1411. Available from https://doi.org/10.1016/j.jbiomech.2015.02.042.

Fithian, D. C., Kelly, M. A., & Mow, V. C. (1990). Material properties and structure-function relationships in the menisci. *Clinical Orthopaedics and Related Research*, *252*, 19–31. Available from https://www.ncbi.nlm.nih.gov/pubmed/2406069.

Freutel, M., Scholz, N. B., Seitz, A. M., Ignatius, A., & Durselen, L. (2015). Mechanical properties and morphological analysis of the transitional zone between meniscal body and ligamentous meniscal attachments. *Journal of Biomechanics*, *48*(8), 1350−1355. Available from https://doi.org/10.1016/j.jbiomech.2015.03.003.

Grassi, A., Zaffagnini, S., Marcheggiani Muccioli, G. M., Benzi, A., & Marcacci, M. (2014). Clinical outcomes and complications of a collagen meniscus implant: A systematic review. *International Orthopaedics*, *38*(9), 1945−1953. Available from https://doi.org/10.1007/s00264-014-2408-9.

Gray, J. C. (1999). Neural and vascular anatomy of the menisci of the human knee. *The Journal of Orthopaedic and Sports Physical Therapy*, *29*(1), 23−30. Available from https://doi.org/10.2519/jospt.1999.29.1.23.

Haemer, J. M., Wang, M. J., Carter, D. R., & Giori, N. J. (2007). Benefit of single-leaf resection for horizontal meniscus tear. *Clinical Orthopaedics and Related Research*, *457*(457), 194−202. Available from https://doi.org/10.1097/BLO.0b013e3180303b5c.

Hauch, K. N., Villegas, D. F., & Haut Donahue, T. L. (2010). Geometry, time-dependent and failure properties of human meniscal attachments. *Journal of Biomechanics*, *43*(3), 463−468. Available from https://doi.org/10.1016/j.jbiomech.2009.09.043.

Hede, A., Larsen, E., & Sandberg, H. (1992). The long term outcome of open total and partial meniscectomy related to the quantity and site of the meniscus removed. *International Orthopaedics*, *16*(2), 122−125. Available from https://doi.org/10.1007/BF00180200.

Herwig, J., Egner, E., & Buddecke, E. (1984). Chemical changes of human knee joint menisci in various stages of degeneration. *Annals of the Rheumatic Diseases*, *43*(4), 635−640. Available from https://doi.org/10.1136/ard.43.4.635.

Higuchi, H., Kimura, M., Shirakura, K., Terauchi, M., & Takagishi, K. (2000). Factors affecting long-term results after arthroscopic partial meniscectomy. *Clinical Orthopaedics and Related Research*, *377*, 161−168. Available from https://doi.org/10.1097/00003086-200008000-00022.

Hutchinson, I. D., Moran, C. J., Potter, H. G., Warren, R. F., & Rodeo, S. A. (2014). Restoration of the meniscus: Form and function. *The American Journal of Sports Medicine*, *42*(4), 987−998. Available from https://doi.org/10.1177/0363546513498503.

Ihn, J. C., Kim, S. J., & Park, I. H. (1993). In vitro study of contact area and pressure distribution in the human knee after partial and total meniscectomy. *International Orthopaedics*, *17*(4), 214−218. Available from https://doi.org/10.1007/BF00194181.

Ishimaru, D., Sugiura, N., Akiyama, H., Watanabe, H., & Matsumoto, K. (2014). Alterations in the chondroitin sulfate chain in human osteoarthritic cartilage of the knee. *Osteoarthritis and Cartilage / OARS, Osteoarthritis Research Society*, *22*(2), 250−258. Available from https://doi.org/10.1016/j.joca.2013.11.010.

Johannsen, A. M., Civitarese, D. M., Padalecki, J. R., Goldsmith, M. T., Wijdicks, C. A., & LaPrade, R. F. (2012). Qualitative and quantitative anatomic analysis of the posterior root attachments of the medial and lateral menisci. *The American Journal of Sports Medicine*, *40*(10), 2342−2347. Available from https://doi.org/10.1177/0363546512457642.

Joshi, M. D., Suh, J. K., Marui, T., & Woo, S. L. (1995). Interspecies variation of compressive biomechanical properties of the meniscus. *Journal of Biomedical Materials Research*, *29*(7), 823−828. Available from https://doi.org/10.1002/jbm.820290706.

Katsuragawa, Y., Saitoh, K., Tanaka, N., Wake, M., Ikeda, Y., Furukawa, H., ... Fukui, N. (2010). Changes of human menisci in osteoarthritic knee joints. *Osteoarthritis and Cartilage / OARS, Osteoarthritis Research Society*, *18*(9), 1133−1143. Available from https://doi.org/10.1016/j.joca.2010.05.017.

Kazi, H. A., Abdel-Rahman, W., Brady, P. A., & Cameron, J. C. (2015). Meniscal allograft with or without osteotomy: A 15-year follow-up study. *Knee Surgery, Sports Traumatology, Arthroscopy: Official Journal of the ESSKA*, *23*(1), 303−309. Available from https://doi.org/10.1007/s00167-014-3291-z.

Kempshall, P. J., Parkinson, B., Thomas, M., Robb, C., Standell, H., Getgood, A., & Spalding, T. (2015). Outcome of meniscal allograft transplantation related to articular cartilage status: Advanced chondral damage should not be a contraindication. *Knee Surgery, Sports Traumatology, Arthroscopy: Official Journal of the ESSKA*, *23*(1), 280−289. Available from https://doi.org/10.1007/s00167-014-3431-5.

Kurzweil, P. R., & Friedman, M. J. (2002). Meniscus: Resection, repair, and replacement. *Arthroscopy*, *18*(2 Suppl 1), 33−39. Available from https://doi.org/10.1016/s0749-8063(02)80003-2.

Kwok, J., Grogan, S., Meckes, B., Arce, F., Lal, R., & D'Lima, D. (2014). Atomic force microscopy reveals age-dependent changes in nanomechanical properties of the extracellular matrix of native human menisci: Implications for joint degeneration and osteoarthritis. *Nanomedicine: Nanotechnology, Biology, and Medicine*, *10*(8), 1777−1785. Available from https://doi.org/10.1016/j.nano.2014.06.010.

LaPrade, C. M., James, E. W., Cram, T. R., Feagin, J. A., Engebretsen, L., & LaPrade, R. F. (2015). Meniscal root tears: A classification system based on tear morphology. *The American Journal of Sports Medicine*, *43*(2), 363−369. Available from https://doi.org/10.1177/0363546514559684.

Lechner, K., Hull, M. L., & Howell, S. M. (2000). Is the circumferential tensile modulus within a human medial meniscus affected by the test sample location and cross-sectional area? *Journal of Orthopaedic Research*, *18*(6), 945−951. Available from https://doi.org/10.1002/jor.1100180614.

Lee, S. J., Aadalen, K. J., Malaviya, P., Lorenz, E. P., Hayden, J. K., Farr, J., ... Cole, B. J. (2006). Tibiofemoral contact mechanics after serial medial meniscectomies in the human cadaveric knee. *The American Journal of Sports Medicine*, *34*(8), 1334−1344. Available from https://doi.org/10.1177/0363546506286786.

Martin Seitz, A., Galbusera, F., Krais, C., Ignatius, A., & Durselen, L. (2013). Stress-relaxation response of human menisci under confined compression conditions. *Journal of the Mechanical Behavior of Biomedical Materials*, *26*, 68−80. Available from https://doi.org/10.1016/j.jmbbm.2013.05.027.

Marzo, J. M., & Gurske-DePerio, J. (2009). Effects of medial meniscus posterior horn avulsion and repair on tibiofemoral contact area and peak contact pressure with clinical implications. *The American Journal of Sports Medicine*, *37*(1), 124−129. Available from https://doi.org/10.1177/0363546508323254.

McDermott, I. D., Masouros, S. D., & Amis, A. A. (2008). Biomechanics of the menisci of the knee. *Current Orthopaedics*, *22*(3), 193−201. Available from https://doi.org/10.1016/j.cuor.2008.04.005.

McKeon, B. P., Zaslav, K. R., Alfred, R. H., Alley, R. M., Edelson, R. H., Gersoff, W. K., ... Kaeding, C. C. (2020). Preliminary results from a US clinical trial of a novel synthetic polymer meniscal implant. *Orthopaedic Journal of Sports Medicine*, *8*(9). Available from https://doi.org/10.1177/2325967120952414, 2325967120952414.

Noyes, F. R., & Grood, E. S. (1976). The strength of the anterior cruciate ligament in humans and Rhesus monkeys. *The Journal of Bone and Joint Surgery. American Volume*, *58*(8), 1074−1082, https://www.ncbi.nlm.nih.gov/pubmed/1002748.

Pauli, C., Grogan, S. P., Patil, S., Otsuki, S., Hasegawa, A., Koziol, J., ... D'Lima, D. D. (2011). Macroscopic and histopathologic analysis of human knee menisci in aging and osteoarthritis. *Osteoarthritis and Cartilage / OARS, Osteoarthritis Research Society*, *19*(9), 1132−1141. Available from https://doi.org/10.1016/j.joca.2011.05.008.

Pena, E., Calvo, B., Martinez, M. A., Palanca, D., & Doblare, M. (2005). Finite element analysis of the effect of meniscal tears and meniscectomies on human knee biomechanics. *Clinical Biomechanics (Bristol, Avon)*, *20*(5), 498−507. Available from https://doi.org/10.1016/j.clinbiomech.2005.01.009.

Petersen, W., & Tillmann, B. (1998). Collagenous fibril texture of the human knee joint menisci. *Anatomy and Embryology (Berl)*, *197*(4), 317−324. Available from https://doi.org/10.1007/s004290050141.

Petersen, W., & Tillmann, B. (1999). Structure and vascularization of the knee joint menisci. *Zeitschrift fur Orthopadie und Ihre Grenzgebiete*, *137*(1), 31−37. Available from https://doi.org/10.1055/s-2008-1037032.

Petty, C. A., & Lubowitz, J. H. (2011). Does arthroscopic partial meniscectomy result in knee osteoarthritis? A systematic review with a minimum of 8 years' follow-up. *Arthroscopy*, *27*(3), 419−424. Available from https://doi.org/10.1016/j.arthro.2010.08.016.

Rangger, C., Klestil, T., Gloetzer, W., Kemmler, G., & Benedetto, K. P. (1995). Osteoarthritis after arthroscopic partial meniscectomy. *The American Journal of Sports Medicine*, *23*(2), 240−244. Available from https://doi.org/10.1177/036354659502300219.

Renstrom, P., & Johnson, R. J. (1990). Anatomy and biomechanics of the menisci. *Clinics in Sports Medicine*, *9*(3), 523−538. Available from https://www.ncbi.nlm.nih.gov/pubmed/2199066.

Samitier, G., Alentorn-Geli, E., Taylor, D. C., Rill, B., Lock, T., Moutzouros, V., & Kolowich, P. (2015). Meniscal allograft transplantation. Part 1: Systematic review of graft biology, graft shrinkage, graft extrusion, graft sizing, and graft fixation. *Knee Surgery, Sports Traumatology, Arthroscopy: Official Journal of the ESSKA*, *23*(1), 310−322. Available from https://doi.org/10.1007/s00167-014-3334-5.

Sandmann, G. H., Adamczyk, C., Grande Garcia, E., Doebele, S., Buettner, A., Milz, S., ... Tischer, T. (2013). Biomechanical comparison of menisci from different species and artificial constructs. *BMC Musculoskeletal Disorders*, *14*, 324. Available from https://doi.org/10.1186/1471-2474-14-324.

Seedhom, B. B., & Hargreaves, D. J. (2016). Transmission of the load in the knee joint with special reference to the role of the menisci. *Engineering in Medicine*, *8*(4), 220−228. Available from https://doi.org/10.1243/emed_jour_1979_008_051_02.

Seitz, A., Kasisari, R., Claes, L., Ignatius, A., & Durselen, L. (2012). Forces acting on the anterior meniscotibial ligaments. *Knee Surgery, Sports Traumatology, Arthroscopy: Official Journal of the ESSKA*, *20*(8), 1488−1495. Available from https://doi.org/10.1007/s00167-011-1708-5.

Seitz, A. M., & Durselen, L. (2019). Biomechanical considerations are crucial for the success of tendon and meniscus allograft integration−a systematic review. *Knee Surgery, Sports Traumatology, Arthroscopy: Official Journal of the ESSKA*, *27*(6), 1708−1716. Available from https://doi.org/10.1007/s00167-018-5185-y.

Seitz, A. M., Lubomierski, A., Friemert, B., Ignatius, A., & Durselen, L. (2012). Effect of partial meniscectomy at the medial posterior horn on tibiofemoral contact mechanics and meniscal hoop strains in human knees. *Journal of Orthopaedic Research: Official Publication of the Orthopaedic Research Society*, *30*(6), 934−942. Available from https://doi.org/10.1002/jor.22010.

Seitz, A. M., Murrmann, M., Ignatius, A., Dürselen, L., Friemert, B., & von Lübken, F. (2021). Neuromapping of the capsuloligamentous knee joint structures. *Arthroscopy, Sports Medicine, and Rehabilitation*, *3*(2), e555−e563. Available from https://doi.org/10.1016/j.asmr.2020.12.009.

Seitz, A. M., Schall, F., Hacker, S. P., van Drongelen, S., Wolf, S., & Durselen, L. (2021). Forces at the Anterior Meniscus Attachments Strongly Increase Under Dynamic Knee Joint Loading. *The American Journal of Sports Medicine*, *49*(4), 994−1004. Available from https://doi.org/10.1177/0363546520988039.

Shemesh, M., Shefy-Peleg, A., Levy, A., Shabshin, N., Condello, V., Arbel, R., & Gefen, A. (2020). Effects of a novel medial meniscus implant on the knee compartments: Imaging and biomechanical aspects. *Biomechanics and Modeling in Mechanobiology*, *19*(6), 2049−2059. Available from https://doi.org/10.1007/s10237-020-01323-6.

Simpson, D. A., Thomas, N. P., & Aichroth, P. M. (1986). Open and closed meniscectomy. A comparative analysis. *The Journal of Bone and Joint Surgery. British Volume*, *68*(2), 301−304. Available from https://doi.org/10.1302/0301-620X.68B2.3754260.

Stein, S., Hose, S., Warnecke, D., Gentilini, C., Skaer, N., Walker, R., ... Durselen, L. (2019). Meniscal Replacement with a silk fibroin scaffold reduces contact stresses in the human knee. *Journal of Orthopaedic Research: Official Publication of the Orthopaedic Research Society*, *37*(12), 2583−2592. Available from https://doi.org/10.1002/jor.24437.

Stein, S. E. C., von Luebken, F., Warnecke, D., Gentilini, C., Skaer, N., Walker, R., ... Duerselen, L. (2019). The challenge of implant integration in partial meniscal replacement: An experimental study on a silk fibroin scaffold in sheep. *Knee Surgery, Sports Traumatology, Arthroscopy: Official Journal of the ESSKA*, *27*(2), 369−380. Available from https://doi.org/10.1007/s00167-018-5160-7.

Sun, Y., Mauerhan, D. R., Honeycutt, P. R., Kneisl, J. S., Norton, H. J., Zinchenko, N., ... Gruber, H. E. (2010). Calcium deposition in osteoarthritic meniscus and meniscal cell culture. *Arthritis Research & Therapy*, *12*(2), R56. Available from https://doi.org/10.1186/ar2968.

Sweigart, M. A., Zhu, C. F., Burt, D. M., DeHoll, P. D., Agrawal, C. M., Clanton, T. O., & Athanasiou, K. A. (2004). Intraspecies and interspecies comparison of the compressive properties of the medial meniscus. *Annals of Biomedical Engineering*, *32*(11), 1569−1579. Available from https://doi.org/10.1114/b:abme.0000049040.70767.5c.

Thompson, W. O., Thaete, F. L., Fu, F. H., & Dye, S. F. (1991). Tibial meniscal dynamics using three-dimensional reconstruction of magnetic resonance images. *The American Journal of Sports Medicine*, *19*(3), 210−215. Available from https://doi.org/10.1177/036354659101900302, discussion 215-6.

Tissakht, M., & Ahmed, A. M. (1995). Tensile stress-strain characteristics of the human meniscal material. *Journal of Biomechanics*, *28*(4), 411−422. Available from https://doi.org/10.1016/0021-9290(94)00081-e.

Vedi, V., Williams, A., Tennant, S. J., Spouse, E., Hunt, D. M., & Gedroyc, W. M. (1999). Meniscal movement. An in-vivo study using dynamic MRI. *The Journal of Bone and Joint Surgery. British Volume*, *81*(1), 37−41. Available from https://doi.org/10.1302/0301-620x.81b1.8928.

Villegas, D. F., & Donahue, T. L. (2010). Collagen morphology in human meniscal attachments: A SEM study. *Connective Tissue Research*, *51*(5), 327−336. Available from https://doi.org/10.3109/03008200903349639.

Vrancken, A. C., Buma, P., & van Tienen, T. G. (2013). Synthetic meniscus replacement: A review. *International Orthopaedics*, *37*(2), 291−299. Available from https://doi.org/10.1007/s00264-012-1682-7.

Walker, P. S., & Erkman, M. J. (1975). The role of the menisci in force transmission across the knee. *Clinical Orthopaedics and Related Research*, *109*, 184−192. Available from https://doi.org/10.1097/00003086-197506000-00027.

Warnecke, D., Balko, J., Haas, J., Bieger, R., Leucht, F., Wolf, N., ... Durselen, L. (2020). Degeneration alters the biomechanical properties and structural composition of lateral human menisci. *Osteoarthritis and Cartilage / OARS, Osteoarthritis Research Society*, *28*(11), 1482−1491. Available from https://doi.org/10.1016/j.joca.2020.07.004.

Zaffagnini, S., Marcheggiani Muccioli, G. M., Lopomo, N., Bruni, D., Giordano, G., Ravazzolo, G., ... Marcacci, M. (2011). Prospective long-term outcomes of the medial collagen meniscus implant vs partial medial meniscectomy: A minimum 10-year follow-up study. *The American Journal of Sports Medicine*, *39*(5), 977−985. Available from https://doi.org/10.1177/0363546510391179.

Zur, G., Linder-Ganz, E., Elsner, J. J., Shani, J., Brenner, O., Agar, G., ... Shterling, A. (2011). Chondroprotective effects of a polycarbonate-urethane meniscal implant: Histopathological results in a sheep model. *Knee Surgery, Sports Traumatology, Arthroscopy: Official Journal of the ESSKA*, *19*(2), 255−263. Available from https://doi.org/10.1007/s00167-010-1210-5.

INTERVERTEBRAL DISC BIOMECHANICS

Fabio Galbusera[1] and Graciosa Quelhas Teixeira[2]

[1]*IRCCS Istituto Ortopedico Galeazzi, Milan, Italy* [2]*Institute of Orthopaedic Research and Biomechanics, Ulm University, Ulm, Germany*

SHAPE AND STRUCTURE

The intervertebral disc is a fibrocartilaginous pad that connects the superior and inferior surfaces of the vertebral bodies, that is, the vertebral endplates (VEPs). Intervertebral discs are present at all spinal levels between C2 and the sacrum, while the disc is notably absent at the level C1−C2. Each disc indeed assumes the name of the vertebral level where it is located, that is, L4−L5 indicates the disc between the L4 and L5 vertebrae. Discs have a roughly elliptical shape covering the whole surface of the endplate, with some variability depending on the vertebral level (Pooni, Hukins, Harris, Hilton, & Davies, 1986). Cervical discs have indeed an elliptical cross-section, thoracic discs tend to be rounder, whereas lumbar discs regain an elliptical shape with a posterior concavity increasing in size proceeding caudally (Fig. 11.1).

Generally, the cross-sectional size of the intervertebral discs increases in the caudal direction, coherently with the dimension of the vertebral bodies and, in turn, with the larger loads to which they are subjected. In the sagittal plane, cervical and lumbar discs show a wedging angle that contributes to the lordotic curvature of these spinal regions; thoracic discs tend to be flatter (Fig. 11.1). The structure of the intervertebral disc is characterized by two regions, namely the nucleus pulposus in the inner part and the annulus fibrosus, which encloses it along its anterior, posterior, and lateral margins (Fig. 11.2). The nucleus pulposus has a jelly consistency and appearance given by its high water content, up to 90%, and is rich in proteoglycans (Natarajan, Williams, & Andersson, 2004; Urban & McMullin, 1985); it also contains small quantities of type II collagen and elastin fibers. Experimental studies showed that 65% of the dry weight of the nucleus pulposus is composed of proteoglycans, whereas 25% shall be attributed to collagen (Eyre, 1979). Although classical papers described the nucleus pulposus as a region with a gelatinous, mucoid structure with low integration with the surrounding structures enclosing it (Coventry, Ghormley, & Kernohan, 1945; Keyes & Compere, 1932), there is now a consensus about the existence of a clear fibrosity in the whole volume of the nucleus, mostly vertically oriented (Wade, Robertson, & Broom, 2011). Despite this fibrosity may appear disordered if analyzed with microscopical techniques, mechanical testing revealed a high degree of organization of the fiber network, which is however convoluted and folded in unloaded conditions. It has been hypothesized that the nucleus fibrosity induces a

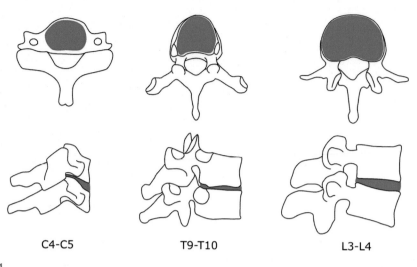

C4-C5   T9-T10   L3-L4

FIGURE 11.1

The shape of the intervertebral discs at the levels C4—C5 (left), T9—T10 (center), and L3—L4 (right) in the transverse (top) and sagittal planes (bottom).

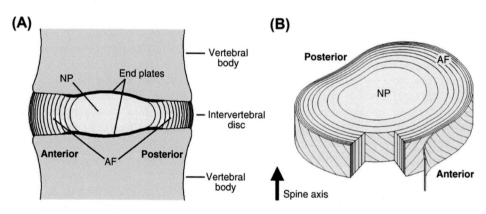

FIGURE 11.2

The structure of a lumbar intervertebral disc, showing the organization and orientation of the collagen fibers: mid-sagittal cross-section (A) and three-dimensional view (B).

From Smith, L. J., Nerurkar, N. L., Choi, K.-S., Harfe, B. D., & Elliott, D. M. (2011). Degeneration and regeneration of the intervertebral disc: Lessons from development. Disease Models & Mechanisms, 4(1), 31—41. https://doi.org/10.1242/dmm.006403.

form of "tethered mobility," which prevents high displacements while not significantly affecting the high deformability and hydrostatic nature of the nucleus (Wade et al., 2011).

The annulus fibrosus is also rich in water, approximately 50%—70% of its volume in lumbar discs, and is characterized by a highly organized macroscopical network of oriented collagen fibers from which it takes its name. Between nucleus and annulus there is no abrupt border, but rather is a transition zone with intermediate composition and properties.

The organization of the annulus fibers has been documented in depth for the lumbar discs, whereas no detailed data exist about their size and orientation in the thoracic and cervical regions of the spine; the following description therefore refers to data measured on lumbar specimens. The annulus fibers are mostly of type I collagen, especially in the outer portion of the annulus in which they constitute approximately 95% of the collagen content (Cassidy, Hiltner, & Baer, 1989); proceeding towards the nucleus, the number of type I fibers progressively decreases, replaced by type II collagen ones.

The fibers are organized in lamellae with variable thickness; in the lumbar spine, this thickness ranges from 80–100 μm in the anterior periphery to 400 μm in the transition zone, with no linear gradient but rather local sudden, sharp variations. In the lateral and posterior regions of the lumbar annulus fibrosus, the lamellar thickness does not show a clear regional variability but rather scattered patterns. While an axial section of a lumbar disc shows a purely circumferential orientation of the fibers, cuts parallel to the craniocaudal axis reveal an alternate orientation with respect to the transverse plane (Fig. 11.2). Such an orientation exhibits a regional variability, with angles ranging from approximately 28 degrees in the outer periphery of the annulus to 43 degrees in the transition zone (Cassidy et al., 1989). Proceeding in the radial direction, the fiber orientation does not vary within every single lamella but rather shows sharp changes at the interface between the lamellae. However, considering each lamella as a whole, the fiber angle exhibits differences between the dorsal and ventral regions of the disc (Marchand & Ahmed, 1990). Studies conducted on cadaver specimens revealed that the regional organization of the fiber networks is not significantly altered by ageing (Holzapfel, Schulze-Bauer, Feigl, & Regitnig, 2005).

CARTILAGINOUS AND VERTEBRAL ENDPLATES

The intervertebral disc is attached to the vertebral bodies by means of the cartilaginous and VEPs. The cartilaginous endplate consists of a layer of hyaline cartilage on the discal side, followed by a subchondral bony plate of about 2-mm thickness connected to the trabecular bone of the vertebra. The cartilaginous endplate has a composition similar to that of articular cartilage but does not show the stratified zonation which is visible in joints. The thickness of the cartilage layer shows a regional variability, being lower in correspondence with the nucleus pulposus and increasing outwards; in general, it ranges between 0.5 and 1 mm, with superior endplates tending to be thicker than inferior ones (Roberts, Menage, & Urban, 1989). The collagen fibers of the cartilaginous endplate do not generally attach directly to the subchondral bony layer (Moore, 2006); anchoring is indeed provided by a layer of calcified cartilage in between the cartilaginous endplate and bone (Wade et al., 2011). Cartilage calcification tends to be thicker in the outer annulus, whereas it decreases in the radial direction, disappearing completely in correspondence of the nucleus pulposus. It has been shown that endplate calcification increases with age, with possible consequences on fluid flow and nutrition of the disc cells (Roberts, Urban, Evans, & Eisenstein, 1996).

The integration between the annulus fibrosus and the cartilaginous endplate is provided by the annulus fibers which insert into the cartilage layer; the depth of the insertion decreases proceeding

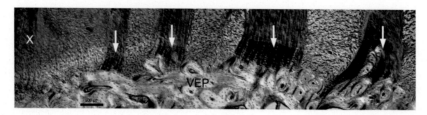

FIGURE 11.3

Photograph of a slice of an ovine specimen showing the integration and interdigitation between the fibers of the annulus fibrosus and the vertebral endplate (VEP). "X" represents the innermost lamella.

From Rodrigues, S. A., Wade, K. R., Thambyah, A., & Broom, N. D. (2012). Micromechanics of annulus–end plate integration in the intervertebral disc. The Spine Journal, 12(2), 143–150. https://doi.org/10.1016/j.spinee.2012.01.003.

from the periphery to the inner annulus (Wade, Galbusera, & Wilke, 2018). When entering the cartilaginous endplate, the fiber bundles subdivide into sub-bundles separated by an extracellular matrix (Rodrigues, Wade, Thambyah, & Broom, 2012; Rodrigues, Thambyah, & Broom, 2015), increasing the contact area and thus enhancing the strength of the integration. In the outer annulus, some fibers protrude into the subchondral bone; this connection is made even more robust by the interdigitation of the fibers in the calcified cartilage layer (Rodrigues et al., 2012; Rodrigues et al., 2015) (Fig. 11.3). Studies on animal models showed that elastin fibers also take part in the connection between annulus fibrosus with endplate; whereas these fibers may not significantly contribute to the strength of the disc-vertebra anchoring, they may facilitate the recovery of the disc height during unloading (Johnson, Caldwell, Berryman, Miller, & Chetty, 1984).

Although the nucleus pulposus has been assumed for decades not to possess any fibrillar connection with the endplate, probably due to limitations in the available experimental and imaging technologies which did not allow detecting them (Inoue & Takeda, 1975), recent studies demonstrated the existence of a structural integration both at the fiber and at the ultrastructural level (Wade et al., 2011). Microstructural analysis conducted on ovine specimens indeed showed that the fibers of the nucleus gather into characteristic nodal insertion points located inside the EP. The function and mechanical relevance of this integration has not been extensively investigated yet.

OSMOTIC SWELLING

The high proteoglycan content of the nucleus pulposus provides it with the capability of attracting water molecules, and therefore to imbibe fluid from the external environment, in other words, the capability of swelling (Urban & McMullin, 1985). Proteoglycans are macromolecules that composed of negatively charged polysaccharide chains with high molecular weight, namely glycosaminoglycans (Malandrino, Galbusera, & Wilke, 2018). Proteoglycans bind cations such as sodium, potassium, and calcium, and are therefore responsible for creating local osmotic effects. Their presence in the intervertebral disc, more precisely the fixed charge density in the glycosaminoglycans, results in an osmotic gradient between disc and surrounding tissues which, in turn, determines the development of an intradiscal pressure even in the absence of any external load. Indeed, it was shown that the osmotic pressure increases sharply in correspondence with an increase of the fixed charge density and is very sensitive to the concentration of glycosaminoglycans (Urban, Maroudas, Bayliss, & Dillon, 1979). However, the equilibrium swelling pressure in the disc, that is, the pressure at which there is no driving force for fluid flow, does not depend only on the tissue composition, but also on its structure and mechanics; it has indeed been shown that the organization and prestress of the collagen fibers play a role in this respect.

The equilibrium swelling pressure was measured in several experimental studies conducted by implanting pressure transducers in cadaver specimens. These investigations revealed pressure values around 0.1−0.25 MPa in healthy lumbar discs and that the pressure is hydrostatic, that is, independent of the direction of insertion of the transducer (Iatridis, MacLean, O'Brien, & Stokes, 2007; Nachemson, 1960). Although the swelling pressure was found to be higher in the nucleus pulposus, the annulus fibrosus also shows a certain swelling capability.

Experimental studies have also been conducted to better understand the effect of osmotic and hydrostatic pressure on the disc tissue with nonuniform results due to a diversity of high loading amplitudes and frequency protocols, as well as stimuli duration. Hydrostatic pressure values, characteristic of healthy discs (up to 0.25 MPa), seemed to have no or little effect on the gene expression profile of rabbit (Kasra et al., 2003), bovine, or human (Neidlinger-Wilke et al., 2006; Wuertz et al., 2007) disc cells, whereas mechanical stimulation in combination with increasing osmolarity (up to 500 mOsm) increased the expression of aggrecan and type II collagen in both nucleus pulposus and annulus fibrosus cells (Wuertz et al., 2007).

High pressure has been shown to affect annulus and nucleus cells cultured in 3D matrices differently. On one hand, dynamic hydrostatic pressure with high loading amplitudes and frequencies (3 MPa and 20 Hz) has been shown to be beneficial to the stimulation of protein synthesis and reduction of protein degradation in 3D cultures of rabbit outer annulus cells (Kasra et al., 2003). On the other hand, human and bovine nucleus cells were shown to decrease gene expression of anabolic proteins such as aggrecan and type II collagen and tended to increase the expression of matrix metalloproteinases MMP1, MMP3, and MMP13 after exposure to hydrostatic pressure at 2.5 MPa and 0.1 Hz (Neidlinger-Wilke et al., 2006).

CELLS AND NUTRITION

The human intervertebral disc is populated by different types of cells: chondrocyte-like cells in the nucleus pulposus, possibly in combination with notochordal cells in immature specimens; chondrocytes in the cartilaginous endplate; and fibroblast-like cells in the annulus fibrosus (Henriksson et al., 2009; Malandrino et al., 2018). While the fibroblast-like cells of the annulus fibrosus and the chondrocytes of the endplate originate from the sclerotomal mesenchyme, the chondrocyte-like cells of the nucleus pulposus derive from the notochord, a structure of mesodermal origin. Although chondrocyte-like cells progressively replace the notochordal cells during skeletal development, small populations of notochordal cells can be retained throughout life (Risbud & Shapiro, 2011). Notochordal cells have been shown to shift into chondrocyte-like cells, under in-vitro standard culture (Kim et al., 2009) or dynamic loading (Purmessur et al., 2013), as well as in vivo after injury (Yang, Leung, Luk, Chan, & Cheung, 2009). Furthermore, while previous studies from Kim et al. (2009) suggested chondrocyte migration from the cartilaginous endplate and the inner annulus fibrosus into the nucleus pulposus as a source of nucleus cells in mature intervertebral discs of rat and rabbit, Henriksson et al. (2009) proposed the existence of stem/progenitor cell niches within the disc (Fig. 11.4).

The avascular nature of the human intervertebral disc determines a challenging environment for its cell population (Fig. 11.5). Since cell nutrition is necessarily based on diffusion phenomena due to the absence of blood supply, the relatively large size of the intervertebral disc, especially in the lumbar region, further contributes to limiting the resources available for cell metabolism. Finite element models have indeed demonstrated that convective transport, that is, the "pumping" effect due to the fluid flow resulting from swelling and daily activities, has a minor effect in enhancing nutrient supply, especially for molecules with low molecular weight such as oxygen (Ferguson, Ito, & Nolte, 2004).

The scarcity of nutrients is reflected by the low cell densities throughout the disc, especially in the regions farthest from the vascular supplies. Indeed, the central nucleus pulposus shows cell densities as low as 4000 cells/mm^3, which increase up to 9000 cells/mm^3 in the outer periphery of the annulus, corresponding in total to approximately 1% of the intervertebral disc volume (Maroudas, Stockwell, Nachemson, & Urban, 1975). This region, as well as the cartilaginous endplate, is supplied by thin capillaries that provide oxygen and glucose. Waste products such as lactate are removed by the combined effect of diffusion and convection, which has a moderate but nonnegligible effect on compounds with larger molecular weight (Ferguson et al., 2004). Studies conducted

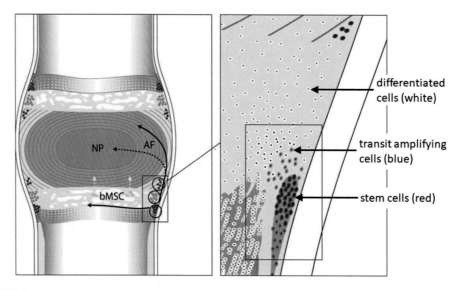

FIGURE 11.4

Schematic overview of the intervertebral disc, stem cell niches (*red*), and hypothetical cellular migration pathways (*arrows*), as well as of niches of transit-amplifying cells (*blue*) and differentiated cells (*white*). AF: annulus fibrosus, NP: nucleus pulposus.

Modified from Henriksson, H., Thornemo, M., Karlsson, C., Hägg, O., Junevik, K., Lindahl, A., & Brisby, H. (2009). Identification of cell proliferation zones, progenitor cells and a potential stem cell niche in the intervertebral disc region: A study in four species. Spine (Philadelphia, PA: 1986), 34(21), 2278–2287. https://doi.org/10.1097/BRS.0b013e3181a95ad2.

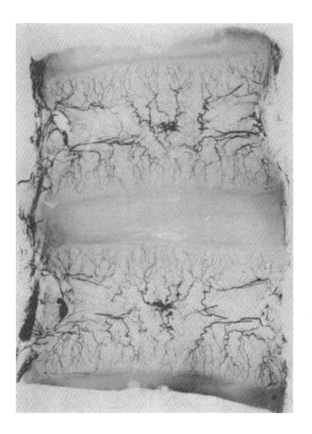

FIGURE 11.5

Blood supply in the vertebral bodies, showing the avascular nature of the intervertebral disc.

From Crock, H. V., & Yoshizawa, H. (1976). The blood supply of the lumbar vertebral column. Clinical Orthopaedics and Related Research, *115, 6–21. https://journals.lww.com/corr/citation/1976/03000/the_blood_supply_of_the_lumbar_vertebral_column.3.*

aspx.

on animal models confirmed that cell densities are indeed associated with and dependent on nutrient availability; small animals show a greater number of cells in the less supplied regions such as the center of the nucleus pulposus (Lotz, 2004).

BIOMECHANICAL RESPONSE OF THE DISCAL TISSUES

The intervertebral disc has been extensively investigated from a biomechanical point of view, both by considering its single components separately and at an organ level. In-vitro tests have been performed to characterize isolated samples of nucleus pulposus and lamellae of the annulus fibrosus as well as on entire discs including VEPs in a variety of loading conditions, both quasi-static and time-dependent.

Since the nucleus pulposus shows mechanical features of both fluids and solids due to the high fluid content (Iatridis et al., 2007; Nachemson, 1960), standard experimental techniques used for other musculoskeletal tissues such as tensile testing are not suitable for its mechanical characterization. Experimental studies focused on its mechanical response under compressive loading, which is the most frequent condition to which the nucleus pulposus is subjected during daily life. The testing methods employed included confined and unconfined compression as well as indentation (Nerurkar, Elliott, & Mauck, 2010); confined should be considered as the most physiological approach, since both indentation and unconfined compression involve lateral expansion and deformations which are not experienced by the nucleus material in vivo. Indeed, testing the nucleus pulposus under unconfined conditions revealed relatively low material properties; on the contrary, under confined conditions the nucleus shows the capability of withstanding high loads, as well as an almost incompressible, evidently time-dependent behavior (Nerurkar et al., 2010). While applying torsion or shear in combination with compression, the nucleus revealed the ability of complete stress relaxation, demonstrating a fluid-like mechanical behavior in this respect (Johannessen & Elliott, 2005).

The complex architecture of the collagen fibers of the annulus fibrosus gives it a distinct anisotropic behavior; experimental tests on annulus samples showed indeed stiffness values one or two orders of magnitude larger in the circumferential direction with respect to the radial and axial directions (Nerurkar et al., 2010). Besides, the experiments showed a clear time-dependent response, which may be attributed to the water content and to the intrinsic viscoelastic properties of the tissue components.

Holzapfel et al. (2005) studied the biomechanical properties of single lamellae of lumbar intervertebral discs, which were assumed to be the elementary units of the annulus fibrosus and were extracted from cadaver specimens by means of a sophisticated preparation technique, which allowed preserving the attachment with the VEP; the lamellae were collected from different locations in the disc and tested in uniaxial tension. The results showed a clear nonlinearity of the tensile properties, higher stiffnesses in ventrolateral specimens with respect to dorsal ones as well as in specimens collected from the outer annulus versus those taken from the inner region (Fig. 11.6). The latter finding has been interpreted as an adaptation phenomenon; since pressurized containers with homogeneous material properties show a stress gradient in the walls, which increases proceeding inwards, a corresponding decreasing gradient in the material properties would mitigate this gradient, making the stress distribution more homogeneous. Interestingly, the samples showed a purely elastic behavior with negligible hysteresis; therefore, the viscoelastic properties that emerge from the testing disc and annulus samples at strain rates comparable to that of the study (0.1−10 mm/min) should not be attributed to the collagen fibers in the lamellae but rather to phenomena related to fluid flow.

The intervertebral disc cells can sense fluid flow, pressure, and deformation of the tissue and, depending on the magnitude and frequency of the stimuli, as well as of the cell type, a different response can be triggered. The cellular responses contribute to homeostatic tissue remodeling. However, when high mechanical stresses are applied, cells sense those stimuli and activate different pathways that contribute to degenerative changes and tissue failure by decreasing the production of matrix components and increasing the secretion of degrading enzymes and proinflammatory factors, as reviewed over time (Molladavoodi, McMorran, & Gregory, 2020; Neidlinger-Wilke et al., 2014; Setton & Chen, 2004, 2006).

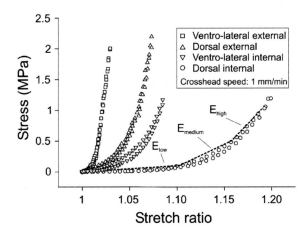

FIGURE 11.6

Stress—stretch responses measured for four single lamellar specimens extracted from the annulus fibrosus of human samples in four different locations (ventrolateral external, dorsal external, ventrolateral internal, and dorsal internal).

From Holzapfel, G. A., Schulze-Bauer, C. A. J., Feigl, G., & Regitnig, P. (2005). Single lamellar mechanics of the human lumbar anulus fibrosus. Biomechanics and Modeling in Mechanobiology, 3*(3), 125–140. https://doi.org/10.1007/s10237-004-0053-8.*

BIOMECHANICS OF THE INTERVERTEBRAL DISC

The intervertebral disc is subjected to several complex loads arising from different postures and motions during daily activities such as walking, running, sitting, carrying loads, flexing, and rotating the trunk. However, most of these activities share a high compressive component; with the exception of relaxed postures such as lying supine, the intervertebral disc is almost always subjected to large compressive loads and is consequently adapted to such a loading condition.

In a healthy intervertebral disc subjected to compression, the high water content of the nucleus pulposus, which is due to its osmotic properties as discussed in the previous paragraphs, determines the development of high hydrostatic pressure. As the nucleus material is incompressible, it tends to bulge outwards determining a tensile stress in the annulus fibers which act as a containment ring (Fig. 11.7). The annulus fibrosus indeed functions as a car tyre in which the inner pressure is contained by a strong fiber-reinforced elastic structure (Boos & Aebi, 2008).

The composite structure of the intervertebral disc is able to effectively induce a relatively homogeneous stress distribution in all of its regions. This hypothesis has been demonstrated in studies in which a needle pressure transducer was pulled along the anteroposterior direction through intervertebral disc specimens subjected to mechanical loads (McNally & Adams, 1992), realizing the so-called "stress profilometry" technique. In healthy specimens under compressive loading, relatively constant stress profiles showing no stress peaks were observed (Fig. 11.8). By aligning the pressure transducer in different directions, the researchers were able to prove that the stress acting in the nucleus pulposus was indeed hydrostatic. As mentioned in the previous paragraphs, the

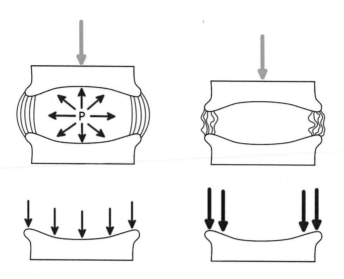

FIGURE 11.7

Under compressive loading, a healthy intervertebral disc (left) develops a hydrostatic pressure in the nucleus pulposus (P) which is transmitted to the annulus fibers, which are loaded in tension; the resulting stresses on the endplates are relatively homogeneous (bottom). In a degenerated disc (right), the low swelling and water content result in lower intradiscal pressure, the collapse of the annulus fibers and, consequently, stress concentrations on the endplates.

spatial gradients observed in the structure and properties of the intervertebral disc (fiber orientation, lamellar thickness and stiffness, proteoglycan and water content, and disc−endplate integration) play a role in determining the homogeneous stress distribution measured by means of stress profilometry.

Intervertebral discs subjected to mechanical loading show an evident time-dependent response, which was investigated in depth both in experimental and numerical studies, and is mostly attributed to the imbibition and exudation of fluids, with the intrinsic viscoelasticity of the disc tissues playing a smaller but nonnegligible role (Costi, Stokes, Gardner-Morse, & Iatridis, 2008). In general terms, the osmotic properties of the disc determine its swelling when not subjected to high mechanical loads, such as during bed rest; intradiscal pressure as well as disc height progressively arise at night (Adams, Dolan, Hutton, & Porter, 1990; Wilke, Neef, Caimi, Hoogland, & Claes, 1999). The onset of compressive loading in the morning determines a fast fluid exudation through the VEPs and, in minor proportions, through the outer periphery of the annulus, determining a loss of disc height and intradiscal pressure. The fluid loss continues throughout the day, progressively slowing down due to the decrease of the pressure gradient between the disc and the external environment. Experimental studies showed that disc specimens subjected to sustained loading show a behavior similar to that observed in vivo and provided additional data not accessible on living subjects such as the progressive changes in the disc stiffness, which increases in compression and decreases in bending (Adams et al., 1990).

Diurnal effects associated with the fluid flow were investigated in depth by means of numerical models. This approach allows for the assessment of the relevance of the various parameters

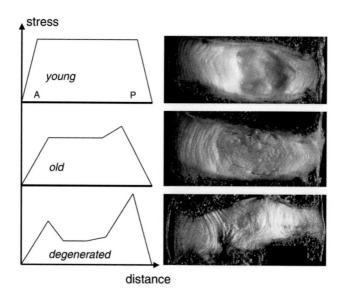

FIGURE 11.8

Schematic representations of stress profiles in the intervertebral disc (top: healthy, young specimen; middle: old specimen; bottom: severely degenerated specimen) measured with needle transducers along the anteroposterior direction.

From Adams, M. A., Dolan, P., & McNally, D. S. (2009). The internal mechanical functioning of intervertebral discs and articular. Matrix Biology, 28(7), 384–389. https://doi.org/10.1016/j.matbio.2009.06.004.

involved such as magnitude and direction of loading, material properties such as stiffness, porosity, and permeability, and the possible presence and extent of cartilage calcification; besides, the use of numerical models eliminates some limitations associated with the use of cadaver specimens such as high interindividual variability and blood clotting occurring postmortem in the endplate pores, which alters the fluid transport properties of the specimens. By the use of numerical models, Schmidt, Shirazi-Adl, Galbusera, and Wilke (2010) performed a detailed analysis of the intradiscal pressures and of the stresses in the annulus fibrosus under the action of a large set of dynamic loads replicating various daily activities and rest periods. The research group was able to determine that the disc is subjected to large stresses especially in the morning, in which a higher risk of injury should be presumed, and that rest periods scattered throughout the day may be effective in reducing this risk.

AGING AND DEGENERATION

The intervertebral discs of elderly subjects show structural and composition differences with respect to those of young, healthy individuals. These changes are commonly labeled as degenerative signs and characterize the condition commonly named "disc degeneration." Nevertheless, the definition of disc degeneration is debated; while several studies considered as degenerated discs even those showing only a moderate loss of height or water content, other authors distinguished discs exhibiting physiological changes associated with aging from those showing major structural failure and disruption, which should be considered as pathologically degenerated (Adams et al., 1990; Wilke et al., 1999).

Cell densities, which are already low in young adults, decrease progressively during adulthood with consequences on the quantity and fragmentation of the proteoglycan macromolecules (Buckwalter, 1995), on cross-linking of the collagen fibrils and, ultimately, on the self-repair capabilities of the tissue (Duance et al., 1998). The decrease of proteoglycan content determines a loss of water retainment and osmotic gradient and, in turn, the water content is partially replaced by fibrotic tissue; experimental studies indeed confirmed a decrease of the swelling pressure from 0.1−0.25 MPa in young subjects to 0.05 in older ones (Urban & McMullin, 1985). Disc dehydration is easily visible on T2-weighted magnetic resonance images that show a clear decrease of the signal intensity in the nucleus pulposus, appearing darker than other structures with high water content such as adipose tissue and spinal cord, and a less evident distinction between the nucleus and the annulus fibrosus (Pfirrmann, Metzdorf, Zanetti, Hodler, & Boos, 2001).

The dehydration of aging and degenerated discs has profound biomechanical implications. In the absence of swelling and thus of intradiscal pressure, compressive loads determine the buckling and collapse of the collagen fibers of the annulus fibrosus (Fig. 11.7), which press on the endplates determining an uneven stress distribution with local peaks; such an effect was confirmed by stress profilometry studies on degenerated discs (Fig. 11.8). It was hypothesized that these stress peaks and concentrations may be associated with discogenic pain, and therefore disc dehydration may indeed be one of the common mechanisms leading to back pain in elderly subjects (Adams, McMillan, Green, & Dolan, 1996). The changes in tissue composition and architecture, as well as the effects in water retainment and load distribution, are associated with cellular alterations, metabolic imbalance, and immune system activation. These are described in more detail in Chapter 6.

As mentioned before, degenerated discs commonly show structural failures such as tears, clefts, and fissures. An analysis of postmortem material indeed showed that the majority of specimens from donors in the third and fourth decades of life already exhibited signs of structural failure (Osti, Vernon-Roberts, Moore, & Fraser, 1992), which should therefore be considered as a physiological phenomenon of the ageing course; older specimens showed however a higher prevalence of tears and fissures as well as a larger average size, demonstrating that the structural failure is implicated in the degenerative cascade. It should be noted that degenerative changes of the intervertebral disc have also been observed in young subjects and, while degeneration is associated with aging, it may indeed occur at any age.

Other degenerative changes involve the VEPs, which calcify and become more sclerotic with aging (Roberts, Menage, & Eisenstein, 1993); this alteration has consequences on the flexibility of the endplates, which may become more prone to fracture, as well as on the nutrition of the disc cells which may be hindered by the decreased permeability and diffusion properties of calcified endplates (Fig. 11.9). It has been observed in symptomatic discs that the impairment of nutrition (in which nutrient concentrations can fall below 0.5 mmol/L glucose) (Kitano et al., 1993), together with low oxygen (under 5% O_2) (Ishihara & Urban, 1999) and an acidic environment (pH levels under 6.8) (Razaq, Wilkins, & Urban, 2003) contribute to a decrease in cell viability and an imbalance of matrix turnover favoring matrix degradation (Natarajan et al., 2004; Urban & McMullin, 1985). The breakdown of macromolecules such as aggrecan reduces the water-binding capacity of

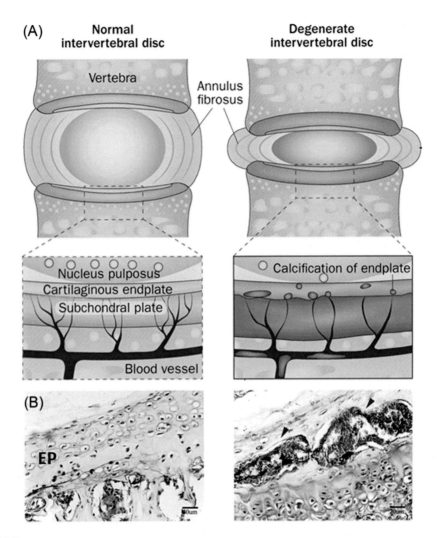

FIGURE 11.9

Schematic showing the nutrient pathways in a normal intervertebral disc and in a degenerated disc with changes such as calcification of cartilaginous endplate, occlusion of marrow spaces (so that they are no longer in contact with cartilage surface), atherosclerosis of vertebral arteries, and reduced capillary density, limiting nutrient transport (Huang et al., 2014) (A). Immunohistochemistry of osterix (a calcification marker), displaying little osterix expression in a healthy endplate (EP), and significantly more (*red arrows*) in degenerated EP (Xiao et al., 2018) (B).

Adapted by permission from Huang, Y.-C., Urban, J. P. G., & Luk, K. D. (2014). Intervertebral disc regeneration: Do nutrients lead the way? Nature Reviews Rheumatology, 10*(9), 561–566, copyright (2014).*

the tissue. Therefore, disc height collapse is another common finding and should be attributed to the loss of the swelling capability and of water content. The nucleus pulposus can be partially replaced by fibrotic tissue, being this reparative fibrosis a way to mechanically stabilize the degenerated disc (Chen et al., 2020). Disc degeneration has large consequences on the flexibility and stability of the motion segment, which are described in detail in Chapter 15.

DISC HERNIATION

A common finding in both the cervical and the lumbar spine, with some sporadic occurrences also in the thoracic region, is the herniation of the intervertebral discs. This disorder consists in the expulsion of a fragment of the nucleus pulposus from the intervertebral space, possibly invading the spinal canal or the foramen and thus compressing the spinal cord and nerves (Dydyk, Ngnitewe Massa, & Mesfin, 2020). Herniation most commonly occurs in early degenerated discs, or even in discs showing no signs of degeneration as a consequence of trauma or overloading. The clinical consequences of disc herniation largely vary among individuals and depend on the location and size of the herniated material; severe pain and disability are relatively common. Disc herniations are commonly asymptomatic and tend to heal spontaneously; symptomatic ones in the lumbar and cervical spine are routinely treated surgically with success. On the contrary, the surgical removal of thoracic herniations, whenever necessary, is technically challenging and associated with high rates of complications.

In comparison with other spinal disorders such as disc degeneration, the biomechanics of disc herniation have been investigated by relatively few studies and many aspects remain still unclear and debated. Disc herniation has been traditionally attributed to a fissure in the annulus fibrosus which undergoes mechanical failure as a consequence of high intradiscal pressure in combination with sudden motion or traumatic events (Adams et al., 1990); the pressurized nucleus material is then expelled through this escape path (Fig. 11.10). While this mechanism of failure is undoubtedly realistic and clinically relevant, recent studies showed that the majority of disc herniations in the lumbar spine, namely 65%, shall be rather associated with the avulsion of the endplate (Rajasekaran, Bajaj, Tubaki, Kanna, & Shetty, 2013) (Fig. 11.10), which has lower mechanical strength in comparison with the annulus fibrosus. Such lesions are most commonly central or paracentral and the lower VEP, which is thinner and supported by a less dense trabecular bone, is involved in the majority of cases.

Several studies have investigated the loads and motions which may be responsible for determining disc herniations, either by annulus rupture or endplate avulsion. Nevertheless, several research questions remain open due to the limitations of the employed experimental methods to the difficulties of obtaining herniations in a reproducible, statistically valid manner. Besides, most experimental studies about disc herniations used animal specimens instead of human ones, taking advantage of the ease of obtaining them in the large quantities necessary for this type of investigation and their low intersubject variability. Nevertheless, the correctness of translating results obtained with animal samples to the bipedal humans, whose discs show distinct anatomies and material properties, remains unclear. The most extensive study in this field was conducted by Berger-Roscher et al. (2017), who tested 30 ovine disc specimens with a device to investigate how different loading

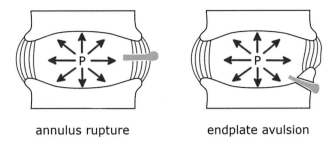

annulus rupture   endplate avulsion

FIGURE 11.10

Schematic representation of the two types of disc herniation: annulus rupture (left) and endplate avulsion (right).

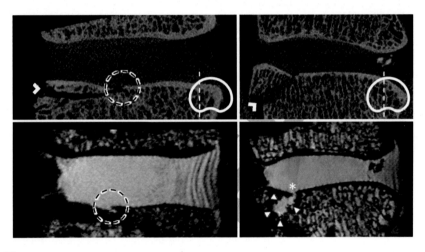

FIGURE 11.11

μCT (top) and ultra-high-field MRI scans of two ovine specimens in which disc herniations were artificially induced. Both specimens show endplate avulsion (*arrow*). In the case on the left the cartilaginous endplate is intact (*circle*), whereas in the case on the right there is rupture of the cartilage layer (*asterisk*) and consequent extrusion of the nucleus in the gap (*triangles*).

From Berger-Roscher, N., Casaroli, G., Rasche, V., Villa, T., Galbusera, F., & Wilke, H.-J. (2017). Influence of complex loading conditions on intervertebral disc failure. Spine, 42(2), E78–E85. https://doi.org/10.1097/BRS.0000000000001699.

combinations influence the mechanism and extent of intervertebral disc failure (Fig. 11.11). The device was designed to be able to apply cyclic rotations around the three main axes as well as axial compression, all with twofold magnitudes with respect to those observed in physiological motions, individually or in different combinations. The authors were able to replicate both the annulus rupture and the endplate avulsion mechanisms and found that the combination of flexion, lateral bending, axial rotation, and axial compression determined the highest risk of disc herniation. In general, flexion and lateral bending seemed to be more associated with failure with respect to axial compression and torsion.

REFERENCES

Adams, M. A., Dolan, P., Hutton, W. C., & Porter, R. W. (1990). Diurnal changes in spinal mechanics and their clinical significance. *The Journal of Bone and Joint Surgery. British Volume*, *72*(2), 266–270. Available from https://doi.org/10.1302/0301-620X.72B2.2138156.

Adams, M. A., McMillan, D. W., Green, T. P., & Dolan, P. (1996). Sustained loading generates stress concentrations in lumbar intervertebral. *Spine (Philadelphia, PA: 1986)*, *21*(4), 434–438. Available from https://doi.org/10.1097/00007632-199602150-00006.

Berger-Roscher, N., Casaroli, G., Rasche, V., Villa, T., Galbusera, F., & Wilke, H.-J. (2017). Influence of complex loading conditions on intervertebral disc failure. *Spine (Philadelphia, PA: 1986)*, *42*(2), E78–E85. Available from https://doi.org/10.1097/BRS.0000000000001699.

Boos, N., & Aebi, M. (2008). *Spinal disorders: Fundamentals of diagnosis and treatment*. Springer, No. 1165.

Buckwalter, J. A. (1995). Aging and degeneration of the human intervertebral disc. *Spine (Philadelphia, PA: 1986)*, *20*(11), 1307–1314. Available from https://doi.org/10.1097/00007632-199506000-00022.

Cassidy, J. J., Hiltner, A., & Baer, E. (1989). Hierarchical structure of the intervertebral disc. *Connective Tissue Research*, *23*(1), 75–88. Available from https://doi.org/10.3109/03008208909103905.

Chen, C., Zhou, T., Sun, X., Han, C., Zhang, K., Zhao, C., ... Zhao, J. (2020). Autologous fibroblasts induce fibrosis of the nucleus pulposus to maintain. *Bone Research*, *8*, 7. Available from https://doi.org/10.1038/s41413-019-0082-7.

Costi, J. J., Stokes, I. A., Gardner-Morse, M. G., & Iatridis, J. C. (2008). Frequency-dependent behavior of the intervertebral disc in response to each of six degree of freedom dynamic loading: solid phase and fluid phase contributions. *Spine (Philadelphia, PA: 1986)*, *33*(16), 1731–1738. Available from https://doi.org/10.1097/BRS.0b013e31817bb116.

Coventry, M. B., Ghormley, R. K., & Kernohan, J. W. (1945). The intervertebral disc: Its microscopic anatomy and pathology: Part I. *The Journal of Bone and Joint Surgery*, *27*(1), 105. Available from https://journals.lww.com/jbjsjournal/Abstract/1945/27010/THE_INTERVERTEBRAL_DISC__ITS_MICROSCOPIC_ANATOMY.11.aspx.

Duance, V. C., Crean, J. K., Sims, T. J., Avery, N., Smith, S., Menage, J., ... Roberts, S. (1998). Changes in collagen cross-linking in degenerative disc disease and scoliosis. *Spine (Philadelphia, PA: 1986)*, *23*(23), 2545–2551. Available from https://doi.org/10.1097/00007632-199812010-00009.

Dydyk, A. M., Ngnitewe Massa, R., & Mesfin, F. B. (2020). Disc herniation. StatPearls Publishing. https://pubmed.ncbi.nlm.nih.gov/28722852.

Eyre, D. R. (1979). Biochemistry of the intervertebral disc. *International Review of Connective Tissue Research*, *8*, 227–291. Available from https://doi.org/10.1016/b978-0-12-363708-6.50012-6.

Ferguson, S. J., Ito, K., & Nolte, L. P. (2004). Fluid flow and convective transport of solutes within the intervertebral. *Journal of Biomechanics*, *37*(2), 213–221. Available from https://doi.org/10.1016/s0021-9290(03)00250-1.

Henriksson, H., Thornemo, M., Karlsson, C., Hägg, O., Junevik, K., Lindahl, A., & Brisby, H. (2009). Identification of cell proliferation zones, progenitor cells and a potential stem cell niche in the intervertebral disc region: A study in four species. *Spine (Philadelphia, PA: 1986)*, *34*(21), 2278–2287. Available from https://doi.org/10.1097/BRS.0b013e3181a95ad2.

Holzapfel, G. A., Schulze-Bauer, C. A. J., Feigl, G., & Regitnig, P. (2005). Single lamellar mechanics of the human lumbar anulus fibrosus. *Biomechanics and Modeling in Mechanobiology*, *3*(3), 125–140. Available from https://doi.org/10.1007/s10237-004-0053-8.

Huang, Y.-C., Urban, J. P. G., & Luk, K. D. (2014). Intervertebral disc regeneration: Do nutrients lead the way? *Nature Reviews Rheumatology*, *10*(9), 561–566.

Iatridis, J. C., MacLean, J. J., O'Brien, M., & Stokes, I. A. F. (2007). Measurements of proteoglycan and water content distribution in human. *Spine (Philadelphia, PA: 1986)*, *32*(14), 1493–1497. Available from https://doi.org/10.1097/BRS.0b013e318067dd3f.

Inoue, H., & Takeda, T. (1975). Three-dimensional observation of collagen framework of lumbar. *Acta Orthopaedica Scandinavica*, *46*(6), 949–956. Available from https://doi.org/10.3109/17453677508989283.

Ishihara, H., & Urban, J. P. (1999). Effects of low oxygen concentrations and metabolic inhibitors on proteoglycan and protein synthesis rates in the intervertebral disc. *Journal of Orthopaedic Research: Official Publication of the Orthopaedic Research Society*, *17*(6), 829–835. Available from https://doi.org/10.1002/jor.1100170607.

Johannessen, W., & Elliott, D. M. (2005). Effects of degeneration on the biphasic material properties of human. *Spine (Philadelphia, PA: 1986)*, *30*(24), E724–E729. Available from https://doi.org/10.1097/01.brs.0000192236.92867.15.

Johnson, E. F., Caldwell, R. W., Berryman, H. E., Miller, A., & Chetty, K. (1984). Elastic fibers in the anulus fibrosus of the dog intervertebral disc. *Acta Anatomica*, *118*(4), 238–242. Available from https://doi.org/10.1159/000145851.

Kasra, M., Goel, V., Martin, J., Wang, S.-T., Choi, W., & Buckwalter, J. (2003). Effect of dynamic hydrostatic pressure on rabbit intervertebral disc cells. *Journal of Orthopaedic Research: Official Publication of the Orthopaedic Research Society*, *21*(4), 597–603. Available from https://doi.org/10.1016/S0736-0266(03)00027-5.

Keyes, D. C., & Compere, E. L. (1932). The normal and pathological physiology of the nucleus pulposus of the. *The Journal of Bone and Joint Surgery*, *14*(4), 897. Available from https://journals.lww.com/jbjsjournal/Abstract/1932/14040/THE_NORMAL_AND_PATHOLOGICAL_PHYSIOLOGY_OF_THE.20.aspx.

Kim, J. H., Deasy, B. M., Seo, H. Y., Studer, R. K., Vo, N. V., Georgescu, H. I., ... Kang, J. D. (2009). Differentiation of intervertebral notochordal cells through live automated. *Spine (Philadelphia, PA: 1986)*, *34*(23), 2486–2493. Available from https://doi.org/10.1097/BRS.0b013e3181b26ed1.

Kitano, T., Zerwekh, J. E., Usui, Y., Edwards, M. L., Flicker, P. L., & Mooney, V. (1993). Biochemical changes associated with the symptomatic human intervertebral. *Clinical Orthopaedics and Related Research*, *293*, 372–377. Available from https://www.ncbi.nlm.nih.gov/pubmed/8339506.

Lotz, J. C. (2004). Animal models of intervertebral disc degeneration: Lessons learned. *Spine (Philadelphia, PA: 1986)*, *29*(23), 2742–2750. Available from https://doi.org/10.1097/01.brs.0000146498.04628.f9.

Malandrino, A., Galbusera, F., & Wilke, H.-J. (2018). *Chapter 6 − Intervertebral disc. Biomechanics of the spine* (pp. 89–103). Academic Press. Available from https://doi.org/10.1016/B978-0-12-812851-0.00006-9.

Marchand, F., & Ahmed, A. M. (1990). Investigation of the laminate structure of lumbar disc anulus fibrosus. *Spine (Philadelphia, PA: 1986)*, *15*(5), 402–410. Available from https://doi.org/10.1097/00007632-199005000-00011.

Maroudas, A., Stockwell, R. A., Nachemson, A., & Urban, J. (1975). Factors involved in the nutrition of the human lumbar intervertebral disc: Cellularity and diffusion of glucose in vitro. *Journal of Anatomy*, *120*(1), 113–130. Available from https://www.ncbi.nlm.nih.gov/pmc/articles/PMC1231728/pdf/janat00375-0115.pdf.

McNally, D. S., & Adams, M. A. (1992). Internal intervertebral disc mechanics as revealed by stress profilometry. *Spine (Philadelphia, PA: 1976)*, *17*(1), 66–73. Available from https://europepmc.org/article/med/1536017.

Molladavoodi, S., McMorran, J., & Gregory, D. (2020). Mechanobiology of annulus fibrosus and nucleus pulposus cells. *Cell and Tissue Research*, *379*(3), 429–444. Available from https://doi.org/10.1007/s00441-019-03136-1.

Moore, R. J. (2006). The vertebral endplate: Disc degeneration, disc regeneration. *European Spine Journal: Official Publication of the European Spine Society, the European Spinal Deformity Society, and the*

European Section of the Cervical Spine Research Society, 15(3), S333−S337. Available from https://doi.org/10.1007/s00586-006-0170-4.

Nachemson, A. (1960). Lumbar intradiscal pressure. Experimental studies on post-mortem material. *Acta orthopaedica Scandinavica Supplementum, 43*, 1−104. Available from https://doi.org/10.3109/ort.1960.31.suppl-43.01.

Natarajan, R. N., Williams, J. R., & Andersson, G. B. J. (2004). Recent advances in analytical modeling of lumbar disc degeneration. *Spine (Philadelphia, PA: 1986), 29*(23), 2733−2741. Available from https://doi.org/10.1097/01.brs.0000146471.59052.e6.

Neidlinger-Wilke, C., Galbusera, F., Pratsinis, H., Mavrogonatou, E., Mietsch, A., Kletsas, D., & Wilke, H.-J. (2014). Mechanical loading of the intervertebral disc: From the macroscopic to the cellular level. *European Spine Journal, 23*(3), S333−S343. Available from https://doi.org/10.1007/s00586-013-2855-9.

Neidlinger-Wilke, C., Würtz, K., Urban, J. P. G., Börm, W., Arand, M., Ignatius, A., ... Claes, L. E. (2006). Regulation of gene expression in intervertebral disc cells by low and high. *European Spine Journal: Official Publication of the European Spine Society, the European Spinal Deformity Society, and the European Section of the Cervical Spine Research Society, 15*(3), S372−S378. Available from https://doi.org/10.1007/s00586-006-0112-1.

Nerurkar, N. L., Elliott, D. M., & Mauck, R. L. (2010). Mechanical design criteria for intervertebral disc tissue engineering. *Journal of Biomechanics, 43*(6), 1017−1030. Available from https://doi.org/10.1016/j.jbiomech.2009.12.001.

Osti, O. L., Vernon-Roberts, B., Moore, R., & Fraser, R. D. (1992). Annular tears and disc degeneration in the lumbar spine. A post-mortem. *The Journal of Bone and Joint Surgery. British Volume, 74*(5), 678−682. Available from https://doi.org/10.1302/0301-620X.74B5.1388173.

Pfirrmann, C. W., Metzdorf, A., Zanetti, M., Hodler, J., & Boos, N. (2001). Magnetic resonance classification of lumbar intervertebral disc. *Spine (Philadelphia, PA: 1986), 26*(17), 1873−1878. Available from https://doi.org/10.1097/00007632-200109010-00011.

Pooni, J. S., Hukins, D. W., Harris, P. F., Hilton, R. C., & Davies, K. E. (1986). Comparison of the structure of human intervertebral discs in the cervical. *Surgical and Radiologic Anatomy: SRA, 8*(3), 175−182. Available from https://doi.org/10.1007/BF02427846.

Purmessur, D., Guterl, C. C., Cho, S. K., Cornejo, M. C., Lam, Y. W., Ballif, B. A., ... Iatridis, J. C. (2013). Dynamic pressurization induces transition of notochordal cells to a mature. *Arthritis Research & Therapy, 15*(5), R122. Available from https://doi.org/10.1186/ar4302.

Rajasekaran, S., Bajaj, N., Tubaki, V., Kanna, R. M., & Shetty, A. P. (2013). ISSLS Prize winner: The anatomy of failure in lumbar disc herniation: An in vivo, multimodal, prospective study of 181 subjects. *Spine, 38*(17), 1491−1500. Available from https://doi.org/10.1097/BRS.0b013e31829a6fa6.

Razaq, S., Wilkins, R. J., & Urban, J. P. G. (2003). The effect of extracellular pH on matrix turnover by cells of the bovine. *European Spine Journal: Official Publication of the European Spine Society, the European Spinal Deformity Society, and the European Section of the Cervical Spine Research Society, 12*(4), 341−349. Available from https://doi.org/10.1007/s00586-003-0582-3.

Risbud, M. V., & Shapiro, I. M. (2011). Notochordal cells in the adult intervertebral disc: New perspective on an old question. *Critical Reviews in Eukaryotic Gene Expression, 21*(1), 29−41. Available from https://doi.org/10.1615/critreveukargeneexpr.v21.i1.30.

Roberts, S., Menage, J., & Eisenstein, S. M. (1993). The cartilage end-plate and intervertebral disc in scoliosis: calcification and other sequelae. *Journal of Orthopaedic Research: Official Publication of the Orthopaedic Research Society, 11*(5), 747−757. Available from https://onlinelibrary.wiley.com/doi/pdf/10.1002/jor.1100110517?casa_token = 9dd16BdHWBcAAAAA:MjK12P-oUTOQLqHqKfdwjhWhjD1sM3-lNjPfdTEz5XLrEm7VMKZPbELDx4pdgvIBJ5f4QSXTlYv7yw.

Roberts, S., Menage, J., & Urban, J. P. (1989). Biochemical and structural properties of the cartilage end-plate and its relation to the intervertebral disc. *Spine*, *14*(2), 166−174. Available from https://doi.org/10.1097/00007632-198902000-00005.

Roberts, S., Urban, J. P., Evans, H., & Eisenstein, S. M. (1996). Transport properties of the human cartilage endplate in relation to its composition and calcification. *Spine*, *21*(4), 415−420. Available from https://doi.org/10.1097/00007632-199602150-00003.

Rodrigues, S. A., Thambyah, A., & Broom, N. D. (2015). A multiscale structural investigation of the annulus-endplate anchorage. *The Spine Journal: Official Journal of the North American Spine Society*, *15*(3), 405−416. Available from https://doi.org/10.1016/j.spinee.2014.12.144.

Rodrigues, S. A., Wade, K. R., Thambyah, A., & Broom, N. D. (2012). Micromechanics of annulus−end plate integration in the intervertebral disc. *The Spine Journal: Official Journal of the North American Spine Society*, *12*(2), 143−150. Available from https://doi.org/10.1016/j.spinee.2012.01.003.

Schmidt, H., Shirazi-Adl, A., Galbusera, F., & Wilke, H.-J. (2010). Response analysis of the lumbar spine during regular daily activities − A finite element analysis. *Journal of Biomechanics*, *43*(10), 1849−1856. Available from https://doi.org/10.1016/j.jbiomech.2010.03.035.

Setton, L. A., & Chen, J. (2004). Cell mechanics and mechanobiology in the intervertebral disc. *Spine (Philadelphia, PA: 1986)*, *29*(23), 2710−2723. Available from https://doi.org/10.1097/01.brs.0000146050.57722.2a.

Setton, L. A., & Chen, J. (2006). Mechanobiology of the intervertebral disc and relevance to disc. *The Journal of Bone and Joint Surgery. American Volume*, *88*(2), 52−57. Available from https://doi.org/10.2106/JBJS.F.00001.

Urban, J. P., & McMullin, J. F. (1985). Swelling pressure of the inervertebral disc: Influence of proteoglycan and collagen contents. *Biorheology*, *22*(2), 145−157. Available from https://doi.org/10.3233/bir-1985-22205.

Urban, J. P., Maroudas, A., Bayliss, M. T., & Dillon, J. (1979). Swelling pressures of proteoglycans at the concentrations found in cartilaginous tissues. *Biorheology*, *16*(6), 447−464. Available from https://doi.org/10.3233/bir-1979-16609.

Wade, K. R., Robertson, P. A., & Broom, N. D. (2011). A fresh look at the nucleus-endplate region: New evidence for significant. *European Spine Journal: Official Publication of the European Spine Society, the European Spinal Deformity Society, and the European Section of the Cervical Spine Research Society*, *20*(8), 1225−1232. Available from https://doi.org/10.1007/s00586-011-1704-y.

Wade, K., Galbusera, F., & Wilke, H.-J. (2018). *Chapter 8 − Vertebral endplates. Biomechanics of the spine* (pp. 125−140). Academic Press. Available from https://doi.org/10.1016/B978-0-12-812851-0.00008-2.

Wilke, H. J., Neef, P., Caimi, M., Hoogland, T., & Claes, L. E. (1999). New in vivo measurements of pressures in the intervertebral disc in daily. *Spine (Philadelphia, PA: 1986)*, *24*(8), 755−762. Available from https://doi.org/10.1097/00007632-199904150-00005.

Wuertz, K., Urban, J. P. G., Klasen, J., Ignatius, A., Wilke, H.-J., Claes, L., & Neidlinger-Wilke, C. (2007). Influence of extracellular osmolarity and mechanical stimulation on gene. *Journal of Orthopaedic Research: Official Publication of the Orthopaedic Research Society*, *25*(11), 1513−1522. Available from https://doi.org/10.1002/jor.20436.

Xiao, Z. F., He, J. B., Su, G. Y., Chen, M. H., Hou, Y., Chen, S. D., & Lin, D. K. (2018). Osteoporosis of the vertebra and osteochondral remodeling of the endplate causes intervertebral disc degeneration in ovariectomized mice. *Arthritis Research & Therapy*, *20*(1), 207. Available from https://doi.org/10.1186/s13075-018-1701-1.

Yang, F., Leung, V. Y. L., Luk, K. D. K., Chan, D., & Cheung, K. M. C. (2009). Injury-induced sequential transformation of notochordal nucleus pulposus. *The Journal of Pathology*, *218*(1), 113−121. Available from https://doi.org/10.1002/path.2519.

FURTHER READING

Adams, M. A., & Roughley, P. J. (2006). What is intervertebral disc degeneration, and what causes it? *Spine (Philadelphia, PA: 1986)*, *31*(18), 2151−2161. Available from https://journals.lww.com/spinejournal/Fulltext/2006/08150/What_is_Intervertebral_Disc_Degeneration,_and_What.24.aspx?casa_token = MhBIZ ZlxVPkAAAAA:nlY7ZJn8_SiXGLBSgi_qWfVSNJV0McmdOe5Dm45rU-huXmrEwYgc-GFSENmaz-qxY IiX7VrWkBUBDs4GfTfViFvzwc0hzQ.

Iatridis, J. C., Weidenbaum, M., Setton, L. A., & Mow, V. C. (1996). Is the nucleus pulposus a solid or a fluid? Mechanical behaviors of the nucleus pulposus of the human intervertebral disc. *Spine (Philadelphia, PA: 1986)*, *21*(10), 1174−1184. Available from https://doi.org/10.1097/00007632-199605150-00009.

Kim, K.-W., Ha, K.-Y., Park, J.-B., Woo, Y.-K., Chung, H.-N., & An, H. S. (2005). Expressions of membrane-type I matrix metalloproteinase, Ki-67 protein. *Spine (Philadelphia, PA: 1986)*, *30*(12), 1373−1378. Available from https://journals.lww.com/spinejournal/Fulltext/2005/06150/Expressions_of_Membrane_Type_I_Matrix.6.aspx?casa_token = 9_Q3_d0hq3oAAAAA:ALx64ym5ygd5PpxcZx7UMX EUkFa_AOZSYFi4Jk_sBH5CyYsoGmDIKhsDwwy40VVLpLgCef9bTKU4f14J05OFTN4WmFa3rLI.

Kim, K.-W., Lim, T.-H., Kim, J. G., Jeong, S.-T., Masuda, K., & An, H. S. (2003). The origin of chondrocytes in the nucleus pulposus and histologic findings. *Spine (Philadelphia, PA: 1986)*, *28*(10), 982−990. Available from https://journals.lww.com/spinejournal/FullText/2003/05150/The_Origin_of_Chondrocytes_in_the_Nucleus_Pulposus.5.aspx?casa_token = VgK84pDc-oIAAAAA:UZeNC3-lEYNXkAhv66l8VDhHP meDBQp5-Chv6CzkjRzx5FYQIk2IIzTkbNtPw5XE7lqsIbu3ZdkFk0ufK7MJU2wjv_oxehw.

Urban, J. P. G., Smith, S., & Fairbank, J. C. T. (2004). Nutrition of the intervertebral disc. *Spine (Philadelphia, PA: 1986)*, *29*(23), 2700−2709. Available from https://doi.org/10.1097/01.brs.0000146499.97948.52.

PART 2

HUMAN JOINTS BIOMECHANICS

BIOMECHANICS OF THE HIP JOINT 12

Fabio Galbusera[1] and Bernardo Innocenti[2]

[1]*IRCCS Istituto Ortopedico Galeazzi, Milan, Italy* [2]*Department of Bio Electro and Mechanical Systems (BEAMS), École Polytechnique de Bruxelles, Université Libre de Bruxelles, Brussels, Belgium*

SKELETAL ANATOMY

The hip joint links the trunk with the lower limbs by providing a highly mobile connection between the pelvis and the femur. Due to this critical location, the hip joint is subjected to high loads associated with both the weight of the upper body and the forces transmitted through the lower limbs from the ground. The joint anatomy, therefore, fulfils two cardinal requirements: high mobility in all directions and high mechanical strength.

The hip joint corresponds perfectly with the definition of a "ball-and-socket" joint, with the femoral head acting as the ball and the acetabulum as the socket (Fig. 12.1). The femoral head has an almost perfect spherical shape with a slightly smaller size in the anteroposterior plane with respect to the frontal one (Anderson & Blake, 1994; Rydell, 1973), and is connected with the diaphysis of the bone through the femoral neck. The neck does not have a perfectly regular shape, but its circular cross-section at the junction with the femoral head progressively becomes oval, with a smaller size in the anteroposterior plane, proceeding toward the shaft (Rydell, 1973). The angle between the femoral shaft and neck shows a certain intersubject variability, with an average of 125 degrees and a standard deviation of 5 degrees (Byrne, Mulhall, & Baker, 2010); pathological cases exceeding this variability range, either on the lower or on the upper side, are defined as coxa-vara and coxa-valga, respectively. The neck—shaft angle does not remain constant throughout life. It shows a tendency to decrease through childhood followed by attaining stability in adulthood (O'Sullivan, Schégl, Varga, Than, & Vermes, 2020; Rydell, 1973). The portion of the femoral head that may come in contact with the acetabulum, typically 60%—70% of the spherical surface (Byrne et al., 2010), is covered by articular cartilage. A small uncovered area at the center of the sphere, the fovea capitis, provides for the femoral insertion of the so-called ligament of the head of the femur or ligamentum teres.

In addition to the neck—shaft angle described above, the femoral neck also exhibits a certain angle (twist) with respect to the shaft and the femoral condyles in the anteroposterior plane (Fig. 12.2). This angle shows a large degree of intersubject variability and is, on average, directed in the anterior direction (namely anteversion) by 15 degrees (Anderson & Blake, 1994). In contrast, a posteriorly directed femoral neck is defined as retroverted.

Human Orthopaedic Biomechanics. DOI: https://doi.org/10.1016/B978-0-12-824481-4.00013-5

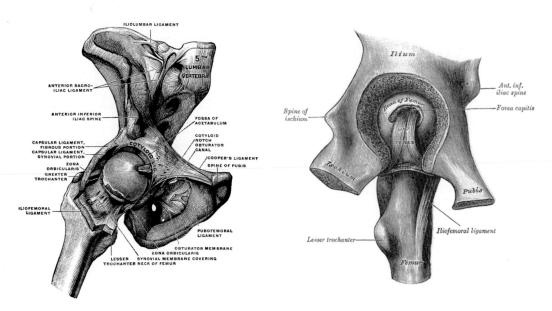

FIGURE 12.1

Section views of the hip joint. Left: the right hip joint from an anterolateral view. Right: the left hip joint view from the pelvis with the acetabular floor removed.

Images reproduced from Grey, H. (1918). Anatomy of the human body, *Lea and Febiger, Philadelphia, PA (public domain).*

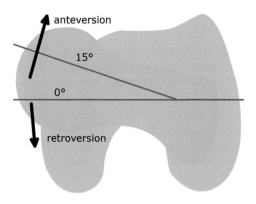

FIGURE 12.2

The anteversion/retroversion angle of the femoral neck.

Anteversion is associated with a biomechanical advantage, that is, it increases the lever arm of muscles involved in hip motion such as the gluteus maximus and piriformis. The anteversion/retroversion angle has also implications in determining the gait pattern of the subject, especially on the rotation of the leg during the swing phase. However, with increase in anteversion angle, the femoral head is twisted anteriorly, increasing the amount of anterior articular surface exposure, and thus predisposing to anterior dislocation. Then, the subject will walk with toe-in gait to restore stability with the assistance of a decreased abductor muscle moment arm, increased demand on hip abductors, and hence an increased joint reaction force; on the other hand, with a retroversion, the femoral head is twisted posteriorly, decreasing the amount of anterior articular surface exposure, and the subject will walk with toe-out gait to restore mobility with the assistance of an increased abductor muscle moment arm, decreased demand on hip abductors, and hence a decreased joint reaction force. On the pelvic side, the femoral head articulates with the acetabulum, a concave surface with a conforming shape. The acetabulum is not formed by a single bone but rather by three bones, the ischium, the ilium, and the pubis, joined in children by the triradiate cartilage, which is ossified when skeletal maturity is achieved. Similar to the femoral head, the anterior, superior, and posterior portions of the acetabulum are covered by articular cartilage, which assumes a crescent moon shape; the acetabular articular surface is, therefore, named a lunate surface. The inner part of the acetabulum not covered by cartilage, the central inferior acetabular fossa, provides for the acetabular insertion of the ligament of the head of the femur.

LIGAMENTS

The hip joint is a synovial joint, that is, its two cartilage-covered articular surfaces are surrounded by a fibrous capsule and a synovial membrane; the joint cavity is filled by synovial fluid secreted by the synovial membrane, which reduces the articular friction. The pelvic attachment of the joint capsule lies lateral to the acetabulum, while on the femur the capsule is attached to the intertrochanteric line anteriorly and on the femoral neck posteriorly, thus effectively covering the whole joint as well as a large part of the neck itself (Anderson & Blake, 1994), and sealing it by containing and preventing the outflow of the intraarticular fluid.

The stability of the hip joint is further enhanced by a strong ligamentous system formed by the iliofemoral, ischiofemoral, and pubofemoral ligaments (Fig. 12.3). The iliofemoral ligament links the anterior—inferior iliac spine with a wide area on the anterior proximal femur in correspondence with the intertrochanteric line. The ligament becomes tensioned in extension and effectively limits the range of motion of the joint while maintaining it under compression; this mechanism is also active during standing and limits the involvement of the hip extensor muscles in achieving equilibrium, thus reducing energy expenditure and fatigue in the bipedal stance (Anderson & Blake, 1994; Johnston, 1973). The ischiofemoral ligament is attached to the anterior aspect of the ischium, and wraps around the joint with its femoral attachment at the posterior intertrochanteric line, limiting the internal rotation in flexion and extension (Martin et al., 2008). The pubofemoral ligament covers the medio-inferior portion of the capsule, and contributes with the iliofemoral ligament to providing stability in extension (Anderson & Blake, 1994), especially controlling external rotation (Martin et al., 2008).

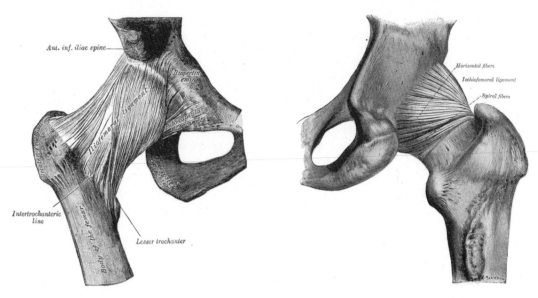

FIGURE 12.3

The ligaments of the hip joint: anterior (left) and posterior (right) views.

Images reproduced from Grey, H. (1918). Anatomy of the human body, *Lea and Febiger, Philadelphia, PA (public domain).*

FEMORAL AXIS

To describe the lower leg kinematics, it is important to define proper anatomical reference axes. Such axes are also necessary in establishing the angular alignment of the lower limb and rotational deformation of individual bones. The most important among them are as follows (Fig. 12.4):

- femoral mechanical axis;
- femoral anatomical axis.

In general, the mechanical axis of a bone is defined as the straight line connecting the joint centers of the proximal and distal joints. It is the axis of the joint force. In the femur, it is the mechanical axis from the center of the femoral head to the center of the distal femur (sometimes the tibial spine).

The anatomical axis is defined as the axis of the bone shaft. In the femur, the anatomical axis is the axis of the femoral shaft determined by the line from the center of the distal femur to the center of the proximal femur. It could be identified by joining the middle points of the two mid-diaphyseal sections.

The mechanical axis is different from the anatomical axis; the angle between the two axes is called the deviation angle. Among a Western population of normal healthy adults, the average deviation angle has been measured at 5.8 degrees (standard deviation 1.9 degrees) (Hsu, Himeno, Coventry, & Chao, 1990).

The mechanical and anatomical axes also exist for the tibia and for the limb. In well-aligned limbs, the mechanical axes of the femur, the tibia, and the limb are aligned. Under pathological conditions (varus or valgus knee), we assist in the medialization or lateralization of the femoral and tibial axes.

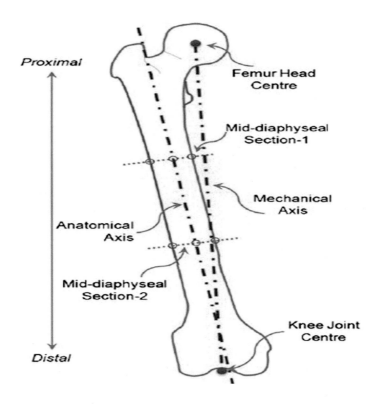

FIGURE 12.4

Mechanical and anatomical axes of the femur.

From Subburaj, K., Ravi, B., & Agarwal, M. (2010). Computer-aided methods for assessing lower limb deformities in orthopaedic surgery planning. Computerized Medical Imaging and Graphics: The Official Journal of the Computerized Medical Imaging Society, *34(4), 277–288. https://doi.org/10.1016/j.compmedimag.2009.11.00.*

FUNCTIONAL ANATOMY OF THE HIP MUSCLES

The complexity of the muscular system acting on the hip joint reflects its broad mobility in all planes, which is actively controlled by individual muscles and stabilized in specific directions (Fig. 12.5). The group most responsible for flexion is the iliopsoas, which consists of the psoas major and the iliacus. The psoas originates on the lateral aspects of the vertebral bodies from T12 to L3, whereas the iliacus originates from the iliac fossa in the pelvis; the two muscles join across the hip joint and insert in the lesser trochanter of the femur. Other muscles that contribute to hip flexion include the sartorius, which is also involved in external rotation and abduction, and rectus femoris (Byrne et al., 2010).

Hip extension is governed in large part by the gluteus maximus, which is the most significant of the gluteal muscles. While being connected to numerous anatomical structures, it mostly originates from the ilium and the lateral aspects of the lower sacrum and coccyx, and it inserts in the femur in correspondence with the linea aspera as well as the fascia lata, a layer of fibrous connective tissue

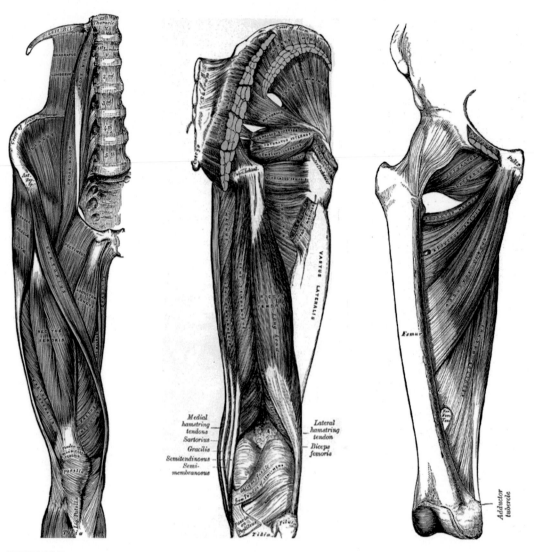

FIGURE 12.5

Muscles of the thigh and hip joint. From left to right: anterior view, posterior view, and anterior view of the deep muscles.

Images reproduced from Grey, H. (1918). Anatomy of the human body, Lea and Febiger, Philadelphia, PA (public domain).

enclosing the thigh. The large size of the human gluteus maximus is unique among mammals, and it is associated with the bipedal posture, and, in particular, with the large forces necessary for standing up from a stooped position. Other muscles such as biceps femoris, semitendinosus, and semimembranosus are also involved in hip extension.

The other gluteal muscles, namely gluteus medius and minimus, are the main abductors of the hip (Anderson & Blake, 1994; Byrne et al., 2010). The gluteus medius originates from the ilium and inserts in the lateral surface of the greater trochanter, and has therefore an oblique line of action effective in hip abduction. The gluteus minimus is located deeper into it and has relatively similar anatomy and function. Both muscles are also important in stabilizing the hip during mono-podial stance and gait. The piriformis, originating in the lateral aspect of the sacrum and inserting in the greater trochanter, is also involved in hip abduction. Adduction is governed and controlled by several muscles, including obturatur externus, adductor longus and brevis, adductor magnus and brevis, gracilis, and pectineus. Most of these muscles are also involved in hip rotation.

LOADS AND STRESSES

As mentioned above, the weight of the upper body and the dynamic loads associated with gait and other daily activities result in high loads acting in the hip joints. A classical way of estimating the joint loads under static conditions, such as in double- or single-leg stance, is by using a free-body diagram and equilibrium equations (Fig. 12.6), which also allow studying in a simple manner the effect of biomechanical variables such as the neck–shaft angle. More sophisticated analyses based

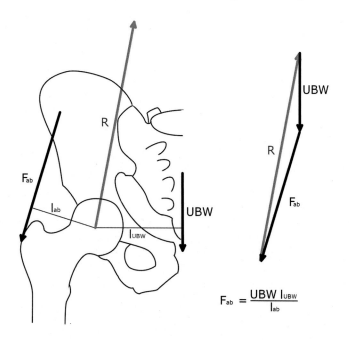

$$F_{ab} = \frac{UBW \; l_{UBW}}{l_{ab}}$$

FIGURE 12.6

The free-body diagram for the estimation of the hip reaction force (R) in single-leg stance. *UBW*, weight of the upper body; F_{ab}, force of the abductors; l_{ab}, moment arm of the abductor force; and l_{UBW}, moment arm of the upper body weight. The force of the abductors is calculated by means of moment equilibrium.

HIP CONTACT FORCE (% BW)

FIGURE 12.7

Peak forces at the hip joint with respect to the body weight (*BW*), measured with instrumented implants during various activities: slow walking, normal walking, fast walking, walking upstairs, walking downstairs, standing up, and sitting down.

on instrumented implants allowed for a detailed study of the loads acting in the joint during various activities such as standing, walking normally as well as with various aids, stumbling, carrying loads, and physiotherapy exercises (Fig. 12.7) (Bergmann et al., 2001). Several studies reported joint forces in the range of two to three times the body weight during normal walking (Bergmann et al., 2001; Davy et al., 1988; English & Kilvington, 1979), with large variability among the tested patients and in general agreement with the results provided by the analytical modeling studies. High values up to 550% of the body weight were recorded during running, and up to 870% during stumbling events (Bergmann, Graichen, & Rohlmann, 1993). It should be noted that such results were measured after total replacement of the hip joint with instrumented implants, after surgical resection of ligaments and joint capsule, which may affect the resulting measured loads. Nevertheless, the studies were able to cast light on the in-vivo load in the hip even during activities with complex dynamics, which would be difficult to replicate with mathematical models, for example, stumbling or specific rehabilitation exercises.

The large loads acting in the joint determine high stresses in the anatomical structures involved, which leads to adaptation of microarchitectures and properties towards fulfilling the challenging mechanical requirements to which they are subjected. An example is the proximal femur, which is subjected to high compression forces as well as bending moments due to the lever arm of the hip joint load with respect to the femoral shaft (Fig. 12.6). Due to the effect of the joint load, in the standing posture, the femoral neck is mostly loaded in compression, with its superior aspect in tension (Fig. 12.8). The stress distribution changes, proceeding in the distal direction; in the proximal shaft, the lateral aspect is tensile, whereas the medial side is compressed (Rudman, Aspden, &

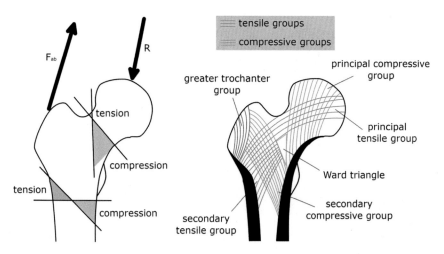

FIGURE 12.8

Schematic representation of the bending moments (left) and of the principal directions of the trabeculae, that is, the trabecular pattern (right) in the proximal femur, under the action of the abductor force (F_{ab}) and hip contact force (R). Areas under tension are depicted in *blue*, and those under compression in *red*.

Meakin, 2006). Based on Wolff's principle according to which bone density increases in correspondence with higher strains, the bony architecture of the proximal femur reflects the stress distribution in it (Fig. 12.8). Its trabecular structure is organized in groups, namely tensile and compressive groups, based on the type of stress to which they are subjected. The principal tensile group is arc-shaped and runs from the greater trochanter to the region below the fovea, being connected with the cortical layer in the superior aspect of the neck; the secondary tensile group follows it distally. The principal compressive group has a vertical orientation and reinforces the femoral head, running from its medial cortex to the femoral neck. The secondary compressive group reinforces the area between the lesser trochanter and the greater trochanter. In between these trabecular groups, the Ward triangle is an area subjected to small stresses and has, consequently, lower reinforcement and mechanical strength; it appears radiolucent in radiographic images and is frequently involved in fragility fractures (Nardi, Ventura, Rossini, & Ramazzina, 2010).

HIP CARTILAGE AND OSTEOARTHRITIS

The articular cartilage of the hip frequently degenerates with aging, resulting in pain and disability commonly requiring medical treatment, and has thus been investigated in depth in numerous studies. Degeneration is characterized by thinning or complete disappearance of the cartilage layer in specific sites or, eventually, in the whole joint, possibly in combination with alterations of the subchondral bone. Hip cartilage degeneration, commonly known as hip osteoarthritis in the medical literature, has been shown to be associated with aging, genetic predisposition, obesity, metabolic disorders, and traumatic events (Houard, Goldring, & Berenbaum, 2013). Indeed, recent studies

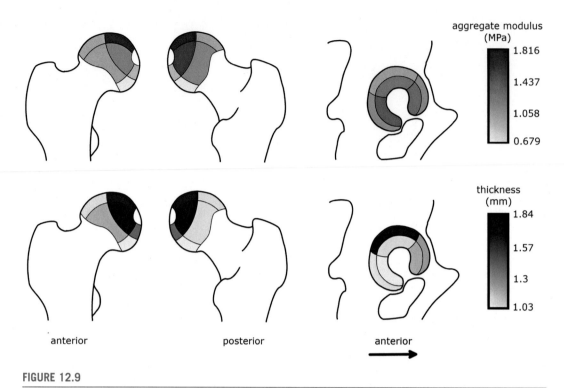

aggregate modulus
(MPa)
1.816
1.437
1.058
0.679

thickness
(mm)
1.84
1.57
1.3
1.03

anterior   posterior   anterior

FIGURE 12.9

Aggregate modulus (top) and thickness (bottom) of the articular cartilage on the femoral head and in the acetabulum.

have shown that osteoarthritis is a complex inflammatory disease that goes well beyond the mere thinning and degeneration of articular cartilage, but also involves bone and synovium and has metabolic and immunological aspects in addition to the purely mechanical, "wear-and-tear" mechanism (Berenbaum, 2013). Nevertheless, the classical hypothesis that cartilage degeneration occurs in locations subjected to high stresses (Afoke, Byers, & Hutton, 1987; Moskowitz, 1987) is well confirmed by literature studies; although this disorder is a subject of investigation from a multifactorial point of view, the association between high local stresses and cartilage loss has been widely demonstrated and remains a critical issue in osteoarthritis research.

Since stress values depend on the acting loads as well as the size, architecture, and material properties of the cartilaginous tissue, their estimation is not trivial; several authors, therefore, studied the local properties and thickness of hip cartilage by means of techniques such as indentation and creep tests on cadaver specimens (Fig. 12.9). Interestingly, experimental results revealed that the properties show a high spatial variability and that early degeneration occurs frequently in sites with significant differences between the cartilage on the femoral and acetabular sides (Athanasiou, Agarwal, & Dzida, 1994). In general, the superior aspect of the femoral head had the highest cartilage thickness in healthy specimens, whereas cartilage was thinner on its inferior side. In the acetabulum, the highest thickness was recorded in the superolateral region, with the other regions showing relatively similar values. The aggregate modulus shows distribution patterns similar to those of thickness on the femoral head, but significantly differs in the acetabulum; here, the medial region showed higher mechanical properties than the lateral one (Fig. 12.9). Differences were also found in terms of permeability and creep-relaxation curves, both between the femoral head and the acetabulum as well as among the different regions (Athanasiou et al., 1994).

A few regions come in contact occasionally during common postures and daily activities and thus not frequently subjected to high loads, such as those on the inferior aspect of the femoral head, which commonly undergo early degeneration (Bullough, Goodfellow, & O'Conner, 1973). This effect has been explained based on the fact that cartilage homeostasis, that is, the dynamic equilibrium between damage and repair mechanisms, requires contact between the two sides for its physiological functioning, whereas regions undergoing habitual contact are able to maintain their mechanical properties over time, and those in occasional contact may progressively lose their characteristics and thus initiate the degenerative cascade.

THE ACETABULAR LABRUM

The bony margin of the acetabulum is covered by a robust fibrocartilaginous structure, the acetabular labrum or cotyloid ligament (Fig. 12.1), which serves in increasing the stability of the joint by enhancing the lock of the spherical surface in the socket while maintaining a certain degree of joint motion due to its nonbony, flexible nature (Hartigan et al., 2018). The biomechanics of the labrum has been investigated in depth since it frequently shows degenerative signs and tears due to pathologies such as femoroacetabular impingement or dysplasia, being therefore associated with pain and disability (Bsat, Frei, & Beaulé, 2016; Philippon, Martin, & Kelly, 2005).

The labrum has a triangular cross-section with sizes ranging from $1-2$ mm to $5-6$ mm, with high regional variability (Fig. 12.10). Its composition and microarchitecture also show spatial differences; in the area in contact with cartilage, the labrum is composed of fibrocartilage, whereas in correspondence with the attachment with the acetabular bone, dense connective tissue replaces the fibrous structure (Bsat et al., 2016). The composition determines a high degree of resistance to fluid flow, giving the labrum the capability of effectively sealing the hip joint by preventing fluid loss even in highly pressurized scenarios (Ferguson, Bryant, Ganz, & Ito, 2003).

The mechanical properties of the acetabular labrum have been extensively investigated with the aim of gaining a deeper understanding of its function and the pathomechanisms of tearing and degeneration. Results showed a marked anisotropy, with relatively high tensile stiffness and strength (Ishiko, Naito, & Moriyama, 2005). Experimental tests in which the acetabular labrum was resected revealed its insignificant function in supporting loads in the healthy hip joint, whereas it had a small but significant contribution (up to 11%) in dysplastic hips (Bsat et al., 2016). However, in general, the labrum appears to be subjected to relatively low strains under physiological conditions, with large safety margins with respect to failure strains (Bsat et al., 2016; Ishiko et al., 2005).

As mentioned above, the labrum has the capability of preventing the outflow of articular fluid from the hip joint, therefore allowing for the development of high pressures, which contributes to an even distribution of the stresses inside the joint. In vitro tests and finite-element models confirmed that the sealing capability of an intact labrum has a marked protective effect on the

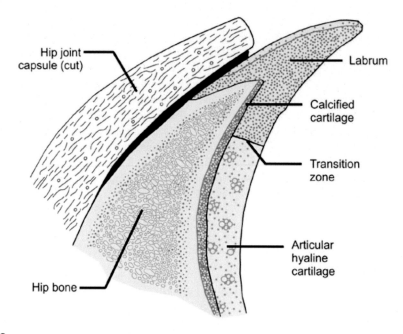

FIGURE 12.10

Section view of the acetabular labrum.

From Lewis, C. L., & Sahrmann, S. A. (2006). Acetabular labral tears. Physical Therapy, *86(1), 110–121.*

cartilage, as it limits its consolidation rate and consequently facilitates an effective lubrication of the joint (Ferguson et al., 2003).

The acetabular labrum has also an evident effect in the stability of the hip; its sealing function is indeed able to enhance the vacuum effect observable between the femoral head and the acetabulum, which prevents dislocation. This function was demonstrated by measuring the force necessary to distract the hip in cadaveric specimens; the presence of a labral tear determined a significant reduction of the distraction loads, especially for small displacements, showing that the labrum is the principal stabilizer of the hip in tension, even more than the joint capsule (Nepple et al., 2014). The same study also demonstrated that labrum reconstruction techniques are effective in restoring the stability of the joint, further confirming that a correct management of the labrum and its disorders has a cardinal importance in degenerative hip surgery.

FRACTURE OF THE FEMORAL NECK

The high loads to which the femoral neck is subjected may result in its fracture, relatively common in elderly subjects with compromised bone quality, that is, osteopenia and osteoporosis, but also occur in young subjects due to traumatic events. The initiation of the fracture is associated with high local strains in the bone, which may undergo microcracking of even catastrophic, immediate failure if the strains exceed the resistance limits of the bone tissue. Whereas the femoral neck is subjected to relatively high strains during daily activities, in the range of 500–2000 microstrains, that is, up to 0.2%, in healthy individuals (Augat, Bliven, & Hackl, 2019; Kersh, Martelli, Zebaze, Seeman, & Pandy, 2018), such values are in the physiological range of bone homeostasis and are therefore beneficial to the tissue (Frost, 1998). However, the same type of loads may result in much higher strains if the bone stiffness and strength are compromised. Osteopenia and osteoporosis indeed affect the quality and material properties of bone, possibly resulting in fractures after minor traumatic events; in the case of severe osteoporosis or in the presence of bone lesions such as metastases, fractures may even occur spontaneously (Cotton, Whitehead, Vyas, Cooper, & Patterson, 1994).

As mentioned above, the physiological loads acting on the hip joint are almost vertical and normally result in a bending moment acting in the femoral neck, with its inferior aspect loaded in compression and the superior surface in tension (Fig. 12.8). Although the vast majority of the volume of the neck is compressed, the presence of an area in tension is most critical in terms of risk of fracture due to the lower strength of cortical bone under tensile loads; besides, since this region is not highly loaded during daily activities, its cortical layer tends to be thinner than in other portions of the neck, especially in older subjects (Mayhew et al., 2005). Indeed, spontaneous fractures occurring under normal physiological loads commonly involve or originate from the superior aspect of the femoral neck (Augat et al., 2019). A biomechanical study showed that, under a standardized loading scenario that replicates physiological loading, the majority of the fractures initiated from the superior region of the proximal neck, with some degree of intersubject variability (Fig. 12.11).

In young subjects with good bone quality, fractures due to physiological, vertical loading are infrequent; other fracture mechanisms associated with different directions of the loads may, however, occur. Sideways falls entail high strains in the greater trochanter and commonly induce fractures of the neck initiating in its superolateral aspect (de Bakker et al., 2009), which are facilitated

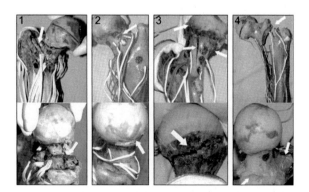

FIGURE 12.11

Some fracture mechanisms observed in an experimental study in which physiological loading was simulated, with the aim of investigating spontaneous fractures. The initiation sites, commonly located on the proximal and superior regions of the femoral neck, are indicated by the *arrows*.

From Cristofolini, L., Juszczyk, M., Martelli, S., Taddei, F., & Viceconti, M. (2007). In vitro replication of spontaneous fractures of the proximal human femur. Journal of Biomechanics, 40(13), 2837–2845.

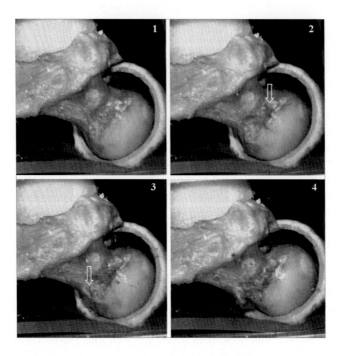

FIGURE 12.12

Fracture mechanism in sideways falls, reproduced in an experimental study. (1) the intact specimen; (2) the fracture initiates in the superior region of the neck (*arrow*), under compression; (3) another crack (*arrow*) appears on the inferior aspect of the neck, loaded in tension; and (4) the fracture involves the whole cross-section of the neck.

From de Bakker, P. M., Manske, S. L., Ebacher, V., Oxland, T. R., Cripton, P. A., & Guy, P. (2009). During sideways falls proximal femur fractures initiate in the superolateral cortex: Evidence from high-speed video of simulated fractures. Journal of Biomechanics, 42(12), 1917–1925. https://doi.org/10.1016/j.jbiomech.2009.05.001.

by the low cortical thickness in this area. Studies showed that fracture risk is increased sixfold by falling sideways rather than forward or backward, and up to 30-fold if the hip directly impacts the ground (Robinovitch, Inkster, Maurer, & Warnick, 2003). This failure mode has been investigated in many experimental and numerical studies in which an impact on the greater trochanter has been simulated. By using high-speed video techniques, researchers demonstrated that the fracture indeed initiates in the superior region of the neck, which is loaded in compression during the impact, whereas the failure of the tensiled inferior region follows shortly afterwards (de Bakker et al., 2009) (Fig. 12.12). Studies based on high-resolution digital image correlation showed that strain concentrations under sideways impact are located in correspondence with large cortical pores, for example, blood vessels (Grassi et al., 2020), which are common on the superior surface of the neck and thus contribute to making it one of the weak links in the hip joint.

The patient-specific risk of undergoing fracture of the femoral neck can be estimated by assessing its bone quality by means of imaging techniques such as dual-energy X-ray absorptiometry (DXA), which uses two X-ray beams at different energy levels to quantitatively measure the local bone mineral density. This value is then used for the calculation of the T-score, that is, the number of standard deviations of the patient-specific bone density with respect to a young, healthy subject, and of the Z-score, which is calculated with respect to the average of age- and sex-matched individuals. T-scores below −2.5 indicate an osteoporotic condition, which is associated with high fracture risk (Kanis, 2002). More refined techniques, employing demographic and clinical information about the patient in addition to the bone mineral density measured with DXA, have been presented and have found wide clinical use. An example is the FRAX, a diagnostic tool developed by the University of Sheffield, which also takes into account information such as smoking, body mass index, and previous fractures, and is able to provide as output the 10-year probability of hip fracture as well as of other common sites of osteoporotic fractures (spine, wrist, shoulder, forearm) (Kanis et al., 2008).

REFERENCES

Afoke, N. Y., Byers, P. D., & Hutton, W. C. (1987). Contact pressures in the human hip joint. *The Journal of Bone and Joint Surgery. British Volume*, *69*(4), 536−541. Available from https://doi.org/10.1302/0301-620X.69B4.3611154.

Anderson, L. C., & Blake, D. J. (1994). The anatomy and biomechanics of the hip joint. *Journal of Back and Musculoskeletal Rehabilitation*, *4*(3), 145−153. Available from https://doi.org/10.3233/BMR-1994-4305.

Athanasiou, K. A., Agarwal, A., & Dzida, F. J. (1994). Comparative study of the intrinsic mechanical properties of the human acetabular and femoral head cartilage. *Journal of Orthopaedic Research: Official Publication of the Orthopaedic Research Society*, *12*(3), 340−349.

Augat, P., Bliven, E., & Hackl, S. (2019). Biomechanics of femoral neck fractures and implications for fixation. *Journal of Orthopaedic Trauma*, *33*(Suppl. 1), S27−S32. Available from https://doi.org/10.1097/BOT.0000000000001365.

Berenbaum, F. (2013). Osteoarthritis as an inflammatory disease (osteoarthritis is not osteoarthrosis!). *Osteoarthritis and Cartilage/OARS, Osteoarthritis Research Society*, *21*(1), 16−21. Available from https://doi.org/10.1016/j.joca.2012.11.012.

Bergmann, G., Deuretzbacher, G., Heller, M., Graichen, F., Rohlmann, A., Strauss, J., ... Duda, G. N. (2001). Hip contact forces and gait patterns from routine activities. *Journal of Biomechanics*, *34*(7), 859−871.

Bergmann, G., Graichen, F., & Rohlmann, A. (1993). Hip joint loading during walking and running, measured in two patients. *Journal of Biomechanics*, *26*(8), 969−990. Available from https://doi.org/10.1016/0021-9290(93)90058-m.

Bsat, S., Frei, H., & Beaulé, P. E. (2016). The acetabular labrum: A review of its function. *Bone & Joint Journal*, *98-B*(6), 730−735. Available from https://doi.org/10.1302/0301-620X.98B6.37099.

Bullough, P., Goodfellow, J., & O'Conner, J. (1973). The relationship between degenerative changes and load-bearing in the human hip. *The Journal of Bone and Joint Surgery. British Volume*, *55*(4), 746−758. Available from https://pubmed.ncbi.nlm.nih.gov/4766179.

Byrne, D. P., Mulhall, K. J., & Baker, J. F. (2010). Anatomy & biomechanics of the hip. *The Open Sports Medicine Journal*, *4*(1), 51−57. Available from https://benthamopen.com/contents/pdf/TOSMJ/TOSMJ-4-51.pdf.

Cotton, D. W., Whitehead, C. L., Vyas, S., Cooper, C., & Patterson, E. A. (1994). Are hip fractures caused by falling and breaking or breaking and falling? *Forensic Science International*, *65*(2), 105−112. Available from https://doi.org/10.1016/0379-0738(94)90265-8.

Davy, D. T., Kotzar, G. M., Brown, R. H., Heiple, K. G., Goldberg, V. M., Heiple, K. G., Jr, ... Burstein, A. H. (1988). Telemetric force measurements across the hip after total arthroplasty. *The Journal of Bone and Joint Surgery. American Volume*, *70*(1), 45−50. Available from https://europepmc.org/article/med/3335573.

de Bakker, P. M., Manske, S. L., Ebacher, V., Oxland, T. R., Cripton, P. A., & Guy, P. (2009). During sideways falls proximal femur fractures initiate in the superolateral cortex: Evidence from high-speed video of simulated fractures. *Journal of Biomechanics, 42*(12), 1917−1925. Available from https://doi.org/10.1016/j.jbiomech.2009.05.001.

English, T. A., & Kilvington, M. (1979). In vivo records of hip loads using a femoral implant with telemetric. *Journal of Biomedical Engineering, 1*(2), 111−115. Available from https://doi.org/10.1016/0141-5425(79)90066-9.

Ferguson, S. J., Bryant, J. T., Ganz, R., & Ito, K. (2003). An in vitro investigation of the acetabular labral seal in hip joint. *Journal of Biomechanics, 36*(2), 171−178. Available from https://doi.org/10.1016/s0021-9290(02)00365-2.

Frost, H. M. (1998). Changing concepts in skeletal physiology: Wolff's Law, the Mechanostat. *American Journal of Human Biology: The Official Journal of the Human Biology Council, 10*(5), 599−605, https://doi.org/10.1002/(SICI)1520-6300(1998)10:5 < 599::AID-AJHB6 > 3.0.CO;2-9.

Grassi, L., Kok, J., Gustafsson, A., Zheng, Y., Väänänen, S. P., Jurvelin, J. S., ... Isaksson, H. (2020). Elucidating failure mechanisms in human femurs during a fall to the side. *Journal of Biomechanics, 106*, 109826. Available from https://doi.org/10.1016/j.jbiomech.2020.109826.

Hartigan, D. E., Perets, I., Meghpara, M. B., Mohr, M. R., Close, M. R., Yuen, L. C., ... Domb, B. G. (2018). Biomechanics, anatomy, pathology, imaging and clinical evaluation of the acetabular labrum: Current concepts. *Journal of ISAKOS, 3*(3), 148−154. Available from https://doi.org/10.1136/jisakos-2017-000159.

Houard, X., Goldring, M. B., & Berenbaum, F. (2013). Homeostatic mechanisms in articular cartilage and role of inflammation in osteoarthritis. *Current Rheumatology Reports, 15*(11), 375. Available from https://doi.org/10.1007/s11926-013-0375-6.

Hsu, R. W., Himeno, S., Coventry, M. B., & Chao, E. Y. (1990). Normal axial alignment of the lower extremity and load-bearing. *Clinical Orthopaedics and Related Research, 255*, 215−227. Available from https://europepmc.org/article/med/2347155.

Ishiko, T., Naito, M., & Moriyama, S. (2005). Tensile properties of the human acetabular labrum-the first report. *Journal of Orthopaedic Research: Official Publication of the Orthopaedic Research Society, 23*(6), 1448−1453. Available from https://doi.org/10.1016/j.orthres.2004.08.025.1100230630.

Johnston, R. C. (1973). Mechanical considerations of the hip joint. *Archives of Surgery, 107*(3), 411−417. Available from https://doi.org/10.1001/archsurg.1973.01350210047015.

Kanis, J. A. (2002). Diagnosis of osteoporosis and assessment of fracture risk. *Lancet, 359*(9321), 1929−1936. Available from https://doi.org/10.1016/S0140-6736(02)08761-5.

Kanis, J. A., Johnell, O., Oden, A., Johansson, H., & McCloskey, E. (2008). FRAX and the assessment of fracture probability in men and women from the USA. Osteoporosis International. *19*(4), 385−397. Available from https://doi.org/10.1007/s00198-007-0543-5.

Kersh, M. E., Martelli, S., Zebaze, R., Seeman, E., & Pandy, M. G. (2018). Mechanical loading of the femoral neck in human locomotion. *Journal of Bone and Mineral Research: The Official Journal of the American Society for Bone and Mineral Research, 33*(11), 1999−2006. Available from https://doi.org/10.1002/jbmr.3529.

Martin, H. D., Savage, A., Braly, B. A., Palmer, I. J., Beall, D. P., & Kelly, B. (2008). The function of the hip capsular ligaments: A quantitative report. *Arthroscopy: The Journal of Arthroscopic & Related Surgery: Official Publication of the Arthroscopy Association of North America and the International Arthroscopy Association, 24*(2), 188−195. Available from https://doi.org/10.1016/j.arthro.2007.08.024.

Mayhew, P. M., Thomas, C. D., Clement, J. G., Loveridge, N., Beck, T. J., Bonfield, W., ... Reeve, J. (2005). Relation between age, femoral neck cortical stability, and hip fracture. *Lancet, 366*(9480), 129−135. Available from https://doi.org/10.1016/S0140-6736(05)66870-5.

Moskowitz, R. W. (1987). Primary osteoarthritis: Epidemiology, clinical aspects, and general. *The American Journal of Medicine, 83*(5A), 5−10. Available from https://doi.org/10.1016/0002-9343(87)90844-8.

Nardi, A., Ventura, L., Rossini, M., & Ramazzina, E. (2010). The importance of mechanics in the pathogenesis of fragility fractures of the femur and vertebrae. *Clinical Cases in Mineral and Bone Metabolism, 7*(2), 130−134. Available from https://pubmed.ncbi.nlm.nih.gov/22460018.

Nepple, J. J., Philippon, M. J., Campbell, K. J., Dornan, G. J., Jansson, K. S., LaPrade, R. F., ... Wijdicks, C. A. (2014). The hip fluid seal−Part II: The effect of an acetabular labral tear repair, resection, and reconstruction on hip stability to distraction. *Knee Surgery, Sports Traumatology, Arthroscopy: Official Journal of the ESSKA, 22*(4), 730−736. Available from https://doi.org/10.1007/s00167-014-2875-y.

O'Sullivan, I. R., Schégl, Á. T., Varga, P., Than, P., & Vermes, C. (2020). Femoral neck-shaft angle and bone age in 4- to 24-year-olds based on 1005 EOS three-dimensional reconstructions. *Journal of Pediatric Orthopaedics. Part B/European Paediatric Orthopaedic Society, Pediatric Orthopaedic Society of North America, 30(4), 337−345*. Available from https://doi.org/10.1097/BPB.0000000000000776.

Philippon, M. J., Martin, R. R., & Kelly, B. T. (2005). A classification system for labral tears of the hip. *Arthroscopy: The Journal of Arthroscopic & Related Surgery: Official Publication of the Arthroscopy Association of North America and the International Arthroscopy Association, 21*(Suppl.), e36.

Robinovitch, S. N., Inkster, L., Maurer, J., & Warnick, B. (2003). Strategies for avoiding hip impact during sideways falls. *Journal of Bone and Mineral Research: The Official Journal of the American Society for Bone and Mineral Research, 18*(7), 1267−1273. Available from https://doi.org/10.1359/jbmr.2003.18.7.1267.

Rudman, K. E., Aspden, R. M., & Meakin, J. R. (2006). Compression or tension? The stress distribution in the proximal femur. *Biomedical Engineering Online, 5*, 12. Available from https://doi.org/10.1186/1475-925X-5-12.

Rydell, N. (1973). Biomechanics of the hip-joint. *Clinical Orthopaedics and Related Research, 92*, 6−15. Available from https://journals.lww.com/corr/Citation/1973/05000/Biomechanics_of_the_Hip_Joint.3.aspx.

BIOMECHANICS OF THE KNEE JOINT

13

Bernardo Innocenti

Department of Bio Electro and Mechanical Systems (BEAMS), École Polytechnique de Bruxelles, Université Libre de Bruxelles, Brussels, Belgium

KNEE FUNCTIONAL ANATOMY

KNEE BONE

From a bony perspective, the knee joint is composed of four bones (Fig. 13.1): femur, tibia, patella, and fibula. While the first three bones interact directly in the knee joint by articulating their articular surfaces, the fibula has the primary function of being the distal attachment site of the lateral

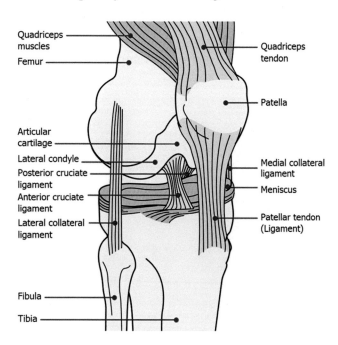

FIGURE 13.1

The knee joint.

Human Orthopaedic Biomechanics. DOI: https://doi.org/10.1016/B978-0-12-824481-4.00004-4

collateral ligament (LCL). The joint between the lateral tibial condyle and the proximal fibula (named the proximal tibiofibular joint) is a synovial plane type joint (arthrodial joint); however, this joint is usually ignored when dealing with the biomechanics of the knee joint.

From a morphological point of view, the femoral condyles could be approximated both by a cylindrical and by a conical surface (or each condyle by a certain spherical surface) (Fig. 13.2); the axis of the cylinder (transcylindrical axis), or the axis that joins the two centers of the sphere, defines the flexion axis, i.e. the axis that describes the flexion—extension movement of the tibio-femoral joint. Looking at the femoral condyles, the medial has an elliptical shape, is more distal than the lateral, has a circular posterior condyle, and has a large posterior offset; the lateral has a more circular (spherical) shape, is less rounded distally, and has a smaller posterior offset (Biscević, Hebibović, & Smrke, 2005). The trochlear groove, which guides the patella during tibial extension—flexion (Elias, Freeman, & Gokcay, 1990) is located in the distal anterior part of the femur. Because of the difference between condylar shapes, the flexion—extension axis is not purely in the medial—lateral direction, i.e., it is not coincident with the transepicondylar axis (TEA) defined by connecting the medial sulcus and lateral prominence but is also offset by a small internal—external rotation, so that in full extension the lateral condylar contact point is usually more anterior than the medial one (Yin et al., 2015). The femoral sulcus and condyles are covered by articular cartilage, which allows proper interaction with the posterior surface of the patella and tibial condyles, respectively.

The proximal tibia (Fig. 13.3) is composed of two plateaus (medial and lateral) separated by the intercondylar eminence, which includes the medial and lateral tibial spines. Morphologically, both the tibial plateaus are concave in the medial—lateral direction; in the anteroposterior direction, the medial tibial plateau is concave, whereas the lateral plateau is convex, increasing lateral mobility compared to the medial (Hashemi et al., 2008). Because of the greater medial force induced by the medial position

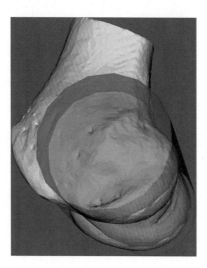

FIGURE 13.2

The cylindrical profile of the condyles.

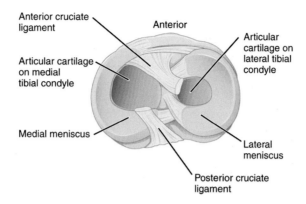

FIGURE 13.3

Anatomy of the proximal tibia.

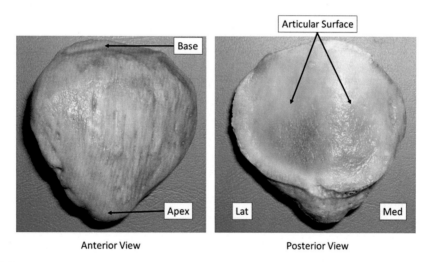

Anterior View   Posterior View

FIGURE 13.4

Patellar bone: anterior (left) and posterior (right) views.

of the femoral head, the medial plateau is larger than the lateral plateau (having up to 50% more surface area), and its articular surface is three times thicker (Yang et al., 2014; Zhang et al., 2018). While the femoral mechanical and anatomical axes are different, they coincide in the tibia. When viewed from the side, the tibia is not orthogonal to the femur, but is characterized by a small posterior slope (descending from anterior to posterior) of around 5.9 and 7.0 degrees for the medial and lateral slopes in female (3.7 and 5.4 degrees in male), respectively. This posterior slope facilitates movement of the lateral side from anterior to posterior, preventing femoral dislocation (Hashemi et al., 2008).

The patella is a hard bone (Fig. 13.4) that articulates with the femur (thigh bone) and covers and protects the anterior articular surface of the knee joint. The patella is a sesamoid bone (completely embedded in the quadriceps tendon) that is approximately triangular in shape, with the apex of the patella pointing downward. The base forms the superior aspect of the bone and provides the attachment area for the quadriceps tendon, and the apex is the lowest part of the patella, and gives attachment to the patellar ligament (Loudon, 2016). The posterior surface, covered by articular cartilage, is divided into two parts: medial and lateral facets, which interact with the femur. The posterior surface of the patella can include up to seven facets, with three on the medial surface and four on the lateral surface. The primary functions of the patello-femoral joint are (Loudon, 2016) as follows:

- protect the distal aspect of the femur from trauma and the quadriceps from frictional wear;
- improve the esthetic appearance of the knee;
- improve the moment arm of the quadriceps;
- decrease the amount of anterior—posterior tibio-femoral shear stress in the knee joint.

SOFT TISSUE ENVELOPE

In addition to the presence of bones, the entire knee is surrounded by soft tissues, called the soft tissue envelope (Figs. 13.1 and 13.5). In terms of ligaments, for the tibio-femoral joint, the four main ligaments are the collaterals and the cruciates, which provide 85% of the restraining force to the anterior tibial displacement at 30 and 90 degrees of knee flexion (Ellison & Berg, 1985), respectively:

- anterior cruciate ligament (ACL);
- posterior cruciate ligament (PCL);
- medial collateral ligaments (MCLs), sometimes called tibial collateral ligaments;
- LCLs, sometimes called fibular collateral ligaments.

In terms of ligaments, the patello-femoral joint comprises the following:

- medial patellofemoral ligament (MPFL);
- lateral and medial retinaculum.

The cruciate ligaments (Figs. 13.1 and 13.3) are intracapsular ligaments (located within the joint), whereas the collateral ligaments are extrinsic ligaments as they are located on the side of the joint. The ACL and PCL are strong, rounded bands that extend from the head of the tibia to the intercondylar notch of the femur. They cross each other like the letter X (reason for the name cruciate). They are called anterior and posterior according to their insertion into the tibia (Fig. 13.3): the ACL attaches to the anterior aspect of the intercondylar area and the PCL attaches to the posterior aspect (Woo, Wu, Dede, Vercillo, & Noorani, 2006). The ACL is most prone to injury out of the four ligaments located in the knee (Siegel, Vandenakker-Albanese, & Siegel, 2012). The cruciate fiber bundles do not function as a simple band of fibers with a constant tension; in fact, they exhibit a different pattern of tension throughout the range of motion. From a functional point of view, the cruciate ligaments guide tibio-femoral kinematics and are responsible for the antero-posterior stability of the joint, as they limit the antero-posterior displacement of the tibia relative to the femur; specifically, the ACL limits the tibia from excessive anterior translation, while the PCL limits posterior translation of the tibia (Woo et al., 2006). In addition, they have sensory endings, which are

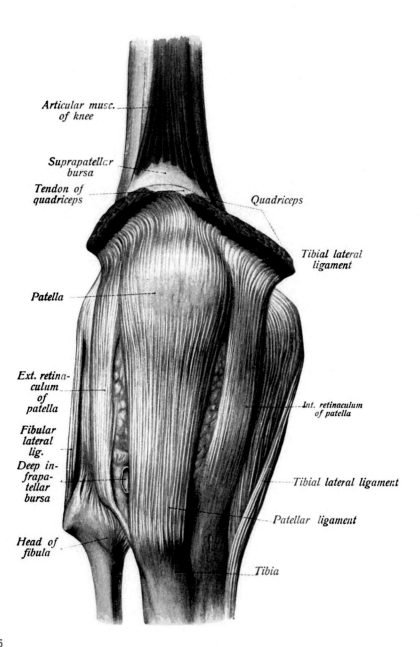

FIGURE 13.5

The soft tissue envelope of the knee joint.

involved in the proprioceptive function of the joint; in fact, an ACL injury not only causes mechanical instability but also leads to a functional deficit in the form of decreased proprioception of the knee joint (Adachi et al., 2002).

Also, the collateral ligaments (Figs 13.1 and 13.5) are responsible for knee stability, primarily limiting the medial—lateral translation of the tibio-femoral joint and, together with the cruciate ligaments, guide knee kinematics under passive conditions (James, LaPrade, & LaPrade, 2015; Warren & Marshall, 1979). Comparing them, the LCL is narrower and less wide than the MCL, has a nearly circular cross-section (the MCL is rather flat), and, mechanically, is more flexible than its medial counterpart, and is, therefore, less susceptible to injury (Wilson, Deakin, Payne, Picard, & Wearing, 2012). The main function of the MCL is to resist varus forces. The MCL could be thought of as consisting of an anterior, middle, and posterior portion; the anterior fibers are stretched in flexion, whereas the posterior fibers are stretched in extension (James et al., 2015; Warren & Marshall, 1979).

The MPFL (Fig. 13.6) originates in the medial aspect of the patella and inserts in the space between the adductor tubercle and the medial femoral epicondyle. Its main function is to prevent lateral displacement of the patella due to the lateral component of the force induced by the quadriceps muscle (Q-angle). The MPFL is the primary stabilizer to lateral displacement of the patella providing approximately 50%—60% of restraining force (Felli et al., 2021).

The lateral (and medial) retinaculum is not a ligament but constituted by fibrous tissue that surrounds the lateral (medial) side of the patellar bone. The main function of the retinaculum is to support the patellar bone position with respect to the femur during the knee motion (Desio, Burks, & Bachus, 1998).

MENISCI

All synovial joints are characterized by the presence of a layer of articular cartilage covering the bony surfaces. The main functions of articular cartilage are to reduce the coefficient of friction between the two surfaces and to increase the contact area due to its elasticity. Since the contact pressure is equal to the ratio of the contact force to the contact area, the cartilage layer allows to reduce the contact pressure in the joint to physiological values, preserving its biological condition. In the knee, since the contact forces are, under certain conditions, very high, the mere presence of the cartilage is not sufficient to reduce the contact pressure to acceptable values. Therefore, it is necessary to add an additional element that increases the contact surface: the menisci (Fig. 13.3).

The menisci are two fibrocartilaginous structures, one placed on the medial and the other on the lateral tibial cartilage. The medial structure has a C shape, is larger and thicker than its lateral counterpart, and is rigidly fixed on the medial tibial cartilage. The lateral meniscus has a round shape, is smaller and thinner than its medial counterpart, and is constrained near the tibial attachments of the ACL and PCL and is more mobile to allow coverage of the lateral femoral condyle throughout the entire range of motion. In addition, because of the tibial shape, the medial meniscus is wider posteriorly than anteriorly (Fig. 13.7).

MUSCLES

The muscles of the knee include the quadriceps (vastus medialis, vastus lateralis, vastus intermedius, and rectus femoris) responsible for knee extension, the medial hamstrings

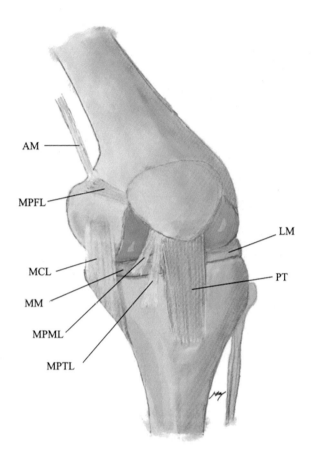

FIGURE 13.6

Schematic drawing of the main medial knee patello-femoral stabilizers. *AM*, adductor magnus; *LM*, lateral meniscus; *MCL*, medial collateral ligament; *MM*, medial meniscus; *MPFL*, medial patello-femoral ligament; *MPML*, medial patella-meniscal ligament; *MPTL*, medial patella-tibial ligament; *PT*, patellar tendon.

From Felli, L., Alessio-Mazzola, M., Lovisolo, S., Capello, A. G., Formica, M., & Maffulli, N. (2020). Anatomy and biomechanics of the medial patellotibial ligament: A systematic review. The Surgeon, 19(5):e168–e174. https://doi.org/10.1016/j.surge.2020.09.005.

(semimembranosus and semitendinosis), and the lateral hamstring (biceps femoris), responsible for the functioning of the knee. These muscles work in coordination to flex, extend, and stabilize the knee joint.

ADDITIONAL SOFT TISSUES

There is also fat in the knee, called infrapatellar fat pad or Hoffa's pad (Fig. 13.7), which is a cylindrical piece of fat situated underneath the patella to maintain proper tension in the ligament and keep it at a certain distance from the femoral bone.

(A)

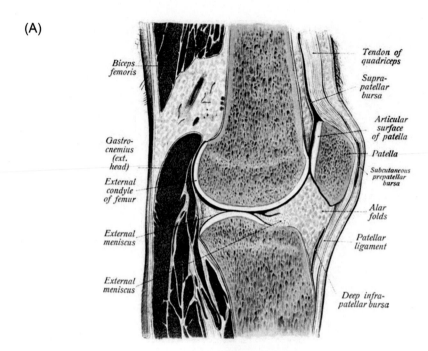

(B)

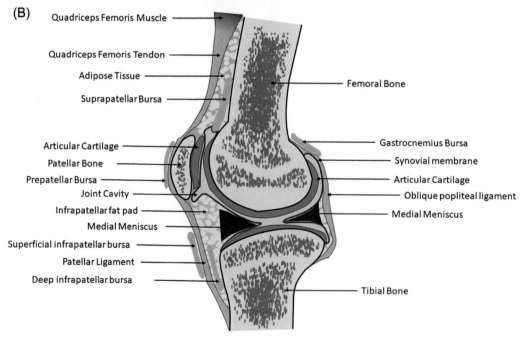

FIGURE 13.7

Sagittal section of right knee joint.

The knee joint is also surrounded by the knee articular capsule (Fig. 13.5), a dual-layered structure that is relatively thin anteriorly and posteriorly and thickened laterally by the collateral ligaments. The outer layer of the knee capsule consists of fibrous connective tissue holding the joint in place, whereas the inner layer consists of a synovial membrane, which secretes synovial fluid into the joint for proper joint lubrication.

THE TIBIO-FEMORAL JOINT: KINEMATICS AND KINETICS
GLOBAL RANGE OF MOTION

From a functional and biomechanical point of view, the complex kinematics of the knee, in terms of tibio-femoral kinematics, is described by a "six degrees of freedom" (6 DOFs) movement considering three rotations and three translations with respect to a series of perpendicular anatomical axes. Although the amplitude of the various movements can vary greatly from subject to subject due to anatomy, age, and training, we can consider the following average ranges of the different movements:

- flexion/extension: -15 to 140 degrees;
- varus/valgus rotations: -6 to 8 degrees;
- internal/external rotations: 25-30 degrees;
- antero-posterior translation: $10-15$ mm;
- medio-lateral translation: $2-4$ mm;
- superior/inferior translation: $2-5$ mm.

Measurement of knee kinematics could be performed using several methods, each characterized by its own pros and cons (e.g., in terms of accuracy, precision, and cost); in general, along with the in-silico study, analysis can be conducted in vivo (e.g., using three-dimensional [3D] motion analysis techniques or fluoroscopic acquisition) or in vitro (robotic simulators). Also, as reported for other joint movements, in vivo measurements are the most relevant, realistic, and interesting as they are directly related to the subject; however, such measurements are also the most complex to perform and usually less accurate than in vitro measurements. For example, looking at the different DOFs calculated, focusing on the knee joint, 3D motion analysis allows us, from the knowledge of the temporal position of the coordinates of the marker placed on the skin, to identify the position and orientation of the bone segments, and from the knowledge of the their relative orientation, the three angles of the knee are determined. In fluoroscopy, the position and orientation of the distal femur and proximal tibia are derived from the superimposition of the contours (calculated as 3D models) on the detected images, obtained from CT or MRI, and this information allows calculation of both the relative rotations and translations of the joint (with an accuracy that depends on the techniques used, i.e., single or dual fluoroscopy).

HISTORICAL KNEE KINEMATICS ANALYSIS: FROM ONE DEGREES OF FREEDOM TO SIX DEGREES OF FREEDOMS

As reported by Bull & Amis (1998) and Pinskerova, Maquet, & Freeman (2000), the motion of the knee joint has been described differently in the literature. Historically, the study of how the femur moves on the tibia began in 1836 with Weber & Weber (1836): they described the 2D motion on the medial side as "cradle-like." Since then, several methods have been used to examine the kinematics of the human knee, and this is still considered a "hot topic" by industry, clinicians, and biomechanical researchers.

The first radiological study was performed by Zuppinger (1904), who reported that the femur rolled back across the tibia during flexion due to the so-called four-bar rigid-link mechanism determined by the two cruciate ligaments. Frankel, Burstein, & Brooks (1971) presented the concept of the instantaneous center of rotation, while Menschik (1974) introduced the four-bar linkage model representing the cruciate ligaments in two dimensions. Grood & Suntay (1983) presented a joint coordinate system providing a geometric description of the 3D rotational and translational motion between two rigid bodies, applied to the knee joint. They introduced the concept of the "floating axis," describing the joint displacements independently of the order in which the component rotations and translations occur. Following this approach, the concept of the helical axis was defined (Blankevoort, Huiskes, & de Lange, 1990), opening the door to a correct scientific description of the kinematics of the knee. However, the complexity of such models was so high that it appeared too difficult to be applied in clinical practice, so the kinematics of the knee was reduced mainly to two rotations and no translations were, initially, considered.

Among the different models available in the literature (in increasing order of complexity: Bull & Amis, 1998; Bobbert, Yeadon, & Nigg, 1992; Ohkoshi, Yasuda, Kaneda, Wada, & Yamanaka, 1991), Bull & Amis has defined the knee joint simply as a 1-DOF hinged joint, considering flexion—extension as the only motion. Such a model has no predictive advantage for the analysis of knee joint structures, although its simplicity allows for its use in larger models, which are not limited to a single joint, e.g., whole body analysis or gait or crash test dummies.

Planar joints, in which the knee moves in parallel to the sagittal plane, with no fixed axis of rotation, have also been widely considered for the knee joint. Such models allow rolling (rotation) if the axis coincides with the contact point, and sliding (translation), either with or without rotation, if the rotation axis is above or below the joint surfaces (Frankel et al., 1971; Gerber & Matter, 1983).

A specific planar description of knee joint motion is known as the "four-bar linkage model" (Bradley, FitzPatrick, Daniel, Shercliff, & O'Connor, 1988; Menschik, 1974; Zavatsky & O'Connor, 1992a, 1992b) (Fig. 13.8), in which the motion is considered in the sagittal plane and, therefore, tibial rotation, medial-lateral translations, and abduction and adduction are neglected. If it is assumed that the cruciate ligaments can each be represented by a rigid connection ("bar") of constant length, then it can be shown that the instantaneous center of rotation must always coincide with the point where the cruciates cross. In this case, the knee is modeled as a mechanism—the path of relative motion is fixed—and so the position of the femur could be derived, relative to the tibia, from the knowledge of only one variable (usually the flexion angle). This model is quite popular because it could be used to explain the relative motion of the femur on the tibia during flexion, coupling flexion, and AP translation; however, it has the major limitation of considering the knee as symmetrical (no difference between the medial and lateral sides), and not considering any internal/external rotation.

Using the ball-and-socket 3-DOF spherical joint, knee motion is represented by three rotations about three general axes intersecting at the center of the joint (as in the hip joint). If the position of the joint center is known, the relative position can also be defined by the coordinates of two other noncollinear points. This model could be adopted in circumstances where joint translations are considered negligible compared to rotation. It has been used for knee in gait analysis (Kettelkamp,

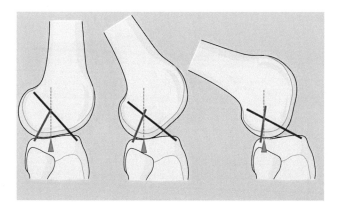

FIGURE 13.8

The four-bar linkage model.

Johnson, Smidt, Chao, & Walker, 1970; Kowalk, Duncan, & Vaughan, 1996), and to calculate muscle force and resultant motion of the lower limb (Apkarian, Naumann, & Cairns, 1989). A special case of such a model is the 2-DOF spherical joint in which only the flexion—extension and the tibial internal—external rotations are considered.

The 6-DOF spatial joint model has no limitations in terms of considering the motion between two bodies. Therefore, the relative motion between two bodies can be described completely as the translation and rotation of the two bodies relative to each other. Various methods have been used to describe motion using the 6-DOF spatial joint, including the one using Grood—Suntay coordinate systems and helical axis approaches in which relative motion is defined in terms of successive displacements rather than successive instantaneous positions (Townsend, Izak, & Jackson, 1977; Wismans, Veldpaus, Janssen, Huson, & Struben, 1980). The main limitation associated with the 6-DOF spatial joint model is its inability to perform spatial description and measurement of motion.

THE GROOD—SUNTAY COORDINATE SYSTEM

Grood & Suntay (1983) described an anatomical coordinate system that has become one of the most widely used coordinate systems to describe knee kinematics (Fig. 13.9). The goal is to ensure, using the anatomical coordinate system, that motion has functional significance for the knee. It is important to note that the Grood—Suntay coordinate system is not a fixed coordinate system: only two segments are fixed, one on the tibial bone and the other on the femoral bone, plus a third floating axis that is always orthogonal to the other two.

The idea of the two researchers was to describe the movement from a functional point of view. According to them:

- the internal—external rotation and the superior—inferior translation are described along the tibial mechanical axis, since the internal—external rotation depends mainly on the tibia and the tibia rotates on its mechanical/anatomical axis;

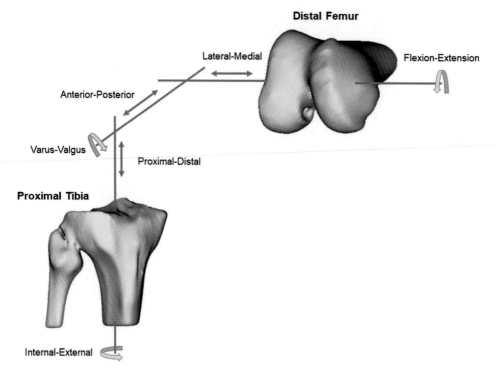

FIGURE 13.9

Joint coordinate system of Grood and Suntay.

From Naendrup, J.-H., Zlotnicki, J. P., Murphy, C. I., Patel, N. K., Debski, R. E., & Musahl, V. (2019). Influence of knee position and examiner-induced motion on the kinematics of the pivot shift. Journal of Experimental Orthopaedics, 6(1), 11. *https://doi.org/ 10.1186/s40634-019-0183-7.*

- flexion—extension rotation and medial—lateral translation are described along the femoral epicondylar axis (flexion/extension axis);
- varus—valgus rotation and antero-posterior translation are described along the third axis (fluctuating axis) obtained from the cross product of the previous tibial and femoral axes.

With this approach, knee motion is described using a functional anatomical coordinate system, with the femoral and tibial axes defined using bony markers and the major axis representing the functional axis. One of the most critical aspects of defining such a coordinate system is related to the accuracy in identifying the bony landmarks; depending on the techniques used, the error on the single knee landmark is larger in the case of standard MRI (average intraobserver variability on landmark up to 3 mm) and reduced in the case of CT imaging (average intraobserver variability on landmark usually less than 1 mm) (Innocenti, Salandra, Pascale, & Pianigiani, 2016; Victor et al., 2009). Another criticism is related to the approximation of the epicondylar axis as flexion/extension axis and not the transcylindrical axis.

MEDIO-LATERAL KNEE KINEMATICS MODEL: MEDIAL PIVOT AND ROLL-BACK KNEE MOTION

As previously reported, the four-bar mechanism used to describe the knee motion is characterized by several limitations. In an attempt to improve the knee model while still keeping the number of DOFs low, Iwaki, Pinskerova, & Freeman (2000) and Pinskerova et al. (2000) proposed an alternative method. The researchers discriminated the movement of the knee by splitting the kinematics analysis in the medial and lateral sides (avoiding the planar motion limitation of the previous model) (Fig. 13.10) and characterizing each condylar movement in the sagittal section by plotting, for different flexion angles, the anteroposterior position of the condylar center (Fig. 13.11). In such a way, it is possible to describe the AP translation of the knee as the average between the medial and lateral AP translations and the internal external rotation by the angle between two consecutive medio-lateral condylar point lines.

Using this model, the researchers, investigating native knee during passive motion (no external muscle load applied), were able to describe the medial pivoting knee motion and roll-back behavior of the knee. In particular, it was clear that:

- The medial condyle is characterized by a different kinematics from the lateral condyle.
- In full extension, the medial condylar center is centered in the medial tibial condyle (little bit posterior), while the lateral condylar point is more anterior.

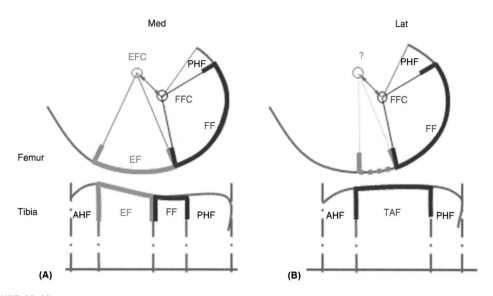

FIGURE 13.10

Sagittal sections through the center of the medial (A) and lateral (B) compartments. *EF*, extended facet, *FF*, flexion facet.

From Pinskerova, A., & Vavrik, P. (2020). Knee anatomy and biomechanics and its relevance to knee replacement. In C. Revière & C. Vendittoli (Eds), Personalized hip and knee joint replacement. Springer. https://doi.org/10.1007/978-3-030-24243-5_14.

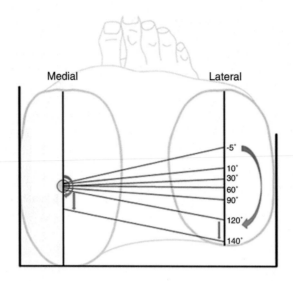

FIGURE 13.11

Kinematics plot of the medial and lateral femoral condylar centers plotted on the tibial plateaus for different flexion angles (−5 to 140 degrees).

From Pinskerova, A., & Vavrik, P. (2020). Knee anatomy and biomechanics and its relevance to knee replacement. In C. Revière & C. Vendittoli (Eds), Personalized hip and knee joint replacement. Springer. https://doi.org/10.1007/978-3-030-24243-5_14.

- At the beginning of the flexion, the medial condyle tries to keep its original position, limiting the AP translation, while the lateral condylar center immediately starts to move from anterior to posterior. This phase, in which the knee motion pivots (rotates) around the medial condyle, is described as "medial pivot knee" (also used for the design of knee arthroplasty): the femur rotates internally during flexion.
- At a higher angle of flexion, from 60 to 90 degrees, which varies from patient to patient, there is a reduction of the femoral internal rotation and both the medial and the lateral condyles start to translate posteriorly, which is called the "knee roll-back": the femur translates posteriorly during flexion.

The reason for this movement could be easily explained by recalling the morphology of the knee. As previously mentioned, the medial condylar side of the knee could easily be approximated as a sphere IN a sphere joint, while the lateral condylar side of the knee, due to the convex antero-posterior shape of the tibia, could be approximated as a sphere ON a sphere joint. Therefore, as soon as a small anteroposterior force is applied, the lateral part immediately starts to move, while the medial part remains stable in its position (medial-pivot); only at high flexion, the lateral part starts to move posteriorly with respect to the medial part, and then both parts move posteriorly (roll-back).

KNEE KINEMATICS IN ACTIVE CONDITIONS

The medial-pivot and roll-back knee motion was observed by different researchers, both in ex vivo and in in vivo knee joints, under passive/static or low-bearing conditions; therefore, it is possible to conclude that the normal kinematics pattern of the human knee, under such conditions, consists of the internal rotation of the tibia relative to the femur with increasing flexion, followed by a posterior translation of the femoral condyles. However, if we analyzed different motor tasks, under different load conditions, the medial-pivot and roll-back motion would disappear and several other patterns could be highlighted. Multicenter ex vivo cadaver studies (Innocenti et al., 2016; J Victor et al., 2009) and in vivo studies on normal volunteers (Desloovere et al., 2010; Scheys et al., 2013) clearly demonstrate a significant inter-individual and activity-dependent variability.

For example, using a robotic simulator, it was possible to verify how during loaded high flexion (squat activity), up to 120/130 degrees, the kinematics model also depends on the soft tissue envelope (Fig. 13.12). Moreover, the results also show that it would be possible to have an anterior translation of the femur (i.e., a paradoxical motion) if the hamstring was not loaded during the movement. Additionally, the presence of the hamstring (single or both lateral and medial) strongly influences the amount of rotation and translation of the tibia (Victor, Labey, Wong, Innocenti, & Bellemans, 2010).

TIBIO-FEMORAL KINETICS

Generally, the forces applied on the knee joint are quite high; for example, the loads transmitted during normal walking vary between two and three times body weight (D'Lima, Townsend, Arms, Morris, & Colwell, 2005) (Taylor & Walker, 2001). This is due partly to the kinetics of acceleration, the high moments generated at the knee, and the simultaneous contraction of multiple muscles (D'Lima, Fregly, Patil, Steklov, & Colwell, 2012). In addition, the presence of knee malalignment (e.g., varus/valgus knee) induces an overload of one of the compartments at the expense of another, which will also lead, over time, to a progression of osteoarthritis (Sharma et al., 1999, 2001).

Knee force measurement is quite complicated in vivo as it is extremely difficult to place load sensors within the joint in a patient. Using 2D or 3D numerical modeling, it is possible to calculate knee force; however, their reliability is highly dependent on several parameters such as surface accuracy, definition of active and passive soft tissues (ligaments and muscles), contact modeling, and handling the problem of muscle redundancy (having more muscles than mechanical DOFs).

An alternative to computational prediction is the direct measurement of knee forces using an instrumented knee prosthesis. Advances in smart implant technology and telemetry systems have made measurement possible in vivo in patients receiving total knee arthroplasty (TKA; Bergmann et al., 2007; D'Lima et al., 2005). Among several groups, one of the pioneers in this field was the group of Prof. Bergmann at Charité University Hospital in Berlin that developed an instrumented telemetric tibial tray that allows measurement of six load components in TKA (Bergmann et al., 2007; D'Lima et al., 2005). With this device, they were able to track force in different knee patients during different daily activities, as reported in their publicly available database at http://www.orthoload.com.

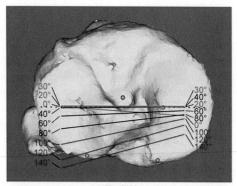

Passive

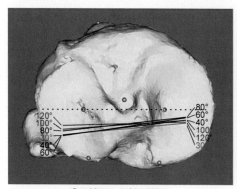

Quadriceps, no hamstrings

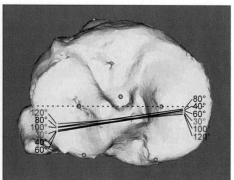

Quadriceps + medial hamstrings

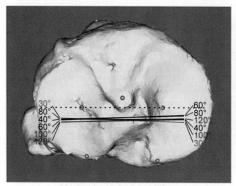

Quadriceps + lateral hamstrings

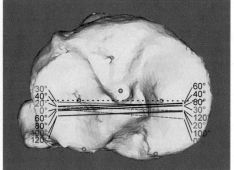

Quadriceps + both hamstrings

FIGURE 13.12

Projection of medial and lateral femoral condyle centers on the horizontal tibial plane for different load cases.

From Victor, J., Labey, L., Wong, P., Innocenti, B., & Bellemans, J. (2010). The influence of muscle load on tibiofemoral knee kinematics. Journal of Orthopaedic Research: Official Publication of the Orthopaedic Research Society, *28(4), 419–428. https://doi. org/10.1002/jor.21019.*

With similar devices, several studies have been performed and knee forces have been reported in vivo for a variety of activities, including recreation and exercise (D'Lima, Steklov, Patil, & Colwell, 2008). Walking on the treadmill at speeds up to 3 mph generated lower peak tibial forces than walking on the laboratory floor. Power walking on the treadmill (at 4 mph) generated higher peak tibial forces. Jogging is a high-impact activity that generated peak forces of 3.6 $\times$ body weight (BW) in two subjects implanted with instrumented distal femoral tumor replacement prostheses (Taylor & Walker, 2001).

THE PATELLO-FEMORAL JOINT
INTRODUCTION: THE PATELLAR FUNCTION

Understanding the biomechanics of the native physiologic patello-femoral joint is an essential prerequisite to investigating the knee biomechanics.

This patello-femoral appears to be an extremely complex joint, in which the forces are a function of both quadriceps tension and flexion angle. Furthermore, they depend on the distance of the joint from the body's barycenter, which justifies why different motor activities, even if characterized by similar tibio-femoral flexion extension angles, can induce extremely different patellar forces.

From a biomechanical point of view, the main function of the patella is to facilitate knee extension by increasing the efficiency of the quadriceps (Bellemans, 2003). This is achieved through the fulcrum function, exerted by the patella, which shifts the muscle line of action anteriorly. This effect produces an increase in the lever arm of the quadriceps muscle moment relative to the center of rotation of the knee. In addition, the patella optimizes the distribution of patello-femoral compressive forces by increasing the contact area during flexion (Aglietti, Insall, Walker, & Trent, 1975; Luyckx et al., 2009). The patella also acts as a guide for the extensor mechanism by centralizing the various traction forces of the different quadriceps heads and transmitting these forces to the patellar tendon (Schindler & Scott, 2011).

Finally, the anatomical shape of the patello-femoral joint provides protection to the extensor apparatus from dislocation (Amis, Senavongse, & Bull, 2006). Obviously, to properly perform these functions, the patella must have proper alignment and slide optimally during the flexion—extension movement of the knee.

PATELLAR KINEMATICS

The relative motion between patella and distal femur, and, consequently, patellar stability, is determined by four main factors: joint geometry, the "active" action of the muscle structures acting on the patella, the passive constraint exerted by the surrounding soft tissues, and the alignment of the lower limb (Amis et al., 2006). Looking at the motion, the patello-femoral joint is a sliding articulation; the patella moves 7 cm caudally during full flexion.

The full trajectory of the bone is, however, three dimensional; also, the AP translation is quite important as the patellar position during flexion allows a change in the AP direction of the patellar tendon force that pulls anteriorly the tibia at a low flexion angle and push posteriorly at a high flexion angle (Victor et al., 2010). The angle between the patellar tendon and the tibial mechanical axis in the sagittal plane was positive between 0 and 65 degrees of flexion and negative from 65 degrees to maximum flexion.

Physiologically, in the frontal plane, due to the Q angle existing between the line of action of the quadriceps muscle force and the patellar tendon, the quadriceps induces, on the patella, a lateral force. This effect is strongly related to the extensor mechanism stability as, during flexion, an increase in the Q angle corresponds to an increase in this lateral force and thus an increased risk of subluxation (Hehne, 1990; Schindler & Scott, 2011). The main antagonist to patellar lateral displacement is the femoral trochlear groove, particularly the depth and inclination of the lateral femoral condyle. When the knee reaches extended flexion, the vector force resulting from quadriceps tension and patellar tendon traction causes the patella to press in the sagittal plane, posteriorly in the trochlear groove, making it less vulnerable to lateral dislocation. On the other hand, this posterior force is more modest in extension, when, simultaneously, the lateral component increases as a result of external rotation of the tibia (screw-home mechanism) (Panni et al., 2011). A dysplastic trochlea, with a flattening of the lateral facet, will, therefore, be unable to withstand this lateral force exerted at low degrees of flexion. Wibeeg (1941) described different forms of patella, not all of which are perfectly articulated with the trochlear groove. In a physiologic patella, the lateral facet is larger than the medial facet, allowing approximately 60% of the total force to be transmitted. On both the patellar and trochlear sides, the articular cartilage and the underlying subchondral bone show different geometries (Shih, Bull, & Amis, 2004), with greater cartilage thickness proximal to the trochlear groove and in the proximal area of the patella. Therefore, care must be taken when evaluating trochlear dysplasia. However, studies in the literature report good correlations between the bone geometry and cartilage thickness in flattened trochleas (Dejour, Walch, Nove-Josserand, & Guier, 1994).

MUSCLE ACTION

Contraction of the quadriceps moves the patella proximally and laterally in extension and guides it within the trochlear groove as the flexion angle increases. Quadriceps muscle force converges on the patella because quadriceps fibers attach to the medial, lateral, and proximal edges of the patella. The orientation of the vastus medialis, as well as that of the vastus lateralis, is not parallel to the anatomical axis of the femur, and, therefore, these structures pull the patella in the medial and lateral directions, respectively. The resultant force vector of the various muscles is, therefore, nearly parallel to the anatomical axis in the frontal plane. However, an insufficiency of the vastus medialis oblique may compromise the force required to move the patella laterally, although this does not necessarily result in lateral subluxation in the absence of other pathologic factors (Senavongse & Amis, 2005). In addition, the vastus medialis oblique exerts an important action of patellar stabilization, as it pulls itself in a posterior direction.

JOINT ALIGNMENT

Patellar biomechanics is not a planar problem so the concept of joint alignment should be considered three-dimensionally and not limited exclusively to a single anatomical view. Any change in alignment from "normal," or "optimal," can produce alterations to the muscular action, and, therefore, to the forces and moments, acting on the patello-femoral joint. Generally, this alteration manifests itself in all three directions; in fact, it is obvious that any change in the sagittal or frontal plane will also affect patellar tracking. This explains why there is a correlation between rotational alignment and anterior patellar pain (Eckhoff, Brown, Kilcoyne, & Stamm, 1997). An increase in femoral anteversion produces femoral internal rotations and thus a distally intrarotated femur. The resulting induced internal rotation of the foot is offset by external tibial rotation, which then increases the effective Q angle. Rotational malalignment can be present at various levels, and when measuring Q angle, the entire alignment of the lower extremity should be considered for ensuring proper tracking, and possibly correction of patellar maltracking (Beaconsfield, Pintore, Maffulli, & Petri, 1994).

PATELLAR SOFT TISSUE ENVELOPE

The medial and lateral patellar retinaculum structures, together with the MPFL, play a key role in maintaining patellar stability (Fig. 13.6). In fact, they act as a static restraint against lateral forces that would induce subluxation of the patella from the trochlear groove into the transverse plane. The MPFL ligament is the primary passive ligamentous restraint of the patello-femoral joint, providing 50%−60% of the resistance to lateral displacement in the small flexion range (0−30 degrees) (Felli et al., 2021). Moreover, the lateral structures of the retinaculum are organized in three layers, in which the superficial layer fuses with the fibers of the ilio-tibial band (ITB). An increase in tension in the ITB reduces the amount of force required to move the patella laterally (Kwak et al., 2000). Lateral retinaculum release is still used in cases of pathological lateral tilt (understood as greater than 20 degrees) (Felli et al., 2021). Nowadays, this practice is, however, subject of much controversy, especially in light of recent studies that have shown that a lateral release can induce an increase in medial tilt, paradoxically making patellar lateral dislocation easier (Cancienne et al., 2019).

PATELLAR CONTACT AREA

In full extension, there is usually no contact between the patella and trochlea. Depending on patellar height and position, trochlear engagement occurs between 10 and 20 degrees of flexion (Smidt, 1973). The patellar contact area is not constant as flexion increases but is initially localized to the inferior lateral margin of the articular surface.

As the flexion angle increases, up to 90 degrees, the contact area shifts from distal to proximal (Leszko et al., 2010). Beyond 30 degrees, the patella enters the interior of the trochlea where it is further stabilized by the strength of the quadriceps and patellar tendon. The contact area extends, like a fascia, from the medial margin of the medial facet to the lateral margin of the lateral facet moving from distal to proximal.

Between 30 and 60 degrees of flexion, the contact area is almost central, shifting, near 90 degrees of flexion, toward the superior pole; beyond 90 degrees of flexion, the patella straddles the medial and lateral condyles, forming two distinct contact areas (Aglietti et al., 1975; Luyckx et al., 2009).

During flexion, the patella undergoes a lateral translation flanked by a slight rotation around the longitudinal axis (Reider, Marshall, & Ring, 1981). Patellar kinematics studies have reported rotations of up to 12−15 degrees relative to the femur, with most of the rotation occurring at more than 50 degrees of knee flexion (Reider et al., 1981; van Kampen & Huiskes, 1990). In addition, the patella tilts in a medial−lateral direction in the axial plane, depending on knee flexion, degree of internal or external rotation, and varus or valgus alignment of the tibio-femoral (van Kampen & Huiskes, 1990). Similarly, the patella undergoes medial displacement, up to 5 mm in the coronal plane, with most of the displacement occurring during the first 30 degrees of knee flexion. In the transverse plane, the patella is perfectly congruent with the trochlea, ensuring its medial and lateral stability (Amis et al., 2006). The flexion–extension angle not only determines the position of the patello-femoral contact but also its contact area. From 20 to 60 degrees of flexion, the contact area increases linearly from 150 to 480 mm^2 (Aglietti et al., 1975; Luyckx et al., 2009), remains constant until approximately 90 degrees of flexion, and then increases again to a value of approximately 360 mm^2 at 120 degrees (Amis et al., 2006; Luyckx et al., 2009).

PATELLO-FEMORAL FORCES

In order to define the forces on the patello-femoral joint, it is necessary to consider them as acting in the sagittal plane (Fig. 13.13). Generally, we can introduce the concept of patello-femoral reaction force (PF) as the force generated by the application, on the patella, of the quadriceps force (QF) and the patellar tendon force (TF). In turn, this force, which depends on the flexion angle, can be divided into an axial compressive force and a tangential force. As bending increases, the value of the force increases and it moves upward. In fact, as the angle between the patellar tendon and the quadriceps becomes more acute, the resulting force vector, corresponding to PF, increases.

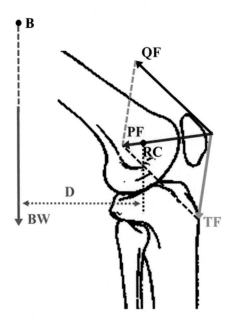

FIGURE 13.13

Diagram of the forces acting on the patello-femoral joint in the sagittal plane. The patello-femoral reaction force (*PF*, in *red*) is obtained from the vector sum of the quadriceps muscle force (*QF*, in *blue*) and the patellar tendon force (*TF*, in *green*). *RC* indicates the center of instantaneous rotation of the knee. *D* is the distance between *RC* and the vertical line passing through the body's barycenter (*B*), where the body weight force (*BW*, in *brown*) is applied.

In addition, the tibial and femoral lever arms increase, requiring greater quadriceps strength to resist the moment exerted by the action of body weight.

One of the most important variables in calculating patellar forces is the distance "D" between the vertical line passing through the body's barycenter (center of gravity) and the instantaneous axis of rotation of the joint (Fig. 13.13). Following a change in posture in the sagittal plane (anteriorly or posteriorly), this distance will change, inducing differences, even large ones, in patellar force.

For example, whenever body weight is shifted posteriorly and away from the joint, the patellar compressive force will increase. Conversely, during normal activities requiring flexion under load, due to hip flexion, the center of gravity is brought forward, shortening the lever arm D. This behavior also justifies the variations in patellar force in the movements during stair climbing (forces 1.8−2.3 BW) and stair descent (forces 2.9−6.0 BW) (Andriacchi, Andersson, Fermier, Stern, & Galante, 1980). Therefore, patello-femoral reaction forces vary greatly and are essentially dependent on the type of activity performed, starting with values from 0.6 BW for walking, increasing to 7.7 BW for jogging up to 20 BW for jumping (Dahlkvist, Mayo, & Seedhom, 1982; Winter, 1983). These forces induce tensile stresses (strains) on the articular cartilage, which can reach even very high values, in conjunction with the small size of the patella and, consequently, its contact areas. As a first approximation, patellar contact pressure can be calculated from the ratio between the applied force and the contact area. Cartilage stress, therefore, also depends on the value of the flexion angle, with values ranging from 1.28 to 12.6 N/mm^2 (Huberti & Hayes, 1984).

REFERENCES

Adachi, N., Ochi, M., Uchio, Y., Iwasa, J., Ryoke, K., & Kuriwaka, M. (2002). Mechanoreceptors in the anterior cruciate ligament contribute to the joint position sense. *Null*, *73*(3), 330−334. Available from https://doi.org/10.1080/000164702320155356.

Aglietti, P., Insall, J. N., Walker, P. S., & Trent, P. (1975). A new patella prosthesis. Design and application. *Clinical Orthopaedics and Related Research*, *107*, 175−187.

Amis, A. A., Senavongse, W., & Bull, A. M. J. (2006). Patellofemoral kinematics during knee flexion-extension: An in vitro study. *Journal of Orthopaedic Research: Official Publication of the Orthopaedic Research Society*, *24*(12), 2201−2211.

Andriacchi, T. P., Andersson, G. B., Fermier, R. W., Stern, D., & Galante, J. O. (1980). A study of lower-limb mechanics during stair-climbing. *The Journal of Bone and Joint Surgery. American Volume*, *62*(5), 749−757.

Apkarian, J., Naumann, S., & Cairns, B. (1989). A three-dimensional kinematic and dynamic model of the lower limb. *Journal of Biomechanics*, *22*(2), 143−155.

Beaconsfield, T., Pintore, E., Maffulli, N., & Petri, G. J. (1994). Radiological measurements in patellofemoral disorders. A review. *Clinical Orthopaedics and Related Research*, *308*, 18−28.

Bellemans, J. (2003). Biomechanics of anterior knee pain. *The Knee*, *10*(2), 123−126.

Bergmann, G., Graichen, F., Rohlmann, A., Westerhoff, P., Bender, A., Gabel, U., ... Heinlein, B. (2007). Loads acting on orthopaedic implants. Measurements and practical applications. *Der Orthopade*, *36*(3), 195−196, 198−200, 202−204.

Biscević, M., Hebibović, M., & Smrke, D. (2005). Variations of femoral condyle shape. *Collegium Antropologicum*, *29*(2), 409−414.

Blankevoort, L., Huiskes, R., & de Lange, A. (1990). Helical axes of passive knee joint motions. *Journal of Biomechanics*, *23*(12), 1219−1229.

Bobbert, M. F., Yeadon, M. R., & Nigg, B. M. (1992). Mechanical analysis of the landing phase in heel-toe running. *Journal of Biomechanics*, *25*(3), 223−234. Available from https://doi.org/10.1016/0021-9290(92)90022-S.

Bradley, J., FitzPatrick, D., Daniel, D., Shercliff, T., & O'Connor, J. (1988). Orientation of the cruciate ligament in the sagittal plane. A method of predicting its length-change with flexion. *The Journal of Bone and Joint Surgery. British Volume*, *70*(1), 94−99.

Bull, A. M., & Amis, A. A. (1998). Knee joint motion: Description and measurement. *Proceedings of the Institution of Mechanical Engineers. Part H, Journal of Engineering in Medicine*, *212*(5), 357−372.

Cancienne, J. M., Christian, D. R., Redondo, M. L., Huddleston, H. P., Shewman, E. F., Farr, J., ... Yanke, A. B. (2019). The biomechanical effects of limited lateral retinacular and capsular release on lateral patellar translation at various flexion angles in cadaveric specimens. *Arthroscopy, Sports Medicine, and Rehabilitation*, *1*(2), e137−e144. Available from https://doi.org/10.1016/j.asmr.2019.09.002.

D'Lima, D. D., Fregly, B. J., Patil, S., Steklov, N., & Colwell, C. W. (2012). Knee joint forces: Prediction, measurement, and significance. *Proceedings of the Institution of Mechanical Engineers. Part H, Journal of Engineering in Medicine*, *226*(2), 95−102.

D'Lima, D. D., Steklov, N., Patil, S., & Colwell, C. W. (2008). The Mark Coventry Award: In vivo knee forces during recreation and exercise after knee arthroplasty. *Clinical Orthopaedics and Related Research*, *466*(11), 2605−2611. Available from https://doi.org/10.1007/s11999-008-0345-x.

D'Lima, D. D., Townsend, C. P., Arms, S. W., Morris, B. A., & Colwell, C. W. (2005). An implantable telemetry device to measure intra-articular tibial forces. *Journal of Biomechanics*, *38*(2), 299−304.

Dahlkvist, N. J., Mayo, P., & Seedhom, B. B. (1982). Forces during squatting and rising from a deep squat. *Engineering in Medicine*, *11*(2), 69−76. Available from https://doi.org/10.1243/EMED_JOUR_1982_011_019_02.

Dejour, H., Walch, G., Nove-Josserand, L., & Guier, C. (1994). Factors of patellar instability: An anatomic radiographic study. *Knee Surgery, Sports Traumatology, Arthroscopy: Official Journal of the ESSKA*, *2*(1), 19−26.

Desio, S. M., Burks, R. T., & Bachus, K. N. (1998). Soft tissue restraints to lateral patellar translation in the human knee. *The American Journal of Sports Medicine*, *26*(1), 59−65.

Desloovere, K., Wong, P., Swings, L., Callewaert, B., Vandenneucker, H., & Leardini, A. (2010). Range of motion and repeatability of knee kinematics for 11 clinically relevant motor tasks. *Gait & Posture*, *32*(4), 597−602. Available from https://doi.org/10.1016/j.gaitpost.2010.08.010.

Eckhoff, D. G., Brown, A. W., Kilcoyne, R. F., & Stamm, E. R. (1997). Knee version associated with anterior knee pain. *Clinical Orthopaedics and Related Research*, *339*, 152−155.

Elias, S. G., Freeman, M. A., & Gokcay, E. I. (1990). A correlative study of the geometry and anatomy of the distal femur. *Clinical Orthopaedics and Related Research*, *260*, 98−103.

Ellison, A. E., & Berg, E. E. (1985). Embryology, anatomy, and function of the anterior cruciate ligament. *The Orthopedic Clinics of North America*, *16*(1), 3−14.

Felli, L., Alessio-Mazzola, M., Lovisolo, S., Capello, A. G., Formica, M., & Maffulli, N. (2021). Anatomy and biomechanics of the medial patellotibial ligament: A systematic review. *The Surgeon*, *19*(5), e168−e174. Available from https://doi.org/10.1016/j.surge.2020.09.005.

Frankel, V. H., Burstein, A. H., & Brooks, D. B. (1971). Biomechanics of internal derangement of the knee. Pathomechanics as determined by analysis of the instant centers of motion. *The Journal of Bone and Joint Surgery. American Volume*, *53*(5), 945−962.

Gerber, C., & Matter, P. (1983). Biomechanical analysis of the knee after rupture of the anterior cruciate ligament and its primary repair. An instant-centre analysis of function. *The Journal of Bone and Joint Surgery. British Volume*, *65*(4), 391−399.

Grood, E. S., & Suntay, W. J. (1983). A joint coordinate system for the clinical description of three-dimensional motions: Application to the knee. *Journal of Biomechanical Engineering, 105*(2), 136−144.

Hashemi, J., Chandrashekar, N., Gill, B., Beynnon, B. D., Slauterbeck, J. R., Schutt, R. C., ... Dabezies, E. (2008). The geometry of the tibial plateau and its influence on the biomechanics of the tibiofemoral joint. *The Journal of Bone and Joint Surgery. American Volume, 90*(12), 2724−2734. Available from https://doi.org/10.2106/JBJS.G.01358.

Hehne, H. J. (1990). Biomechanics of the patellofemoral joint and its clinical relevance. *Clinical Orthopaedics and Related Research, 258*, 73−85.

Huberti, H. H., & Hayes, W. C. (1984). Patellofemoral contact pressures. The influence of q-angle and tendofemoral contact. *The Journal of Bone and Joint Surgery American Volume, 66*(5), 715−724.

Innocenti, B., Salandra, P., Pascale, W., & Pianigiani, S. (2016). How accurate and reproducible are the identification of cruciate and collateral ligament insertions using MRI? *The Knee, 23*(4), 575−581. Available from https://doi.org/10.1016/j.knee.2015.07.015.

Iwaki, H., Pinskerova, V., & Freeman, M. A. (2000). Tibiofemoral movement 1: The shapes and relative movements of the femur and tibia in the unloaded cadaver knee. *The Journal of Bone and Joint Surgery. British Volume, 82*(8), 1189−1195.

James, E. W., LaPrade, C. M., & LaPrade, R. F. (2015). Anatomy and biomechanics of the lateral side of the knee and surgical implications. *Sports Medicine and Arthroscopy Review, 23*(1), 2−9. Available from https://doi.org/10.1097/JSA.0000000000000040.

Kettelkamp, D. B., Johnson, R. J., Smidt, G. L., Chao, E. Y., & Walker, M. (1970). An electrogoniometric study of knee motion in normal gait. *The Journal of Bone and Joint Surgery. American Volume, 52*(4), 775−790.

Kowalk, D. L., Duncan, J. A., & Vaughan, C. L. (1996). Abduction-adduction moments at the knee during stair ascent and descent. *Journal of Biomechanics, 29*(3), 383−388.

Kwak, S. D., Ahmad, C. S., Gardner, T. R., Grelsamer, R. P., Henry, J. H., Blankevoort, L., ... Mow, V. C. (2000). Hamstrings and iliotibial band forces affect knee kinematics and contact pattern. *Journal of Orthopaedic Research, 18*(1), 101−108. Available from https://doi.org/10.1002/jor.1100180115.

Leszko, F., Sharma, A., Komistek, R. D., Mahfouz, M. R., Cates, H. E., & Scuderi, G. R. (2010). Comparison of in vivo patellofemoral kinematics for subjects having high-flexion total knee arthroplasty implant with patients having normal knees. *The Journal of Arthroplasty, 25*(3), 398−404. Available from https://doi.org/10.1016/j.arth.2008.12.007.

Loudon, J. K. (2016). Biomechanics and pathomechanics of the patellofemoral joint. *International Journal of Sports Physical Therapy, 11*(6), 820−830.

Luyckx, T., Didden, K., Vandenneucker, H., Labey, L., Innocenti, B., & Bellemans, J. (2009). Is there a biomechanical explanation for anterior knee pain in patients with patella alta?: Influence of patellar height on patellofemoral contact force, contact area and contact pressure. *The Journal of Bone and Joint Surgery. British Volume, 91*(3), 344−350. Available from https://doi.org/10.1302/0301-620X.91B3.21592.

Menschik, A. (1974). Mechanik des Kniegelenkes. 1. *Zeitschrift Fur Orthopadie Und Ihre Grenzgebiete, 112* (3), 481−495.

Ohkoshi, Y., Yasuda, K., Kaneda, K., Wada, T., & Yamanaka, M. (1991). Biomechanical analysis of rehabilitation in the standing position. *The American Journal of Sports Medicine, 19*(6), 605−611.

Panni, A. S., Cerciello, S., Maffulli, N., Di Cesare, M., Servien, E., & Neyret, P. (2011). Patellar shape can be a predisposing factor in patellar instability. *Knee Surgery, Sports Traumatology, Arthroscopy: Official Journal of the ESSKA, 19*(4), 663−670. Available from https://doi.org/10.1007/s00167-010-1329-4.

Pinskerova, A., & Vavrik, P. (2020). Knee anatomy and biomechanics and its relevance to knee replacement. In C. Revière & C. Vendittoli (Eds.), Personalized Hip and Knee Joint Replacement. Springer. Available from https://doi.org/10.1007/978-3-030-24243-5_14

Pinskerova, V., Maquet, P., & Freeman, M. A. (2000). Writings on the knee between 1836 and 1917. *The Journal of Bone and Joint Surgery. British Volume*, *82*(8), 1100−1102.

Reider, B., Marshall, J. L., & Ring, B. (1981). Patellar tracking. *Clinical Orthopaedics and Related Research*, *157*, 143−148.

Scheys, L., Leardini, A., Wong, P. D., Van Camp, L., Callewaert, B., Bellemans, J., . . . Desloovere, K. (2013). Three-dimensional knee kinematics by conventional gait analysis for eleven motor tasks of daily living: Typical patterns and repeatability. *Journal of Applied Biomechanics*, *29*(2), 214−228.

Schindler, O. S., & Scott, W. N. (2011). Basic kinematics and biomechanics of the patello-femoral joint. Part 1: The native patella. *Acta Orthopaedica Belgica*, *77*(4), 421−431.

Senavongse, W., & Amis, A. A. (2005). The effects of articular, retinacular, or muscular deficiencies on patellofemoral joint stability: A biomechanical study in vitro. *The Journal of Bone and Joint Surgery. British Volume*, *87*(4), 577−582.

Sharma, L., Lou, C., Felson, D. T., Dunlop, D. D., Kirwan-Mellis, G., Hayes, K. W., . . . Buchanan, T. S. (1999). Laxity in healthy and osteoarthritic knees. *Arthritis and Rheumatism*, *42*(5), 861−870.

Sharma, L., Song, J., Felson, D. T., Cahue, S., Shamiyeh, E., & Dunlop, D. D. (2001). The role of knee alignment in disease progression and functional decline in knee osteoarthritis. *Journal of the American Medical Association*, *286*(2), 188−195.

Shih, Y.-F., Bull, A. M. J., & Amis, A. A. (2004). The cartilaginous and osseous geometry of the femoral trochlear groove. *Knee Surgery, Sports Traumatology, Arthroscopy: Official Journal of the ESSKA*, *12*(4), 300−306.

Siegel, L., Vandenakker-Albanese, C., & Siegel, D. (2012). Anterior cruciate ligament injuries: Anatomy, physiology, biomechanics, and management. *Clinical Journal of Sport Medicine: Official Journal of the Canadian Academy of Sport Medicine*, *22*(4), 349−355. Available from https://doi.org/10.1097/JSM.0b013e3182580cd0.

Smidt, G. L. (1973). Biomechanical analysis of knee flexion and extension. *Journal of Biomechanics*, *6*(1), 79−92.

Taylor, S. J., & Walker, P. S. (2001). Forces and moments telemetered from two distal femoral replacements during various activities. *Journal of Biomechanics*, *34*(7), 839−848.

Townsend, M. A., Izak, M., & Jackson, R. W. (1977). Total motion knee goniometry. *Journal of Biomechanics*, *10*(3), 183−193.

van Kampen, A., & Huiskes, R. (1990). The three-dimensional tracking pattern of the human patella. *Journal of Orthopaedic Research*, *8*(3), 372−382. Available from https://doi.org/10.1002/jor.1100080309.

Victor, J., Van Doninck, D., Labey, L., Innocenti, B., Parizel, P. M., & Bellemans, J. (2009). How precise can bony landmarks be determined on a CT scan of the knee? *The Knee*, *16*(5), 358−365. Available from https://doi.org/10.1016/j.knee.2009.01.001.

Victor, Jan, Labey, L., Wong, P., Innocenti, B., & Bellemans, J. (2010). The influence of muscle load on tibiofemoral knee kinematics. *Journal of Orthopaedic Research: Official Publication of the Orthopaedic Research Society*, *28*(4), 419−428. Available from https://doi.org/10.1002/jor.21019.

Warren, L. F., & Marshall, J. L. (1979). The supporting structures and layers on the medial side of the knee: An anatomical analysis. *The Journal of Bone and Joint Surgery. American Volume*, *61*(1), 56−62.

Weber, W., & Weber, F. (1836). Mechanik der menschlichen Gehwerkzeuge (in German). Göttingen, Dieterich.

Wibeeg, G. (1941). Roentgenographs and anatomic studies on the femoropatellar joint: With special reference to chondromalacia patellae. *Null*, *12*(1−4), 319−410. Available from https://doi.org/10.3109/17453674108988818.

Wilson, W. T., Deakin, A. H., Payne, A. P., Picard, F., & Wearing, S. C. (2012). Comparative analysis of the structural properties of the collateral ligaments of the human knee. *Journal of Orthopaedic & Sports Physical Therapy*, *42*(4), 345−351. Available from https://doi.org/10.2519/jospt.2012.3919.

Winter, D. A. (1983). Moments of force and mechanical power in jogging. *Journal of Biomechanics*, *16*(1), 91−97. Available from https://doi.org/10.1016/0021-9290(83)90050-7.

Wismans, J., Veldpaus, F., Janssen, J., Huson, A., & Struben, P. (1980). A three-dimensional mathematical model of the knee-joint. *Journal of Biomechanics*, *13*(8), 677−685. Available from https://doi.org/10.1016/0021-9290(80)90354-1.

Woo, S. L.-Y., Wu, C., Dede, O., Vercillo, F., & Noorani, S. (2006). Biomechanics and anterior cruciate ligament reconstruction. *Journal of Orthopaedic Surgery and Research*, *1*, 2.

Yang, B., Song, C., Yu, J., Yang, Y., Gong, X., Chen, L., . . . Wang, J. (2014). Intraoperative anthropometric measurements of tibial morphology: Comparisons with the dimensions of current tibial implants. *Knee Surgery, Sports Traumatology, Arthroscopy: Official Journal of the ESSKA*, *22*(12), 2924−2930. Available from https://doi.org/10.1007/s00167-014-3258-0.

Yin, L., Chen, K., Guo, L., Cheng, L., Wang, F., & Yang, L. (2015). Identifying the functional flexion-extension axis of the knee: An in-vivo kinematics study. *PloS One*, *10*(6), e0128877. Available from https://doi.org/10.1371/journal.pone.0128877.

Zavatsky, A. B., & O'Connor, J. J. (1992a). A model of human knee ligaments in the sagittal plane. Part 1: Response to passive flexion. *Proceedings of the Institution of Mechanical Engineers. Part H, Journal of Engineering in Medicine*, *206*(3), 125−134.

Zavatsky, A. B., & O'Connor, J. J. (1992b). A model of human knee ligaments in the sagittal plane. Part 2: Fibre recruitment under load. *Proceedings of the Institution of Mechanical Engineers. Part H, Journal of Engineering in Medicine*, *206*(3), 135−145.

Zhang, Y., Chen, Y., Qiang, M., Zhang, K., Li, H., Jiang, Y., . . . Jia, X. (2018). Comparison between three-dimensional CT and conventional radiography in proximal tibia morphology. *Medicine*, *97*(30), e11632. Available from https://doi.org/10.1097/MD.0000000000011632.

Zuppinger, H. (1904). Die aktive flexion im unbelasteten Kniegelenk:Züricher Habil Schr. Wiesbaden: Bergmann.

BIOMECHANICS OF THE SPINE

14

Fabio Galbusera
IRCCS Istituto Ortopedico Galeazzi, Milan, Italy

ANATOMY

The spine is composed of several types of tissues such as bones, namely the vertebrae, intervertebral discs, ligaments, and synovial joints known as zygapophyseal or facet joints. The spine can be subdivided into three regions: cervical, that is, the neck; thoracic or dorsal; and lumbar, that is, the low back. Whereas all vertebrae, with the exception of the most cranial vertebra of the cervical spine, share a common structure, their size and appearance vary considerably among the different spinal regions.

Each vertebra is composed of the vertebral body, with an approximate oval or bean-like shape and constituting the most ventral part; two pedicles, one for each side, protruding from the posterior aspect of the vertebral body and with an approximately cylindrical shape; the laminae, connecting the dorsal aspect of the pedicle and forming, together with the posterior wall of the body and the pedicles, the spinal canal; the zygapophyses or articular processes, which articulate with the adjacent vertebrae forming the zygapophyseal or facet joints; the spinous process, protruding posteriorly; and the transverse processes, protruding laterally and acting as important muscle insertion sites (Fig. 14.1).

The cervical spine is composed of seven vertebrae, identified with progressive numbers (C1−C7) (Fig. 14.2). The occiput, that is, the outer surface of the occipital bone articulating with C1, is frequently named C0. Intervertebral discs are present at the levels between C2−C3 and C6−C7, whereas the levels C0−C1 and C1−C2, which constitute the so-called upper cervical spine, do not possess a disc. Indeed, the vertebrae C1 and C2 have a peculiar, specialized anatomy aimed at enhancing the mobility of the head while maintaining sufficient robustness. C1, also known as atlas, has a ring shape and articulates, in its inner side, with a protrusion of the C2 vertebra named odontoid process or dens. C1 has two cranial facets, which articulate with the condyles of the occiput, forming the atlanto-occipital joint, and two caudal facets in contact with corresponding facets on C2. Whereas the superior structures of C2, also known as axis, possess these specialized anatomical features, its inferior aspect has a more conventional anatomy and is connected to C3 through an intervertebral disc and zygapophyseal joints. Vertebrae of the lower cervical spine, that is, the C3−C7 region, have the conventional structure with vertebral body, pedicles, laminae, and facet joints; their vertebral bodies have a characteristic saddle-like shape and possess hook-shaped processes on their superior surface, which articulate with the cranial vertebra forming the

Human Orthopaedic Biomechanics. DOI: https://doi.org/10.1016/B978-0-12-824481-4.00019-6

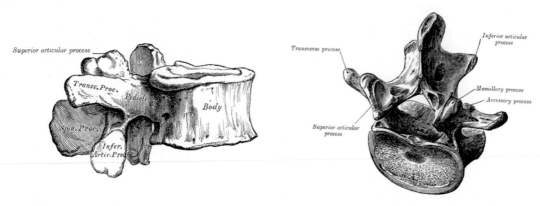

FIGURE 14.1

The anatomy of a vertebra, from anterolateral (left) and posterosuperior (right) views.

Reproduced from Gray, H. (1918). Anatomy of the human body. Lea and Febiger. (public domain).

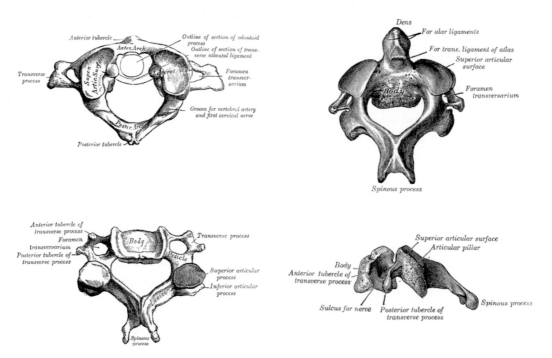

FIGURE 14.2

Cervical vertebrae. Top: C1, the atlas (left), and C2, the axis (right). Bottom: a vertebra of the lower cervical spine from superior (left) and lateral (right) views.

Reproduced from Gray, H. (1918). Anatomy of the human body. Lea and Febiger. (public domain).

uncovertebral or Luschka joints. Articular facets are flat and perpendicular to the sagittal plane. The transverse processes present a hole, the foramen intertransversarium, which hosts the vertebral artery and vein. The cervical spine has a lordotic shape, that is, it has a curvature in the sagittal plane with the concavity in the dorsal direction.

The thoracic vertebrae, T1–T12, are all connected to the ribs through the costovertebral joints (Fig. 14.3). These 12 vertebrae have a relatively similar shape, with heart-shaped bodies with size slightly increasing in the caudal direction and prominent transverse processes that provide the articulation sites with the tubercles of the ribs, whereas the head of each rib is directly connected with the vertebral body. Intervertebral discs are present at all levels. The spinous processes are long, triangular-shaped, and pointing downward. Articular processes are almost flat and vertical, and directed backward and slightly lateralward. The thoracic spine has a kyphotic curvature, that is, its concavity in the sagittal plane points to the ventral direction.

The lumbar spine consists of five vertebrae, L1–L5, and has a lordotic concavity in healthy individuals. Its vertebrae have a larger size compared to the other spinal regions, allowing for a

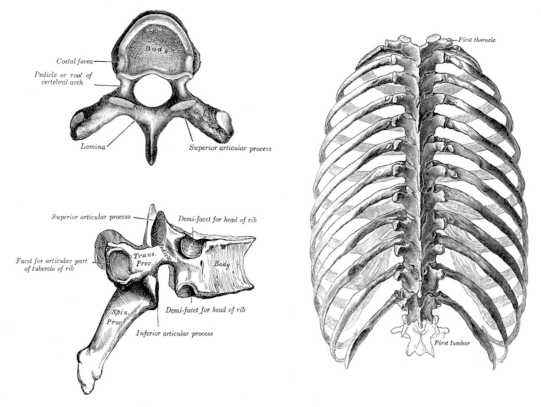

FIGURE 14.3

Left: thoracic vertebrae from superior (top) and lateral (bottom) views. Right: the thoracic spine and the rib cage from a posterior view.

Reproduced from Gray, H. (1918). Anatomy of the human body. Lea and Febiger. (public domain).

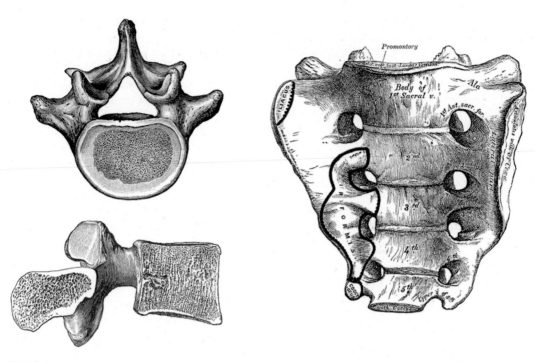

FIGURE 14.4

Left: lumbar vertebrae from superior (top) and lateral, midsagittally sectioned (bottom) views. Right: the sacrum from an anterior view.

Reproduced from Gray, H. (1918). Anatomy of the human body. Lea and Febiger. (public domain).

higher resistance to the body weight, which increases proceeding in the caudal direction (Fig. 14.4). Lumbar vertebrae have a large, bean-shaped body connected to the adjacent vertebrae by means of thick intervertebral discs. Spinous processes are thick and robust and have an approximately square appearance. The articular processes are not flat, but present a curved surface; those on the superior processes are concave, and directed backward and medialward; inferior facets are convex, and directed forward and lateralward. Transverse processes are long and relatively thin. Intervertebral discs are again present at all levels.

The sacrum articulates with the L5 vertebra and the pelvis, thus connecting the rest of spine with the lower limbs (Fig. 14.4). It consists of five fused vertebrae, S1–S5, and has an approximately triangular shape determined by the medial fused vertebral bodies and the two alae (wings), which articulate with the pelvis forming the sacroiliac joints. Nerve roots exit the sacral canal through the anterior and posterior sacral foramina. The cranial surface of the sacrum articulates with the L5 vertebra with a rather conventional spinal anatomy, featuring an intervertebral disc and zygapophyseal joints. Caudally, the sacrum is connected to the coccyx, a complex of rudimentary vertebrae, which is the remnant of a vestigial tail.

Vertebrae and discs are strengthened by a complex ligamentous system (Fig. 14.5). The anterior and posterior longitudinal ligaments are, respectively, connected to the ventral and dorsal surfaces

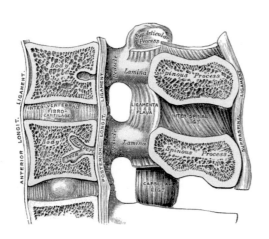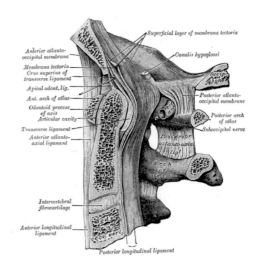

FIGURE 14.5

Left: ligaments of the thoracolumbar spine. Right: ligaments of the upper cervical spine.

Reproduced from Gray, H. (1918). Anatomy of the human body. Lea and Febiger. (public domain).

of vertebral bodies and intervertebral discs, acting as a containment ring to limit disc bulging in flexion—extension. The posterior elements are reinforced by the flaval or yellow ligament, which connects the ventral side of the laminae of adjacent vertebrae and is rich in elastin; the supraspinous ligament, connecting the tips of the spinous processes; the interspinous ligament, a thin layer connecting the superior and inferior aspects of adjacent spinous processes and blending into the yellow ligament and supraspinous ligament ventrally and dorsally, respectively; the capsular ligaments, envelopes enclosing the zygapophyseal joints; and the intertransverse ligaments, which connect the adjacent transverse ligaments. In the cervical spine, the nuchal ligament is located dorsally with respect to the supraspinous ligaments and connect the occiput to the spinous process of C7. The upper cervical spine possesses a specific ligament system, which includes the alar ligament, the transverse ligament, and the anterior atlantoaxial and posterior atlantoaxial ligaments, which accommodate and strengthen the peculiar anatomy of this region. Other ligaments are also present in the costovertebral joints and in the sacropelvic complex.

FLEXIBILITY AND MOBILITY

Despite being a robust structure able to support the weight of trunk, upper limbs, and head as well as the loads generated by the muscles of the upper body, the human spine has considerable flexibility, which allows for a wide range of motions and activities. The various spine regions have distinct flexibilities, associated with the specific anatomy of vertebrae and intervertebral discs and with the mobility requirements of the various body parts.

The cervical spine allows for a high mobility of the head with respect to the trunk, which is fundamental to humans due to the frontal position of both eyes, which permits a wide binocular vision but restricts the field of view. Indeed, in vitro tests conducted on cadaver specimens demonstrated that the cervical spine has a large range of motion in flexion—extension, lateral bending, and axial rotation (Fig. 14.6) (Panjabi et al., 2001). In particular, the design of the C1—C2 level allows for a rotation of up to 55 degrees, accounting for more than half of the physiological range of motion of the cervical spine altogether. Flexion—extension is maximal at C0—C1 and C1—C2, which show a range of motion up to 30 degrees (Anderst, Donaldson, Lee, & Kang, 2015). The lower cervical spine exhibits a stiffer behavior; its flexibility tends to increase proceeding in the caudal direction, reaching a maximum of 20 degrees at C5—C6 (Anderst et al., 2015). The shape and orientation of the facet joints determine a motion coupling in lateral bending and axial rotation; in other words, if a pure moment in lateral bending is applied to a cervical spine specimen, the resulting motion would feature a component in axial rotation in addition to the one in lateral bending, and vice versa. Such a phenomenon has been deeply investigated in vivo and in vitro by determining the helical

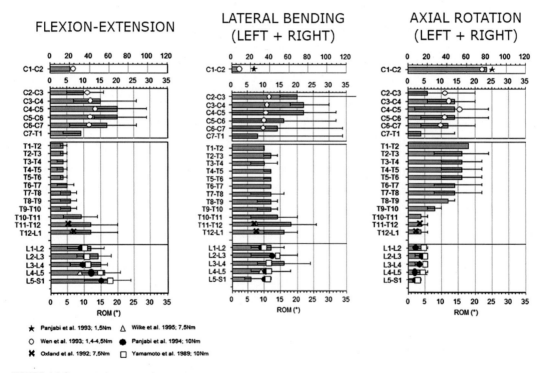

FIGURE 14.6

Segmental flexibility of the spine, as measured by various in vitro studies (Oxland et al., 1992; Panjabi et al., 1993; Wen et al., 1993; Wilke et al., 1995; Yamamoto, Panjabi, Crisco, & Oxland, 1989).

From Wilke, H. -J., Geppert, J., & Kienle, A. (2011). Biomechanical in vitro evaluation of the complete porcine spine in comparison with data of the human spine. European Spine Journal, 20(11), 1859–1868. https://doi.org/10.1007/s00586-011-1822-6.

axes of motion with imaging techniques or motion analysis systems (Anderst et al., 2015; Bogduk & Mercer, 2000; Penning, 1978).

In general, the thoracic spine exhibits a lower flexibility with respect to the cervical and lumbar counterparts (Fig. 14.6). Studies highlighted the role of facet joint orientation and rib cage in determining the stiffer behavior (Wilke, Herkommer, Werner, & Liebsch, 2017). In vitro testing on cadaver specimens demonstrated that the range of motion tends to decrease proceeding caudally, whereas in vivo measurements reported the opposite behavior, likely due to an increase in the loads to which the lower thoracic spine is subjected in the physiological conditions (Willems, Jull, & Ng, 1996).

Cadaver tests under controlled laboratory conditions showed that the lumbar spine presents a relatively large flexibility in flexion—extension, especially in the most caudal segments (Yamamoto et al., 1989) (Fig. 14.6); the latter finding is not corroborated by in vivo data (Pearcy, 1985) and may be attributed to the resection of muscles and spinopelvic ligaments, in particular the iliolumbar ligament, which is performed during the preparation of the specimens. In lateral bending, the lumbar spine exhibits a higher flexibility in the upper region, whereas the range of motion strongly decreases in the most caudal segments. Axial rotation is largely limited by the orientation of the facet joints, and does not show a marked dependence on the spinal level (Yamamoto et al., 1989).

LOADS

The mechanical loads acting in the spine result from the simultaneous effect of the body weight and of the trunk muscles, which act to maintain the equilibrium in various postures and to allow for body motion. Indeed, even in relaxed positions such as standing or sitting, the anterior location of the center of gravity of the upper body requires the activation of the extensor muscles; the combined action of weight and muscle forces determine high loads in vertebrae and intervertebral discs. As a matter of fact, the vertical orientation of the trabecules in the vertebral bone tissue clearly demonstrates the compressive nature of the load most commonly acting in the spine. However, other postures such as forward flexion, or performing dynamic activities such as walking, running, or lifting loads, alter the loading environment, introducing bending and torsional moments or traction.

In the 1950s, Alf Nachemson and colleagues quantified for the first time the compressive loads in the spine in various postures by inserting a pressure-sensitive needle in the spine of healthy volunteers, namely in the nucleus pulposus of a lumbar intervertebral disc (Nachemson, 1959). The authors did not report absolute intradiscal pressure values, but normalized the signal of the transducer with respect to the measurement in the standing position. The tests revealed substantial increases of the pressure, between 50% and 220%, with the subject in the forward-bending position or carrying loads. More recent studies deepened the investigations by performing measurements on degenerated discs (Sato, Kikuchi, & Yonezawa, 1999) and during a large set of activities and tasks (Wilke et al., 1999). Due to the improvements in the sensor technology, the latter study also reported absolute values for the pressure, which ranged between 0.08 MPa (lying supine) and 2.3 MPa (lifting a 20 kg load with rounded back and straight knees) (Fig. 14.7). To our knowledge, intradiscal pressures in the cervical and thoracic spines in vivo were assessed only in two studies,

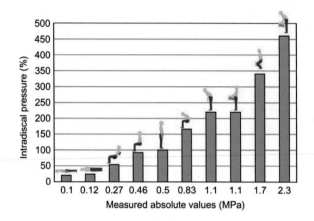

FIGURE 14.7

The intradiscal pressure in the lumbar spine measured with a needle transducer in various postures.

From Wilke, H. -J., Volkheimer, D., Galbusera, F., & Wilke, H. -J. (2018). Chapter 4 — Basic biomechanics of the lumbar spine. In Biomechanics of the spine (pp. 51–67). Academic Press. https://doi.org/10.1016/B978-0-12-812851-0.00004-5.

which reported values around 1 MPa in the cervical region (Hattori, Oda, & Kawai, 1981) and in the same range of the lumbar pressures in the thoracic region (Polga et al., 2004).

An alternative technique for the estimation of the in vivo loads is based on the use of telemeterized implants instrumented with strain gauges and able to assess the mechanical loads which they are subjected to. This technique has been introduced by the research group at Charité, Berlin University of Medicine, for the hip joint, and has been expanded by the same group to spine fixators (Rohlmann, Bergmann, & Graichen, 1997) and vertebral body replacements (Rohlmann et al., 2007) for the study of the spinal loads. With respect to intradiscal pressure measurements, instrumented implants allow for a direct assessment of forces and moments in the three directions rather than a single pressure value, but are affected by the surgical procedure and the implants themselves, which may alter the loading environment.

DEGENERATION

Aging is associated with degenerative changes in the spinal tissues such as intervertebral discs, ligaments, and facet joints, which can, in turn, result in pain and dysfunction. The link between spinal degeneration and pain is, however, not fully understood; subjects showing clear signs of degeneration in a radiological examination may not be symptomatic, whereas patients suffering from severe pain and disability may not show evident degenerative changes. As a matter of fact, a certain identification of the source of pain is not possible in a large portion of back pain patients.

The intervertebral disc is arguably the tissue that undergoes the most evident changes (Fig. 14.8). As mentioned in Chapter 12, the healthy disc has the capability of swelling, that is, imbibing fluid and developing an inner pressure even in the absence of external loading, due to the

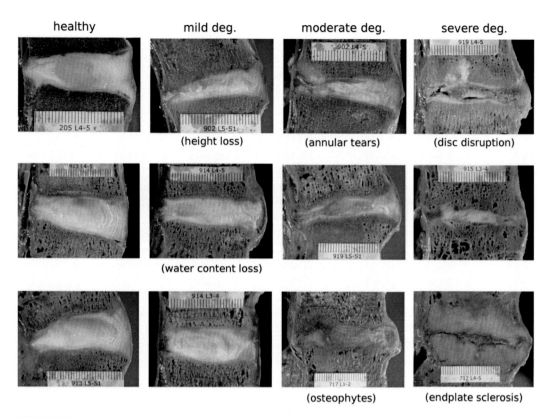

FIGURE 14.8

The degenerative changes observable in degenerated intervertebral discs.

From Galbusera, F., van Rijsbergen, M., Ito, K., Huyghe, J. M., Brayda-Bruno, M., & Wilke, H. -J. (2014). Ageing and degenerative changes of the intervertebral disc and their. European Spine Journal, *23(Suppl 3), S324–S332. https://doi.org/10.1007/s00586-014-3203-4.*

presence of proteoglycans (Urban & McMullin, 1985). In degenerative conditions, the cell population decreases, leading to a lower and more fragmented proteoglycan content, which. in turn. determines a reduction of the capability of imbibing water. Degenerated discs have therefore a lower water content with respect to healthy ones, which is reflected in the so-called "black disc" appearance in T2-weighted MRI scans. The intradiscal pressure decreases from 0.2 MPa, as recorded in healthy specimens, to approximately 0.05 MPa (Urban & McMullin, 1985); such a pressure loss has frequently an impact on the disc height, and narrowing of the intervertebral space is indeed a common clinical finding. Structural damage, that is, tears and fissures in both the nucleus pulposus and the annulus fibrosus, are also frequently observed. Degenerative changes also affect the vertebral endplates, that is, the osteocartilaginous layers connecting vertebral bodies and intervertebral discs, which may show calcification and sclerosis (Wilke et al., 2006). Osteophytes, that is, heterotopic bony beaks and spurs, may appear around the edges of the vertebral body. In combination with the relatively frequent calcification of the ligaments, especially of the flaval and of the posterior longitudinal ligaments, posterior osteophytes can pinch and compress spinal cord and nerve roots determining the so-called spinal stenosis, which is associated with pain in the back and lower limbs as well as with neurological deficits. The latter condition is relatively easy to diagnose and is frequently treated surgically, decompressing the neurological structures by removing part of the ossified ligaments and laminae.

Facet joints also frequently undergo degenerative changes, typically in the form of osteoarthritis, and have been indicated as a frequent cause of chronic back pain. Such a degenerative process is characterized by a loss of the joint space, cartilage erosion and degeneration, loss of synovial fluid, and growth of bone spurs. Studies showed that the facet joint and intervertebral disc degeneration are associated with each other; increasing facet joint osteoarthritis may have a negative effect of the loads in the intervertebral disc, facilitating the degenerative cascade (Bashkuev, Reitmaier, & Schmidt, 2019).

The impact of disc and facet joint degeneration on the spinal biomechanics has been investigated in several studies, based on both in vitro testing of cadaver specimens and numerical models. Contradicting results were, however, reported; whereas several studies reported increased flexibility of degenerated specimens, especially in the early stages of degeneration (Fujiwara et al., 2000), other papers described a trend toward a stiffening with the progression of the degenerative cascade (Kettler, Rohlmann, Ring, Mack, & Wilke, 2011), coherently with the hypothesis of a natural history tending to a spontaneous fusion (reviewed in Galbusera et al., 2014). As a matter of fact, spinal degeneration is highly variable among individuals, and both cases of degenerative hypermobility and reduced flexibility are frequently seen in the clinical practice.

SAGITTAL ALIGNMENT AND DEGENERATIVE DEFORMITIES

As mentioned above, the spine exhibits an S-shape in the sagittal plane, which results from the lordotic and kyphotic curvatures of its regions. Such a shape provides stability by decreasing the risk of buckling and minimizes the energy requirement when keeping the standing posture (Le Huec, Charosky, Barrey, Rigal, & Aunoble, 2011); this optimal condition is frequently named "sagittal balance." Degenerative changes can, however, negatively impact this balance; degenerated lumbar intervertebral discs commonly lose the wedging angle, which determines the lumbar lordosis, resulting in a hypolordotic or even kyphotic lumbar curvature. Although thoracic intervertebral discs are in general less affected by degenerative changes than the lumbar counterparts, the decrease in bone quality and density common in the elderly subjects may in some cases lead to wedging of the thoracic vertebrae, resulting in an increased thoracic kyphosis. A combination of these alterations to the physiological curvatures determines a forward shift of the trunk and of its center of gravity, and, in turn, an increase of the muscle activity necessary to maintain equilibrium. The consequences of such a condition, commonly named sagittal imbalance and considered a degenerative deformity of the spine, are fatigue, functional disability, and impairment in activities such as walking, and excessive cervical strain to gain a sufficient field of view.

Many subjects affected by mild or moderate degenerative alterations of the sagittal curvature of the spine employ compensatory mechanisms in order to achieve satisfactory and asymptomatic

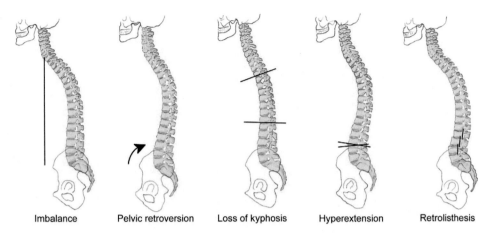

| Imbalance | Pelvic retroversion | Loss of kyphosis | Hyperextension | Retrolisthesis |

FIGURE 14.9

The compensatory mechanisms employed by subjects in order to regain an appropriate sagittal alignment of the spine after a loss of the lumbar lordosis: pelvic retroversion, loss of thoracic kyphosis, hyperextension, and retrolisthesis.

From Ottardi, C., Luca, A., Galbusera, F., Galbusera, F., & Wilke, H. -J. (2018). Chapter 21 – Sagittal imbalance. In Biomechanics of the spine *(pp. 379–391). Academic Press. https://doi.org/10.1016/B978-0-12-812851-0.00021-5.*

posture and gait; this condition is named "compensated imbalance" (Barrey, Roussouly, Perrin, & Le Huec, 2011) (Fig. 14.9). The most common of these compensatory mechanisms is the retroversion of the pelvis; indeed, pelvic retroversion can be observed in the majority of elderly subjects, and does not have in general any negative consequences. By rotating the pelvis backward, the subject can shift the weight of the trunk to the posterior direction, thus reducing the muscular stress and energy expenditure. In more severe cases, a more vertical alignment of the spine is obtained by flexing the knees; such compensation, however, requires the activation of the muscles of the lower limb, and is therefore commonly associated with fatigue and gait impairments. Another compensatory mechanism is the hyperlordosis of the nondegenerated lumbar segments, which can, however, induce foraminal stenosis and overload of the facet joints. Whereas patients with mild or moderate compensated sagittal imbalance may have sufficient functionality and health-related quality of life without any treatment, cases with severe alterations of the sagittal alignment frequently undergo surgical intervention aimed at the restoration of a correct curvature and balance.

The sagittal alignment of the spine in elderly subjects can be effectively investigated by means of planar radiographs in the standing posture (Fig. 14.10). Several radiological parameters describing the sagittal alignment can be assessed on such images; the most commonly measured are the sagittal vertical axis (SVA), that is, the distance between the plumb line from the center of C7 and the postero-superior corner of the sacral endplate (Jackson & McManus, 1994); the pelvic incidence (PI), that is, the angle between the line orthogonal to the sacral endplate and the line passing through the hip joint center and the center of the sacral endplate (Duval-Beaupere, Schmidt, & Cosson, 1992); the sacral slope (SS), that is, the slope of the sacral endplate with respect to the horizontal line; and the pelvic tilt (PT), that is, the angle between the line passing through the hip joint

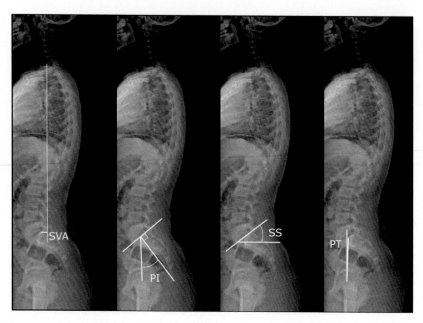

FIGURE 14.10

The radiological parameters describing the sagittal alignment: sagittal vertical axis (SVA), pelvic incidence (PI), sacral slope (SS), and pelvic tilt (PT).

center and the center of the sacral endplate and the vertical line. In recent years, a good deal of attention has been paid to the PI, since it is the only parameter that does not depend on the posture but only on the spinopelvic anatomy of the subject. Radiological studies showed that healthy patients with a large PI also exhibit a large lumbar lordosis and in general a more curved appearance of the spine in the sagittal plane (Roussouly, Gollogly, Berthonnaud, & Dimnet, 2005), and therefore assumes a key role in the preoperative planning of surgeries for the restoration of the spinal alignment. As a matter of fact, the target lumbar lordosis, that is, the lordosis to be achieved in standing after surgical correction, is normally calculated based on the value of PI (Le Huec, Saddiki, Franke, Rigal, & Aunoble, 2011).

Degenerative changes can also determine deformities of the spine in the coronal plane, that is, degenerative scoliosis. This type of deformity may arise from an asymmetrical pattern of degenerative changes, for example, facet osteoarthritis more pronounced on one side than on the other, resulting in a progressive deformity. In contrast with pediatric and adolescent scoliosis, degenerative coronal deformities mostly involve the lumbar spine only, whereas alterations in the coronal alignment of the thoracic spine are relatively rare. Degenerative scoliosis is commonly associated with pain, discomfort, and disability; however, a recent study observed a relatively high prevalence of scoliosis (27%) in a large cohort of healthy elderly subjects (Bassani et al., 2019), indicating that asymptomatic cases are relatively frequent.

CONGENITAL, PEDIATRIC, AND ADOLESCENT SCOLIOSIS

Spinal deformities do not affect only elderly individuals and are not only associated with degenerative changes; they are also observed, even if with a rather low prevalence, in children, adolescents, and young adults. Scoliosis is the most frequent type of spinal deformity in young subjects, and is classified as congenital when present at birth, infantile when appearing before 3 years of age, juvenile between 3 and 9 years, and adolescent between 10 and 18 years. Scoliosis appearing after 18 years of age is classified as an adult deformity, and corresponds in most cases with the degenerative deformities of the elderly subjects described in the previous section. Many cases of adult scoliosis indeed originate in the adolescent age, with the scoliotic curves restarting their progression in the fifth or sixth decade when degenerative changes appear; in this case, the pathology is commonly named de novo scoliosis. Whereas being asymptomatic in several mild and moderate cases, scoliosis is commonly associated with pain, functional disabilities, cosmetic issues, limited quality of life, and cardiac and respiratory disorders in the most severe cases involving major deformities of the rib cage.

Scoliosis is a spinal deformity most evident in the coronal plane; indeed, the diagnostic criterion is based on the maximal angle between the vertebral endplates in frontal radiographs of the spine in the standing posture as assessed by using the Cobb method (Cobb, 1948). If this angle exceeds a threshold value of 10 degrees, the diagnosis of scoliosis is confirmed. However, several papers highlighted the three-dimensional nature of scoliosis; indeed, vertebrae often present a rotation in the axial plane, which is observable in coronal images by assessing the position of the pedicles (Perdriolle & Vidal, 1987).

Congenital scoliosis develops as a consequence of an incomplete formation of a vertebra during gestation, that is, hemivertebra, or lack of separation of the vertebrae during their development. In some individuals, these two mechanisms are simultaneously observable, resulting in severe deformities and health consequences. Congenital defects are also commonly associated with compensatory curves aimed at keeping an acceptable global alignment of the spine.

Infantile and juvenile scoliosis share several characteristics, and are therefore grouped under the name early onset scoliosis. This spinal disorder is commonly associated with a fast progression of the scoliotic curves, which may have severe consequences on the development of lungs, which takes place at the same time (Yang, Andras, Redding, & Skaggs, 2016). Its treatment is particularly challenging because it should aim at correcting the deformity or limiting its progression while allowing for a physiological growth of the trunk; multiple surgeries are frequently performed, resulting in a high rate of complications.

Adolescent scoliosis is relatively more frequent in comparison with cases with an earlier age of onset, having a prevalence of 1%−3% of the general population (Lowe et al., 2000). It is more common in female subjects, and may have a very different appearance in the affected individuals (Fig. 14.11). Indeed, classification schemes of adolescent curve types based on planar radiographs have been introduced. The King−Moe scheme has been widely used for decades and relies on the coronal radiographic projection, and subdivides the possible deformity patterns into five types (King, Moe, Bradford, & Winter, 1983). This classification has been criticized in recent years due to the lack of consideration of the sagittal profile and of the three-dimensional aspect of scoliosis, that is, the vertebral rotation in the transverse plane. Nowadays, adolescent scoliosis is classified

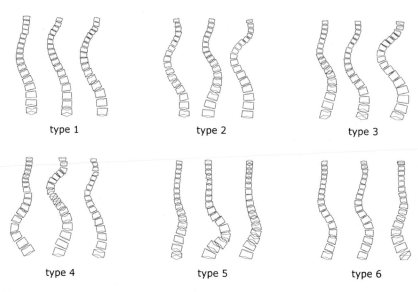

FIGURE 14.11

Examples of the coronal appearance of the thoracolumbar spine of adolescent subjects with different types of scoliotic deformities, classified based on the Lenke scheme.

based on the Lenke scheme, which takes into account the coronal appearance of the spine as well as the thoracic and lumbar sagittal curvature (Lenke et al., 2002). The flexibility of the spine in lateral bending, also assessed on coronal radiographs, is taken into account as well in order to distinguish the structural curves, that is, the curves that do not tend to be reduced in bending, from the nonstructural or compensatory curves, which are more flexible in bending.

A word that is commonly associated with scoliosis is "idiopathic," meaning that its cause is unknown. Indeed, in pediatric and adolescent subjects, only 15% of the cases of scoliosis can be attributed to a specific cause, which might be a development defect or a neuromuscular disorder such as muscular dystrophy (Lowe et al., 2000). In some cases, scoliosis is iatrogenic, that is, it develops as a consequence of a surgical intervention such as a pneumonectomy. Recent studies indicated that the etiopathogenesis of scoliosis is likely multifactorial, including a genetic component and an involvement of the remodeling mechanism of bone tissue (Stokes, Spence, Aronsson, & Kilmer, 1996). Curve progression has been described as a vicious circle, in which a coronal wedging or curvature induces an asymmetry of loads and stresses in the spine, which triggers an asymmetric growth resulting in a progression of the curve. However, the pathogenesis of idiopathic scoliosis remains an open and active field of research due to the lack of evidence and contradictory observations reported in the literature (Kouwenhoven & Castelein, 2008).

Scoliosis can be treated either conservatively or surgically. Physical therapy is commonly prescribed; both general physiotherapy and scoliosis-specific exercises are commonly performed, being normally tolerable by the patient and having no side effects. However, evidence about their efficacy is rather poor (Romano et al., 2013). Bracing aims at limiting the progression of the curve and at guiding a correct, physiological development of the spine shape by applying corrective forces on the trunk. Brace treatment is normally prescribed for skeletally immature subjects, and corrects both the coronal shape of the spine by applying lateral forces to it and the rib hump, which is commonly associated with the deformity. Braces should be worn for 2−5 years for several hours per day (18−23 hours); therefore, they are associated with patient discomfort and compliance issues. Similar to physical therapy, evidence about the efficacy of bracing is insufficient; although it is not clear whether bracing improves the health-related quality of life, its effect in reducing or stopping the curve progression has been consistently observed in clinical studies (Negrini et al., 2016).

The surgical correction of scoliosis is performed on more severe, progressive curves with Cobb angle greater than 40 degrees; whereas it is most commonly done on patients with a certain degree of skeletal maturity, the surgical treatment of cases of early-onset scoliosis is not infrequent. The aim of surgery is correcting the deforming by manually displacing and rotating the vertebrae in the scoliotic region, and fixing them by means of instrumentation, described in detail in Chapter 22. The selection of the region to be instrumented and of the target of correction is typically based on the type and severity of the curves (Lenke et al., 2002). The surgery is highly invasive but is associated with high success rates and patient satisfaction; postoperative complications such as implant failure or loss of correction are, however, rather frequent.

TRAUMA AND FRACTURES

Vertebral fractures due to domestic falls, traffic accidents, and sports are relatively frequent in young, middle-aged, and elderly individuals. The injury mechanism is described as high-energy trauma when the accident involves high values of forces and moments applied to vertebrae, sufficient to cause mechanical failure also in good-quality bone, as found in young and healthy subjects. In elderly, osteoporotic subjects as well as in patients suffering from bone disorders such as osteogenesis imperfecta, vertebral fractures may also result from low-energy trauma; in some cases, the single traumatic event triggering the fracture cannot even be identified.

Although vertebral fractures may involve any region of the spine depending on the injury mechanisms, some vertebrae appear to be more frequently involved than others, especially in high-energy trauma. The region around the thoracolumbar junction, approximately between T10 and L2, is indeed the most common location of vertebral fractures (Lindtner, Schmoelz, Galbusera, & Wilke, 2018). Being a transitional zone between two regions with distinct mobility patterns, that is, the thoracic spine, which is stiffened by the presence of the ribs and facet joints limiting flexion−extension and the more flexible lumbar spine, the thoracolumbar junction may be subjected to higher stresses leading to a higher incidence of fractures.

Different mechanical loads during the injury events may have distinct fracture patterns, and, in turn, distinct instability and clinical consequences. Such fracture patterns have been described by Magerl and coworkers, who proposed a classification scheme, which gained widespread popularity (Magerl, Aebi, Gertzbein, Harms, & Nazarian, 1994). In this scheme, fractures are classified into three main types: compression, distraction, and rotational injuries, based on their appearance on radiographs and CT scans, which are further subdivided into 25 subgroups. Compression injuries were attributed to axial compression, were the most frequent in the examined pool of cases, and did not involve the posterior elements; distraction injuries resulted from hyperflexion or hyperextension and may have involved the posterior elements involving the ligaments; rotational injuries involved both the posterior and anterior elements and were attributed to torsional moments. The classification scheme was later improved by adding information about the possible neurological injury and two clinical modifiers regarding the condition of the posterior tension band and comorbidities, and by eliminating subgroups that were difficult to assess; the resulting AOSpine thoracolumbar spine injury classification system is nowadays considered the standard tool for the description of spinal fractures and for planning their treatment (Vaccaro et al., 2013).

In general, the choice of the treatment for a specific vertebral fracture depends on its degree of stability and on its neurological consequences. There is consensus about the need for stabilization of unstable fractures, and for the decompression of the spinal canal and foramina in the case of impingement of neural structures. The surgical stabilization consists of the implantation of pedicle screws and posterior rods, and can span one or more vertebral levels depending on the spinal region and on the degree of instability. Anterior implants aimed at restoring the support capability of the vertebral bodies, as well as cement augmentation aimed at improving screw purchase in osteoporotic bone, are also used in some cases; such fixation techniques are described in detail in Chapter 22. In several cases of stable fractures, the conservative treatment is sufficient, although elderly patients may benefit from the injection of bone cement, that is, vertebroplasty, in terms of early recovery and improved quality of life.

REFERENCES

Anderst, W. J., Donaldson, W. F., 3rd, Lee, J. Y., & Kang, J. D. (2015). Three-dimensional intervertebral kinematics in the healthy young adult. *Journal of Biomechanics*, *48*(7), 1286−1293. Available from https://doi.org/10.1016/j.jbiomech.2015.02.049.

Barrey, C., Roussouly, P., Perrin, G., & Le Huec, J.-C. (2011). Sagittal balance disorders in severe degenerative spine. Can we identify the compensatory mechanisms? *European Spine Journal: Official Publication of the European Spine Society, the European Spinal Deformity Society, and the European Section of the Cervical Spine Research Society*, *20*(Suppl. 5), 626−633. Available from https://doi.org/10.1007/s00586-011-1930-3.

Bashkuev, M., Reitmaier, S., & Schmidt, H. (2019). Relationship between intervertebral disc and facet joint degeneration: A probabilistic finite element model study. *Journal of Biomechanics*, *26*, 109518. Available from https://doi.org/10.1016/j.jbiomech.2019.109518.

Bassani, T., Galbusera, F., Luca, A., Lovi, A., Gallazzi, E., & Brayda-Bruno, M. (2019). Physiological variations in the sagittal spine alignment in an asymptomatic elderly population. *The Spine Journal: Official Journal of the North American Spine Society*, *19*(11), 1840−1849. Available from https://doi.org/10.1016/j.spinee.2019.07.016.

Bogduk, N., & Mercer, S. (2000). Biomechanics of the cervical spine. I: Normal kinematics. *Clinical Biomechanics*, *15*(9), 633−648. Available from https://doi.org/10.1016/s0268-0033(00)00034-6.

Cobb, J. (1948). Outline for the study of scoliosis. *Instructional Course Lectures*, *5*, 261−275. Available from https://ci.nii.ac.jp/naid/10027964210/.

Duval-Beaupere, G., Schmidt, C., & Cosson, P. H. (1992). A barycentremetric study of the sagittal shape of spine and pelvis: The conditions required for an economic standing position. *Annals of Biomedical Engineering*, *20*(4), 451−462, https://idp.springer.com/authorize/casa?redirect_uri = https://link.springer.

com/content/pdf/10.1007/BF02368136.pdf&casa_token = Kp3GNz7PiUsAAAAA:Z04rja8DbwRGCm1AZ QIrAn8xS9Y8wdKpkTumTdU9Mcl-X9uv7WpdzCAZYTJVtuMyTrxmLSo9XRzvCf8.

Fujiwara, A., Lim, T. H., An, H. S., Tanaka, N., Jeon, C. H., Andersson, G. B., & Haughton, V. M. (2000). The effect of disc degeneration and facet joint osteoarthritis on the segmental flexibility of the lumbar spine. *Spine (Philadelphia, Pa.: 1986)*, *25*(23), 3036–3044. Available from https://doi.org/10.1097/00007632-200012010-00011.

Galbusera, F., van Rijsbergen, M., Ito, K., Huyghe, J. M., Brayda-Bruno, M., & Wilke, H.-J. (2014). Ageing and degenerative changes of the intervertebral disc and their impact on spinal flexibility. *European Spine Journal: Official Publication of the European Spine Society, the European Spinal Deformity Society, and the European Section of the Cervical Spine Research Society*, *23*(Suppl. 3), S324–S332. Available from https://doi.org/10.1007/s00586-014-3203-4.

Hattori, S., Oda, H., & Kawai, S. (1981). Cervical intradiscal pressure in movements and traction of the cervical. *Zeitschrift fur Orthopadie und ihre Grenzgebiete*, *119*, 568–569. Available from https://ci.nii.ac.jp/naid/10030713617/.

Jackson, R. P., & McManus, A. C. (1994). Radiographic analysis of sagittal plane alignment and balance in standing. *Spine (Philadelphia, Pa.: 1986)*, *19*(14), 1611–1618. Available from https://doi.org/10.1097/00007632-199407001-00010.

Kettler, A., Rohlmann, F., Ring, C., Mack, C., & Wilke, H.-J. (2011). Do early stages of lumbar intervertebral disc degeneration really cause. *European Spine Journal: Official Publication of the European Spine Society, the European Spinal Deformity Society, and the European Section of the Cervical Spine Research Society*, *20*(4), 578–584. Available from https://doi.org/10.1007/s00586-010-1635-z.

King, H. A., Moe, J. H., Bradford, D. S., & Winter, R. B. (1983). The selection of fusion levels in thoracic idiopathic scoliosis. *The Journal of Bone and Joint Surgery. American Volume*, *65*(9), 1302–1313. Available from http://www.srf-india.org/wp-content/uploads/2017/02/pub022.pdf.

Kouwenhoven, J.-W. M., & Castelein, R. M. (2008). The pathogenesis of adolescent idiopathic scoliosis: Review of the literature. *Spine (Philadelphia, Pa.: 1986)*, *33*(26), 2898–2908. Available from https://doi.org/10.1097/BRS.0b013e3181891751.

Le Huec, J. C., Charosky, S., Barrey, C., Rigal, J., & Aunoble, S. (2011). Sagittal imbalance cascade for simple degenerative spine and consequences: Algorithm of decision for appropriate treatment. *The Spine Journal: Official Journal of the North American Spine Society*, *20*(Suppl. 5), 699–703. Available from https://doi.org/10.1007/s00586-011-1938-8.

Le Huec, J. C., Saddiki, R., Franke, J., Rigal, J., & Aunoble, S. (2011). Equilibrium of the human body and the gravity line: The basics. *European Spine Journal: Official Publication of the European Spine Society, the European Spinal Deformity Society, and the European Section of the Cervical Spine Research Society*, *20*(5), 558–563. Available from https://doi.org/10.1007/s00586-011-1939-7.

Lenke, L. G., Betz, R. R., Clements, D., Merola, A., Haher, T., Lowe, T., . . . Blanke, K. (2002). Curve prevalence of a new classification of operative adolescent. *Spine (Philadelphia, Pa.: 1986)*, *27*(6), 604–611. Available from https://doi.org/10.1097/00007632-200203150-00008.

Lindtner, R. A., Schmoelz, W., Galbusera, F., & Wilke, H.-J. (2018). *Chapter 22 — Biomechanics of vertebral fractures and their treatment. Biomechanics of the spine* (pp. 395–407). Academic Press. Available from https://doi.org/10.1016/B978-0-12-812851-0.00022-7.

Lowe, T. G., Edgar, M., Margulies, J. Y., Miller, N. H., Raso, V. J., Reinker, K. A., & Rivard, C. H. (2000). Etiology of idiopathic scoliosis: Current trends in research. *The Journal of Bone and Joint Surgery. American Volume*, *82*(8), 1157–1168. Available from https://doi.org/10.2106/00004623-200008000-00014.

Magerl, F., Aebi, M., Gertzbein, S. D., Harms, J., & Nazarian, S. (1994). A comprehensive classification of thoracic and lumbar injuries. *European Spine Journal: Official Publication of the European Spine Society,* the European Spinal Deformity Society, and the European Section of the Cervical Spine Research Society, *3*(4), 184–201. Available from https://doi.org/10.1007/bf02221591.

Nachemson, A. (1959). Measurement of intradiscal pressure. *Acta Orthopaedica Scandinavica, 28*, 269–289. Available from https://doi.org/10.3109/17453675908988632.

Negrini, S., Minozzi, S., Bettany-Saltikov, J., Chockalingam, N., Grivas, T. B., Kotwicki, T., . . . Zaina, F. (2016). Braces for idiopathic scoliosis in adolescents. *Spine (Philadelphia, Pa.: 1986), 41*(23), 1813–1825. Available from https://doi.org/10.1097/BRS.0000000000001887.

Oxland, T. R., Lin, R. M., & Panjabi, M. M. (1992). Three-dimensional mechanical properties of the thoraco-lumbar junction. *Journal of Orthopaedic Research, 10*(4), 573–580. Available from https://doi.org/10.1002/jor.1100100412.

Panjabi, M. M., Crisco, J. J., Vasavada, A., Oda, T., Cholewicki, J., Nibu, K., & Shin, E. (2001). Mechanical properties of the human cervical spine as shown by three-dimensional load-displacement curves. *Spine (Philadelphia, Pa.: 1986), 26*(24), 2692–2700. Available from https://doi.org/10.1097/00007632-200112150-00012.

Panjabi, M. M., Oda, T., Crisco, J. J., 3rd, Dvorak, J., & Grob, D. (1993). Posture affects motion coupling patterns of the upper cervical spine. *Journal of Orthopaedic Research, 11*(4), 525–536. Available from https://doi.org/10.1002/jor.1100110407.

Pearcy, M. J. (1985). Stereo radiography of lumbar spine motion. *Acta Orthopaedica Scandinavica. Supplementum, 212*, 1–45. Available from https://doi.org/10.3109/17453678509154154.

Penning, L. (1978). Normal movements of the cervical spine. *American Journal of Roentgenology, 130*(2), 317–326. Available from https://doi.org/10.2214/ajr.130.2.317.

Perdriolle, R., & Vidal, J. (1987). Morphology of scoliosis: Three-dimensional evolution. *Orthopedics, 10*(6), 909–915. Available from http://search.proquest.com/openview/6b3f1f6f1ad342727b6d2795aef27764/1.pdf?pq-origsite = gscholar&cbl = 47931&casa_token = HJDKmUKIe3IAAAAA:cJ41adCZ5VsaD0969YGIUwabw6PH93CC9bCxncfuBv4HBN00fVDJDnQQjeHNTnbHwgqLyVjJ.

Polga, D. J., Beaubien, B. P., Kallemeier, P. M., Schellhas, K. P., Lew, W. D., Buttermann, G. R., & Wood, K. B. (2004). Measurement of in vivo intradiscal pressure in healthy thoracic intervertebral discs. *Spine (Philadelphia, Pa.: 1986), 29*(12), 1320–1324. Available from https://doi.org/10.1097/01.brs.0000127179.13271.78.

Rohlmann, A., Bergmann, G., & Graichen, F. (1997). Loads on an internal spinal fixation device during walking. *Journal of Biomechanics, 30*(1), 41–47. Available from https://doi.org/10.1016/s0021-9290(96)00103-0.

Rohlmann, A., Gabel, U., Graichen, F., Bender, A., & Bergmann, G. (2007). An instrumented implant for vertebral body replacement that measures loads. *Medical Engineering & Physics, 29*(5), 580–585. Available from https://doi.org/10.1016/j.medengphy.2006.06.012.

Romano, M., Minozzi, S., Zaina, F., Saltikov, J. B., Chockalingam, N., Kotwicki, T., . . . Negrini, S. (2013). Exercises for adolescent idiopathic scoliosis: A Cochrane systematic review. *Spine (Philadelphia, Pa.: 1986), 38*(14), E883–E893. Available from https://doi.org/10.1097/BRS.0b013e31829459f8.

Roussouly, P., Gollogly, S., Berthonnaud, E., & Dimnet, J. (2005). Classification of the normal variation in the sagittal alignment of the human lumbar spine and pelvis in the standing position. *Spine (Philadelphia, Pa.: 1986), 30*(3), 346–353. Available from https://doi.org/10.1097/01.brs.0000152379.54463.65.

Sato, K., Kikuchi, S., & Yonezawa, T. (1999). In vivo intradiscal pressure measurement in healthy individuals and in patients with ongoing back problems. *Spine (Philadelphia, Pa.: 1986), 24*(23), 2468–2474. Available from https://doi.org/10.1097/00007632-199912010-00008.

Stokes, I. A., Spence, H., Aronsson, D. D., & Kilmer, N. (1996). Mechanical modulation of vertebral body growth. Implications for scoliosis. *Spine (Philadelphia, Pa.: 1986), 21*(10), 1162–1167. Available from https://doi.org/10.1097/00007632-199605150-00007.

Urban, J. P., & McMullin, J. F. (1985). Swelling pressure of the inervertebral disc: Influence of proteoglycan and collagen contents. *Biorheology, 22*(2), 145–157. Available from https://doi.org/10.3233/bir-1985-22205.

Vaccaro, A. R., Oner, C., Kepler, C. K., Dvorak, M., Schnake, K., Bellabarba, C., ... Vialle, L. (2013). AOSpine thoracolumbar spine injury classification system: Fracture description, neurological status, and key modifiers, & AOSpine Spinal Cord Injury & Trauma Knowledge Forum*Spine (Philadelphia, Pa.: 1986)*, *38*(23), 2028–2037. Available from https://doi.org/10.1097/BRS.0b013e3182a8a381.

Wen, N., Lavaste, F., Santin, J. J., & Lassau, J. P. (1993). Three-dimensional biomechanical properties of the human cervical spine in vitro. *I. Analysis of normal.*

Wilke, H.-J., Herkommer, A., Werner, K., & Liebsch, C. (2017). In vitro analysis of the segmental flexibility of the thoracic spine. *PLoS One*, *12*(5), e0177823. Available from https://doi.org/10.1371/journal.pone.0177823.

Wilke, H. J., Neef, P., Caimi, M., Hoogland, T., & Claes, L. E. (1999). New in vivo measurements of pressures in the intervertebral disc in daily. *Spine (Philadelphia, Pa.: 1986)*, *24*(8), 755–762. Available from https://doi.org/10.1097/00007632-199904150-00005.

Wilke, H.-J., Rohlmann, F., Neidlinger-Wilke, C., Werner, K., Claes, L., & Kettler, A. (2006). Validity and interobserver agreement of a new radiographic grading system. *European Spine Journal: Official Publication of the European Spine Society, the European Spinal Deformity Society, and the European Section of the Cervical Spine Research Society*, *15*(6), 720–730, https://idp.springer.com/authorize/casa? redirect_uri = https://link.springer.com/article/10.1007/s00586-005-1029-9&casa_token = gmEgzXs9NyEAA AAA:fRgrSCLw3vDiB7dDkYZcotIi75D3HyzWz8ZucNpaVV3W8H_knDwRJPqlprge3KUJRqO1Tsc-kjC8HLk.

Wilke, H. J., Wolf, S., Claes, L. E., Arand, M., & Wiesend, A. (1995). Stability increase of the lumbar spine with different muscle groups. A biomechanical in vitro study. *Spine*, *20*(2), 192–198. Available from https://doi.org/10.1097/00007632-199501150-00011.

Willems, J. M., Jull, G. A., & Ng, J. K.-F. (1996). An in vivo study of the primary and coupled rotations of the thoracic spine. *Clinical Biomechanics*, *11*(6), 311–316. Available from https://doi.org/10.1016/0268-0033(96)00017-4.

Yamamoto, I., Panjabi, M. M., Crisco, T., & Oxland, T. (1989). Three-dimensional movements of the whole lumbar spine and lumbosacral joint. *Spine (Philadelphia, Pa.: 1986)*, *14*(11), 1256–1260. Available from https://doi.org/10.1097/00007632-198911000-00020.

Yang, S., Andras, L. M., Redding, G. J., & Skaggs, D. L. (2016). Early-onset scoliosis: A review of history, current treatment, and future directions. *Pediatrics*, *137*(1). Available from https://doi.org/10.1542/peds.2015-0709.

BIOMECHANICS OF THE SHOULDER JOINT

15

Paolo Dalla Pria

Waldemar Link GmbH & Co. KG, Hamburg, Germany

SKELETAL ANATOMY

The movements of the upper limb are very sophisticated thanks to a complex chain of joints in which the shoulder girdle has an important role. The shoulder girdle is composed of the clavicle and the scapula, which articulates with the head of the humerus (Fig. 15.1).

The shoulder girdle is connected to the axial skeleton through the sternoclavicular joint, which is a synovial saddle joint connecting the sternum with the clavicle. The clavicle is then connected to the scapula through the acromion, a large apophysis of the scapula which is the highest part of the shoulder (from ancient Greek ἄκρος (ákros), "highest" + ὦμος (ômos), "shoulder"). The acromioclavicular joint is a plane synovial joint (Fig. 15.2).

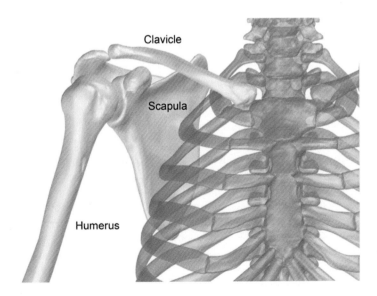

Clavicle

Scapula

Humerus

FIGURE 15.1

The shoulder girdle. The shoulder girdle consists of the clavicle and scapula, the bones which connect to the arm.

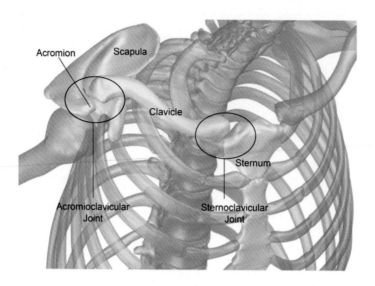

FIGURE 15.2

The shoulder girdle. The shoulder girdle is connected to the axial skeleton through the sternoclavicular joint.

The scapula is a thin, flat, triangular, spade-shaped bone (from Latin scapula which is from scabere, "dig") placed on the posterior and superior wall of the thorax, which serves as a site of attachment for muscles. The posterior flat surface is divided into two parts by a thin cortical transversal wall, called the scapular spine (Fig. 15.3).

The lateral extension of the scapular spine is called acromion, which is the most stressed part of the scapular bone. In the anterior part of the scapula, next to the glenoid, another prominence called the coracoid process, shaped like a raven beak (from ancient Greek κόραξ (kórax), "raven"), is a site of attachment for muscles and ligaments.

The lateral part of the scapula is the glenoid. The shape of the glenoid vault is like an inverted comma, with a height in the coronal plane of about 30−40 mm and a width of approximately 22−30 mm in the transverse plane (Churchill, Brems, & Kotschi, 2001; Rockwood, 2009) (Fig. 15.3).

The widest movements of the arm are reached around the center of the humeral head through the glenohumeral joint, a "ball and socket" joint (like the hip joint) with the humeral head acting as the ball and the glenoid fossa acting as the socket. The very large range of motion of the arm is possible due to the very low mechanical constraint between the humeral head and the glenoid. Unlike the hip joint, in the shoulder the "ball" articulates against—and not within—the "socket." The radial mismatch between head and glenoid is about 0.7 (low congruency) in the majority of people (McPherson, Friedman, An, Chokesi, & Dooley, 1997; Randelli & Gambrioli, 1986; Saha, 1971), and the constraint of the head in the socket is very low because the arc of curvature of the glenoid in the coronal plane is only about 66 degrees and in the transverse plane is only about 46 degrees (McPherson et al., 1997). In the majority of people, the glenoid bone is superomedially tilted to enhance the elevation of the arm, from 5 to 15 degrees (Basmajian & Bazant, 1959; Rockwood, 2009). In the prosthetic glenohumeral joint, this anatomical feature of the glenoid bone can be counterproductive because it does not promote the stability of the implant.

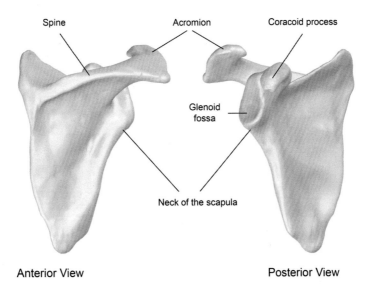

Spine   Acromion   Coracoid process

Glenoid
fossa

Neck of the scapula

Anterior View   Posterior View

FIGURE 15.3

The scapular bone. The glenoid fossa is coated by cartilage and articulates with the humeral head. The large processes (coracoid and acromion) are the site of the muscles of the upper humerus.

The humeral head is about one-third of a sphere, and its diameter ranges from 37 to 57 mm (average 46 mm) (Boileau & Walch, 1997; Rockwood, 2009). The real shape is ellipsoidal, rather than spherical. The long axis lies on the frontal plane, and the ratio between long and short axis depends on the diameter of the head: the larger the head, the larger the ratio between the axes (Humphrey, Sears, & Curtin, 2016; Iannotti, Gabriel, Schneck, Evans, & Misra, 1992).

The head of the humerus is superomedially oriented (like the head of the femur), and the neck-shaft angle, equivalent to the cervico-diaphyseal angle of the femur, is about 130 degrees. The humeral head is retroverted about 30 degrees to better match the scapular surface, with the scapula being 30 degrees anteverted (Rockwood, 2009) (Fig. 15.4).

On the sagittal plane, there is normally a posterior offset of the center of the humeral head with respect to the axis of the humerus. This is not surprising in the long bones; for example, the center of the femoral head is offset with respect to the diaphyseal axis. In the femur such offset is evident due to the length of the femoral neck, while in the humerus the neck is very short. On the sagittal plane, the distance between the center of the humeral head and the axis of the humerus is about 5 mm (range 2−12 mm) (Humphrey, Sears, & Curtin, 2016; Iannotti, Gabriel, Schneck, Evans, & Misra, 1992; Zhang et al., 2016) . Unrelated to the offset between humeral axis and center of the head, two different necks are defined in the humerus. The anatomical neck is defined as the base level of the humeral head and is relevant for the resection level of the humeral head in a shoulder arthroplasty, while the surgical neck is located at the end of the proximal metaphysis and represents the transition between the metaphysis and the diaphysis. The surgical neck of the humerus is the most common fracture site on the proximal humerus. Just under the anatomical neck two

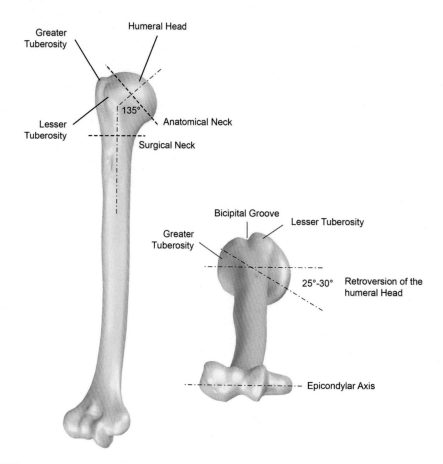

FIGURE 15.4

Geometrical parameters of the humerus. The neck-shaft angle is defined as the angle between the diaphyseal axis of the humerus and the axis perpendicular to the anatomical neck. The retroversion angle is defined as the angle on the transversal plane between the transepicondylar axis of the distal humerus and the axis of symmetry of the humeral head.

tuberosities are present, the lesser tuberosity, located in the anterior part of the humerus, and the greater tuberosity, located laterally. The tuberosities are divided by the bicipital groove and sulcus, which hosts the tendon of the long head of the biceps muscle.

SOFT TISSUES

The intrinsic mechanical instability of the glenohumeral joint is counterbalanced by a complex interaction of muscles and ligaments.

CAPSULE AND LIGAMENTS

The glenoid is surrounded by an annular ligament called the labrum, whose function is similar to the menisci of the knee joint (Fig. 15.5). Because the radius of curvature of the glenoid surface is larger than the radius of the humeral head (Humphrey et al., 2016; Iannotti et al., 1992), the labrum absorbs the peripheral loadings of the humeral head and allows some degree of translation of the head (Graichen et al., 2000; Lin et al., 2007; McMahon et al., 1995). Similar to a gasket, the labrum also maintains the synovial fluid between the humeral head and the glenoid in order to reduce friction.

Like all joints lubricated by synovial fluid, the glenohumeral joint is coated by an elastic and sealed sleeve, called the capsule. The capsule is evenly attached to the rim of the glenoid in the scapular side, while the humeral attachment is irregular, starting laterally on the rim of the anatomical neck, next to the greater tuberosity and ending medially at the level of the surgical neck. This means that a small medial portion of the humerus is intraarticular, even though not covered by the cartilage. The shape of the capsule allows a long stretching of the fibrous tissue, and this is necessary due to the very large range of motion of the glenohumeral joint.

However, the intrinsic mechanical instability of the glenohumeral joint cannot be fully counterbalanced by the capsule alone, which is therefore reinforced by three glenohumeral ligaments (superior, medial, and inferior) that connect the anterior base of the anatomical neck with the anterior surface of the glenoid vault and at a more external layer humerus and scapula are bridged by the coracohumeral ligament, which starts from the coracoid up to the greater tuberosity (Fig. 15.6).

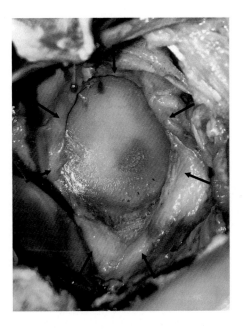

FIGURE 15.5

The labrum. The labrum is an annular ligament surrounding the glenoid vault.

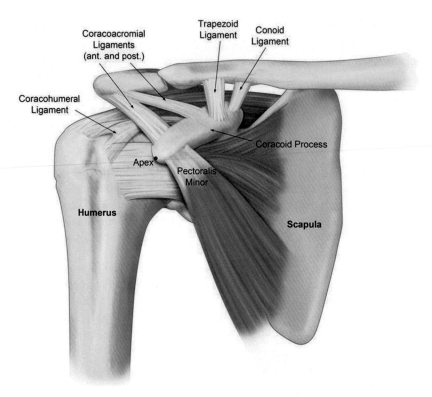

FIGURE 15.6

Ligaments of the shoulder joint. The three glenohumeral ligaments connect humerus and glenoid vault. The coracoid process is connected to the greater tuberosity of the humerus through the coracohumeral ligament, to the anterior surface of the acromion through the coracoacromial ligament and to the anterior surface of the clavicle through the coracoclavicular ligament.

From Chahla, J., Marchetti, D. C., Moatshe, G., Ferrari, M. B., Sanchez, G., Brady, A. W., ... Provencher, M. T. (2018). Quantitative assessment of the coracoacromial and the coracoclavicular ligaments with 3-dimensional mapping of the coracoid process anatomy: A cadaveric study of surgically relevant structures. Arthroscopy - Journal of Arthroscopic and Related Surgery, 34*(5), 1403–1411.* *https://doi.org/10.1016/j.arthro.2017.11.033.*

Two other reinforcing ligaments connect the coracoid process to the surrounding structures: the coracoacromial ligament, a strong elastic roof between the superior surface of the coracoid and the anterior surface of the acromion, and the coracoclavicular ligament, divided into two ligaments called the trapezoid ligament and the conoid ligament, starting respectively from the middle and the posterior part of the coracoid up to the anterior surface of the clavicle (Fig. 15.6). It can be observed that all the reinforcing ligaments are located anteriorly, while no ligaments are present in the posterior part of the humerus. This is due to two reasons: the motion of the arm has a high prevalence in anterior, and the glenohumeral joint is posteriorly protected by the scapular spine. Posterior instability of the shoulder is a rare condition that represents about 10% of shoulder instability (Bäcker, Galle, Maniglio, & Rosenwasser, 2018).

MUSCLES

The intrinsic muscles of the shoulder connect the scapula and/or clavicle to the humerus. They are the deltoid and the rotator cuff muscles. The extrinsic muscles are partially related to the anatomy of the shoulder and affect the function (biceps, triceps, pectoralis major, coracobrachialis, pectoralis minor, teres major, latissimus dorsi, trapezius, levator scapulae, subclavius, sternocleidomastoid). Only the intrinsic muscles will be described.

The deltoid

The deltoid muscle is the largest and most important of the glenohumeral muscles (Fig. 15.7). It consists of three major sections having different tasks (Rockwood, 2009):

- the anterior deltoid, originating from the lateral third of the clavicle is responsible for flexion and medial rotation of the arm;
- the middle third of the deltoid, originating from the acromion is responsible for the abduction of the arm (up to 90 degrees);
- the posterior deltoid, originating from the spine of the scapula is responsible for extension and lateral rotation of the arm.

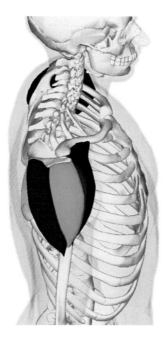

FIGURE 15.7

The deltoid muscle. The deltoid muscle consists of three major sections: anterior (red), middle (green), and posterior (blue).

From (n.d.-a). Available at: https://it.m.wikipedia.org/wiki/File:Deltoid_muscle_top10.png.

In the lateral side of the humeral shaft, a tubercle can be seen. This is the deltoid tubercle, the humeral insertion of the deltoid muscle.

The rotator cuff

The rotator cuff is composed of four muscles (supraspinatus, infraspinatus, teres minor, subscapularis) (Fig. 15.8).

which appear separate superficially but in their deeper regions they are associated with each other, with the capsule underneath, and with the tendon of the long head of the biceps. The deltoid muscle and the muscles of the cuff show an interesting symbiosis and opposition. The muscles of the cuff cooperate with the deltoid to achieve the same movements and, at the same time, the direction of the forces are in opposition in order to maintain the stability of the joint (Rockwood, 2009).

- The supraspinatus muscle lies on the superior portion of the scapula (as the name suggests, above the scapular spine) and is attached to the humerus in the lateral part of the greater tuberosity. It keeps the humeral head in contact with the glenoid vault and avoids the superior shifting of the arm. The supraspinatus has the most prominent elevator moment arms during early abduction in both the coronal and scapular planes (30 degrees anterior to the frontal plane) as well as in flexion (Hik & Ackland, 2018).
- The infraspinatus lies below the scapular spine and its humeral attachment is posterior to the supraspinatus, on the greater tuberosity. It is a powerful external rotator of the humerus and

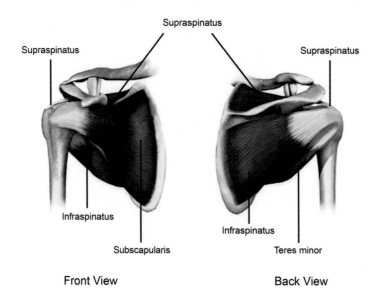

FIGURE 15.8

The cuff of the rotator muscles. The rotator cuff is composed of four muscles: supraspinatus, infraspinatus, teres minor, subscapularis.

Modified from (n.d.-b). Available at: https://www.unionpt.com/seattle-physical-therapy-for-rotator-cuff-injuries/.

accounts for as much as 60% of external rotation force (Colachis, Strohm, & Brechner, 1969) and plays a role as a stabilizer against posterior subluxation (Bäcker et al., 2018).

- The teres minor lies below the infraspinatus, both on the scapular and on the humeral sides. It improves the extrarotation of the humerus (accounts for up to 45% of external rotation force) and is an important stabilizer of the joint, being a strong depressor of the head (Crouch, Plate, Li, & Saul, 2013).
- The subscapularis is the anterior portion of the rotator cuff and equivalent of the infraspinatus on the anterior side of the shoulder. Characterized by a large surface of attachment on the scapular side, its humeral attachment is located on the lesser tuberosity, which is separated from the greater tuberosity by the bicipital groove. The subscapularis functions as an internal rotator and a passive stabilizer to prevent anterior subluxation and serves to depress the humeral head resisting the shear force of the deltoid (Inman, Saunders, & Abbott Leroy, 1996).

FUNCTIONAL ANATOMY

The range of motion of the shoulder complex is very large and multiplanar. The arm can move up to 180 degrees in elevation (upward lateral movement of humerus out to the side, away from the body, in the plane of the scapula); up to 170 degrees in flexion (straight anteriorly) and extension (straight posteriorly) and in horizontal adduction and abduction (anterior and posterior rotation in the horizontal plane); up to 150 degrees in internal and external rotation (Steindler, 1955). In the normal activities of daily living, such a wide range of motion is not required. Namdari et al. (2011) used an electromagnetic tracking system to record 10 functional tasks of the American Shoulder and Elbow Score on healthy volunteers, and they found that the average shoulder motions required to perform the 10 functional tasks were flexion 121 degrees, extension 46 degrees, abduction 128 degrees, external rotation 59 degrees (with arm 90 degrees abducted), and internal rotation 102 degrees (arm at the side). This motion, which represents the composite motion of several joints, occurs primarily in the glenohumeral joint; extreme positions require rotation of the sternoclavicular and acromioclavicular joints.

The glenohumeral joint is the most important joint of the shoulder girdle and has the widest angular range of motion. During the motion, before the extreme limits, the head remains centered in the glenoid vault but thanks to the low mechanical congruency between head and glenoid the head can translate of few millimeters in superoinferior and anteroposterior direction, pushing against the inner rim of the labrum (Graichen et al., 2000). Both glenohumeral and hip joints are ball and socket joints, but the kinematics of the glenohumeral joint is more similar to the knee joint. The hip joint, due to its high intrinsic stability, allows only the spinning of the femoral head, that is, only three rotational degrees of freedom are allowed. The glenohumeral joint is a low congruency joint; therefore, rolling and sliding of the humeral head are also permitted, similar to the motion of the femoral condyles against the tibial condyles, and the glenoid labrum acts like the tibial menisci. The glenohumeral joint has five degrees of freedom even though the translations of the humeral head are quite limited.

During the first 30 degrees of abduction, the glenohumeral joint is mainly responsible for the movement of the arm, then the acromioclavicular and the sternoclavicular joints start moving together. The clavicle is highly movable and affects large movements in elevation and flexion. During the elevation of the arm, the clavicle rotates about 30−35 degrees upwards at the sternoclavicular joint with the maximum at about 130 degrees of elevation and 10 degrees forward up to 40 degrees of elevation. Then, no change takes place during the next 90 degress and at the terminal arc of elevation an additional 15−20 degrees of forward rotation occurs (Inman et al., 1996).

The acromioclavicular joint is a plane joint, and its motion is limited. Three rotations are possible: anteroposterior rotation of the clavicle on the scapula, superoinferior rotation, and anterior and posterior axial rotation. Of these, anteroposterior rotation of the clavicle with respect to the acromion is approximately three times as great as superoinferior rotation (Collins, Tencer, Sidles, & Matsen, 1992). Superoinferior rotation of the clavicle is quite limited at the acromioclavicular joint.

Thanks to the mutual action of the mentioned joints, both the arms can be elevated up to 150 degrees simultaneously on the scapular plane, up to the impingement with the acromion. A further elevation up to 180 degrees is possible after an extrarotation of the arms, moving the greater tuberosity behind the acromion. However, if only one arm is abducted, the maximum elevation on the scapular plane is about 180 degrees because an additional movement of about 30 degrees is given by the flexion of the trunk.

During the arm elevation, the joints of the shoulder girdle move simultaneously and not sequentially. Because the glenohumeral joint is the most important for daily activities, it is of high interest to know the kinematical relationship between the glenohumeral joint and the others, which are grouped together as scapulothoracic joints. The ratio between the angular movement of the glenohumeral joint and the scapulothoracic joints of the shoulder girdle is called scapulohumeral rhythm. The sum of the two angular movements is the abduction angle (in the scapular plane), or more generally it is called the humerothoracic angle.

The scapulohumeral rhythm allows the muscles crossing the joints to operate at the optimal strain and to rotate the glenoid upward during high degrees of abduction, supporting some weight of the upper limb and reducing the effort of the muscles. In the past, several authors studied the scapulohumeral rhythm and the results are quite similar, although with some differences (Freedman & Munro, 1966; Inman et al., 1996; Nobuhara, 1987; Poppen & Walker, 1978; Reeves, Jobbins, & Flowers, 1972). It was demonstrated that the average scapulohumeral rhythm is about 2:1 during the elevation from 0 to 180 degrees, that is, for each two degrees of glenohumeral elevation there is one degree of scapular upward rotation. Almost all studies demonstrated that in the first range of motion, up to 30 degrees, only the glenohumeral joint is active, while the influence of the scapulothoracic joints is negligible. In the previous studies, the scapulohumeral rhythm was considered linear along the whole arc of motion, at least after 30 degrees of abduction, but more recent studies (Ruiz Ibán et al., 2020; Scibek & Carcia, 2012; Zdravkovic, Alexander, Wegener, Spross, & Jost, 2020) revealed a variable rhythm, according to the elevation angle. For example, according to Scibek & Carcia (2012), the scapula contributes 2.5% of total motion for the first 30 degrees of shoulder elevation, between 20% and 38% for 30−90 degrees of shoulder elevation, and 53% for 90−120 degrees of shoulder elevation. Zdravkovic et al. (2020) observed that patients with rotator cuff arthropathy show an altered and predominantly scapular motion pattern. This altered pattern of scapular position does not change after reverse total shoulder arthroplasty (rTSA). The pattern of scapular position after rTSA was found to differ between operated shoulders and nonoperated, contralateral shoulders. In particular, there was more upward rotation after surgery (Bruttel, Spranz, Wolf, & Maier, 2020; Kim, Lim, Lee, Kovacevic, & Cho, 2012). This phenomenon is probably due to the change of the movement patterns in pathological conditions, which persist after the implant of a prosthesis (like in the knee joint, where the automatic external rotation of the leg is lost due to the arthritis).

Most of the alterations/pathologies have specific patterns that are reflected in a well-defined and recognizable deviation from normality. A quick noninvasive examination therefore can be used to support diagnosis and to easily check for appropriateness of the treatment through an objective quantification of the compensation adopted. Shoulder kinematic and scapulohumeral rhythm tracking are therefore key factors to consider for identifying proper joint functionality and alterations. Its implementation into daily practice can be easily done to support clinical decisions in a noninvasive quick and effective approach.

New systems have been developed to record and evaluate the scapulohumeral rhythm in real time using wearable sensors positioned on the body and a receiver is able to reliably translate these sensors' data into the relative movement between trunk, arm, and scapula (Ruiz Ibán et al., 2020). The most advanced software can quickly compare the recorded data with the physiological scapulohumeral rhythms (available in the form of normality bands), allowing for easy and immediate identification of the alteration of the scapulohumeral rhythm (Fig. 15.9).

Soft tissues are fundamental to proper functioning of the shoulder motion, and a shoulder surgery may be successful only if the soft tissues are fully reconstructed and well-functioning. In the complex chain of joints and soft tissues, there is a position of the arm where the joint tissues are under the

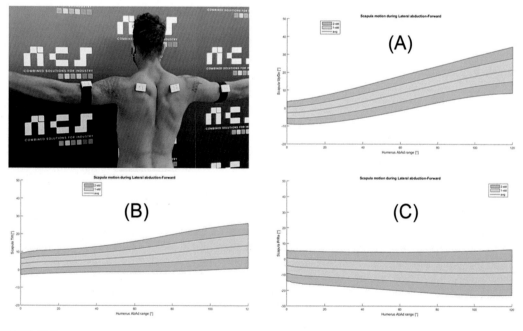

FIGURE 15.9

The SHoW Motion 3D kinematic tracking system (NCS Lab, Carpi, Italy). Scapular angles against the total shoulder abduction. (A) Upward/downward rotation. (B) Anterior/posterior tilt. (C) Internal/external rotation. The green shade areas are the prediction bands for 1 SD (light green) and 2 SD (dark green).

least amount of stress and the joint capsule has its greatest laxity. It is called the resting position of the joint, and it is also regarded as the position of minimal congruence between joint surfaces, allowing the greatest passive separation between articular surfaces. The resting position of a joint is generally considered to be the position of maximal mobility, and it is used by clinicians to evaluate the passive range of motion. In the glenohumeral, joint the resting position is generally considered to be located at a position in neutral rotation between 55 and 70 degrees of shoulder abduction in the scapular plane. The glenohumeral joint has both rotational and translational degrees of freedom, and there are two different resting positions: the mean resting position determined by rotational movement was located at about 50 degrees of abduction, while the mean resting position determined by translational movement was located at about 24 degrees of abduction (Lin et al., 2007).

STABILITY OF THE GLENOHUMERAL JOINT

The stability of the glenohumeral joint can be static (passive) or dynamic (active).

STATIC STABILITY

Static stability takes place in rest conditions or when the movements are limited with minimal effort of the muscles. In this condition, the labrum surrounding the glenoid vault avoids the subdislocation of the head. The labrum does not play a significant role in the anterior stability, but in the absence of the labrum there is a higher tendency of the inferior instability of the head (Halder, Kuhl, Zobitz, Larson, & An, 2001). This can be an issue after the implant of an anatomical prosthesis because the labrum is never saved. Another important passive stability factor is the negative pressure in the joint, which tends to keep the humeral head against the glenoid. In general, all the soft tissues surrounding the glenohumeral joint help to maintain the stability.

DYNAMIC STABILITY

Dynamic stability occurs when the effort of the muscles is relevant. In almost all movements of the arm, the deltoid is active and the force generated by this powerful muscle always has a higher component towards the diaphyseal axis of the humerus and a lower component far from the body (Fig. 15.10).

At low abduction angles (<30 degrees), the humeral axis and the glenoid surface are almost parallel; therefore, there is no bone support for the humeral head. Starting an elevation movement from this position, the force component of the deltoid along the humeral axis will act as a destabilizer for the joint, pushing the humerus upward, while at higher angles of abduction the arm is pushed against the glenoid and the deltoid will work as a stabilizer (Fig. 15.11).

In this variable scenario, the acromion also plays a role. The deltoid fibers are attached to the lateral surface of the acromion, and the acromion can be laterally short or long. The farther the distance between the lateral rim of the acromion and the center of the humeral head, the more destabilizing the deltoid muscle at the start of abduction because of the higher lateralizing component of the deltoid force. In contrast, the deltoid always works as an anterior stabilizer because the three sections of the muscle wrap the humeral head, preventing anterior dislocation.

The destabilizing effect of the deltoid is compensated mainly by the teres major, latissimus dorsi, and pectoralis major, which work as depressors of the humeral head. The infraspinatus, teres

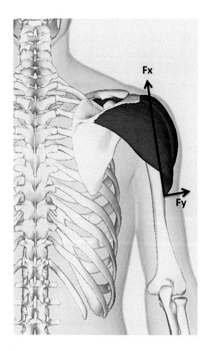

FIGURE 15.10

Deltoid action. At the beginning of the arm elevation the deltoid force pushes the humerus up and a lower perpendicular component tends to move the humerus apart from the joint.

From (n.d.-c). Available at: https://www.physio-pedia.com/File:Deltoid_action.png.

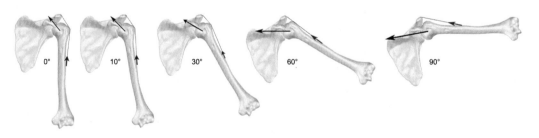

FIGURE 15.11

Stabilizing effect of the deltoid. At low abduction angles, the deltoid acts as a destabilizer for the joint, pushing the humerus upward, while at higher angles the arm is pushed against the glenoid and the deltoid works as a stabilizer.

minor, and subscapularis muscles are weak humeral depressors (Hik & Ackland, 2018) and their contribution is complex because they act as either an elevator or a depressor depending on joint position. The supraspinatus plays an important role in the stabilization of the glenohumeral joint (Terrier, Reist, Vogel, & Farron, 2007), acting as a humeral head depressor when the arm is fully abducted and externally rotated, and in this position the long head of the biceps muscle also acts as a humeral head depressor (Dalton & Snyder, 1989) (Fig. 15.8).

The mutual interaction of the supraspinatus and the deltoid is very interesting at the start of arm elevation. They are equally responsible for generating torque during arm flexion and abduction, and these movements can take place even if either the deltoid or the supraspinatus is missing (Hecker et al., 2020; Werthel, Bertelli, & Elhassan, 2017). In case of supraspinatus paralysis, a significantly larger deltoid force is required to initiate abduction, although only a small increase in the deltoid muscle force is required to achieve maximum abduction, and the translation of the humeral head does not change significantly with respect to a healthy condition of the cuff (Apreleva, Parsons, Warner, Fu, & Woo, 2000). However, the supraspinatus is a roof of the humeral head and prevents its superior shifting. In severe cases of lesion of the supraspinatus, even if the elevation of the arm can still take place, the balance of the glenohumeral joint is irreparably compromised and the destabilizing effect of the deltoid prevails at the beginning of abduction; the humeral head squeezes the bursa subacromialis and pushes against the acromial vault, causing pain and deficiency of the movements.

GLENOHUMERAL FORCES

The first detailed analysis of the forces acting on the glenohumeral joint on the scapular plane was done by Poppen & Walker (1978). They took into consideration all the active muscles during the isometric abduction of an extended arm recording their activity through electromyography. Lever arm and direction of the fibers were then measured graphically in the X-rays. Assuming a scapulo-humeral rhythm of 2:1, the equations of the equilibrium of forces and moments were then solved per several angles of abduction. However, there were more unknowns than permissible equations: three equilibrium equations contain four variables (force of the deltoid, force of the supraspinatus, horizontal and vertical components of the glenohumeral reaction force); hence, the solution was statically indeterminate and the authors assumed as an optimization function that the relative force in a given muscle is proportional to its cross-sectional area times the integrated electromyographic signal. They found that the resultant force reached a maximum of 0.89 times body weight (BW) at 90 degrees of abduction, while the shearing component up the face of the glenoid was a maximum of 0.42 times BW at 60 degrees of abduction. With arm elevated to 90 degrees and the elbow flexed to 90 degrees, the forces were reduced by 30% (about 62% of the BW).

Starting from the work of Poppen and Walker, in 1988 Dul developed a two-dimensional mathematical model to quantify shoulder muscle load, joint load, and endurance time in work situations, therefore assuming the elbow flexed at 90 degrees (Dul, 1988). To solve the equilibrium equations, the minimum fatigue optimization technique was selected. This technique is based upon two premises: (1) each muscle exhibits an endurance time-output characteristic which decreases continuously with increasing force, and (2) the system seeks to maximize the minimum endurance time of the muscles working in parallel, and hence the system minimizes muscular fatigue. According to Dul, at 87 degrees the joint force reaches a maximum force level of 43% of the BW, which is less than the value found by Poppen and Walker. Movements with the elbow flexed are less tiresome than with the arm fully extended and, as a consequence, also the glenohumeral reaction force is lower.

Applying numerical systems offers several advantages in comparison to the analysis on cadavers; therefore, Karlsson and Petersen developed a numerical system of 46 variables to simulate a three-dimensional model of the shoulder assuming the bones as rigid bodies and considering the glenohumeral joint a ball and socket joint (Karlsson & Peterson, 1992). There are only six equations in the force and moment equilibrium system, and 23 internal forces were considered. To find a unique solution, the authors used the optimization criterion based on minimizing the sum of squared muscle tensions. Simulating the elevation of the arm on the scapular plane with a hand load of 1 kg, a contact force about 0.8 times the BW in 60−90 degrees elevation was obtained.

A very interesting application of this computer model was used by Anglin, Wyss, & Pichora (2000) to determine the glenohumeral contact forces for tasks which are demanding of the shoulder but which would commonly be performed by elderly people. The functional tasks chosen were using the arms to stand up from and sit down into a chair, walking with a cane, lifting a 5 kg box to shoulder height with both hands, and lifting a 10 kg suitcase. Average contact forces ranged from 1.3 to 2.4 times BW (930−1720 N). Anglin analyzed the direction of loading in all planes which is relevant in the dynamic stress tests of the implants because eccentrical loading on the prosthetic glenoid can be cause of loosening.

Starting from the CT scan of a cadaver shoulder, Terrier et al. (2007) developed a three-dimension FEM model with six muscles: middle deltoid, anterior deltoid, posterior deltoid, supraspinatus, subscapularis, and infraspinatus combined with teres minor. Even with only six muscles, there were more unknown forces than available equilibrium equations. This indeterminate in the muscular forces was characterized by relating each muscle force to the middle deltoid force. Unlike other models, the humeral head and glenoid are not treated as a highly congruent ball and socket joint; on the contrary, the natural translation of the head is permitted. Considering this fact the authors compared a normal shoulder and the same without supraspinatus, simulating a common pathological condition. For the normal shoulder, the glenohumeral force was maximal at 82 degrees of abduction and corresponded to 81% of the BW (608 N). The middle deltoid force was maximal at 75 degrees of abduction and reached 25% of BW (190 N). Without supraspinatus, the maximum glenohumeral force was 8% higher at 70 degrees of abduction, while the maximum muscle forces were 30% higher at 64 degrees of abduction.

Numerical models offer the chance to modify the input parameters and to simulate different conditions. On the other hand, all the numerical models have to consider some optimization function to address the problem of the indeterminate between the number of the equilibrium equations and the number of the chosen muscles because the real mutual function between muscles is unknown. Apreleva et al. (2000) used a dynamic shoulder testing apparatus equipped with a force−moment sensor to directly measure reaction forces connected to fresh-frozen, full upper extremities obtained from human cadavers. The objective of this work was to measure the magnitude and direction of glenohumeral joint reaction forces under simulated active loading conditions using four different muscle force combinations in which the relative contributions of the deltoid and supraspinatus muscles to glenohumeral abduction were varied (Equal Force generated by supraspinatus and deltoid, Supraspinatus Dominant, Deltoid Dominant, and Supraspinatus Paralysis). Assuming that the upper extremity is approximately 5.2% of total BW (Poppen & Walker, 1978), the reaction force magnitudes were normalized to the weight of each upper extremity and expressed as a percentage of total BW. Expressed in this manner, the magnitude of the normalized joint reaction forces at maximum abduction for the Equal Force condition was 44% ± 8% of BW. The Supraspinatus Dominant condition resulted in an 8% increase of the reaction force from the Equal Force condition (48% ± 8% of BW). The Deltoid Dominant and Supraspinatus Paralysis conditions resulted respectively in 7% (41% ± 7% BW) and 18% (36% ± 7% BW) decreases in the reaction force from the Equal Force condition. When the reaction force at maximum abduction for the Supraspinatus Dominant condition was compared with that of the Supraspinatus Paralysis condition, a 25% decrease in joint compression was observed in the latter case. The same conclusion about the influence of the supraspinatus on the glenohumeral force reaction was later published by Terrier et al. (2007) who used a numerical model.

All numerical and physical models require some assumptions and simplifications because the real mutual function between muscles is unknown and the only way to know the realistic loads acting in vivo in a human joint is applying strain gauges in the implanted prosthesis. To measure forces and moments, Bergmann et al. (2011) transformed commercially available humeral prostheses, which are equipped with a nine-channel telemetry, six strain-gages, and an inductive power supply. The inner electronics are connected to the antenna by a heart-pacemaker feedthrough. The loads are monitored in real time and stored with the subject's video images for detailed analyses. In vivo glenohumeral joint loads during forward flexion and abduction were measured in six patients operated due to osteoarthritis with a still well-functioning rotator cuff. Abduction up to 90 degrees causes a glenohumeral force of about 0.81 times BW, while 0.73 times BW is caused by a forward elevation. For elevation angles of less than 90 degrees, the forces agreed with many previous model-based calculations, while at higher elevation angles the measured loads still rose in contrast to the analytical results (Fig. 15.12).

It was also observed that when the exercises are performed at a higher speed, the peak forces decrease.

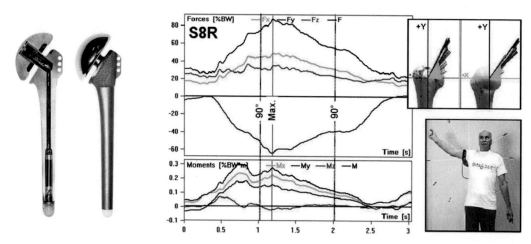

FIGURE 15.12

In vivo measurements of forces and moments acting on the glenohumeral joint. Left: shoulder prosthesis provided of telemetrized strain-gages. Right: abduction >90 degrees. Upper diagrams = resultant force F and components in % of body weight (%BW). Lower diagrams = resultant moment M and components in %BWm. Elevation angles are indicated. Vector plots = force directions in frontal and sagittal plane during whole motion.

Modified from Bergmann, G., Graichen, F., Bender, A., Rohlmann, A., Halder, A., Beier, A., & Westerhoff, P. (2011). In vivo gleno-humeral joint loads during forward flexion and abduction. Journal of Biomechanics, 44(8), 1543–1552. *https://doi.org/10.1016/j.jbiomech.2011.02.142.*

PATHOLOGIES

The normal biomechanics of the shoulder can be affected by several pathologies and traumatic events. Like in all human joints, osteoarthritis is the main responsible of the deterioration of the sliding surfaces. Osteoarthritis can be primitive or caused by a previous trauma (fracture sequelae). A typical effect of the glenohumeral osteoarthritis is wear and flattening of the humeral head and often an inferior osteophyte of the head is generated. Bone deformations occur to extend the articular surface in order to reduce the pressure between the surfaces and to limit the freedom of the glenohumeral movement in an attempt to prevent pain. In a more advanced stage of disease, the glenoid surface is also worn and deformed. In such a pathological condition, the stability of the glenohumeral joint is compromised and weakening of the muscles and eccentric erosion of the glenoid can take place. If the supraspinatus muscle is affected by the abnormal loads, the humerus rises up pushing against the acromion (eccentrical osteoarthritis). A similar effect in absence of osteoarthritis, called cuff tear arthropathy, occurs when the supraspinatus thickness decreases up to the rupture.

REFERENCES

Anglin, C., Wyss, U. P., & Pichora, D. R. (2000). Glenohumeral contact forces. *Proceedings of the Institution of Mechanical Engineers, Part H: Journal of Engineering in Medicine, 214*(6), 637−644. Available from https://doi.org/10.1243/0954411001535660.

Apreleva, M., Parsons, I. M., Warner, J. J. P., Fu, F. H., & Woo, S. L. Y. (2000). Experimental investigation of reaction forces at the glenohumeral joint during active abduction. *Journal of Shoulder and Elbow Surgery, 9*(5), 409−417. Available from https://doi.org/10.1067/mse.2000.106321.

Bäcker, H. C., Galle, S. E., Maniglio, M., & Rosenwasser, M. P. (2018). Biomechanics of posterior shoulder instability − Current knowledge and literature review. *World Journal of Orthopedics, 9*(11), 245−254. Available from https://doi.org/10.5312/wjo.v9.i11.245.

Basmajian, J. V., & Bazant, F. J. (1959). Factors preventing downward dislocation of the adducted shoulder joint. *The Journal of Bone & Joint Surgery, 41*, 1182−1186. Available from https://doi.org/10.2106/00004623-195941070-00002.

Bergmann, G., Graichen, F., Bender, A., Rohlmann, A., Halder, A., Beier, A., & Westerhoff, P. (2011). In vivo gleno-humeral joint loads during forward flexion and abduction. *Journal of Biomechanics, 44*(8), 1543−1552. Available from https://doi.org/10.1016/j.jbiomech.2011.02.142.

Boileau, P., & Walch, G. (1997). The three-dimensional geometry of the proximal humerus. *Journal of Bone and Joint Surgery − Series B, 79*(5), 857−865. Available from https://doi.org/10.1302/0301-620X.79B5.7579.

Bruttel, H., Spranz, D. M., Wolf, S. I., & Maier, M. W. (2020). Scapulohumeral rhythm in patients after total shoulder arthroplasty compared to age-matched healthy individuals. *Gait and Posture, 82*, 38−44. Available from https://doi.org/10.1016/j.gaitpost.2020.08.111.

Churchill, R. S., Brems, J. J., & Kotschi, H. (2001). Glenoid size, inclination, and version: An anatomic study. *Journal of Shoulder and Elbow Surgery, 10*(4), 327−332. Available from https://doi.org/10.1067/mse.2001.115269.

Colachis, S. C., Strohm, B. R., & Brechner, V. L. (1969). Effects of axillary nerve block on muscle force in the upper extremity. *Archives of Physical Medicine and Rehabilitation, 50*(11), 647−654.

Collins, D., Tencer, A., Sidles, J., & Matsen, F. (1992). Edge displacement and deformation of glenoid components in response to eccentric loading. The effect of preparation of the glenoid bone. *Journal of Bone and Joint Surgery − Series A, 74*(4), 501−507. Available from https://doi.org/10.2106/00004623-199274040-00005.

Crouch, D. L., Plate, J. F., Li, Z., & Saul, K. R. (2013). Biomechanical contributions of posterior deltoid and Teres minor in the context of axillary nerve injury: A computational study. *Journal of Hand Surgery*, *38*(2), 241−249. Available from https://doi.org/10.1016/j.jhsa.2012.11.007.

Dalton, S. E., & Snyder, S. J. (1989). Glenohumeral instability. *Bailliere's Clinical Rheumatology*, *3*(3), 511−534. Available from https://doi.org/10.1016/S0950-3579(89)80006-8.

Dul, J. (1988). A biomechanical model to quantify shoulder load at the work place. *Clinical Biomechanics*, *3*(3), 124−128. Available from https://doi.org/10.1016/0268-0033(88)90057-5.

Freedman, L., & Munro, R. R. (1966). Abduction of the arm in the scapular plane: Scapular and glenohumeral movements. A roentgenographic study. *The Journal of Bone and Joint Surgery. American Volume*, *48*(8), 1503−1510. Available from https://doi.org/10.2106/00004623-196648080-00004.

Graichen, H., Stammberger, T., Bonel, H., Englmeier, K.-H., Reiser, M., & Eckstein, F. (2000). Glenohumeral translation during active and passive elevation of the shoulder − A 3D open-MRI study. *Journal of Biomechanics*, *33*(5), 609−613. Available from https://doi.org/10.1016/S0021-9290(99)00209-2.

Halder, A. M., Kuhl, S. G., Zobitz, M. E., Larson, D., & An, K. N. (2001). Effects of the glenoid labrum and glenohumeral abduction on stability of the shoulder joint through concavity-compression: An in vitro study. *Journal of Bone and Joint Surgery − Series A*, *83*(7), 1062−1069. Available from https://doi.org/10.2106/00004623-200107000-00013.

Hecker, A., Aguirre, J., Eichenberger, U., Rosner, J., Schubert., Sutter, R., ... Bouaicha, S. (2020). Deltoid muscle contribution to shoulder flexion and abduction strength: An experimental approach. *Journal of Shoulder and Elbow Surgery/American Shoulder and Elbow Surgeons*, *30*(2), e60−e68. Available from https://doi.org/10.1016/j.jse.2020.05.023.

Hik, F., & Ackland, D. C. (2018). The moment arms of the muscles spanning the glenohumeral joint: A systematic review. *Journal of Anatomy*, *234*(1), 1−15. Available from https://doi.org/10.1111/joa.12903.

Humphrey, C. S., Sears, B. W., & Curtin, M. J. (2016). An anthropometric analysis to derive formulae for calculating the dimensions of anatomically shaped humeral heads. *Journal of Shoulder and Elbow Surgery*, *25*(9), 1532−1541. Available from https://doi.org/10.1016/j.jse.2016.01.032.

Iannotti., Gabriel, J., Schneck, S., Evans, B., & Misra, S. (1992). The normal glenohumeral relationships. An anatomical study of one hundred and forty shoulders. *The Journal of Bone and Joint Surgery. American Volume*, *74*(4), 491−500.

Inman, V. T., Saunders, J. Bd. M., & Abbott Leroy, C. (1996). Observations of the function of the shoulder joint. *Clinical Orthopaedics and Related Research*, *330*, 3−12. Available from https://doi.org/10.1097/00003086-199609000-00002.

Karlsson, D., & Peterson, B. (1992). Towards a model for force predictions in the human shoulder. *Journal of Biomechanics*, *25*(2), 189−199. Available from https://doi.org/10.1016/0021-9290(92)90275-6.

Kim, M. S., Lim, K. Y., Lee, D. H., Kovacevic, D., & Cho, N. Y. (2012). How does scapula motion change after reverse total shoulder arthroplasty? − A preliminary report. *BMC Musculoskeletal Disorders*, *13*, 210. Available from https://doi.org/10.1186/1471-2474-13-210.

Lin, H. T., Hsu, A. T., Chang, G. L., Chien, J. R. C., An, K. N., & Fong, C. S. (2007). Determining the resting position of the glenohumeral joint in subjects who are healthy. *Physical Therapy*, *87*(12), 1669−1682. Available from https://doi.org/10.2522/ptj.20050391.

McMahon, P. J., Debski, R. E., Thompson, W. O., Warner, J. J. P., Fu, F. H., & Woo, S. L. Y. (1995). Shoulder muscle forces and tendon excursions during glenohumeral abduction in the scapular plane. *Journal of Shoulder and Elbow Surgery*, *4*(3), 199−208. Available from https://doi.org/10.1016/S1058-2746(05)80052-7.

McPherson, E. J., Friedman, R. J., An, Y. H., Chokesi, R., & Dooley, R. L. (1997). Anthropometric study of normal glenohumeral relationships. *Journal of Shoulder and Elbow Surgery*, *6*(2), 105−112. Available from https://doi.org/10.1016/S1058-2746(97)90030-6.

Namdari, S., Yagnik, G., Ebaugh., Nagda., Ramsey, M., Jr, & Mehta. (2011). Defining functional shoulder range of motion for activities of daily living. *Journal of Shoulder and Elbow Surgery/American Shoulder and Elbow Surgeons*, *21*(9), 1177−1183. Available from https://doi.org/10.1016/j.jse.2011.07.032, Epub.

Nobuhara. (1987). *The shoulder: Its function and clinical aspects*. Tokyo: Igaku-Shoin.

Poppen, N. K., & Walker, P. S. (1978). Forces at the glenohumeral joint in abduction. *Clinical Orthopaedics and Related Research*, *135*, 165−170. Available from https://doi.org/10.1097/00003086-197809000-00035.

Randelli, M., & Gambrioli, P. L. (1986). Glenohumeral osteometry by computed tomography in normal and unstable shoulders. *Clinical Orthopaedics and Related Research*, *208*, 151−156.

Reeves, B., Jobbins, B., & Flowers, M. (1972). Biomechanical problems in the development of a total shoulder endoprosthesis. *The Journal of Bone and Joint Surgery*, *54*, 193.

Rockwood, C. (2009). *The shoulder*. Elsevier.

Ruiz Ibán, M., Paniagua Gonzalez, A., Muraccini, A., Asenjo Gismero, C., Varini, A., Berardi, A., & Mantovani, A. (2020). Evaluation of a novel portable three-dimensional scapular kinematics assessment system with inter and intraobserver reproducibility and normative data for healthy adults. *Journal of Experimental Orthopaedics*, *7*(1), 31. Available from https://doi.org/10.1186/s40634-020-00238-6, PMID: 32405717.

Saha, A. K. (1971). Dynamic stability of the glenohumeral joint. *Acta Orthopaedica*, *42*(6), 491−505. Available from https://doi.org/10.3109/17453677108989066.

Scibek, J. S., & Carcia, C. R. (2012). Assessment of scapulohumeral rhythm for scapular plane shoulder elevation using a modified digital inclinometer. *World Journal of Orthopedics*, *3*(6), 87−94. Available from https://doi.org/10.5312/wjo.v3.i6.87.

Steindler, A. (1955). *Kinesiology of the human body under normal and pathological conditions*. Springfield, IL: Charles C Thomas Pub Ltd.

Terrier, A., Reist, A., Vogel, A., & Farron, A. (2007). Effect of supraspinatus deficiency on humerus translation and glenohumeral contact force during abduction. *Clinical Biomechanics*, *22*(6), 645−651. Available from https://doi.org/10.1016/j.clinbiomech.2007.01.015.

Werthel, J. D., Bertelli, J., & Elhassan, B. T. (2017). Shoulder function in patients with deltoid paralysis and intact rotator cuff. *Orthopaedics and Traumatology: Surgery and Research*, *103*(6), 869−873. Available from https://doi.org/10.1016/j.otsr.2017.06.008.

Zdravkovic, V., Alexander, N., Wegener, R., Spross, C., & Jost, B. (2020). How do scapulothoracic kinematics during shoulder elevation differ between adults with and without rotator cuff arthropathy? *Clinical Orthopaedics and Related Research*, *478*(11), 2640−2649. Available from https://doi.org/10.1097/CORR.0000000000001406.

Zhang, Q., Shi, L. L., Ravella, K. C., Koh, J. L., Wang, S., Liu, C., . . . Wang, J. (2016). Distinct proximal humeral geometry in Chinese population and clinical relevance. *Journal of Bone and Joint Surgery*, *98*(24), 2071−2081. Available from https://doi.org/10.2106/JBJS.15.01232.

BIOMECHANICS OF THE ANKLE JOINT

16

Luc Labey

Department of Mechanical Engineering, KU Leuven, Geel, Belgium

PRELIMINARY DEFINITIONS

Before discussing the biomechanics of the ankle joint, we first need to introduce some basic terminology. Fig. 16.1 shows a schematic view of the distal part of a right lower leg and foot with two orthogonal coordinate systems superimposed. It identifies the main planes, directions, and axes of rotation that are typically used when talking about human foot and ankle biomechanics. The definitions below follow the recommendation by the International Society of Biomechanics and are inspired by the convention used for the description of knee kinematics as proposed by Grood and Suntay (Grood & Suntay, 1983; Wu et al., 2002). The latter means that the three translations and rotations that are necessary to fully describe the 3D kinematics of the joint are defined with respect to three nonorthogonal axes of which one is fixed to the proximal segment of the joint, the second is fixed to the distal segment, and the third is a so-called floating axis, which is perpendicular to the first two. A big advantage of such an approach is that large joint displacements are independent of the order in which the translations and rotations occur (Grood & Suntay, 1983; Wu et al., 2002).

Only four bony landmarks are needed to define all planes and axes, which are subsequently used for the description of ankle and foot kinematics. All four are identified on the proximal segment, the tibia + fibula in this case (considered as one, almost nondeformable, body), and none on the other bones of the ankle or foot. Therefore, it is also necessary to define a neutral configuration of the ankle and foot with respect to the tibia + fibula. This is an unfortunate drawback of this approach, which is difficult to circumvent given the fact that the distal bones of the ankle joint complex do not dispose of easily identifiable external landmarks. With current medical imaging techniques, this issue could be solved, however.

The four landmarks are the two bony prominences on the distal tibia and fibula called the malleoli (one medially MM and one laterally LM). The other two are the most medial (MC) and most lateral (LC) points on the borders of the proximal tibial condyles. Starting from these four, one can easily define IM, the point in the center between MM and LM, and IC, the point midway between MC and LC.

Based on these six points, two planes and two axes are defined:

1. The torsional plane contains MM, LM, and IC.
2. The frontal plane contains IM, MC, and LC.

Human Orthopaedic Biomechanics. DOI: https://doi.org/10.1016/B978-0-12-824481-4.00036-6

305

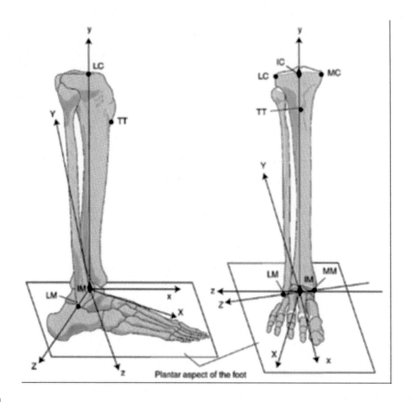

FIGURE 16.1

Definitions of landmarks and coordinate axes according to ISB. ISB, International Society of Biomechanics.

From Wu, G., Siegler, S., Allard, P., Kirtley, C., Leardini, A., Rosenbaum, D., Whittle, M., D'Lima, D., Cristofolini, L., & Witte, H. (2002). ISB recommendation on definitions of joint coordinate system of various joints for the reporting of human joint motion—part I: Ankle, hip, and spine. Journal of Biomechanics, *35(4), 543–548.*

3. The Z axis of the tibia + fibula contains MM and LM and points to the right (from medial to lateral in a right leg, from lateral to medial in a left leg). This axis also lies in the torsional plane.
4. The long axis of the tibia + fibula contains IM and IC and points upwards. It is also the intersection line of the frontal and torsional planes.

The sagittal plane is further defined as the plane perpendicular to the frontal plane and containing the long axis of the tibia + fibula. The transverse plane is a plane that is perpendicular to both the frontal and sagittal planes (and thus also to the torsional plane).

The neutral configuration of the ankle and foot with respect to the tibia + fibula is such that the plantar face of the foot (the horizontal plane when standing upright) is perpendicular to both the frontal and the sagittal planes (and thus parallel to the transverse plane). Furthermore, the line perpendicular to the frontal plane and the long axis of the second metatarsal should be parallel. In this neutral configuration, the axis and planes defined above can also be used to identify a coordinate system fixed to the heel bone (calcaneus).

The coordinate system for the tibia + fibula consists of an origin O and three orthogonal axes X, Y, and Z:

1. The origin O is situated at the point IM.
2. The Z axis is already defined above.
3. The X axis is perpendicular to the torsional plane and points anteriorly.
4. The Y axis is perpendicular to Z and X and points upwards. This axis is thus also lying in the torsional plane.

The coordinate system for the calcaneus consists of an origin o and three orthogonal axes x, y, and z. As mentioned before, it is defined in the neutral configuration of the ankle and foot:

1. The origin o coincides with O.
2. The y axis is equal to the long axis of the tibia + fibula.
3. The x axis is perpendicular to the frontal plane and points anteriorly.
4. The z axis is perpendicular to x and y and points to the right. This axis is also lying in the frontal plane.

Finally, the different translational and rotational components of motion in the ankle and foot are defined.

The first component is defined with respect to the Z axis of the tibia + fibula (with a unit vector $\vec{e_1}$). Translation with respect to this axis is called medial (negative for a right leg) or lateral (positive for a right leg) shift. Rotation around this axis is called dorsiflexion (positive or counterclockwise rotation, the toes are pulled up) or plantar flexion (negative or clockwise rotation, the toes point downwards).

The second component is defined with respect to the y axis of the calcaneus (with a unit vector $\vec{e_3}$). Translation with respect to this axis corresponds to compression (positive) or distraction (negative). Rotation around this axis is described as internal (the toes moving medially) or external (the toes moving laterally) rotation. For a right leg, internal rotation is positive and external rotation is negative.

The third component is defined with respect to the floating axis that is at all times (even during motion) perpendicular to both Z and y. It has a unit vector $\vec{e_2} = \vec{e_3} \times \vec{e_1}$. Defined like this, the positive sense of $\vec{e_2}$ is towards anterior. Translation with respect to this axis corresponds to the anterior (positive) or posterior (negative) drawer. Rotation around this axis is called inversion (the big toe moves upward, the foot goes into a varus position) or eversion (the big toe moves downward, the foot goes into a valgus position). For a right leg, inversion is positive and eversion is negative.

ANATOMY AND MORPHOLOGY OF THE HUMAN ANKLE JOINT

In nature, and certainly also in biomechanics, form and function are closely interrelated. The function of a joint, its motion, and the force it withstands are reflected in the size and shape of its bones and the insertion sites and direction of its ligaments and vice versa. It is therefore instructive to open this chapter with a description of the anatomy and morphology of the human ankle joint. From this, much information can be gained to better understand the human ankle kinematics and kinetics, which will be covered afterwards. Despite the fact that we commonly consider the human ankle as a single joint, it is also often studied and described as a structure, called the ankle joint complex. This should have become clear already from the definitions above, where motion is described with respect to axes fixed to tibia + fibula on the one hand and to the calcaneus on the other, disregarding the real ankle bone, which is the talus. Indeed, what is usually perceived as the movement of the ankle cannot unambiguously be attributed to the motion of one bone with respect to another, at least when observing it from outside. The full motion of the ankle joint complex is in fact a combination of rotation and translation over different articular surfaces at once. This is different from the situation in most other joints in the human body.

BONES AND JOINTS

Fig. 16.2 shows a frontal (A) and lateral (B) X-ray of the human ankle and midfoot as well as a top and bottom drawing (C). The bones of the lower leg (tibia to the left and fibula to the right in panel A) are clearly visible. Below those two bones, the talus (the most proximal of the hindfoot bones) is shown. Strictly speaking, the ankle joint is defined as the joint between tibia + fibula on the one hand and talus on the other. This joint is also called the talocrural joint (Manganaro & Alsayouri, 2021).

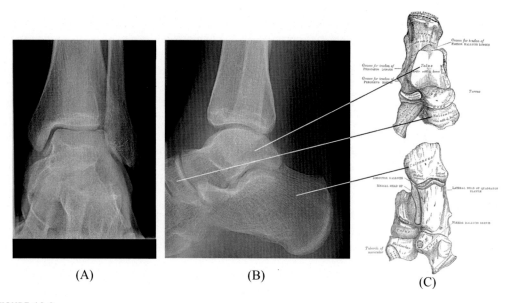

(A)   (B)   (C)

FIGURE 16.2

The bones of the ankle.

Adapted from Wikimedia Commons and Gray, H. (1918). Anatomy of the human body. Lea and Febiger, Philadelphia.

THE TALUS

The shape of the talus is quite irregular, looking somewhat like a saddle, and it is composed of a head, neck, and body (Fig. 16.3). It serves as a linkage between the lower leg, the calcaneus, and the proximal bones of the midfoot, somewhat like a ball in a roller bearing. As such, it contributes to various movements of the foot and ankle.

The shape of the articular interface of the talus with the tibia + fibula can be approximated quite well by a truncated cone (Khan & Varacallo, 2021). As a result, its relative motion with respect to the lower leg is close to a rotation around the axis of this cone, certainly in weight-bearing conditions. The talocrural joint can thus be approximated by a hinge joint with a fixed axis of rotation. Rotation and translation along all other axes are very limited by the fact that the talus fits snugly in the so-called mortise formed by the tibia and fibula. The axis of rotation is however tilted with

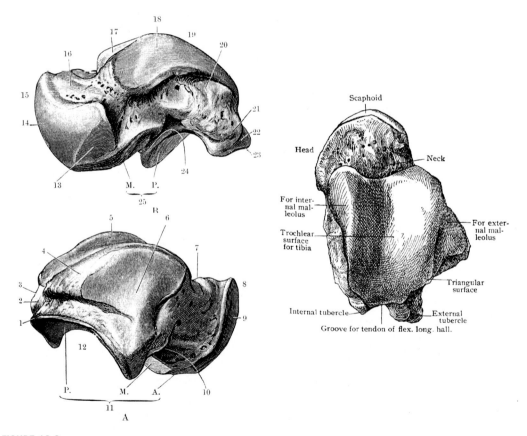

FIGURE 16.3

The talus.

Modified from Cunningham, D. J. (1903). Manual of practical anatomy (Vol. 1). J. B. Lippincott.

respect to the major body axes. It runs from posterolateral to anteromedial and from superomedial to inferolateral. Rotation around this axis thus results in a combination of the three rotation components described above. The proximal articular surface of the talus is somewhat wider anteriorly than posteriorly (Brockett & Chapman, 2016). Consequently, the joint is more stable in dorsiflexion than in plantar flexion.

The talus is completely encapsulated by other bones and tissues. Its motion is thus almost impossible to measure by conventional clinical techniques such as motion capture in a gait lab. In addition, no tendons are directly attached to the talus. Its motion is therefore not actively controlled but is purely the result of interaction with the other bones and tissues surrounding it.

The talus transfers the weight of the body to the foot and thus withstands quite a heavy loading which is distributed over a relatively small articular surface, certainly when compared to the size of the articular surface in the knee and the hip.

THE CALCANEUS

The calcaneus (or heel bone) is the biggest bone in the foot, and it is situated distally and posteriorly to the talus (Fig. 16.4). Together, both bones form the hindfoot. The calcaneus is more or less shaped like a beam with its long axis extending approximately parallel to the midline of the foot. Its main function is to transfer most of the body weight from the lower extremity to the ground during the heel-contact phase of gait. Furthermore, the calcaneus provides leverage for the Achilles tendon during walking or running.

The joint between the calcaneus and talus is also called the subtalar joint. The articular surface of the subtalar joint is in fact a combination of three separate facets: two of them are situated anteriorly and one posteriorly. The posterior facet is the largest. It is concave on the talus and convex on the calcaneus, curved around an axis, which runs from superoanterior to inferoposterior (Gupton, Özdemir, & Terreberry, 2021). Despite the complex shape of the articular surface, the talus and calcaneus can still move slightly with respect to each other certainly when loaded. Taking into account the fact that the main load on the calcaneus will typically lead to an extension moment with respect to the talus, the main component of its motion will be rotation and sliding over its posterior contact area with the talus, while the more anterior contact areas will show a tendency to separate. This motion leads to inversion–eversion combined with an internal–external rotation of the calcaneus with respect to the talus (Sammarco, 2004). The subtalar joint is therefore sometimes referred to as the "steering wheel" in human locomotion.

The groove between the two anterior facets and the posterior facet is called the calcaneal sulcus. Together with the corresponding talar sulcus, a canal is formed. This tarsal sinus is a quite large space that contains neurovascular structures and ligaments, which will be discussed later.

Since the talus is not accessible (and not visible) from outside, the externally visible kinematics of the ankle joint complex as measured in motion capture settings is often described as a combination of the motion in the talocrural joint and in the subtalar joint.

In front of the talus and calcaneus, the two most proximal bones of the midfoot are located: the navicular bone on the medial side and the cuboid bone on the lateral side.

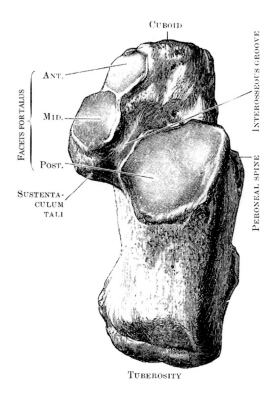

FIGURE 16.4

The calcaneus. Top view: the right calcaneus bone.

Modified from Cunningham, D. J. (1903). Manual of practical anatomy *(Vol. 1). J. B. Lippincott.*

THE NAVICULAR BONE

The navicular bone's name stems from the fact that it resembles a small boat lying on its side when looking from the medial side.

Posteriorly, the navicular has a strongly concave semispherical-shaped articular surface where it articulates with the head of the talus (Fig. 16.5) (Prapto & Dreyer, 2021). Inferiorly to this area also the calcaneus is present. The combined navicular and anterior and middle calcaneal articular surfaces form an acetabulum-like structure in which the head of the talus articulates. Thus, the three bones form the talocalcaneonavicular (TCN) joint. This joint can be described as a ball-and-socket joint (named the coxa pedis), similar to the hip. The TCN joint is further stabilized by the plantar calcaneonavicular ligament (also called the spring ligament) on the medial side and the lateral calcaneonavicular ligament on the lateral side. On the lateral side, the navicular articulates with the cuboid and anteriorly with the three cuneiform bones of the foot.

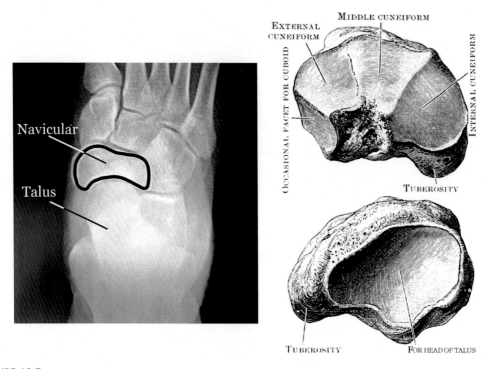

FIGURE 16.5

The navicular bone. The right navicular bone, anterior (top) and posterior (bottom) views.

From Prapto, D. & Dreyer, M. A. (2021). Anatomy, bony pelvis and lower limb, navicular bone. Available at: https://www.ncbi.nlm.nih.gov/books/NBK547675/. Drawings from Cunningham, D. J. (1903). Manual of practical anatomy (Vol. 1). J. B. Lippincott.

THE CUBOID

As its name indicates, the cuboid is more or less shaped like a cube (Fig. 16.6) (Gill & Vilella, 2021). This laterally situated bone of the midfoot has five articular surfaces. It articulates with the calcaneus posteriorly, over a saddle-shaped surface. On the medial aspect, the cuboid has two joint surfaces: one with the navicular and one with the most lateral cuneiform. Anteriorly, there are two articular surfaces with the fourth and fifth metatarsals. The cuboid has prominence on its plantar surface, which is called the cuboid tuberosity.

The combination of the talonavicular (TN) and the calcaneocuboid (CC) joints is also referred to as the transverse tarsal (or Chopart) joint. This joint complex facilitates motion between the hindfoot and the midfoot, and it contributes to the inversion and eversion motion of the foot (Bonnel, Teissier, Colombier, Toullec, & Assi, 2013).

Finally, the cuboid also has a role in stabilizing the static and rigid lateral sides of the foot.

All these bones consist of a thin outer layer of dense, cortical bone that encloses an interior of porous trabecular bone where the trabeculae follow the lines of the principal stresses. The articular

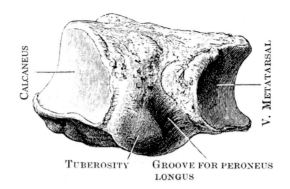

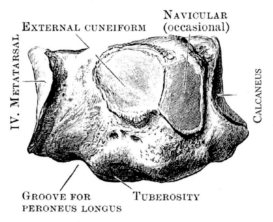

FIGURE 16.6

The cuboid. The right cuboid bone, lateral (top) and medial (bottom) views.

Modified from Cunningham, D. J. (1903). Manual of practical anatomy *(Vol. 1). J. B. Lippincott.*

surfaces of the bones are covered with a 1–1.5 mm thick layer of cartilage. This soft, viscoelastic tissue serves to generate a low-friction interface, thus enabling smooth, effortless motion.

MAJOR LIGAMENTS

The bones in a joint provide stability and strength when pushed together by a combination of external loads and muscle forces. However, they do not provide stability whenever they are pulled apart. To keep the joint from falling apart under these circumstances, bands of soft tissue are present that hold different bones together while still allowing motion. These bands are called ligaments (Watanabe et al., 2012). Because of their function, the insertion sites of these ligaments on the bones should be such that the ligaments either remain isometric or slacken during the movement of the bones.

The entire ligamentous apparatus in the ankle structure is quite complex as can be seen in Fig. 16.7. Rather than discussing each ligament, which would lead to a long list of ligaments with their respective insertion sites, lengths and orientation, we will present the most relevant ones from a functional biomechanical point of view. They are also listed in Table 16.1.

The ankle joint is stabilized on the medial side by the medial collateral ligamentous assembly. This assembly consists of three superficial and one deep band and has roughly the shape of a Greek capital delta, hence its name deltoid ligament (Harper, 1987). This ligament connects the tibia to the calcaneus by a superficial tibiocalcaneal band (sTC), to the talus by both the superficial posterior tibiotalar band (sTT) and the deep anterior tibiotalar band (dTT), and to the navicular by the superficial tibionavicular band (sTN). All of them stabilize against eversion and distraction of the talus (and thus the foot), although the sTC is the primary stabilizer against these motion components. The sTN and dTT are stabilizers against the anterior drawer and external rotation of the talus, while the sTT band prevents the posterior drawer and internal rotation of the talus. The talus is further also directly connected to the calcaneus medially by the much thinner talocalcaneal ligament (mTC).

The lateral or fibular collateral ligament is much weaker than the medial one. It is also composed of different bands, three in this case, roughly in the shape of a capital T, tilted anteriorly. The fibula is connected to the calcaneus by the calcaneofibular band (CF, the vertical bar of the T) and to the talus by both a posterior and an anterior talofibular band (pTF and aTF, respectively, the horizontal bar of the T). The CF is the primary stabilizer against ankle inversion, external rotation, and distraction at the subtalar joint. However, since this band is thinner and weaker than the deltoid, it typically suffers injury much more easily in an inversion ankle sprain. The CF also resists the posterior drawer of the calcaneus. The aTF originates at the lateral malleolus and inserts anteriorly on the talus. It restricts anterior drawer, internal rotation, and inversion of the talus. The pTF arises on the lateral malleolus and inserts posteriorly on the talus. It resists the posterior drawer and external rotation of the talus. Also laterally, there is a direct link between talus and calcaneus by

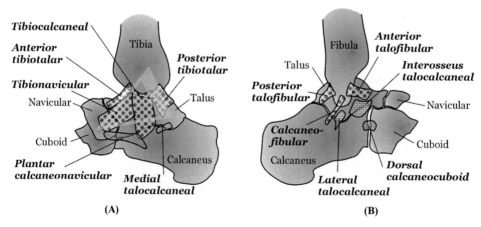

FIGURE 16.7

Medial (A) and lateral (B) views of the ankle with its ligaments.

Modified from Cunningham, D. J. (1903). Manual of practical anatomy (Vol. 1). J. B. Lippincott.

Table 16.1 Overview of the most important ankle ligaments.

Name	Abbreviation	Translation			Rotation		Inversion	Eversion
		Distraction	Anterior	Posterior	Internal	External		
Deltoid: superficial tibiocalcaneal	sTC	x						x
Deltoid: superficial tibiotalar	sTT	x		x	x			x
Deltoid: deep tibiotalar	dTT	x	x			x		x
Deltoid: superficial tibionavicular	sTN	x	x			x		x
Medial talocalcaneal	mTC							
Lateral collateral: calcaneofibular	CF	x		x		x	x	
Lateral collateral: posterior talofibular	pTF			x		x		
Lateral collateral: anterior talofibular	aTF		x		x		x	
Lateral talocalcaneal	lTC	x		x			x	
Interosseous talocalcaneal	iTC							
Plantar calcaneonavicular	pCN							
Chopart: lateral calcaneonavicular	lCN							
Chopart: medial calcaneocuboid	mCC							

the lateral talocalcaneal ligament which is situated close to and parallel with the CF. This ligament is however not always present.

Apart from the ligaments on either side of the ankle, the calcaneus has also some other ligaments attached to it which are worth mentioning.

Firstly, the interosseous talocalcaneal ligament (iTC) passes through the tarsal canal and connects the talus to the calcaneus. It stabilizes the posterior talocalcaneal joint.

Furthermore, and as already mentioned above, the calcaneus and the navicular form a "socket" for the head of the talus. The plantar calcaneonavicular ligament, otherwise known as the spring ligament, is in fact a group of ligaments that hold this socket together. It is visible at the medial side, somewhat more plantar than and almost parallel to the sTN.

The bifurcate ligament, also known as the Chopart ligament, is visible from the lateral side of the foot and situated anteriorly to the iTC. It is a Y-shaped ligament that consists of two components. These are the lateral calcaneonavicular ligament and the medial calcaneocuboid ligament. Together, these components stabilize the cuboid and navicular against the anterior drawer and internal rotation.

Finally, the plantar aponeurosis and long plantar ligament support the arch of the foot and attach to the calcaneus as well.

MUSCLES AND TENDONS

Due to the fact that the ankle joint can be considered as a hinge joint, the most important muscles and tendons that control its movement can be grossly divided into two groups. The muscles that make the ankle and foot rotate downwards are called the plantar flexors, and the muscles that raise the ankle and foot are called the dorsiflexors. Since muscles can only pull, the plantar flexors are situated at the posterior side of the lower leg while the dorsiflexors are located anteriorly. Functionally, during walking and running, humans (but also animals) need much more torque to provide plantar flexion and thus the plantar flexors are more powerful and have a larger cross-section than the dorsiflexors.

However, most of the tendons linking the muscles to the ankle and foot bones are attached medially or laterally from the central axis of the foot. Activation of the muscles will therefore typically result in plantar flexion or dorsiflexion combined with either or both inversion−eversion or internal−external rotation. Because no muscles attach to the talus, none of them acts exclusively on the ankle in the strict sense.

THE POSTERIOR COMPARTMENT

In the posterior compartment, we can make a distinction between several layers of plantar flexors. Two muscles, together called the triceps surae, combine distally in the Achilles tendon and connect to a tubercle at the posterior side of the calcaneus. These two muscles are the two-headed gastrocnemius at the surface and the soleus that lies underneath. The gastrocnemius attaches proximally on the lower part of the femur (it is thus also a knee flexor), and the soleus attaches proximally at the top of the tibia. Sometimes, the plantaris muscle is also considered as part of the triceps surae. This is a thin, long muscle with similar attachments as the gastrocnemius. The triceps surae is the main plantar flexor of the ankle and foot.

Still posteriorly, but in the layer below the soleus, one can find three more muscles that are also plantar flexors, though this is not their primary function. From medial to lateral, these are the flexor digitorum longus, the tibialis posterior, and the flexor hallucis longus. As their names indicate, the flexor digitorum longus and flexor hallucis longus are flexors of the toes. The first one flexes all the toes except the big toe, while the second one flexes the big toe. Superiorly, the former is connected to the posterior face of the tibia and the latter to the posterior face of the fibula. They both cross the medial side of the calcaneus and attach to the most distal bones of the toes or the big toe, respectively. Because of their location, they will also tend to plantar flex and invert the ankle when they contract. The tibialis posterior is the stronger of the three muscles. It originates from the posterior faces of both the tibia and fibula (and even their interosseous membrane), runs again medially across the calcaneus, and passes into the tibialis posterior tendon which inserts on the plantar faces of both the navicular and the cuboid (among other tarsal bones). This muscle mainly stabilizes the medial arch of the foot and leads to an inversion of the foot, but it is also a plantar flexor.

THE ANTERIOR COMPARTMENT

The dorsiflexor muscles are both lower in number and in strength compared to the plantar flexors. To begin with, there is no equivalent of the triceps surae on this side of the ankle joint. The dorsiflexors are in fact mirror images of the second layer of plantar flexors. All dorsiflexors are situated laterally on the lower leg. They consist of the extensor digitorum longus, the extensor hallucis longus, and the tibialis anterior.

The extensor digitorum longus and hallucis longus are both mainly attached on the anterior face of the fibula, with the first more superiorly and the second more inferiorly. The extensor digitorum longus then runs more or less straight downwards. The muscle belly passes into a tendon at the bottom of the fibula and crosses the ankle joint at the lateral side, slightly medially from the lateral malleolus. More or less at the level of the cuboid, the tendon splits into four separate units that are each attached to the distal toe bones. The extensor hallucis longus runs downwards and slightly medially. It crosses the ankle centrally between the two malleoli where the muscle belly also passes into a tendon. This tendon attaches to the most distal bone of the big toe. Finally, the tibialis anterior is the biggest of these three dorsiflexor muscles. It attaches to the lateral face of the tibia, just in front of the joint between the tibia and fibula. From there it runs downwards along and in front of the tibia. It crosses the ankle joint just laterally from the medial malleolus. Its tendon then attaches to the first metatarsal bone.

From a functional point of view, these three muscles serve to pull up the toes and dorsiflex the foot. The extensor digitorum longus furthermore everts the foot, while the extensor hallucis longus can assist in both eversion and inversion (depending on foot position) and the tibialis anterior inverts the foot.

Apart from these three most important dorsiflexion muscles in the anterior compartment, there is another smaller muscle called the peroneus tertius. It is more or less parallel to the extensor digitorum longus, but it attaches to the fibula at a lower position and attaches at the fifth metatarsal. Apart from its dorsiflexion function, it serves also to evert the foot.

THE LATERAL COMPARTMENT

The lateral compartment consists of two muscles: the peroneus longus and peroneus brevis. They both originate from the lateral surface of the fibula. The peroneus longus attaches at the top, while the brevis attaches in the middle. From there they run downwards along the fibula to cross the ankle laterally just behind and underneath the lateral malleolus. Both tendons finally insert into the base of the first metatarsal bone—the longus at its lateral side and the brevis in the tuberosity. Their main function is eversion of the foot, but they are also (weak) plantar flexors.

KINEMATICS OF THE HUMAN ANKLE JOINT

The kinematics of a joint is at least partially determined by the shape of its mating surfaces. The forces on the joint (either external forces, or forces in ligaments or muscles) only determine the movement according to degrees of freedom that the geometry does not define. It is therefore appropriate to start the discussion of the kinematics of the ankle joint in the passive situation when it is not loaded by external forces.

THE RANGE OF MOTION OF THE ANKLE

Several investigators have tried to determine the range of motion of the ankle joint in the past. In most cases, this has been done in vivo and using footplates fixed to a goniometer. As a consequence, the range of motion thus obtained is a combination of motion in the talocrural and the subtalar joint. Moreover, the description of the range of motion in terms of its different components such as plantar flexion−dorsiflexion and inversion−eversion does not exactly match with the components as defined in the preliminary definitions.

Myburgh, Vaughan, & Isaacs (1984) determined the ankle range of motion in 12 healthy, young volunteers using an ankle goniometer, which registered motion of a footplate with respect to a fixed and a moving axis. During the measurements, the knee joint was extended with the leg positioned horizontally. The fixed axis of the device was perpendicular to the long axis of the clamped lower leg and oriented anteriorly. Rotation around this axis was described as an inversion−eversion range of motion. The second axis was always perpendicular to the previous axis and oriented mediolaterally. Rotation around this axis was described as plantar flexion−dorsiflexion. The authors did not explicitly mention how the foot was moved, by the volunteers themselves or by another person. However, they compared their results with those in another paper where the volunteers moved their feet by exerting maximal muscle force. They found the following average ranges of motion:

- plantar flexion: 45.8−48.7 degrees;
- dorsiflexion: 25−26.8 degrees;

- inversion: 27.3−33.3 degrees in the neutral position and 33−38 degrees in the plantar flexed position;
- eversion: 14.5−19.8 degrees in the neutral position and 7.3−11.6 degrees in the plantar flexed position.

In another paper by Bok, Lee, & Lee (2013), ranges of motion of the ankle were determined in three age groups of 20 healthy volunteers each. As mentioned in the paper above, the measurements were performed using an ankle goniometer, which registered the motion of a footplate. This time, however, the knee was flexed at 90 degrees during the measurements (which may have an effect on the plantar flexors). No information was provided regarding the definition of the measurement axes. Both the range of motion and the strength at the ankle joint were determined by the maximal contraction of muscles by the participants. Results were as follows:

- plantar flexion: 37.4 degrees (age 20−40 years) down to 32.1 degrees (age > 65 years);
- dorsiflexion: 20.9 degrees (age 20−40 years) down to 15.4 degrees (age > 65 years);
- inversion: 30.8 degrees (age 20−40 years) down to 24.9 degrees (age > 65 years) in the neutral position;
- eversion: 32.9 degrees (age 20−40 years) down to 24.3 degrees (age > 65 years) in the neutral position.

The values are typically somewhat lower compared to the previous paper, but still within one standard deviation. Only the eversion range of motion seems rather high. It is also quite clear that the range of motion in the ankle joint diminishes with aging.

If one wants to discriminate between the talocrural and subtalar ranges of motion, this can only be done based on medical imaging. Since medical imaging techniques have only recently advanced to the point that motion measurements are possible, very few studies are already available with information about ankle range of motion.

One paper by Tuijthof et al. (2009) used a 3D CT stress test to quantify the range of motion in both the talocrural and subtalar joints in 20 healthy volunteers. The subjects were lying supine with their knee extended and the right foot fixed to a radiolucent footplate. After a scan with the foot in the neutral position, eight more scans were made, while the foot was loaded and pulled in eight different extreme positions with a subjective specific tolerable load (average value of 60.9 N) applied through a cable. After segmenting the lower tibia + fibula, talus, and calcaneus, the pose change of the ankle bones was determined and described by a finite helical axis (FHA). The three rotation components of the FHAs were determined with respect to a coordinate system fixed to the talus (and defined by its principal axes). These three components were then considered to describe the true plantar flexion−dorsiflexion, inversion−eversion, and internal−external rotation, although this definition obviously does not match the ISB definitions.

The results show that the total plantar flexion−dorsiflexion range of motion is situated mostly in the talocrural joint (63.3 degrees around the talocrural FHA), while the subtalar joint contributes very little (4.1 degrees around the subtalar FHA). The inversion−eversion range of motion seems to be more equally distributed between the two joints. The talocrural joint contributes on average 23.7 degrees and the subtalar joint 37.3 degrees. The full inversion−eversion range of motion from this study (56.6 degrees) seems to be closer to the range found in the previous paper (63.7 degrees) than that of the first one (41.8−53.1 degrees).

Moreover, it was also clear that the different rotation components are coupled. It is not possible to obtain pure plantar flexion−dorsiflexion or pure inversion−eversion motion. This can of course be expected (and has been already mentioned above) due to the shape of the articular surfaces. Coupled motion is less in plantar flexion−dorsiflexion because it takes place almost exclusively in the talocrural joint and this rotation component is defined as the rotation around the principal Y axis. Nevertheless, the full 63.3-degree plantar flexion−dorsiflexion talocrural rotation around the FHA can be attributed to 57.4-degree pure plantar flexion−dorsiflexion (around the Y axis) combined with 23.4-degree inversion−eversion and 12.9-degree internal−external rotation of the talus. Macroscopic inversion−eversion is always accompanied by an almost equal amount of internal−external rotation in both joints.

KINEMATICS DURING GAIT

Gait is the most important daily activity of human beings. It is therefore of paramount importance to also understand the kinematics of the ankle joint during this motor task. Most information in this respect is based on studies using 3D motion capture techniques. Unfortunately, these datasets are both inaccurate and incomplete. Firstly, motion capture is inaccurate because of a number of different error sources which are, alas, unavoidable. Two of the most important ones are soft tissue artifacts and difficulties in correctly identifying the movement axes. Secondly, motion capture of the ankle can only provide incomplete data because the talus is not accessible from outside. Marker-based measurements of the ankle can therefore only provide information on the kinematics of the calcaneus with respect to the tibia + fibula.

Before giving a quantitative overview of the findings from these studies, it makes sense to start with a description of the different phases in a typical gait pattern, which are relevant from the perspective of the ankle and foot. Most authors discriminate between three phases during stance, which are weight acceptance (heel strike), single limb support, and propulsion (push-off). During each of them, the foot and ankle serve a different purpose and behave differently.

During weight acceptance, the CC and TN joints (commonly designated as the transverse tarsal joint) need to make sure that the foot behaves as a flexible structure. This is necessary to be able to adapt to differences in terrain and to dissipate the mechanical energy released during the heel strike. One of the strategies to enable this is the eversion of the subtalar joint. As a consequence, the axes of the CC and the TN joints become parallel, which leads to flexibility. The transverse tarsal joint is unlocked.

Later, during stance and push-off, the foot becomes gradually more and more stable and stiff. This is needed to use the plantar flexion moment from the gastrocnemius and soleus to generate a ground reaction force that propels the body. To this end, the foot is inverted which leads to a change in orientation of the CC and TN joint axes. They are no longer parallel and thus the midfoot loses its mobility. The transverse tarsal joint is locked.

The results of 3D motion capture measurements are usually represented as illustrated in Fig. 16.8. The data set is obtained from (Burg, 2014). They were collected from 15 healthy volunteers with an average age of 56.3 years who underwent gait analysis while walking barefoot at a self-selected speed. The marker set included a lower leg segment and a multisegment foot model

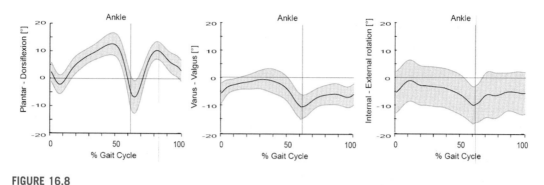

FIGURE 16.8

AJC gait kinematics. Kinematics of the ankle joint complex during gait. Ajc, Ankle joint complex

(hindfoot, midfoot, forefoot, and hallux). Marker trajectories were recorded using a Vicon system with 10 cameras at a sampling rate of 200 Hz.

The graphs clearly show the plantar flexion–dorsiflexion excursion of the ankle joint complex during gait, which is much smaller than the full range of motion described in the previous paragraphs. During the majority of the stance phase, the ankle is dorsiflexed. It moves quickly into plantar flexion just after the heel contact (forced by the external moment exerted by the ground reaction force) and just before push-off (by the plantar flexion moment produced by the plantar flexor muscles). During the first half of the swing phase, the ankle and foot quickly move into dorsiflexion again, to prevent the toes from hitting the ground when the foot swings forward. The other movement components never exceed much more than 10 degrees from the neutral ankle and foot position.

To be able to discriminate between talocrural and subtalar kinematics during gait, 3D motion capture techniques are not helpful. Fortunately, more recent developments have enabled us to obtain some information on this aspect. Again, this requires the use of medical imaging technologies. One recent paper (de Asla, Wan, Rubash, & Li, 2006) used a combination of fluoroscopy and MRI scans to obtain talocrural and subtalar kinematics during a weight-bearing gait-like motor task in five healthy subjects. The data showed that the full plantar flexion–dorsiflexion range of motion of 7.5 degrees during the first half of the stance phase was primarily situated in the talocrural joint (9.1 degrees) and much less in the subtalar joint (−0.9 degrees). During the second half of the stance phase, however, the situation was opposite: 8.5 degrees of the full range of 13.4 degrees took place in the subtalar joint and 4.4 degrees in the talocrural joint. The two other components are smaller, particularly during the first half of the stance phase. From mid-stance up to toe-off, inversion–eversion and internal–external rotation are clearly taking place primarily in the subtalar joint.

KINETICS OF THE HUMAN ANKLE JOINT

Apart from the kinematics, the forces in the ankle joint during functional motor tasks should be also discussed. Again, gait is by far the most important daily activity and the discussion will therefore be devoted to information during this activity. Most of the information is again obtained from gait analyses in motion capture labs where kinematics as well as ground reaction forces are measured, sometimes also combined with muscle activity.

EXTERNAL LOADS ON THE ANKLE JOINT

Ground reaction forces recorded during gait in a motion capture lab are usually considered to be opposite and equal to the external forces experienced by the ankle joint itself. The data obtained by Burg (2014) show typical values and courses for the three components of the ground reaction force expressed with respect to body weight (Fig. 16.9).

The force in the vertical direction is the largest. It shows two peaks, one shortly after the heel strike and one just before push-off, which are almost equal in value and slightly larger than body weight. The anteroposteriorly directed force is almost 10 times smaller and also shows two peaks at the same instances as the vertical force. The first peak is in the posterior direction and serves to decelerate the body immediately after the heel strike. The second peak is directed anteriorly and serves to accelerate the body during push-off. The mediolateral force component is the smallest of the three. It is directed medially for a very short time immediately after the heel strike, due to the shift of support from one leg to the other. Thereafter, it is directed laterally during the rest of the stance phase apart from a very short period of time just before push-off.

Combining the magnitude and direction of the ground reaction force vector with its point of application on the foot allows for the calculation of the moment exerted on the ankle joint. The graph is shown in Fig. 16.10. It expresses the moment as a function of body weight. The moment exerted by the muscles (plantar and dorsiflexors) can be considered to be equal and opposite to the shown moment. Thus, while the moment of the ground reaction force leads to plantar flexion immediately after the heel contact and to dorsiflexion in the rest of the stance phase as shown in Fig. 16.9, the muscles will need to counteract this external moment. Immediately after the heel contact, the dorsiflexors will thus slow down the plantar flexion movement of the foot. Afterwards, the

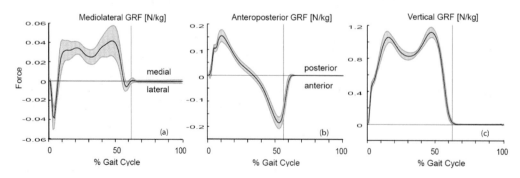

FIGURE 16.9

Gait GRF. Ground reaction force components on the ankle joint during gait. GRF, Ground reaction force.

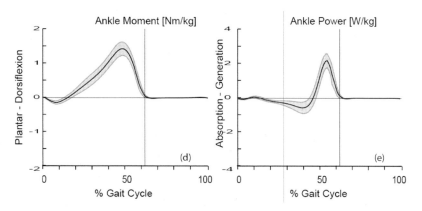

FIGURE 16.10

Gait moment and power. Ankle moment and power during gait.

plantar flexors will start contracting, first to slow down the dorsiflexion of the foot and then to bring the ankle and foot quickly in plantar flexion and propel the body forward.

Based on the plantar flexion–dorsiflexion moments of the muscles and the rotation speed of the ankle (the derivative with respect to time of the graph in Fig. 16.8), the power delivered by the plantar and dorsiflexor muscles can also be calculated. Negative power means that the muscles dissipate mechanical energy, while positive power means that the muscles produce mechanical energy (both potential energy and kinetic energy). Fig. 16.10 shows that the ankle is a joint where energy is mainly produced, contrary to what happens in the knee joint during walking.

MUSCLE FORCES AND JOINT CONTACT FORCES IN THE ANKLE JOINT

A final word then on the muscle and joint contact forces working in the ankle joint. As it is impossible to really measure forces in the muscles or in the articular surface, information about these parameters during gait can only be derived by combining data from motion capture measurements with mathematical models and simulations. Nowadays, quite complex and advanced musculoskeletal models are available including large numbers of muscles, tendons, and ligaments. They range from rigid body mechanics (in which bones are considered undeformable) up to finite element models.

A full discussion of these models would lead us too far in this chapter. A quite old, very basic, and static approximation will suffice to give an indication of the order of magnitude of the internal forces in the ankle. It has been worked out by Brewster, Chao, & Stauffer (1974) already, but the findings are still largely valid (although more advanced models have tweaked the values somewhat).

It is rather easy to see from the derivation that the force in the Achilles tendon can reach a maximal value up to four times body weight during the stance phase close to push-off. At the same time, the contact force on the tibiotalar articular surface reaches a maximal value of five times body weight. The latter value is larger than the maximal contact forces in the hip and knee joint during walking. Taking into account that the contact surface in the ankle joint is smaller than in those two joints, it is easy to see that the average contact pressure on the articular cartilage is higher in the ankle than in the hip and knee joint.

REFERENCES

Bok, S.-K., Lee, T. H., & Lee, S. S. (2013). The effects of changes of ankle strength and range of motion according to aging on balance. *Annals of Rehabilitation Medicine, 37*(1), 10.

Bonnel, F., Teissier, P., Colombier, J. A., Toullec, E., & Assi, C. (2013). Biometry of the calcaneocuboid joint: Biomechanical implications. *Foot and Ankle Surgery, 19*(2), 70–75.

Brewster, R. C., Chao, E. Y., & Stauffer, R. N. (1974). *Force analysis of the ankle joint during the stance phase of gait*. 27th ACEMB Alliance for Engineers. Philiadelphia.

Brockett, C. L., & Chapman, G. J. (2016). Biomechanics of the ankle. *Orthopaedics and Trauma, 30*(3), 232–238.

Burg, J. (2014). *Biomechanics of the foot-ankle complex before and after total ankle arthroplasty: An in vivo and in vitro analysis*. Leuven: KU Leuven.

de Asla, R. J., Wan, L., Rubash, H. E., & Li, G. (2006). Six DOF in vivo kinematics of the ankle joint complex: Application of a combined dual-orthogonal fluoroscopic and magnetic resonance imaging technique. *Journal of Orthopaedic Research, 24*(5), 1019–1027.

Gill, M., & Vilella, R. C. (2021). *Anatomy, Bony Pelvis and Lower Limb, Foot Cuboid Bone*. Available at: https://www.ncbi.nlm.nih.gov/books/NBK549912/.

Grood, E. S., & Suntay, W. J. (1983). A joint coordinate system for the clinical description of three-dimensional motions: Application to the knee. *Journal of Biomechanical Engineering, 105*(2), 136–144.

Gupton, M., Özdemir, M., & Terreberry, R. R. (2021). *Anatomy, Bony Pelvis and Lower Limb, Calcaneus*. Available at: https://www.ncbi.nlm.nih.gov/books/NBK519544/.

Harper, M. C. (1987). Deltoid ligament: An anatomical evaluation of function. *Foot & Ankle, 8*(1), 19–22.

Khan, I. A., & Varacallo, M. (2021). *Anatomy, Bony Pelvis and Lower Limb, Foot Talus*. Available at: https://www.ncbi.nlm.nih.gov/books/NBK541086/.

Manganaro, D., & Alsayouri, K. (2021). *Anatomy, Bony Pelvis and Lower Limb, Ankle Joint*. Available at: https://www.ncbi.nlm.nih.gov/books/NBK545158/.

Myburgh, K. H., Vaughan, C. L., & Isaacs, S. K. (1984). The effects of ankle guards and taping on joint motion before, during, and after a squash match. *The American Journal of Sports Medicine, 12*(6), 441–446.

Prapto, D., & Dreyer, M. A. (2021). *Anatomy, Bony Pelvis and Lower Limb, Navicular Bone*. Available at: https://www.ncbi.nlm.nih.gov/books/NBK547675/

Sammarco, V. J. (2004). The talonavicular and calcaneocuboid joints: Anatomy, biomechanics, and clinical management of the transverse tarsal joint. *Foot and Ankle Clinics, 9*(1), 127–145.

Tuijthof, G. J. M., Zengerink, M., Beimers, L., Jonges, R., Maas, M., van Dijk, C. N., & Blankevoort, L. (2009). Determination of consistent patterns of range of motion in the ankle joint with a computed tomography stress-test. *Clinical Biomechanics, 24*(6), 517–523. Available from https://doi.org/10.1016/j.clinbiomech.2009.03.004.

Watanabe, K., Kitaoka, H. B., Berglund, L. J., Zhao, K. D., Kaufman, K. R., & An, K.-N. (2012). The role of ankle ligaments and articular geometry in stabilizing the ankle. *Clinical Biomechanics, 27*(2), 189–195. Available from https://doi.org/10.1016/j.clinbiomech.2011.08.015.

Wu, G., Siegler, S., Allard, P., Kirtley, C., Leardini, A., Rosenbaum, D., ... Witte, H. (2002). ISB recommendation on definitions of joint coordinate system of various joints for the reporting of human joint motion—Part I: Ankle, hip, and spine. *Journal of Biomechanics, 35*(4), 543–548.

BIOMECHANICS OF WRIST AND ELBOW

Emmannuel J. Camus[1,2,3,4], **Fabian Moungondo**[5] and **Luc Van Overstraeten**[5]

[1]*IMPPACT - Lille Sud Clinic, Lesquin, France* [2]*Functional Anatomy Laboratory, Université Libre de Bruxelles, Brussels, Belgium* [3]*IMPPACT - Val de Sambre Clinic, Maubeuge, France* [4]*IMPPACT - Clinic of Cambresis, Cambrai, France* [5]*Department of Orthopedics and Traumatology, Faculty of Medicine, Erasme University Hospital, Université Libre de Bruxelles, Brussels, Belgium*

THE WRIST

Wrist biomechanics is the study of wrist and carpal bone movements, wrist motors, stabilizers, and means of controlling movements, and the distribution of loads that pass from the hand to the forearm. The wrist is an articular group that joins the hand to the forearm. It gathers the different articular spaces between the forearm bones (radius and ulna), the carpal bones, and the palm bones (the metacarpals).

The eight carpal bones are organized into two rows of four: the proximal and the distal row. The proximal row includes the scaphoid, lunate, triquetrum, and pisiform. The first three articulate directly with the forearm bones. The joint between the forearm bones and the proximal row is the radiocarpal joint. The distal row includes the trapezium, trapezoid, capitate, and hamate. They articulate directly with the proximal row. The joint between the proximal and distal rows is the midcarpal joint. The five metacarpals articulate directly with the bones of the distal carpal row: the first metacarpal with the trapezium, the second metacarpal with the trapezoid, the third metacarpal with the capitate, and the fourth and fifth metacarpals with the hamate. The joints between the carpus and the metacarpals are the carpometacarpal joints. With this architecture, the bony continuity between the hand, and thus the metacarpals and the forearm, is ensured by all the bones of the carpus, except for the pisiform. The pisiform is articulated in front of the triquetrum. It is integrated into the thickness of the terminal tendon of the flexor carpi ulnaris and mobilizes with it. Although it has its place in the description of the carpus, its kinematics allows it to be considered outside the movement of the proximal and distal rows (Fig. 17.1).

Among the movements that take place inside the carpus, some can be considered as functional movements, because they allow the hand to be oriented in space, and others as adaptive movements, because the bony surfaces of the carpus are not simple geometric surfaces, and the fluidity of the movements requires a malleability of the two carpal rows.

All the motor muscles of the hand are inserted at the base of the metacarpals. No tendon inserts on the carpal rows. Two flexor tendons for the wrist, the flexor carpi radialis and the flexor carpi ulnaris, insert respectively on the volar surfaces of the second and fifth metacarpals (Fig. 17.2).

Human Orthopaedic Biomechanics. DOI: https://doi.org/10.1016/B978-0-12-824481-4.00037-8

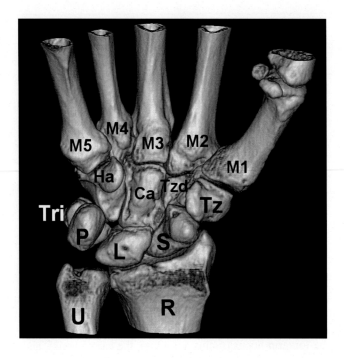

FIGURE 17.1

Wrist bones. *R*, Radius; *U*, ulna; *S*, scaphoid; *L*, lunate; *P*, pisiformis; *Tri*, triquetrum (behind P); *Tz*, trapezium; *Tzd*, trapezoid; *Ca*, capitate; *Ha*, hamate.

Three extensor tendons for the wrist, the extensor carpi radialis longus, the extensor carpi radialis brevis, and the extensor carpi ulnaris, insert on the base of the second, third, and fifth metacarpals, respectively (Fig. 17.2). The motor tendons of the fingers, flexors and extensors of the long fingers, long abductor, short extensor, long extensor of the thumb, can have an accessory role of mobilization of the wrist, but which is not systematic, as there is independence between the movements of the fingers and those of the wrist. In most actions, the finger flexors are strongest during wrist extension, and the finger extensors are strongest during wrist flexion.

However, all the motor tendons of the wrist and fingers, distributed all around the wrist, can have a global role of cohesion of the carpus when they are put under tension, by coming to push on the bony reliefs to prevent the carpus from subluxing forward or backwards, and to allow it to remain permanently centered under the distal end of the two bones of the forearm. This phenomenon was named by Kapandji: "the tendon encagement of the carpus" (Kapandji, Martin-Bouyer, & Verdeille, 1991).

WRIST MOVEMENTS

They are classically described in relation to the anatomical planes, starting from the anatomical reference position.

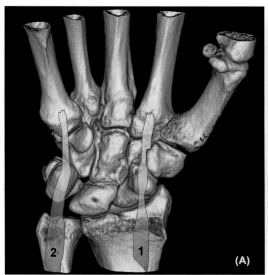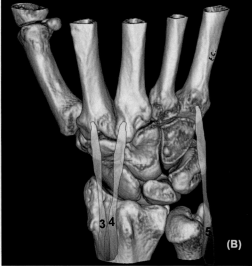

FIGURE 17.2

Motor tendons of the wrist. (A) Flexor carpi: radialis (1) and ulnaris (2) tendons. (B) Extensor carpi: radialis longus (3), brevis (4), and carpi ulnaris (5) tendons.

1. *In the sagittal plane*, the movements of flexion, carrying the palm forward and upward, and extension, carrying the palm backwards and downward, are described.
   a. *Wrist flexion* is controlled by the flexor carpi radialis and the flexor carpi ulnaris. This movement includes flexion of the proximal row, combined with flexion of the distal row. However, the range of flexion is not equal for each carpal bone due to the deformability of both rows (Camus, Millot, Lariviere, Raoult, & Rtaimate, 2004). The amplitude of flexion, as well as extension, is 70 to 80° in active motion, and 90° or even more in passive motion.
   b. *Wrist extension* is produced by the extensor carpi radialis longus and brevis, and extensor carpi ulnaris, with a minor involvement of the extensor digitorum. This movement combines the extension of the two rows.
2. In the *volar plane*, the movements defined as radial inclination and ulnar inclination are defined as follows:
   a. *Radial inclination* of the wrist is controlled by the radial muscles: the extensor carpi radialis longus and the flexor carpi radialis. In the frontal plane, the two rows bend radially, and their inclination values adding up (Camus, Millot, Larivière, Rtaimate, & Raoult, 2008). This movement involves a flexion of the proximal carpal row according to the forearm, and an extension of the distal row according to the proximal row. The latter compensates proximal row flexion, to keep the hand in the volar plane. This combination of movements results in a decrease in the height of the carpus on its radial edge, and an increase in its ulnar edge, and thus in the inclination of the hand. For this change in carpal height to occur, it involves shearing between the two carpal rows on the radial edge of the wrist.

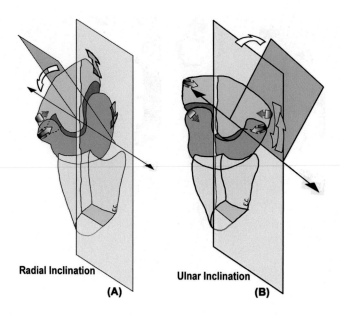

Radial Inclination

(A)

Ulnar Inclination

(B)

FIGURE 17.3

Movements of the carpal rows in radial/ulnar inclination. (A) Radial inclination of the wrist. Association of radial inclination of both rows, flexion/pronation of proximal row and extension/supination of distal row, so the proximal row crosses radially and volarly the distal row. (B) Ulnar inclination of the wrist. Association of ulnar deviation of both rows, extension/supination of the proximal row and flexion/pronation of the distal row, so the proximal row crosses ulnarly and volarly the distal row. Yellow: radius. Orange: proximal row. Green: distal row.

The scaphoid, representing the proximal row, passes in front of the trapezium, which represents the distal row. This shearing implies that passively the flexion of the proximal row is done with a pronation, and the extension of the distal row is done with a supination (Kapandji et al., 1991) (Fig. 17.3). Its amplitude of motion is 5° to 15°.

b. *Ulnar inclination* of the wrist is controlled by the flexor carpi ulnaris and extensor carpi ulnaris muscles. In the volar plane, the two rows bend ulnarly, their inclination values adding up. This movement involves an extension of the proximal carpal row, and a flexion of the distal row beyond the proximal row. This combination of movements results in an increase in height of the carpus on its radial edge, a decrease in height on its ulnar edge, and the ulnar tilt of the hand. In this movement, the hamate, located distal to the triquetrum in the neutral position, comes to position behind it, which in reaction shifts the triquetrum forward, implying a joint supination of the proximal carpal row as it extends, and a pronation of the distal, as it flexes (Fig. 17.3). Its amplitude of motion is 20° to 50°.

c. When the wrist successively flexes, inclines radially, extends, inclines ulnarly, but also in reverse succession, the movement is described as circumduction (Camus et al., 2008; Kapandji, 1998).

d. Functionally, there is a movement that is more frequent in daily activities than the movements described in the anatomical reference planes. This movement occurs in an *oblique plane* between two extreme positions: one represented by the combination of extension and radial inclination of the wrist, and the other represented by the combination of flexion and ulnar inclination of the wrist. The transition from one position to the other, which is widely used in repetitive movements, is called the "dart-throwing motion" (DTM). If this ludic activity is a good example, this movement is very much used in many daily manual activities using tools with an alternative gesture, such as hammering. In DTM, radiocarpal joint motion is minimal, most of the movement occurs in the midcarpal joint (Crisco et al., 2005; Moritomo et al., 2007).

e. In the *axial plane*, the hand can rotate 180 degrees along the axis of the forearm. However, this movement, called *prono-supination*, is not purely produced in the wrist, but in the forearm, both in the proximal and distal radioulnar joints. The radius, which has a longitudinal S-shape, behaves like a crank. Its rotation around its longitudinal axis allows its distal end, which is laterally offset, to rotate around the distal end of the ulna, passing forward and then positioning itself medial to it, which describes the pronation. The opposite movement is called supination (Kapandji, 2005a, 2005b).

MEANS OF CARPAL STABILITY

The congruence of the joint is inversely proportional to its stability. The interlocking of the bony forms is not sufficient to ensure joint stability on its own. The wrist is therefore stabilized by a complex system of ligaments (Nanno et al., 2009), completed by the tonus of the various periarticular tendons.

Many ligaments are included in the thickness of the joint capsule and schematically connect the forearm with the proximal row, or the proximal and distal rows. Some ligaments are purely intraarticular and schematically connect the bones of the same row. The ligaments between the forearm and the carpus are named extrinsic, the purely carpal ligaments are named intrinsic (Camus et al., 2004) (Fig. 17.4).

The ligaments have a fibrous structure that is not very extensible, resistant to traction and torsion. They therefore act as a mechanical brake when the movement reaches its extreme amplitude. But they are also rich in mechanoreceptors, which induces proprioceptive control (Esplugas, Garcia-Elias, Lluch, & Llusá Pérez, 2016; Hagert, Forsgren, & Ljung, 2005; Hagert, Garcia-Elias, Forsgren, & Ljung, 2007; Mataliotakis et al., 2009).

The periarticular tendons have a role of direct control of the position of the hand by their motor action according to the classical agonist/antagonist principle. The movement is allowed by the contraction of the agonists with the release of the antagonists. The simultaneous and isometric contraction of the agonists and antagonists allows the hand to be held securely in place, regardless of its position. Indirectly, the muscle contraction tightens the tendons, which stabilizes the carpus by brace effect. The flexor carpi radialis prevents anterior subluxation of the scaphoid tubercle and the trapezium. The flexor carpi ulnaris, reinforced by the pisiform, prevents anterior subluxation of the triquetrum and hamate. The radial extensors prevent dorsal subluxation of the proximal pole of the scaphoid and trapezoid. The ulnar extensor prevents dorsal subluxation of the triquetrum and hamate.

The combination of the action of the periarticular tendons and the interosseous ligaments forces the carpus to mobilize in a coherent manner. The loss of this stabilization system can lead to serious intrarow instabilities (carpal instability dissociative—CID), such as scapholunate dissociation,

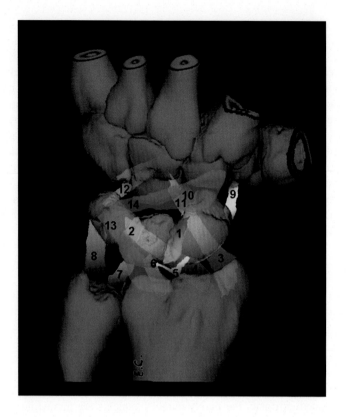

FIGURE 17.4

Main wrist ligaments. Green: intrinsic scapholunate and lunotriquetral ligaments (1–2). Orange shades: volar ligaments. Radioscaphocapitate (radiocarpal part) (3). Long radiolunate (4). Radioscapholunate (Testut) (5). Short radiolunate (6). Ulnolunate (7). Ulnotriquetral (8). Scaphotrapezial (9). Scaphocapitate (10). Radioscaphocapitate (midcarpal part) (11). Triquetrohamatocapitate (12). Blue shades: dorsal ligaments. Dorsal radiocarpal (13). Dorsal midcarpal (14).

or extrarow instabilities, such as radiocarpal dislocation or midcarpal instability (carpal instability nondissociative—CIND), or to a more severe degree, as a combined instability, such as perilunate dislocation of the carpus.

CARPAL LOADS

The carpus is subjected to many loads that are mostly axial, transmitted through the metacarpals towards the two forearm bones. The highest loads are due to the grip force. They come from the strength of the flexor fingers tendons and the motor tendons of the wrist when the latter is locked in a fixed position (Fig. 17.5).

These tendons exert an upward longitudinal vector that produces a global load from the base of each metacarpal. These stresses are then transmitted to the distal row of the carpus, but the

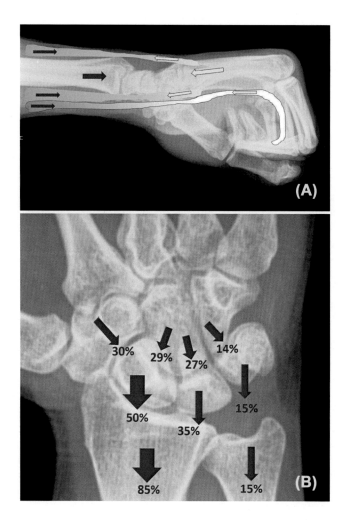

FIGURE 17.5

Origin and distribution of carpal loads. (A) Red arrows: muscles contraction and forearm bones reaction. Yellow arrows: tendons traction and hand bones reaction. (B) Distribution of carpal loads (in %).

architecture of the midcarpal joint results in a significant concentration of pressure stresses to the scaphoid, since the scaphoid can receive pressure from the first three rays, through the trapezium, trapezoid, and capitate. The lunate receives some loads from the capitate and sometimes the hamate, and the triquetrum essentially receives some loads from the hamate. These stresses are then transmitted to the forearm, about 85% to the radius, and 15% to the triangular fibro-cartilage complex (TFCC) and ulna (Camus, 2013; Camus, Aimar, Van Overstraeten, Schuind, & Innocenti, 2020; Camus, Van Overstraeten, & Schuind, 2021) (Fig. 17.5).

However, in the forearm, this proportion is reversed thanks to the interosseous membrane, whose main oblique fibers inserted on the radius, and directed more distally towards the ulna, transmit more than half of this upward stress towards the proximal ulna and its articulation with the humerus. It may be noted that the proximal end of the radius at the elbow, which articulates with the capitulum of the humerus, as well as the distal end of the ulna at the wrist, which articulates with the TFCC and the carpus, withstand axial rotation under load, but are relatively relieved when pressure is transmitted from the hand to the humerus, thus decreasing resistance to prono-supination.

THE ELBOW

The elbow is a complex structure that allows flexion of the forearm on the arm and also pronation and supination of the forearm in order to provide the human ability to interact both with himself and with the environment. Combination of mobility and stability is the main challenge that is met by this very congruent joint which has to make possible subtle and accurate motions as well as weight-bearing. To complete this task, there is not a single joint but a complex made of three joints enclosed in a single joint capsule. Flexion and extension of the elbow are provided by the ulnohumeral and radiohumeral joints, whereas pronation and supination motions involve the radiohumeral joint and the radioulnar joint proximally at the elbow but also distally out of this joint. Forearm rotation, by the bias of proximal and distal radioulnar joints, is thereby a direct link between elbow and wrist function and is highly dependent on the antebrachial frame integrity.

ANATOMY

Bones

The distal extremity of the humerus presents two condyles with different shapes of articular surfaces. Medially the trochlea is intended to articulate with the proximal ulna in a diarthrodial joint, and laterally the almost spherical capitellum is designed to meet the proximal radius in a spheroid joint. These condyles are flexed anteriorly by 30 degrees and tilted laterally at 6−8 degrees, while a medial rotation of 5−7 degrees gives a joint configuration of a joint placed with some offset compared to the upper part of the humerus (Alcid, Ahmad, & Lee, 2004).

The proximal ulna has two articular surfaces: the greater sigmoid notch oriented anteriorly, which articulates with the trochlea, and the lesser sigmoid notch oriented laterally, which articulate with the radial head. Both proximal and distal extremities of the greater sigmoid notch have a beak shape process named olecranon and coronoid process, respectively. The lesser sigmoid is based on the lateral side of the coronoid process. Compared to the proximal ulna, the distal ulna demonstrates a varus angulation of 11−28 degrees at ∼85 mm to the olecranon (Windisch, Clement, Grechenig, Tesch, & Pichler, 2007).

The proximal extremity of the radius, the radial head, has an almost cylindrical shape. The proximal surface of the radial head has a dish shape called the fovea radialis that is articulated with the capitellum, while the circumference of the head is articulated with the ulna lesser sigmoid notch also called radial notch (Captier, Canovas, Mercier, Thomas, & Bonnel, 2002). Between the fovea radialis and the circumference of the radial head a beveled shape of the radial head is intended to articulate with the humerus at a region between the trochlea and the capitulum named the zona conoidea (Jeon et al., 2012). At the radial neck, the radial head is angulated by 6−28 degrees with the radius diaphysis (Van Riet et al., 2004). The radial head shape is not always circular but sometimes elliptical and Captier et al. note that is a correlation between the angulation of the radial neck and the circular aspect of the radial head with neck angle being greater in the case of circular shape.

Ligaments

Two main ligamentous complexes provide the elbow integrity medially and laterally.

The annular ligament is attached at the anterior and posterior edges of the ulnar lesser sigmoid notch. This ligament stabilizes the radial head into the lesser sigmoid notch during forearm rotation.

Laterally, the radial collateral ligament joins the inferior surface of the lateral epicondyle (Olsen, Vaesel, Søjbjerg, Helmig, & Sneppen, 1996) to the proximal radius at the level of the annular ligament (Morrey & An, 1985) and the ulnar collateral ligament joins the inferior surface of the lateral epicondyle to the proximal ulna on a tubercle of the supinator crest named the crista supinatoris (Fig. 17.6). According to Seki et al., these three ligaments should be considered as the elements of a single structure, the lateral collateral ligament complex that is a three-dimensional Y shape structure made of a superior band, a posterior band, and an anterior band (Seki, Olsen, Jensen, Eygendaal, & Søjbjerg, 2002). Further investigation of this theory shows that the radial

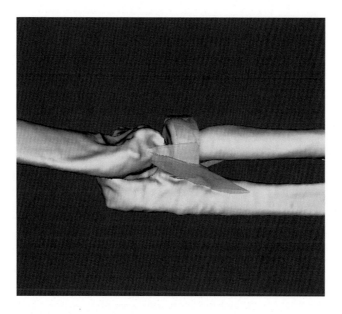

FIGURE 17.6

Anatomy of the lateral ligaments of the elbow. While the origin of the lateral ligaments is at the same point at the inferior surface of the lateral epicondyle, the lateral ulnar collateral ligament is inserted distally at the level of the crista supinatoris (*red arrow*) and the radial collateral ligament is inserted distally on the radial head through the annular ligament (*arrowhead*).

FIGURE 17.7

Anatomy of the medial collateral ligament complex. The origins of both anterior and oblique bundles are located at the inferior surface of the epitrochlear. The anterior bundle is inserted distally on the sublime tubercle of the proximal ulna (*arrow*), and the posterior bundle is inserted distally at the medial border of the olecranon (*arrowhead*).

head plays an important role in the maintenance of the strain of these ligaments by supporting the three-dimensional "y" shape of this complex (Olsen et al., 1996).

Medially the medial collateral complex is composed of three elements: an anterior or oblique bundle, a posterior bundle, and a transverse segment (Morrey & An, 1985). Both the anterior and posterior bundles originate from the inferior surface of the epitrochlear (O'Driscoll, Jaloszynski, Morrey, & An, 1992). While the anterior bundle is inserted on the medial aspect of the coronoid process, the so-called sublime tubercle, the fan-shaped posterior bundle is inserted at the olecranon medial side (Fig. 17.7). The transverse segment is made of horizontally oriented fibers between the olecranon and the coronoid process (Morrey & An, 1985).

ELBOW STABILITY

The elbow is a very congruent joint, and the bony structures play an important role in its stability. In extension, 80% of the valgus stresses are resisted by the proximal portion of the sigmoid notch and 65% of the varus stresses are resisted by the distal part of the sigmoid notch, while the coronoid process is essential to prevent posterior subluxation of the elbow (King, Morrey, & An, 1993).

The elbow stability is further enhanced by ligaments and muscles around the joint. These structures may be classified as dynamic and static stabilizers (Bryce & Armstrong, 2008). Dynamic stabilizers are the muscles crossing the elbow the triceps, brachialis, and anconeus whose contractions induce joint coaptation. Static stabilizers can be divided into primary and secondary stabilizers. The secondary stabilizers are the radial head, the joint capsule, and the insertion of the flexors and extensors tendons. The primary stabilizers are the ulnohumeral joint, the medial collateral ligament (mainly the oblique bundle), and the lateral collateral ligament.

While the radial head is the secondary constraint to resist the valgus stresses, the primary valgus stabilizer of the elbow is the anterior bundle of the medial collateral ligament (Morrey, Tanaka, & An, 1991). On the other side, the primary varus stabilizer is the lateral collateral ligament complex (Olsen et al., 1996). Through this complex, the radial head plays a role in the varus stability of the elbow by preserving its tridimensional shape (Morrey & An, 1985). If the medial and lateral ligaments of the elbow remain intact, a radial head removal would not have a major effect on the valgus stability of the joint but it will induce a relaxation of this complex and thus varus instability will appear.

FORCE TRANSMISSION AT THE ELBOW

There is no consensus about the force transmission through the elbow. According to Halls and Travill (1964), with the elbow in complete extension, 60% of force applied to the forearm pass through the radiocapitellar joint and 40% through the ulnohumeral joint. This assertion is not shared by Chantelot et al. (2008), who described that no more than 25% of the load passed through the radiocapitellar joint and instead, 60% of the load passed from the ulnohumeral joint through the coronoid process. These authors also observed that the forearm rotation had an influence on the amount of load passing through the radiocapitellar joint, the neutral being the position of the forearm with the higher radiocapitellar load and the supination showing the lower load transfer. The relation between elbow rotation and the radiohumeral contact was also described in other studies. Following a combined in vivo and cadaver study, the contact between the bevel edge of the radial head and the humeral zona conoidea was described by Jeon et al. (2012) as maximal in supination, these authors state that supination is the position of higher stability of the elbow. In a cadaver study, Hwang et al. have observed the opposite, with the radiocapitellar contact decreasing from pronation to supination while the contact pressure was increasing. These authors suggest the pronation is a protective position of the radiocapitellar joint because the joint contact area is higher (Hwang et al., 2018). According to Captier et al. (2002), the contact between the radial head and the zona conoidea mainly depends on the radial head shape and when the radial head is not circular, they described the contact occurs in the neutral forearm position. In a cadaver study, we previously observed that the radiocapitellar contact areas were increased with supination (Hwang et al., 2018). More recently, in a similar study design, we observed mainly in this forearm position an additional contact area corresponding to the contact between the bevel edge of the radial head and the zona conoidea (Moungondo et al., 2010).

ELBOW MOTION

Elbow flexion occurs at the ulnohumeral and radiohumeral joints. It is performed around an axis passing through the center of the trochlea and the center of the capitellum and the anteroinferior aspect of the medial epicondyle (Bryce & Armstrong, 2008). The axis of flexion−extension is a loose hinge that allows 5 degrees of internal and external rotation during the elbow flexion. The normal range of motion of the elbow is approximately 140 degrees of flexion.

The trochlear groove is the central depression of the trochlea. Its orientation is oblique distally and laterally resulting in a lateral deviation of the forearm in extension of the elbow. The angle between the ulna and the humerus in this position is called the Carrying angle (10−15 degrees) (Stroyan & Wilk, 1993).

Pronation and supination occur at the proximal and distal radioulnar joint. During this movement, the radius turns around itself and around the ulna following an axis passing through the center of radial and the ulnar head. The main movement of the radius is a rotation, while the main movement of the ulna is a translation (Nakamura, Yabe, Horiuchi, & Yamazaki, 1999).

At the elbow, the radial head shows some anterior and posterior translation during forearm rotation. Captier et al. (2002) observed that in noncircular radial head, there was a translation of the center of the fovea radialis anteriorly in pronation and dorsally in supination.

REFERENCES

Alcid, J. G., Ahmad, C. S., & Lee, T. Q. (2004). Elbow anatomy and structural biomechanics. *Clinics in Sports Medicine, 23*(4), 503−517. Available from https://doi.org/10.1016/j.csm.2004.06.008, vii.

Bryce, C. D., & Armstrong, A. D. (2008). Anatomy and biomechanics of the elbow. *The Orthopedic Clinics of North America, 39*(2), 141−154. Available from https://doi.org/10.1016/j.ocl.2007.12.001, v.

Camus, E. (2013). *Carpal biomechanics: Application to ligamentous injuries. Carpal ligament surgery: Before arthritis* (pp. 19−37). France: Springer-Verlag Vol. 9782817803791. Available from https://doi.org/10.1007/978-2-8178-0379-1_2.

Camus, E. J., Aimar, A., Van Overstraeten, L., Schuind, F., & Innocenti, B. (2020). Lunate loads following different osteotomies used to treat Kienböck's disease: A 3D finite element analysis. *Clinical Biomechanics, 78.* Available from https://doi.org/10.1016/j.clinbiomech.2020.105090.

Camus, E. J., Millot, F., Lariviere, J., Raoult, S., & Rtaimate, M. (2004). Kinematics of the wrist using 2D and 3D analysis: Biomechanical and clinical deductions. *Surgical and Radiologic Anatomy, 26*(5), 399−410. Available from https://doi.org/10.1007/s00276-004-0260-0.

Camus, E. J., Millot, F., Larivière, J., Rtaimate, M., & Raoult, S. (2008). Le carpe à double cupule: Illustration de la géométrie variable du carpe. *Chirurgie de la Main, 27*(1), 12−19. Available from https://doi.org/10.1016/j.main.2007.10.005.

Camus, E. J., Van Overstraeten, L., & Schuind, F. (2021). Lunate biomechanics: Application to Kienböck's disease and its treatment. *Hand Surgery and Rehabilitation, 40*(2), 117−125. Available from https://doi.org/10.1016/j.hansur.2020.10.017.

Captier, G., Canovas, F., Mercier, N., Thomas, E., & Bonnel, F. (2002). Biometry of the radial head: Biomechanical implications in pronation and supination. *Surgical and Radiologic Anatomy: SRA, 24*(5), 295−301. Available from https://doi.org/10.1007/s00276-002-0059-9.

Chantelot, C., Wavreille, G., Dos Remedios, C., Landejerit, B., Fontaine, C., & Hildebrand, H. (2008). Intra-articular compressive stress of the elbow joint in extension: An experimental study using Fuji films. *Surgical and Radiologic Anatomy: SRA, 30*(2), 103−111. Available from https://doi.org/10.1007/s00276-007-0297-y.

Crisco, J. J., Coburn, J. C., Moore, D. C., Akelman, E., Weiss, A. P. C., & Wolfe, S. W. (2005). In vivo radio-carpal kinematics and the dart thrower's motion. *Journal of Bone and Joint Surgery - Series A, 87*(12 I), 2729−2740. Available from https://doi.org/10.2106/JBJS.D.03058.

Esplugas, M., Garcia-Elias, M., Lluch, A., & Llusá Pérez, M. (2016). Role of muscles in the stabilization of ligament-deficient wrists. *Journal of Hand Therapy*, *29*(2), 166−174. Available from https://doi.org/10.1016/j.jht.2016.03.009.

Hagert, E., Forsgren, S., & Ljung, B. O. (2005). Differences in the presence of mechanoreceptors and nerve structures between wrist ligaments may imply differential roles in wrist stabilization. *Journal of Orthopaedic Research*, *23*(4), 757−763. Available from https://doi.org/10.1016/j.orthres.2005.01.011.

Hagert, E., Garcia-Elias, M., Forsgren, S., & Ljung, B. O. (2007). Immunohistochemical analysis of wrist ligament innervation in relation to their structural composition. *Journal of Hand Surgery*, *32*(1), 30−36. Available from https://doi.org/10.1016/j.jhsa.2006.10.005.

Halls, A. A., & Travill, A. (1964). Transmission of pressures across the elbow joint. *The Anatomical Record*, *150*, 243−247. Available from https://doi.org/10.1002/ar.1091500305.

Hwang, J.-T., Kim, Y., Bachman, D. R., Shields, M. N., Berglund, L. J., Fitzsimmons, A. T., ... O'Driscoll, S. W. (2018). Axial load transmission through the elbow during forearm rotation. *Journal of Shoulder and Elbow Surgery/American Shoulder and Elbow Surgeons*, *27*(3), 530−537. Available from https://doi.org/10.1016/j.jse.2017.10.003.

Jeon, I.-H., Chun, J.-M., Lee, C.-S., Yoon, J.-O., Kim, P.-T., An, K.-N., ... Shin, H.-D. (2012). Zona conoidea of the elbow: Another articulation between the radial head. *The Journal of Bone and Joint Surgery. British Volume*, *94*(4), 517−522. Available from https://doi.org/10.1302/0301-620X.94B4.27842.

Kapandji, A. (1998). Le carpe considere comme "un tout" ou la conception holistique du carpe. *Main*, *3*(4), 417−428.

Kapandji, A. I. (2005a). La pronosupination. *Physiologie Articulaire*, *1*, 104−145.

Kapandji, A. I. (2005b). Le mouvement de circumduction. *Physiologie Articulaire*, *1*, 152−153.

Kapandji, A. I., Martin-Bouyer, Y., & Verdeille, S. (1991). Etude du carpe au scanner à trois dimensions sous contraintes de prono-supination. *Annales de Chirurgie de la Main*, *10*(1), 36−47. Available from https://doi.org/10.1016/S0753-9053(05)80036-5.

King, G. J., Morrey, B. F., & An, K. N. (1993). Stabilizers of the elbow. *Journal of Shoulder and Elbow Surgery/American Shoulder and Elbow Surgeons*, *2*(3), 165−174. Available from https://doi.org/10.1016/S1058-2746(09)80053-0.

Mataliotakis, G., Doukas, M., Kostas, I., Lykissas, M., Batistatou, A., & Beris, A. (2009). Sensory innervation of the subregions of the scapholunate interosseous ligament in relation to their structural composition. *Journal of Hand Surgery*, *34*(8), 1413−1421. Available from https://doi.org/10.1016/j.jhsa.2009.05.007.

Moritomo, H., Apergis, E. P., Herzberg, G., Werner, F. W., Wolfe, S. W., & Garcia-Elias, M. (2007). 2007 IFSSH committee report of wrist biomechanics committee: Biomechanics of the so-called dart-throwing motion of the wrist. *Journal of Hand Surgery*, *32*(9), 1447−1453. Available from https://doi.org/10.1016/j.jhsa.2007.08.014.

Morrey, B. F., & An, K. N. (1985). Functional anatomy of the ligaments of the elbow. *Clinical Orthopaedics and Related Research*, *201*, 84−90. Available from https://europepmc.org/article/med/4064425.

Morrey, B. F., Tanaka, S., & An, K. N. (1991). Valgus stability of the elbow. A definition of primary and secondary. *Clinical Orthopaedics and Related Research*, *265*, 187−195. Available from https://europepmc.org/article/med/2009657.

Moungondo, F., El Kazzi, W., van Riet, R., Feipel, V., Rooze, M., & Schuind, F. (2010). Radiocapitellar joint contacts after bipolar radial head arthroplasty. *Journal of Shoulder and Elbow Surgery/American Shoulder and Elbow Surgeons*, *19*(2), 230−235. Available from https://doi.org/10.1016/j.jse.2009.09.015.

Nakamura, T., Yabe, Y., Horiuchi, Y., & Yamazaki, N. (1999). In vivo motion analysis of forearm rotation utilizing magnetic resonance. *Clinical Biomechanics*, *14*(5), 315−320. Available from https://doi.org/10.1016/s0268-0033(98)90091-2.

Nanno, M., & Viegas, S. F. (2009). Three-dimensional computed tomography of the carpal ligaments. *Seminars in Musculoskeletal Radiology*, *13*(1), 3−17. Available from https://doi.org/10.1055/s-0029-1202240.

O'Driscoll, S. W., Jaloszynski, R., Morrey, B. F., & An, K. N. (1992). Origin of the medial ulnar collateral ligament. *Journal of Hand Surgery. American Volume*, *17*(1), 164−168. Available from https://doi.org/10.1016/0363-5023(92)90135-c.

Olsen, B. S., Vaesel, M. T., Søjbjerg, J. O., Helmig, P., & Sneppen, O. (1996). Lateral collateral ligament of the elbow joint: Anatomy and kinematics. *Journal of Shoulder and Elbow Surgery / American Shoulder and Elbow Surgeons*, *5*(2 Pt 1), 103−112. Available from https://doi.org/10.1016/s1058-2746(96)80004-8.

Seki, A., Olsen, B. S., Jensen, S. L., Eygendaal, D., & Søjbjerg, J. O. (2002). Functional anatomy of the lateral collateral ligament complex of the elbow: Configuration of Y and its role. *Journal of Shoulder and Elbow Surgery/American Shoulder and Elbow Surgeons*, *11*(1), 53−59. Available from https://doi.org/10.1067/mse.2002.119389.

Stroyan, M., & Wilk, K. E. (1993). The functional anatomy of the elbow complex. *The Journal of Orthopaedic and Sports Physical Therapy*, *17*(6), 279−288. Available from https://doi.org/10.2519/jospt.1993.17.6.279.

Van Riet, R. P., Van Glabbeek, F., Neale, P. G., Bimmel, R., Bortier, H., Morrey, B. F., ... An, K. N. (2004). Anatomical considerations of the radius. *Clinical Anatomy (New York, N.Y.)*, *17*(7), 564−569. Available from https://doi.org/10.1002/ca.10256.

Windisch, G., Clement, H., Grechenig, W., Tesch, N. P., & Pichler, W. (2007). The anatomy of the proximal ulna. *Journal of Shoulder and Elbow Surgery/American Shoulder and Elbow Surgeons*, *16*(5), 661−666. Available from https://doi.org/10.1016/j.jse.2006.12.008.

BIOMECHANICS AND DESIGN OF ORTHOPAEDIC DEVICES

BIOMATERIALS AND BIOCOMPATIBILITY

18

Ludovica Cacopardo
Research Center 'E. Piaggio', University of Pisa, Pisa, Italy

BIOMATERIALS: DEFINITIONS

A **biomaterial** is formally defined as "a material designed to take a form that can direct, through interactions with living systems, the course of any therapeutic or diagnostic procedure" (Ghasemi-Mobarakeh, Kolahreez, Ramakrishna, & Williams, 2019; Williams, 2019). More generally, biomaterials are those materials that can be placed in a biological environment promoting a desired response (e.g., promote tissue growth or repair) or without causing adverse effects (e.g., inflammation, cytotoxic, or thrombogenic effects).

Biomaterials are intrinsically related with the concept of biocompatibility, which can be defined as "the ability of a material to perform with an appropriate host response in a specific application" (Ghasemi-Mobarakeh et al., 2019; Williams, 2019). This means that biocompatibility is not only a property of the material itself but of the material–host system. Indeed, the material behavior depends on several environmental conditions (e.g., temperature, pH, and mechanical loading) and its performance is clearly dependent on the site of implant and application. This means that a given material can be classified as "bio"-material or not, depending on the desired performance. On the other hand, the biological response can be influenced by several factors related to specific pathophysiologic conditions.

Depending on the application, a biomaterial can be referred to as bioinert, if it has no (or at least low) effect on the body, or as bioactive, in the case of kindling a specific (desired) biological response. In orthopedics, bioinert materials are generally resistant to chemical corrosion, thermal decomposition, and biodegradation, thus ensuring implant durability and preventing side effects. However, materials able to favor the integration between bone and implant (osteointegration) are also desired. Bioactive materials may form chemical bonds with bones or promote *osteoconduction* when they support bone growth (e.g., thanks to a porous and cell adhesive microstructure), and *osteoinduction*, when they are able to stimulate osteogenesis (i.e., stem cell differentiation into mature bone cells) (Costa, Bartolomeu, Alves, Silva, & Miranda, 2019; Ghasemi-Mobarakeh et al., 2019; Ibrahim et al., 2017; Wang et al., 2018). In the case of temporary implants, controlled *biodegradation*, obtained through specific microstructures or with the presence of a particular chemical element, is desirable to allow implant resorbing rather than its surgical removing after it has fulfilled its function (Li et al., 2016; Prakasam et al., 2017).

Human Orthopaedic Biomechanics. DOI: https://doi.org/10.1016/B978-0-12-824481-4.00038-X

BIOMATERIAL CLASSES AND PROPERTIES

In (bio)engineering, knowledge of material properties is fundamental for product design. The material selection process has to consider technical, safety, and economical specifications, which are based on both intrinsic and extrinsic material properties (Fig. 18.1). Intrinsic properties are those properties that do not depend on the amount of matter considered. These include bulk and local (e.g., surface) physicochemical properties, such as atomic number, chemical structure, and density (ρ), and mechanical properties, such as the elastic modulus (E), yield stress, and fracture stress. Extrinsic properties are related to dimensional factors, such as size and shape, and take into account both economic factors, such as material cost and market availability, and manufacturing-related properties, such as production cost, finishing, and texture (Leo, 2008).

Material performance typically depends on a combination of properties, which can be compared thanks to graphics known as Ashby charts. On the basis of the specific application, performance indexes can be defined and the optimal material can be selected to maximize these indexes. Fig. 18.1B shows a comparison between the elastic modulus and density of a different material class. The dashed lines refer to a constant E/ρ ratio, also known as *specific elastic modulus*. In engineering applications, a material with a high E/ρ ratio is often desirable since it indicates a light and resistant material (Ashby & Cebon, 1993; Ferro & Bonollo, 2019). Even though an intrinsic "ideal" material is not always available, extrinsic properties such as shape or manufacturing process may contribute to the desired material performance. For example, a hollow cylinder optimizes bending resistance, weight, and costs (less material amount) with respect to a solid one. These concepts have been excellently adopted by nature which designed our long bones as hollow cylinders.

As illustrated in Fig. 18.2, the main material classes include:

- Metals: characterized by the metallic bond (i.e., an electron cloud shared between atoms), which contributes to the good thermal and electrical conductivity and to malleability (atoms are able to slide over one another without breaking the lattice structure). Thus, they present good

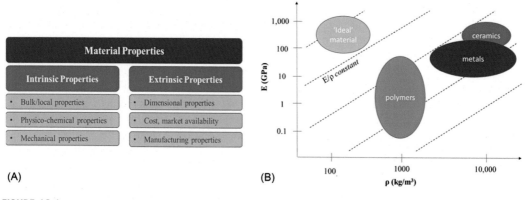

(A)

(B)

FIGURE 18.1

Material comparison and classification. (A) Material property classification; (B) Ashby chart comparing material density and elastic modulus.

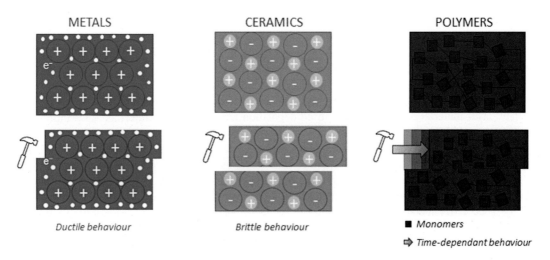

METALS   CERAMICS   POLYMERS

Ductile behaviour   *Brittle behaviour*   ■ *Monomers*

⇨ *Time-dependant behaviour*

FIGURE 18.2

Schematic illustration of the structure and mechanical behavior of the main material classes. Metals are composed of positively charged atoms and an electronic cloud, which is responsible for their ductile behavior in response to mechanical stimuli. Ceramics are characterized by bonds between atoms with different charges and, consequently, by fragile behavior. Finally, polymers are constituted by chains of repeated monomers able to slide upon loading, resulting in a time-dependent or (viscoelastic) mechanical response.

mechanical resistance and their (room-temperature) failure typically occurs by ductile fracture. However, particularly in wet environments, metals are often subject to chemical reactions responsible for their degradation (as corrosion).

- Ceramics: hard and brittle materials characterized by ionic and covalent bonds between atoms of metal and nonmetal elements. In this case, atom sliding is not favored and material fracture is typically fragile. In contrast, they have good resistance to wear and good chemical stability. Other properties depend on atom types and on the different crystalline structures. In particular, the microstructure and the resulting material properties are strongly affected by the manufacturing process (e.g., purity of the powder, duration of the thermal steps and maximum temperature, which is typically higher with respect to the other material classes).

- Polymers: formed by a chain of repeated functional units (monomers). Generally, they present lower elastic moduli and density of the previous two material classes, but their properties also depend on chain lengths and cross-links. Interchain bonds can be of weak nature, in the case of thermoplastic polymers, or of covalent nature, in the case of thermosetting polymers. Additionally, elastomers present both weak and covalent cross-links, which contribute to their resilience (i.e., the ability to undergo large deformations absorbing energy and then recover the original shape releasing energy). Chain entanglement sliding and water content (in the case of hydrophilic polymers such as gels) are also responsible for their viscoelastic behavior, which means that mechanical properties are time-dependent (Cacopardo, Guazzelli, & Ahluwalia, 2021).

Manufacturing processes can be generally classified as (1) formative, based on casting or injection of a liquid or powder form material in a mold or on material modeling thanks to both compressive (e.g., rolling, extrusion, and forging) or tensile forming; (2) subtractive, when the desired shape is obtained by a controlled material-removal process (e.g., turning, drilling, milling, laser cutting, and electrical discharge machining); (3) additive (also known as rapid prototyping or solid freeform fabrication), which allow the formation of a three-dimensional object thanks to a layer-by-layer material deposition (e.g., fused deposition modeling, laser sintering, 3D printing, and stereolithography).

The choice of the manufacturing strategies is typically related to the raw material properties, the desired final specifications (surface properties, tolerances, etc.), and the related costs. For example, the temperature required to liquefy the material or make it malleable significantly affects the fabrication process. To get an idea, melting temperatures can vary around 2000°C–4000°C for ceramics, 600°C–1800°C for metals, and 100°C–300°C for thermoplastic polymers. Thermosetting polymers cannot be melted after curing, which can occur at a temperature between 20°C and 200°C. Moreover, considering the number of parts, manufacturing costs are typically constant for additive techniques. Thus, they are mainly advantageous for small production volume or in the case of complex geometries. Differently, subtractive and formative techniques have a high initial cost related to the initial tool setup but allow a significant cost reduction for large production volumes.

Orthopedic implants can be thus based on different material types and fabricated by means of different manufacturing processes depending on the required performance. In addition, they need to meet safety requirements to minimize acute or chronic adverse body reactions. Acute responses generate as a consequence of the insertion of a foreign material in the body: they involve an inflammatory process which, in normal conditions, resolves within a week (Anderson, Rodriguez, & Chang, 2008). Chronic reactions are related to metal ions (corrosion) and particulate wear *debris* release over time, which cause *metallosis* (when ion concentration exceeds the toxicity threshold) and chronic inflammation. The latter can induce the formation of granulation tissue and of a *fibrotic capsule*, leading to implant failure. Bacterial colonization of the implant surface can also occur, leading to the formation of *biofilm* infections, which can be resolved only with the surgical removal of the implant. From a mechanical point of view, besides wear and fatigue implant failure, the difference in bone and implant elastic moduli is responsible for the *stress shielding* phenomenon, that is, loads are primarily sustained from the implant causing bone resorption and possibly implant mobilization (Hallab & Jacobs, 2009; Ibrahim et al., 2017; Williams, 2019).

BIOMATERIALS FOR ORTHOPEDIC DEVICES

Depending on the specific application and duration of the orthopedic implant, materials belonging to the metal, ceramic, and polymer classes have been employed. Examples are shown in Table 18.1.

Metals are extensively used for both load-bearing (e.g., knee or hip prosthesis) and nonbearing (e.g., screws or maxillofacial plates) applications. They generally have high elastic modulus (100–200 GPa) and yielding stresses (300–1000 GPa) and good fatigue resistance. The main issues are related to the high densities (4–17 g/cm^3), to corrosion in contact with biological fluids, which

Table 18.1 Common biomaterials used in orthopedics.

	Example	ρ (g/cm³)	E (GPa)	Advantages	Disadvantages	Main Applications
Metals	Steel alloys	7.9	190	• Good mechanical resistance to fatigue.	• Corrosion and ion releasing; • High density and elastic modulus; • Poor bioactivity	• Load bearing implant parts (e.g. stems or femoral heads and cups); • Short term implant (e.g. fracture plates, screws, nails).
	Cobalt alloys	8.9	210			
	Titanium alloys	4.5	110			
	Bulk Tantalum	16.6	186			
Ceramics	Allumina	3.9	380	• Wear resistance; • Biocompatibility	• Brittle behaviour; • Very high density and elastic modulus	• Joint implant components (e.g. femoral heads and cups); • Implant coatings; • Cranial plates • Bone or tissue engineering scaffolds fillers
	Zirconia	5.6	200			
	Hydroxyapatite	3.2	120			
	Bioglass	2.7	44			
Polymers	PMMA	1.2	2.2	• Elastic modulus and density comparable with bone; • Easy manufacturing.	• Poor fatigue resistance; • Wear and debris production	• Bone cement; • Joint implant component (e.g acetabular cups); • Bone fracture fixation; • Tissue engineering scaffolds.
	UHWPE	0.9	1.2			
	PEEK	1.3	3-4			

can lead to metal ion poisoning, and to low bioactivity, which can lead to implant mobilization. We can distinguish them as follows (Hudecki et al., 2019; Ibrahim et al., 2017):

- *Stainless steels (SSs)*, that is, iron-based alloys such as 316L-SS (containing Cr, Ni, Mo, Mn, Si, Cu, and C) and ASTM-F2229 (nickel-free) steel. In dry conditions, this material presents high corrosion resistance thanks to the formation of a chemically inert chromium oxide film, which efficiency is enhanced in the presence of molybdenum. The austenitic structure (face-centered cubic) is the material phase that better resists in corrosive media, with the additional advantage of not being ferromagnetic. However, in corrosive media, such as in the body, this property is strongly reduced. In addition, they show poor fatigue and wear resistance. Thus, these alloys are mainly used for temporary implants, such as fracture plates, screws, or hip nails. The main advantage is related to their low cost and acceptable biocompatibility.
- *Cobalt (CoCrMo) alloys* (e.g., ASTM-F75), characterized by good wear and corrosion resistance which is related to the formation of a chromium oxide layer, even in wet conditions. Despite the low fatigue resistance in liquid and the high cost of raw material and manufacturing

processes, they are widely used for long-term joint implants (e.g., prosthesis stem). Other alloys such as ASTM F90 (CoCrWNi), which contain tungsten and nickel for improving mechanical strength (i.e., stabilizing the austenitic structure at low temperature), are instead used only for short-term implants, such as fracture fixation devices because of the reduced corrosion resistance and the consequent release of toxic nickel ions.

- *Titanium alloys* (e.g., Ti6Al4V), with an elastic modulus closer to bone and a significantly lower density with respect to the other two alloy classes. These alloys typically present a biphasic structure (hexagonal compact [HC] and body-centered cubic [BCC], packing), thus combining the advantages of both phases: malleability for the HC and suitability to thermal treatment to improve mechanical properties for the BCC. They also present high corrosion resistance, thanks to the formation of a titanium oxide passive layer, and bioactive properties, which promote osteointegration. However, from a mechanical point of view, they show poor fatigue and bending resistance. Thus, their use for long-term load-bearing applications is limited. In addition, for the presence of elements with different specific weights and melting points, they require advanced and expensive fabrication processes. Ti-based amorphous phase alloys (i.e., without a crystalline structure) have been developed improving fatigue strength from 250 to 1500 MPa. In this case, the main issue is related to the limited fabrication size.

- *Porous metals*, fabricated with a wide range of techniques enabling the formation of multiscale porosities to improve both primary and secondary implant stabilization. With respect to bulk materials, the porous structure is typically responsible for lower elastic moduli, which prevents stress shielding (Figs. 18.3 and 18.4). For example, *Trabecular* Metal (Zimmer Biomet) is obtained by chemical vapor deposition of *tantalum* on a carbon foam, resulting in a material composed of 99% metal and 1% carbon with an 80% porosity with 430−650 μm pore size. Although not particularly resistant with respect to corrosion, the good osteointegration level obtained thanks to porosity, the low elastic modulus (3 GPa) and the wear resistance allow the

Trabecular Titanium ®

FIGURE 18.3

Trabecular titanium. Trabecular titanium (hip cup).

From Marin, E., et al. "Characterization of cellular solids in Ti6Al4V for orthopaedic implant applications: trabecular titanium."
Journal of the Mechanical Behavior of Biomedical Materials *3.5 (2010): 373–381. Available at: https://doi.org/10.1016/j.*
jmbbm.2010.02.001. Epub 2010 Feb 16.

Trabecular metal ®

FIGURE 18.4

Trabecular metal. Trabecular metal (hip cup).

From Sporer, S. M., & Paprosky, W. G. (2006). The use of a trabecular metal acetabular component and trabecular metal augment for severe acetabular defects. Journal of Arthroplasty, 21(6), 83–86. https://doi.org/10.1016/j.arth.2006.05.008.

common use of this material in total joint arthroplasty. On the contrary, bulk tantalum is rarely employed because of the elevated manufacturing cost (melting point temperature equal to 2996°C) and the high density with respect to the other metals. *Trabecular Titanium* (Lima Corporate) is produced by electron beam melting (EBM), which allows the realization of a regular three-dimensional hexagonal cell structure. The pore percentage can be modulated between 60% and 90% (with a mean pore size of 640 μm), obtaining monolithic parts with variable porosity which optimize surface and bulk properties. Compared to bulk titanium, the material presents a lower elastic modulus (1.2 GPa) and compression failure typically occurs on 45-degree planes with respect to the loading direction, corresponding to the weakest plane of the hexagonal structure. Similarly, *Tri-Por* (Adler Ortho) devices are fabricated with EBM techniques from titanium, CoCrMo, and SS powders. As an example, *Fixa Ti-Por* acetabular cups, obtained from Ti6Al4V powders, are characterized by a 65% constant porosity with a pore size of 700 μm. These cups present excellent performance, with lower aseptic loosening cases with respect to standard cups (Castagnini et al., 2019; Ibrahim et al., 2017; Matassi et al., 2013; Regis et al., 2015; Sporer & Paprosky, 2006).

- *Oxinium (oxide zirconium*—Smith&Nephew), a novel alloy of niobium and zirconium (Zr−2.5 Nb). After high-temperature treatment in air, a zirconium oxide layer (with a thickness in the order of micrometers) forms upon the material surface, thus combining the benefits of both metals and ceramics. The superior wear performance, the low friction, and the good corrosion resistance, without the ceramic brittleness, make oxinium an excellent choice for bearing components of orthopedic implants. The main drawback of this material is the expensive cost (Hernigou et al., 2007).

Ceramics are widely used for orthopedic implants due to their excellent wear resistance and biocompatibility. The main issue is their brittle nature, which limits their use in load-bearing applications. Moreover, ceramic materials have relatively high densities (3−6 g/cm^3) and low traction resistance. Nevertheless, they are often used for cranial plates and spinal spacers, as bone defect fillers, to replace parts of load-bearing metallic implants or as bioactive coatings. These materials can be classified as (Hudecki et al., 2019; Ibrahim et al., 2017; Merola & Affatato, 2019; Soto-Gutierrez et al., 2010):

- Bioinert: typically metal oxides such as *alumina* (Al_2O_3) and *zirconia* (ZrO_2), which have been widely used for the femoral head in hip prosthesis thanks to manufacturing process able to produce highly spherical and smooth components. Zirconia presents not only a higher fracture resistance with respect to alumina but also a lower wear resistance and contains traces of undesired radioactive elements such as uranium and thorium. These materials can also be fabricated in the presence of foaming agents to increase their porosity. However, material strength decreases exponentially with the increase of the porosity volume fraction. To date, Biolox (Ceramtec), made up of 82% alumina, 17% zirconia (plus smaller percentages of other oxides such as chromium and strontium oxide), is the most used ceramic in joint replacements. Yttrium-stabilized zirconia nanoparticles prevent crack propagation and improve the mechanical strength of the matrix, thus combining the high wear resistance of alumina with a superior mechanical behavior compared to pure alumina.
- Bioactive ceramics, which promote the formation of chemical bonds between the implant and bone tissue. For example, synthesized *hydroxyapatite* [HA − $Ca_{10}(PO_4)_6(OH)_2$] is cheap, has a chemical composition similar to that of a bone and presents a good porosity, which further improves osteointegration allowing cell migration into the pores. *Bioglass 45S5* (45% SiO_2, 24.5% CaO, 24.5% Na_2O, and 6% P_2O_5) typically forms a carbonated hydroxyapatite layer, which shows an even more efficient osteointegration than simple HA. However, manufacturing issues are related to the glass-phase maintaining, avoiding material crystallization.
- Biodegradable ceramic: basically *calcium phosphates*, in which solubility is dependent on the Ca/P ratio.

Polymers have lower densities ($1-1.5$ g/cm^3) and elastic moduli ($1-4$ GPa) with respect to metal and ceramics; thus, their employ for load-bearing application is limited since they can undergo plastic deformation especially under cyclic loading. However, they are easily fabricated in different forms and their properties can be modulated varying the cross-linking degree or with the combination with specific fillers. Polymers that find application in the orthopedic field include (Hudecki et al., 2019; Cacopardo, Guazzelli, & Ahluwalia, 2021; Cacopardo et al., 2020; Navarro et al., 2008; Wang et al., 2018):

- *Polymethyl-methacrylate* (*PMMA*), an acrylic bone cement that guarantees implant primary fixation. The main disadvantages are the lack of a secondary biological fixation, the presence of residual monomer that may enter the circulation, the exotherm reaction that may produce tissue necrosis, and the shrinking after polymerization that can produce gaps between cement and implant. Ceramic radiopacifier particles ($BaSO_4$ or ZrO_2) are often included for diagnostic purposes, despite the introduction of discontinuities and the consequent worsening of the mechanical performance. In addition to prosthesis anchorage, PMMA also finds application in vertebroplasty and kyphoplasty, reinforcing the vertebra body.
- *Ultra-high molecular weight polyethylene* (*UHMWPE*), widely used for joint implant components such as acetabular cups or tibial and patellar elements, due to the low friction, good mechanical resistance, and biocompatibility. The high molecular weight is obtained thanks to the catalyst that allows for the formation of longer chains. As an example, low-density PE has a

molecular weight of $\sim 3 \times 10^3$ g/mol, a density of ~ 0.9 g/cm^3, and a crystallinity percentage around 60%, while the correspondent values for UHMWPE are around 2×10^6 g/mol, 0.9 g/cm^3, and 80%. Although lower with respect to standard PE, the formation of wear debris is the main side effect of this material in such applications. In order to improve wear performance, *highly crosslinked polyethylene (HXLPE)* has been produced as PE exposure to high-dose gamma-radiation which induce the formation of crosslinks in the amorphous phase. The irradiation is also responsible for the formation of free radicals in the crystalline phase, which may induce oxidative degradation. Strategies to overcome this issue include high-temperature annealing, remelting, and vitamin-E doping. This last method seems to be the most promising, since it does not affect the material mechanical properties and its antioxidant effect prevents free-radical formation during sterilization procedures with ionizing radiations (Baker, Bellare, & Pruitt, 2003; Oral & Muratoglu, 2011; ZIMMER TECHNICAL MEMO, 2014).

Polyetherether ketone (PEEK), which has good mechanical resistance to plastic deformation with respect to other polymers. The elastic moduli comparable with bone tissue and the chemical resistance make this material a good choice for spinal implants. However, its bioinertness may limit implant success and some attempts to improve PEEK bioactivity (e.g., incorporating hydroxyapatite or tricalcium phosphate) concomitantly affect the mechanical performance. Therefore, while its orthopedic use is still limited, PEEK is widely used in traumatology for bone fracture fixation as a radiopaque alternative to metals (Kurtz & Devine, 2007).

- Polyurethanes (PUs) are elastomeric polymers that present a mechanical strength comparable to UHMWPE and PEEK. To increase their resistance to degradation in vivo, different PU variants, such as polyether urethane, polyether urethane urea, and polycarbonate urethane, have been developed. These *elastomers are often used in combination with metallic components to preserve spinal implant flexibility* (St. John, 2014).
- *Medical grade silicone elastomers*, mainly used for small joint implants due to their flexibility that matches with the mobility range of these articulations. The main drawbacks are possible fracture, subluxation, loosening, and abrasion.
- *Biodegradable polymers*, such as polyglycolic acid (PGA), polylactic acid (PLA), polycaprolactone (PCL), and chitosan, widely used for the fixation of different bone fractures, osteotomies, and reconstructive surgery. Their biodegradability is mainly related to hydrolysis of the polymer chains, and can be controlled by varying polymer concentration, crystallinity, and molecular weight. In the case of PGA and PLA, the final degradation products are lactic acid and glycolic acid monomers used in the metabolic pathways.
- *Hydrogels*, composed of highly hydrophilic polymers such as polyhydroxy-ethyl-methacrylate, polyethylene glycol, and hyaluronic acid are used for the repair of cartilage, ligaments, tendons, and intervertebral disc (nucleus pulposus).

Composite materials also find different applications in the orthopedic field, with the advantage of combining the benefits of different materials. Usually, a reinforcement phase (e.g., ceramic powder or carbon fibers) are dispersed in the polymeric matrix. For instance, carbon fiber−reinforced PEEK has been used to fabricate bone plates with a bone-like elastic modulus and an improved fatigue resistance (Steinberg et al., 2013). Similarly, biodegradable materials for orthopedic tissue engineering, such as polypropylene fumarate, have been reinforced with ceramic materials such as calcium carbonate and calcium phosphate (Temenoff & Mikos, 2000).

Other composites are based on metal or polymeric matrix with bioactive fillers such as hydroxy-apatite or on the combination of a ceramic matrix with a metallic phase. As an example, PEEK-HA composites have been developed for load-bearing implants. Despite the increase in bioactivity, an increase of brittle behavior with increasing HA concentration (typically between 30% and 40% v/v) has to be considered (Abu Bakar et al., 2003). In addition to the properties of the base materials and their relative volume fractions, the performance of composites is affected by the shape, size, and distribution of the filler and by the interaction at the interface between the two materials (Ghasemi-Mobarakeh et al., 2019; Mattei et al., 2020; Merola & Affatato, 2019; Paital & Dahotre, 2008; Saad et al., 2018; Shuai et al., 2020).

Finally, self-healing materials have been recently introduced to improve orthopedic implant life-time. These materials are able to repair microdamages in situ, avoiding defect accumulation and implant failure. For example, stainless steel implants coated with cerium-niobium oxide (Ce-Nb$_2$O$_3$) are capable of self-repairing during electrochemical corrosion due to the ability of Ce ions to migrate toward the defects, while PMMA with self-repair properties has been developed due to the incorporation of microcapsules, which release crack-healing agents (Brochu, Craig, & Reichert, 2011; Celestine, Sottos, & White, 2015; Katta & Nalliyan, 2019).

BIOTRIBOLOGY

Biotribology is defined as the science of the interaction between surfaces in relative motions in the human body. In the case of articular bearings, such as hip or knee, the most used material combinations can be classified as (Fig. 18.5):

- Soft-on-hard couplings, that is, metal-on-plastic (MoP) and ceramic-on-plastic (CoP). The polymeric material is typically used for the cup, in hip joints, or as a plastic spacer in the tibial component, in knee joints, while the hard material is respectively employed for the hip femoral head and for the knee femoral component.

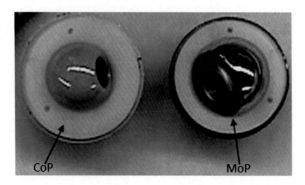

FIGURE 18.5

Material couplings. Examples of material couplings for hip prosthesis.

From (N.d.). Available at: https://doi.org/10.1016/j.matdes.2003.11.011.

- Hard-on-hard couplings, that is, ceramic-on-ceramic (CoC), metal-on-metal (MoM) and ceramic-on-metal (CoM).

Because of surface roughness, friction and wear typically occur. In engineering as well as in nature, lubrication is introduced to reduce frictional forces minimizing the contact between surface asperities. The parameter λ, defined as the ratio between lubricant film thickness and combined roughness of the articulating surfaces, identifies three lubrication regimes: boundary ($\lambda < 1$), mixed ($1 < \lambda < 3$), and fluid film lubrication ($\lambda > 3$) (Di Puccio & Mattei, 2015; Stewart, 2010). Typically, MoM bearings, with surface roughness around $0.01-0.05\ \mu m$, are known to operate in a mixed lubrication regime. In this case, the design of components with larger diameters, maintaining a low radial clearance, improves their performance minimizing wear. Other strategies to improve the tribological behavior of metals include thermal oxidation treatments, which improve surface wettability and thus lubrication. Differently, because of the high roughness of polymeric surfaces ($0.1-2\ \mu m$), MoP couplings are considered to work in a boundary lubrication regime. Therefore, surface asperities inevitably come into contact with a resulting important wear amount. On the contrary, ceramics present very fine surface finishing (roughness around $0.001-0.005\ \mu m$) and higher wettability properties. Hence, CoC contacts generally operate in a fluid film lubrication regime with low wear (Rieker, 2016; Di Puccio & Mattei, 2015; Wang et al., 2018).

In particular, wear can be classified as (1) *adhesive*, when local welding between surface asperities occurs causing their rupture during motion; (2) *abrasive*, in the presence of hard entrapped particles (third body wear), or asperities (two-body wear), which scratch the softer surface; (3) *fatigue-related*, in the case of cycling loading of the subsurface material which is subject to microcrack and particles detaching. Table 18.2 reports volumetric wear rates for different material couplings, highlighting that soft-on-hard contacts are generally the most affected by wear. At a microscale, abrasion is the main wear cause flanked by adhesion, while at macroscale, fatigue wear is also responsible for delamination. Moreover, when subject to multidirectional sliding, polymeric chains usually align along with principal molecular orientation. This behavior is responsible for anisotropic wear since wear resistance results higher along the direction of chain orientation. Nevertheless,

Table 18.2 Wear rates reported in vitro for typical articular bearing couplings (Mc = million cycles).

			V (mm³/Mc)
Soft-on-hard contact	MoP	CoCr-XLPE	6.7
		CoCrMo-XLPE	4.1
	CoP	Biolox®-XLPE	2.0
		Alumina-XLPE	3.4
Hard-on-hard contact	MoM	CoCrMo-CoCrMo	0.1-0.6
	CoC	Biolox®-Biolox®	0.05-0.1
		Alumina-Alumina	0.03-0.74
	CoM	Biolox®-CoCrMo	0.02-0.87

MoP and CoP couplings show lower friction coefficients (0.062 and 0.056, respectively) with respect to MoM (0.12). CoC bearings present the best performance in terms of wear and friction (0.04). Despite that, they are still affected by a higher fracture risk and by the squeaking phenomena, which is known to be a cause of great discomfort for the prostheses wearers (Rieker, 2016; Di Puccio & Mattei, 2015; Merola & Affatato, 2019).

SURFACE FUNCTIONALIZATION

In metals, the combination of wear and corrosion is one of the main drawbacks. Passive layers, typically 2—5 nm thick, are hindered by the surface contacts. In the case of contact between different metals, galvanic corrosion may also occur. Additionally, the poor bioactivity of metals and other biomaterials is likely responsible for poor implant fixation. Therefore, surface treatments are often employed to improve chemical, mechanical, and biological properties of the material interface with bone (Fig. 18.6).

The surface-free energy and chemical composition are responsible for material surface interaction with the aqueous environment in the body. Higher energies are related to higher affinity with polar molecules, such as water. Since hydrophilic surfaces typically form a water layer, which prevents protein adsorption, surface *chemical modifications* are typically performed to increase their hydrophobicity. However, nonspecific protein adsorption, which may also assume a different form from their native configuration, can also provoke undesired biological reactions. Thus, surface functionalization with biomolecules is often required to obtain good cell adhesion and osteointegration.

Surface topology is also fundamental for implant success. At the microscale, surface roughness enhances mechanical grip and stability, while at lower scales, it affects cell and protein interactions. Thus, *roughening treatments* such as grit blasting, acid etching, or plasma treatment can be performed.

In addition, bioactive coatings can be considered to reduce corrosion and mechanical failure, and to avoid fibrotic tissue formation. The most widely used coating techniques include physical vapor deposition, plasma spray, laser cladding, electrochemical deposition, and solution-based methods (Di Puccio and Mattei, 2015).

Intriguing solutions include *antimicrobial coatings* to prevent infections and biofilm formation. For example, antibiotic or antiseptic drugs can be incorporated into the implant surface, allowing the local administration of the substance with a known pharmacokinetics. Moreover, nanosilver coatings are often employed for their well-known antimicrobic activities related to the ability of binding thiol groups, thus damaging cell membranes and nucleic acids. Thus, silver nanoparticles are typically incorporated into titanium surfaces or polymeric to control the release profile. In particular, galvanic currents, which carry out a further bactericidal action, occur when the particles are

FIGURE 18.6

Surface functionalization strategies. Schematic illustration of the different surface functionalization strategies.

embedded into a titanium matrix, which act as anode and cathode. Finally, photoactive coatings can be used for their oxidizing power which is responsible for cell damage. Photocatalysts such as titanium oxide are employed as photocatalytic agents that are activated after ultraviolet light exposure (Eltorai et al., 2016).

ADDING "SMARTNESS" TO ORTHOPEDIC IMPLANTS

Smart materials are able to respond to an external stimulus (i.e., a change in a physical parameter) doing something "intelligent" or useful (i.e., performing a task in a controlled and repeatable manner) without the need for an external sensor, actuator, and control logics. More specifically, these materials are able to exchange energy between multiple physical domains (e.g., mechanical, chemical, thermal, magnetic, and electric), which becomes coupled. The two main classes of smart materials that find application in the orthopedic field are shape memory alloys (SMAs) and piezoelectric materials, which respectively present thermomechanical and electromechanical coupling. As shown in Table 18.3, SMAs typically present higher maximum strains and slower responses with respect to piezoelectric materials, which in turn are capable of smaller but faster strains (Leo, 2008).

SMAs are metallic materials that show wide strains (and stresses) in response to an environmental temperature change. This behavior is known as the shape memory effect: if the material is deformed in its martensitic state (low temperature) and then heated, an austenitic phase change occurs and the material recovers its original shape generating large deformations (Fig. 18.7A). SMAs also show superelastic behavior: if the material is deformed in its austenitic phase, a stress-induced martensitic state forms (Fig. 18.7B). As the stress is removed, the material immediately recovers its austenitic shape. Alloys such as Nitinol have been used in several biomedical applications for their excellent corrosion resistance and biocompatibility. For example, *bone plates and stables* for fracture fixation take advantage of both shape memory and superplastic effects thanks to Nitinol alloys with austenitic transition temperatures around 37°C. In the first case, the stable is inserted at a martensitic temperature and once positioned in the body, it recovers the original shape. In the second case, the stable is maintained in an austenitic phase and deformed generating stress-induced martensite before implantation. Thus, as soon as the stress applied during insertion is removed, it recovers the original austenitic shape. In both cases, stables are designed to exert a dynamic force between bone fragments. Adaptable shape memory implants can also be employed during bone healing processes allowing progressive implant stiffness reduction when the bone increases its load-bearing capacity. For example, the bending stiffness can be modulated by varying

Class	Material	ρ (g/cm³)	E (GPa)	Maximum strain (%)	Response time
SMA	Nitinol	6.4	23-27 (martensite) 84-89 (austenite)	8	s
Piezoeletric	PZT	7.9	90-69	0.1	µs
	BT	6.0	67	0.1	µs

Table 18.3 Smart materials in orthopedics.

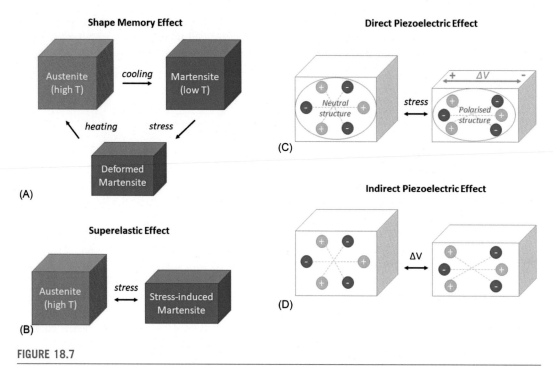

FIGURE 18.7

Shape memory and superelastic materials. Schematic illustration of the shape memory (A), superelastic (B), direct (C), and indirect (D) piezoelectric effect.

implant geometry (i.e., the moment of inertia decreases with decreasing cross-sectional area). In this case, Nitinol alloys with higher austenitic transition temperatures can be used to trigger shape change on demand using induction devices (Biscardini, Mazzolai, & Tuissi, 2010; Russell, 2009; Takale & Chougule, 2019; Tarniţč et al., 2010).

Piezoelectric materials are dielectric ceramics characterized by an anisotropic lattice structure. When deformed, electric dipole moments generate resulting in a global polarization of the material. The direct piezoelectric effect is related to charge generation in response to a mechanical deformation, whereas the indirect effect occurs when an electrical stimulus causes a mechanical deformation (Fig. 18.7C and D). Despite showing only short-term biocompatibility (because of the possible release of Pb ions over time), lead zirconate titanate (PZT) is one of the most used piezoelectric materials in the orthopedic field. For example, it has been used for intra- and postoperative *sensing* in knee joint surgery. The advantage with respect to the classic sensor such as load cells is that it does not require an external source of energy. Indeed, the same piezoelectric element can be used for energy harvesting to power the electronics needed for signal conditioning and data transmission. This integrated circuit is generally embedded in the UHMWPE bearing, thus limiting possible toxic effects. Another interesting application of piezoelectric materials relies on *mimicking bone electrical microenvironment*. Indeed, bone tissues present a spontaneous piezoelectric activity, which stimulates and directs cell activity during bone remodeling. To this end, biocompatible barium titanate (BT) particles can be incorporated into bone cement mixtures or into polymeric bone scaffolds (Jeong et al., 2017; Koju et al., 2018; Platt et al., 2005; Safaei et al., 2018).

BONE TISSUE ENGINEERING AND PERSONALIZED ORTHOPEDIC MEDICINE

Tissue engineering is defined as "the practice of combining cells and scaffolds to develop biological substitutes that restore, maintain, and improve tissue function." In the orthopedic field, tissue engineering products find a wide application, and several materials and fabrication strategies have been investigated to obtain scaffolds with controlled microstructure and mechanical properties. Besides biocompatibility, osteoconduction, and osteoinduction, the main specifications for bone scaffolds regard controlled biodegradation to allow their gradual replacement with newly deposited extracellular matrix (ECM), mechanical properties able to bear weight during the healing period, and a porous structure to promote cell migration inside the scaffolds (Fig. 18.8A). The most successful materials in bone tissue engineering are polymer/ceramic composites, which actually mimic natural bone composition. Thus, calcium phosphates and HA have been combined with both natural and synthetic polymers. Natural polymers can be classified as proteins, such as collagen and gelatin, and polysaccharides, for example, alginate and chitosan. Their mechanical stability is generally improved with chemical (glutaraldehyde and genepin), enzymatic (transglutaminase), or photo-cross-linking strategies which however may reduce the intrinsic biocompatibility of these materials. On the other hand, the synthetic polymers are mainly polyesters such as PGA, PLA, PCL, and polyethylene terephthalate, which present good mechanical properties and biocompatibility. However, both polysaccharides and synthetic polymers are not cell adhesive "per se." Thus, their surface is generally functionalized with bioactive coatings containing bioadhesive molecules such as the arginyl-glycyl-aspartic acid (RGD) sequence (Ghassemi et al., 2018; Mattei et al., 2015, 2020).

Moreover, tissue-derived scaffolds are often used with the advantage of maintaining the mechanical properties and architecture of native ECM. The main issue is related to the material source. Autograft (generally derived from the patient ribs or iliac crest) are highly osteogenic and present low immunological response. However, drawbacks are associated with additional surgery and with the limited size. In contrast, allograft and xenograft, respectively derived from other

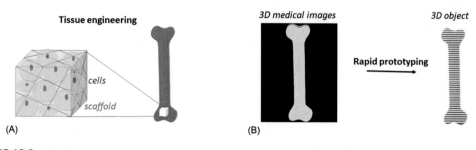

(A) (B)

FIGURE 18.8

Tissue engineering and rapid prototyping. Schematic illustration of tissue engineering (A) and rapid prototyping (B) applications in orthopedics.

human donors or from animals, can be subjected to severe immune responses. For this reason, *decellularization* techniques have been developed to obtain acellular scaffolds which can be recolonized from patient cells after implantation, thus reducing adverse host reactions (Ghassemi et al., 2018; Wang et al., 2018).

Finally, rapid prototyping strategies such as *3D printing*, *fused deposition modeling*, *stereolithography*, *EBM*, and *laser sintering* are emerging as powerful tools for personalized orthopedic applications. These techniques belong to additive manufacturing methods and are typically based on a layer-by-layer material deposition, allowing the fabrication of a 3D object starting from three-dimensional medical images or from computer-aided design models (Fig. 18.8B). A further advantage is the possibility to obtain internal structures with controlled porosity and filling density, thus improving osteointegration and reducing stress-shielding effects. For example, 3D-printed trabecular metals and PEEK implants have been used for hip and knee arthroplasty and for preoperative surgical training models (Eltorai, Nguyen, & Daniels, 2015; Trauner, 2018; Wong, 2016). *4D printing* is also becoming popular to fabricate custom devices capable of adapting their structure on the basis of physiological bone remodeling processes. The difference of this technology with respect to standard rapid prototyping strategies is the use of smart materials. For instance, Nitinol or shape memory polymers have been used to realize implants able to change shape upon heating or cooling (Haleem, Javaid, & Vaishya, 2018).

REFERENCES

Abu Bakar, M. S., Cheng, M. H. W., Tang, S. M., Yu, S. C., Liao, K., Tan, C. T., ... Cheang, P. (2003). Tensile properties, tension-tension fatigue and biological response of polyetheretherketone-hydroxyapatite composites for load-bearing orthopedic implants. *Biomaterials*, *24*(13), 2245−2250. Available from https://doi.org/10.1016/S0142-9612(03)00028-0.

Anderson, J. M., Rodriguez, A., & Chang, D. T. (2008). Foreign body reaction to biomaterials. *Seminars in Immunology*, *20*(2), 86−100. Available from https://doi.org/10.1016/j.smim.2007.11.004.

Ashby, M. F., & Cebon, D. (1993). Materials selection in mechanical design. *Journal De Physique*, *3*(7), 1−9. Available from https://doi.org/10.1051/jp4:1993701.

Baker, D. A., Bellare, A., & Pruitt, L. (2003). The effects of degree of crosslinking on the fatigue crack initiation and propagation resistance of orthopedic-grade polyethylene. *Journal of Biomedical Materials Research − Part A*, *66*(1), 146−154. Available from https://doi.org/10.1002/jbm.a.10606.

Biscardini, A., Mazzolai, G., & Tuissi, A. (2010). Enhanced nitinol properties for biomedical applications. *Recent Patents on Biomedical Engineering*, *1*, 180−196.

Brochu, A. B. W., Craig, S. L., & Reichert, W. M. (2011). Self-healing biomaterials. *Journal of Biomedical Materials Research − Part A*, *96*(2), 492−506. Available from https://doi.org/10.1002/jbm.a.32987.

Cacopardo, L., Mattei, G., & Ahluwalia, A. (2020). A new load-controlled testing method for viscoelastic characterisation through stress-rate measurements. *Materialia*, *9*, 100552.

Cacopardo, L., Guazzelli, N., & Ahluwalia, A. (2021). Characterising and engineering biomimetic materials for viscoelastic mechanotransduction studies. *Tissue Engineering - Part B*. Available from https://doi.org/10.1089/ten.TEB.2021.0151PDF/EPUB.

Castagnini, F., Bordini, B., Stea, S., Calderoni, P. P., Masetti, C., & Busanelli, L. (2019). Highly porous titanium cup in cementless total hip arthroplasty: Registry results at eight years. *International Orthopaedics*, *43*(8), 1815−1821. Available from https://doi.org/10.1007/s00264-018-4102-9.

Celestine, A. D. N., Sottos, N. R., & White, S. R. (2015). Autonomic healing of PMMA via microencapsulated solvent. *Polymer*, *69*, 241−248. Available from https://doi.org/10.1016/j.polymer.2015.03.072.

Costa, M. M., Bartolomeu, F., Alves, N., Silva, F. S., & Miranda, G. (2019). Tribological behavior of bioactive multi-material structures targeting orthopedic applications. *Journal of the Mechanical Behavior of Biomedical Materials*, *94*, 193−200. Available from https://doi.org/10.1016/j.jmbbm.2019.02.028.

Di Puccio, F., & Mattei, L. (2015). Biotribology of artificial hip joints. *World Journal of Orthopaedics*, *6*(1), 77−94. Available from https://doi.org/10.5312/wjo.v6.i1.77.

Eltorai, A. E. M., Haglin, J., Perera, S., Brea, B. A., Ruttiman, R., Garcia, D. R., ... Daniels, A. H. (2016). Antimicrobial technology in orthopedic and spinal implants. *World Journal of Orthopaedics*, *7*(6), 361−369. Available from https://doi.org/10.5312/wjo.v7.i6.361.

Eltorai, A. E. M., Nguyen, E., & Daniels, A. H. (2015). Three-dimensional printing in orthopedic surgery. *Orthopedics*, *38*(11), 684−687. Available from https://doi.org/10.3928/01477447-20151016-05.

Ferro, P., & Bonollo, F. (2019). Materials selection in a critical raw materials perspective. *Materials and Design*, *177*, 107848. Available from https://doi.org/10.1016/j.matdes.2019.107848.

Ghasemi-Mobarakeh, L., Kolahreez, D., Ramakrishna, S., & Williams, D. (2019). Key terminology in biomaterials and biocompatibility. *Current Opinion in Biomedical Engineering*, *10*, 45−50. Available from https://doi.org/10.1016/j.cobme.2019.02.004.

Ghassemi, T., Shahroodi, A., Ebrahimzadeh, M. H., Mousavian, A., Movaffagh, J., & Moradi, A. (2018). Current concepts in scaffolding for bone tissue engineering. *Archives of Bone and Joint Surgery*, *6*(2), 90−99. Available from https://doi.org/10.22038/abjs.2018.26340.1713.

Haleem, A., Javaid, M., & Vaishya, R. (2018). 4D printing and its applications in Orthopaedics. *Journal of Clinical Orthopaedics and Trauma*, *9*(3), 275−276. Available from https://doi.org/10.1016/j.jcot.2018.08.016.

Hallab, N. J., & Jacobs, J. J. (2009). Biologic effects of implant debris. *Bulletin of the NYU Hospital for Joint Diseases*, *67*(2), 182−188.

Hernigou, P., Mathieu, G., Poignard, A., Manicom, O., Filippini, P., & Demoura, A. (2007). Oxinium, a new alternative femoral bearing surface option for hip replacement. *European Journal of Orthopaedic Surgery and Traumatology*, *17*(3), 243−246. Available from https://doi.org/10.1007/s00590-006-0180-2.

Hudecki, A., Kiryczyński, G., & Łos, M. J. (2019). *Biomaterials, definition, overview. Stem cells biomater regenerative medicine* (pp. 85−98). Academic Press.

Ibrahim, M. Z., Sarhan, A. A. D., Yusuf, F., & Hamdi, M. (2017). Biomedical materials and techniques to improve the tribological, mechanical and biomedical properties of orthopedic implants − A review article. *Journal of Alloys and Compounds*, *714*, 636−667. Available from https://doi.org/10.1016/j.jallcom.2017.04.231.

Jeong, C. K., Han, J. H., Palneedi, H., Park, H., Hwang, G. T., Joung, B., ... Lee, K. J. (2017). Comprehensive biocompatibility of nontoxic and high-output flexible energy harvester using lead-free piezoceramic thin film. *APL Materials*, *5*(7). Available from https://doi.org/10.1063/1.4976803.

Katta, P. P. K., & Nalliyan, R. (2019). Corrosion resistance with self-healing behavior and biocompatibility of Ce incorporated niobium oxide coated 316L SS for orthopedic applications. *Surface and Coatings Technology*, *375*, 715−726. Available from https://doi.org/10.1016/j.surfcoat.2019.07.042.

Koju, N., Sikder, P., Gaihre, B., & Bhaduri, S. B. (2018). Smart injectable self-setting monetite based bioceramics for orthopedic applications. *Materials*, *10*(7), 1258. Available from https://doi.org/10.3390/ma11071258.

Kurtz, S. M., & Devine, J. N. (2007). PEEK biomaterials in trauma, orthopedic, and spinal implants. *Biomaterials*, *28*(32), 4845−4869. Available from https://doi.org/10.1016/j.biomaterials.2007.07.013.

Leo, D. J. (2008). *Engineering analysis of smart material systems. Engineering analysis of smart material systems* (pp. 1−556). John Wiley and Sons. Available from https://doi.org/10.1002/9780470209721.

Li, X., Chu, C., & Chu, P. K. (2016). Effects of external stress on biodegradable orthopedic materials: A review. *Bioactive Materials*, *1*(1), 77−84. Available from https://doi.org/10.1016/j.bioactmat.2016.09.002.

Matassi, F., Botti, A., Sirleo, L., Carulli, C., & Innocenti, M. (2013). Porous metal for orthopedics implants. *Clinical Cases in Mineral and Bone Metabolism: Clinical and Experimental, 10*(2), 111−115. Available from https://doi.org/10.11138/ccmbm/2013.10.2.111.

Mattei, G., Cacopardo, L., & Ahluwalia, A. (2020). *Engineering gels with time − Evolving viscoelasticity. Materials (Basel)* (13, pp. 1−14).

Mattei, G., Ferretti, C., Tirella, A., Ahluwalia, A., & Mattioli-Belmonte, M. (2015). Decoupling the role of stiffness from other hydroxyapatite signalling cues in periosteal derived stem cell differentiation. *Scientific Reports, 5.* Available from https://doi.org/10.1038/srep10778.

Merola, M., & Affatato, S. (2019). Materials for hip prostheses: A review of wear and loading considerations. *Materials, 12*(3), 495. Available from https://doi.org/10.3390/ma12030495.

Navarro, M., Michiardi, A., Castaño, O., & Planell, J. A. (2008). Biomaterials in orthopaedics. *Journal of the Royal Society Interface, 5*(27), 1137−1158. Available from https://doi.org/10.1098/rsif.2008.0151.

Oral, E., & Muratoglu, O. K. (2011). Vitamin E diffused, highly crosslinked UHMWPE: A review. *International Orthopaedics, 35*(2), 215−223. Available from https://doi.org/10.1007/s00264-010-1161-y.

Paital, S. R., & Dahotre, N. B. (2008). Review of laser based biomimetic and bioactive Ca-P coatings. *Materials Science and Technology, 24*(9), 1144−1161. Available from https://doi.org/10.1179/174328408X341825.

Platt, S. R., Farritor, S., Garvin, K., & Haider, H. (2005). The use of piezoelectric ceramics for electric power generation within orthopedic implants. *IEEE/ASME Transactions on Mechatronics, 10*(4), 455−461. Available from https://doi.org/10.1109/TMECH.2005.852482.

Prakasam, M., Locs, J., Salma-Ancane, K., Loca, D., Largeteau, A., & Berzina-Cimdina, L. (2017). Biodegradable materials and metallic implants − A review. *Journal of Functional Biomaterials, 8*(4), 44. Available from https://doi.org/10.3390/jfb8040044.

Regis, M., Marin, E., Fedrizzi, L., & Pressacco, M. (2015). Additive manufacturing of Trabecular Titanium orthopedic implants. *MRS Bulletin, 40*(2), 137−144. Available from https://doi.org/10.1557/mrs.2015.1.

Rieker, C. B. (2016). Tribology of total hip arthroplasty prostheses. *EFORT Open Reviews, 1,* 52−57. Available from https://doi.org/10.1302/2058-5241.1.000004.

Russell, S. M. (2009). Design considerations for nitinol bone staples. *Journal of Materials Engineering and Performance, 18*(5−6), 831−835. Available from https://doi.org/10.1007/s11665-009-9402-1.

Saad, M., Akhtar, S., & Srivastava, S. (2018). Composite polymer in orthopedic implants: A review. *Materials Today: Proceedings, 5*(9), 20224−20231. Available from https://doi.org/10.1016/j.matpr.2018.06.393.

Safaei, M., Meneghini, R. M., & Anton, S. R. (2018). Energy harvesting and sensing with embedded piezo-electric ceramics in knee implants. *IEEE/ASME Transactions on Mechatronics, 23*(2), 864−874. Available from https://doi.org/10.1109/TMECH.2018.2794182.

Shuai, C., Liu, G., Yang, Y., Qi, F., Peng, S., Yang, W., . . . Qian, G. (2020). A strawberry-like Ag-decorated bar-ium titanate enhances piezoelectric and antibacterial activities of polymer scaffold. *Nano Energy, 74,* 104825.

Soto-Gutierrez, A., Navarro-Alvarez, N., Yagi, H., Nahmias, Y., Yarmush, M. L., & Kobayashi, N. (2010). Engineering of an hepatic organoid to develop liver assist devices. *Cell Transplantation, 19*(6−7), 815−822. Available from https://doi.org/10.3727/096368910X508933.

Sporer, S. M., & Paprosky, W. G. (2006). The use of a trabecular metal acetabular component and trabecular metal augment for severe acetabular defects. *Journal of Arthroplasty, 21*(6), 83−86. Available from https://doi.org/10.1016/j.arth.2006.05.008.

St. John, K. R. (2014). The use of polyurethane materials in the surgery of the spine: A review. *Spine Journal, 14*(12), 3038−3047. Available from https://doi.org/10.1016/j.spinee.2014.08.012.

Steinberg, E. L., Rath, E., Shlaifer, A., Chechik, O., Maman, E., & Salai, M. (2013). Carbon fiber reinforced PEEK Optima-A composite material biomechanical properties and wear/debris characteristics of CF-PEEK composites for orthopedic trauma implants. *Journal of the Mechanical Behavior of Biomedical Materials, 17,* 221−228. Available from https://doi.org/10.1016/j.jmbbm.2012.09.013.

Stewart, T. D. (2010). Tribology of artificial joints. *Orthopaedics and Trauma*, *24*(6), 435−440. Available from https://doi.org/10.1016/j.mporth.2010.08.002.

Takale, A. M., & Chougule, N. K. (2019). Effect of wire electro discharge machining process parameters on surface integrity of Ti49.4Ni50.6 shape memory alloy for orthopedic implant application. *Materials Science and Engineering C*, *97*, 264−274. Available from https://doi.org/10.1016/j.msec.2018.12.029.

Tarniţč, D., Tarniţč, D. N., Hacman, L., Copiluş, C., & Berceanu, C. (2010). In vitro experiment of the modular orthopedic plate based on nitinol, used for human radius bone fractures. *Romanian Journal of Morphology and Embryology*, *51*(2), 315−320. Available from http://www.rjme.ro/RJME/resources/files/510210315320.pdf.

Temenoff, J. S., & Mikos, A. G. (2000). Injectable biodegradable materials for orthopedic tissue engineering. *Biomaterials*, *21*(23), 2405−2412. Available from https://doi.org/10.1016/S0142-9612(00)00108-3.

Trauner, K. B. (2018). The emerging role of 3D printing in arthroplasty and orthopedics. *Journal of Arthroplasty*, *33*(8), 2352−2354. Available from https://doi.org/10.1016/j.arth.2018.02.033.

Wang, C., Wang, S., Yang, Y., Jiang, Z., Deng, Y., Song, S., . . . Chen, Z. G. (2018). Bioinspired, biocompatible and peptide-decorated silk fibroin coatings for enhanced osteogenesis of bioinert implant. *Journal of Biomaterials Science, Polymer Edition*, *29*(13), 1595−1611. Available from https://doi.org/10.1080/09205063.2018.1477316.

Williams, D. F. (2019). Enabling biomaterials based on the principles of biocompatibility mechanisms. *Frontiers in Bioengineering and Biotechnology*, *7*, 1−10. Available from https://doi.org/10.3389/fbioe.2019.00255.

Wong, K. C. (2016). 3D-printed patient-specific applications in orthopedics. *Orthopedic Research and Reviews*, *8*, 57−66. Available from https://doi.org/10.2147/ORR.S99614.

ZIMMER TECHNICAL MEMO. (2014). Vivacit-E® vitamin E highly crosslinked polyethylene long-term performance for high demand patients. Zimmer Tech. Memo. Available at: https://www.zimmerbiomet.com/content/dam/zimmer-biomet/medical-professionals/hip/vivacit-e/vivacite-white-paper.pdf.

HIP PROSTHESIS: BIOMECHANICS AND DESIGN 19

Edoardo Bori[1], Fabio Galbusera[2] and Bernardo Innocenti[1]

[1]*Department of Bio Electro and Mechanical Systems (BEAMS), École Polytechnique de Bruxelles, Université Libre de Bruxelles, Brussels, Belgium* [2]*IRCCS Istituto Ortopedico Galeazzi, Milan, Italy*

INTRODUCTION

Over the centuries, total hip arthroplasty (THA) has gone from being one of the first surgical procedures, characterized by a significant amount of issues and thus poor long-term outcomes, to be one of the most successful and frequently performed ones.

The hip replacement operation is mostly required in case the joint is affected by severe osteoarthritis (OA), which is known to lead to pain and different levels of limitations in the joint functions (Pivec, Johnson, Mears, & Mont, 2012); this intervention is especially necessary when the patient turns refractory to conservative treatments (No Author, 2000). This disease, therefore, currently represents a serious issue for public health, with symptoms present in 9% of men and 11% of women (Oliveria, Felson, Reed, Cirillo, & Walker, 1995; Zhang & Jordan, 2008), and is considered to be the main reason for 93% of total hip arthroplasties performed. Other pathologies leading to the necessity of hip replacement are osteonecrosis (2%), femoral neck fracture (2%), developmental dysplasia of the hip (2%), and inflammatory arthritis (1%) (Pivec et al., 2012).

The greatest number of procedures belongs to patients aged 65 years and older (65%) (Pivec et al., 2012), but the number of younger patients is expected to increase to 50% of all arthroplasties by 2030 (Kurtz et al., 2009); since age is considered to be one of the major risk factors leading to its development (Arden & Nevitt, 2006), the number of people affected is growing proportionally to the rapid aging of the world's population. (The number of people older than 60 years is expected to double from 11% to 22% by 2050 [Gerland et al., 2014]). It is also to be noted that, nowadays, patients' expectations concerning the quality of life have changed significantly: many patients desire to lead an active lifestyle in old age as well, thus requiring smooth functioning of their joints or, in case they have undergone an arthroplasty, the prosthesis implanted (Mahomed et al., 2002). Consequently, this factor, together with the increasing incidence of OA, represents one of the main reasons for the increasing demand of THA (Birrell, Johnell, & Silman, 1999; Fear, Hillman, Chamberlain, & Tennant, 1997).

Nowadays, more than 500,000 THAs are performed yearly in the United Kingdom and United States, and more than one million worldwide, and, as stated previously, the number is expected to double within the next 10 years (Pivec et al., 2012); with ever-growing successful clinical outcomes, the survival rate has gone up to more than 95% at 10-year follow-up and more than 80% at

Human Orthopaedic Biomechanics. DOI: https://doi.org/10.1016/B978-0-12-824481-4.00032-9

25-year follow-up (Kurtz, Ong, Lau, Mowat, & Halpern, 2007; Kurtz et al., 2009; National Joint Registry for England and Wales, 2010). For these reasons, THA can be considered to be one of the most successful surgical procedures performed and has thus been correctly identified as the "operation of the century" (Learmonth, Young, & Rorabeck, 2007).

IMPLANT OPERATION

In order to better understand the function of all the components of such prostheses, it is useful to understand how the surgery itself is performed.

Based on the wide variety of models available nowadays to address different patients' pathological situations, different procedures are implemented according to the prosthesis design; however, the standard primary implant surgery generally follows the steps listed below.

— After performing the deep incisions and exposing the joint, the surgeon opens the hip capsule to expose the femoral head while taking care of subcutaneous tissues and collecting bone debris and blood during the procedure followed by dislocating the femoral head (Fig. 19.1A). Then, the surgeon performs the neck osteotomy with an oscillating saw, removing the femoral head. It is to be noted that different incision options for THA are adopted according to the surgeon's preferences and the type of prosthesis to implant.

— The next focus of the operation is the acetabulum, which represents the "socket" portion of the hip joint: in most cases where THA is required, the acetabulum is also affected by the pathology and, therefore, has to be reshaped to restore its original center and related functionality. Reaming is thus performed to remove the osteophytes and articular cartilage until the deepened socket again assumes a regular hemispherical shape (Fig. 19.1B); at this point, the size of the component to be inserted is evaluated via dedicated trials, and the surgeon impacts the optimal shell into the cavity, typically with 15−20 degrees of anteversion. According to the fixation technique chosen, drilling can be performed to anchor the shell with two to three cancellous screws (no vascular and neurological structure should be damaged during such a procedure) or cement can be used (Fig. 19.1C). A polyethylene liner is selected according to the shell size and the surgeon's preferences for femoral head sizes and geometrical orientations, in order to best fit the patient and thus prevent dislocation; the surgeon then seats the polyethylene liner into the shell and locks it into place.

— After completing the procedure on the acetabulum, the femur is once again addressed: a hollow box osteotome is inserted in the femoral canal to create a pilot hole for femoral reaming instrumentation, and incremental sizes of femoral broaches and reamers are pushed into the medullary canal until reaching the predetermined implant size (Fig. 19.1D). If the chosen fixation method involves the use of cement, reaming and broaching are further performed in order to obtain a wider canal to host the cement mantle. Different trial femoral stems and heads are then implanted (Fig. 19.1E), and then for each of them, the surgeon performs a specific maneuver on the joint: this test is necessary to check for eventual dislocation and to evaluate different sizes of these components, in order to select the optimal ones to be finally applied. The new femoral head is then relocated in the acetabulum cavity (Fig. 19.1F), and the surgeon proceeds with the suturing.

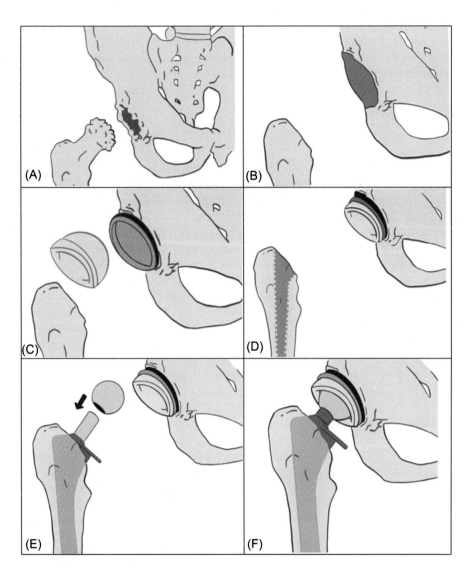

FIGURE 19.1

Implant operation steps: (A) joint dislocation; (B) neck osteotomy and acetabulum reaming; (C) acetabular cup and liner implant; (D) femoral reaming; (E) femoral stem and head implant; and (F) joint relocation.

SCORES

In order to quantify the outcomes of an implant and thus its effectiveness, different scores have been described in the literature and are currently used in the medical community: Harris Hip Score, Hip Disability and Osteoarthritis Outcome Score, Oxford Hip Score, Lequesne Index of Severity for Osteoarthritis of the Hip, American Academy of Orthopedic Surgeons Score, and Hip and Knee Questionnaire and Forgotten Joint Score (Nilsdotter & Bremander, 2011). These scores are important to measure improvementafter treatment, to facilitate communication withother physicians, and to offer patients a more accu-rate prognosis before surgery. However, not all have been properly validated and many of them assessonly few aspects of hip function (Longo et al., 2019).

HISTORY OF PROSTHESIS

The first operation ever performed took place in 1891, with an ivory ball and socket joint implanted by Themistocles Gluck (Brand, Mont, & Manring, 2011), who used nickel-plated screws to fix it to the bones; he furthermore worked on the use of a mixture of plaster of Paris and powdered pumice with resins in order to provide fixation (Gomez & Morcuende, 2005a). The variety of implants he developed (who also designed wrist, elbow, shoulder, ankle, and knee implants) had great results in the short term, but all these prostheses failed in the long term due to chronic infection issues (mostly related to the materials used to reconstruct the joint and the tendons) (Eynon-Lewis, Ferry, & Pearse, 1992).

After 10 years, Gluck's models were followed by the works of John Benjamin Murphy, who introduced the concept of interpositional hip arthroplasty; at the core of this approach, as an interposition between the acetabular cup and the femoral head, is the use of a layer of material to bear the stress and facilitate the relative motion between the two prosthetic components. While Murphy, for this purpose, adopted the "Fascia Lata," pig bladder was used by the French surgeon Foedre, which was first applied by William Steven Baer at the Johns Hopkins Hospital (Baer, 1909, 1918); Sir Robert Jones, instead, used gold foils to cover the reconstructed femoral heads, and was the first one to record smooth functionality of the joint of an implanted patient at 21-year follow-up (Jones, 1929).

In 1923, Marius Smith-Petersen designed a glass mold to be placed between the femoral head and the acetabulum to "guide nature's repair" of the joint. The first version was, however, made of a very fragile material, and some of the implanted molds broke afterward; for this reason, after having tried different other materials as celluloid, Bachelite, and Pyrex, in 1937, he used the Austenal (later renamed Vitallium, an alloy of 65% cobalt, 30% chromium, and 5% molybdenum), which was originally developed for dental applications: it was the first example of CoCrMo alloy ever used in THA and the first device able to provide predictable results in interpositional hip arthroplasty (Smith-Petersen, 1948).

In 1938, Philip Wiles performed the first THA using stainless steel components: they were precisely fitted and fixed to the bone with screws and bolts (Wiles, 1958), but no actual satisfactory result was obtained.

Robert and Jean Judet, in 1948, were then the first proponents of the introduction of acrylic as a material for hip implants with their prosthesis (Judet & Judet, 1950); such a model was, however, highly susceptible to wear and thus prone to failure. Some of the design features adopted by the Judet brothers, nonetheless, were incorporated in 1950 in New York by Frederick Röeck Thompson (who used Vitallium for his curved stem models [Thompson, 1952]) and, independently from Thompson, in 1952 by Harold R. Böhlman and Austin Moore in the United Kingdom, who further developed a long-stemmed straight prosthesis made of CoCrMo alloy and introduced the concept of fenestrated stems to allow bone ingrowth (Moore & Bohlman, 1943). These were the first hip arthroplasty implants to be distributed on a wide scale, and they are actually still used for the replacement of the femoral head and neck (especially following femoral neck fractures in the elderly) (Gomez & Morcuende, 2005a).

Kenneth McKee, who had studied with Wiles in London, developed different prosthesis designs in the 1950s and tried the use of dental acrylic cement for fixation; he used the Thompson prosthesis on the femoral side and articulated it with a three-claw—type cup that was screwed into the acetabulum, but this implant was characterized by high failure rates mainly due to component loosening (McKee & Watson-Farrar, 1966). Peter Ring, active in England in the same period, developed a cementless model based on the Austin Moore stem, with a metal-on-metal articulation: surprisingly good results were achieved with his early models (up to 97% of implant survival rate at 17-year follow-up) (Ring, 1971).

Both the McKee—Farrar and the Ring models were anyway abandoned in the 1970s, with the advent of the actual father of the modern THA: Sir John Charnley. Charnley was aware that one of the main achievements to improving the prosthesis performances was to guarantee the proper lubrication at the articular interface of the joint (Charnley, 1961), in order to reduce frictional resistance to motion; he furthermore realized how necessary it was to minimize the torque transmitted from the metal femoral head to the socket. Therefore, he initially worked on lubrication of animal and artificial joints (Charnley, 1960) and, to address both the issues he was focused on, introduced a 22.2 mm femoral head, way smaller than the ones available at the time (thus reducing friction force and torque) but still adequate to avoid a high incidence of dislocation (Charnley, 1961).

His studies on the optimal insert bearing material led him to adopt, first, the polytetrafluoroethylene (with poor results in terms of preventing wear) and, finally, ultra-high molecular weight polyethylene (UHMWPE): his model was then characterized by the use of a small femoral head, polymethylmethacrylate as bone-prosthesis cement, and UHMWPE as a bearing surface, thus representing the first THA, as we know today (Gomez & Morcuende, 2005b).

Nowadays, years after the introduction of the Charnley implant, surgeons have access to a large variety of options developed by different device manufacturers and different implant designs to address hip pathologies and eventual revision of previous implants: the clinician, thus, has to ponder which is the optimal solution considering the expected longevity, level of activity and expectations of the patient, the bone dimensions and quality, and the availability and service life of the implant.

In the following section, a deeper analysis of the characteristics and features of different implant components will be addressed.

FEMORAL COMPONENT

The femoral component replaces the resected arthritic or necrotic head and neck, and its stem is usually composed of cobalt—chromium or titanium alloys.

In order to obtain the optimal joint restoration respecting the patient's anatomy, different dimensional parameters are customizable (see Fig. 19.2):

— stem length;
— medial offset: the horizontal distance from the middle of the femoral head to a line through the distal stem;

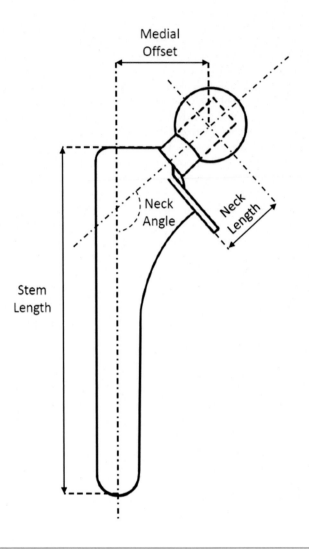

FIGURE 19.2

The femoral component.

—   neck length: the distance from the center of the femoral head to the base of the implant;
—   head diameter;
—   design of the neck, in terms of orientation (anteversion or retroversion) and neck angle with the stem axis.

The interface between head and stems is often standardized to allow for a vast series of possible combinations, and femoral heads can be in different sizes and made of different materials in order to be optimally coupled with the acetabular component in terms of dimensions (and thus the range of motion [RoM]) and friction (related to eventual wear); increasing the head diameter, however, leads to a reduction of the polyethylene liner thickness and may further lead to polyethylene wear.

It is to be noted that the variety of stems is not only in terms of dimensions but also in terms of overall shape and features. Concerning the proximal region of the femoral component, a feature that may be present is the collar: the rationale of this design is to help distribute the stresses in the proximal femur, thanks to this additional contact surface. However, when this feature is present, particular attention is paid to the implant procedure in order to guarantee optimal contact between bone and collar, as even slight imprecisions may lead to insufficient load transmission and increased stem tilting effect (Jeon et al., 2011), leading to unsatisfactory performances.

Turning the attention to the distal region of the prosthesis and addressing the stem shape, it is to be noted that, together with the traditional straight stems, anatomic stems are considered one of the main categories currently used. These stems, thanks to their length and curved shape, aim to provide improved stability by following the patient medullary canal anatomy (hence the name "anatomic stems"): for this reason, their performances are strictly related to the level of congruence among prosthesis and canal curvatures, hence requiring a more personalized and patient-specific approach.

In order to provide a similar level of congruence but preventing the cost associated with a patient-specific implant, the modular approach was introduced as an alternative to these monoblock (aka "monolithic") models: different components for prosthesis stem and neck are thus assembled to obtain an implant able to optimally fulfill the surgeon's and patient's needs, furthermore allowing adjustments intraoperatively in case required. Subsidence issues can then be addressed and anteversion degrees can be adapted to patient anatomy in order to reduce dislocation risk, just by interchanging components. It is, nonetheless, important to highlight that this approach brings a series of drawbacks, mostly related to the concept of modularity itself: it introduces a series of interfaces and contact surfaces that bring all the consequent complications in terms of fretting, absent in the monoblock designs.

Therefore, several options could characterize a hip prosthesis in terms of features & uses, among them the most relevant are: prosthesis use (primary or revision), the stem fixation (cemented or press-fit), the stem shape (straight, anatomical/bowed, tapered or modular), the stem interface (porous or non-porous/smooth) and the system type (modular or monolithic).

ACETABULAR COMPONENT

The acetabular component occupies the reamed socket and is usually composed of two parts: the shell and liner. The acetabular liner, as seen before, is usually made of UHMWPE; in order to reduce wear and to facilitate sterilization, this material is often treated with radiation to trigger cross-linking, but particular attention has to be paid to prevent oxidation during and after the process. (Some liners can be impregnated with antioxidants, such as Vitamin E, to further decrease free radical breakdown after implantation). Different material combinations for liner and femoral head components, involving the use of ceramic or metal to adjust the bearing surface properties, are also available to meet different requirements (see Table 19.1).

The standard liner design is characterized by a hemispherical cavity, allowing a maximal RoM; liners used in revision cases (and thus designed to address a higher risk of dislocation) can, instead,

Table 19.1 Interface material combinations.

Materials	Advantages	Disadvantages
Metal on polyethylene	Most frequently used, cost-effective, predictable lifespan	Aseptic loosening can occur due to polyethylene debris
Metal on metal	Reduced wear if compared to polyethylene (increased lifespan), larger femoral head (lower dislocation rate, higher range of motion)	Metallosis potential carcinogenic effect of metal ions
Ceramic on ceramic	Low-friction, low-debris generation, inert material	High cost, requires expertise in its insertion to avoid damages, might generate noise during movement

Table 19.2 Acetabular cups.

System type	Implant interface	Use	Fixation	Shape
Two pieces: shell (metal) + liner (polyethylene, metal, ceramic)	Nonporous (smooth)	Primary	Cemented	Hemisperical
One piece: polyethylene	Porous	Revision	Press-fit bone screws	Treated

be dual-mobility liners (also referred to as "bi-polar head," featuring the possibility of the liner to rotate inside the acetabular cup) or constrained liners (in which the femoral head is held captive within the socket to decrease dislocation risks) (Hoskins, Bingham, Hatton, & de Steiger, 2020).

The shell is generally made of a porous-coated alloy, and a locking mechanism is present to allow the liner to be rigidly fixed in place; it is to be noted that the locking is compatible with different liner cavity sizes. (The smaller the diameter of the cavity, the thicker the liner.) The design is usually hemispherical in order to limit bone resection, and screws can then be used to secure the fixation of the shell component to the bone and encourage cancellous bone ingrowth. (This solution is often adopted as the standard in the case of revision implant.)

To provide further stability, threaded cups are used to improve the fixation to the underlying bone, but, on the other hand, this design is more invasive and therefore cannot be used in the case of low bone availability.

See Table 19.2 for an overview on the acetabular cup component.

FIXATION APPROACHES

The design of the components has to comply, in addition to patient anatomy, to the fixation approach intended for the prosthesis: stability of the implant depends on that, and this is one of the main prerequisites of implant success. For this reason, great attention has been paid to this topic, and, consequently, significant improvements have been made over the years. Different approaches in term of fixation and relative designs are available, and the optimal one has to be chosen according to the surgeon's preferences and the patient's needs.

CEMENTED

Cement can be applied to both stem and acetabular cups or to any one of them, as shown in Fig. 19.3A–C.

Cemented stems have smooth, textured, and coated surfaces, designed to secure the fixation to the layer of cement that is applied. These stems occupy only 80% of the medullary canal reamed during the operation, and the remaining space is occupied by the cement mantle (Fig. 19.3B).

A similar concept applies to the acetabular cup shell, but it is to be noted that a cemented cup is often made of a single polyethylene component, featuring backside grooves to host the cement mantle (Fig. 19.3C). This solution, however, is often discarded because of the high risks of aseptic loosening due to the shear stress, which is particularly high in the acetabulum, and the susceptibility of the cement to this solution; press-fit approach (with a metal shell and a liner) is thus more appropriate.

CEMENTLESS OR PRESS-FIT

This approach was developed in the mid-1970s, as issues concerning cemented fixation had started to emerge: loosening, bone loss, and general signs of damages in the cement layer were the main reasons leading to research on fixation methods not involving cement, thus using femoral stems and acetabular cups able to fit tightly into the bare medullary canal and in the reamed acetabulum. This technique, as it requires the surgeon to strongly insert the prosthesis into the bone to provide proper fixation, is referred to as "press-fit" (Fig. 19.3D).

It is to be noted that, with this approach, the bone remains in direct contact with the prosthetic component: for this reason, this kind of components features a porous surface or coating in order to facilitate bone ingrowth and thus fixation. With a similar aim, coated implants were designed and equipped with a calcium phosphate–based material (hydroxyapatite): this material is the same as found in the inorganic phase of bone (and is thus biocompatible) and therefore greatly promotes bone ingrowth and improves implant fixation, providing a biological bond at the interface.

The stems belonging to this category should therefore be able to provide immediate stability once implanted, and this fixation can be done in different regions of the bone based on the type of operation (primary or revision) being performed considering the patient bone quality.

Press-fit can be performed at the metaphysis (the standard for primary hip replacements), at the meta-diaphysis (in the case of osteoporotic patient, for which rigid fixation does not depend only on the softer and pathologic metaphyseal bone), or at the diaphysis (only in the most extreme cases for primary implants but the standard for revision hip replacements as the metaphyseal bone is not available anymore).

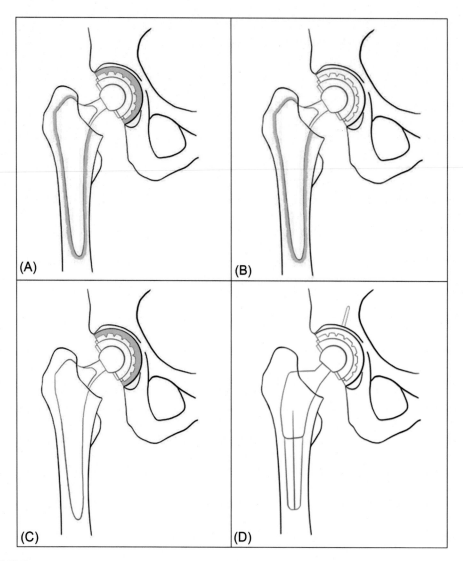

FIGURE 19.3

Fixation approaches: (A) cemented stem and cup; (B) cemented stem and press-fitted cup; (C) press-fitted stem and cemented cup; and (D) press-fitting.

GEOMETRY

Different stem designs are available based on the region where fixation is required (see Fig. 19.4).

When metaphyseal fixation is sought, flat tapered designs or dual taper ones can be used.

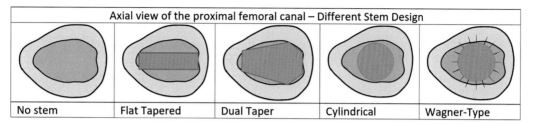

FIGURE 19.4

Femoral stem cross-sections.

The flat tapered stems, also known as "single-wedge" or "blade," achieve early fixation by wedging a rectangular-shaped stem into the medullary circular-like cavity: this geometry mismatch guarantees good torsional stability, and, therefore, fixation occurs by medial−lateral engagement or along the length of the stem called three-point fixation. Moreover, the minimal distal contact between implant and diaphysis prevents proximal bone resorption (stress shielding).

The dual taper stems, also called "elliptical," achieve early fixation by completely filling the metaphyseal canal and circumferentially engaging the femoral cortex proximally; the implant, therefore, presents tapers in all the planes and, consequently, occupies more volume compared to a flat tapered design. It is to be noted that this kind of designs are also used in the case of cemented options.

When diaphyseal fixation is required, cylindrical and Wagner-type stems are suitable.

The rationale behind using cylindrical stems is to achieve early fixation by engaging not only the metaphysis (when available) but also diaphysis for bone ingrowth; since no taper is present in these designs, the cylindrical model engages the diaphyseal cortical bone. Poor metaphyseal bone quality (or even the complete lack of it, as in the case of revision prosthesis) is the main reason behind using these models, which, however, involve drawbacks such as nonuniformity of bone ingrowth along the stem length. Being the bone-prosthesis engaging mostly distally, this region receives the majority of the stress and generates "stress shielding" in the proximal metaphysis, which leads to further bone loss in the proximal area: for such reasons, this design is preferred for revision surgery (in which the metaphysis is already compromised and thus stress-shielding effects are negligible).

Wagner-type stems follow a similar rationale but, unlike the cylindrical stems, present a tapered cylindrical shape in order to provide axial stability; moreover, this is combined with the presence of splines, to grip the cortex circumferentially, thus providing rotational stability.

When dealing with press-fit acetabular cups, bone screws may be used as an additional means of providing further stability of the implant, in addition to improving adhesion and fixation due to the surface being porous or with a specific pattern (i.e., meshed or threaded cups); then, there is a category of cups, the "expansion cups," allowing radial compression before the implantation, and they expand in the reamed cavity once inserted by the surgeon, providing further fixation.

HYBRID FIXATION

Together with the aforementioned approaches, press-fit porous acetabular shells and cemented femoral stems (Fig. 19.3C) are also preferred by surgeons, and it is considered as the gold standard for total hip replacement, avoiding the issues related to the use of cement in the acetabular cup but maintaining its advantages for stem fixation.

Different philosophies, however, exist when it comes to making a decision on which kind of fixation is the best to adopt: in the 1980s, the trend was that of using cementless implants, while in the 1990s cemented implants regained popularity. Currently, the shared choice of adopting the press-fit approach is a consequence of the proven extended service life of these porous implants; however, cement is still an option of choice in case patient bone quality does not allow reaching proper stability relying on press-fit alone.

LATEST DESIGNS

As can be clearly understood from the above discussion, a wide variety of approaches are nowadays available for THA, and they may follow different "philosophies" even if aiming at the same result. As an example of this variety, the German Zweymuller stem presents a squared, sharp, long, and straight double tapered design aiming at guaranteeing the diaphyseal three-point fixation with the cortical bone; the ABG design, on the other hand, has round, curved, and without sharp edges, and this stem is also relatively shorter and relies on a proximal hydroxyapatite coating to build the load on the metaphyseal area. Despite following diametrically opposite philosophies, both these models yield good results, along with other designs falling "in between" these two categories; an example of this category is the Corail stem: this French model is straight (as the Zweymuller one) but also fully coated in hydroxyapatite in order to progressively transfer stresses to the surrounding cancellous bone along the entire length (opposite of Zweymuller but similar to ABG).

In light of the latest innovations and trends in the design of THA, it is worth mentioning that the growing interest is in bone preservation both in terms of bone removal during the surgery and in terms of long-term resorption: these concepts led to the development of prosthesis aimed at avoiding stress shielding and maintaining the highest amount of bone possible, as the "mini-stems" and the "neck preserving stems." The mini-stems, as the name goes, is a category of stems characterized by very short length, which relies on metaphyseal stability and aims at avoiding considerable variations in stress distribution compared to a physiological configuration; neck-preserving stems, instead, allow to reduce the amount of bone required to be removed for the implant of the prosthesis, thanks to their curved shape, which allows them to be inserted in the femur with a more proximal cut if compared to traditional stems.

Based on this concept, resurfacing prosthesis represents the most commonly used bone-conserving option: this type of prosthesis is composed of two components, namely a femoral and an acetabular one, joined in a metal-on-metal bearing articulation, which aims at substituting the pathological contact surface of the hip joint. This category of hip prosthesis represents an interesting solution for maintaining the original configuration to the extent possible; however, it requires a skilled surgeon and good selectivity for the patient, as it is suitable for selective cases.

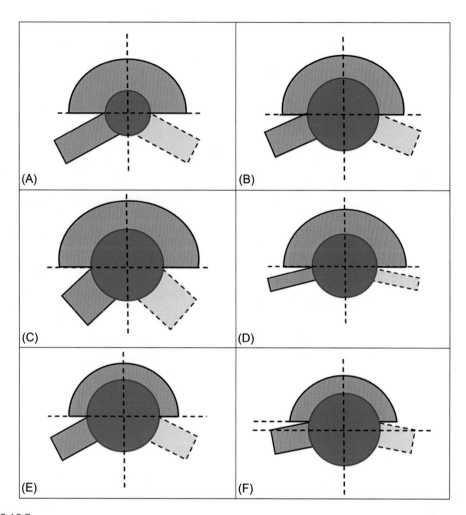

FIGURE 19.5

Design features and influences on range of motion: (A) small-sized femoral head; (B) big-sized femoral head; (C) thick neck component; (D) thin neck component; (E) hemispherical acetabular cup entrance plane; and (F) subhemispherical acetabular cup entrance plane.

KINEMATICS AND KINETICS

All the models available today with different features, therefore, aim at restoring joint functionality and, consequently, gaining proper kinematics and kinetics of the hip. It is fundamental for the optimal joint performances that the positioning of the implant is made respecting the patient anatomy, as malpositioning issues are one of the main causes of kinetics and kinematics deviations from the physiological configuration, which eventually leads to failure.

Kinematics performances, therefore, depend on implant positioning, but also on the bearing surfaces in terms of geometry: the size of the head is proportional to the RoM (see Fig. 19.5A and B), but it is of note that this does not always represent the active or passive one guaranteed to the patient, as the latter strongly depends on component orientation, muscular and soft tissue situation, and patient characteristics. In addition, other ways of increasing RoM include using thinner neck components (see Fig. 19.5C and D) and subhemispherical acetabular cup entrance planes (see Fig. 19.5E and F). However, it should be kept in mind that the increase of RoM via increasing the head size is often associated with the risks of impingement or dislocation issues of the prosthesis, which are important causes of failure: during the operation, while inserting the big head in the acetabulum, the forces required to relocate the joint increase, and this may lead to weakening or damage of soft tissue structures, factors leading to dislocation issues.

These geometrical features have also kinetics implications: in details, the use of large heads produces higher friction moments, also depending on the material used for the bearing surfaces, with metal-on-metal being a particularly bad choice for wear and metallosis issues, which have to be supported by the fixation of the prosthesis in order not to lead to aseptic loosening in the cup (a significant cause of failure for this kind of implants).

Thinner necks, on the other hand, may lead to fatigue issues: the neck of the prosthesis is the component that bears the highest cyclical loads during daily tasks (walking, climbing stairs, etc.) so its mechanical properties should be adjusted accordingly.

Whereas the kinetics of the acetabular cup and its interaction with the bone are mainly influenced by the head size, the stem shape represents the main factor influencing stress distribution in bone metaphysis and diaphysis; as described previously, its features deeply affect the way the prosthesis interacts with the bone, and for this reason, special attention should be paid to choosing the appropriate length and geometry: an incorrect stress distribution may lead to bone remodeling due to stress shielding and, thus, to higher risks of aseptic loosening.

REFERENCES

Arden, N., & Nevitt, M. (2006). Osteoarthritis: Epidemiology. *Best Practice & Research. Clinical Rheumatology*, 20(1), 3–25. Available from https://doi.org/10.1016/j.berh.2005.09.007.

Baer, W. (1909). Preliminary report of animal membrane in producing mobility in ankylosed joints. *The American Journal of Orthopedic Surgery*, 7, 1.

Baer, W. (1918). Arthroplasty with the aid of animal membrane. *The American Journal of Orthopedic Surgery*, 16(1), 171.

Birrell, F., Johnell, O., & Silman, A. (1999). Projecting the need for hip replacement over the next three decades: Influence of changing demography and threshold for surgery. *Annals of the Rheumatic Diseases*, 58(9), 569–572. Available from https://doi.org/10.1136/ard.58.9.569.

Brand, R. A., Mont, M. A., & Manring, M. M. (2011). Biographical Sketch: Themistocles Gluck (1853–1942). *Clinical Orthopaedics & Related Research*, 469(6), 1525–1527. Available from https://doi.org/10.1007/s11999-011-1836-8.

Charnley, J. (1960). The lubrication of animal joints in relation to surgical reconstruction by arthroplasty. *Annals of the Rheumatic Diseases*, 19, 10.

Charnley, J. (1961). Arthroplasty of the hip a new operation. *The Lancet*, 277(7187), 1129–1132. Available from https://doi.org/10.1016/S0140-6736(61)92063-3.

Eynon-Lewis, N. J., Ferry, D., & Pearse, M. F. (1992). Themistocles Gluck: An unrecognised genius. *British Medical Journal (Clinical Research ed.)*, *305*(6868), 1534−1536. Available from https://doi.org/10.1136/bmj.305.6868.1534.

Fear, J., Hillman, M., Chamberlain, M. A., & Tennant, A. (1997). Prevalence of hip problems in the population aged 55 years and over: Access to specialist care and future demand for hip arthroplasty. *Rheumatology*, *36*(1), 74−76. Available from https://doi.org/10.1093/rheumatology/36.1.74.

Gerland, P., Raftery, A. E., ev ikova, H., Li, N., Gu, D., Spoorenberg, T., ... Wilmoth, J. (2014). World population stabilization unlikely this century. *Science (New York, N.Y.)*, *346*(6206), 234−237. Available from https://doi.org/10.1126/science.1257469.

Gomez, P. F., & Morcuende, J. A. (2005a). A historical and economic perspective on Sir John Charnley, Chas F. Thackray Limited, and the early arthoplasty industry. *The Iowa Orthopaedic Journal*, *25*, 30−37. Available from http://www.pubmedcentral.nih.gov/articlerender.fcgi?artid = PMC1888784.

Gomez, P. F., & Morcuende, J. A. (2005b). Early attempts at hip arthroplasty−1700s to 1950s. *The Iowa Orthopaedic Journal*, *25*, 25−29. Available from http://www.pubmedcentral.nih.gov/articlerender.fcgi?artid = PMC1888777.

Hoskins, W., Bingham, R., Hatton, A., & de Steiger, R. N. (2020). Standard, large-head, dual-mobility, or constrained-liner revision total hip arthroplasty for a diagnosis of dislocation: An analysis of 1,275 revision total hip replacements. *The Journal of Bone and Joint Surgery. American Volume*, *102*(23), 2060−2067. Available from https://journals.lww.com/jbjsjournal/subjects/hip/Fulltext/2020/12020/Standard,_Large_Head,_Dual_Mobility,_or.6.aspx.

Jeon, I., Bae, J.-Y., Park, J.-H., Yoon, T.-R., Todo, M., Mawatari, M., & Hotokebuchi, T. (2011). The biomechanical effect of the collar of a femoral stem on total hip arthroplasty. *Computer Methods in Biomechanics and Biomedical Engineering*, *14*(1), 103−112. Available from https://doi.org/10.1080/10255842.2010.493513.

Jones, R. L. (1929). *Orthopaedic*. Baltimore: Wm Wood.

Judet, J., & Judet, R. (1950). The use of an artificial femoral head for arthroplasty of the hip joint. *The Journal of Bone and Joint Surgery. British Volume*, *32-B*(2), 166−173. Available from https://doi.org/10.1302/0301-620X.32B2.166.

Kurtz, S., Ong, K., Lau, E., Mowat, F., & Halpern, M. (2007). Projections of primary and revision hip and knee arthroplasty in the United States from 2005 to 2030. *The Journal of Bone & Joint Surgery*, *89*(4), 780−785. Available from https://doi.org/10.2106/JBJS.F.00222.

Kurtz, S. M., Lau, E., Ong, K., Zhao, K., Kelly, M., & Bozic, K. J. (2009). Future young patient demand for primary and revision joint replacement: National projections from 2010 to 2030. *Clinical Orthopaedics & Related Research*, *467*(10), 2606−2612. Available from https://doi.org/10.1007/s11999-009-0834-6.

Learmonth, I. D., Young, C., & Rorabeck, C. (2007). The operation of the century: Total hip replacement. *The Lancet*, *370*(9597), 1508−1519. Available from https://doi.org/10.1016/S0140-6736(07)60457-7.

Mahomed, N. N., Liang, M. H., Cook, E. F., Daltroy, L. H., Fortin, P. R., Fossel, A. H., & Katz, J. N. (2002). The importance of patient expectations in predicting functional outcomes after total joint arthroplasty. *The Journal of Rheumatology*, *29*(6), 1273−1279. Available from http://www.ncbi.nlm.nih.gov/pubmed/12064846.

McKee, G. K., & Watson-Farrar, J. (1966). Replacement of arthritic hips by the McKee-Farrar prosthesis. *The Journal of Bone and Joint Surgery, British Volume*, *48*(2), 245−259. Available from http://www.ncbi.nlm.nih.gov/pubmed/5937593.

Moore, A. T., & Bohlman, H. R. (1943). The classic. Metal hip joint. A case report. *Clinical Orthopaedics and Related Research*, *176*, 3−6. Available from http://www.ncbi.nlm.nih.gov/pubmed/6342894.

Nilsdotter, A., & Bremander, A. (2011). Measures of hip function and symptoms: Harris Hip Score (HHS), Hip Disability and Osteoarthritis Outcome Score (HOOS), Oxford Hip Score (OHS), Lequesne Index of Severity for Osteoarthritis of the Hip (LISOH), and American Academy of Orthopedic Surgeons (AAOS) Hip and Knee Questionnaire. *Arthritis Care & Research*, *63*(S11), S200−S207. Available from https://doi.org/10.1002/acr.20549.

Oliveria, S. A., Felson, D. T., Reed, J. I., Cirillo, P. A., & Walker, A. M. (1995). Incidence of symptomatic hand, hip, and knee osteoarthritis among patients in a health maintenance organization. *Arthritis & Rheumatism*, *38*(8), 1134−1141. Available from https://doi.org/10.1002/art.1780380817.

Pivec, R., Johnson, A. J., Mears, S. C., & Mont, M. A. (2012). Hip arthroplasty. *The Lancet*, *380*(9855), 1768−1777. Available from https://doi.org/10.1016/S0140-6736(12)60607-2.

No Author. (2000). Recommendations for the medical management of osteoarthritis of the hip and knee: 2000 update. American College of Rheumatology Subcommittee on Osteoarthritis Guidelines. *Arthritis & Rheumatism*, *43*(9), 1905−1915. Available from https://doi.org/10.1002/1529-0131(200009)43:9 < 1905:: AID-ANR1 > 3.0.CO;2-P.

Ring, P. A. (1971). Replacement of the hip joint. *Annals of the Royal College of Surgeons of England*, *48*(6), 344−355. Available from http://www.pubmedcentral.nih.gov/articlerender.fcgi?artid = PMC2387876.

Smith-Petersen, M. N. (1948). Evolution of mould arthroplasty of the hip joint. *The Journal of Bone and Joint Surgery. British Volume*, *30B*(1), 59−75. Available from http://www.ncbi.nlm.nih.gov/pubmed/18864947.

Thompson, F. R. (1952). Vitallium intramedullary hip prosthesis, preliminary report. *New York State Journal of Medicine*, *52*(24), 3011−3020. Available from http://www.ncbi.nlm.nih.gov/pubmed/13013552.

Wiles, P. (1958). The surgery of the osteo-arthritic hip. *British Journal of Surgery*, *45*(193), 488−497. Available from https://doi.org/10.1002/bjs.18004519315.

Zhang, Y., & Jordan, J. M. (2008). Epidemiology of osteoarthritis. *Rheumatic Disease Clinics of North America*, *34*(3), 515−529. Available from https://doi.org/10.1016/j.rdc.2008.05.007.

FURTHER READING

Longo, U. G., Ciuffreda, M., Candela, V., Berton, A., Maffulli, N., & Denaro, V. (2019). Hip scores: A current concept review.. *Br Med Bull*, *131*(1), 81−96. doi:10.1093/bmb/ldz026. Available from https://doi.org/10.1093/bmb/ldz026.

National Joint Registry for England and Wales. (2010). *7th Annual Report*. National Joint Registry: Hemel Hempstead.

KNEE PROSTHESIS: BIOMECHANICS AND DESIGN

20

Bernardo Innocenti

Department of Bio Electro and Mechanical Systems (BEAMS), École Polytechnique de Bruxelles,
Université Libre de Bruxelles, Brussels, Belgium

INTRODUCTION AND GENERAL CONCEPTS

The development of knee osteoarthritis, by the degeneration of the articular cartilage, dramatically reduces the quality of life in a patient as the joint contact, instead of cartilage-on-cartilage, is achieved through bone-on-bone contact, which is extremely painful and, in addition, is generally associated with swelling, stiffness in the knee, and related decreased mobility. Among the various treatment options available to relieve pain and restore mobility in the patient, the surgical insertion of a knee replacement is a successful option.

A generic total knee arthroplasty (TKA) consists of different parts (Fig. 20.1):

- a metallic femoral component, attached to the femoral bone;
- a polyethylene (PE) tibial insert (or liner or spacer), placed between the tibial and femoral components, attached mechanically to the tibial component;
- a tibial component (tibial tray), attached to the tibial bone, usually made of metal;
- an (optional) patellar component (patellar button), also in polyethylene, that could be used to replace the articular cartilage of the patellar if required clinically.

To properly investigate TKA and its relative design features, it is important to know the following preliminary concepts:

- CR (cruciate retaining) and PS (posterior stabilized) design implants;
- cemented and press-fit implant;
- fix-bearing and mobile-bearing;
- alignment and balancing;
- primary and revision TKA;
- total and partial knee arthroplasty.

CRUCIATE RETAINING AND POSTERIOR STABILIZED IMPLANTS

The concepts of CR and PS are related to the possibility of eventually retaining or resecting the cruciate ligaments during the implantation of a TKA. The anterior and posterior cruciate ligaments (ACL

Human Orthopaedic Biomechanics. DOI: https://doi.org/10.1016/B978-0-12-824481-4.00015-9

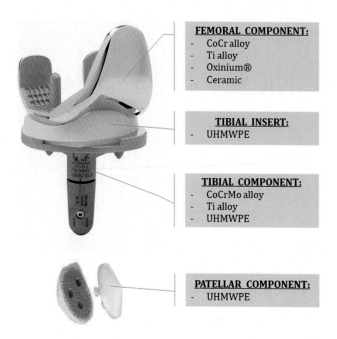

FEMORAL COMPONENT:
- CoCr alloy
- Ti alloy
- Oxinium®
- Ceramic

TIBIAL INSERT:
- UHMWPE

TIBIAL COMPONENT:
- CoCrMo alloy
- Ti alloy
- UHMWPE

PATELLAR COMPONENT:
- UHMWPE

FIGURE 20.1

Generic total knee arthroplasty components and materials.

Courtesy Waldemar Link GmbH & Co. KG.

and PCL, respectively, see also Chapter 13) are intracapsular ligaments, and, therefore, they could produce an obstacle during the surgical placement of the implant. However, the cruciates are responsible for the biomechanical stability of the knee joint, limiting anteroposterior translation, and, thus, the replaced knee implant should be able to perform such function if the cruciates are retained.

Surgically, the surgeon can decide between two different options (Fig. 20.2):

- PS TKA, in which both the cruciates are removed;
- CR TKA, in which the PCL is kept and only the ACL is removed.

With a PS design, the surgeon cuts both the ACL and the PCL, so the knee alone will lose the AP stability function provided by the cruciate ligaments, and this must be provided mechanically by the design of the implant itself. In detail, the PS TKA features a Post/Cam system (Figs. 20.2 and 20.3), with a metal cam positioned between the two femoral condyles, which engage, during flexion, with a polyethylene post placed in the insert, limiting the femoral AP motion.

With a CR design, the surgeon cuts only the ACL, as its position is rather central in the tibial bone, but retains the PCL, which then becomes the only cruciate responsible for stability. From a design standpoint, the condyles of the CR femoral component are separated, allowing the PCL to function properly during knee flexion; likewise, the polyethylene insert has a posterior slot for correct PCL motion.

In general, the tibial baseplate has the same design for both PS and CR versions of the TKA with the advantage of reducing the inventory cost.

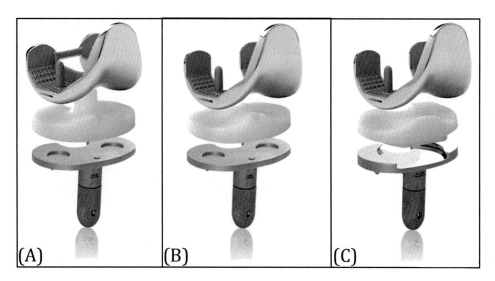

FIGURE 20.2

TKA design features: (A) PS TKA fix-bearing, (B) CR TKA fix-bearing, and (C) CR TKA mobile-bearing. *CR*, cruciate retaining; *PS*, posterior stabilized; *TKA*, total knee arthroplasty.

Courtesy Waldemar Link GmbH & Co. KG.

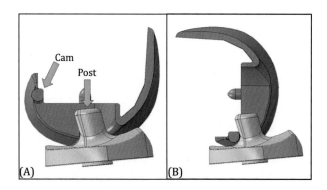

FIGURE 20.3

Overview of the Post and Cam system in a section of a PS TKA (only femoral component and tibial insert): (A) full extension configuration and (B) 90 degrees of flexion. *PS*, Posterior stabilized; *TKA*, total knee arthroplasty.

Surgically, while the cutting of the tibia during a PS implant operation is performed by cutting the entire proximal bone, the use of a CR implant requires the surgeon to carefully prepare the tibial attachment of the PCL (bone island) avoiding any resection, and, therefore, the procedure is slightly more delicate. An additional surgical advantage of the PS design, compared to CR, is related to the greater opening (distraction) of the knee joint and thus a better view of the posterior side of the bones. In fact, the intact PCL (present in TKA CR surgery) exerts a constraint to the opening of the joint space (Nowakowski, Majewski, Müller-Gerbl, & Valderrabano, 2012), reducing the overall field of view. However, due to the presence of the Cam, the femoral preparation in a PS design required an additional step (PS box preparation) in comparison to a CR implant.

Post/Cam designs differ among the various models of TKA, and several studies have clearly demonstrated that these features strongly influence the kinematics of the knee joint (Arnout et al., 2015; Fitzpatrick, Clary, Cyr, Maletsky, & Rullkoetter, 2013; Innocenti & Bori, 2020; Nakayama et al., 2005; Walker, Sussman-Fort, Yildirim, & Boyer, 2009). In particular, along with the shape, it is important to consider the following parameters:

- the engagement angle (the initial contact angle, during flexion, between the Post and the Cam);
- the position (height) of the Post/Cam contact, as a high value could increase the risk of femoral loosening; sometimes this concept is expressed in terms of jumping distance, as the distance between the contact point and the tip of the post; so, a lower jumping distance means a high risk of femoral luxation;
- the direction of the Post/Cam contact force, which should be exerted along a proximal—distal direction to avoid lift-off and further prevent femoral loosening;
- the Post contact area and associated contact pressure, with the latter being as low as possible to avoid any plasticization and permanent deformation of the post geometry.

There is a continuing debate about the superiority of CR vs PS TKA; theoretically, in patients with a nonfunctional PCL, a PS TKA system has to be used. In patients with a functional PCL, the decision on which design should be used depends largely on the preference, experience, and training of the surgeon (van den Boom, Brouwer, van den Akker-Scheek, Bulstra, & van Raaij, 2009). Looking at registry data, PS and CR are both valid and common options, and according to the data collected from different countries, the percentage use of cruciate-retaining implants varies from 40% to 90% (Heckmann et al., 2019). Clinically, there are no relevant differences between CR and PS TKA in terms of clinical, functional, and radiological outcomes, and complications, while PS TKA is superior to CR TKA in respect of range of motion (ROM), although whether this superiority matters or not in clinical practice still needs further investigation and longer follow-up (Jiang et al., 2016).

Recently, as a potential alternative to PS and CR implants, the bicruciate retaining (BCR) TKA has been introduced with the aim of reproducing more natural kinematics and superior proprioception while improving postoperative function and stability. In this design, both the ACL and the PCL are preserved. In more details, the tibial component (Fig. 20.4) consists of two separated trays that are connected anteriorly, in which the medial and lateral tibial inserts are placed (also allowing the possible selection of different shapes and heights). The surgical implantation of such a device is quite challenging, and, currently, the usefulness of such implants is debatable, and the literature has not shown clear indications and guidelines for the value and use of this implant (Boese, Ebohon, Ries, & De Faoite, 2021; Osmani, Thakkar, Collins, & Schwarzkopf, 2017).

CEMENTED AND PRESS-FIT IMPLANT

Similar to total hip arthroplasty, it is also possible to use bone cement (PMMA, polymethyl methacrylate) for TKA to fix the implant to the bone (cemented implant) or use a press-fit implant.

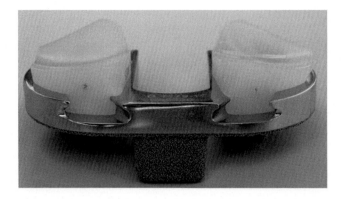

FIGURE 20.4

Bicruciate retaining tibial plate and inserts.

From Nowakowski, A. M., Stangel, M., Grupp, T. M., & Valderrabano, V. (2013). Comparison of the primary stability of different tibial baseplate concepts to retain both cruciate ligaments during total knee arthroplasty. Clinical Biomechanics (Bristol, Avon), 28*(8), 910–915. https://doi.org/10.1016/j.clinbiomech.2013.08.008.*

Actually, "cement" is a misleading term because the word cement is used to describe a substance that binds two things together. However, PMMA acts as a spatial filler that creates a tight space that holds the implant against the bone and thus acts as a "grout." Bone cements do not have intrinsic adhesive properties, but, instead, rely on a tight mechanical fit between the irregular bone surface and the prosthesis (Vaishya, Chauhan, & Vaish, 2013).

The ideal fixation of a TKA is still debated, and cemented TKA continues to be the gold standard for primary TKA; however, patient demographics are changing to include younger, obese, and more active patients who have demonstrated higher failure rates with cemented TKAs, leading to increased use of press-fit TKA in recent years (Miller et al., 2018). Some authors have also proposed a "hybrid" fixation technique, which consists of a press-fit femoral component and a cemented tibial component.

The cementing technique is multifactorial and includes preparation of the bone before cementation; where, when, and how to apply the cement; and the curing and stabilization phase after installation (Refsum et al., 2019). Cementing technique is very important for the fixation of TKA components, and several techniques are available; some studies have shown that a cement penetration depth (cement mantle) of 3–5 mm is optimal, since thicker cement layers (5–10 mm) increase the risk for thermal damage during cement polymerization, while lower values lead to higher radiolucency and lower tensile strength, which is associated with early implant micromotion (Vaninbroukx, Labey, Innocenti, & Bellemans, 2009; Vanlommel et al., 2011).

To promote bone ingrowth, press-fit implants could incorporate a porous coating or a surface mesh or hydroxyapatite layer, and fixation devices (pins, screws), which can reduce stress conditions and micromotion at the bone/metal interfaces. Press-fit prostheses are more expensive than cemented implants, although the prices vary among companies (Aprato et al., 2016).

There are many theoretical benefits of using press-fit knee prostheses: shorter operating room time, preservation of bone stock, ease of revision, and elimination of complications associated with cemented fixation like third-body wear and retained loose fragments. However, to ensure good primary stability of the implant, the bone resections must be performed accurately, avoiding any gaps between the host bone and the components, while in cemented TKAs, the cement mantle can easily fill small defects in resections without affecting the stability (Aprato et al., 2016).

FIXED- AND MOBILE-BEARING TOTAL KNEE ARTHROPLASTY

The concept of fixed- and mobile-bearing TKA is related to the relative motion that might exist between the polyethylene insert and the tibial component. Whereas in a fixed-bearing design (Fig. 20.2A and B) the insert is fixed to the tibial tray, in a mobile-bearing design (Fig. 20.2C), the insert could rotate along a vertical axis.

"Mobile-bearing" (MB) designs were introduced in the 1970s to reduce the wear and the consequent loosening of the implant reported, in the same time as "Fix-Bearing" (FB) devices (Hamelynck, 2006; Huang, Liau, & Cheng, 2007).

The two main features of these designs are as follows:

- A more congruent articular surface, which would increase the contact area and reduce polyethylene contact stresses and, consequently, polyethylene insert wear.
- The insert is free to axially rotate; therefore, the MB design induces fewer rotational forces on the tibial implant, theoretically reducing implant loosening.

In addition, MB designs reduce the shear force transmitted to the bone because they do not constrain the prosthesis rotationally; this is beneficial for cemented implants, as bone cement has very low shear strength. A study examining the influence of mobile-bearing design on torque and torsional stress across the entire proximal tibia in primary and revision TKAs (Small et al., 2013) showed that FB tibias generated up to 13.7 times the torsional moment of the rotating platform in primary designs.

However, the introduction of a mobile insert doubles the interactions as the insert will move with respect to the tibial component and femoral component, thus doubling the potential wear surfaces; to avoid the eventual backside wear, i.e., the wear generated between the tibial component and the insert, MB tibial baseplates are usually made of a highly polished chromium–cobalt alloy. FB implants, in order to grant a suitable locking junction in the tibial component, are necessarily made of titanium alloy and, thus, although well finished, they are unable to provide an ideal smooth surface for PE (Callaghan et al., 2001).

Clinically, the survival rate and patient's satisfaction is similar between these two models. Many studies comparing the clinical results of mobile- and fixed-bearing prostheses have shown no difference between the two prostheses in terms of ROM, clinical score, long-term survival rate, or radiographic findings (Poirier, Graf, & Dubrana, 2015; Smith, Jan, Mahomed, Davey, & Gandhi, 2011). A few studies have reported either superior (Price et al., 2003) or inferior (Aglietti, Baldini, Buzzi, Lup, & De Luca, 2005) results of mobile-bearing knees. No statistically significant difference was found between the FB and MB design PS TKA in terms of patellofemoral pain and function (Feczko et al., 2017).

All the revision TKAs are FB TKAs, however, An MB TKA could be eventually selected in the case of a revision of a unicompartmental knee arthroplasty [UKA].

IMPLANT ALIGNMENT AND BALANCING

The shape of each TKA implant is designed following a certain philosophy; however, its in-vivo performance is strongly related to the position and orientation of the implant that the surgeon chooses and the relative management of the soft tissue envelope. Although extremely important, this topic is still highly debatable and a clear, standardized approach has not yet been established.

The goal of TKA is to obtain symmetrical and balanced flexion and extension gaps (Daines & Dennis, 2014), defined as the space between the femoral and tibial bones after bony cuts in flexion and in extension (Laskin, 1995).

Theoretically, there are two main approaches, depending on whether the surgeon prioritizes the ligaments and the relative knee tension (tension gap) or the bone cuts (measured resection).

Following the "tension gap" approach, the surgeon chooses, using a ligament tensioner, the proper ligament tension and positions the implant, performing the bone cuts, accordingly. In detail, the femoral component is placed parallel to the resected proximal tibia with each collateral ligament equally tensioned to achieve a rectangular flexion gap (Daines & Dennis, 2014).

Following a "measured resection" approach, the bone cuts are initially made regardless of soft tissue tension; in detail, bone landmarks, such as the transepicondylar, anterior–posterior, or posterior condylar axes, are used by the surgeon to determine the proper rotation of the femoral component, and subsequent gap (soft tissue) balancing is performed to properly stabilize the joint (Daines & Dennis, 2014).

Regarding bone cuts, several alignment strategies are currently available (Hirschmann, Becker, Tandogan, Vendittoli, & Howell, 2019; Lustig et al., 2021; also see Fig. 20.5):

- mechanical alignment;
- anatomical alignment;
- kinematic alignment;
- inverse kinematic alignment.

If the surgeon decides to use the "mechanical alignment" (A), which was introduced by Freeman in the 1970s, the tibial and femoral components are positioned orthogonally to the tibial and femoral mechanical axes. This approach enables achieving a neutral lower limb alignment with a hip–knee–ankle angle of 180 degrees. Such an approach was primarily focused on allowing a proper distribution of the load from the implant to the bone (Lustig et al., 2021).

The execution of the cuts following the mechanical alignment, due to the oblique joint line, induces an asymmetric (mid-lateral) thickness in the resected bony parts. To avoid this issue, in the 1980s, Hungerford and Krackow introduced the concept of "anatomical alignment," with the goal of improving TKA functionality closely mimicking the alignment of the native knee joint line. To duplicate the population's mean orientation of the native joint line (3-degree femoral valgus and 3-degree tibial varus), they suggested cutting the bones at an inclination of 3 degrees to the mechanical axis (Rivière et al., 2017). This value was chosen regardless of the patient actual joint line, so, even with this approach, the resection could result to be nonsymmetrical between the medial and lateral sides.

The "kinematic alignment" (B), first proposed in 2006 by Howell, is a patient-specific technique, aimed at the restoration of the native joint line alignment of each patient (Howell, Howell, Kuznik, Cohen, & Hull, 2013). As the TKA is mainly a resurfacing procedure, the kinematic alignment technique aims at realigning the joint line of the components with the actual patient-specific

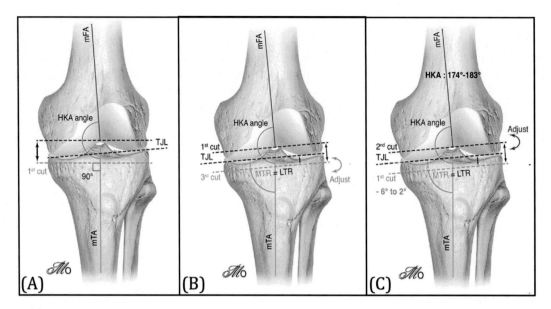

FIGURE 20.5

(A) Mechanical alignment, (B) kinematic alignment, and (C) inverse kinematic alignment. The order of the cut is also reported in the figure. *HKA*, Hip−knee−ankle angle; *LTR*, lateral tibial resection; *mFA*, mechanical femoral axis; *mTA*, mechanical tibial axis; *MTR*, medial tibial resection; *TJL*, tibial joint line.

Modified from Lustig, S., Sappey-Marinier, E., Fary, C., Servien, E., Parratte, S., & Batailler, C. (2021). Personalized alignment in total knee arthroplasty: Current concepts. SICOT-J, 7, 19. https://doi.org/10.1051/sicotj/2021021.

joint lines. With such an approach, the femoral and tibial bone resection thicknesses, checked with caliper measurements, are equal, and they should match the thickness of the components (after compensating for articular cartilage wear). This alignment strategy is a "femur-first" approach, which means that the surgeon must prepare the femur initially and then the tibia accordingly, with flexion and extension gaps balanced with tibial resection.

Following the principle of the "inverse kinematic alignment" (C), the surgeon initially performs the tibial cut ("tibial-first" approach) with similar medial and lateral resections after correcting for wear, maintaining the prearticular tibial joint line obliquity. The gap balancing is then performed by adjusting the femoral posterior and distal resections (Lustig et al., 2021).

An additional alignment, named "functional alignment," could also be added to the list as an updated kinematic alignment approach achieved using advanced technology as patient-specific implants and 3D printed cutting blocks that are designed preoperatively for the purpose of achieving a more accurate kinematic alignment in TKA than the one achieved intraoperatively (Lustig et al., 2021).

Several studies have attempted to compare the different alignments in terms of operation time, length of hospital stay, postoperative complications, clinical and radiographic outcomes, and patient kinematics (Luo et al., 2020; Yoon, Han, Jee, & Shin, 2017); however, no conclusive results could be finalized and long-term unbiased multicenter follow-up studies have yet to be performed to clearly demonstrate the superiority of one approach over the other (Hirschmann et al., 2019; Lustig et al., 2021;

Rivière et al., 2017). In addition, these different surgical alignment approaches are usually performed using TKA components that are, however, designed according to the traditional mechanical alignment, and this discrepancy could therefore affect clinical outcomes and make difficult their interpretation.

Moreover, to be able to reach a reliable agreement, the problem should be analyzed by coupling to the clinical findings the following biomechanical aspects:

- the tibial and femoral bone—prosthesis interface and the relative stress distribution;
- the mediolateral tibiofemoral interaction (both in terms of contact forces and polyethylene stress);
- the soft tissue tension;
- the kinematics of the tibiofemoral joint;
- the kinetics and kinematics of the patellofemoral joint.

PRIMARY AND REVISION TOTAL KNEE ARTHROPLASTYS

A primary TKA is the first implant used by a surgeon in a patient. During the primary TKA procedure, the surgeon opens the knee joint to have access to the damaged bone and he have a perfect overview of the main bony landmarks, ligament structure, and joint line and position. This information is fundamental for the proper implant placement and alignment. However, although TKA is one of the most successful orthopedic procedures with reproducible long-term results, this device can fail for a number of reasons (Baldini et al., 2015).

If a knee replacement fails, it is necessary to undergo a second surgery: the revision total knee replacement. In this procedure, the surgeon removes some or all of the parts of the original primary prosthesis and replaces them with new ones. The revision of a revision TKA is called second revision.

Although both procedures have the same goal (i.e., to relieve pain and to improve function), it is important to note that revision surgery is different from primary total knee replacement. It is a longer, more complex procedure that requires extensive planning and sometimes performed in two stages, bone defects are managed with bone grafts or prosthetic augmentation, and specialized implants and instruments are used to achieve a good result (Tigani, Comitini, Leonetti, & Affatato, 2015).

Comparing revision TKA with primary TKA, we have the following:

- Revision TKA surgeries are fewer in number compared to primary TKA surgeries. As an example, the American (http://connect.ajrr.net/2020-ajrr-annual-report) and the Australian (https://aoanjrr.sahmri.com/) registries reported 8.2% and 9.5% of revision TKA compared to primary TKA.
- Revision TKA surgery incurs a higher cost as the patient length of stay is longer (mean length of stay of 1.9 for primary TKA compared to 3.3 for revision TKA [American registry https://aoanjrr.sahmri.com/]).
- Revision TKA implants are, in general, more expensive (due to also the presence of additional features as wedges and stems) (Bhandari, Smith, Miller, & Block, 2012).
- Their revision rate is higher. Survival rate of primary TKA varies between 90% and 95% at 15-year follow-up, while, unfortunately, the survival rate of revision TKA varies between 71% and 86% at 10-year follow-up (Rosso et al., 2019).

- The incidence of revision TKA is likely to increase, particularly in young patients for infection (Schwartz, Farley, Guild, & Bradbury, 2020), and even if the incidence rate of primary TKAs is expected to increase, the annual total number of revision procedures is forecast to increase even more rapidly in the coming years (Klug et al., 2020; Lustig et al., 2021).

TOTAL KNEE ARTHROPLASTY AND PARTIAL KNEE REPLACEMENT

The TKA is called "total" because it replaces all three articulations of the knee joint with artificial material: the medial and lateral tibiofemoral joints and the patellofemoral joint (see Chapter 13).

However, along with TKA, orthopedic companies also make available other types of implants to surgeons, called "early intervention devices" or "partial knee replacement," which are only related to one specific knee articulation, in particular (Fig. 20.6):

- unicompartmental (or monocompartmental) knee arthroplasty (UKA), which replaces only either the medial (most common) or lateral tibiofemoral condylar joint;
- patellofemoral joint (PFJ) replacement, which replaces the PFJ.

The advantages of such devices, compared to TKA, are as follows:

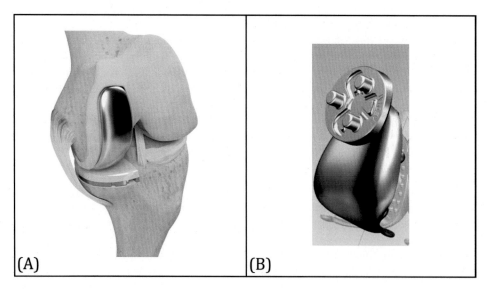

(A)   (B)

FIGURE 20.6

Partial knee replacement: (A) unicompartmental knee arthroplasty of the medial compartment; and (B) patellofemoral joint replacement.

Courtesy Waldemar Link GmbH & Co. KG.

- They could be adopted in the case of a minor knee defect, avoiding the use of a more "aggressive treatment," such as TKA.
- They replace only the damaged area, usually reducing the skin incision and often resulting in a faster recovery with less bleeding.
- In the case of UKA, both the ACL and the PCL are preserved, so this device improves knee proprioception and allows for more "normal" knee kinematics.

In comparison with TKA, the number of UKAs and PFJs is rather low; for example, the 2020 American registry (http://connect.ajrr.net/2020-ajrr-annual-report) reported, of all primary knee arthroplasties, a percentage of 4.1% for UKA (both medial and lateral) and 0.4% for PFJ, and the Swedish registry (http://myknee.se/en) reported 10.7% for UKA ans 0.4% for PFJ. (This number is evaluated during a 10-year period.) This small number is strongly related to a minor indication for partial replacement compared to total replacement but also to the small number of surgeons that implant such devices; for example, the American registry reports that in 2019 only 9.2% of all surgeons who performed primary knee arthroplasty procedures preferred PFJs, and only 27.2% performed medial and/or lateral UKAs.

The primary concern with UKA (and PFJ) is the reportedly higher revision rate when compared to TKA: revisions of medial UKA, lateral UKA, and PFJ occur at an annual rate of 2.18-, 2.31-, and 3.57-fold that of TKA, respectively (Chawla et al., 2017). However, the revision rate is drastically reduced with experienced surgeons making a proper patient selection (Sun & Su, 2018). In well-selected patients treated by experienced surgeons performing a proper implant positioning, both TKR and PKR are effective, have similar clinical outcomes, and result in a similar incidence of reoperations, revision, and complications (Siman et al., 2017). Based on recent UK multicentric, pragmatic, randomized, controlled effectiveness trial studies (Beard et al., 2019, 2020), UKA demonstrates a similar incidence of reoperations and complications to TKA, lower costs and better cost-effectiveness, greater health benefits (measured using quality-adjusted life-years), and lower health-care costs. However, the researchers mention the limitation that the surgeons who performed partial knee replacement were relatively experienced with the procedure.

HISTORY OF TOTAL KNEE PROSTHESIS DESIGN
INTRODUCTION

Historically, several (and different) types of prostheses were used for TKA, and their evolution, which has a history of almost 50 years, involves cycles of failure and further development.

Knowledge of the early implants, their purpose, and the reason for their failure (and the approach that was used to solve the issues) is essential to the design of an innovative (and theoretically better) device.

It is important to note that the reason for most of the mistakes committed in the past was mainly related/influenced by the lack of knowledge of biomechanics, metallurgy, biology, and other fields involved in the medical device development and production processes.

In order to develop an effective device, it is important to highlight that the design should always be developed in accordance with the current state of the art of all fields related to this subject.

THE FIRST HINGED DESIGNS

The German surgeon Themistocles Gluck (1853−1942) is considered one of the fathers of TKA with a first preliminary TKA proposed in 1891. The device (Fig. 20.7) is a hinge implant that links the femoral component with the tibial component with a mechanical constraint, allowing the main knee motion: the flexion−extension.

During that period, this design was relatively functional and efficient as it was able to integrate several advantages:

- The entire TKA consisted of only three parts: a femoral component, a tibial component, and a central peg that was used to connect the components.
- It was made of ivory, which was considered a good biomaterial at that time (as it was inert, thus potentially safer).
- The surgery was considered feasible (at the time) because it consisted of the resection of the distal femoral bone and of the tibial proximal bone and insertion of the two components.

Although it was a revolutionary concept and the procedures appeared successful over the short term, these implants often failed due to high infection rates, poor metallurgy, and inadequate fixation (Song, Seon, Moon, & Hyoun, 2013).

Despite the failure of Gluck's hinged TKA, the idea of physically constraining the knee joint with a hinge constraint was considered a correct approach; thus, following this philosophy, in 1950, Dr. Waldius developed a hinged TKA, initially made of acrylic, and later (1958) manufactured in Vitallum (a corrosion-free CoCr alloy), intended for the treatment of rheumatoid patients. This design was coupled with an intramedullary stem that allows the alignment of the implant with the patient leg, and it was technically easy to perform because all ligaments and soft tissues could be removed

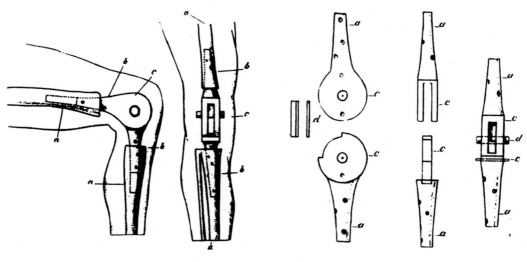

FIGURE 20.7

Gluck's ivory total knee hinge arthroplasty.

due to the mechanical and structural stability of the prosthesis. This design was in use until the early 1970s along with several other hinged designs as the Shiers and the Stanmore (Ranawat, 2002).

Unfortunately, the long-term results of these devices were not extremely good. All devices allowed only flexion/extension movements and internal/external rotation was not considered as a relevant motion; thus, the forces (and relative stress) on the implant (and on the bone) were quite high and induced early loosening or stem fracture. In addition, the implant was produced in only one size and only a few instruments were available, and metal debris, produced due to metal-on-metal contact, was an additional issue with such designs (Manning, Chiang, & Freiberg, 2005).

In an attempt to compensate for the high revision rate of previous designs, the Guepar prosthesis, introduced in 1969, was designed with the aim of minimal bone resection, joint stability, valgus alignment, preservation of motion, preservation of patellar tracking, and a damping effect in extension due to the presence of a silicone bumper present on the anterior tibia (Manning et al., 2005). A 5−8 year follow-up study of such design in 101 cases reported an overall revision rate of 6% (Aubriot, Deburge, & Genet, 2014).

THE FIRST CONDYLAR IMPLANTS

The first condylar implants were (total or partial) knee resurfacing prosthesis of the tibiofemoral joint. As an alternative to the hinged implants, the advantages of these designs were the reduced bone resection and the possible preservation of the soft tissue.

The first femoral resurfacing design was developed by Dr. Campbell in 1940, who, after obtaining good results for mold arthroplasty in the hip joint, used a similar approach to cover the articular distal femoral condyles; this design failed due to the absence of any anchorage, making it extremely unstable. Later, in 1941, Dr. Smith and Dr. Petersen developed a femoral resurfacing prosthesis that, compared to the previous design, featured a longer stem in order to ensure proper bone position and stability. This design failed mainly due to poor metallurgy and consequent material degeneration; later, in order to solve the flaws of the previous design, Dr. Platt in 1955 introduced a design made of stainless steel with a reduced stem length, similar to a UNI femoral component (Song et al., 2013).

Other similar design approaches (and results) were also developed for the tibia resurfacing prosthesis, starting from 1958 with the McIntosh hemiarthroplasty, which was primarily aimed at correcting varus−valgus alignment, made of acrylic material, and fixed to the tibia using surface irregularities, to the one developed by McKeever, still in 1958, which was fixed to the tibia with a flange; both designs failed due to poor fixation (Song et al., 2013). From a design standpoint, these prostheses were quite close to a UKA; however, they were used simultaneously for both plateaus, creating another problem related to the incorrect load distribution among the medial and lateral sides induced by any malalignment introduced surgically during implant positioning.

ANATOMICAL AND FUNCTIONAL APPROACHES

The development of the condylar-resurfacing TKA was based on two different design philosophies: the anatomical approach or the functional approach (Dall'Oca et al., 2017).

The anatomical approach, which was also the one followed by the first generation of resurfacing implants, is based on the concept that the correct function of a certain device used to replace the joint is achieved copying/mimicking the anatomy of the joint itself. Therefore, following this approach, the cruciate ligaments and most of the soft tissue were preserved, the implant surfaces (resurfacing) were designed to mimic the anatomy, and any potential conflict among soft tissues and articular surfaces were avoided.

An example of such design was the polycentric knee introduced by Gunston in 1968, which consists of a highly conforming implant with separate medial and lateral stainless steel femoral components articulating with plastic tibial runners. This design TKA failed mainly due to the extremely low contact area (and therefore high contact pressure, as the menisci were not included in the articular surfaces), low rotational freedom, and quite difficult surgical technique due to the preservation of the cruciate ligaments (Ranawat, 2002).

Following the anatomical approach, the Kodama–Yamamoto in 1970 was designed with an anatomical femoral component coupled with a minimally constrained single-piece PE tibial component, placed uncemented, shaped with a central cutout (horseshoe) for the preservation of both cruciate ligaments, and the UCI knee (University of California at Irvine) by Waugh & Smith in 1971 with a duplication of femoral condyles and tibial plateaus using casting techniques, avoiding cruciate ligament resection (Dall'Oca et al., 2017; Hamilton, 1982).

The clinical results of the designs developed with anatomical approaches were not very good, and their revision rate was also quite high. The main issues were as follows:

- The design, aimed at following the bone anatomy in some cases, was extremely difficult to develop in those days.
- The contact pressure and force, due to the limited contact area, were quite high, inducing failures due to fatigue, wear, and mechanical fractures.
- The surgeon complained about the difficulty in the identification of the proper position, i.e., the implant was difficult to position accurately and the surgery was considered too difficult for most surgeons.
- Most surgeons considered cruciate ligament resection necessary to correct deformity.

Therefore, these issues have led to a shift from the anatomical approach to the functional approach: the function of the joint was obtained without being anatomical but by providing in the design specific features that allowed the proper joint function. Avoiding to be anatomical, the biomechanics of the knee joint was simplified by the resection of the condyles and the cruciate ligaments.

Following this approach, the total condylar prosthesis was designed in 1970 with a chrome–cobalt femoral shield and two components in polyethylene (tibial and patellar). Learning from the previous issues, such design was developed keeping in view a high congruency of the tibiofemoral joint: two symmetrical femoral condyles with a smaller ray of curvature interfaced with a congruent tibial base. This enables increasing the contact area and a reduction of the contact pressure, which eventually formed the base for the development of the first mobile-bearing TKA. This concept, nevertheless, showed an anterior translation of the femur during flexion, low levels of ROM, and high anterior polyethylene debris (Dall'Oca et al., 2017).

From bicondylar to tricompartmental prosthesis andthe introduction of the Post/Cam

In the early 1970s, three types of condylar prostheses were developed, which ushered in an era of modern knee arthroplasty:

- the duocondylar prosthesis (Ranawat & Shine, 1973);
- the geometric prosthesis (Coventry, Finerman, Riley, Turner, & Upshaw, 1972);
- the anatomic prosthesis (Townley, 1972).

The condylar prosthesis, even if they preserved both the anterior and the posterior cruciate ligaments, provided stability to the knee joint, and used bone cement for fixation to bone. However, these designs showed, during the first experiences, a high rate of complications due to early loosening of fixation, component mobilization or malfunctioning, and infection due to complicated surgery.

Therefore, to avoid these issues, the total (tricompartmental) condylar prostheses were introduced as a design that required the removal of the anterior and posterior cruciate ligaments. Moreover, the femoral component, which was made of chrome−cobalt alloy, had a symmetric femur with a flat patellar trochlear groove. An aspect of this prosthetic model was a posteriorly decreasing radius of curvature, and, thus, the components resulted in being perfectly congruent in extension and partially congruent in flexion. The tibial prosthesis, equipped with a stem to improve tibial fixation, was completely made of polyethylene, and had a prominence in the mid-surface to improve the anteroposterior and mediolateral stability. Also, the patellar prosthesis was all-polyethylene, with a central peg, to be fixed with bone cement.

Early total condylar prostheses did not show any roll-back, and the tibial posterior offset was high, reducing the mobility range when the flexion gap was not balanced. According to early clinical reports, the average flexion angle was 90−100 degrees (Song et al., 2013).

To improve knee kinematics and also add additional stability to the replaced joint, Insall in 1982 (Insall, Lachiewicz, & Burstein, 1982) introduced the concept of posteriorly stabilized condylar prosthesis, modifying the total condylar design, adding a Post−Cam system to the design, and, thus, laying the foundation for modern knee arthroplasty.

DESIGN OF A TOTAL KNEE REPLACEMENT
WHY DESIGN A TOTAL KNEE ARTHROPLASTY TODAY?

Currently, there are several TKA designs available on the market, each with their own features, options, and properties, so it begs the question: "Why are new designs needed?", which is quite reasonable and should be answered appropriately. Also, looking at the different national registries, the success rate of TKA is quite high (in general, up to 95% at 10-year follow-up), so another question arises: "Why should the design of a new TKA change?", which is also an additional aspect that needs to be explained.

Among the different reasons, one main aspect should be pointed out: as highlighted by Noble already in 2005 (Noble et al., 2005), despite the high success rate of TKA, almost 20% of patients reported dissatisfaction after surgery, i.e., they are not satisfied with the final outcome of the surgery mainly because of different types of difficulties faced in performing a certain activity due to the presence of the implant. To provide some numbers, overall, 52% of the patients who underwent TKAs reported some degree of limitation in doing functional activities, compared with 22% of subjects without previous knee disorders, and only about 40% of the functional deficits present after a TKA appear to be related to the normal physiologic effects of aging (Noble et al., 2005).

This is the driving force behind surgeons and engineers making continuous effort toward designing new implants: to improve patient lifestyle, in terms of better clinical follow-up and restoration of normal knee function after TKA surgery.

TOTAL KNEE ARTHROPLASTY DESIGN OBJECTIVES, CRITERIA, AND DIRECTIONS

The design of a TKA is translational (as it involves knowledge from several fields as, but not limited to medical and engineering) and multifactorial, and different design approaches, strategies, and directions could be adopted.

In general, most engineering designs involve a multitude of considerations, and it is a challenge for the engineer to recognize all of them in right proportion. Although no single checklist can be adequate or complete, it may be helpful to report the list of the major categories subdivided among the major fields involved:

- surgical/clinical:
  o easy to implant;
  o replicate anatomical factor;
  o fit for the different patient populations and morphologies;
  o preservation of the soft tissue structure;
- mechanical:
  o material biocompatibility;
  o minimization of the joint contact pressure;
  o minimization of the fixation and shear forces;
  o reduced wear and micromotions;
- industrial:
  o manufacturability;
  o reasonable production and inventory cost.

Furthermore, looking at the reasons for TKA failure of the early designs (such as implant loosening and osteolysis, polyethylene wear, component fracture, joint instability/dislocation, joint stiffness, pain, infection, extensor mechanism dysfunction), the engineers should target the following main objectives:

- a correct joint biomechanics, both in terms of motion (kinematics) and force (kinetics);
- a proper joint stability (sometimes named functional stability (Innocenti, Bori, & Paszicsnyek, 2021));
- fit with patient anatomy (and eventually correct the deformity);
- good interaction with the surrounding materials, both in terms of artificial−biological contact (avoiding stress shielding, soft tissue damage, inflammation, and bacterial adhesion) and artificial−artificial contact (reduced wear, correct lubrication, roughness, etc).

Lastly, it is important to note that, as for the design of any medical device, there are three design directions that must be followed. The design of the prosthesis should be:

- functional, meaning that the design should allow a proper joint function as in native conditions;
- durable, meaning that the function of the design should be guaranteed not only on the day of surgery, but throughout the entire life of the implant without any catastrophic change in performance over time;
- robust (in relation to both the patient and the surgeon), meaning that each implant design should fit the different anatomies, morphologies, and related features characterizing different patients, together with the expectations of different surgeons (in terms of skills, experience level, etc) and related surgical techniques.

In any implant, these aims, objectives, and directions are not achieved only with a good implant design but also, inextricably, through a good correlation with well-designed surgical tools and related instruments (not considered in this chapter for brevity) and, possibly, with additional sensors providing the surgeon with additional intraoperative information.

FEMORAL COMPONENT DESIGN

Several aspects must be considered when designing the femoral component of a TKA.

In general, comparing the medial to the lateral condylar shape, it is possible to design femoral components to be the same, for symmetrical condyles, or different, for asymmetrical condyles. The condyles could be separated (CR design) or connected by a metallic Cam (PS design), and the Cam itself could be symmetric or asymmetric (to promote internal/external rotation during flexion as in some medial-pivot knee design). In addition, similar to the tibial implant, in terms of implant—bone interface, the metal surface could be treated with porous metal or some coating (as hydroxyapatite) to enhance bone ingrowth.

Among other aspects, one of the main design features that characterize a femoral component is the sagittal shape of the condyle, which could be single curved or J-curved (Fig. 20.8).

With the aim of reproducing the anatomical shape of the femoral condyle (which is not perfectly circular), the J-curved condyles consist of a multiradius (MR) design where the condylar curvature is different, from a large radius in extension to a smaller radius in flexion. Usually, the decrease in the radius during flexion is achieved with three different radii. However, the variation in radius has been considered by surgeons as a possible risk factor for mid-flexion instability, inducing, due to a change in ligament isometry, possible joint laxity. To avoid this phenomenon, the single-radius (SR) design has been introduced in which the condylar radius is kept constant up to 90 degrees (even up to 130 degrees in some designs). Sometimes, this design is also coupled with a medial congruent tibial insert allowing a medial ball-in-socket joint motion, the so-called medial-pivot knee (MP). Recently, the MR design has been further improved by introducing the concept of the gradually reduced (GR) design (Gradius™ curve) with the idea of changing continuously the radius for all flexion angles. In other words, the gradually reducing femoral radius was designed with the aim of providing a smoother transition from stability to rotational freedom through a patient's ROM.

However, although some designs are characterized by some biomechanical advantages and better knee kinematics, this has not translated into better clinical outcomes, and there is still insufficient evidence to clinically support SR, MP, or GR TKA over the traditional MR TKA design (Ng,

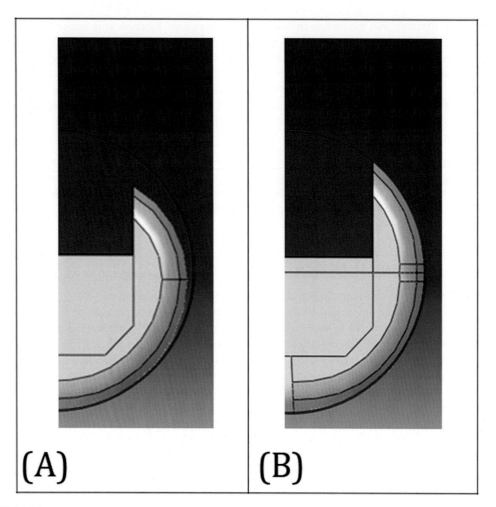

FIGURE 20.8

Femoral condylar design: (A) multiradius design and (B) single-radius design.

Bloch, & James, 2019). Moreover, the link between the radius-of-curvature of the femoral component and mid-flexion instability was analyzed clinically and biomechanically, with contradictory results (Vajapey et al., 2020), and some authors (Stoddard, Deehan, Bull, McCaskie, & Amis, 2013), reporting no statistically significant difference in mid-flexion stability of the knee between SR and MR TKAs, suggested that mid-flexion instability may be due to unrecognized ligament laxity during surgery rather than an implant design feature.

Because anterior knee pain following TKA is one of the main reasons for implant revision, which is equally observed in knees with resurfaced and nonresurfaced patellae, the geometry of the femoral trochlear is also considered extremely important: It is strongly related to extensor mechanism function and joint kinematics, and any defect in this aspect could induce patellofemoral complications following TKA (Kulkarni, Freeman, Poal-Manresa, Asencio, & Rodriguez, 2000). Analyzing the shape of the trochlear groove in different designs (Qin et al., 2018), recent TKAs have been designed following a more anatomic trochlear geometry than earlier TKA models by the same manufacturers. The sulcus angles remain 3−6 degrees greater (shallower) in prosthetic trochlear compartments than in healthy knees, and they exceed radiologic signs of trochlear dysplasia by 2−5 degrees.

TIBIAL COMPONENT DESIGN

For the design of the tibial component shape, apart from the features related to the press-fit/cemented implant and MB/FB design, there are two main options: symmetrical or asymmetrical baseplate.

A symmetrical design has the advantage of reducing the inventory cost (since there is no difference between left and right designs); however, compared with the asymmetric design, the coverage of the proximal tibial surface is reduced, the selection of the correct implant size is more complicated (due to the difference in medial and lateral AP dimensions of the tibial bone), and it is also more difficult to place the baseplate with the correct rotational implant alignment (along the mediolateral direction). This issue is especially relevant for FB design as it could induces a wrong rotational position of the insert. A possible advantage of the symmetric design could be minimizing the risk, fortunately to quite an extent, of a possible error in selecting the correct side of the implant. Conversely, an asymmetric tibial baseplate allows better coverage, as it is designed with a larger medial AP dimension, and an easier placement in tibial rotation.

The distal part of the tibia features a central stem, which allows correct axial positioning, and to improve the internal−external rotational constraint, the surface could present additional pegs or flange.

A special design of the tibial baseplate is the all-polyethylene tibial component in which the tibial component and the tibial insert are a single block, made of polyethylene, cemented to the tibial bone (Fig. 20.9).

The main advantage of using of an all-polyethylene tibial component over a metal-backed modular component is the lower cost and the possibility of using it in patients with metal allergy or sensitivity (coupled with a femoral component in ceramic material or in Oxinium or covered with a ceramic coating). Further advantages are the absence of problems with the locking mechanism and backside wear. Disadvantages include lack of modularity, limiting intraoperative options, and no option for the removal or exchange of the insert due to infection or late implant failures (Gioe & Maheshwari, 2010).

TIBIAL INSERT DESIGN

The medial and lateral compartments of the polyethylene tibial insert could be symmetrical or nonsymmetrical and could have different levels of congruity with the femoral component.

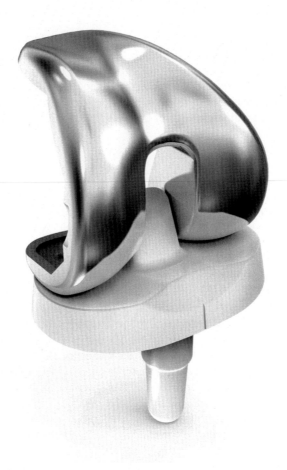

FIGURE 20.9

TKA with an all-polyethylene tibial implant.

Courtesy Waldemar Link GmbH & Co. KG.

In a symmetrical insert, the medial and lateral compartments are concave in the anteroposterior direction, whereas in an asymmetrical insert, the lateral surface is less concave (or flat or even slight convex). In some design, to promote native (passive) medial-pivot motion, the tibial insert is characterized by a more congruent medial side (medial-pivot knee) or a flatter lateral side (lateral-sliding knee).

In the frontal view, the thickness of the insert could be constant (equal between the medial and lateral compartments) or asymmetric. A frontal asymmetry of the insert could be introduced to mimic the average native joint line (of 3 degrees), increasing the lateral insert thickness. With such an insert, it is possible to have an oblique joint line with a mechanically aligned tibial cut (Fig. 20.5).

The insert is characterized by a size (the same as the tray) and a thickness that are chosen to correctly stabilize the joint (avoiding ligament laxity or overtension); surgically, the tibial cut corresponds to the total height (given by the sum of the tibial tray plus the height of the insert).

Some CR inserts are designed with an ultracongruent shape. The advantage of using this insert is to provide additional AP stability of the joint still keeping the posterior cruciate ligament (and therefore avoiding the use of PS design). Such an insert enables, in a mobile-bearing device, stabilizing the joint even with PCL deficiency (Innocenti, 2020) and can be considered as a safe alternative to the well-established PS design if necessary (Innocenti & Bori, 2020).

PATELLAR COMPONENT DESIGN

Patella components can be cemented or noncemented, inlayed or onlayed, metal-backed or all-polyethylene, and are available in various morphologies. According to the national registry, the most commonly used patellar component (Fig. 20.1) is fully made in polyethylene and is cemented on the patellar bone.

From a design point of view, it could be symmetrical or asymmetrical, presenting in this case a medial patellar facet to improve patellar stability and reduce patellar tilting. Patellar components can be generally classified as onset designs (oval dome) or inset designs. Inset designs are smaller and have hemispheric domes compared with the oval domes associated with onset designs. In terms of bone−implant interface, the inset design is characterized by one central peg, while the onset has three pegs. Moreover, the inset design has a roughly biconvex shape in the sagittal plane (Fig. 20.10). Potential advantages of an inset design include increased contact at the bone−cement−prosthesis interface without additional weakening of the patellar remnant (Ezzet et al., 2001).

ADDITIONAL TOTAL KNEE ARTHROPLASTY DESIGN ASPECTS
IMPLANT SIZE

Recent anatomical studies have shown that the size and shape of the femur and tibia in the knee vary significantly between different individuals, particularly between males and females. For this reason, many different designs and sizes of TKAs are currently available on the market, in order to provide the surgeons (and therefore the patients) with a variety of appropriate devices to choose from based on their needs.

The size of the implant is a critical factor, and, in parallel with the design itself, it has a fundamental impact on the performance of the implant. Therefore, great care must be taken while deciding on the implant size (Innocenti & Bori, 2020). Industrially, it is also particularly relevant the proper decision on the number of size of the implant. On one hand, a large number of size is associated with a high cost, but a closer fit with patient's anatomy, while, on the other hand, a small number of size increases the risk of having an implant that does not fit well with the shape of the bone.

Note that it is not always possible to decide on the proper size for every patient unless custom-made devices are used: when the patient's geometry is between two sizes, the surgeons must go for oversizing or undersizing the implant. Usually, oversizing should be avoided because it induces soft tissue envelope damage, patient pain, and related detrimental biomechanical functions. On the other hand, studies on implant undersizing have reported possible problems with ligament laxity or anterior cortical notching but overall acceptable results (Innocenti & Bori, 2020).

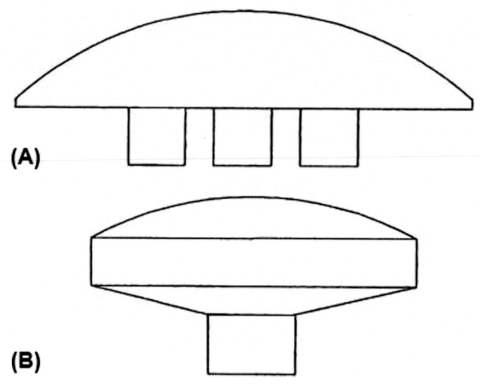

FIGURE 20.10

Patellar designs: (A) oval dome onset patella with three pegs for fixation to bone; and (B) hemispherical inset patella with a central peg for fixation to bone.

From Ezzet, K. A., Hershey, A. L., D'Lima, D. D., Irby, S. E., Kaufman, K. R., & Colwell, C. W. (2001). Patellar tracking in total knee arthroplasty: Inset vs onset design. The Journal of Arthroplasty, 16(7), 838–43.

DESIGN OF UNICOMPARTMENTAL KNEE ARTHROPLASTY

UKA is the typical solution for monocompartmental osteoarthritis (OA) (Fig. 20.6A). If well implanted, under the right indications, the knee after UKA can reproduce accurately the intact knee motion (Suggs et al., 2006), and national registries reported excellent results at 10 years of follow-up with modern designs. However, some cases of failure are also described in the literature.

In addition to wear, four major postoperative problems are reported (Hansen, Ong, Lau, Kurtz, & Lonner, 2019; Kim, Lee, Lee, & Kim, 2016):

- loosening of the prosthesis component (due to stress shielding) in the bone;
- malpositioning of prosthetic components (failure in the fixation of the implant due to excessive bone stress and increased strains in the soft tissues);
- medial knee pain (due to bone overload, component malalignment, and soft tissue tensioning);
- OA progression in the lateral side (due to altered stress pattern in the bone/cartilage).

Therefore, when designing a UKA, it is crucial to investigate the biomechanical effects that such design will induce in the knee joint such as implant kinematics, bone stresses, and ligament strains, as these are fundamental to guaranteeing proper implant (and joint) performance.

From a design point of view, UKA could be fix-bearing or mobile-bearing (Fig. 20.11). Usually, in MB UKA, the insert is congruent, while in FB UKA, the insert is almost flat (Mittal, Meshram, Kim, & Kim, 2020). To minimize the friction force between insert and baseplate, MB UKA requires a highly polished cobalt—chromium tibial tray. FB UKA, on the other hand, is designed with an all-polyethylene tibia or a metal-backed titanium tibial tray as they require no polishing. The modulus of elasticity of these trays is much closer to the one of the tibial bone.

Despite the difference in design and their mechanics, clinical outcomes and revision rates of primary FB and MB UKAs are reported as equal. Swedish (http://myknee.se/en/) and Finnish (https://www.thl.fi) arthroplasty registries comparing both bearing designs of UKA suggested no conclusive advantage of one bearing design over the other in terms of prosthesis survival (Mittal et al., 2020).

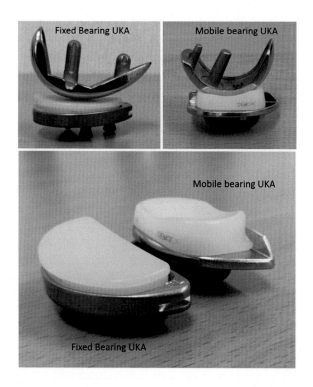

FIGURE 20.11

UKA design: fix-bearing and mobile-bearing. The difference in the insert shapes between the two designs is also reported. *UKA*, unicompartmental knee arthroplasty.

Data from Mittal, A., Meshram, P., Kim, W. H., & Kim, T. K. (2020). Unicompartmental knee arthroplasty, an enigma, and the ten enigmas of medial UKA. Journal of Orthopaedics and Traumatology: Official Journal of the Italian Society of Orthopaedics and Traumatology, 21(1), 15. doi:10.1186/s10195-020-00551-x.

From a biomechanical point of view, the insertion of a UKA alters, in one compartment of the knee, the material stiffness. In fact, in a healthy knee, the contact interface, both on the medial and lateral sides, consists of different layers (from superior to inferior): the femoral cortical bone, the femoral articular cartilage, menisci, the tibial articular cartilage, and the tibial cortical bone. However, in a knee with a UKA, the artificial contact interface, usually the medial side, changes, and it consists of the following different layers: the metal of the femoral component, the polyethylene insert, and the metal of the tibial component, with an increase in stiffness (between metal and cortical bone) of around 10 times.

One study reported that after implantation, due to the increased stiffness in the medial side, UKA resulted in a slight valgus alignment configuration (Innocenti et al., 2014), and the valgus alignment was more pronounced when the thickness of the insert was increased (overstuffing) along with an increase in load on the medial side. Because of this, some surgeons prefer to understuff the UKA during surgery.

From a kinematic point of view, a study (Heyse et al., 2014) has demonstrated that, qualitatively, native kinematics were reproduced after UKA, in different specimens, replicating different motor tasks; the mismatch in stiffness between the lateral native joint and the medial artificial UKA may lead to three consequences:

- The femur becomes more superior than in the native knee.
- The tibia tilts toward a more valgus position.
- The PCL remains more taught and the femoral medial condyle is held more posteriorly.

Additionally, the performance of UKA is strongly related to implant position, and any alteration could induce a big change in bone stress, both average and maximum (Innocenti, Pianigiani, Ramundo, & Thienpont, 2016). Moreover, a malaligned position alters the contact point position and the poly stress, especially for high varus/valgus configurations.

DESIGN OF REVISION TOTAL KNEE ARTHROPLASTY: CONDYLAR CONSTRAINT KNEE AND HINGED DESIGN, STEM, AND AUGMENT

The increasing number of patients undergoing primary TKA has been accompanied by a similar increase in the number of revision TKAs. The primary goals of revision are to increase joint stability, usually, due to a lack of soft tissue tension, identify a proper knee joint line (as most of the bony references are lost after primary TKA removal), and stable fixation of the implants.

In general, two different levels of constraint for TKA can be considered for revision surgeries: semiconstrained (or condylar constrained) and constrained (hinged) implants.

The condylar constraint knee (CCK) design is a PS design in which the Post is bigger and thicker to fit the intercondylar femoral space. This design confers different levels of varus−valgus and rotational stability depending on the Post−Cam engagement design (different for each manufacturer). In general, a bigger Post allows a reduction of the mediolateral joint translation and of the internal−external rotation.

With the hinged design (Fig. 20.12A), the femur is mechanically constrained, with a mechanical hinge joint, to the tibia. Typically, a hinged joint allows only flexion/extension and internal/external

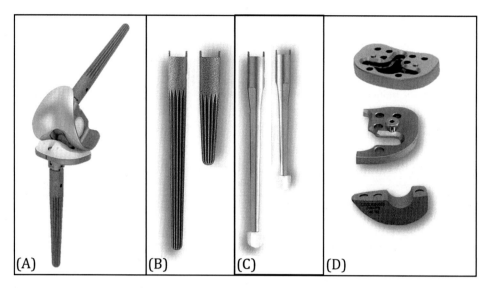

FIGURE 20.12

Revision options: (A) hinged TKA, (B) press-fit stems, (C) cemented stems, and (D) tibial and femoral augments. *TKA*, total knee arthroplasty.

Courtesy Waldemar Link GmbH & Co. KG.

rotation (rotating hinged design); however, fixed hinged designs are also available in which only the flexion—extension rotation is allowed. This design is primarily used in the case of patellar deficit. Different hinged designs are currently available on the market, characterized by different positions of the axis of rotation, different degrees of mobility, and different femoral shapes (e.g., some have a large posterior condylar thickness requiring a large bone loss).

Hinged TKAs are indicated in the cases of severe deformities, primary TKA revision, revision of a revision TKA, collateral ligament deficiency, and inadequate bone support.

To improve implant-bone stability, usually, revision implants are coupled with cemented or press-fit stems (Fig. 20.12B and C). Stems with different features are provided by different manufacturers, which can change their shape (being straight or bowed), length (short/long), interaction with bone (cemented/press-fit), and can be massive or present a slot.

Moreover, bone loss and subsequent defects are often encountered in the revision of TKA, and augments are a common solution for treating them (Fig. 20.12D). Nowadays, a variety of options are available for surgeons, in terms of material, thickness, and shapes (usually related to the TKA design).

An engineer should consider all of these design aspects very carefully, as they may alter the biomechanics of implant fixation. Some studies showed that in the case of cemented stems, both long and short stems offer good stability with little implant micromotions and acceptable stress levels in bone. The press-fit stems show more implant micromotions, higher stress levels, and more stress peaks in the bone. They also reported a higher concentration of bone stress in the case of a massive stem, while the presence of a slot reduced the bone stress value (El-Zayat et al., 2016, Bori, Armaroli, & Innocenti, 2021). Another study comparing CCK and hinged designs has shown that different levels of constraint in revision arthroplasty were always associated with different biomechanical outputs. Rotating hinged implants are characterized by higher tibial stress, especially in the region close to the stem tip; condylar implants, instead, increase the proximal tibial stress and therefore implant micromotions, as a result of the presence of the Post−Cam mechanism (Andreani, Pianigiani, Bori, Lisanti, & Innocenti, 2020). Finally, one study reported that the use of any type of augment induces a change in bone stresses, especially localized in the region close to the bone cut, and the stiffness of the augment material should be as close to that of the bone as possible, suggesting a porous metal as a better option to treat bone defect than solid metal (Innocenti, Fekete, & Pianigiani, 2018).

REFERENCES

Aglietti, P., Baldini, A., Buzzi, R., Lup, D., & De Luca, L. (2005). Comparison of mobile-bearing and fixed-bearing total knee arthroplasty: A prospective randomized study. *The Journal of Arthroplasty*, *20*(2), 145−153.

Andreani, L., Pianigiani, S., Bori, E., Lisanti, M., & Innocenti, B. (2020). Analysis of biomechanical differences between condylar constrained knee and rotating hinged implants: A numerical study. *The Journal of Arthroplasty*, *35*(1), 278−284. Available from https://doi.org/10.1016/j.arth.2019.08.005.

Aprato, A., Risitano, S., Sabatini, L., Giachino, M., Agati, G., & Massè, A. (2016). Cementless total knee arthroplasty. *Annals of Translational Medicine*, *4*(7), 129. Available from https://doi.org/10.21037/atm.2016.01.34.

Arnout, N., Vanlommel, L., Vanlommel, J., Luyckx, J. P., Labey, L., Innocenti, B., ... Bellemans, J. (2015). Post-cam mechanics and tibiofemoral kinematics: A dynamic in vitro analysis of eight posterior-stabilized total knee designs. *Knee Surgery, Sports Traumatology, Arthroscopy : Official Journal of the ESSKA*, *23*(11), 3343−3353. Available from https://doi.org/10.1007/s00167-014-3167-2.

Aubriot, J.-H., Deburge, A., & Genet, J.-P. (2014). GUEPAR hinge knee prosthesis. *Orthopaedics & Traumatology, Surgery & Research: OTSR*, *100*(1), 27−32. Available from https://doi.org/10.1016/j.otsr.2013.12.012.

Baldini, A., Castellani, L., Traverso, F., Balatri, A., Balato, G., & Franceschini, V. (2015). The difficult primary total knee arthroplasty: A review. *The Bone & Joint Journal*, *97-B*(10 Suppl A), 30−39. Available from https://doi.org/10.1302/0301-620X.97B10.36920.

Beard, D. J., Davies, L. J., Cook, J. A., MacLennan, G., Price, A., Kent, S., ... Campbell, M. K. (2020). Total vs partial knee replacement in patients with medial compartment knee osteoarthritis: The TOPKAT RCT. *Health Technology Assessment (Winchester, England)*, *24*(20), 1−98. Available from https://doi.org/10.3310/hta24200.

Beard, D. J., Davies, L. J., Cook, J. A., MacLennan, G., Price, A., Kent, S., ... Warner, K. (2019). The clinical and cost-effectiveness of total vs partial knee replacement in patients with medial compartment osteoarthritis (TOPKAT): 5-year outcomes of a randomised controlled trial. *The Lancet*, *394*(10200), 746−756. Available from https://doi.org/10.1016/S0140-6736(19)31281-4.

Bhandari, M., Smith, J., Miller, L. E., & Block, J. E. (2012). Clinical and economic burden of revision knee arthroplasty. *Clinical Medicine Insights. Arthritis and Musculoskeletal Disorders*, *5*, 89−94. Available from https://doi.org/10.4137/CMAMD.S10859.

Boese, C. K., Ebohon, S., Ries, C., & De Faoite, D. (2021). Bi-cruciate retaining total knee arthroplasty: A systematic literature review of clinical outcomes. *Archives of Orthopaedic and Trauma Surgery*, *141*(2), 293−304. Available from https://doi.org/10.1007/s00402-020-03622-0.

Bori, E., Armaroli, F., & Innocenti, B. (2021). Biomechanical analysis of femoral stems in hinged total knee arthroplasty in physiological and osteoporotic bone. *Computer Methods and Programs in Biomedicine, 213*, 106499. Available from https://doi.org/10.1016/j.cmpb.2021.106499. 34763174.

Callaghan, J. J., Insall, J. N., Greenwald, A. S., Dennis, D. A., Komistek, R. D., Murray, D. W., ... Dorr, L. D. (2001). Mobile-bearing knee replacement: Concepts and results. *Instructional Course Lectures, 50*, 431–449.

Chawla, H., van der List, J. P., Christ, A. B., Sobrero, M. R., Zuiderbaan, H. A., & Pearle, A. D. (2017). Annual revision rates of partial vs total knee arthroplasty: A comparative *meta*-analysis. *The Knee, 24*(2), 179–190. Available from https://doi.org/10.1016/j.knee.2016.11.006.

Coventry, M. B., Finerman, G. A., Riley, L. H., Turner, R. H., & Upshaw, J. E. (1972). A new geometric knee for total knee arthroplasty. *Clinical Orthopaedics and Related Research, 83*, 157–162.

Daines, B. K., & Dennis, D. A. (2014). Gap balancing vs. measured resection technique in total knee arthroplasty. *Clinics in Orthopedic Surgery, 6*(1), 1–8. Available from https://doi.org/10.4055/cios.2014.6.1.1.

Dall'Oca, C., Ricci, M., Vecchini, E., Giannini, N., Lamberti, D., Tromponi, C., & Magnan, B. (2017). Evolution of TKA design. *Acta Bio-Medica : Atenei Parmensis, 88*(2S), 17–31. Available from https://doi.org/10.23750/abm.v88i2-S.6508.

El-Zayat, B. F., Heyse, T. J., Fanciullacci, N., Labey, L., Fuchs-Winkelmann, S., & Innocenti, B. (2016). Fixation techniques and stem dimensions in hinged total knee arthroplasty: A finite element study. *Archives of Orthopaedic and Trauma Surgery, 136*(12), 1741–1752.

Ezzet, K. A., Hershey, A. L., D'Lima, D. D., Irby, S. E., Kaufman, K. R., & Colwell, C. W. (2001). Patellar tracking in total knee arthroplasty: Inset vs onset design. *The Journal of Arthroplasty, 16*(7), 838–843.

Feczko, P. Z., Jutten, L. M., van Steyn, M. J., Deckers, P., Emans, P. J., & Arts, J. J. (2017). Comparison of fixed and mobile-bearing total knee arthroplasty in terms of patellofemoral pain and function: A prospective, randomised, controlled trial. *BMC Musculoskeletal Disorders, 18*(1), 279. Available from https://doi.org/10.1186/s12891-017-1635-9.

Fitzpatrick, C. K., Clary, C. W., Cyr, A. J., Maletsky, L. P., & Rullkoetter, P. J. (2013). Mechanics of post-cam engagement during simulated dynamic activity. *Journal of Orthopaedic Research : Official Publication of the Orthopaedic Research Society, 31*(9), 1438–1446. Available from https://doi.org/10.1002/jor.22366.

Gioe, T. J., & Maheshwari, A. V. (2010). The all-polyethylene tibial component in primary total knee arthroplasty. *The Journal of Bone and Joint Surgery. American Volume, 92*(2), 478–487. Available from https://doi.org/10.2106/JBJS.I.00842.

Hamelynck, K. J. (2006). The history of mobile-bearing total knee replacement systems. *Orthopedics, 29*(9 Suppl.), S7–S12.

Hamilton, L. R. (1982). UCI total knee replacement. A follow-up study. *The Journal of Bone and Joint Surgery. American Volume, 64*(5), 740–744.

Hansen, E. N., Ong, K. L., Lau, E., Kurtz, S. M., & Lonner, J. H. (2019). Unicondylar knee arthroplasty has fewer complications but higher revision rates than total knee arthroplasty in a study of large United States databases. *The Journal of Arthroplasty, 34*(8), 1617–1625. Available from https://doi.org/10.1016/j.arth.2019.04.004.

Heckmann, N., Ihn, H., Stefl, M., Etkin, C. D., Springer, B. D., Berry, D. J., & Lieberman, J. R. (2019). Early results from the American joint replacement registry: A comparison with other national registries. *The Journal of Arthroplasty, 34*(7S), S125–S134. Available from https://doi.org/10.1016/j.arth.2018.12.027, e1.

Heyse, T. J., El-Zayat, B. F., De Corte, R., Chevalier, Y., Scheys, L., Innocenti, B., ... Labey, L. (2014). UKA closely preserves natural knee kinematics in vitro. *Knee Surgery, Sports Traumatology, Arthroscopy : Official Journal of the ESSKA, 22*(8), 1902–1910. Available from https://doi.org/10.1007/s00167-013-2752-0.

Hirschmann, M. T., Becker, R., Tandogan, R., Vendittoli, P.-A., & Howell, S. (2019). Alignment in TKA: What has been clear is not anymore!. *Knee Surgery, Sports Traumatology, Arthroscopy : Official Journal of the ESSKA, 27*(7), 2037−2039. Available from https://doi.org/10.1007/s00167-019-05558-4.

Howell, S. M., Howell, S. J., Kuznik, K. T., Cohen, J., & Hull, M. L. (2013). Does a kinematically aligned total knee arthroplasty restore function without failure regardless of alignment category? *Clinical Orthopaedics and Related Research, 471*(3), 1000−1007. Available from https://doi.org/10.1007/s11999-012-2613-z.

Huang, C.-H., Liau, J.-J., & Cheng, C.-K. (2007). Fixed or mobile-bearing total knee arthroplasty. *Journal of Orthopaedic Surgery and Research, 2*, 1.

Innocenti, B. (2020). High congruency MB insert design: Stabilizing knee joint even with PCL deficiency. *Knee Surgery, Sports Traumatology, Arthroscopy: Official Journal of the ESSKA, 28*(9), 3040−3047. Available from https://doi.org/10.1007/s00167-019-05764-0.

Innocenti, B., Bilgen, Ö. F., Labey, L., van Lenthe, G. H., Sloten, J. V., & Catani, F. (2014). Load sharing and ligament strains in balanced, overstuffed and understuffed UKA. A validated finite element analysis. *The Journal of Arthroplasty, 29*(7), 1491−1498. Available from https://doi.org/10.1016/j.arth.2014.01.020.

Innocenti, B., & Bori, E. (2020). Change in knee biomechanics during squat and walking induced by a modification in TKA size. *Journal of Orthopaedics, 22*, 463−472. Available from https://doi.org/10.1016/j.jor.2020.10.006.

Innocenti, B., Fekete, G., & Pianigiani, S. (2018). Biomechanical analysis of augments in revision total knee arthroplasty. *Journal of Biomechanical Engineering*. Available from https://doi.org/10.1115/1.4040966.

Innocenti, B., Pianigiani, S., Ramundo, G., & Thienpont, E. (2016). Biomechanical effects of different varus and valgus alignments in medial unicompartmental knee arthroplasty. *The Journal of Arthroplasty, 31*(12), 2685−2691. Available from https://doi.org/10.1016/j.arth.2016.07.006.

Innocenti, B., Bori, E., & Paszicsnyek, T. (2021). Functional stability: an experimental knee joint cadaveric study on collateral ligaments tension. *Archives of Orthopaedic and Trauma Surgery*. Available from https://doi.org/10.1007/s00402-021-03966-1. 34046716.

Insall, J. N., Lachiewicz, P. F., & Burstein, A. H. (1982). The posterior stabilized condylar prosthesis: A modification of the total condylar design. Two to four-year clinical experience. *The Journal of Bone and Joint Surgery. American Volume, 64*(9), 1317−1323.

Jiang, C., Liu, Z., Wang, Y., Bian, Y., Feng, B., & Weng, X. (2016). Posterior cruciate ligament retention vs posterior stabilization for total knee arthroplasty: A meta-analysis. *PLoS One, 11*(1), e0147865. Available from https://doi.org/10.1371/journal.pone.0147865.

Kim, K. T., Lee, S., Lee, J. I., & Kim, J. W. (2016). Analysis and treatment of complications after unicompartmental knee arthroplasty. *Knee Surgery & Related Research, 28*(1), 46−54. Available from https://doi.org/10.5792/ksrr.2016.28.1.46.

Klug, A., Gramlich, Y., Rudert, M., Drees, P., Hoffmann, R., Weißenberger, M., & Kutzner, K. P. (2020). The projected volume of primary and revision total knee arthroplasty will place an immense burden on future health care systems over the next 30 years. *Knee Surgery, Sports Traumatology, Arthroscopy : Official Journal of the ESSKA, 29*(10), 3287−3298. Available from https://doi.org/10.1007/s00167-020-06154-7.

Kulkarni, S. K., Freeman, M. A., Poal-Manresa, J. C., Asencio, J. I., & Rodriguez, J. J. (2000). The patellofemoral joint in total knee arthroplasty: Is the design of the trochlea the critical factor? *The Journal of Arthroplasty, 15*(4), 424−429.

Laskin, R. S. (1995). Flexion space configuration in total knee arthroplasty. *The Journal of Arthroplasty, 10*(5), 657−660.

Luo, Z., Zhou, K., Peng, L., Shang, Q., Pei, F., & Zhou, Z. (2020). Similar results with kinematic and mechanical alignment applied in total knee arthroplasty. *Knee Surgery, Sports Traumatology, Arthroscopy: Official Journal of the ESSKA, 28*(6), 1720−1735. Available from https://doi.org/10.1007/s00167-019-05584-2.

Lustig, S., Sappey-Marinier, E., Fary, C., Servien, E., Parratte, S., & Batailler, C. (2021). Personalized alignment in total knee arthroplasty: Current concepts. *SICOT-J, 7*, 19. Available from https://doi.org/10.1051/sicotj/2021021.

Manning, D. W., Chiang, P. P., & Freiberg, A. A. (2005). *Hinge implants* (pp. 219−236). New York: Springer. Available from https://doi.org/10.1007/0-387-27085-X_20.

Miller, A. J., Stimac, J. D., Smith, L. S., Feher, A. W., Yakkanti, M. R., & Malkani, A. L. (2018). Results of cemented vs cementless primary total knee arthroplasty using the same implant design. *The Journal of Arthroplasty, 33*(4), 1089−1093. Available from https://doi.org/10.1016/j.arth.2017.11.048.

Mittal, A., Meshram, P., Kim, W. H., & Kim, T. K. (2020). Unicompartmental knee arthroplasty, an enigma, and the ten enigmas of medial UKA. *Journal of Orthopaedics and Traumatology : Official Journal of the Italian Society of Orthopaedics and Traumatology, 21*(1), 15. Available from https://doi.org/10.1186/s10195-020-00551-x.

Nakayama, K., Matsuda, S., Miura, H., Iwamoto, Y., Higaki, H., & Otsuka, K. (2005). Contact stress at the post-cam mechanism in posterior-stabilised total knee arthroplasty. *The Journal of Bone and Joint Surgery. British Volume, 87*(4), 483−488.

Ng, J. W. G., Bloch, B. V., & James, P. J. (2019). Sagittal radius of curvature, trochlea design and ultracongruent insert in total knee arthroplasty. *EFORT Open Reviews, 4*(8), 519−524. Available from https://doi.org/10.1302/2058-5241.4.180083.

Noble, P. C., Gordon, M. J., Weiss, J. M., Reddix, R. N., Conditt, M. A., & Mathis, K. B. (2005). Does total knee replacement restore normal knee function? *Clinical Orthopaedics and Related Research, 431*, 157−165.

Nowakowski, A. M., Majewski, M., Müller-Gerbl, M., & Valderrabano, V. (2012). Measurement of knee joint gaps without bone resection: "physiologic" extension and flexion gaps in total knee arthroplasty are asymmetric and unequal and anterior and posterior cruciate ligament resections produce different gap changes. *Journal of Orthopaedic Research : Official Publication of the Orthopaedic Research Society, 30*(4), 522−527.

Osmani, F. A., Thakkar, S. C., Collins, K., & Schwarzkopf, R. (2017). The utility of bicruciate-retaining total knee arthroplasty. *Arthroplasty Today, 3*(1), 61−66. Available from https://doi.org/10.1016/j.artd.2016.11.004.

Poirier, N., Graf, P., & Dubrana, F. (2015). Mobile-bearing vs fixed-bearing total knee implants. Results of a series of 100 randomised cases after 9 years follow-up. *Orthopaedics & Traumatology, Surgery & Research : OTSR, 101*(4 Suppl), S187−S192. Available from https://doi.org/10.1016/j.otsr.2015.03.004.

Price, A. J., Rees, J. L., Beard, D., Juszczak, E., Carter, S., White, S., ... Murray, D. W. (2003). A mobile-bearing total knee prosthesis compared with a fixed-bearing prosthesis. A multicentre single-blind randomised controlled trial. *The Journal of Bone and Joint Surgery. British Volume, 85*(1), 62−67.

Qin, J., Chen, D., Xu, Z., Shi, D., Dai, J., & Jiang, Q. (2018). Evaluation of the effect of the sulcus angle and lateral to medial facet ratio of the patellar groove on patella tracking in aging subjects with stable knee joint. *BioMed Research International, 2018*, 4396139. Available from https://doi.org/10.1155/2018/4396139.

Ranawat, C. S. (2002). History of total knee replacement. *Journal of the Southern Orthopaedic Association, 11*(4), 218−226.

Ranawat, C. S., & Shine, J. J. (1973). Duo-condylar total knee arthroplasty. *Clinical Orthopaedics and Related Research, 94*, 185−195.

Refsum, A. M., Nguyen, U. V., Gjertsen, J.-E., Espehaug, B., Fenstad, A. M., Lein, R. K., ... Furnes, O. (2019). Cementing technique for primary knee arthroplasty: A scoping review. *Null, 90*(6), 582−589. Available from https://doi.org/10.1080/17453674.2019.1657333.

Rivière, C., Iranpour, F., Auvinet, E., Howell, S., Vendittoli, P.-A., Cobb, J., & Parratte, S. (2017). Alignment options for total knee arthroplasty: A systematic review. *Orthopaedics & Traumatology, Surgery & Research : OTSR, 103*(7), 1047−1056. Available from https://doi.org/10.1016/j.otsr.2017.07.010.

Rosso, F., Cottino, U., Dettoni, F., Bruzzone, M., Bonasia, D. E., & Rossi, R. (2019). Revision total knee arthroplasty (TKA): Mid-term outcomes and bone loss/quality evaluation and treatment. *Journal of Orthopaedic Surgery and Research, 14*(1), 280. Available from https://doi.org/10.1186/s13018-019-1328-1.

Schwartz, A. M., Farley, K. X., Guild, G. N., & Bradbury, T. L. (2020). Projections and epidemiology of revision hip and knee arthroplasty in the United States to 2030. *The Journal of Arthroplasty, 35*(6S), S79−S85. Available from https://doi.org/10.1016/j.arth.2020.02.030.

Siman, H., Kamath, A. F., Carrillo, N., Harmsen, W. S., Pagnano, M. W., & Sierra, R. J. (2017). Unicompartmental knee arthroplasty vs total knee arthroplasty for medial compartment arthritis in patients older than 75 years: Comparable reoperation, revision, and complication rates. *The Journal of Arthroplasty, 32*(6), 1792−1797. Available from https://doi.org/10.1016/j.arth.2017.01.020.

Small, S., Rogge, R., Malinzak, R., Archer, D., Oja, J., & Berend, M. (2013). *Digital Image Correlation Analysis of Tibial Loading in Rotating Platform Total Knee Arthroplasty* (Poster 1733). Available at: https://www.ors.org/Transactions/59/PS2-099/1733.html

Smith, H., Jan, M., Mahomed, N. N., Davey, J. R., & Gandhi, R. (2011). Meta-analysis and systematic review of clinical outcomes comparing mobile bearing and fixed bearing total knee arthroplasty. *The Journal of Arthroplasty, 26*(8), 1205−1213. Available from https://doi.org/10.1016/j.arth.2010.12.017.

Song, E., Seon, J., Moon, J., & Hyoun, Y. (2013). The evolution of modern total knee prostheses. *Arthroplasty - Update*. Available from https://doi.org/10.5772/54343, IntechOpen.

Stoddard, J. E., Deehan, D. J., Bull, A. M. J., McCaskie, A. W., & Amis, A. A. (2013). The kinematics and stability of single-radius vs multi-radius femoral components related to mid-range instability after TKA. *Journal of Orthopaedic Research: Official Publication of the Orthopaedic Research Society, 31*(1), 53−58. Available from https://doi.org/10.1002/jor.22170.

Suggs, J. F., Li, G., Park, S. E., Sultan, P. G., Rubash, H. E., & Freiberg, A. A. (2006). Knee biomechanics after UKA and its relation to the ACL−A robotic investigation. *Journal of Orthopaedic Research : Official Publication of the Orthopaedic Research Society, 24*(4), 588−594.

Sun, X., & Su, Z. (2018). A meta-analysis of unicompartmental knee arthroplasty revised to total knee arthroplasty vs primary total knee arthroplasty. *Journal of Orthopaedic Surgery and Research, 13*(1), 158. Available from https://doi.org/10.1186/s13018-018-0859-1.

Tigani, D., Comitini, S., Leonetti, D., & Affatato, S. (2015). *14 - Revision total knee arthroplasty (TKA)* (pp. 229−242). Woodhead Publishing. Available from https://doi.org/10.1533/9781782420385.3.229.

Townley, C. O. (1972). The anatomic total knee resurfacing arthroplasty. *Clinical Orthopaedics and Related Research, 192*, 82−96.

Vaishya, R., Chauhan, M., & Vaish, A. (2013). Bone cement. *Journal of Clinical Orthopaedics and Trauma, 4*(4), 157−163. Available from https://doi.org/10.1016/j.jcot.2013.11.005.

Vajapey, S. P., Pettit, R. J., Li, M., Chen, A. F., Spitzer, A. I., & Glassman, A. H. (2020). Risk factors for mid-flexion instability after total knee arthroplasty: A systematic review. *The Journal of Arthroplasty, 35*(10), 3046−3054. Available from https://doi.org/10.1016/j.arth.2020.05.026.

van den Boom, L. G. H., Brouwer, R. W., van den Akker-Scheek, I., Bulstra, S. K., & van Raaij, J. J. A. M. (2009). Retention of the posterior cruciate ligament vs the posterior stabilized design in total knee arthroplasty: A prospective randomized controlled clinical trial. *BMC Musculoskeletal Disorders, 10*, 119. Available from https://doi.org/10.1186/1471-2474-10-119.

Vaninbroukx, M., Labey, L., Innocenti, B., & Bellemans, J. (2009). Cementing the femoral component in total knee arthroplasty: Which technique is the best? *The Knee, 16*(4), 265−268. Available from https://doi.org/10.1016/j.knee.2008.11.015.

Vanlommel, J., Luyckx, J. P., Labey, L., Innocenti, B., De Corte, R., & Bellemans, J. (2011). Cementing the tibial component in total knee arthroplasty: Which technique is the best? *The Journal of Arthroplasty, 26*(3), 492−496. Available from https://doi.org/10.1016/j.arth.2010.01.107.

Walker, P. S., Sussman-Fort, J. M., Yildirim, G., & Boyer, J. (2009). Design features of total knees for achieving normal knee motion characteristics. *The Journal of Arthroplasty*, *24*(3), 475−483. Available from https://doi.org/10.1016/j.arth.2007.11.002.

Yoon, J.-R., Han, S.-B., Jee, M.-K., & Shin, Y.-S. (2017). Comparison of kinematic and mechanical alignment techniques in primary total knee arthroplasty: A meta-analysis. *Medicine*, *96*(39), e8157. Available from https://doi.org/10.1097/MD.0000000000008157.

SPINAL IMPLANTS: BIOMECHANICS AND DESIGN

21

Fabio Galbusera

IRCCS Istituto Ortopedico Galeazzi, Milan, Italy

INSTRUMENTED SPINE SURGERY

Spinal disorders such as spinal instability and spinal tuberculosis have been known for centuries, and external fixators and supports have been in use since the XV century (De Kunder et al., 2018). The introduction of radiography, general anesthesia, and antisepsis in the XIX century allowed the treatment of spinal diseases by performing a surgical stabilization. The first surgical intervention in which a device was implanted dates back to 1891, when a cervical fracture was stabilized by wrapping a silver wire around the spinous processes of C6 and C7. This surgery also opened the history of implant-related complications and revision surgery: The wire underwent postoperative loosening and a reoperation was necessary (De Kunder et al., 2018). Surgical stabilization was also performed in the lumbar spine in the following decades; in 1911, Albee and Hibbs independently treated two cases of Pott's disease by implanting a bone graft using a posterior approach, securing the graft in place by means of sutures (Strömqvist, 1993). The technique for lumbar fusion was then refined by Watkins in 1953 (Watkins, 1953), who implanted autologous bone grafts in contact with decorticated laminae and applied transverse processes for the treatment of degenerative disorders and spinal stenosis, with good clinical success.

Despite negative experiences such as the implantation of steel rods in 1908 by Lange, the idea of using metal instrumentation in combination with bone grafting, aiming at achieving solid primary stability and bony fusion in the months after the operation, was not abandoned; in the 1940s, fixation of the facet joints by means of transarticular screws was reported by several surgeons. In 1962, Harrington introduced the use of instrumentation for the treatment of spinal deformities, which arguably constitutes the most challenging testing ground for spinal implants. Although several complications have been reported, many scoliotic patients who underwent correction by means of Harrington rods are still doing well decades after the surgery (Mariconda, Galasso, Barca & Milano, 2005).

Nowadays, hundreds of innovative implantable devices for the treatment of spinal disorders, including degeneration, deformities, trauma, and tumors, are available on the market; surgeons have a variety of options available, including implants for posterior spinal fixation, interbody fusion, intervertebral disk arthroplasty, and dynamic stabilization. Besides the selection of the instrumentation and of the surgical strategy, the clinical outcome of spine surgery is invariably dependent on the pathology and on the clinical condition of the subject; however, the clinical

Human Orthopaedic Biomechanics. DOI: https://doi.org/10.1016/B978-0-12-824481-4.00023-8

success of instrumented spine surgery can be judged as generally good. Intraoperative, perioperative, and postoperative complications also associated with implants are, however, not infrequent, especially after major surgeries.

PEDICLE SCREW FIXATION

Posterior fixation of the spine by means of pedicle screws and rods is nowadays the cornerstone of spine surgery. Pedicle screws for the fixation of thoracolumbar fractures were introduced in the 1960s by Roy-Camille, and were discussed in a publication a few years later (Roy-Camille, Saillant & Mazel, 1986). Other applications of posterior fixation, such as degenerative disorders, spondylolisthesis, and tumors, soon followed in the following years (Fig. 21.1). Decades of clinical experience demonstrated that pedicle screw fixation is safe and effective for the achievement of primary stability, and for promoting fusion in combination with bone grafts.

Pedicle screws are implanted through a posterior approach and achieve stability and purchase, as the name suggests, by the interference of their thread with the bone tissue in the pedicle. The implantation can be performed both with free-hand techniques and by using guiding technologies such as fluoroscopy and navigation systems. Proper placement was found to be crucial in order to achieve sufficient primary stability, as well as to avoid neurological and vascular lesions; as a matter of fact, pedicle screws implanted free-hand have been found to be relatively frequently misplaced even in experienced hands (Castro et al., 1996). The standard, free-hand implantation technique involves the exposure of the posterior elements, the identification of the entry point, drilling of the pilot hole, checking the pedicle wall with a ball-tipped probe, tapping the whole, and, finally, implanting the screw (Defino, Galbusera & Wilke, 2020). A careful exposure of the entry point is fundamental; in some cases, part of the facet joints is removed in order to allow better visibility of the anatomy. The entry point and the trajectory are variable from level to level (Fig. 21.2); a high level of attention, experience, and knowledge of anatomy are necessary for the achievement of accurate placement. The use of fluoroscopy equipment or more sophisticated navigation devices

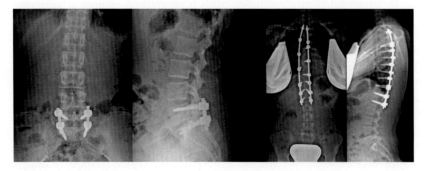

FIGURE 21.1

Two examples of posterior fixation with pedicle screws and rods: for the treatment of symptomatic degenerative disk disease (left) and for the correction of adolescent idiopathic scoliosis (right).

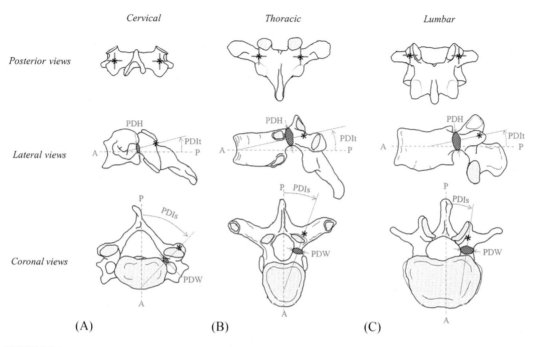

FIGURE 21.2

Entry points and trajectories for pedicle screws in the cervical (A), thoracic (B), and lumbar spine (C). *PDIs/t*: pedicle inclination in the sagittal/transverse plane; *PDW/H*: pedicle width/height; *A*: anterior; *P*: posterior.

From La Barbera, L. (2018). Fixation and fusion. In Biomechanics of the spine *(pp. 301–327). Elsevier. https://www.sciencedirect. com/science/article/pii/B9780128128510000173.*

has been proven to be effective in improving the rate of misplaced screws (Costa et al., 2011); such guiding systems have allowed the development of percutaneous techniques in which the spine is only minimally exposed, which reduces the muscle damage and blood loss, facilitating a shorter and easier recovery.

Whereas the original screws used by Roy-Camille and colleagues were posteriorly connected to plates, most of the current solutions feature tulip-shaped connectors, which allow assembling two rods, most commonly having a circular cross-section and a diameter of 5–7 mm. Rods are usually contoured in order to match the physiological shape of the region to be fixed; they can either be provided already contoured by the manufacturer or be bent in the operating room by the surgeon, allowing for the fitting of any patient-specific shape based on the clinical needs. Based on the design of the tulips, pedicle screws can be classified as monoaxial, i.e., having a rigid connection between the screw shaft and the tulip, and polyaxial, i.e., allowing for a certain degree of rotation of the tulip with respect to the shaft (Fig. 21.3). Polyaxial screws permit an easier alignment and fitting of the rods in the tulips, which are especially useful in the case of complex deformities; on the other hand, monoaxial screws provide higher primary stability and are superior for correcting the vertebral rotation in the transverse plane in deformity surgeries (Defino et al., 2020).

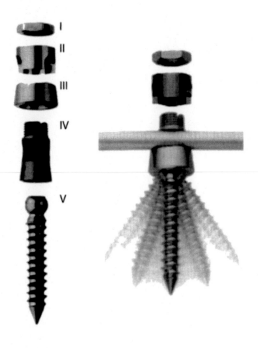

FIGURE 21.3

An example of a polyaxial screw from the Universal Spine System (USS). The assembled screw allows for a rotation of 36 degrees (right). *I*: nut; *II*; collar; *III*: locking ring; *IV*: rod–screw connector; *V*: screw.

From Aebi, M., Thalgott, J. S., Webb, J. K., Aebi, M., Thalgott, J. S., & Webb, J. K. (1998). Modular stabilization system: The universal spine system. In AO ASIF principles in spine surgery (pp. 123–196). Springer Berlin Heidelberg. https://doi.org/10.1007/ 978-3-642-58824-2_8.

Pedicle screws and rods determine major changes in the biomechanics of the levels undergoing fixation. Indeed, the aim of fixation is achieving primary stability, and the range of motion of the implanted levels is therefore severely reduced. In vitro testing revealed that the largest stabilization is achieved in flexion–extension and lateral bending, whereas a certain degree of flexibility is preserved in axial rotation (Cripton, Jain, Wittenberg & Nolte, 2000). Posterior fixation also induces a shift of the center of rotation in flexion–extension toward the spinal canal (Jahng, Kim & Moon, 2013).

In general, pedicle screws are subjected to complex loads, including pull-out, i.e., traction in the axial direction, shear, and bending moments, which can undermine their stability after the implantation (La Barbera, 2018). The primary stability is indeed influenced by several factors such as bone quality, accuracy of the screw placement, diameter of the pilot hole, material and diameter of the posterior rods, screw diameter, length, and thread profile. Pitch, thread depth, and shape were indeed found to be associated with the pull-out strength (Bianco, Aubin, Mac-Thiong, Wagnac & Arnoux, 2016). Primary stability is also affected by the screw trajectory, which is, however, largely dictated by the geometry of pedicles and vertebral body; pedicle screw triangulation, i.e., the use of more convergent trajectories, was found to provide higher resistance to pull-out and better stability of the fixation (Ruland, McAfee, Warden & Cunningham, 1991).

The load-sharing pattern is also strongly altered by posterior fixation. Since pedicle screws and rods determine an increase in the stiffness of the posterior spine but do not affect the anterior structures, posterior fixation tends to induce a posterior shift of the compressive load acting in the spine at the implanted levels. Such an alteration has been demonstrated in vitro; implanted specimens showed decreased intradiscal pressure under physiological loading (Cripton et al., 2000). Whereas this effect can be beneficial in some clinical conditions such as in the treatment of vertebral fractures, the decreased anterior load may induce stress-shielding effects in cases in which the target is an anterior fusion, leading to delayed ossification or pseudoarthrosis.

INTERBODY CAGES

Anterior accesses were not employed in the early days of spine surgery; the first attempts were made in the 1930s after careful study and tests on cadavers (De Kunder et al., 2018). The first anterior spinal surgery was a case of traumatic spondylolisthesis at L5-S1, in which a tibial autograft was implanted through a transperitoneal approach with a good clinical outcome, thus constituting the first case of interbody fusion. Anterior approaches did not, however, gain widespread popularity in the following years due to the technical difficulties and risk of complications. The successive interbody fusion technique, reported in 1944, was indeed performed through a posterior access; Briggs and Milligan described a series of degenerative patients in which a bone graft was posteriorly implanted between adjacent vertebrae (Briggs & Milligan, 1944). The technique was named posterior lumbar interbody fusion and was refined by Cloward in 1953 (Cloward, 1953), but did not become popular in the successive years due to its technically challenging nature, the risk of pseudoarthrosis, and morbidity at the donor site. The interest toward interbody fusion rose again in the 1980s, when an alternative technique with lower surgical invasivity, the transforaminal lumbar interbody fusion, was introduced by Harms and Rolinger (1982).

Interbody cages, i.e., devices to be implanted in between vertebral bodies, providing a mechanical support to bone grafts and chips, were introduced in the 1980s as a means of facilitating bony fusion and limiting complications such as fragmentation of the grafts and pseudoarthrosis. The first cage, a simple perforated metal cylinder filled with bone chips, was developed for veterinary use and was implanted in the cervical spine of horses, with very good outcomes; implantation in the spine of human patients soon followed (Kuslich, Ulstrom, Griffith, Ahern & Dowdle, 1998).

Nowadays, hundreds of interbody cage models are available on the market and are widely used stand-alone or in combination with posterior fixation. Implants designed for several surgical accesses such as posterior, transforaminal, anterior, extraforaminal, and lateral approaches have been developed and are successfully used (Fig. 21.4). As an alternative to the metal construction, several manufacturers employ polyetheretherketone (PEEK), which provides better compatibility with CT and MR imaging and a stiffness more similar to that of the adjacent bone tissue. Cages may also be implanted in combination with bone fusion enhancers such as recombinant human bone morphogenetic protein-2 (rhBMP-2). Clinical studies showed clear superiority of instrumented interbody fusion with respect to interbody fusion with bone grafts only or posterolateral fusion, in terms of preservation of the disk height, reduced incidence of pseudoarthrosis, and improved fusion rates (Eck, Hodges & Humphreys, 2007).

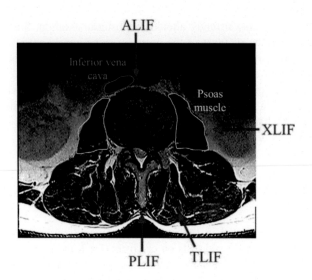

FIGURE 21.4

The surgical accesses for the most common types of interbody cages. *ALIF*: anterior lumbar interbody fusion; *PLIF*: posterior lumbar interbody fusion; *TLIF*: transforaminal lumbar interbody fusion; *XLIF*: extreme lateral lumbar interbody fusion.

From La Barbera, L. (2018). Fixation and fusion. In Biomechanics of the spine (pp. 301–327). Elsevier. https://www.sciencedirect. com/science/article/pii/B9780128128510000173.

Numerous in vitro studies have been conducted to investigate the biomechanics of thoracolumbar interbody fusion. Several papers highlighted the low stabilizing potential of stand-alone interbody cages, especially in extension (Schmoelz & Keiler, 2015), whereas the facet joints, the posterior tension band, and the cage itself may avoid an iatrogenic destabilization in flexion, lateral bending, and axial rotation, and the lack of anterior structures that maintain tension, except for portions of the intervertebral disk and the anterior longitudinal ligament whenever preserved, may lead to instability and separation at the implant—bone interface in extension. Anterior lumbar interbody cages, including locking screws or other anchoring systems, however, suffer from such a limitation to a less extent (Chen, Chiang, Lin, Lin & Hung, 2013), although the effectiveness of the anchoring structure specifically depends on the applied technology (Nagaraja & Palepu, 2017). Lateral cages such as XLIF (extreme lumbar interbody fusion) also do not show instability in extension, due to the largely preserved anterior intervertebral disk (Fogel, Parikh, Ryu & Turner, 2014).

Although biomechanical studies and several clinical papers showed that stand-alone interbody cages are safe and effective, this type of fixation is most commonly performed in combination with pedicle screws and rods to facilitate the so-called circumferential fusion (Fig. 21.5), which supposedly prevents the risk of destabilization. The scientific literature seems to prove that posterior stabilization is effective in enhancing the stiffness of segments subjected to interbody fusion (Rapoff, Ghanayem & Zdeblick, 1997).

A complication that is commonly associated with instrumented interbody fusion is subsidence, i.e., the dislocation of the cage into the adjacent vertebra, most commonly the caudal one. Factors

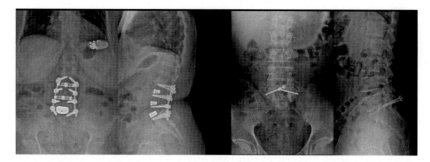

FIGURE 21.5

Two examples of interbody fixation with cages for the treatment of a degenerative disorder. Left: three-level fixation with large lateral interbody cages at L3-L4 and L4-L5, and with an anterior cage at L5-S1, supplemented with pedicle screws and rods. Right: interbody fixation with a PEEK cage supplemented by transarticular screws.

associated with the risk of subsidence are bone quality, the footprint of the cage on the vertebral endplates, the presence of keels, threads or spikes, which may provoke the local failure of bone but also stimulate remodeling and deposition of new tissue, and the surgical preparation of the endplate (La Barbera, 2018).

CERVICAL FIXATION

Besides thoracolumbar fixation, the implantation of posterior instrumentation such as pedicle screws and rods in the lower cervical spine is common and successful. In addition to pedicles, the prominent lateral masses also offer solid anchoring sites for fixation devices; lateral mass and transarticular screws are indeed frequently implanted, especially concomitant with decompression surgeries in which laminae and spinous processes are resected. Several techniques of lateral mass screw placement have been described, with each having its unique entry point and screw trajectory. The implantation of these screws requires careful planning and technical execution due to the risk of violation of nerve roots and vertebral arteries (Xu, Haman, Ebraheim & Yeasting, 1999).

The anatomy of the neck, however, also offers a safe and easy anterior access to the spine (Fig. 21.6). Anterior plates and interbody cages are more frequently employed in cervical surgery with respect to the thoracolumbar region, in which the ventral access is technically demanding and associated with a relatively high risk of complications. Anterior cervical discectomy and fusion (ACDF), i.e., the removal of the intervertebral disk through the anterior access and the implantation of a bone graft possibly in combination with a cage or a plate, has become the solution of choice for the treatment of cervical disk herniation, trauma, instability, radiculopathy, and myelopathy (Rhee, Park, Yang & Riew, 2005).

Stand-alone anterior cervical cages revealed the same limitations of anterior lumbar interbody fusion, i.e., possible instability in extension, which can be mitigated by locking screws and anchoring systems (Cunningham, Hu, Zorn & McAfee, 2010). An alternative technique is the combination of an anterior cage with an anterior plate, which effectively prevents its dislocation and provides a

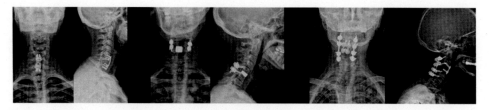

FIGURE 21.6

Three examples of anterior cervical discectomy and fusion (ACDF) achieved with different instrumentation strategies. Left: a two-level fixation with anterior cages and an anterior plate fixed with anchoring screws to the vertebral bodies. Middle: one-level instrumentation with an anterior cage and posterior fixation. Right: one-level anterior fixation with a PEEK cage and a plate secured with four screws, which was later supplemented by posterior instrumentation of the adjacent levels for the treatment of their early degeneration. *PEEK*: polyetheretherketone.

higher degree of stability (Stein et al., 2014) (Fig. 21.6). Such a solution is, however, associated with peculiar postoperative complications determined by the contact of the plate with the surrounding soft tissues; neck pain, hoarseness, and especially dysphagia have been frequently reported (Shao et al., 2015). Zero-profile plates, which aim at offering a thin and smooth anterior surface blending with the bony surface, seem to be effective in limiting the incidence of these complications.

Due to its peculiar anatomy, the surgical fixation of the upper cervical spine is performed by means of specific techniques. One of the most frequent indications for surgery is the traumatic fracture of the odontoid process of C2, which can be treated by the anterior or transoral implantation of one or two screws (Böhler, 1982). Posterior fixation can also be performed in the case of nonunion or when the implantation of the odontoid screws is unsafe or not feasible. The earliest posterior technique for C1-C2 fixation involved the implantation of a bone graft between the spinous process of the axis and the posterior arch of the atlas and its anchoring with a wire cerclage; this technique was then improved by additional fixation by means of transarticular screws (Magerl, Seemann, Kehr & Weidner, 1987). The current approaches, in addition to the anterior fixation with one or two screws, which remains the golden standard whenever feasible, include the use of rods secured to C1 and C2 by means of pedicle screws, lateral mass screws, or claws. Posterior fixation is also used in the case of trauma or instability of the occipitocervical joint and is achieved by means of dedicated plate, screw, and rod systems commonly involving the entire cervical spine. Biomechanical studies showed a substantial equivalence between the different technical solutions (Richter et al., 2000).

INSTRUMENTATION FOR DEFORMITY CORRECTION

With the exception of a few unsuccessful attempts in the first half of the XX century, sometimes with catastrophic results, spinal deformities have been treated with plaster casts, braces, and devices able to apply traction and posture correction, with limited clinical success, long treatment times, and patients' discomfort (Mohan & Das, 2003). In the 1940s and 1950s, repeated waves of poliomyelitis epidemics increased the incidence of severe kyphoscoliotic deformities, especially in

North America; such patients were the first to be treated with modern spinal instrumentation by Harrington. After attempting with transarticular screws, Harrington focused on the use of rods fitted with hooks able to be anchored to the laminae, the so-called laminar or sublaminar hooks (Harrington, 1962). The indications of Harrington rods soon expanded to the treatment of idiopathic scoliosis, neuromuscular deformities, fractures, and spondylolisthesis, with evident clinical success in the short and intermediate terms (Mohan & Das, 2003). The most common technique for the treatment of scoliosis with Harrington instrumentation involved the implantation of a rod on the concave side of the scoliotic curve applying a traction force, and therefore reducing the deformity, at the upper and lower limits of the curve by means of the hooks. In some cases, a compression rod with hooks fitted in the opposite direction in order to apply a compressive force was implanted on the convex side. After the surgery, the patients had to wear plaster casts and braces until bony fusion was achieved.

Long-term studies highlighted the limitations of Harrington rods, and triggered further advances in implant and surgical techniques. As a matter of fact, many treated patients developed the so-called flat-back syndrome due to the fact that the rods were not able to restore a correct sagittal alignment of the spine, which is associated with back pain, discomfort and functional loss. The subsequent techniques indeed focused on including more anchoring sites, which would allow for the use of contoured instrumentation replicating a correct sagittal profile. In the 1970s, Luque introduced contoured rods fixed to multiple anchoring sites by means of sublaminar wires (Luque, 1982) (Fig. 21.7), with good clinical results but also frequent neurological complications.

The introduction of pedicle screws, able to provide a solid anchoring point to which mechanical loads in any direction could be safely applied, opened the way to a long series of innovations, which is still ongoing. Indeed, the Cotrel−Dubousset (CD) fixation system introduced in the early 1980s can still be considered the foundation of all deformity implants currently available on the market. The CD instrumentation system was based on stainless steel rods, which could be contoured by the surgeon in order to fit the target shape of the spine, both in the coronal plane and in the sagittal plane. Anchoring implants such as laminar and pedicle hooks, as well as pedicle screws, could be fitted and secured in any position of the rod (Lenke, Bridwell, Baldus, Blanke & Schoenecker, 1992). The CD system allowed for employing the derotation technique, in which a three-dimensional correction is achieved by converting the coronal curvature into a sagittal one; in this way, a restoration of the physiological sagittal alignment of the spine could be obtained, as well as a reduction of the rib deformity by decreasing the rotation in the transverse plane of the thoracic vertebrae. The Universal Spine System (USS) employed a different principle to achieve a similar three-dimensional correction (Laxer & Others, 1994). With this system, a contoured rod is first solidly fixed to the cranial and caudal extremities of the fusion region, then a forceps is used to fit other anchoring devices located in between the extremities to the rods themselves. Similar to the CD, the USS system allows for the use of both hooks and pedicle screws. The posterior instrumentation systems currently available on the market share many similarities with the CD and the USS systems, although they also allow for newer correction techniques such as the direct vertebral rotation, which aims at enhancing the three-dimensional correction by reducing the local rotation in the transverse plane at the apex of the scoliotic curve (Lee, Suk & Chung, 2004). Current surgeries tend to favor the use of pedicle screws instead of hooks in the thoracic spine, due to their capability of supporting higher mechanical loads and thus allowing for a greater correction (Mohan & Das, 2003).

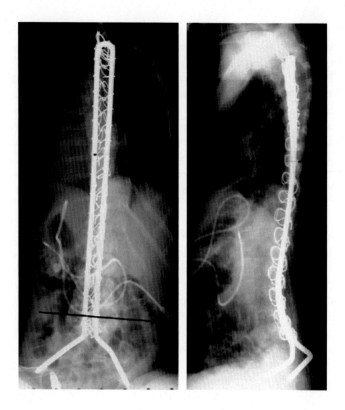

FIGURE 21.7

The correction of a case of neuromuscular scoliosis by means of the Luque—Galveston technique, including pelvic fixation and sublaminar wires.

From Tsirikos, A. I., & Spielmann, P. (2007). Spinal deformity in paediatric patients with cerebral palsy. Current Orthopaedics, 21 (2), 122—134. https://doi.org/10.1016/j.cuor.2007.01.001.

In parallel to methods and technologies for posterior instrumentation, anterior techniques for deformity correction have been developed and are still in widespread use (Dwyer, Newton & Sherwood, 1969; Kaneda, Shono, Satoh & Abumi, 1996), especially in the German-speaking countries. Anterior techniques are based on one or two bicortical screws in the vertebral bodies, solidly connected to each other by means of plates and cables. The main advantage of anterior instrumentation is the shorter fusion length necessary to achieve the correction of the deformity; implant loosening and failure are, however, not uncommon (Kaneda, et al., 1996). Recent studies have introduced a novel technique named anterior vertebral body tethering for the treatment of adolescent idiopathic scoliosis in skeletally immature patients, which employs lateral bicortical screws in the vertebral bodies on the convex side of the curvature connected by a flexible cable. The technique is based on the concept of growth modulation; the growing spine creates a tension in the cable, which applies a progressively increasing correction force to the scoliotic curve while allowing for motion preservation and further growth (Samdani et al., 2015). Early clinical results are promising, but studies also reported high revision rates (Newton et al., 2018). Adult deformities commonly involve the loss of lumbar lordosis and a progressive spinal ankylosis; specific techniques have been introduced in order to achieve the restoration of a physiological sagittal balance in such degenerative scenarios. Adult deformity correction surgeries frequently involve osteotomies, i.e., the removal of pieces of bone aiming at gaining flexibility to facilitate the correction of the deformity. Osteotomies were introduced in 1922, when vertebral column resection was first performed for the treatment of severe scoliosis (Meredith & Vaccaro, 2014). The Smith—Petersen osteotomy, consisting of the resection of facet joints or of the fusion mass in the case of ankylosis, was first described in 1945 (Smith-Petersen, Larson & Aufranc, 1945), and is still used for the restoration of lumbar lordosis. However, the technique most commonly employed for the correction of severe adult sagittal deformities is pedicle subtraction osteotomy, which was introduced in the 1980s (Thomasen, 1985), in which the posterior elements, the pedicles, and a wedge of vertebral body are resected in order to gain a local increase of the lordotic curvature up to 30 degrees (Fig. 21.8). Due to the removal of large quantities of bone and a strong reduction of the remaining cross-sectional area of the spine, this procedure involves a significant iatrogenic destabilization requiring concomitant posterior fixation. After studies reported frequent mechanical failures of the instrumentation due to the excessive loading associated with the destabilization, stronger fixation techniques involving accessory rods in addition to the primary ones have been introduced (Luca et al., 2017). With such techniques appear to be able to reduce the risk of implant failures, research in this field is underway at a rapid pace and further advances are expected in the near future.

SACROPELVIC FIXATION

As mentioned in the previous paragraphs, long thoracolumbar fixation is frequently performed for the treatment of several spinal disorders such as adolescent idiopathic scoliosis, sagittal imbalance in adult patients, as well as adult scoliosis; the length of the fusion zone is determined in the preoperative planning phase based on the radiological characteristics of the deformity and on the stiffness of the deformed regions. In some cases, the fusion length involves the lumbosacral junction, i.e., the level L5-S1, the successful fixation of which has, however, proven to be challenging to achieve. However, studies reported high rates of nonunion, pseudoarthrosis (Lee et al., 2004), and mechanical failure of the implants as well as their loosening, especially of S1 pedicle screws and of the posterior rods in the L5-S1 region (Nguyen et al., 2019). Other studies reported painful early degeneration of the sacroiliac joints due to overloading and high motion (Ha, Lee & Kim, 2008).

Beginning in the 1980s, surgeons attempted to include the pelvis in the posterior instrumentation in order to improve the stability of the lumbosacral junctions and reduce complications. The first spinopelvic fixation technique was the Galveston Iliac Rod introduced by Allen and Ferguson, an L-shaped rod inserted into the ilium through the posterior superior iliac spine (Allen & Ferguson, 1984) (Fig. 21.7). Although this technique allowed reducing the rates of nonunions at L5-S1, painful loosening of the rod extremity inserted in the ilium was not infrequent (Lombardi, Shillingford, Lenke & Lehman, 2018).

Nowadays, two techniques for pelvic fixation in combination with thoracolumbar instrumentation are commonly in use (Fig. 21.9). Iliac screws are inserted in the ilium and solidly linked to the

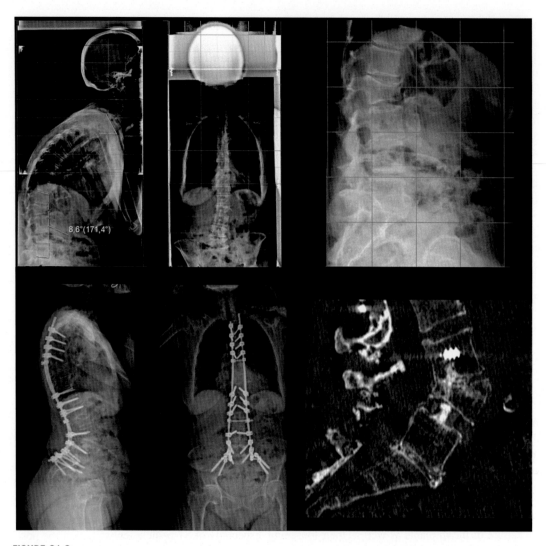

FIGURE 21.8

Pedicle subtraction osteotomy (PSO) for the correction of the loss of lumbar lordosis. First row: preoperative radiographs of the subject, showing a flat lumbar spine as well as the increase of the thoracic kyphosis. Second row: the same patient after correction by means of PSO at L4 and long instrumentation of the whole thoracolumbar spine, including sacropelvic fixation, in which the improvement in the sagittal balance is evident.

From Ottardi, C., Luca, A., Galbusera, F., Galbusera, F., & Wilke, H.-J. (2018). Chapter 21 — Sagittal imbalance. In Biomechanics of the spine (pp. 379–391). Academic Press. https://doi.org/10.1016/B978-0-12-812851-0.00021-5.

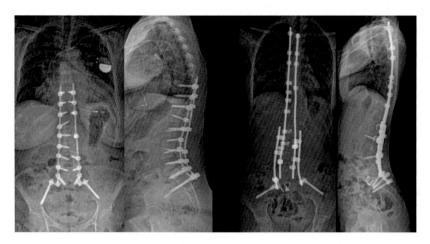

FIGURE 21.9

The most common instrumentation techniques for pelvic fixation: iliac screws (left) and S2 alar-iliac screws (right).

posterior rods by means of metal connectors; they provide a more solid fixation with respect to Galveston rods since they possess a thread and can be inserted in more advantageous locations, allowing for a better purchase. In patients in which the risk of pseudoarthrosis and implant failure is supposedly higher, constructs including multiple iliac screws can provide a biomechanical advantage at the cost of dealing with high technicalities and increased risk of infection (Shen, Harper, Foster, Marks & Arlet, 2006). Despite their generally good clinical success, iliac screws have been found to be associated with a nonnegligible risk of infection due to the large soft tissue dissection necessary for their implantation, and with the risk of violating the superior gluteal artery and sciatic nerve (Lombardi et al., 2018).

S2 alar-iliac (S2AI) screws have been introduced in 2007 and achieve a solid sacropelvic fixation by passing through the sacroiliac joints; the optimal entry point and trajectory have been described in detail (Kebaish, 2010). With respect to iliac screws, they require a lower degree of soft tissue dissection and are less prominent, reducing the risk of patient discomfort; furthermore, they do not require the use of connectors but can be linked directly to the posterior rods.

A new frontier for sacropelvic fixation in combination with long thoracolumbar instrumentation has recently been opened by the use of minimally invasive implants, which are commonly used for the treatment of dysfunction of the sacroiliac joint (Duhon et al., 2016). Recent studies hypothesized that such implants can supplement or replace iliac and S2AI screws, enhancing the fixation and limiting the risk of perioperative and postoperative complications (Galbusera et al., 2020).

ARTIFICIAL DISKS

Fixation and fusion is the treatment of choice for patients suffering from symptomatic spinal degeneration requiring surgical treatment, and is generally believed to be associated with good clinical results. Nevertheless, a recent large-scale study indicated that degenerative spine surgery delivers significantly poorer patient-reported outcomes if compared with hip and knee joint replacement (Mannion et al., 2018); persistent pain and low patient satisfaction are also not infrequent. In addition, complications such as pseudoarthrosis and chronic pain at the donor site in the case of autologous bone grafting have been reported after at least 10% of the surgeries (Turner et al., 1992).

Early degeneration of intervertebral disks, facet joints, and ligaments adjacent to the fixation, supposedly determined by the overload and increased motion necessary to compensate for the loss of mobility, has also been advocated in several papers as a frequent adverse effect of fixation and fusion (Lee et al., 2009). Although this issue has been highlighted and investigated in several papers, it remains controversial; several authors hypothesized that the progression of degeneration observed after the fixation could be independent of the surgery itself, and the hypermobility at the adjacent segments has never been clearly demonstrated (Malakoutian et al., 2015). Nevertheless, there is consensus about the fact that adjacent segment degeneration exists to some extent, although its understanding from a basic science point of view is poor.

In the 1960s, the concept of replacing a degenerated intervertebral disk with an implant preserving physiological mobility stimulated pioneering research in the field. Two notable attempts are the implantation of a testicular prosthesis inside a lumbar intervertebral disk by Nachemson (1962) and of steel balls by Fernström (1966). Both devices had, however, no clinical success; whereas the testicular prostheses underwent fragmentation due to the high loads, the steel balls frequently subsided into the adjacent vertebral bodies. The first successful mechanical intervertebral disk, the Charité Artificial Disk (DePuy Spine, Inc., Raynham, MS, USA), which was released in the 1980s and had been on the market until 2012, was designed to be implanted in the lumbar spine through an anterior approach and consisted of two titanium endplates articulating with a polymeric insert. This three-component design was described as unconstrained since it did not impose any constraints on the degrees of freedom of the motion segment nor to the axes of rotation. Currently, two artificial disks, also known as lumbar total disk replacements (TDRs), are approved by the US Food and Drug Administration (FDA) for use in the United States: ProDisc-L Total Disk Replacement (Synthes Spine, Inc., West Chester, PA, USA) and activL Artificial Disk (Aesculap Implant Systems, LLC., Center Valley, PA, USA). Several non-FDA-approved implants are available in other countries, including the European Union. The current indications of lumbar TDR include degenerative disk disease, spondylolisthesis, stenosis, degenerative scoliosis, segmental instability, discogenic pain, and failed back surgery (Büttner-Janz, Guyer & Ohnmeiss, 2014).

Current lumbar TDRs impose different degrees of kinematic constraint to the operated segment, based on their design (Fig. 21.10). As mentioned above, unconstrained designs such as Charité feature three distinct components and do not restrain any degree of freedom. The ball-in-socket semi-constrained design (such as ProDisc-L) guides the axis of rotation to pass through the center of its spherical articulating surfaces. The stiffness and allowed range of motion are therefore the same in flexion—extension and lateral bending, whereas axial rotation is not limited. Other designs, such as the ActivL artificial disk, possess an additional degree of freedom, the anteroposterior shear translation, which is achieved by allowing a translational motion of the polymeric insert. Nevertheless, all these design concepts are radically different from the anatomy and physiology of the intervertebral disk and aim only at replicating its mobility from a global point of view. In vitro tests revealed ranges of motion substantially higher than those of physiological motion segments (Wilke et al., 2012), although in vivo observations based on medical imaging could not confirm such findings

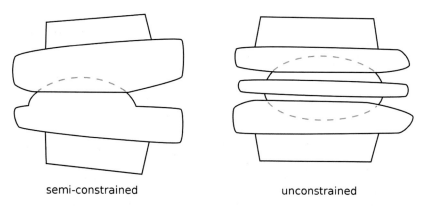

semi-constrained   unconstrained

FIGURE 21.10

Schematic representation of the two most common designs for artificial disks: semiconstrained (left) and unconstrained (right).

From Galbusera, F., Wilke, H.-J., Galbusera, F., & Wilke, H.-J. (2018). Chapter 18 - Motion preservation. In Biomechanics of the spine *(pp. 329–342). Academic Press. https://doi.org/10.1016/B978-0-12-812851-0.00018-5.*

(Guyer et al., 2016). Experimental and clinical studies, however, agree in describing an increase of the segmental lordosis (Cakir, Richter, Käfer, Puhl & Schmidt, 2005), the long-term consequences of which remain unclear.

In recent years, artificial disks going beyond the concept of mechanical articulating surfaces and aiming at replicating the deformable structure of the native intervertebral disk have been introduced in the market, with contrasting, sometimes catastrophic clinical results. Acroflex (AcroMed Corp., DePuy Spine Inc., Raynham, MA, USA) employed a polyolefin rubber core, which was able to mimic the mechanical behavior of the disk, including its shock absorption capability and its stiffness. Although the device was tested for biological and biomechanical compatibility prior to clinical trials, fragmentation and fatigue failure occurred frequently after the implantation, resulting in its withdrawal from the market. More recent Cadisc-L (Ranier Technology, Cambridge, UK) had a sophisticated core made of polyurethane-polycarbonate with a functionally graded stiffness from nucleus to annulus. Although preclinical biomechanical testing was favorable as the artificial disk showed a behavior very similar to that of the natural disk (McNally, Naylor & Johnson, 2012), several implants underwent mechanical failure under implantation, again resulting in market withdrawal and legal investigations. The only TDR currently available on the market that takes inspiration from the real anatomy and physiology of the intervertebral disk is the M6-L Artificial Lumbar Disk (Spinal Kinetics, Sunnyvale, CA, USA), which features a deformable polymeric core surrounded by a fiber jacket resembling the annulus fibrosus, enclosed in a polymer sheath and two keeled titanium endplates for bone fixation.

Artificial disks have also been implanted in the cervical spine since the first days of motion preservation surgery. Fernström also inserted steel balls in cervical disks in 1966, resulting in a high rate of subsidence (Fernström, 1966). The first mechanical cervical disk prosthesis that gained a certain degree of clinical success was the Cummins–Bristol disk, a prototype introduced in 1989 consisting of two articulating metal plates fixed to the anterior aspect of the vertebral bodies by means of anchoring screws. The TDR was tested on 18 subjects with mixed results (Baaj, Uribe, Vale, Preul & Crawford, 2009; Cummins, Robertson & Gill, 1998). The prototype was then refined, and finally found its way to the market with the name PRESTIGE Cervical Disk System (Medtronic Disk, Minneapolis, MN, USA). In the following years, the device was implanted in thousands of patients with satisfactory clinical results and gained the FDA marketing approval in the United States. In 2020, seven cervical TDRs approved by the FDA are available on the US market, whereas several other devices are commercialized in Europe. Cervical artificial disks can, in general, be classified, similar to their lumbar counterparts, into semiconstrained and unconstrained (Fig. 21.10); a cervical version of the biomimetic M6-L, the M6-C Artificial Cervical Disk (Spinal Kinetics, Sunnyvale, CA, USA), is also available on the market and is approved for use in the United States by the FDA. The semiconstrained ProDisc-C Total Disk Replacement (Synthes Spine, Inc., West Chester, PA, USA) is also available in a cervical version, and shares many characteristics in terms of design and materials with the lumbar one. In contrast, the PCM Cervical Disk System (NuVasive Manufacturing LLC, Fairborn, OH, USA) has been designed to mimic specifically the biomechanics of a cervical motion segment, and employs a saddle-like design replicating the shape of the cervical vertebral endplates, which aims at guiding the center of rotation in its physiological location. In vitro testing confirmed its effectiveness in achieving a physiological motion pattern in flexion−extension (Cunningham et al., 2010). In general, cervical TDR appears to be a safe and effective alternative to ACDF in select patients, effective in reducing adjacent segment degeneration and the need for revision surgeries (Buckland, Baker, Roach & Spivak, 2016). The complications most commonly reported include migration, subsidence, and, especially, heterotopic ossification, which has, however, marginal clinical consequences in many patients (Murtagh & Castellvi, 2014).

DYNAMIC STABILIZATION AND OTHER MOTION-PRESERVING IMPLANTS

Besides artificial disks, other implants aimed at stabilizing lumbar motion segments while preserving a certain flexibility have been presented and are currently available on the market. The concept of the so-called dynamic stabilization arose from the hypothesis of hypermobility in the early stages of disk degeneration, as described by Kirkaldy-Willis and Farfan (1982); flexible implants aimed at restoring a physiological flexibility in this hypermobility stage could supposedly help in reducing pain and discomfort, and possibly in slowing down the progression of the degenerative cascade. Dynamic stabilization may also provide advantages with respect to fixation and fusion in terms of early degeneration of the adjacent segments, due to the lower loss of mobility and therefore supposedly smaller overloading of the neighboring disks. Currently, indications for dynamic stabilization of the lumbar spine include painful moderate disk degeneration, mild osteoarthritis of the facet joints, low-grade spondylolisthesis, and spinal stenosis (Prud'homme et al., 2015). In the case of spinal deformities, tumors, and fractures, dynamic stabilization is not indicated. Although the clinical outcome in select patients is generally good, its superiority to fixation and fusion has not been clearly proven (Prud'homme et al., 2015).

Dynamic stabilization was pioneered in the early 1990s by the Graf ligament, a couple of tension bands fixed to specially designed pedicle screws, which was used in patients suffering from chronic low back pain attributed to disk degeneration, in most cases in association with prophylactic foraminal decompression (Gomleksiz, Sasani, Oktenoglu & Ozer, 2012; Grevitt et al., 1995). Indeed, since the device was able to limit the flexibility in flexion but had no effect in extension, foraminal stenosis and radiculopathy were frequently observed in subjects in which decompression was not performed. To overcome such limitations, the successive implantation of the Dynesys Dynamic Neutralization System(Zimmer Biomet, Warsaw, IN, USA) integrates a polymeric spacer active in limiting the flexibility in extension, in addition to a cord reducing flexion. In vitro testing demonstrated that this implant was able to reduce the spinal flexibility by at least 50%, and to restore a stiffness close to the physiological levels in artificially destabilized specimens (Schmoelz et al., 2003). Clinical studies reported good outcomes, the most frequent complication being screw loosening (Kocak, Cakir, Reichel & Mattes, 2010). Several other implants are available on the market nowadays, either based on full-metal flexible designs or based on the use of polymeric components.

In addition to pedicle-anchored dynamic stabilization systems, other motion-preserving designs not replacing the whole intervertebral disk have been introduced with various degrees of clinical success. Nucleus replacement based on polymeric implants or injectable materials, which could be implanted mini-invasively, has been used to treat symptomatic disk degeneration that could not be reduced by conservative treatment, especially in early degenerated disks with an intact annulus fibrosus and facet joints in good conditions. Such implants are still employed in a relatively low number of cases due to the restrictive indications, but may find a larger use in the future in combination with regeneration therapies now under development and preclinical testing.

Interspinous spacers aim at maintaining a minimum distance between adjacent spinous processes and at reducing the range of motion in extensions, and are used in the case of spinal stenosis alone or in combination with surgical decompression. These implants became very popular in the 2000s due to the possibility of minimally invasive implantation and the immediate pain relief in several patients; their use, however, decreased in recent years due to high reoperation rates associated with device migration, fracture of the spinous processes, and recurrent radiculopathy (Bowers, Amini, Dailey & Schmidt, 2010). Total facet joint replacement systems are available as well, and indicated in the case of severe facet osteoarthritis alone or in combination with stenosis or degenerative spondylolisthesis; although the small amount of data available are encouraging (Anekstein et al., 2015), longer clinical studies are necessary to foster a wider adoption of this technology.

FATIGUE FAILURE AND LOOSENING OF SPINAL IMPLANTS

As described in the previous paragraphs, postoperative mechanical complications such as implant breakage, disassembly, and loosening are relatively frequent in spine surgery, especially after major interventions involving iatrogenic destabilization such as osteotomies; they are, however, not infrequent in smaller surgeries in which such a catastrophic outcome could not be expected. Whereas implant breakage and coupling failures require a revision surgery to replace the broken part and in some cases to elongate the fixation length, implant loosening has frequently limited consequences on the clinical outcome and can even go unnoticed in mild cases if not actively searched (Galbusera et al., 2015). However, reoperation after symptomatic implant loosening is also relatively common.

Implant breakage is most commonly due to fatigue failure, and induced by repetitive loads arising from daily activities such as walking, running, lifting, and carrying loads as well as doing sports. In addition to high cyclic loads, risk factors for implant breakage include a low stability of the anatomical structures, either due to pathological changes or due to surgical resections; the implant size, especially the cross-sectional area of the parts subjected to high stress; the presence of stress concentrators, such as material defects, connections with other devices, decreases in cross-section, threads, and notches; the surface finish and coating; the fatigue properties of the material; the surgical configuration (e.g., the choice of using accessory rods or not in osteotomies) (Luca et al., 2017); the success of the surgery, for example, the achievement of a bony fusion rather than the occurrence of pseudoarthrosis; and possible damages during the implantation, such as notches created while manually contouring posterior rods to match the target curvature of the spine (Nguyen et al., 2019) or due to electrocautery (Almansour et al., 2019). Design choices are often associated with compromises, which may impact the fatigue life of the device; for example, reducing the thickness of a cervical plate in order to decrease the risk of dysphagia may induce a higher risk of fatigue fracture.

In the case of spinal fixation, the achievement of an osseous fusion in the months after the surgery is a key risk factor for the fatigue life of the implants. If fusion is successfully obtained, the load acting on the bony mass would progressively increase over time, decreasing the stresses in the devices and, consequently, reducing the chances of failure. Conversely, the development of a pseudoarthrosis or nonunion would only minimally affect the implant loads and stresses, increasing the risk of fatigue failure (Pihlajamäki, Myllynen & Böstman, 1997).

Assessing the fatigue life is a key issue in the marketing approval of any spinal implant by the relevant regulatory bodies, which involves the verification of the mechanical safety of the device by means of preclinical testing, in addition to the requirements of patient safety and effectiveness assessed with clinical trials (Kienle et al., 2018). Such preclinical tests are conducted in laboratory environments using material testing systems, following national or international standards defined by organizations such as the American Society for Testing and Materials (ASTM) and the International Organization for Standardization (ISO). Standard tests for spinal implants favor the aspects of repeatability and reproducibility over the application of realistic loads and motions, which are complex and challenging to replicate with standard laboratory equipment.

Two notable standard tests commonly used to assess the fatigue life of spinal fixation systems are the ASTM F1717 and the ISO 12189 (Fig. 21.11). The two standards mimic a two-level fixation and share many similarities in terms of the geometry of the polymeric blocks to be used as vertebrae surrogates, and the type, direction, and point of application of the applied load, as well the boundary conditions. However, the standards differ in a few key aspects. First, the ASTM F1717 simulates a vertebrectomy model with no anterior support, which maximizes the stresses in the implants and acts as a "worst case scenario," whereas in the ISO 12189, calibrated springs are used to replicate the stiffness and load-sharing contribution of healthy intervertebral disks. Second, whereas in the ASTM F1717 there is no predefined value for the applied load and the maximum safe load, which is determined by the user, the ISO 12189 prescribes a fixed load amplitude. Such differences prevent a direct comparability of the results of the two tests, and demonstrate the distinct viewpoints of ASTM and ISO with respect to determining implant safety. In fact, the ASTM generally prefers not to provide acceptance criteria to judge the success of a test, leaving this responsibility to the user and regulatory bodies, while the ISO directly provides such success criteria (Graham et al., 2014).

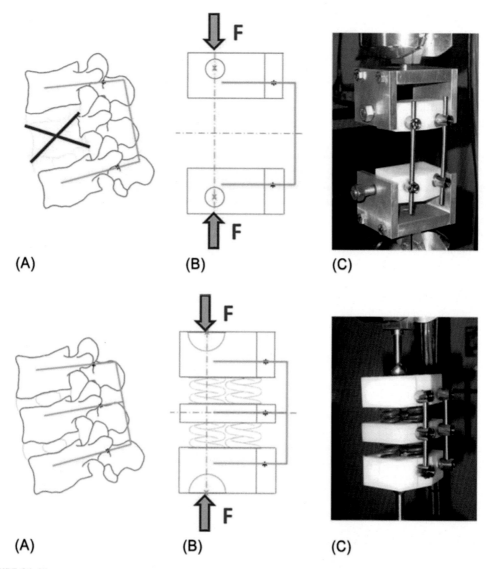

FIGURE 21.11

Schematic representation of the ASTM F1717 and ISO 12189 standards to assess the fatigue life of posterior instrumentation systems: (A) anatomical sketch, (B) sketch, and (C) photographs of the loading set-up.

From Kienle, A., Graf, N., Villa, T., La Barbera, L., Galbusera, F., & Wilke, H.-J. (2018). Chapter 13 - Standard testing. In Biomechanics of the spine (pp. 223–238). Academic Press. https://doi.org/10.1016/B978-0-12-812851-0.00013-6.

Implant loosening is diagnosed on planar radiographs or CT scans by assessing the existence of a radiolucent halo around the edges of the device, which is not observable in implants solidly connected to bone, or of the so-called "double halo" in which the radiolucent zone is contained in a visible, dense bony rim. Studies showed that radiolucency and double halos in pedicle screws are associated with lower extraction torques and pull-out strength in comparison with osseointegrated implants (Sanden, Olerud, Petren-Mallmin, Johansson & Larsson, 2004). A literature review showed that pedicle screw loosening is relatively infrequent in healthy, nonelderly subjects, whereas it has higher prevalence in osteoporotic bone, although clinical data supporting such statements are not abundant (Galbusera et al., 2015). Implants specially designed to reduce the risk of loosening in patients at risk by augmenting expandable screws and fenestrated screws with the injection of bone cement appear to be effective, but limited data are available. Pedicle-anchored dynamic stabilization systems, as mentioned in the previous paragraphs, showed a higher risk of loosening due to the loads sustained by the posterior instrumentation and by the bone—implant interface, which do not decrease in the months after the fixation since fusion is not the target of the surgery (Fig. 21.12).

Any standard tests defined by entities such as ASTM and ISO for the investigation of the loosening behavior of spinal implants do not currently exist. Indeed, the ASTM F543 standard addresses the mobilization of a metallic bone screw in a polymeric foam acting as a bone surrogate, but the applied load, a purely axial pull-out force, differs substantially from the spinal loads acting *in vivo*, which determine the loosening of the implants. As a matter of fact, most experimental

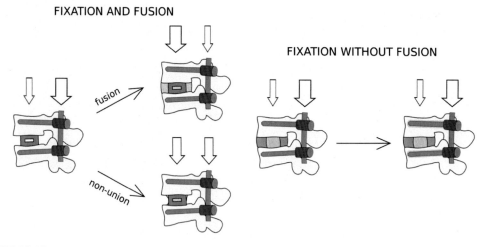

FIGURE 21.12

In fixation and fusion (left), if a solid fusion is achieved, the loads on the posterior fixation and on the bone—implant interface decrease over time, whereas in the case of nonunion, the posterior implants continue to support a substantial part of the load. The target of posterior dynamic stabilization implants is to preserve mobility over time, resulting in higher loads at the bone—implant interface and therefore higher risk of loosening.

From Galbusera, F., Volkheimer, D., Reitmaier, S., Berger-Roscher, N., Kienle, A., & Wilke, H.-J. (2015). Pedicle screw loosening: A clinically relevant complication? European Spine Journal, 24(5), 1005—1016. https://doi.org/10.1007/s00586-015-3768-6.

studies employed such pull-out tests to investigate the loosening behavior of pedicle screws, with consequent limitations on the translation of the results into clinical practice. Other types of testing, for example, the toggle test in which a rod connected to the pedicle screw is pulled out in a more physiological direction (Santoni et al., 2009), have been proposed but did not find a widespread use yet. Besides, while loosening is clinically the result of repetitive loads associated with walking, running, etc., most tests are based on the application of a quasi-static load to failure. As a matter of fact, the experimental investigation of the loosening behavior of spinal implants is an open research field offering large room for improvement.

REFERENCES

Allen, B. L., & Ferguson, R. L. (1984). A Pictorial guide to the galveston lri pelvic fixation technique. *Journal of Pediatric Orthopaedics*, *4*(2), 267. Available from https://journals.lww.com/pedorthopaedics/Citation/1984/03000/A_Pictorial_Guide_To_the_Galveston_Lri_Pelvic.47.aspx.

Almansour, H., Sonntag, R., Pepke, W., Bruckner, T., Kretzer, J. P., & Akbar, M. (2019). Impact of electrocautery on fatigue life of spinal fusion constructs—An in vitro biomechanical study. *Materials*, *12*(15), 2471. Available from https://doi.org/10.3390/ma12152471.

Anekstein, Y., Floman, Y., Smorgick, Y., Rand, N., Millgram, M., & Mirovsky, Y. (2015). Seven years follow-up for total lumbar facet joint replacement (TOPS) in the management of lumbar spinal stenosis and degenerative spondylolisthesis. *European Spine Journal: Official Publication of the European Spine Society, the European Spinal Deformity Society, and the European Section of the Cervical Spine Research Society*, *24*(10), 2306–2314. Available from https://doi.org/10.1007/s00586-015-3850-0.

Baaj, A. A., Uribe, J. S., Vale, F. L., Preul, M. C., & Crawford, N. R. (2009). History of cervical disc arthroplasty. *Neurosurgical Focus*, *27*(3), E10. Available from https://doi.org/10.3171/2009.6.FOCUS09128.

Bianco, R.-J., Aubin, C.-E., Mac-Thiong, J.-M., Wagnac, E., & Arnoux, P.-J. (2016). Pedicle screw fixation under nonaxial loads: A cadaveric study. *Spine (Philadelphia, PA: 1986)*, *41*(3), E124–E130. Available from https://doi.org/10.1097/BRS.0000000000001200.

Böhler, J. (1982). Anterior stabilization for acute fractures and non-unions of the dens. *The Journal of Bone and Joint Surgery. American Volume*, *64*(1), 18–27. Available from https://europepmc.org/abstract/med/7033229.

Bowers, C., Amini, A., Dailey, A. T., & Schmidt, M. H. (2010). Dynamic interspinous process stabilization: Review of complications. *Neurosurgical Focus*, *28*(6), E8. Available from https://doi.org/10.3171/2010.3.FOCUS1047.

Briggs, H., & Milligan, P. R. (1944). Chip fusion of the low back following exploration of the spinal canal. *The Journal of Bone and Joint Surgery*, *26*(1), 125–130. Available from http://citeseerx.ist.psu.edu/viewdoc/download?doi = 10.1.1.978.6512&rep = rep1&type = pdf.

Buckland, A. J., Baker, J. F., Roach, R. P., & Spivak, J. M. (2016). Cervical disc replacement—emerging equivalency to anterior cervical. *International Orthopaedics*, *40*(6), 1329–1334. Available from https://idp.springer.com/authorize/casa?redirect_uri = https://link.springer.com/article/10.1007/s00264-016-3181-8&casa_token = PZatfOKBsg8AAAAA:tAbEQol5mOjMlNnkvw448OFQcRIJIncA-f1SbNNiWmMkj5QzfIK0i9_YYA2ty1joZ9_Vsus6uZquR89jc5k.

Büttner-Janz, K., Guyer, R. D., & Ohnmeiss, D. D. (2014). Indications for lumbar total disc replacement: Selecting the right patient. *International Journal of Spine Surgery*, 8, 12. Available from https://doi.org/10.14444/1012.

Cakir, B., Richter, M., Käfer, W., Puhl, W., & Schmidt, R. (2005). The impact of total lumbar disc replacement on segmental and total lumbar. *Clinical Biomechanics*, *20*(4), 357–364. Available from https://doi.org/10.1016/j.clinbiomech.2004.11.019.

Castro, W. H., Halm, H., Jerosch, J., Malms, J., Steinbeck, J., & Blasius, S. (1996). Accuracy of pedicle screw placement in lumbar vertebrae. *Spine (Philadelphia, PA: 1986)*, *21*(11), 1320−1324. Available from https://doi.org/10.1097/00007632-199606010-00008.

Chen, S.-H., Chiang, M.-C., Lin, J.-F., Lin, S.-C., & Hung, C.-H. (2013). Biomechanical comparison of three stand-alone lumbar cages—a three-dimensional finite element analysis. *BMC Musculoskeletal Disorders*, *14*(1), 281. Available from https://link.springer.com/article/10.1186/1471-2474-14-281.

Cloward, R. B. (1953). The treatment of ruptured lumbar intervertebral discs by vertebral body. *Journal of Neurosurgery*, *10*(2), 154−168. Available from https://doi.org/10.3171/jns.1953.10.2.0154.

Costa, F., Cardia, A., Ortolina, A., Fabio, G., Zerbi, A., & Fornari, M. (2011). Spinal navigation: standard pre-operative versus intraoperative computed tomography data set acquisition for computer-guidance system: radiological and clinical study in 100 consecutive patients. *Spine*, *36*(24), 2094−2098.

Cripton, P. A., Jain, G. M., Wittenberg, R. H., & Nolte, L. P. (2000). Load-sharing characteristics of stabilized lumbar spine segments. *Spine*, *25*(2), 170−179. Available from https://doi.org/10.1097/00007632-200001150-00006.

Cummins, B. H., Robertson, J. T., & Gill, S. S. (1998). Surgical experience with an implanted artificial cervical joint. *Journal of Neurosurgery*, *88*(6), 943−948. Available from https://doi.org/10.3171/jns.1998.88.6.0943.

Cunningham, B. W., Hu, N., Zorn, C. M., & McAfee, P. C. (2010). Biomechanical comparison of single- and two-level cervical arthroplasty versus arthrodesis: effect on adjacent-level spinal kinematics. *The Spine Journal: Official Journal of the North American Spine Society*, *10*(4), 341−349. Available from https://doi.org/10.1016/j.spinee.2010.01.006.

De Kunder, S. L., Rijkers, K., Caelers, I. J. M. H., De Bie, R. A., Koehler, P. J., & Van Santbrink, H. (2018). Lumbar interbody fusion: A historical overview and a future perspective. *Spine (Philadelphia, PA: 1986)*, *43*(16), 1161−1168. Available from https://doi.org/10.1097/BRS.0000000000002534.

Defino, H., Galbusera, F., & Wilke, H.-J. (2020). *Lumbar Spine Online Textbook*. Wheelessonline.com. http://www.wheelessonline.com/ISSLS/section-11-chapter-8-pedicle-screw-fixation-and-design/.

Duhon, B. S., Bitan, F., Lockstadt, H., Kovalsky, D., Cher, D., Hillen, T., & Study Group, S. I. F. I. (2016). Triangular titanium implants for minimally invasive sacroiliac joint. *International Journal of Spine Surgery*, *10*, 13. Available from https://doi.org/10.14444/3013.

Dwyer, A. F., Newton, N. C., & Sherwood, A. A. (1969). 26 An anterior approach to scoliosis: A preliminary report. *Clinical Orthopaedics and Related Research*, *62*, 192. Available from https://journals.lww.com/corr/Citation/1969/01000/26_An_Anterior_Approach_to_Scoliosis__A.27.aspx.

Eck, J. C., Hodges, S., & Humphreys, S. C. (2007). Minimally invasive lumbar spinal fusion. *The Journal of the American Academy of Orthopaedic Surgeons*, *15*(6), 321−329. Available from https://doi.org/10.5435/00124635-200706000-00001.

Fernström, U. (1966). Arthroplasty with intercorporal endoprothesis in herniated disc and in painful disc. *Acta Chirurgica Scandinavica. Supplementum*, *357*, 154−159. Available from https://ci.nii.ac.jp/naid/10027252337/.

Fogel, G. R., Parikh, R. D., Ryu, S. I., & Turner, A. W. L. (2014). Biomechanics of lateral lumbar interbody fusion constructs with lateral and posterior plate fixation: laboratory investigation. *Journal of Neurosurgery. Spine*, *20*(3), 291−297. Available from https://thejns.org/spine/view/journals/j-neurosurg-spine/20/3/article-p291.xml.

Galbusera, F., Casaroli, G., Chande, R., Lindsey, D., Villa, T., Yerby, S., ... Brayda-Bruno, M. (2020). Biomechanics of sacropelvic fixation: A comprehensive finite element comparison of three techniques. *European Spine Journal: Official Publication of the European Spine Society, the European Spinal Deformity Society, and the European Section of the Cervical Spine Research Society*, *29*(2), 295−305. Available from https://doi.org/10.1007/s00586-019-06225-5.

Galbusera, F., Volkheimer, D., Reitmaier, S., Berger-Roscher, N., Kienle, A., & Wilke, H.-J. (2015). Pedicle screw loosening: A clinically relevant complication? *European Spine Journal: Official Publication of the*

European Spine Society, the European Spinal Deformity Society, and the European Section of the Cervical Spine Research Society, 24(5), 1005–1016. Available from https://doi.org/10.1007/s00586-015-3768-6.

Gomleksiz, C., Sasani, M., Oktenoglu, T., & Ozer, A. F. (2012). A short history of posterior dynamic stabilization. *Advances in Orthopedics, 2012,* 629698. Available from https://doi.org/10.1155/2012/629698.

Graham, J. H., Anderson, P. A., & Spenciner, D. B. (2014). Letter to the editor in response to Villa T, La Barbera L, Galbusera F "comparative analysis of international standards for the fatigue testing of posterior spinal fixation systems". *Spine Journal, 14*(12), 3067–3068. Available from https://doi.org/10.1016/j.spinee.2014.07.026.

Grevitt, M. P., Gardner, A. D., Spilsbury, J., Shackleford, I. M., Baskerville, R., Pursell, L. M., ... Mulholland, R. C. (1995). The Graf stabilisation system: Early results in 50 patients. *European Spine Journal: Official Publication of the European Spine Society, the European Spinal Deformity Society, and the European Section of the Cervical Spine Research Society, 4*(3), 169–175. Available from https://doi.org/10.1007/bf00298241, discussion 135.

Guyer, R. D., Pettine, K., Roh, J. S., Dimmig, T. A., Coric, D., McAfee, P. C., & Ohnmeiss, D. D. (2016). Five-year follow-up of a prospective, randomized trial comparing two lumbar total disc replacements. *Spine (Philadelphia, PA: 1986), 41*(1), 3–8. Available from https://doi.org/10.1097/BRS.0000000000001168.

Ha, K.-Y., Lee, J.-S., & Kim, K.-W. (2008). Degeneration of sacroiliac joint after instrumented lumbar or lumbosacral fusion: A prospective cohort study over five-year follow-up. *Spine (Philadelphia, PA: 1986), 33* (11), 1192–1198. Available from https://doi.org/10.1097/BRS.0b013e318170fd35.

Harms, J., & Rolinger, H. (1982). A one-stager procedure in operative treatment of spondylolistheses: Dorsal traction-reposition and anterior fusion (author's transl). *Zeitschrift fur Orthopadie und Ihre Grenzgebiete, 120*(3), 343–347. Available from https://europepmc.org/abstract/med/7113376.

Harrington, P. R. (1962). Treatment of scoliosis. Correction and internal fixation by spine instrumentation. *The Journal of Bone and Joint Surgery. American Volume* (44-A), 591–610. Available from http://citeseerx.ist.psu.edu/viewdoc/download?doi = 10.1.1.828.5421&rep = rep1&type = pdf.

Jahng, T.-A., Kim, Y. E., & Moon, K. Y. (2013). Comparison of the biomechanical effect of pedicle-based dynamic. *The Spine Journal: Official Journal of the North American Spine Society, 13*(1), 85–94. Available from https://doi.org/10.1016/j.spinee.2012.11.014.

Kaneda, K., Shono, Y., Satoh, S., & Abumi, K. (1996). New anterior instrumentation for the management of thoracolumbar and lumbar scoliosis. Application of the Kaneda two-rod system. *Spine (Philadelphia, PA: 1986), 21*(10), 1250–1261, discussion 1261-2. Available from https://doi.org/10.1097/00007632-199605150-00021.

Kebaish, K. M. (2010). Sacropelvic fixation: Techniques and complications. *Spine (Philadelphia, PA: 1986), 35*(25), 2245–2251. Available from https://doi.org/10.1097/BRS.0b013e3181f5cfae.

Kienle, A., Graf, N., Villa, T., La Barbera, L., Galbusera, F., & Wilke, H.-J. (2018). *Chapter 13 - Standard testing. Biomechanics of the spine* (pp. 223–238). Academic Press. Available from https://doi.org/10.1016/B978-0-12-812851-0.00013-6.

Kirkaldy-Willis, W. H., & Farfan, H. F. (1982). Instability of the lumbar spine. *Clinical Orthopaedics and Related Research, 165,* 110–123. Available from https://journals.lww.com/clinorthop/Citation/1982/05000/Instability_of_the_Lumbar_Spine.15.aspx.

Kocak, T., Cakir, B., Reichel, H., & Mattes, T. (2010). Screw loosening after posterior dynamic stabilization—review of the literature. *Acta Chirurgiae Orthopaedicae et Traumatologiae Cechoslovaca, 77*(2), 134–139. Available from https://pdfs.semanticscholar.org/be72/e9dacda30099c01ce9155947a12f39ce664b.pdf.

Kuslich, S. D., Ulstrom, C. L., Griffith, S. L., Ahern, J. W., & Dowdle, J. D. (1998). The Bagby and Kuslich method of lumbar interbody fusion. History, techniques, and 2-year follow-up results of a United States prospective, multicenter trial. *Spine, 23*(11), 1267–1278. Available from https://doi.org/10.1097/00007632-199806010-00019, discussion 1279.

La Barbera, L. (2018). *Fixation and fusion. Biomechanics of the spine* (pp. 301–327). Elsevier. Available from https://www.sciencedirect.com/science/article/pii/B9780128128510000173.

Laxer, E., & Others. (1994). A further development in spinal instrumentation. *European Spine Journal, 3*(6), 347–352. Available from https://idp.springer.com/authorize/casa?redirect_uri = https://link.springer.com/content/pdf/10.1007/BF02200149.pdf&casa_token = qGvgws1f0IUAAAAA:xEBB4NFEsuxkPR9uQy7IsQwW6VDG8gpv3qFdtbGPxWO6a3ZejsqMfdGeGn1xIu_7a_r2YAEt_EXF1R7giII.

Lee, C. S., Hwang, C. J., Lee, S.-W., Ahn, Y.-J., Kim, Y.-T., Lee, D.-H., & Lee, M. Y. (2009). Risk factors for adjacent segment disease after lumbar fusion. *European Spine Journal: Official Publication of the European Spine Society, the European Spinal Deformity Society, and the European Section of the Cervical Spine Research Society, 18*(11), 1637–1643. Available from https://doi.org/10.1007/s00586-009-1060-3.

Lee, S.-M., Suk, S.-I., & Chung, E.-R. (2004). Direct vertebral rotation: A new technique of three-dimensional deformity. *Spine (Philadelphia, PA: 1986), 29*(3), 343–349. Available from https://doi.org/10.1097/01.brs.0000109991.88149.19.

Lenke, L. G., Bridwell, K. H., Baldus, C., Blanke, K., & Schoenecker, P. L. (1992). Cotrel-Dubousset instrumentation for adolescent idiopathic scoliosis. *The Journal of Bone and Joint Surgery. American Volume, 74*(7), 1056–1067. Available from https://europepmc.org/abstract/med/1522092.

Lombardi, J. M., Shillingford, J. N., Lenke, L. G., & Lehman, R. A. (2018). Sacropelvic fixation: When, why, how? *Neurosurgery clinics of North America, 29*(3), 389–397. Available from https://dl.uswr.ac.ir/bitstream/Hannan/83636/1/2018%20NCNA%20Volume%2029%20Issue%203%20July%20%282%29.pdf.

Luca, A., Ottardi, C., Sasso, M., Prosdocimo, L., La Barbera, L., Brayda-Bruno, M., ... Villa, T. (2017). Instrumentation failure following pedicle subtraction osteotomy: The role of rod material, diameter, and multi-rod constructs. *European Spine Journal: Official Publication of the European Spine Society, the European Spinal Deformity Society, and the European Section of the Cervical Spine Research Society, 26*(3), 764–770. Available from https://doi.org/10.1007/s00586-016-4859-8.

Luque, E. R. (1982). Segmental spinal instrumentation for correction of scoliosis. *Clinical Orthopaedics and Related Research, 163*, 192–198. Available from https://europepmc.org/abstract/med/7067252.

Magerl, F., Seemann, P.-S., Kehr, P., & Weidner, A. (1987). Stable posterior fusion of the atlas and axis by transarticular screw fixation. *Cervical spine I: Strasbourg* (1985, pp. 322–327). Vienna: Springer. Available from https://doi.org/10.1007/978-3-7091-8882-8_59.

Malakoutian, M., Volkheimer, D., Street, J., Dvorak, M. F., Wilke, H.-J., & Oxland, T. R. (2015). Do in vivo kinematic studies provide insight into adjacent segment. *European Spine Journal: Official Publication of the European Spine Society, the European Spinal Deformity Society, and the European Section of the Cervical Spine Research Society, 24*(9), 1865–1881. Available from https://doi.org/10.1007/s00586-015-3992-0.

Mannion, A. F., Impellizzeri, F. M., Leunig, M., Jeszenszy, D., Becker, H.-J., Haschtmann, D., ... Fekete, T. F. (2018). Eurospine 2017 full Paper Award: Time to remove our rose-tinted. *European Spine Journal: Official Publication of the European Spine Society, the European Spinal Deformity Society, and the European Section of the Cervical Spine Research Society, 27*(4), 778–788. Available from https://doi.org/10.1007/s00586-018-5469-4.

Mariconda, M., Galasso, O., Barca, P., & Milano, C. (2005). Minimum 20-year follow-up results of Harrington rod fusion for idiopathic. *European Spine Journal: Official Publication of the European Spine Society, the European Spinal Deformity Society, and the European Section of the Cervical Spine Research Society, 14*(9), 854–861. Available from https://doi.org/10.1007/s00586-004-0853-7.

McNally, D., Naylor, J., & Johnson, S. (2012). An in vitro biomechanical comparison of CadiscTM-L with natural lumbar. *European Spine Journal: Official Publication of the European Spine Society, the European Spinal Deformity Society, and the European Section of the Cervical Spine Research Society, 21*(5), S612–S617. Available from https://doi.org/10.1007/s00586-012-2249-4.

Meredith, D. S., & Vaccaro, A. R. (2014). History of spinal osteotomy. *European Journal of Orthopaedic Surgery & Traumatology, 24*(1), S69–S72. Available from https://doi.org/10.1007/s00590-013-1406-8.

Mohan, A. L., & Das, K. (2003). History of surgery for the correction of spinal deformity. *Neurosurgical Focus*, *14*(1), e1. Available from https://doi.org/10.3171/foc.2003.14.1.2.

Murtagh, R., & Castellvi, A. E. (2014). Motion preservation surgery in the spine. *Neuroimaging Clinics of North America*, *24*(2), 287–294. Available from https://doi.org/10.1016/j.nic.2014.01.008.

Nachemson, A. (1962). Some mechanical properties of the lumbar intervertebral discs. *Bulletin of the Hospital for Joint Diseases*, *23*, 130–143. Available from https://www.ncbi.nlm.nih.gov/pubmed/13937019.

Nagaraja, S., & Palepu, V. (2017). Integrated Fixation Cage Loosening Under Fatigue Loading. *International Journal of Spine Surgery*, *11*, 20. Available from https://doi.org/10.14444/4020.

Newton, P. O., Kluck, D. G., Saito, W., Yaszay, B., Bartley, C. E., & Bastrom, T. P. (2018). Anterior spinal growth tethering for skeletally immature patients with scoliosis: A retrospective look two to four years postoperatively. *The Journal of Bone and Joint Surgery. American Volume*, *100*(19), 1691–1697. Available from https://doi.org/10.2106/JBJS.18.00287.

Nguyen, J. H., Buell, T. J., Wang, T. R., Mullin, J. P., Mazur, M. D., Garces, J., … Smith, J. S. (2019). Low rates of complications after spinopelvic fixation with iliac screws in 260 adult patients with a minimum 2-year follow-up. *Journal of Neurosurgery. Spine*, *2019*, 1–9. Available from https://doi.org/10.3171/2018.9.SPINE18239.

Pihlajamäki, H., Myllynen, P., & Böstman, O. (1997). Complications of transpedicular lumbosacral fixation for non-traumatic disorders. *The Journal of Bone and Joint Surgery. British Volume*, *79*(2), 183–189. Available from https://pdfs.semanticscholar.org/d8ef/dbec33d0bb58a2b2c5d86e7da3e053e49e46.pdf.

Prud'homme, M., Barrios, C., Rouch, P., Charles, Y. P., Steib, J.-P., & Skalli, W. (2015). Clinical outcomes and complications after pedicle-anchored dynamic or hybrid lumbar spine stabilization: A systematic literature review. *Journal of Spinal Disorders and Techniques*, *28*(8), E439–E448. Available from https://journals.lww.com/jspinaldisorders/fulltext/2015/10000/Clinical_Outcomes_and_Complications_After.4.aspx?casa_token = sc4Qyb9G-UAAAAAA:npLsMecmfFMvqzwSnZD3k1Zf_LCap5q8uA9AEzc6IuvNTcfl-cjZKzKo2VhLu5E9VmrY1UZ2rBuWUPv49WY149nyfvtIsaxSnw.

Rapoff, A. J., Ghanayem, A. J., & Zdeblick, T. A. (1997). Biomechanical comparison of posterior lumbar interbody fusion cages. *Spine (Philadelphia, PA: 1986)*, *22*(20), 2375–2379. Available from https://doi.org/10.1097/00007632-199710150-00010.

Rhee, J. M., Park, J.-B., Yang, J.-Y., & Riew, D. K. (2005). Indications and techniques for anterior cervical plating. *Neurology India*, *53*(4), 433–439. Available from https://doi.org/10.4103/0028-3886.22609.

Richter, M., Wilke, H.-J., Kluger, P., Neller, S., Claes, L., & Puhl, W. (2000). Biomechanical evaluation of a new modular rod-screw implant system for posterior instrumentation of the occipito-cervical spine: in-vitro comparison with two established implant systems. *European Spine Journal: Official Publication of the European Spine Society, the European Spinal Deformity Society, and the European Section of the Cervical Spine Research Society*, *9*(5), 417–425. Available from https://idp.springer.com/authorize/casa?redirect_uri = https://link.springer.com/content/pdf/10.1007/s005860000173.pdf&casa_token = YKiiCfq4BycAAAAA:1lHmODwnNAIwn07h2zL-p7BVl2KbLptjhJwgNhKhSMUJ_vc7irXFyu0Udt9SgCV47RG0WxREumViXg.

Roy-Camille, R., Saillant, G., & Mazel, C. (1986). Internal fixation of the lumbar spine with pedicle screw plating. *Clinical Orthopaedics and Related Research*, *203*, 7–17. Available from https://europepmc.org/abstract/med/3955999.

Ruland, C. M., McAfee, P. C., Warden, K. E., & Cunningham, B. W. (1991). Triangulation of pedicular instrumentation. A biomechanical analysis. *Spine (Philadelphia, PA: 1986)*, *16*(6), S270–S276. Available from https://doi.org/10.1097/00007632-199106001-00019.

Samdani, A. F., Ames, R. J., Kimball, J. S., Pahys, J. M., Grewal, H., Pelletier, G. J., & Betz, R. R. (2015). Anterior vertebral body tethering for immature adolescent idiopathic. *European Spine Journal: Official Publication of the European Spine Society, the European Spinal Deformity Society, and the European Section of the Cervical Spine Research Society*, *24*(7), 1533–1539. Available from https://doi.org/10.1007/s00586-014-3706-z.

Sanden, B., Olerud, C., Petren-Mallmin, M., Johansson, C., & Larsson, S. (2004). The significance of radiolucent zones surrounding pedicle screws. Definition of screw loosening in spinal instrumentation. *The Journal of Bone and Joint Surgery. British Volume*, *86*(3), 457−461. Available from https://pdfs.semanticscholar.org/2b6a/f9f779c32df7752fb0d6e4e0e1a32873e4d0.pdf.

Santoni, B. G., Hynes, R. A., McGilvray, K. C., Rodriguez-Canessa, G., Lyons, A. S., Henson, M. A. W., ... Puttlitz, C. M. (2009). Cortical bone trajectory for lumbar pedicle screws. *The Spine Journal: Official Journal of the North American Spine Society*, *9*(5), 366−373. Available from https://doi.org/10.1016/j.spinee.2008.07.008.

Schmoelz, W., & Keiler, A. (2015). Intervertebral cages from a biomechanical point of view. *Der Orthopade*, *44*(2), 132−137. Available from https://europepmc.org/abstract/med/25595216.

Schmoelz, W., Huber, J. F., Nydegger, T., Dipl-Ing Claes, L., & Wilke, H. J. (2003). Dynamic stabilization of the lumbar spine and its effects on adjacent. *Journal of Spinal Disorders & Techniques*, *16*(4), 418−423. Available from https://doi.org/10.1097/00024720-200308000-00015.

Shao, H., Chen, J., Ru, B., Yan, F., Zhang, J., Xu, S., & Huang, Y. (2015). Zero-profile implant vs conventional cage-plate implant in anterior. *Journal of Orthopaedic Surgery and Research*, *10*, 148. Available from https://doi.org/10.1186/s13018-015-0290-9.

Shen, F. H., Harper, M., Foster, W. C., Marks, I., & Arlet, V. (2006). A novel "Four-Rod Technique" for lumbo-pelvic reconstruction: Theory and technical considerations. *Spine (Philadelphia, PA: 1986)*, *31*(12), 1395. Available from https://doi.org/10.1097/01.brs.0000219527.64180.95.

Smith-Petersen, M. N., Larson, C. B., & Aufranc, O. E. (1945). Osteotomy of the spine for correction of flexion deformity in rheumatoid. *The Journal of Bone and Joint Surgery*, *27*(1), 1. Available from https://journals.lww.com/jbjsjournal/Abstract/1945/27010/OSTEOTOMY_OF_THE_SPINE_FOR_CORRECTION_OF_FLEXION.1.aspx.

Stein, M. I., Nayak, A. N., Gaskins, R. B., 3rd, Cabezas, A. F., Santoni, B. G., & Castellvi, A. E. (2014). Biomechanics of an integrated interbody device vs ACDF anterior. *The Spine Journal: Official Journal of the North American Spine Society*, *14*(1), 128−136. Available from https://doi.org/10.1016/j.spinee.2013.06.088.

Strömqvist, B. (1993). Posterolateral uninstrumented fusion. *Acta Orthopaedica Scandinavica. Supplementum*, *251*, 97−99. Available from https://doi.org/10.3109/17453679309160134.

Thomasen, E. (1985). Vertebral osteotomy for correction of kyphosis in ankylosing spondylitis. *Clinical Orthopaedics and Related Research*, *194*, 142−152. Available from https://europepmc.org/abstract/med/3978906.

Turner, J. A., Ersek, M., Herron, L., Haselkorn, J., Kent, D., Ciol, M. A., & Deyo, R. (1992). Patient outcomes after lumbar spinal fusions. *JAMA: The Journal of the American Medical Association*, *268*(7), 907−911. Available from https://jamanetwork.com/journals/jama/article-abstract/399285.

Watkins, M. B. (1953). Posterolateral fusion of the lumbar and lumbosacral spine. *The Journal of Bone and Joint Surgery. American Volume*, *35-A*(4), 1014−1018. Available from https://journals.lww.com/jbjsjournal/Citation/1953/35040/POSTEROLATERAL_FUSION_OF_THE_LUMBAR_AND.24.aspx.

Wilke, H.-J., Schmidt, R., Richter, M., Schmoelz, W., Reichel, H., & Cakir, B. (2012). The role of prosthesis design on segmental biomechanics. *European Spine Journal: Official Publication of the European Spine Society, the European Spinal Deformity Society, and the European Section of the Cervical Spine Research Society*, *21*(5), 577−584. Available from https://idp.springer.com/authorize/casa?redirect_uri = https://link.springer.com/article/10.1007/s00586-010-1552-1&casa_token = d8DK_HfoQB8AAAAA:Mu0iWz7Q1-5JuRFnUJIs5nn4ECjfM3tyNmC6gMg9SvDnH8YpqSjLHRAdwoQWpuLi6mjzIMolpzbfamCoTw.

Xu, R., Haman, S. P., Ebraheim, N. A., & Yeasting, R. A. (1999). The anatomic relation of lateral mass screws to the spinal nerves. A comparison of the Magerl, Anderson, and An techniques. *Spine (Philadelphia, PA: 1986)*, *24*(19), 2057−2061. Available from https://doi.org/10.1097/00007632-199910010-00016.

SHOULDER PROSTHESIS: BIOMECHANICS AND DESIGN

22

Paolo Dalla Pria

Waldemar Link GmbH & Co. KG, Hamburg, Germany

EVOLUTION OF THE SHOULDER ARTHROPLASTY

The shoulder prosthesis has an important role in the human imagination. The very first shoulder prosthesis has been described in Greek mythology. The infamous Tantalus (whose name was taken for the chemical element: Ta) offered his son Pelops as food for the gods and by mistake Demeter consumed Pelops' left shoulder. Demeter then created an ivory replacement for Pelops and Zeus punished Tantalus (see *Tantalus' punishment*).

The very first joint arthroplasty implanted in a human being was also a shoulder prosthesis. In 1893, the French surgeon Jules Pean implanted a rubber and platinum prosthesis to replace a gleno-humeral joint destroyed by tuberculosis (Pean, 1894) (Fig. 22.1). The patient was not as lucky as Pelops because due to an infection the implant was retrieved after 2 years. In the early decades of the 20th century, some other attempts were made using ivory (König, 1914), transplant of fibula (Albee, 1929), and acrylic resin (Boron & Sevin, 1951). The first humeral prosthesis made in metal (CoCr alloy) was designed by Krueger (1951). In this implant, the anatomical posterior offset of the humeral head was considered. Such a peculiar design feature was then abandoned for many years and reintroduced only a couple of decades ago. Even more recent than Krueger's, the implant designed by Charles Neer (Neer, Brown, & McLaughlin, 1953) to treat the fresh proximal fractures is often mentioned as the first modern shoulder prosthesis. Neer is considered the father of modern shoulder surgery, developing advanced surgical techniques, and updated designs of the implants, and he was the inventor of the polyethylene glenoid implant. Until the 1980s, the shoulder arthro-plasties were simple straight cemented stems available in different diameters and in different humeral head diameters (these kinds of implants are called the first-generation implants). Few improvements were done over the years, introducing modularity of the humeral head, i.e., the chance to use different diameters or heights of the metal head with stems of different sizes, lengths, and materials (second-generation prostheses). For many years, surgeons have been aware that an anatomical replacement of the glenohumeral joint may be successful only if the rotator cuff is healthy, otherwise problems of instability or malfunction occur. During the 1970s and the 1980s, instability was approached by increasing the constraint between humerus and glenoid. High-retention implants were proposed by several authors, taking inspiration from hip arthroplasty (lipped polyethylene glenoids, constrained liners of metal-backed glenoids) (Fig. 22.2) or using the revolutionary concept of reversing the anatomy of the joint.

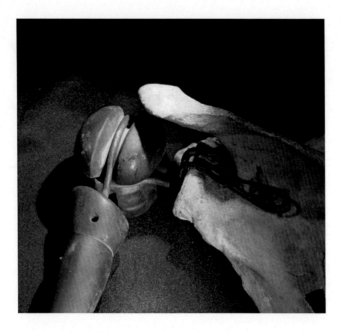

FIGURE 22.1

Pean's prosthesis made of rubber and platinum. The shoulder prosthesis designed by Jules Pean in 1893 is the very first joint prosthesis implanted in a human being.

From (N.d.-a). Available at: https://commons.wikimedia.org/wiki/File:Mock_up_of_artificial_shoulder.jpg

The lipped and constrained glenoid implants were abandoned over the years due to the high rate of failure (Nam et al., 2010; Nwakama, Cofield, Kavanagh, & Loehr, 2000; Wiedel, 1990) and the reverse anatomy designs were also abandoned. Rupture and cuff tear arthropathies were then managed using large humeral heads or bipolar heads (Boyer, Huten, & Alnot, 2006; Diaz-Borjon, Yamakado, Pinilla, Keith, & Worland, 2007). During the 1990s, the exact restoration of the joint anatomy influenced the design of new implants and the constraint of the glenohumeral joint was considered more as a potential risk of failure, rather than a means to achieve a satisfactory outcome, due to the common belief that the higher the glenohumeral constraint, the higher the risk of glenoid loosening. In the new systems more adaptability was introduced through modular components, like the choices of the neck resection angle and the retroversion angle of the stem, the offsets of the humeral head, and the height of the implant. In some models, the uncemented fixation of the stems was also improved. Moreover, new designs of uncemented glenoid components were proposed to bypass the use of cement. All these systems are included in the third generation of shoulder arthroplasties. However, the improvements on the anatomical constructs of the implants can be effective in the reproduction of the native anatomy only if the muscles of the rotator cuff are well-functioning, and the treatment of cuff tear degenerations was still unsolved. The French surgeon Paul Grammont spent much effort in finding an effective solution for the treatment of the non-reparable supraspinatus. The increase of the lever arm of the deltoid muscle was beneficial in improving the active motion because the deltoid muscle is mainly responsible for the motion of the arm, and the longer the lever arm, the lower the muscular effort required to elevate

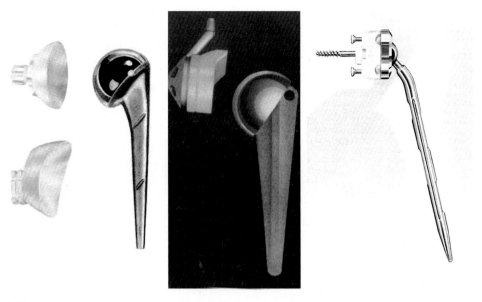

FIGURE 22.2

High-retention total anatomical prostheses. The St. Georg cemented TSA (Waldemar Link GmbH & Co. KG) (left). The UHMWPE glenoids are lipped to avoid migration of the humeral head. The English-McNab uncemented TSA (Biomet) (middle). The Buchholz-Zippel fully constrained TSA (Waldemar Link GmbH & Co. KG) (right). *TSA*, total shoulder arthroplasty.

the arm. At the same time, restoration of the joint stability was also required and eventually he proposed a reverse shoulder arthroplasty (RSA) characterized by the medialization of the center of rotation (Baulot, Sirveaux, & Boileau, 2011; Grammont, Trouilloud, Laffay, & Deries, 1987) (Fig. 22.3).

For several years, RSA has become very popular and currently it is used in over 80% of replacement cases (Australian Orthopaedic Association, 2020). This dramatic change over the years is due to several aspects. Nowadays, shoulder arthroplasty is rarely used in younger patients and, on the other hand, in elderly patients reverse arthroplasty is the preferred choice because it is a quick option and is easier and more reliable than other choices, especially in fresh traumas, where the cuff is often broken or not well-functioning (Dey Hazra et al., 2021; Lo, Rizkalla, Montemaggi, & Krishnan, 2021; Luciani, Farinelli, Procaccini, Verducci, & Gigante, 2019; Schwarz et al., 2021; Urch, Dines, & Dines, 2016). Cuff degeneration is an important factor to be considered in the design of total shoulder arthroplasties. Over time, because of cuff degeneration a well-fixed and well-positioned anatomical total shoulder arthroplasty (TSA) is not functioning well, which leads to stiffness and pain of the glenohumeral joint. In such a situation, replacement with a total RSA is the preferred choice. The retrieval of a well-fixed hemi-shoulder arthroplasty (HSA) or TSA to implant an RSA can be very demanding, while a conversion without removing well-integrated metal parts is safer and easier. The first conversion from TSA to RSA was done in 2002 in Australia. In a modular implant (Randelli system, Lima-Lto), a few components were replaced without damaging the humeral stem and the metal glenoid component, both uncemented (Fig. 22.4).

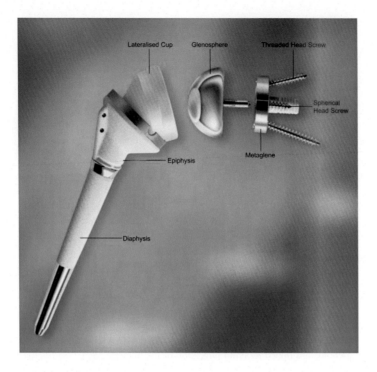

FIGURE 22.3

The Delta III reverse shoulder arthroplasty (DePuy). The Delta III is the first modern reverse prosthesis according to Grammont's design.

From (N.d.-b). Available at: https://www.yumpu.com/en/document/read/31680952/delta-cta-depuy-shoulderdoccouk

The prosthesis had not been designed to host a reverse construct; therefore, custom-made parts were produced. Present modern systems already consider the chance of an easier conversion from HSA/TSA to RSA, replacing just a few components, reducing at a minimum the damage to the bone—implant interface. At the present time, several kinds of implants are available in the market to address different indications and pathologies. Most of the anatomical and reverse TSAs are provided of a humeral stem which can be short or long. However, in some models, called "stemless," the humeral implant is fixed through a cage in the upper metaphysis after the cut of the humeral head (Fig. 22.5).

The least invasive humeral prosthesis is the resurfacing of the native bone articular surface through a metal head with fixation occurring through one or more pins into the epiphyseal region of the humerus. Implants with lower invasiveness like resurfacing and stemless arthroplasties require well-preserved anatomy of the proximal humerus. Resurfacing and stemless implants are expected to save the humeral bone stock in case of revision, and hence a new low invasiveness implant might be used. In case of revision of previous stems or diaphyseal fractures, long stems are necessary and if the upper extremity is lost (resorption and pseudoarthritis) or has to be resected (tumor), special implants provided of modular cylindrical segments are used to reconstruct the upper part of the humerus (Fig. 22.6).

In these peculiar cases, the fixation of the stem can be either cemented or uncemented.

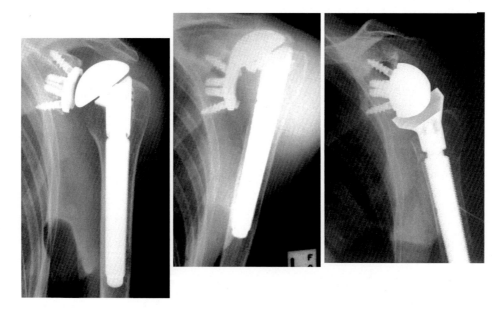

FIGURE 22.4

The first conversion from anatomical to reverse total shoulder arthroplasty (2002). Well-functioning cuff (left). Supraspinatus muscle tear (middle). Conversion to RSA (right). Special custom-made components were designed to leave in situ the glenoid metal-back and the humeral stem.

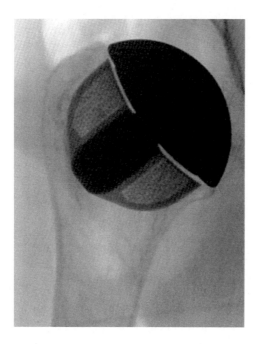

FIGURE 22.5

Stemless prosthesis (Embrace, Waldemar Link GmbH Co. KG). The stemless humeral prosthesis is a cage fixed in the upper metaphysis after the cut of the humeral head. The final prosthetic head and the cage are assembled through a conical connection.

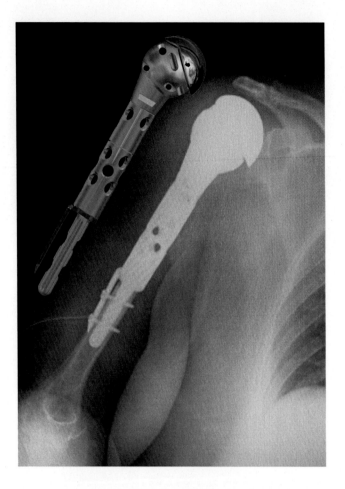

FIGURE 22.6

Humeral mega-prosthesis (Mega-C UL, Waldemar Link GmbH & Co. KG). A mega-prosthesis is used when a large resection of the humerus is required. Modular cylindrical segments are chosen to restore the original length of the resected bone.

BIOMECHANICS OF THE SHOULDER PROSTHESIS

BIOMECHANICS OF THE ANATOMICAL IMPLANTS

Anatomical implants can be either partial or total. Total implants (humeral and scapular implants) are nowadays preferred (Fig. 22.7) and partial humeral implants (HSA) are limited to some fresh traumas, degenerative illness related only to the humeral head, or if the scapular bone is not able to host a glenoid implant. Unpaired cuff muscles are accepted only in the latter case, otherwise the anatomical implants always require a well-functioning rotator cuff. The aim of the anatomical

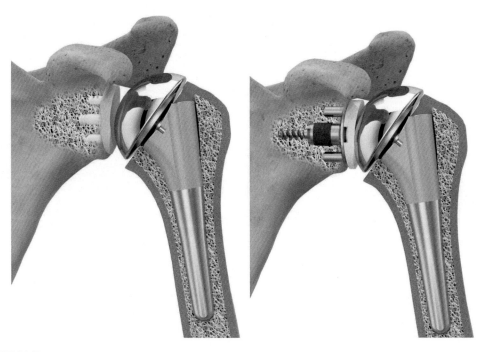

FIGURE 22.7

Anatomical total shoulder arthroplasty (aTSA). An anatomical total shoulder prosthesis is composed of a humeral implant and a scapular implant. Cemented glenoid implants are made of UHMWPE and provided of spikes (left) or a keel. Uncemented glenoid implants are often composed of a metal-back and a UHMWPE liner (right). Additional screws are often used to improve primary fixation.

implants is to reproduce as much as possible the native non-pathological anatomy of the glenohumeral joint, and the joint may work only if the muscles are healthy and powerful.

Diameter, thickness, and position of the humeral bone head should be precisely reproduced by the prosthetic one. In the prosthetic systems, the range of the head diameter can vary from 38 mm up to 54 mm with a step of 2−3 mm between sizes. Each diameter is often available in two or three heights. There is an offset between the axis of the humerus and the center of the head on the sagittal plane (the head is posterior to the humeral axis) and in all modern systems the heads can be placed at various offsets on the plane of the anatomical neck of the humerus to cover the resected surface as accurately as possible (Fig. 22.8). The same offsets are also used to modify the position of the head on the frontal plane, giving the chance to displace the head cranially or caudally. The correct choice of diameter and height of the head and placement on the resected surface of the humerus are very important to allow the glenohumeral motion because an unwanted overlapping between resected humeral surface and prosthetic head will lead to an overstretching of the surrounding muscles. If the diameter of the head is too small, the joint can be unstable, while if too large, the joint can be stiff and painful. If the cuff is healthy and there is some doubt on the head size, the smaller size is often preferred to avoid any unwanted overstretching of the surrounding

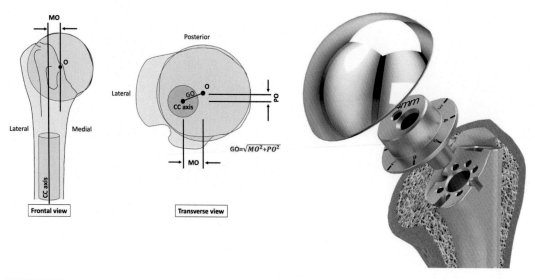

FIGURE 22.8

Offset of the humeral head. The prosthetic head can be offset with respect to the cervical axis of stem to cover the resected surface as accurately as possible.

Modified from Barth, J., Garret, J., Boutsiadis, Sautier, E., Geais, L., & Bothorel, H., Shoulder Friends Institute, Godenèche A.
(2018). Is global humeral head offset related to intramedullary canal width? A computer tomography morphometric study. Journal of
Experimental Orthopaedics, 5(1), 35. https://doi.org/10.1186/s40634-018-0148-2. PMID: 30209642; PMCID: PMC6135727.

muscles. Conversely, a larger head was preferred in case of cuff tear arthropathy (CTA) when RSA was not available. Nowadays, when RSA is not a feasible solution, special anatomical heads are chosen that are characterized by a superolateral extension to reduce the pressure of the rim of the head against the acromion (Fig. 22.9).

In anatomical arthroplasty, the height of the implant has the same relevance as the head diameter and for the same reasons. If the implant is too short, problems of instability may occur and if too high, the supraspinatus muscle can be damaged due to the pressure of the head and due to the overstretching of its tendon on the greater tuberosity. It has been hypothesized that overstretching on the greater tuberosity may lead to bone resorption of the superolateral humeral bone. This phenomenon was mainly observed in cases of fresh trauma and can be explained by the fact that in fresh traumas it is quite difficult to place the prosthesis and the fragments (greater and lesser tuberosities) at the right level because the anatomical landmarks are missing or not well defined. In this regard, it is interesting to observe that the posterior offset of the humeral head is often ignored in cases of tuberosities' fracture because the original offset of the humeral cross-section at the level of the anatomical neck is not recognized and in the majority of these cases no posterior offset is chosen.

Another important geometrical parameter to be considered to reach the right fine balance between shape and muscles is the retroversion of the humeral head with respect to the humeral axis. During elective surgeries such a value (standard range between 0 and 30 degrees) is easily

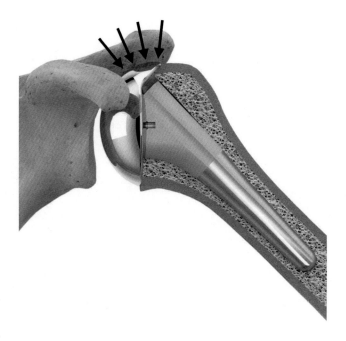

FIGURE 22.9

CTA (cuff tear arthropathy) head. The CTA head is a special prosthetic head characterized by a superolateral additional extension to prevent painful impingement between the metal rim and the acromion when the supraspinatus muscle is torn.

found, while in trauma cases it is chosen at about 20−25 degrees. If the retroversion angle is too small, the head can damage the subscapularis muscle, which passes in front of the head and if too large, the prosthesis can dislocate posteriorly. Sometimes, in trauma surgeries when the tuberosities are split and the posterior offset is unknown, the joint balance is obtained by skilled surgeons increasing the retroversion. This works because both retroversion and posterior offset displace the head posteriorly. All the geometrical parameters related to the head are checked during the surgery, up to the end and just before the assembly of the final components. Moreover, any change can also be done after the final assembly, or in case of revision. The height of the implant and retroversion are less forgiving geometrical parameters because the removal of the stem can be very demanding.

BIOMECHANICS OF THE REVERSE IMPLANTS

Cuff muscles can deteriorate over the years, or traumatic events can severely damage the muscles, the tendons, and the bone zones where the muscles are attached (greater and lesser tuberosity). In the choice of treatment, surgical or conservative, the supraspinatus muscle plays an important role. If broken or torn with little chance of repair, the humerus will shift upwards, impinging against the *subacromial bursa* and the acromion. RSA is designed to let possible a satisfactory restoration of the glenohumeral motion also in case of relevant deficiency of the rotator cuff muscles. In the old

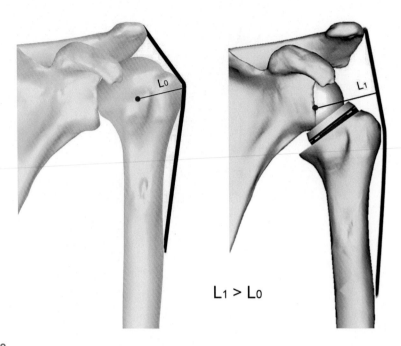

$$L_1 > L_0$$

FIGURE 22.10

Grammont's principle. Medialization of the center of rotation of the glenohumeral joint increases the lever arm of the deltoid muscle, reducing the effort to elevate the arm.

models of reverse implants, the center of rotation of the glenohumeral joint was kept as much as possible in the native anatomical position, while in the modern RSA, introduced by Paul Grammont, the center of rotation is placed closer to the glenoid (medialization of the center of rotation) to increase the lever arm of the deltoid muscle, reducing in such a way the effort to elevate the arm (Fig. 22.10).

The deltoid muscle wraps the proximal humerus; therefore, the increase of its lever arm with respect to the center of rotation is not constant. The middle deltoid takes the most advantage from the medialization of the center of rotation because its lever arm can be increased from about 35 mm up to 60 mm (Hamilton et al., 2015; Kontaxis & Johnson, 2009; Masjedi & Johnson, 2010; Terrier, Reist, Merlini, & Farron, 2008). The medialization of the center of rotation is beneficial also for the anterior deltoid, while the lever arm of the posterior deltoid is minimally affected and this explains why poor extrarotation is reached in case of deficiency of the infraspinatus and the teres minor muscles. The reduction of the deltoid effort in the arm elevation ranges from 30% to 67% (Ackland, Roshan-Zamir, Richardson, & Pandy, 2010; Dalla Pria, 2002; Giles, Langohr, Johnson, & Athwal, 2015; Kontaxis & Johnson, 2009; Terrier et al., 2008). The deltoid muscle in the glenohumeral joint works as an abductor/elevator, similarly to the abductor muscles of the hip joint and also in the hip prosthesis, the increase of the lever arm of the abductors is sometimes performed but it is obtained in the opposite way compared with the shoulder. In the shoulder, the attachment of the deltoid on the lateral rim of the acromion is fixed and the center of rotation is

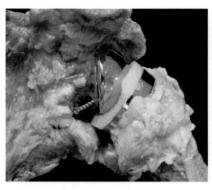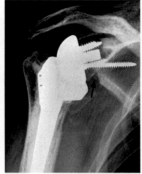

FIGURE 22.11

The scapular notch. Scapular notches observed from cadaveric specimen (left), radiographic studies (middle) and predicted with mathematical models (right).

From Kontaxis, A., & Johnson, G. R. (2009). The biomechanics of reverse anatomy shoulder replacement — A modelling study. Clinical Biomechanics, 24(3), 254–260. https://doi.org/10.1016/j.clinbiomech.2008.12.004.

medially displaced, while in the hip the center of rotation is fixed in the acetabulum and the abductors are lateralized increasing the horizontal offset of the neck of the stem. The medialization of the center of rotation has not only the beneficial effect of increasing the lever arm of the deltoid muscle, but also reduces the lever arm of the forces acting on the bone—implant scapular interface, reducing in this way the torque transmitted by the implant to the hosting bone. This beneficial feature of the Grammont prosthesis is considered fundamental for the survival of the glenoid components and in the modern literature, it is often stated that the pre-Grammont reverse prostheses used to fail due to the lever arm between the center of the joint and the scapula. However, Pupello (2016) reviewed all the published reports of the pre-Grammont reverse implants and he did not find any correlation between failure and position of the center of rotation. His conclusion is that there are current misconceptions on the RSA principles. Unfortunately, the medialization of the center of rotation also involves the medialization of the humerus, and the medial rim of the prosthetic socket articulating with the glenosphere can impinge against the scapular bone during some movements (adduction, intra- or extra-rotation), depending also on the mutual orientation between scapular and humeral components. The continuous scratching between humeral liner and scapula can lead to abnormal wear of the polyethylene, which causes a localized bone resorption of the scapular bone known as a *scapular notch* (Boileau, Watkinson, Hatzidakis, & Balg, 2005; Kim, Kim, Jang, & Yoo, 2021; Kontaxis & Johnson, 2009; Nyffeler, Werner, Simmen, & Gerber, 2004; Sirveaux et al., 2004) (Fig. 22.11).

It occurs because the RSA is a highly congruent joint to provide static stability and a stable fulcrum (Baulot et al., 2011; Grammont et al., 1987; Werthel et al., 2019) and the angle between the axis of the stem and the axis perpendicular to the humeral liner (NSA, *neck-shaft angle*, Fig. 22.12) is in some models quite high (155 degrees in the original Grammont design) to avoid the dislocation of the humeral component and the glenosphere.

Its occurrence is dependent on several factors, but it seems to be related to the glenoid morphology because the same implant may result in notching in a short-neck glenoid and no notching in a

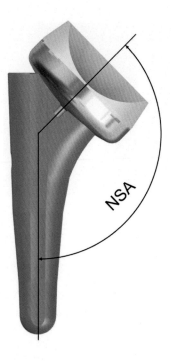

FIGURE 22.12

The neck-shaft angle (NSA). The NSA is the angle between the axis of the stem and the axis perpendicular to the humeral liner.

long-neck glenoid. The scapular notch is commonly considered as the contact between the humeral cup and the scapular bone during the adduction movement. However, several types of impingement exist in RSA. Using 12 fresh frozen shoulder specimens, Lädermann analyzed the impingements between scapula and humeral implant and scapular notching seems to be caused by more movements or combination of movements than previously considered, and in particular by movements of flexion/extension and internal/external rotations with the arm at the side (Lädermann et al., 2015). Scapular notch was very frequent in the last decade and responsible for more than 50% of the complications related to the RSA (Lévigne et al., 2011; Scarlat, 2013; Simovitch, Flurin, Wright, Zuckerman, & Roche, 2019; Zumstein, Pinedo, Old, & Boileau, 2011).

ADVANCED DESIGN CONCEPTS OF THE REVERSE SHOULDER ARTHROPLASTY

In summary, the concepts attributed to Grammont are (1) medialization of the center of rotation to increase the lever arm of the deltoid muscle, (2) distalization of the humerus to stretch deltoid fibers, (3) high NSA, and (4) high congruency of the joint to improve stabilization (Boileau et al., 2005; Sirveaux et al., 2004; Werthel et al., 2019), even if the RSAs of the previous generation were also characterized by high congruency, and the distalization of the humerus is an intrinsic consequence of the RSA design. The original Grammont concepts are still valid and still used in some modern implants, though medialized designs of both the glenosphere and the humerus may lead to scapular notching, instability, poor restoration of internal and external rotation, and loss of contour of the shoulder (Werthel et al., 2019). Over the years, several changes and new solutions have been introduced in RSA design to optimize its use and minimize the drawbacks.

SIZE OF THE GLENOSPHERE

The original diameter of the Grammont RSA was 36 mm. This diameter is still in use in many systems and it is preferred for small shoulders. Larger diameters are offered in all systems, but there is not a common standard in the market. According to the producer, available diameters are 38, 40, 42, 44, and 46 mm. Height and sex are highly correlated with a surgeon's choice of glenosphere size, but no recommendation can be made for surgeons to select a particular glenosphere size based on a patient's height (Schoch et al., 2019). RSA is often defined as a constrained (Goutallier, Postel, Zilber, & Van Driessche, 2003) or semiconstrained implant (Sayana, Kakarala, Bandi, & Wynn-Jones, 2008; Werthel et al., 2019) due to the high congruency between glenosphere and humeral liner. These definitions are not fully correct because almost all the available RSAs are not provided with a constraint mechanism able to avoid the dislocation of the joint. While in the total hip arthroplasty, a large femoral head prevents the risk of dislocation of the joint, in the RSA the glenosphere diameter has minimal effect on the joint stability (Gutiérrez, Keller, Levy, Lee, & Luo, 2008). This is because the depth of the humeral liner is quite low and similar for all the diameters, while in the acetabular liners the depth is almost always the radius of the head. However, a larger glenosphere tends to lateralize the humerus, and the fibers of the deltoid are more wrapping around the tuberosities and the residual muscles of the cuff are then tightened. It was shown that patients treated with larger glenospheres had significant improvements in active forward elevation and active external rotation when compared with smaller glenospheres (Mollon, Mahure, Roche, & Zuckerman, 2016; Müller et al., 2018). Moreover, the larger the glenosphere, the wider the potential range of motion of the joint and the lower the risk of the scapular notch.

ECCENTRICITY OF THE GLENOSPHERE

Eccentrical glenospheres are often used as an alternative to larger diameters to prevent the scapular notch or to increase the joint tension. The sphere is placed eccentrically towards the inferior side of the glenoid (Fig. 22.13). In this way, the liner of the humeral component has to accomplish a larger movement in adduction before impinging the scapular bone and the humerus is distalized, increasing the joint tension and the lever arm of the deltoid muscle. In a comparison between eccentrical 38 mm glenospheres and concentrical 42 mm glenospheres it was found that small glenospheres with eccentricity fared slightly worse than large glenospheres regarding the scapular notch development, even though no significant differences were noted (Torrens, Miquel, Martínez, & Santana, 2019). Inferior and medial positioning of the glenosphere also serves to decrease acromial stress, thought to be primarily due to increased deltoid mechanical advantage. The greatest effect magnitudes are seen at lower abduction angles, where the humerus is more frequently positioned (Wong, Langohr, Athwal, & Johnson, 2016).

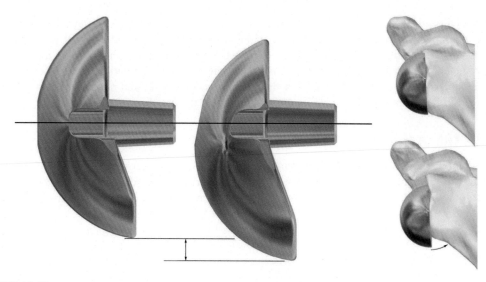

FIGURE 22.13

Glenosphere eccentricity. In the eccentric glenosphere, the spherical portion is placed eccentrically towards the inferior side of the glenoid to increase the range of motion in adduction and prevent contact between the humeral reverse liner and the scapula.

GLENOSPHERE MEDIALIZATION/LATERALIZATION

In the original Grammont design, the full medialization of the center of rotation combined with a small concentric glenosphere led to several cases of instability and the scapular notch (Boileau et al., 2005; Sirveaux et al., 2004; Werthel et al., 2019) and for these reasons in many modern systems the glenosphere is less medialized than in the Grammont design. The distance between the center of the glenosphere and the glenoid surface is called *lateralization* of the glenosphere. The term *lateralization* may cause some confusion because the center of rotation of the reverse joint remains medialized compared with the anatomical center of rotation of the native glenohumeral joint but it is a relative term based on comparison with a traditional Grammont-style prosthesis. The lateralization of the glenosphere can be reached in three ways (Fig. 22.14): the glenosphere is more extended than a half-sphere; there is some offset between the scapular baseplate and the center of the glenosphere; a graft of bone is interposed between the baseplate and the glenoidal bone.

The latest is called BIO-RSA (Bony Increased Offset Reverse Shoulder Arthroplasty), and it was introduced by Boileau to reduce the scapular notch, improve the stability, and stretch the residual cuff muscles without increasing the lever arm of the forces acting on the baseplate (Boileau, Moineau, Roussanne, & O'Shea, 2011). Progressive lateralization of the glenosphere results in improved stability of the joint and a lower risk of the scapular notch (Ferle, Pastor, Hagenah, Hurschler, & Smith, 2019). The range of motion of the shoulder can be improved by some lateralization of the glenosphere (Lädermann et al., 2019) and, if excessive, might negatively affect the glenohumeral abduction range of motion (Ferle et al., 2019; Werner, Chaoui, & Walch, 2017).

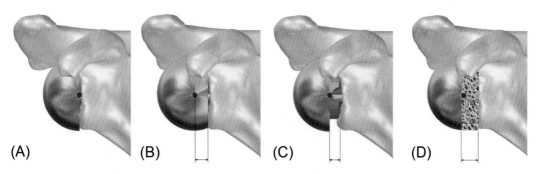

FIGURE 22.14

Lateralization of the glenosphere. (A) Glenosphere fully medialized. The center of rotation lies on the glenoidal surface. (B) The center of rotation is lateralized through the extension of the spherical surface of the glenosphere. (C) The center of rotation is lateralized through the thickening of the baseplate. (D) The center of rotation is lateralized through the interposition of a bone graft between the baseplate and the scapula.

A direct comparison between conventional Grammont design and the same implant where the BIO-RSA was applied was published by some authors. According to Collin (Collin et al., 2018), at the 2-year follow-up, BIO-RSA does not lead to a clinically significant improvement in range of motion, outcome scores, or change in scapular notching compared with a traditional RSA. However, some improvements in the outcome score and in forward flexion were recorded. Interestingly, Greiner (Greiner, Schmidt, Herrmann, Pauly, & Perka, 2015) found that BIO-RSA shows a trend toward improved external rotation, with a statistically significant improvement in external rotation in patients with an intact teres minor muscle and no difference between BIO-RSA and conventional Grammont type if the teres minor muscle was affected by degenerative changes. No differences were found by Kirzner, Paul and Moaveni (2018), who observed that the BIO-RSA technique was associated with an increase in scapular stress fracture rate when compared to the standard RSA (16.7% vs. 9.1%); however, this was not found to be significant. An increase of the stress on the scapular spine has to be considered when the glenosphere is lateralized. Moreover, bending moments at the implant interface increase with lateralization (except for the BIO-RSA) and moment arms of the deltoid consistently decrease requiring higher effort to the deltoid muscle in abduction (Costantini, Choi, Kontaxis, & Gulotta, 2015; Helmkamp et al., 2018; Henninger et al., 2011).

NECK-SHAFT ANGLE

The NSA is the angle between the axis of the stem and the axis perpendicular to the humeral liner (Fig. 22.12). In the original Grammont design, the NSA was 155 degrees. In modern systems available in the market, different NSAs are available, ranging from 135 to 155 degrees. Some advanced systems also offer the chance to choose the most appropriate NSA during the surgery (Fig. 22.15). In theory, a large NSA can reduce the risk of superior dislocation of the humerus, but the larger the angle, the easier the impingement of the humeral liner against the scapular bone (scapular notch) and decreasing the NSA demonstrated through a significant increase in impingement-free range of

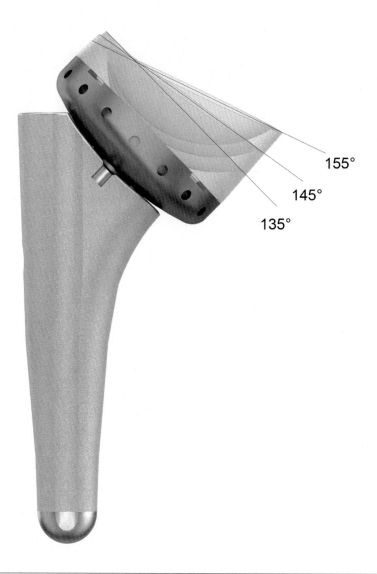

FIGURE 22.15

Neck-shaft angle (NSA). If multiple angles of the reverse liners are available, the most appropriate NSA can be chosen during the surgery.

motion (Nelson et al., 2018; Werner et al., 2017). However, the correlation between NSA and range of motion is model-related and cannot be assumed the same for all RSAs. A computational analysis of a Grammont-style implant found no correlation between NSA and range of motion, although it was found that decreasing NSA by 5 degrees lowered the inferior and superior impingement points (Roche et al., 2009). Decreasing the NSA can also positively affect the stability of the joint: configurations with an NSA of 135 degrees showed significantly higher dislocation forces than configurations with an NSA of 145 or 155 degrees at 30 degrees of abduction in 30 degrees of external rotation (Ferle et al., 2019). A systematic review of the influence of TSA in RSA to determine if a difference exists between RSA prostheses with a 135 versus 155 degrees humeral component inclination angle with respect to dislocation rates and scapular notching rates (Erickson, Frank, Harris, Mall, & Romeo, 2015) found that the rate of scapular notching was significantly higher with the 155 degrees prosthesis than with the 135 degrees prosthesis with a lateralized glenosphere, with no difference in dislocation rates between prostheses.

INLAY/ONLAY

In the original Grammont and almost all the previous pre-Grammont RSAs, the center of rotation of the joint was located partially into the metaphysis of the humerus. The first modern RSAs (De Puy Delta III, DJO Reverse, Lima-LTO SMR) were designed according to this solution. Only the SMR was offered as a system comprising both TSA and RSA and easily convertible from aTSA to RSA through a partial replacement of some components. The demand of having comprehensive systems, rather than aTSA- and RSA-specific models, led to a new solution. The humeral stem can host on top of the cut surface of the humerus either the anatomical head or a modular metal tray supporting the reverse polyethylene liner. The two designs are called *inlay* and *onlay* (Fig. 22.16). A humeral inlay implant is set into the metaphysis of the humerus. An onlay implant uses a reverse humeral tray that sits on top of the cut surface of the humerus (Wright & Murthi, 2021).

Even if the onlay concept was initially introduced to make surgeries and conversions from aTSA to RSA easier and less invasive, several studies demonstrated some important biomechanical benefits. The onlay tray lateralizes the humerus regardless of the position of the center of the glenosphere. Increasing humeral lateralization decreases deltoid forces required for active abduction (Giles et al., 2015) because a large moment arm of the deltoid muscle can be achieved. This decrease is greatest in the middle deltoid of the lateralized humerus for abduction and flexion and in the rotator cuff muscles under both internal rotation and external rotation (Liou et al., 2016). The lower the deltoid force, the lower the joint reaction forces in abduction and flexion and the lower the stresses acting on the glenoid baseplate. Lateralizing the humerus improves the function of the residual cuff muscles and the wrapping of the deltoid around the proximal humerus which prevents dislocation of the reverse joint. In some systems, the reverse tray is designed with an offset and can be dialed around the axis of the conical coupling between the tray and humeral stem reaching different positions on the resection plane (similarly to the offset of the anatomical heads).

Offset direction was found to have an effect on extra-articular impingement and muscle moment arms for all activities. Overall, impingement-free range of motion was maximized using a posterolateral tray offset and muscle moment arms were maximized using a medial tray offset (Berhouet et al., 2014; Glenday, Kontaxis, Roche, & Sivarasu, 2019). Onlay designs may provide better active external rotation, extension, and adduction. No significant differences were found in abduction and forward flexion (Beltrame et al., 2019). Both the lateralization of the glenosphere and the lateralization of the humerus increase the stresses on the scapular spine and also in the onlay design scapular spine fractures can occur (Ascione et al., 2018). At the present time, many different RSA systems are available in the market with large differences between them. In all systems, the thickness of the polyethylene liner can be chosen during the surgery and almost all systems offer different glenosphere diameters and different offsets of lateralization and eccentricities. Also NSA and offset of

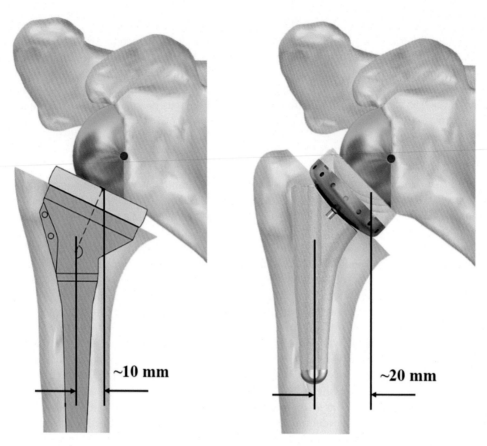

FIGURE 22.16

Inlay and onlay designs of the reverse humeral component. A humeral inlay implant (left) is set into the metaphysis of the humerus and the glenosphere can intersect the plane of resection. An onlay implant (right) uses a reverse humeral tray that sits on top of the cut surface of the humerus and the glenosphere always lies above the plane of resection.

lateralization of the onlay trays can be changed according to the surgeon's preferences. An interesting comparison (Werthel et al., 2019) between 22 systems revealed a very large range of lateralization of the glenosphere and the humeral axis with respect to the original Grammont design. The offset between the glenoid surface and the axis of the humeral component ranges from 13.1 mm in the Grammont design (Delta III, De Puy) up to 34.5 mm in the most lateralized RSA (Arrow, FH Ortho).

CONCLUSION

Scapular notch was the most commonly reported complication following the procedure, with rates previously ranging from 44% to 96% (Lévigne et al., 2011; Scarlat, 2013; Zumstein, Pinedo, Old,

& Boileau, 2011). Biomechanical and clinical studies have demonstrated that modern lateralized implants (from either the glenosphere or humeral side) and techniques have decreased the rate down to 0% to 19% in recent years (Cuff, Clark, Pupello, & Frankle, 2012; Valenti, Sauzières, Katz, Kalouche, & Kilinc, 2011). A large systematic review compared implants with a medialized vs. a lateralized center of rotation RSA (Helmkamp et al., 2018). The lateralized center of rotation prosthesis showed increased postoperative external rotation and decreased scapular notching. However, no difference in any patient-reported outcomes was found.

REFERENCES

Ackland, D. C., Roshan-Zamir, S., Richardson, M., & Pandy, M. G. (2010). Moment arms of the shoulder musculature after reverse total shoulder arthroplasty. *Journal of Bone and Joint Surgery - Series A*, *92*(5), 1221−1230. Available from https://doi.org/10.2106/JBJS.I.00001.

Albee. (1929). Restoration of shoulder function in cases of loss of head and upper portion of humerus. *Gynecologic Surgery*, 32.

Ascione, F., Kilian, C. M., Laughlin, M. S., Bugelli, G., Domos, P., Neyton, L., ... Walch, G. (2018). Increased scapular spine fractures after reverse shoulder arthroplasty with a humeral onlay short stem: An analysis of 485 consecutive cases. *Journal of Shoulder and Elbow Surgery*, *27*(12), 2183−2190. Available from https://doi.org/10.1016/j.jse.2018.06.007.

Australian Orthopaedic Association. (2020). *National Joint Replacement Registry (AOANJRR)*, 293. Available from https://aoanjrr.sahmri.com/annual-reports-2020.

Baulot, E., Sirveaux, F., & Boileau, P. (2011). *Grammont's idea: The story of paul grammont's functional surgery concept and the development of the reverse principle*, . *Clinical orthopaedics and related research* (vol. 469, pp. 2425−2431). New York LLC: Springer 9. Available from https://doi.org/10.1007/s11999-010-1757-y.

Beltrame, A., Di Benedetto, P., Cicuto, C., Cainero, V., Gisonni, R., & Causero, A. (2019). Onlay vs inlay humeral steam in reverse shoulder arthroplasty (RSA): Clinical and biomechanical study. *Acta Biomedica*, *90*, 54−63. Available from https://doi.org/10.23750/abm.v90i12-S.8983.

Berhouet, J., Kontaxis, A., Gulotta, L., Warren, R., Craig, E., Dines, D., & Dines, J. (2014). Effects of the humeral tray component positioning for onlay reverse shoulder arthroplasty design: A biomechanical analysis. *Journal of Shoulder and Elbow Surgery / American Shoulder and Elbow Surgeons*, *24*(4), 569−577. Available from https://doi.org/10.1016/j.jse.2014.09.022, Epub.

Boileau, P., Moineau, G., Roussanne, Y., & O'Shea, K. (2011). Bony increased-offset reversed shoulder arthroplasty: Minimizing scapular impingement while maximizing glenoid fixation. *Clinical Orthopaedics & Related Research*, *469*(9), 2558−2567. Available from https://doi.org/10.1007/s11999-011-1775-4.

Boileau, P., Watkinson, D. J., Hatzidakis, A. M., & Balg, F. (2005). Grammont reverse prosthesis: Design, rationale, and biomechanics. *Journal of Shoulder and Elbow Surgery*, *14*((1)), S147−S161. Available from https://doi.org/10.1016/j.jse.2004.10.006.

Boron, R., & Sevin, L. (1951). Prothèse acrylique de l'épaule. *La Presse médicale*, *59*(71), 1480.

Boyer, P., Huten, D., & Alnot, J. Y. (2006). La prothèse humérale bipolaire dans l'épaule rhumatoïde avec rupture de coiffe irréparable: Résultats à plus de 5 ans de recul. *Revue de Chirurgie Orthopedique et Reparatrice de l'Appareil Moteur*, *92*(6), 556−566. Available from https://doi.org/10.1016/s0035-1040(06)75913-x.

Collin, P., Liu, X., Denard, P. J., Gain, S., Nowak, A., & Lädermann, A. (2018). Standard vs bony increased-offset reverse shoulder arthroplasty: A retrospective comparative cohort study. *Journal of Shoulder and Elbow Surgery*, *27*(1), 59−64. Available from https://doi.org/10.1016/j.jse.2017.07.020.

Costantini, M. S., Choi, D., Kontaxis, A., & Gulotta, L. V. (2015). The effects of progressive lateralization of the joint center of rotation of reverse total shoulder implants. *Journal of Shoulder and Elbow Surgery / American Shoulder and Elbow Surgeons, 24*(7), 1120−1128. Available from https://doi.org/10.1016/j.jse.2014.11.040, Epub.

Cuff, D., Clark, R., Pupello, D., & Frankle, M. (2012). Reverse shoulder arthroplasty for the treatment of rotator cuff deficiency: A concise follow-up, ata minimum of five years, of a previous report. *Journal of Bone and Joint Surgery - Series A, 94*(21), 1996−2000. Available from https://doi.org/10.2106/JBJS.K.01206.

Dalla Pria, P. (2002). *La spalla: Protesi e particolari problematiche.* Biomeccanica. VI Congress Società Italiana Chirurgia Spalla e Gomito (SICSeG), Triest (Italy).

Dey Hazra, R., Blach Robert, M., Ellwein, A., Katthagen, J., Lill, H., & Jensen, G. (2021). Latest trends in the current treatment of proximal humeral fractures − An analysis of 1162 cases at a level-1 trauma centre with a special focus on shoulder surgery. *Zeitschrift Für Orthopädie Und Unfallchirurgie.* Available from https://doi.org/10.1055/a-1333-3951.

Diaz-Borjon, E., Yamakado, K., Pinilla, R., Keith, P., & Worland, R. L. (2007). Shoulder replacement in end-stage rotator cuff tear arthropathy: 5- to 11-year follow-up analysis of the bi-polar shoulder prosthesis. *Journal of Surgical Orthopaedic Advances, 16*(3), 123−130.

Erickson, B., Frank, R., Harris, J., Mall, N., & Romeo, A. A. (2015). The influence of humeral head inclination in reverse total shoulder arthroplasty: A systematic review. *Journal of Shoulder and Elbow Surgery / American Shoulder and Elbow Surgeons, 24*(6), 988−993. Available from https://doi.org/10.1016/j.jse.2015.01.001, Epub.

Ferle, M., Pastor, M. F., Hagenah, J., Hurschler, C., & Smith, T. (2019). Effect of the humeral neck-shaft angle and glenosphere lateralization on stability of reverse shoulder arthroplasty: A cadaveric study. *Journal of Shoulder and Elbow Surgery, 28*(5), 966−973. Available from https://doi.org/10.1016/j.jse.2018.10.025.

Giles, J. W., Langohr, G. D. G., Johnson, J. A., & Athwal, G. S. (2015). Implant design variations in reverse total shoulder arthroplasty influence the required deltoid force and resultant joint load. *Clinical Orthopaedics and Related Research, 473*(11), 3615−3626. Available from https://doi.org/10.1007/s11999-015-4526-0.

Glenday, J., Kontaxis, A., Roche, S., & Sivarasu, S. (2019). Effect of humeral tray placement on impingement-free range of motion and muscle moment arms in reverse shoulder arthroplasty. *Clinical Biomechanics, 62*, 136−143. Available from https://doi.org/10.1016/j.clinbiomech.2019.02.002.

Goutallier, D., Postel, J. M., Zilber, S., & Van Driessche, S. (2003). Shoulder surgery: From cuff repair to joint replacement. An update. *Joint, Bone, Spine: Revue du Rhumatisme, 70*(6), 422−432. Available from https://doi.org/10.1016/j.jbspin.2003.08.003.

Grammont, P., Trouilloud, P., Laffay, J., & Deries, X. (1987). Etude et realisation d'une nouvelle prothèse d'épaule. *Rhumatologie, 39*, 407−418.

Greiner, S., Schmidt, C., Herrmann, S., Pauly, S., & Perka, C. (2015). Clinical performance of lateralized vs non-lateralized reverse shoulder arthroplasty: A prospective randomized study. *Journal of Shoulder and Elbow Surgery, 24*(9), 1397−1404. Available from https://doi.org/10.1016/j.jse.2015.05.041.

Gutiérrez, S., Keller, T. S., Levy, J. C., Lee, W. E., & Luo, Z. P. (2008). Hierarchy of stability factors in reverse shoulder arthroplasty. *Clinical Orthopaedics and Related Research, 466*(3), 670−676. Available from https://doi.org/10.1007/s11999-007-0096-0.

Hamilton, M. A., Diep, P., Roche, C., Flurin, P. H., Wright, T. W., Zuckerman, J. D., & Routman, H. (2015). Effect of reverse shoulder design philosophy on muscle moment arms. *Journal of Orthopaedic Research, 33*(4), 605−613. Available from https://doi.org/10.1002/jor.22803.

Helmkamp, J. K., Bullock, G. S., Amilo, N. R., Guerrero, E. M., Ledbetter, L. S., Sell, T. C., & Garrigues, G. E. (2018). The clinical and radiographic impact of center of rotation lateralization in reverse shoulder arthroplasty: A systematic review. *Journal of Shoulder and Elbow Surgery, 27*(11), 2099−2107. Available from https://doi.org/10.1016/j.jse.2018.07.007.

Henninger, H., Barg, A., Anderson, A., Bachus, K., Burks, R., & Tashjian, R. Z. (2011). Effect of lateral offset center of rotation in reverse total shoulder arthroplasty: A biomechanical study. *Journal of Shoulder and Elbow Surgery / American Shoulder and Elbow Surgeons, 21*(9), 1128−1135. Available from https://doi.org/10.1016/j.jse.2011.07.034, Epub.

Kim, S., Kim, I., Jang, M., & Yoo, J. (2021). Complications of reverse shoulder arthroplasty: A concise review. *Clinics in Shoulder and Elbow, 24*(1), 42−52. Available from https://doi.org/10.5397/cise.2021.00066.

Kirzner, N., Paul, E., & Moaveni, A. (2018). Reverse shoulder arthroplasty vs BIO-RSA: Clinical and radiographic outcomes at short term follow-up. *Journal of Orthopaedic Surgery and Research, 13*(1), 256. Available from https://doi.org/10.1186/s13018-018-0955-2.

König, F. (1914). Ober die Implantation von Elfenbein zum Ersatz von Knochen und Gelenkenden. *Bruns' Beitrage fur Klinische Chirurgie, 85.*

Kontaxis, A., & Johnson, G. R. (2009). The biomechanics of reverse anatomy shoulder replacement − A modelling study. *Clinical Biomechanics, 24*(3), 254−260. Available from https://doi.org/10.1016/j.clinbiomech.2008.12.004.

Krueger, F. (1951). A vitallium replica arthroplasty on shoulder: A case report of aseptic necrosis of the proximal end of the humerus. *Surgery, 30,* 1005−1011.

Lädermann, A., Gueorguiev, B., Charbonnier, C., Stimec, B. V., Fasel, J. H. D., Zderic, I., ... Walch, G. (2015). Scapular notching on kinematic simulated range of motion after reverse shoulder arthroplasty is not the result of impingement in adduction. *Medicine (United States), 94*(38), e1615. Available from https://doi.org/10.1097/MD.0000000000001615.

Lädermann, A., Tay, E., Collin, P., Piotton, S., Chiu, C. H., Michelet, A., & Charbonnier, C. (2019). Effect of critical shoulder angle, glenoid lateralization, and humeral inclination on range of movement in reverse shoulder arthroplasty. *Bone and Joint Research, 8*(8), 378−386. Available from https://doi.org/10.1302/2046-3758.88.BJR-2018-0293.R1.

Lévigne, C., Garret, J., Boileau, P., Alami, G., Favard, L., & Walch, G. (2011). *Scapular notching in reverse shoulder arthroplasty: Is it important to avoid it and how? Clinical orthopaedics and related research* (vol. 469, pp. 2512−2520). New York LLC: Springer Issue 9. Available from https://doi.org/10.1007/s11999-010-1695-8.

Liou, W., Yang, Y., Petersen-Fitts, G., Lombardo., Stine, S., & Sabesan, V. J. (2016). Effect of lateralized design on muscle and joint reaction forces for reverse shoulder arthroplasty. *Journal of Shoulder and Elbow Surgery / American Shoulder and Elbow Surgeons, 26*(4), 564−572. Available from https://doi.org/10.1016/j.jse.2016.09.045, Epub.

Lo, E., Rizkalla, J., Montemaggi, M., & Krishnan, S. G. (2021). Clinical and radiographic outcomes of cementless reverse total shoulder arthroplasty for proximal humeral fractures. *Journal of Shoulder and Elbow Surgery / American Shoulder and Elbow Surgeons, 30*(8), 1949−1956. Available from https://doi.org/10.1016/j.jse.2020.11.009, Epub.

Luciani, P., Farinelli, L., Procaccini, R., Verducci, C., & Gigante, A. (2019). Primary reverse shoulder arthroplasty for acute proximal humerus fractures: A 5-year long term retrospective study of elderly patients. *Injury, 50*(11), 1974−1977. https://doi.org/10.1016/j.injury.2019.09.019. Epub

Masjedi, M., & Johnson, G. R. (2010). Reverse anatomy shoulder replacement: Comparison of two designs. *Proceedings of the Institution of Mechanical Engineers, Part H: Journal of Engineering in Medicine, 224*(9), 1039−1049. Available from https://doi.org/10.1243/09544119JEIM759.

Mollon, B., Mahure, S. A., Roche, C. P., & Zuckerman, J. D. (2016). Impact of glenosphere size on clinical outcomes after reverse total shoulder arthroplasty: An analysis of 297 shoulders. *Journal of Shoulder and Elbow Surgery, 25*(5), 763−771. Available from https://doi.org/10.1016/j.jse.2015.10.027.

Müller, A. M., Born, M., Jung, C., Flury, M., Kolling, C., Schwyzer, H. K., & Audigé, L. (2018). Glenosphere size in reverse shoulder arthroplasty: Is larger better for external rotation and abduction strength? *Journal of Shoulder and Elbow Surgery*, *27*(1), 44−52. Available from https://doi.org/10.1016/j.jse.2017.06.002.

Nam, D., Kepler, C. K., Neviaser, A. S., Jones, K. J., Wright, T. M., Craig, E. V., & Warren, R. F. (2010). Reverse total shoulder arthroplasty: Current concepts, results, and component wear analysis. *Journal of Bone and Joint Surgery - Series A*, *92*(2), 23−35. Available from https://doi.org/10.2106/JBJS.J.00769.

Neer, C. S., Brown, T. H., & McLaughlin, H. L. (1953). Fracture of the neck of the humerus with dislocation of the head fragment. *The American Journal of Surgery*, *85*(3), 252−258. Available from https://doi.org/10.1016/0002-9610(53)90606-0.

Nelson, R., Lowe, J. T., Lawler, S. M., Fitzgerald, M., Mantell, M. T., & Jawa, A. (2018). Lateralized center of rotation and lower neck-shaft angle are associated with lower rates of scapular notching and heterotopic ossification and improved pain for reverse shoulder arthroplasty at 1 year. *Orthopedics*, *41*(4), 230−236. Available from https://doi.org/10.3928/01477447-20180613-01.

Nwakama, A. C., Cofield, R. H., Kavanagh, B. F., & Loehr, J. F. (2000). Semiconstrained total shoulder arthroplasty for glenohumeral arthritis and massive rotator cuff tearing. *Journal of Shoulder and Elbow Surgery*, *9*(4), 302−307. Available from https://doi.org/10.1067/mse.2000.106467.

Nyffeler, R. W., Werner, C. M. L., Simmen, B. R., & Gerber, C. (2004). Analysis of a retrieved Delta III total shoulder prosthesis. *Journal of Bone and Joint Surgery - Series B*, *86*(8), 1187−1191. Available from https://doi.org/10.1302/0301-620X.86B8.15228.

Pean, J. (1894). Des moyens prosthetiques destines a obtenir la reparation de parties osseuses. *Gaz Hop Paris*, *67*, 291.

Pupello, D. (2016). Origins of reverse shoulder arthroplasty and common misconceptions. In. In M. Frankle, S. Marberry, & D. Pupello (Eds.), *Reverse shoulder arthroplasty* (pp. 3−18). Springer International Publishing. Available from https://doi.org/10.1007/978-3-319-20840-4.

Roche, C., Flurin, P. H., Wright, T., Crosby, L. A., Mauldin, M., & Zuckerman, J. D. (2009). An evaluation of the relationships between reverse shoulder design parameters and range of motion, impingement, and stability. *Journal of Shoulder and Elbow Surgery*, *18*(5), 734−741. Available from https://doi.org/10.1016/j.jse.2008.12.008.

Sayana, M., Kakarala, G., Bandi, S., & Wynn-Jones. (2008). Medium term results of reverse total shoulder replacement in patients with rotator cuff arthropathy. *Irish Journal of Medical Science*, *178*(2), 147−150. Available from https://doi.org/10.1007/s11845-008-0262-8, Epub.

Scarlat, M. M. (2013). Complications with reverse total shoulder arthroplasty and recent evolutions. *International Orthopaedics*, *37*(5), 843−851. Available from https://doi.org/10.1007/s00264-013-1832-6.

Schoch, B., Vasilopoulos, T., LaChaud, G., Wright., Roche, C., King, J., & Werthel, J. D. (2019). Optimal glenosphere size cannot be determined by patient height. *Journal of Shoulder and Elbow Surgery / American Shoulder and Elbow Surgeons*, *29*(2), 258−265. Available from https://doi.org/10.1016/j.jse.2019.07.003, Epub.

Schwarz, A., Hohenberger, G., Sauerschnig, M., Niks, M., Lipnik, G., Mattiassich, G., . . . Plecko, M. (2021). Effectiveness of reverse total shoulder arthroplasty for primary and secondary fracture care: Mid-term outcomes in a single-centre experience. *BMC Musculoskeletal Disorders*, *22*(1), 48. Available from https://doi.org/10.1186/s12891-020-03903-0.

Simovitch, R., Flurin, P. H., Wright, T. W., Zuckerman, J. D., & Roche, C. (2019). Impact of scapular notching on reverse total shoulder arthroplasty midterm outcomes: 5-year minimum follow-up. *Journal of Shoulder and Elbow Surgery*, *28*(12), 2301−2307. Available from https://doi.org/10.1016/j.jse.2019.04.042.

Sirveaux, F., Favard, L., Oudet, D., Huquet, D., Walch, G., & Molé, D. (2004). Grammont inverted total shoulder arthroplasty in the treatment of glenohumeral osteoarthritis with massive rupture of the cuff. Results of a multicentre study of 80 shoulders. *Journal of Bone and Joint Surgery - Series B, 86*(3), 388−395. Available from https://doi.org/10.1302/0301-620X.86B3.14024.

Terrier, A., Reist, A., Merlini, F., & Farron, A. (2008). Simulated joint and muscle forces in reversed and anatomic shoulder prostheses. *Journal of Bone and Joint Surgery - Series B, 90*(6), 751−756. Available from https://doi.org/10.1302/0301-620X.90B6.19708.

Torrens, C., Miquel, J., Martínez, R., & Santana, F. (2019). Can small glenospheres with eccentricity reduce scapular notching as effectively as large glenospheres without eccentricity? A prospective randomized study. *Journal of Shoulder and Elbow Surgery / American Shoulder and Elbow Surgeons, 29*(2), 217−224. Available from https://doi.org/10.1016/j.jse.2019.09.030, Epub.

Urch, E., Dines, J. S., & Dines, D. M. (2016). Emerging indications for reverse shoulder arthroplasty. *Instructional Course Lectures, 65*, 157−169.

Valenti, P., Sauzières, P., Katz, D., Kalouche, I., & Kilinc, A. S. (2011). *Do less medialized reverse shoulder prostheses increase motion and reduce notching? Clinical orthopaedics and related research* (vol. 469, pp. 2550−2557). New York LLC: Springer 9. Available from https://doi.org/10.1007/s11999-011-1844-8.

Werner, B. S., Chaoui, J., & Walch, G. (2017). The influence of humeral neck shaft angle and glenoid lateralization on range of motion in reverse shoulder arthroplasty. *Journal of Shoulder and Elbow Surgery, 26*(10), 1726−1731. Available from https://doi.org/10.1016/j.jse.2017.03.032.

Werthel, J. D., Walch, G., Vegehan, E., Deransart, P., Sanchez-Sotelo, J., & Valenti, P. (2019). Lateralization in reverse shoulder arthroplasty: A descriptive analysis of different implants in current practice. *International Orthopaedics, 43*(10), 2349−2360. Available from https://doi.org/10.1007/s00264-019-04365-3.

Wiedel, J. D. (1990). Dissociation of the glenoid component in the Macnab/English total shoulder arthroplasty. *Journal of Arthroplasty, 5*(1), 15−18. Available from https://doi.org/10.1016/S0883-5403(06)80004-5.

Wong, M. T., Langohr, G. D. G., Athwal, G. S., & Johnson, J. A. (2016). Implant positioning in reverse shoulder arthroplasty has an impact on acromial stresses. *Journal of Shoulder and Elbow Surgery, 25*(11), 1889−1895. Available from https://doi.org/10.1016/j.jse.2016.04.011.

Wright, M. A., & Murthi, A. M. (2021). Offset in reverse shoulder arthroplasty: Where, when, and how much. *The Journal of the American Academy of Orthopaedic Surgeons, 29*(3), 89−99. Available from https://doi.org/10.5435/JAAOS-D-20-00671.

Zumstein, M. A., Pinedo, M., Old, J., & Boileau, P. (2011). Problems, complications, reoperations, and revisions in reverse total shoulder arthroplasty: A systematic review. *Journal of Shoulder and Elbow Surgery, 20*(1), 146−157. Available from https://doi.org/10.1016/j.jse.2010.08.001.

DEVICES FOR TRAUMATOLOGY: BIOMECHANICS AND DESIGN 23

Pankaj Pankaj

Computational Biomechanics, School of Engineering, Institute for Bioengineering, The University of Edinburgh, United Kingdom

ORTHOPEDIC TRAUMA AND ITS TREATMENT

MECHANICAL PROPERTIES OF BONE

There are two main types of bones, cortical bone (also known as compact bone) and trabecular bone (also known as cancellous or spongy bone). Cortical bone is a dense, solid mass with microscopic channels. Trabecular bone, on the other hand, has much higher porosity compared to cortical bone and comprises a lattice of plates and rods known as the trabeculae. In terms of mechanical properties both these bone types are anisotropic (properties vary with direction) and heterogeneous (properties vary with location) (Donaldon, 2011; Donaldson et al., 2011; Pankaj, 2013). Bone's mechanical behavior is also known to be time dependent (Manda, Wallace, Xie, Levrero-Florencio, & Pankaj, 2017), which means that deformation is not instantaneous on load application. Bone's ability to sustain loads deteriorates with age and in particular with a disease called osteoporosis, which causes bone to become highly porous. Consequently, how bone deforms and fails depends on how loads are applied, where they are applied, the rate at which they are applied, and the individual who experiences them. It is apparent that the stresses levels that result bone yielding or failure vary with bone quality. Interestingly, however, studies have shown that the strains that cause failure are relatively uniform (Bayraktar et al., 2004; Levrero-Florencio et al., 2016). This implies that bone of poor quality reaches the failure threshold or fracture strains at lower loads (or lower stresses), whereas attainment of identical strains requires much higher loads for good-quality bone. Fractures are generally a result of trauma that can be caused by a simple fall from a standing position in osteoporotic bone; healthy bone requires larger forces such as a blow imparted in a contact sport (e.g., football) or in an automobile accident.

DEVICE MATERIALS FOR SURGICAL APPROACHES

Once fracture occurs, it needs to be treated. The traditional conservative method of treatment requires fracture reduction (or restoring fractured bone to its initial position and alignment) and stabilizing it through splints and plaster, sometimes combined with traction. Lorenz Böhler (1885–1973), a surgeon based in Vienna, was the chief proponent of this approach. Development

Human Orthopaedic Biomechanics. DOI: https://doi.org/10.1016/B978-0-12-824481-4.00033-0

of aseptic techniques, antibiotics, and surgical methods has led to the adoption of different surgical approaches to treating fracture.

Devices used in surgical approaches employ a variety of materials. An ideal material should be nontoxic to the body, inert, and corrosion-proof. It should also have very high strength and fatigue resistance. Unfortunately, there is no perfect material. Fracture fixation implants are most commonly made of metals such as stainless steel, titanium, and cobalt-chromium alloys (Thakur, 2020). Stainless steel (e.g., grade 316 L) is used extensively in fracture fixation implants. It is less expensive, has good corrosion resistance, and is bioinert in comparison to other biocompatible metals. Its Young's modulus is around 200 GPa, tensile yield strength around 500 MPa, and it has good ductility. Titanium alloy, such as Ti-6AL-4V, is also widely used in the manufacture of fracture fixation implants. It has a Young's modulus of around 110 GPa tensile strength of around 900 MPa. The ductility of titanium alloys is considerably lower than that of stainless steel.

Cobalt-based alloys are also used for making fracture fixation implants. They have a good corrosion resistance and a Young's modulus of around 250 GPa. Their fabrication, however, is expensive (Ganesh, Ramakrishna, & Ghista, 2005). Bioabsorbable polymers have also emerged as implants for fracture fixation. They provide stability during the healing process and are then slowly resorbed. Despite this apparent advantage, they have not gained popularity due to apprehensions about their provision of initial fixation stability and their ability to maintain strength as healing progresses (Thakur, 2020). Their acceptance, however, appears to be increasing.

TREATMENT OBJECTIVES

A fractured bone is not capable of carrying physiological loads on its own and fracture fixation devices aim to redirect load and shield the bone from undesirable motion during the process of healing (Claes, 2021; Gaston & Simpson, 2007; McKibbin, 1978). Introduction of an implant alters the natural load distribution in the host bone. This redistribution of load also results in some unwanted effects: stress shielding and strain concentration in the bone at the bone—implant interface. Stress shielding is defined as the reduction in the stress carried by the bone through sharing it with stiff implants. In the absence of loading stimulus to the bone, which diminishes continued remodeling, the bone becomes less dense and weaker. Stress shielding has received considerable attention in research (e.g. Uhthoff, Poitras, & Backman, 2006). Strain concentrations in the bone arise because loads are transferred from the bone to the devices via the small bone—implant interface areas, for example, from the bone surface in contact with a screw. These concentrations can take the bone beyond its yield capacity which can result in implant loosening. These stress—strain concentrations have received relatively little attention in literature (MacLeod & Pankaj, 2014; Pankaj & Xie, 2019; Xie, Manda, & Pankaj, 2018).

For any fixation device, there are three key clinical requirements and consequent mechanical demands arising from them (MacLeod & Pankaj, 2014). First, the device must promote healing by providing appropriate level of "stability." Stability is a term used by clinicians to describe relative motion between the bone fragments at the fracture site or interfragmentary motion (IFM). Complete absence of IFM is often termed absolute stability, while small IFM is called relative stability (Thakur, 2020). Correct amount of IFM is crucial for healing (Claes, 2021; Gaston & Simpson, 2007); this aspect is further discussed in the following section. In case of a post-surgery fracture gap, the device takes most of the load and consequently IFM is strongly related to the stiffness of the device.

The second key requirement is that the device should not fail and continue to carry loads in the fracture healing duration. This implies that the stresses sustained by the implant due to physiological loading process are below their failure limits. In general, implant failure occurs when healing is delayed and implants continue to carry cyclic loads and fail due to fatigue (Vallier, Hennessey, Sontich, & Patterson, 2006). Life span due to fatigue reduces significantly with a small increase in the magnitude of cyclic loads. Failure of devices is rarely due to a single event with excessive loading.

Lastly, devices should not cause excessive strain concentrations in the bone at the bone—implant interface (as mentioned earlier); these can cause loosening which may be followed by infection (Fragomen & Rozbruch, 2007; Moroni, Vannini, Mosca, & Giannini, 2002). Large strains in the bone also cause patient discomfort and can result in permanent damage to the bone (Turner, 2006). These requirements are not independent—faster healing helped by optimum IFM will prevent device loosening and reduction in the number of cycles of loading the device has to carry thereby reducing fatigue strength demand.

FRACTURE HEALING

There are two types of fracture healing: direct healing, which is also called primary healing, and indirect healing, which is also termed as secondary healing. Primary healing uses the body's osteoclast-driven remodeling mechanism, requires the fracture to be completely reduced, and conditions close to absolute stability. Direct healing is slower (Perren, 2002; Woo et al., 1984) and requires a more invasive surgical procedure, which consequently entails higher risk of delayed healing and infection (Perren, 2002; Strauss, Schwarzkopf, Kummer, & Egol, 2008). As a result, there has been a shift towards more "biological" fixation, which aims to reduce the detrimental impact of the implant on the bone, utilize indirect healing and improve vascularization (Perren, 2002). Indirect healing is a more natural form of fracture repair. It requires a certain amount of IFM for progression of callus formation and also, therefore, sometimes referred to as callus healing. Callus formed in the medullary cavity is called medullary callus and that formed on the external surface of the bone periosteal callus. Callus between bone ends is referred to as gap callus. This chapter does not discuss the biology of bone healing, but focuses on the suitable mechanical environment required for it to occur. Another term used in fracture healing literature is interfragmentary strain (IFS), which is the ratio of IFM to the initial fracture gap. Healing is known to occur based on the original fracture gap and IFS (Claes, 2021; Perren, 2002; Perren et al., 1973). Studies in sheep models have shown that one to three osteotomy gaps heal under compressive IFS of 7%—31% and large gaps do not heal. Claes and Meyers (2020) suggest that compressive IFS stimulates greater vessel formation than tensile or shearing strain. The primary message from these studies is that, for secondary healing via callus formation, some flexibility is required in the fixation device to permit optimum IFM/IFS.

IMPLANTS USED FOR FRACTURE FIXATION

In this chapter, we consider three primary categories of fixation devices: (1) external fixators; (2) plates and screws; and (3) intramedullary devices—rods or nails. The choice of device depends on the fracture location and its type and extent. Factors such as patient's age and weight also play a role. With increasing aging population, the importance of bone quality in treatment also needs to be considered; this is often not taken into account in a robust manner. While thumb rules exist, currently there appears to be no standardized accepted methodology for selecting a fixation device and its configuration for different fracture types and bone properties to achieve the best possible clinical outcome. Studies have shown that the way a device is selected and configured by different surgeons varies significantly even for similar fractures (Saving, Ponzer, Enocson, Mellstrand Navarro, & Blank, 2018) and the choice has a surgeon bias (Ansari, Adie, Harris, & Naylor, 2011). While this chapter focuses on the fulfillment of mechanical demands, a surgeon also needs to consider the effect of treatment on soft tissues, nerves, and blood vessels, which is not discussed here.

EXTERNAL FIXATORS

External fixation is commonly used for high-energy injuries that cause open fractures, which also require extensive soft tissue repair. External fixation comprises use of screws, pins, or wires to restrain the bone on both sides of the fracture site. These are attached to an external frame made of stiff rings or bars. Wires are always supported on both sides of the bone they traverse, while pins can be supported on only one side or both sides. There are two main types of external fixators: pin fixators and ring fixators.

PIN FIXATORS

Pin fixators readily stabilize diaphyseal fractures and permit wound access for management of soft tissue injuries. Pins may be applied in varying configurations varying from unilateral uniplanar configurations (Fig. 23.1) to bilateral and multiplanar configurations (Fragomen & Rozbruch, 2007; Grubor & Borrelli, 2020). As mentioned above, IFS plays an important role in fracture healing,

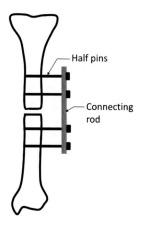

Half pins

Connecting rod

FIGURE 23.1

Unilateral and uniplanar fixator.

which depends on bone-fixator stiffness. Therefore fixator stiffness has been extensively investigated over several decades (Behrens & Johnson, 1989; Fragomen & Rozbruch, 2007). Behrens and Johnson (1989) tested different configurations in not only compression (i.e., load being applied in along the axis of the bone) but also in bending and torsion. Their study showed that unilateral and uniplanar configurations had the lowest stiffness.

Unilateral configuration, shown in Fig. 23.1, that uses screws/pins rigidly connected to a stiff connection rod on a single side of a limb (also called half-pins) is very commonly used. These devices are fairly unobtrusive and are often better accepted than devices that encircle the whole limb, particularly by children (Gordon et al., 2003). Fragomen and Rozbruch (2007) discussed the parameters that influence the stiffness of unilateral fixators and suggested that reducing pin separation on either side of the fracture and decreasing the distance between the bar and the bone increase the stiffness. As expected, increasing the stiffness of the external bar, pin numbers and diameter also increase the stiffness.

While stiffness has been extensively investigated, the influence of bone quality on fixation configuration and device materials has received relatively less attention in research. Strain localization at the implant−bone interface is the cause of the most common complications—loosening and infection, which are often severe enough to require implant removal (Moroni et al., 2002). Pin design, placement, and thermal injury during predrilling have all been shown to promote loosening (Wikenheiser, Markel, Lewallen, & Chao, 1995). Dynamic loading, induced by daily activity, also increases the risk of pin loosening by the repeated action of alternating loads (Pettine, Chao, & Kelly, 1993). Infection of pin sites can act in combination with mechanical loosening to destabilize fixation (Moroni et al., 2002). It is not straightforward to investigate loosening using in vitro experiments as it is not readily possible to find bone with similar geometries but having varying bone qualities to evaluate the effect of different configurations. Moreover, there is no apparent way to measure strains at the bone−pin interface. Computational modeling approaches present a viable alternative. However, modeling too presents a range of challenges: (1) how to model bone−pin/screw interaction—does one assume the two surfaces to be tied (Shahar & Shani, 2004) or sliding (Huiskes, Chao, & Crippen, 1985; MacLeod, Pankaj, & Simpson, 2012); (2) how to model bone's nonlinear material behavior—do you assume it to be elastic (Karunratanakul, Schrooten, & Van Oosterwyck, 2010; MacLeod, Simpson, & Pankaj, 2016), elasto-plastic (Donaldson et al., 2012a), or use other complex nonlinear models; (3) how does one model bone's time-dependent behavior (Xie et al., 2018)—most studies assume time-independent properties; and (4) should bone's anisotropy and inhomogeneity be included—most fixation studies assume bone to be isotropic and homogeneous (Karunratanakul et al., 2010; Shahar & Shani, 2004).

Donaldon (2011) and Donaldson et al. (2012a) considered loosening of unilateral fixator in a tibial midshaft in which they included a frictional bone−implant interface; a strain-based plasticity criterion (Pankaj & Donaldson, 2013) to describe the nonlinear behavior of bone; orthotropic and inhomogeneous bone properties (Donaldson et al., 2011); and change in bone geometry with age (Russo et al., 2006). They found that bone yielding (an indicator of loosening) was concentrated at the pin entrance sites (cortex close to the connecting rod). Bone yielding initiated on the periosteal side of the half-pin threads and in case of poor-quality bone propagated through the entire cortical thickness (Fig. 23.2). Peri-implant yielding at the opposite cortex was only observed with the old bone, and in these cases, only small regions of yielded bone were observed. They also observed that a couple was generated between the far and near half-pins such that they experienced tensile

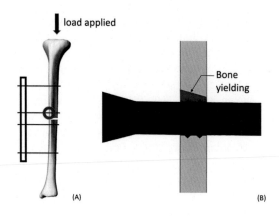

FIGURE 23.2

(A) Unilateral fixation subjected to axial load. (B) Bone yielding at the near cortex for poor-quality bone.

Modified from Donaldson, F. E., Pankaj, P., & Simpson, A. H. R. W. (2012a). Bone properties affect loosening of half-pin external fixators at the pin-bone interface. Injury, *43(10), 1764–1770. https://doi.org/10.1016/j.injury.2012.07.001.*

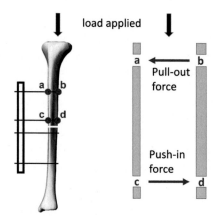

FIGURE 23.3

Couple generated to prevent rotation of the bone fragment in unilateral fixation.

Modified from Donaldson, F. E., Pankaj, P., & Simpson, A. H. R. W. (2012a). Bone properties affect loosening of half-pin external fixators at the pin-bone interface. Injury, *43(10), 1764–1770. https://doi.org/10.1016/j.injury.2012.07.001.*

and compressive axial forces, respectively, to resist the rigid body rotation of the bone fragment (Fig. 23.3). These pull-out and push-in forces between the far and near half-pins resulted in considerable load transfer in the radial direction. In old bone, this couple (Fig. 23.3) in conjunction with yielding of the full cortical thickness (Fig. 23.2) can cause loss of screw purchase resulting in device loosening. The largest yielding occurred in the pins closest to the fracture site and the amount of yielding decreased, expectedly, with increasing number of pins. There were two particularly interesting findings in these computational studies. First, they showed that while bone quality strongly influenced loosening it had a relatively small effect on IFM, which increased only slightly with poor bone quality. This feature was also observed later for locking plates (MacLeod et al., 2016). This implies that IFM can be fairly accurately estimated without including bone quality. Second, the studies showed that, in comparison to stainless steel pins, the flexibility of titanium pins results in much higher yielding at bone−implant interface. Therefore the general notion that titanium is always better than stainless steel needs to be reconsidered.

RING FIXATORS

The ring fixator was developed in 1954 by the Russian surgeon Gavriil Ilizarov. The Ilizarov method of fracture fixation makes use of pretensioned, thin wires (1.5−1.8 mm in diameter), which transfix the bone, supported by circular rings and stiff longitudinal bars (Fig. 23.4). Ilizarov fixators are also extensively used in the corrective treatment of nonunions, posttrauma residual misalignment, and limb deformities (by distraction osteogenesis).

As with other fixations, Ilizarov fixators have been extensively investigated for their stiffness (Fleming, Paley, Kristiansen, & Pope, 1989; Podolsky & Chao, 1993; Zamani & Oyadiji, 2008). It is the wire components rather than the frame that are critical for stiffness. On application of loads, the wires sag, behaving like cables rather than beams. As the sagging increases, so does the tension in the wires, resulting in a nonlinear load-displacement response, that is, increasing stiffness. Relatively larger IFS is caused on application of small loads, which stimulates osteogenesis by indirect healing, while their increasing stiffness on application of higher loads prevents excessive IFS, which is deleterious to bone formation. The rapid fracture healing times that have been reported with the Ilizarov technique have been attributed to these mechanical qualities (Fleming et al., 1989). This nonlinear response can be readily derived using basic principles of mechanics (Zamani & Oyadiji, 2008). It can be shown that fixator stiffness increases by increasing wire pretension, using wires with a higher Young's modulus, increasing wire diameter, and using more wires on either side of the fracture. It has been reported that small amounts of wire slippage at their ring

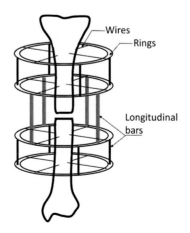

FIGURE 23.4

Ilizarov ring fixator.

attachments ($\sim$0.5 mm) could result in complete loss of pretension (Zamani & Oyadiji, 2008), therefore design of clamps to prevent this from happening is of great importance. Clearly, there is a limit to the amount of pretension that can be applied as too large tension can result in wire yielding and breakage. Wire yielding has been reported to occur under walking loads at the clamps and at the periosteum. In metal wire yielding is followed by work-hardening resulting in strains that reduce pretension (Hillard, Harrison, & Atkins, 1998).

While the work on fixator stiffness and failure of wires has been significant, relatively little work has been done on bone yielding and failure at the wire—implant interface. Concerns that thin wires will cut through poor-quality bone like cheese cutters have been unfounded. Ilizarov wires are associated with a lower rate of loosening than external fixation using half-pins (Ali, Burton, Hashmi, & Saleh, 2003; Board, Yang, & Saleh, 2007). Donaldon (2011) and Donaldson et al. (2012b) undertook a finite element analysis in which they used anisotropic (Donaldson et al., 2011) and strain-based plasticity for bone (Pankaj & Donaldson, 2013) and sliding interaction between wires and bone. Bone yielding was used as an estimate for loosening. As for unilateral fixators, they found that IFM increased only slightly for poor bone quality, that is, it was largely independent of bone quality. Increasing wire tension reduced bone yielding and yielding increased dramatically with poorer bone quality. They found that Ilizarov wire fixation produces axial displacements while retaining lateral stability due to the fact that wires are installed in crossed pairs. Unilateral fixation, on the other hand, was found to induce cantilever-like bone fragment displacement and thus relies on screw pull-out resistance for lateral stability. An interesting observation was that under Ilizarov fixation the yielded bone remained concentrated separately at the periosteum and endosteum (Fig. 23.5), superior and inferior to the wire, respectively. Since a substantial proportion

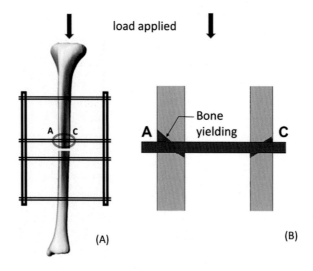

FIGURE 23.5

(A) Ilizarov ring fixator subjected to axial load. (B) Bone yielding pattern.

Modified from Donaldson, F. E., Pankaj, P., & Simpson, A. H. R. W. (2012b). Investigation of factors affecting loosening of Ilizarov ring-wire external fixator systems at the bone-wire interface. Journal of Orthopaedic Research, 30(5), 726–732. https://doi.org/10.1002/jor.21587.

of the cortical thickness does not yield around Ilizarov wires, this remaining bone continues to provide stability. This contrasts with unilateral fixators where yielding penetrates the full cortical thickness (for poor bone quality) (Fig. 23.2), which as discussed earlier, would result in the loss of screw thread purchase and half-pin loosening especially when accompanied by push-in and pull-out forces. This explains why Ilizarov fixators have fared better than unilateral fixators.

INTERNAL FIXATION—PLATES AND SCREWS

Two types of plate fixations are used most commonly: dynamic compression plates (DCP) (Fig. 23.6A) and the relatively recently developed locking compression plates (LCP) (Fig. 23.6B). The screw design is different and the manner in which they work is significantly different (MacLeod, 2014; MacLeod, Simpson, & Pankaj, 2015) (Fig. 23.7). The DCP uses compression screws that tighten the plate to the bone (Figs. 23.6A and 23.7A), whereas the screw in LCP is "locked" within the plate and remains at a fixed angle to it (Figs. 23.6B and 23.7B).

DCP may also have angled screw holes in the plate to induce a preload between fracture fragments during screw insertion. "Dynamic" compression is induced between bone fragments either by using a tensioning device or generated by the screws within the holes during insertion. This compression aids in fracture reduction and in preventing any gaps between fractured bone ends. Thus compression plating aims to achieve direct bone healing using "absolute stability" with minimal IFM. Absolute stability of all fragments may be too difficult to achieve for cases where the fracture pattern is segmental or comminuted; in this case "relative stability" is the goal, where the primary fragments are secured and the remaining fragments or debris are untouched. In all cases, as

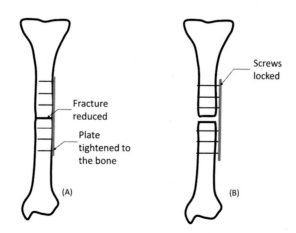

FIGURE 23.6

(A) DCP. Fracture is reduced and plate tightened to the bone with an aim to achieve absolute stability. (B) LCP behaves like unilateral external fixation. The fracture may not be completely reduced and there may be a gap between the plate and the bone. *DCP*, dynamic compression plate; *LCP*, locking compression plate.

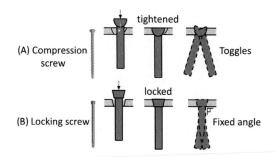

FIGURE 23.7

(A) Compression screw tightens the plate to the bone and can toggle within the plate. (B) Locking screw is locked within the plate and remains at a fixed angle.

From MacLeod, A. R. (2014). Modelling and optimising the mechanical behaviour of fractures treated with locking plates *(Ph.D. thesis). The University of Edinburgh.*

discussed earlier, unless all fracture displacement is almost completely avoided, indirect bone healing ensues.

In LCP, the plate is not tightened against the bone and is generally offset from it, helping to preserve the periosteum (Mushtaq, Shahid, Asif, & Maqsood, 2009; Perren, 2002). This is also known as bridge plating. The mechanical functioning of LCP is similar to a unilateral fixator, which is internal rather than external.

LCP has been extensively presented as a superior performer in comparison to DCP in older poorer quality bone (Gardner, Helfet, & Lorich, 2004; Kim, Ayturk, Haskell, Miclau, & Puttlitz, 2007; Yanez, Carta, & Garces, 2010). Failure and fatigue strength and, ultimately, failure load for locked plating has been reported to be higher (Seide et al., 2007; Uhl et al., 2008; Zehnder, Bledsoe, & Puryear, 2009). Some other studies have, however, shown that compression screws perform better than locked plating in healthy bone as their pull-out strength increases with bone density (Kim et al., 2007). The reasons for the differences involve multiple factors. The preloads involved in compression screw tightening induce strains at the screw–bone interface even before physiological loads are applied, whereas locking screws have negligible screw tightening preloads if any. During physiological loading, however, the DCP allows for frictional load transfer at the plate–bone interface; locked plating, on the other hand, transfers all physiological loads via the screw–bone interface. Therefore the local environment around compression screws in a DCP is influenced by a mix of effects, whereas the mechanics related to locking screws is more straightforward—similar to a unilateral external fixation device.

Locking and compression screws have been compared using laboratory experiments (Kim et al., 2007; Seide et al., 2007; Uhl et al., 2008; Yanez et al., 2010); while these tests are able to compare the performance, they cannot fully explain the mechanics associated with the difference.

MacLeod et al. (2015) examined the stress–strain response around the screws using finite element simulation; this is not possible in a laboratory experiment. They considered both DCP and LCP with and without fracture gaps in a tibia fracture model. The DCP is not generally used as a bridge plate; however, complete reduction of the fracture may not always be achieved. For DCP models, screw tightening preloads and interfragmentary compression along with frictional interface at the bone–implant interface were included. An axial load was then applied to the bone. The study showed that locking plates produce lower tensile strains around screw-hole locations in osteoporotic bone than the DCPs. In healthy bone, there was a much smaller difference between the two types of plating. In both types of plating, an incomplete fracture reduction or a fracture gap resulted in increased strain concentration around screw holes. These results support clinical conclusions that simple fractures in healthy bone should be treated with compressive reduction and absolute stability (Leahy, 2010). Compression plated osteoporotic bone, however, results in regions of very high tensile strain, even with complete fracture reduction. While strains due to weight-bearing may be reduced by the addition of more screw, the strains caused by screw fastening in DCPs will be present regardless of the number of screws used. As for external fixators, appropriate IFM is a key requirement for locking plates. Working length, also known as the bridging span, defined as the distance between the two innermost screws on either side of the fracture is one of the most important parameters for regulating IFM (MacLeod & Pankaj, 2018). For comminuted fractures with a fracture gap, the implant carries all the load (prior to healing) and, therefore, a larger working length means larger IFM and larger stresses experienced by the plate (Fig. 23.8A). For nearly reduced fractures or those with a relatively small fracture gap, load sharing (between the bone and the implant) comes into play earlier for a large working length (less stiff system) inducing lower stresses in the plate in comparison to a small working length where load sharing will be initiated only after considerable stresses have been experienced by the plate (Fig. 23.8B). This difference between load sharing and nonload sharing systems is poorly understood, and some studies have

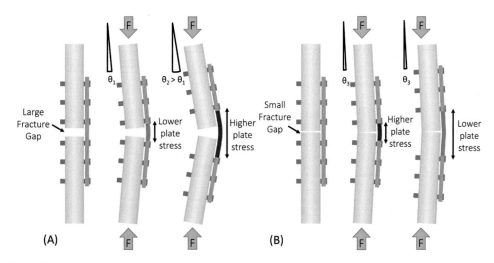

FIGURE 23.8

Influence of working length and fracture gap on plate stress in a bone-plate system under applied axial load where: (A) there is a large fracture gap and interfragmentary contact cannot occur (the plate is supporting all applied loads), resulting in larger plate stresses and larger angular deformation for a system with larger working length; and (B) there is a small fracture gap in which load sharing is enabled for a larger working length resulting in lower plate stresses in comparison to a shorter working length where there is no load sharing.

From MacLeod, A. R., & Pankaj, P. (2018). Pre-operative planning for fracture fixation using locking plates: device configuration and other considerations. Injury, 49, S12–S18. https://doi.org/10.1016/S0020-1383(18)30296-1.

attributed small working length to higher plate stresses even for cases with a fracture gap and non-load sharing arrangement (Gardner et al., 2004; Gautier & Sommer, 2003; Smith, Ziran, Anglen, & Stahel, 2007). Thus device performance varies with fracture pattern. It was found that larger working lengths produced the lowest plate stresses for small fracture gaps (1 mm), and the highest plate stresses for larger fracture gaps (6 mm).

The stiffness of the bone-plate system also depends on the plate material. Unlike external fixators where the external frame is relatively stiff and the IFM arises primarily due to the bending or sagging of pins or wires, in locking plates it is plate bending that contributes most to IFM. It is apparent that titanium, with a lower Young's modulus than steel, will produce greater IFM. Also the geometry of the plate, particularly the bending stiffness, influences the IFM.

The bone-plate offset in locking plates is another important parameter that influences stiffness and, therefore, IFM. This parameter is important in unilateral fixators as well; however, the effect on IFM for them is due to the change in the cantilevered length of the pins. As discussed above, IFM for locking plates is generated largely due to plate bending which is subjected to both bending and axial forces, so larger offsets imply increased bending forces. IFM generated by plate bending is considerably asymmetric with the near cortex (cortex close to the plate) experiencing much smaller IFM in comparison to the far cortex. Techniques that utilize bending of the locking screws to increase the effective bone-plate distance have been described in a number of studies using "far-cortical locking screws" (Bottlang & Feist, 2011), "dynamic locking screws," (Dobele et al., 2010) and "near cortical slots" (Gardner, Nork, Huber, & Krieg, 2010). These techniques have been shown to generate increased motion at the fracture site; however, they also produce larger strains at the screw−bone interface due to the larger lever arm (MacLeod & Pankaj, 2014; Moazen et al., 2013) (Fig. 23.9).

As with external fixators, screw loosening is reported as one of the most frequent complications in locked plating and is commonly attributed to an incorrect choice of screw configuration. MacLeod et al. (2016) used finite element simulations to evaluate screw placement to reduce

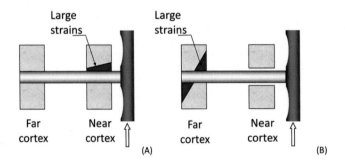

FIGURE 23.9

Large interface strain pattern for (A) locking plate and (B) far-cortical locking. The applied loading direction is shown. Far-cortical locking generates a more uniform IFM at the fracture gap; the screw, however, is only supported at the far cortex. *IFM*, interfragmentary motion.

Modified from MacLeod, A. R., & Pankaj, P. (2014). Computer simulation of fracture fixation using extramedullary devices: An appraisal. In B. Doyle, K. Miller, A. Wittek, and P. M. F. Nielsen (Eds.), Computational biomechanics for medicine: Fundamental science and patient-specific applications *(Vol. 9781493907458, pp. 87−99). Springer New York. https://doi.org/10.1007/978-1-939-0745-8_7.*

loosening risk in healthy and poor-quality bone. Strain levels within the bone were used as indicators of regions that may undergo loosening. They found that, in healthy bone under axial loading, the most important variables influencing strain levels at the bone—screw interface were the size of the bridging span (working length) and the plate stiffness. Larger working lengths were found to not only increase IFM and stresses within the plate, but also increase strains at the interface. Increasing the number of screws beyond three on either side of the fracture was found to have minimal influence on interfacial strain predictions regardless of the position of the screws. This was because the two screws closest to the fracture, on either side of the fracture, had the largest interfacial strain values around them in all cases (Fig. 23.10). As with external fixators, reduced bone quality was found to have little influence on IFM and plate stress but it substantially alters the strains experienced at the bone—implant interface (Fig. 23.10). They also found that for the same working length the manner in which rest of the screws were employed had a significant effect on interfacial strains in osteoporotic bone in comparison to healthy bone. Interfacial strain levels in osteoporotic bone were found to be lowest when using a two-hole spacing between screws on either side of the fracture (i.e., screw closest to the fracture site the next one on the same fracture fragment). In healthy bone, the influence of this spacing was much smaller.

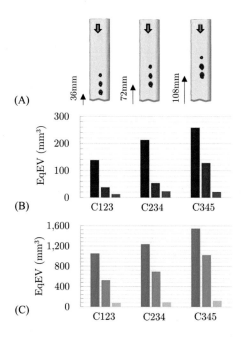

FIGURE 23.10

Predicted volumes of bone above 0.02% equivalent strain (EqEV) for different working lengths. (A) Screw arrangements C123, C234, and C345. EqEV values at different screw locations for (B) healthy bone and (C) osteoporotic bone. Load of 250 N is applied from above and the fracture is located below.

From MacLeod, A. R., Simpson, A. H. R. W., & Pankaj, P. (2016). Age-related optimization of screw placement for reduced loosening risk in locked plating. Journal of Orthopaedic Research, *34(11), 1856–1864. https://doi.org/10.1002/jor.23193.*

INTRAMEDULLARY NAILING

German orthopedic surgeon Gerhard Küntscher initiated intramedullary nails for fixation of femoral fractures; an intramedullary nail acts as an internal splint, allowing a fractured bone to heal (Fig. 23.11). Fractures of the tibia, femur, and humerus are often treated with intramedullary nailing. The stiffness of intramedullary nails can be altered by changing material used in their construction, their cross-sectional shape, and the length over which they are applied. Both stainless steel (316L) and titanium alloy (Ti-6AL-4V) discussed earlier have been used in the fabrication of intramedullary nails. Since the Young's modulus of elasticity of titanium alloy is much closer to that of

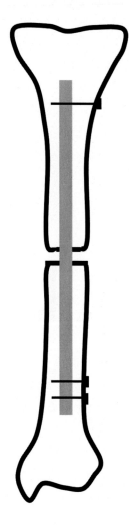

FIGURE 23.11

Locked intramedullary nail.

cortical bone, it appears to be a more suitable material for intramedullary nails (Woods & Della Rocca, 2020); however, clinical results with stainless steel have also been good (Clavert, Gicquel, & Giacomelli, 2010). Küntscher's original design was stainless steel nail with a V-shaped cross-section, which was later replaced by the more universally recognized cloverleaf-shaped nail and several design changes followed (Woods & Della Rocca, 2020). As bones are not straight (e.g., the femur), nails need to be flexible to conform to the endosteal surface. A good conformity also provides a frictional fit to maintain fracture reduction. Solid nails, cannulated nails, and slotted nails have all been used for fracture fixation. Cannulated nails have a much larger torsional stiffness in comparison to slotted nails; and solid nails are even stiffer.

Küntscher employed medullary reaming to insert a nail with a larger cross-section for increased bone–nail contact. Reaming the medullary channel for insertion of an intramedullary nail causes destruction of vessels and reduces the endosteal blood supply (Claes, 2021). To prevent these negative effects of reaming, unreamed intramedullary nails and approaches requiring minimal reaming were developed. The less invasive unreamed nailing spares the perfusion in the early phase of bone healing but does not lead to better bone healing in the latter phase (Claes, 2021). Shear and tensile IFMs permitted by loosely fitting unreamed nailing override the advantages of preserving perfusion in the early phase (Claes, 2021).

It is possible to treat simple and minimally comminuted midshaft fractures of long bones without using any interlocking screws. The locked nail, however, is now the popular implant for diaphyseal fracture fixation (Thakur, 2020). The distal and proximal parts of the fracture are reduced and aligned using screw insertions at the two ends (termed as static locking) of the nail to interlock it. Interlocking contributes to the reduction in IFM and induces rotational stability. When screws are used only on one end of the nail, the fixation is called dynamic locking. Clearly for dynamic locking to remain stable a significant proportion of the medullary canal needs to be in contact with the nail. Also dynamic interlocking with excessive postop weight-bearing can result in shortening.

EFFECT OF HEALING ON DEVICE CHOICE AND CONFIGURATION

With the progression of indirect healing and callus formation, a device that was bearing all the load will start sharing it with the bone. Early healing results in early load sharing, while delayed healing will cause the device to continue carrying physiological loads for a longer period. Clinical studies on fracture treatments with locking plates have shown that that more flexible bone-plate systems (Henderson et al., 2011; Lujan et al., 2010) lead to faster callus formation. On the other hand, lab studies with load bearing plates indicate that stiffer constructs perform better (Chao, Conrad, Lewis, Horodyski, & Pozzi, 2013; Hoffmeier, Hofmann, & Mückley, 2011; Schmidt, Penzkofer, Bachmaier, & Augat, 2013; Stoffel, Dieter, Stachowiak, Gächter, & Kuster, 2003). MacLeod (2014) and MacLeod and Pankaj (2018) evaluated the influence of callus formation on plate stresses using finite element simulations for a short working length (stiffer bone–implant system) and large working length (a more flexible system). To represent healing, callus was assumed to have varying properties with stiffness increasing with time, based on existing studies (Fig. 23.12A) (Comiskey, Macdonald, McCartney, Synnott, & O'Byrne, 2010; Steiner, Claes, Ignatius, Simon, & Wehner, 2014). With callus formation, the bone starts sharing the load and IFM decreases as do the

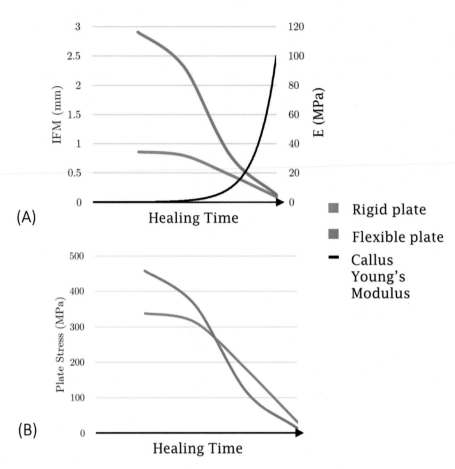

FIGURE 23.12

The influence of rigid and flexible locking plates on (A) the IFM produced at the fracture site for a given callus Young's modulus; (B) the resulting plate stresses. A flexible plate can become a load sharing plate earlier than a rigid plate. *IFM*, interfragmentary movement.

From MacLeod, A. R., & Pankaj, P. (2018). Pre-operative planning for fracture fixation using locking plates: Device configuration and other considerations. Injury, 49, *S12–S18. https://doi.org/10.1016/S0020-1383(18)30296-1.*

plate stresses for both the stiff and flexible systems. However, they decrease faster for the more flexible plate despite starting higher (Fig. 23.12B). Thus a flexible system becomes a load sharing system faster with reduced plate stresses. Early load sharing of a flexible system implies that the implant does not need to carry cycles of relatively high stresses which reduces the chances of fatigue. It can be inferred that, clinically, the reason longer working lengths reduce the risk of plate failure in the medium to long term is because of increased load sharing at the fracture site (MacLeod & Pankaj, 2018). Clearly, in vitro tests do not incorporate this effect of healing and therefore their predictions are only good for the prehealing period.

BOUNDARY CONDITIONS

The aim of undertaking in vitro experiments or numerical simulations of the mechanical behavior of bone-fixator systems is to understand how they are likely to behave in an in vivo scenario. In the human body, bone is supported in complex ways—at the joints and by muscles and ligaments—and experiences a wide range of forces caused by varied physiological activities. These loading and boundary conditions cannot be replicated in lab experiments and are not easy to incorporate in computational models (Pankaj, 2013; Phillips, Pankaj, Howie, Usmani, & Simpson, 2007). In fact, the need to validate computational modeling has resulted in developing these models to simulate lab experiments rather than in vivo conditions. In the evaluation of the behavior of bone-fixator systems of long bones subjected to loads, fully restrained boundary conditions (also called clamped or potted conditions) are often used in experimental and numerical work to provide stability to the bone (Bottlang, Doornink, Fitzpatrick, & Madey, 2009; Yanez, Cuadrado, Carta, & Garces, 2012) although they have been criticized in literature (Speirs, Heller, Duda, & Taylor, 2007). Another approach involves load application through a universal joint that permits rotation but restrains translation (Hoffmeier et al., 2011; Stoffel et al., 2003). In numerical simulations, it is possible to clamp one end and apply a load at the other end, which has no displacement/rotational restraints; this is difficult to achieve in experiments in which a loading jack generally has a fixed direction of travel. Fundamental mechanics shows that boundary conditions influence not only stiffness or IFM but the entire mechanical response. This is reflected in the wide range of stiffness values predicted in the literature for the same fixator type (Bottlang et al., 2010; Hoffmeier et al., 2011; Stoffel et al., 2003). This influence of boundary conditions is not frequently discussed in the evaluation of bone-fixator systems. Some recent studies (MacLeod & Pankaj, 2014; MacLeod, Simpson, & Pankaj, 2018) evaluated bone-locking plate constructs under different boundary conditions (Fig. 23.13). They found that for the configuration shown in Fig. 23.13, the predicted stiffness can vary by over seven times; this difference can be significantly higher for other configurations. These studies on locking plates also showed that the screw location around which largest strains occur also changes with boundary conditions; the screw farthest from the fracture is critical for conditions shown in Fig. 23.13A and B (Bottlang et al., 2010), whereas the screw closest to the fracture is critical for the condition shown in Fig. 23.13C (Donaldson et al., 2012a; MacLeod & Pankaj, 2014; Oni, Capper, & Soutis, 1993).

TIME-DEPENDENT PROPERTIES OF BONE

Experiments on bone-fixator systems have shown that fixator loosening under cyclic loading is a function of the number of cycles (Born, Karich, Bauer, von Oldenburg, & Augat, 2011). Theoretically, time-independent material models of bone are unable to explain this as the response does not change with an increasing number of cycles. Some recent studies have attempted to explain this using time-dependent behavior of bone (Pankaj & Xie, 2019; Xie et al., 2018). Xie et al. (2018) considered the influence of cyclic loading in an idealized unicortical bone—screw system (Fig. 23.14A and B). In this the screw was subjected to 500 cycles of lateral loads (Fig. 23.15C) with loading frequency $f = 1$ Hz followed by 1000 s recovery. The trabecular bone

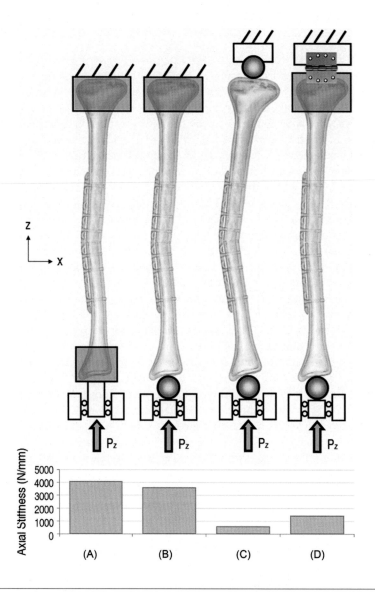

FIGURE 23.13

Examples of typical loading conditions employed in studies and the resulting axial stiffness of the construct: (A) fully restrained proximally and distally (Yanez et al., 2012); (B) fully restrained proximally pinned distally (Bottlang et al., 2010); (C) pinned proximally and distally (Hoffmeier et al., 2011; Stoffel et al., 2003); and (D) hinged proximally and pinned distally which could be used as an alternative to the other conditions (MacLeod & Pankaj, 2014, 2018).

Modified from MacLeod, A. R., & Pankaj, P. (2014). Computer simulation of fracture fixation using extramedullary devices: An appraisal. In B. Doyle, K. Miller, A. Wittek, and P. M. F. Nielsen (Eds.), Computational biomechanics for medicine: Fundamental science and patient-specific applications *(Vol. 9781493907458, pp. 87–99). Springer New York. https://doi.org/10.1007/978-1-4939-0745-8_7.*

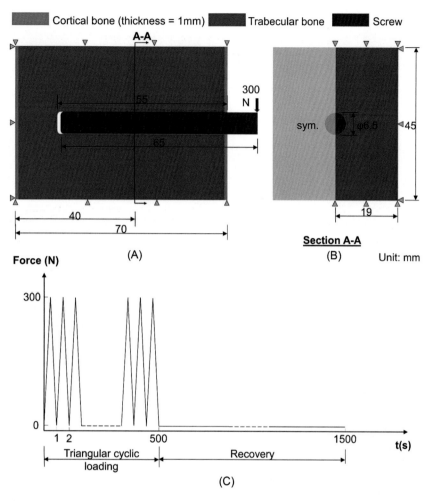

FIGURE 23.14

Idealized geometry of the bone–screw system showing symmetry surface with location of load application (A); section A-A (B); load application, each model was subjected to 500 cycles of triangular load of 300 N amplitude followed by 1000 s of recovery (C).

From Xie, S. Manda, K., & Pankaj, P. (2018). Time-dependent behaviour of bone accentuates loosening in the fixation of fractures using bone-screw systems. Bone and Joint Research, 7(10), 580–586. https://doi.org/10.1302/2046-3758.710.BJR-2018-0085.R1.

was modeled as time-dependent nonlinear viscoelastic viscoplastic material-based constitutive models developed from experimental studies (Manda et al., 2017; Manda, Xie, Wallace, Levrero-Florencio, & Pankaj, 2016; Xie et al., 2017). The study (Xie et al., 2018) examined the accumulation of strain at the bone–screw interface with increasing number of cycles and after recovery. The principal strain contours (Fig. 23.15) showed that the strain experienced by bone around the screw

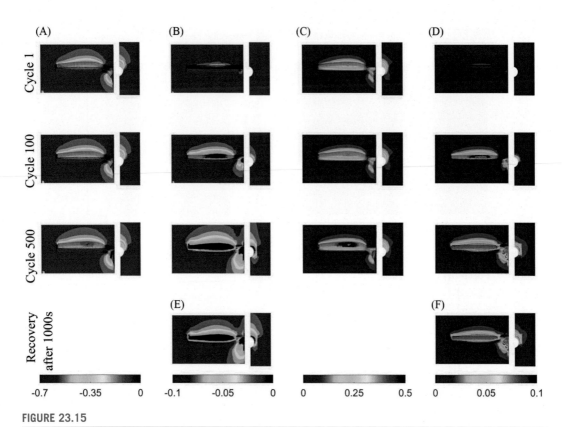

FIGURE 23.15

Compressive (A, B, and E) and tensile (C, D, and F) strain (%) contours from the symmetry surface and section A-A. Three representative cycles are shown: the strain accumulation with increasing cycle number when the load is at its peak (A and C); at the time points when the load is zero (B and D); and recovery after 1000 s (E and F).

From Xie, S. Manda, K., & Pankaj, P. (2018). Time-dependent behaviour of bone accentuates loosening in the fixation of fractures using bone-screw systems. Bone and Joint Research, 7(10), 580– 586. https://doi.org/10.1302/2046-3758.710.BJR-2018-0085.R1.

increases with increasing number of cycles and not all strain is recovered even after 1000 s of recovery. This increase in strain with number of loading cycles and the presence of residual strain indicates that the mechanical environment at the bone—implant interface will change as physiological activities are undertaken by the patient and will accentuate implant loosening. It has also been shown that the strain—displacement environment at the bone—implant interface is also frequency dependent (Xie, Manda, & Pankaj, 2019). Low-frequency cyclic loading results in larger interfacial strains in the first few cycles as slow loading rate gives more time to bone to deform. With increasing cycle numbers, higher loading frequency starts playing an important role. At higher frequencies, the loading—unloading time is shorter and the bone is loaded again by the next cycle before it can recover from its last loading cycle resulting in accumulation of separation between the bone and the implant. In all cases, this accumulation is larger for low-density poor-quality bone.

While these studies include time-dependent mechanical behavior, they do not consider the biological response of bone that can be supportive or disadvantageous for the bone-fixator system over time.

REFERENCES

Ali, A. M., Burton, M., Hashmi, M., & Saleh, M. (2003). Treatment of displaced bicondylar tibial plateau fractures (OTA-41C2&3) in patients older than 60 years of age. *Journal of Orthopaedic Trauma*, *17*(5), 346−352. Available from https://doi.org/10.1097/00005131-200305000-00005.

Ansari, U., Adie, S., Harris, I. A., & Naylor, J. M. (2011). Practice variation in common fracture presentations: A survey of orthopaedic surgeons. *Injury*, *42*(4), 403−407. Available from https://doi.org/10.1016/j.injury.2010.11.011.

Bayraktar, H. H., Morgan, E. F., Niebur, G. L., Morris, G. E., Wong, E. K., & Keaveny, T. M. (2004). *Comparison of the elastic and yield properties of human femoral trabecular and cortical bone tissue.* *Journal of Biomechanics*, *37*(1), 27−35. Available from https://doi.org/10.1016/s0021-9290(03)00257-4.

Behrens, F., & Johnson, W. (1989). Unilateral external fixation. Methods to increase and reduce frame stiffness. *Clinical Orthopaedics and Related Research*, *241*, 48−56. Available from https://www.ncbi.nlm.nih.gov/pubmed/2924479.

Board, T. N., Yang, L., & Saleh, M. (2007). Why fine-wire fixators work: An analysis of pressure distribution at the wire-bone interface. *Journal of Biomechanics*, *40*(1), 20−25. Available from https://doi.org/10.1016/j.jbiomech.2005.12.005.

Born, C. T., Karich, B., Bauer, C., von Oldenburg, G., & Augat, P. (2011). Hip screw migration testing: First results for hip screws and helical blades utilizing a new oscillating test method. *Journal of Orthopaedic Research: Official Publication of the Orthopaedic Research Society*, *29*(5), 760−766. Available from https://doi.org/10.1002/jor.21236.

Bottlang, M., & Feist, F. (2011). Biomechanics of far cortical locking. *Journal of Orthopaedic Trauma*, *25* (Suppl 1), S21−S28. Available from https://doi.org/10.1097/BOT.0b013e318207885b.

Bottlang, M., Doornink, J., Fitzpatrick, D. C., & Madey, S. M. (2009). Far cortical locking can reduce stiffness of locked plating constructs while retaining construct strength. *The Journal of Bone and Joint Surgery. American Volume*, *91*(8), 1985−1994. Available from https://doi.org/10.2106/JBJS.H.01038.

Bottlang, M., Doornink, J., Lujan, T. J., Fitzpatrick, D. C., Marsh, L., Augat, P., ... Madey, S. M. (2010). Effects of construct stiffness on healing of fractures stabilized with locking plates. *Journal of Bone and Joint Surgery-American Volume*, *92A*, 12−22. Available from https://doi.org/10.2106/jbjs.j.00780.

Chao, P., Conrad, B. P., Lewis, D. D., Horodyski, M., & Pozzi, A. (2013). Effect of plate working length on plate stiffness and cyclic fatigue life in a cadaveric femoral fracture gap model stabilized with a 12-hole 2.4 mm locking compression plate. *BMC Veterinary Research*, *9*, 125. Available from https://doi.org/10.1186/1746-6148-9-125.

Claes, L. E. (2021). Improvement of clinical fracture healing − What can be learned from mechano-biological research? *Journal of Biomechanics*, *115*, 110148. Available from https://doi.org/10.1016/j.jbiomech.2020.110148.

Claes, L. E., & Meyers, N. (2020). The direction of tissue strain affects the neovascularization in the fracture-healing zone. *Medical Hypotheses*, *137*, 109537. Available from https://doi.org/10.1016/j.mehy.2019.109537.

Clavert, J. M., Gicquel, P., & Giacomelli, M. C. (2010). Stainless steel or titanium? In P. Lascombes (Ed.), *Flexible Intramedullary Nailing in Children*. Berlin: Springer.

Comiskey, D. P., Macdonald, B. J., McCartney, W. T., Synnott, K., & O'Byrne, J. (2010). The role of interfragmentary strain on the rate of bone healing-a new interpretation and mathematical model. *Journal of Biomechanics*, *43*(14), 2830−2834. Available from https://doi.org/10.1016/j.jbiomech.2010.06.016.

Dobele, S., Horn, C., Eichhorn, S., Buchholtz, A., Lenich, A., Burgkart, R., ... Stockle, U. (2010). The dynamic locking screw (DLS) can increase interfragmentary motion on the near cortex of locked plating constructs by reducing the axial stiffness. *Langenbeck's Archives of Surgery/Deutsche Gesellschaft fur Chirurgie*, *395*(4), 421−428. Available from https://doi.org/10.1007/s00423-010-0636-z.

Donaldon, F. E. (2011). *On incorporating bone microstructure in macro-finite-element models* (Ph.D. thesis). The University of Edinburgh.

Donaldson, F. E., Pankaj, P., Cooper, D. M., Thomas, C. D., Clement, J. G., & Simpson, A. H. R. W. (2011). Relating age and micro-architecture with apparent-level elastic constants: A micro-finite element study of female cortical bone from the anterior femoral midshaft. *Proceedings of the Institution of Mechanical Engineers, Part H*, *225*(6), 585−596. Available from https://doi.org/10.1177/2041303310395675.

Donaldson, F. E., Pankaj, P., & Simpson, A. H. R. W. (2012a). Bone properties affect loosening of half-pin external fixators at the pin-bone interface. *Injury*, *43*(10), 1764−1770. Available from https://doi.org/10.1016/j.injury.2012.07.001.

Donaldson, F. E., Pankaj, P., & Simpson, A. H. R. W. (2012b). Investigation of factors affecting loosening of Ilizarov ring-wire external fixator systems at the bone-wire interface. *Journal of Orthopaedic Research: Official Publication of the Orthopaedic Research Society*, *30*(5), 726−732. Available from https://doi.org/10.1002/jor.21587.

Fleming, B., Paley, D., Kristiansen, T., & Pope, M. (1989). A biomechanical analysis of the Ilizarov external fixator. *Clinical Orthopaedics and Related Research* (241), 95−105, Retrieved from. Available from https://www.ncbi.nlm.nih.gov/pubmed/2924484.

Fragomen, A. T., & Rozbruch, S. R. (2007). The mechanics of external fixation. *HSS Journal: The Musculoskeletal Journal of Hospital for Special Surgery*, *3*(1), 13−29. Available from https://doi.org/10.1007/s11420-006-9025-0.

Ganesh, V. K., Ramakrishna, K., & Ghista, D. N. (2005). Biomechanics of bone-fracture fixation by stiffness-graded plates in comparison with stainless-steel plates. *Biomedical Engineering Online*, *4*, 46. Available from https://doi.org/10.1186/1475-925X-4-46.

Gardner, M. J., Helfet, D. L., & Lorich, D. G. (2004). Has locked plating completely replaced conventional plating? *American Journal of Orthopedics (Belle Mead, NJ)*, *33*(9), 439−446. Available from https://doi.org/10.1016/j.injury.2010.02.022.

Gardner, M. J., Nork, S. E., Huber, P., & Krieg, J. C. (2010). Less rigid stable fracture fixation in osteoporotic bone using locked plates with near cortical slots. *Injury*, *41*(6), 652−656. Available from https://doi.org/10.1016/j.injury.2010.02.022.

Gaston, M. S., & Simpson, A. H. R. W. (2007). Inhibition of fracture healing. *The Journal of Bone and Joint Surgery. British Volume*, *89*(12), 1553−1560. Available from https://doi.org/10.1302/0301-620X.89B12.19671.

Gautier, E., & Sommer, C. (2003). Guidelines for the clinical application of the LCP. *Injury*, *34*(Suppl 2), B63−B76. Available from https://doi.org/10.1016/j.injury.2003.09.026.

Gordon, J. E., Schoenecker, P. L., Oda, J. E., Ortman, M. R., Szymanski, D. A., Dobbs, M. B., & Luhmann, S. J. (2003). A comparison of monolateral and circular external fixation of unstable diaphyseal tibial fractures in children. *Journal of Pediatric Orthopaedics. Part B/European Paediatric Orthopaedic Society, Pediatric Orthopaedic Society of North America*, *12*(5), 338−345. Available from https://doi.org/10.1097/01.bpb.0000078262.58527.38.

Grubor, P., & Borrelli, J. (2020). Biomechanics of external fixators for fracture fixation: Uniplanar, multiplanar, and circular frames. In B. D. Crist (Ed.), *Essential biomechanics for orthopedic trauma* (pp. 45−60). Switzerland AG: Springer Nature.

Henderson, C. E., Lujan, T. J., Kuhl, L. L., Bottlang, M., Fitzpatrick, D. C., & Marsh, J. L. (2011). 2010 mid-America orthopaedic association physician in training award: Healing complications are common after locked plating for distal femur fractures. *Clinical Orthopaedics and Related Research*, *469*(6), 1757−1765. Available from https://doi.org/10.1007/s11999-011-1870-6.

Hillard, P. J., Harrison, A. J., & Atkins, R. M. (1998). The yielding of tensioned fine wires in the Ilizarov frame. *Proceedings of the Institution of Mechanical Engineers, Part H, 212*(1), 37–47. Available from https://doi.org/10.1243/0954411981533809.

Hoffmeier, K. L., Hofmann, G. O., & Mückley, T. (2011). Choosing a proper working length can improve the life-span of locked plates: A biomechanical study. *Clinical Biomechanics, 26*(4), 405–409, Retrieved from. Available from http://www.sciencedirect.com/science/article/B6T59-51SMSGH-2/2/68c822c94c62366ff519f13c7393648a.

Huiskes, R., Chao, E. Y., & Crippen, T. E. (1985). Parametric analyses of pin-bone stresses in external fracture fixation devices. *Journal of Orthopaedic Research: Official Publication of the Orthopaedic Research Society, 3*(3), 341–349. Available from https://doi.org/10.1002/jor.1100030311.

Karunratanakul, K., Schrooten, J., & Van Oosterwyck, H. (2010). Finite element modelling of a unilateral fixa-tor for bone reconstruction: Importance of contact settings. *Medical Engineering & Physics, 32*(5), 461–467. Available from https://doi.org/10.1016/j.medengphy.2010.03.005.

Kim, T., Ayturk, U. M., Haskell, A., Miclau, T., & Puttlitz, C. M. (2007). Fixation of osteoporotic distal fibula fractures: A biomechanical comparison of locking vs conventional plates. *The Journal of Foot & Ankle Surgery, 46*(1), 2–6. Available from https://doi.org/10.1053/j.jfas.2006.09.009.

Leahy, M. (2010). When locking plates fail. *AAOS Now, 4*(5), 9.

Levrero-Florencio, F., Margetts, L., Sales, E., Xie, S., Manda, K., & Pankaj, P. (2016). Evaluating the macro-scopic yield behaviour of trabecular bone using a nonlinear homogenisation approach. *Journal of the Mechanical Behavior of Biomedical Materials, 61*, 384–396. Available from https://doi.org/10.1016/j.jmbbm.2016.04.008.

Lujan, T. J., Henderson, C. E., Madey, S. M., Fitzpatrick, D. C., Marsh, J. L., & Bottlang, M. (2010). Locked plating of distal femur fractures leads to inconsistent and asymmetric callus formation. *Journal of Orthopaedic Trauma, 24*(3), 156–162. Available from https://doi.org/10.1097/BOT.0b013e3181be6720.

MacLeod, A. R. (2014). *Modelling and optimising the mechanical behaviour of fractures treated with locking plates* (Ph.D. thesis). The University of Edinburgh.

MacLeod, A. R., & Pankaj, P. (2014). Computer simulation of fracture fixation using extramedullary devices: An appraisal. In B. Doyle (Ed.), *Computational biomechanics for medicine: Fundamental science and patient-specific applications* (pp. 87–99). New York: Springer Science + Business Media.

MacLeod, A. R., & Pankaj, P. (2018). Pre-operative planning for fracture fixation using locking plates: Device configuration and other considerations. *Injury, 49*(Suppl 1), S12–S18. Available from https://doi.org/10.1016/S0020-1383(18)30296-1.

MacLeod, A. R., Pankaj, P., & Simpson, A. H. R. W. (2012). Does screw-bone interface modelling matter in finite element analyses? *Journal of Biomechanics, 45*(9), 1712–1716. Available from https://doi.org/10.1016/j.jbiomech.2012.04.008.

MacLeod, A. R., Simpson, A. H. R. W., & Pankaj, P. (2015). Reasons why dynamic compression plates are inferior to locking plates in osteoporotic bone: A finite element explanation. *Computer Methods in Biomechanics and Biomedical Engineering, 18*(16), 1818–1825. Available from https://doi.org/10.1080/10255842.2014.974580.

MacLeod, A. R., Simpson, A. H. R. W., & Pankaj, P. (2016). Age-related optimization of screw placement for reduced loosening risk in locked plating. *Journal of Orthopaedic Research: Official Publication of the Orthopaedic Research Society, 34*(11), 1856–1864. Available from https://doi.org/10.1002/jor.23193.

MacLeod, A. R., Simpson, A. H. R. W., & Pankaj, P. (2018). Experimental and numerical investigation into the influence of loading conditions in biomechanical testing of locking plate fracture fixation devices. *Bone & Joint Research, 7*(1), 111–120. Available from https://doi.org/10.1302/2046-3758.71.BJR-2017-0074.R2.

Manda, K., Wallace, R. J., Xie, S., Levrero-Florencio, F., & Pankaj, P. (2017). Nonlinear viscoelastic charac-terization of bovine trabecular bone. *Biomechanics and Modeling in Mechanobiology, 16*(1), 173–189. Available from https://doi.org/10.1007/s10237-016-0809-y.

Manda, K., Xie, S., Wallace, R. J., Levrero-Florencio, F., & Pankaj, P. (2016). Linear viscoelasticity — bone volume fraction relationships of bovine trabecular bone. *Biomechanics and Modeling in Mechanobiology*, *15*(6), 1631−1640. Available from https://doi.org/10.1007/s10237-016-0787-0.

McKibbin, B. (1978). The biology of fracture healing in long bones. *The Journal of Bone and Joint Surgery. British Volume*, *60-B*(2), 150−162. Available from https://doi.org/10.1302/0301-620X.60B2.350882.

Moazen, M., Mak, J. H., Jones, A. C., Jin, Z., Wilcox, R. K., & Tsiridis, E. (2013). Evaluation of a new approach for modelling the screw-bone interface in a locking plate fixation: A corroboration study. *Proceedings of the Institution of Mechanical Engineers, Part H*, *227*(7), 746−756. Available from https://doi.org/10.1177/0954411913483259.

Moroni, A., Vannini, F., Mosca, M., & Giannini, S. (2002). State of the art review: Techniques to avoid pin loosening and infection in external fixation. *Journal of Orthopaedic Trauma*, *16*(3), 189−195. Available from https://doi.org/10.1097/00005131-200203000-00009.

Mushtaq, A., Shahid, R., Asif, M., & Maqsood, M. (2009). Distal tibial fracture fixation with locking compression plate (LCP) using the minimally invasive percutaneous osteosynthesis (MIPO) Technique. *European Journal of Trauma and Emergency Surgery*, *35*(2), 159−164. Available from https://doi.org/10.1007/s00068-008-8049-1.

Oni, O. O. A., Capper, M., & Soutis, C. (1993). A finite element analysis of the effect of pin distribution on the rigidity of a unilateral external fixation system. *Injury*, *24*(8), 525−527. Available from https://doi.org/10.1016/0020-1383(93)90028-5.

Pankaj, P. (2013). Patient-specific modelling of bone and bone-implant systems: The challenges. International. *Journal for Numerical Methods in Biomedical Engineering*, *29*(2), 233−249. Available from https://doi.org/10.1002/cnm.2536.

Pankaj, P., & Donaldson, F. E. (2013). Algorithms for a strain-based plasticity criterion for bone. *International Journal for Numerical Methods in Biomedical Engineering*, *29*(1), 40−61. Available from https://doi.org/10.1002/cnm.2491.

Pankaj, P., & Xie, S. (2019). The risk of loosening of extramedullary fracture fixation devices. *Injury, 50, Suppl 1*, S66−S72. Available from https://doi.org/10.1016/j.injury.2019.03.051.

Perren, S. M. (2002). Evolution of the internal fixation of long bone fractures. The scientific basis of biological internal fixation: Choosing a new balance between stability and biology. *The Journal of Bone and Joint Surgery. British Volume*, *84*(8), 1093−1110. Available from https://doi.org/10.1302/0301-620x.84b8.13752.

Perren, S. M., Allgower, M., Cordey, J., & Russenberger, M. (1973). Developments of compression plate techniques for internal fixation of fractures. *Progress in Surgery*, *12*, 152−179. Available from https://doi.org/10.1159/000394905.

Pettine, K. A., Chao, E. Y. S., & Kelly, P. J. (1993). Analysis of the external fixator pin-bone interface. *Clinical orthopaedics and related research* (Issue 293), 18−27. Available from https://doi.org/10.1097/00003086-199308000-00004.

Phillips, A. T. M., Pankaj, P., Howie, C. R., Usmani, A. S., & Simpson, A. H. (2007). Finite element modelling of the pelvis: Inclusion of muscular and ligamentous boundary conditions. *Medical Engineering & Physics*, *29*(7), 739−748. Available from https://doi.org/10.1016/j.medengphy.2006.08.010.

Podolsky, A., & Chao, E. Y. S. (1993). Mechanical performance of Ilizarov circular external fixators in comparison with other external fixators. *Clinical Orthopaedics and Related Research* (Issue 293), 61−70. Available from https://doi.org/10.1097/00003086-199308000-00009.

Russo, C. R., Lauretani, F., Seeman, E., Bartali, B., Bandinelli, S., Di Iorio, A., ... Ferrucci, L. (2006). Structural adaptations to bone loss in aging men and women. *Bone*, *38*(1), 112−118. Available from https://doi.org/10.1016/j.bone.2005.07.025.

Saving, J., Ponzer, S., Enocson, A., Mellstrand Navarro, C., & Blank, R. D. (2018). Distal radius fractures—Regional variation in treatment regimens. *PLoS One*, *13*(11), e0207702. Available from https://doi.org/10.1371/journal.pone.0207702.

Schmidt, U., Penzkofer, R., Bachmaier, S., & Augat, P. (2013). Implant material and design alter construct stiffness in distal femur locking plate fixation: A pilot study. *Clinical Orthopaedics and Related Research, 471*(9), 2808−2814. Available from https://doi.org/10.1007/s11999-013-2867-0.

Seide, K., Triebe, J., Faschingbauer, M., Schulz, A. P., Püschel, K., Mehrtens, G., & Jürgens, C. (2007). Locked vs. unlocked plate osteosynthesis of the proximal humerus − A biomechanical study. *Clinical Biomechanics, 22*(2), 176−182. Available from https://doi.org/10.1016/j.clinbiomech.2006.08.009.

Shahar, R., & Shani, Y. (2004). Fracture stabilization with type II external fixator vs. type I external fixator with IM pin: Finite element analysis. *Veterinary and Comparative Orthopaedics and Traumatology, 17*(2), 91−96. Available from https://doi.org/10.1055/s-0038-1636480.

Smith, W. R., Ziran, B. H., Anglen, J. O., & Stahel, P. F. (2007). Locking plates: Tips and tricks. *Journal of Bone and Joint Surgery - Series A, 89*(10), 2298−2307. Available from https://doi.org/10.2106/00004623-200710000-00028.

Speirs, A. D., Heller, M. O., Duda, G. N., & Taylor, W. R. (2007). Physiologically based boundary conditions in finite element modelling. *Journal of Biomechanics, 40*(10), 2318−2323. Available from https://doi.org/10.1016/j.jbiomech.2006.10.038.

Steiner, M., Claes, L., Ignatius, A., Simon, U., & Wehner, T. (2014). Disadvantages of interfragmentary shear on fracture healing−mechanical insights through numerical simulation. *Journal of Orthopaedic Research: Official Publication of the Orthopaedic Research Society, 32*(7), 865−872. Available from https://doi.org/10.1002/jor.22617.

Stoffel, K., Dieter, U., Stachowiak, G., Gächter, A., & Kuster, M. S. (2003). Biomechanical testing of the LCP − How can stability in locked internal fixators be controlled? *Injury, 34*(2), SB11−SB88. Available from https://doi.org/10.1016/j.injury.2003.09.021.

Strauss, E. J., Schwarzkopf, R., Kummer, F., & Egol, K. A. (2008). The current status of locked plating: The good, the bad, and the ugly. *Journal of Orthopaedic Trauma, 22*(7), 479−486. Available from https://doi.org/10.1097/BOT.0b013e31817996d6.

Thakur, A. J. (2020). *The elements of fracture fixation.* Elsevier.

Turner, C. H. (2006). Bone strength: Current concepts. *Annals of the New York Academy of Sciences, 1068*, 429−446. Available from https://doi.org/10.1196/annals.1346.039.

Uhl, J. M., Seguin, B., Kapatkin, A. S., Schulz, K. S., Garcia, T. C., & Stover, S. M. (2008). Mechanical comparison of 3.5 mm broad dynamic compression plate, broad limited-contact dynamic compression plate, and narrow locking compression plate systems using interfragmentary gap models. *Veterinary Surgery, 37*(7), 663−673. Available from https://doi.org/10.1111/j.1532-950X.2008.00433.x.

Uhthoff, H. K., Poitras, P., & Backman, D. S. (2006). Internal plate fixation of fractures: Short history and recent developments. *Journal of Orthopaedic Science: Official Journal of the Japanese Orthopaedic Association, 11*(2), 118−126. Available from https://doi.org/10.1007/s00776-005-0984-7.

Vallier, H. A., Hennessey, T. A., Sontich, J. K., & Patterson, B. M. (2006). Failure of LCP condylar plate fixation in the distal part of the femur. A report of six cases. *Journal of Bone and Joint Surgery - Series A, 88*(4), 846−853. Available from https://doi.org/10.2106/JBJS.E.00543.

Wikenheiser, M. A., Markel, M. D., Lewallen, D. G., & Chao, E. Y. (1995). Thermal response and torque resistance of five cortical half-pins under simulated insertion technique. *Journal of Orthopaedic Research: Official Publication of the Orthopaedic Research Society, 13*(4), 615−619. Available from https://doi.org/10.1002/jor.1100130418.

Woo, S. L., Lothringer, K. S., Akeson, W. H., Coutts, R. D., Woo, Y. K., Simon, B. R., & Gomez, M. A. (1984). Less rigid internal fixation plates: Historical perspectives and new concepts. *Journal of Orthopaedic Research: Official Publication of the Orthopaedic Research Society, 1*(4), 431−449. Available from https://doi.org/10.1002/jor.1100010412.

Woods, J. C., & Della Rocca, G. J. (2020). *Biomechanics of intramedullary nails relative to fracture fixation and deformity correction. Essential Biomechanics for Orthopedic Trauma* (pp. 221−235). Springer Science and Business Media LLC. Available from https://doi.org/10.1007/978-3-030-36990-3_16.

Xie, S., Manda, K., & Pankaj, P. (2018). Time-dependent behaviour of bone accentuates loosening in the fixation of fractures using bone-screw systems. *Bone & Joint Research*, *7*(10), 580−586. Available from https://doi.org/10.1302/2046-3758.710.BJR-2018-0085.R1.

Xie, S., Manda, K., & Pankaj, P. (2019). Effect of loading frequency on deformations at the bone-implant interface. *Proceedings of the Institution of Mechanical Engineers, Part H*, *233*(12), 1219−1225. Available from https://doi.org/10.1177/0954411919877970.

Xie, S., Manda, K., Wallace, R. J., Levrero-Florencio, F., Simpson, A. H. R. W., & Pankaj, P. (2017). Time dependent behaviour of trabecular bone at multiple load levels. *Annals of Biomedical Engineering*, *45*(5), 1219−1226. Available from https://doi.org/10.1007/s10439-017-1800-1.

Yanez, A., Carta, J. A., & Garces, G. (2010). Biomechanical evaluation of a new system to improve screw fixation in osteoporotic bones. *Medical Engineering & Physics*, *32*(5), 532−541. Available from https://doi.org/10.1016/j.medengphy.2010.02.014.

Yanez, A., Cuadrado, A., Carta, J. A., & Garces, G. (2012). Screw locking elements: A means to modify the flexibility of osteoporotic fracture fixation with DCPs without compromising system strength or stability. *Medical Engineering & Physics*, *34*(6), 717−724. Available from https://doi.org/10.1016/j.medengphy.2011.09.015.

Zamani, A. R., & Oyadiji, S. O. (2008). Analytical modelling of kirschner wires in Ilizarov circular external fixators using a tensile model. *Proceedings of the Institution of Mechanical Engineers, Part H*, *222*(6), 967−976. Available from https://doi.org/10.1243/09544119JEIM373.

Zehnder, S., Bledsoe, J. G., & Puryear, A. (2009). The effects of screw orientation in severely osteoporotic bone: A comparison with locked plating. *Clinical Biomechanics*, *24*(7), 589−594. Available from https://doi.org/10.1016/j.clinbiomech.2009.04.008.

REGENERATION AND REPAIR OF LIGAMENTS AND TENDONS

24

Rocco Aicale[1,2], Nicola Maffulli[1,2,3,4] and Francesco Oliva[1,2]

[1]*Department of Trauma and Orthopaedic Surgery, Surgery and Dentistry, University of Salerno School of Medicine, Salerno, Italy* [2]*Department of Musculoskeletal Disorders, Faculty of Medicine and Surgery, University of Salerno, Baronissi, Salerno, Italy* [3]*School of Pharmacy and Bioengineering, Keele University, Stoke-on-Trent, United Kingdom* [4]*Centre for Sports and Exercise Medicine, Queen Mary University of London, London, United Kingdom*

INTRODUCTION

Tendon and ligament injuries remain a clinical challenge. They account for 45% of the 32 million musculoskeletal injuries each year in the United States (Butler, Juncosa, & Dressler, 2004), with rates rising due to contact sports participation and the aging population. The current management options fail to restore the preinjury structural and biochemical properties of tendons and ligaments, resulting in scar tissue formation and fibrosis. Consequently, the principal elements of tissue engineering, such as cells, scaffolds, and bioactive molecules, have been explored to improve healing. At present, no tissue engineered construct thus far has achieved complete regeneration.

Studies regarding tendon fetal healing show restoration of the native structural and functional tissue properties, with no scar formation (Favata et al., 2006). Unfortunately, the understanding of tendon biology and healing is incomplete, and strategies are still lagging behind increasing demands (Chisari, Rehak, Khan, & Maffulli, 2019).

To treat tendon diseases supporting regeneration, cell-based therapy and tissue engineering are considered as potential approaches to reproduce a safe and successful long-term outcome for full microarchitecture and biomechanical tissue recovery. The first step is the in vitro model, which is characterized by some fundamentals as follows:

- identify and/or compare the tenogenic plasticity of different stem/progenitor cell source,
- define and drive cell mechanism and environmental conditions leading tenogenesis,
- control stepwise signaling molecules and pathways,
- direct stem cell precommitment before transplantation (reducing tumorigenic risks with embryonic stem cells, unwilling differentiation path of mesenchymal stem cells (MSCs) or to increase tissue integration),
- test teno-inductive properties of new scaffolds,
- validate biomechanical teno-inductive stimuli.

Disorders of soft tissue, in the United Kingdom, only have a prevalence of 18 cases for 1000 individuals per year, and require a specialist consultation in 40% of the cases (Aicale, Tarantino, & Maffulli, 2017).

Human Orthopaedic Biomechanics. DOI: https://doi.org/10.1016/B978-0-12-824481-4.00030-5

In the general population, tendons may undergo traumatic and degenerative processes. In particular, the tendons most vulnerable to overload are those of rotator cuff, long head of the biceps, the extensors and flexors of the wrist, the posterior tibial tendon, and the patellar and Achilles tendons (Sharma, & Maffulli, 2005b). However, one-third of patients do not practice intense physical activity (Waldecker, Hofmann, & Drewitz, 2012).

A healthy tendon is fibroelastic, constituted by white tissue, composed of scanty cells (mainly tenoblasts and tenocytes) and extracellular matrix (ECM) (Kannus, 2000). Chondrocytes, synovial, vascular endothelial, and smooth muscle cells represent only 5%−10% of tendon cell population (Sharma, & Maffulli, 2005b). Tenoblasts are immature spindle-shaped cells, with metabolically active cytoplasmic organelles. They change over time, becoming elongated when they mature into tenocytes (Kannus, 2000). Tenocytes are described as having a lower nucleus-to-cytoplasm ratio and a relatively reduced metabolic activity, and are able to synthesize collagen and ECM (De Albornoz, Aicale, Forriol, & Maffulli, 2018; O'Brien, 1997).

Collagen type I and type III, proteoglycans (such as lumican, decorin, aggrecan, fibromodulin, and versican), and glycosaminoglycans (chondroitin and dermatan sulfate) are the main elements that compose the tendon ECM (Chuen et al., 2004; McNeilly, Banes, Benjamin, & Ralphs, 1996).

Among the cell populations present in tendons are stem/progenitor cells (TSPCs) (Bi et al., 2007; Ruzzini et al., 2014), which contain MSCs characterized by the universal criteria defining stem cells, such as clonogenicity, multipotency, and self-renewal. Both TSPCs and differentiated cells are located in an adequate microenvironment of biophysical and biochemical signals, in which mechanical forces and ECM topography are the dominating elements that govern cellular processes and tissue function (Li, Xiao, & Liu, 2017; Spanoudes, Gaspar, Pandit, & Zeugolis, 2014).

Tenoblasts are round cells with ovoid nuclei, contained mainly in the endotenon (Chuen et al., 2004), and are immature cells that give rise to tenocytes. Indeed, they are the main cells population in young tendons (Bi et al., 2007). TSPCs have been recently characterized in tendon tissue of several species (Lui, 2013; Mienaltowski, Adams, & Birk, 2013), and represent 1%−4% of tendon resident cells. They exhibit the same characteristics as adult MSCs (Mienaltowski et al., 2013). TSPCs can be sorted on CD44 positivity (Ruzzini et al., 2014), and express MSC markers Stro 1 and CD146 and tenogenic markers a-smooth muscle actin (a-Sma) and tenomodulin (Tnmd) (Zhang & Wang, 2010; Zhang, Pan, Liu, & Wang, 2010). During life, their number and self-renewal potential decrease (Ruzzini et al., 2014) explaining the low ability of adult tendons to spontaneous healing.

Tendon injuries can be acute or chronic, and caused by intrinsic [age (Riley et al., 2002), body structure (Franceschi et al., 2014; Schwellnus, Jordaan, & Noakes, 1990), nutrition (Longo, Ronga, & Maffulli, 2009), metabolic diseases (Oliva et al. 2016; Oliva, Misiti, & Maffulli, 2014), genetics (Maffulli, Khan, & Puddu, 1998)] or extrinsic [excessive and improper (Maffulli, Sharma, & Luscombe, 2004) loading disuse (Yasuda & Hayashi, 1999) and external damage (Bisaccia, Aicale, Tarantino, Peretti, & Maffulli, 2019; Van der Linden et al., 2003)] factors, alone or in combination (Aicale et al., 2017). In acute trauma, extrinsic factors predominate, whilst in chronic injuries intrinsic factors also play a role. These factors are associated with the onset of overload pathology of tendons, though there is not a specific cause−effect relationship.

At microscopic evaluation, the abnormal tendon tissues present a noninflammatory process (Khan, Cook, Bonar, Harcourt, & Åstrom, 1999). However, inflammation is associated with tendon ruptures (Maffulli, Barrass, & Ewen, 2000). In tendinopathy, there is a disordered collagen fibers arrangement with loss of hierarchical structure (Maffulli et al., 2004), increased vascularization (Benazzo, Stennardo, Mosconi, Zanon, & Maffulli, 2001), and poor healing tendency. Blood vessels have a random orientation demonstrating angioblastic features, sometimes at right angles to collagen fibers (Zafar, Mahmood, & Maffulli, 2009). Six different subcategories of collagen degeneration have been described, but the most common are the mucoid or lipoid varieties. The areas of altered collagen fiber structure and increased interfibrillar ground substance exhibit an increased signal at magnetic resonance imaging (MRI) (Aicale et al., 2017; Zafar et al., 2009), and are hypoechoic on ultrasound (US) (Paavola, Paakkala, Kannus, & Järvinen, 1998).

As mentioned above, the etiopathogenesis of tendinopathy remains unclear, but it is currently considered multifactorial, and an interaction between intrinsic and extrinsic factors has been postulated (Maffulli et al., 2004). Changes in training pattern, poor technique, previous injuries, and environmental factors, such as training on hard, slippery, or slanting surfaces, are extrinsic factors that may predispose an athlete to tendinopathies (Aicale, Tarantino, & Maffulli, 2018; Maffulli et al., 2004). Drugs such as fluoroquinolones (i.e., ciprofloxacin) and corticosteroids have been implicated as risk factors in tendinopathy (Bisaccia et al., 2019; Parmar & Hennessy, 2007). Imbalance in matrix metalloproteinase activity in response to repeated injury or mechanical strain may result in tendinopathy (Magra & Maffulli, 2005a, 2005b, 2006; Magra, Hughes, El Haj, & Maffulli, 2007). Metabolic disorders with a genetic component seem to play a role (Aicale, Tarantino, Maccauro, Peretti, & Maffulli, 2019; de Oliveira, Lemos, de Castro Silveira, da Silva, & de Moraes, 2011; Maffulli, Via, & Oliva, 2015; Oliva, Berardi, Misiti, & Maffulli, 2013; Oliva, Via, & Maffulli, 2012).

Diagnostic imaging tools, such as plain radiography, US and MRI, may be helpful to verify a clinical suspicion or to exclude other musculoskeletal disorders (Maffulli et al., 2017). However, treatment of tendinopathies lacks strong evidence-based support, involving tendinopathies as a pathology with a high risk for long-term morbidity (Kader, Saxena, Movin, & Maffulli, 2002). Primarily, management is conservative, and many patients show good outcomes (Aicale, Bisaccia, Oliviero, Oliva, & Maffulli, 2020). However, if conservative management fails, surgery is recommended after 3–6 months of conservative management (Maffulli & Peretti, 2020; Sayana & Maffulli, 2007).

In human, regeneration tissue development is a dramatic achievement, with limited knowledge about signals implicated in the regenerative processes. With the term "regeneration," commonly, we indicate the postnatal restoration of a structure damaged or lost, which may occur through several mechanisms to produce functional structures similar to the original ones before injury (Bely & Nyberg, 2010). After injury, tissues healing occurs through three stages, interdependent and overlapping with each other: an inflammatory response, followed by a fibroblastic/proliferative phase, and, a prolonged remodeling phase (Mendes et al., 2018; Nichols, Best, & Loiselle, 2019). In many cases, tissue healing is associated with a profibrotic response that leads to scar formation (Atala, Irvine, Moses, & Shaunak, 2010), with excessive and disorganized deposition of the ECM (Nichols et al., 2019).

In this scenario, cells have a primary role in maintaining the normal tendon ECM dynamics. Therefore, biologics strategies and therapies are commanding attention, through the combination of cells and biomaterials (Loebel & Burdick, 2018).

TISSUE ENGINEERING FOR COMMON TENDON AND LIGAMENT INJURIES

Tendons anchor muscle to bone, transmitting their forces. The most common injuries involve the rotator cuff tendons, Achilles tendon, and flexor tendons of the hand. These three tendons present a great challenge to repair, given their unique anatomy, function, biomechanical properties, healing capacities, and rehabilitation approaches (Cipollaro et al., 2020; De Albornoz et al., 2018; Gott et al., 2011).

In cases of tendon injury, surgical management may be commonly used to repair or replace tendon with autografts, allografts, xenografts, or prosthetic devices (Liu et al., 2011). However, clinical outcomes remain unsatisfactory from several limitations, including donor site morbidity, high failure rates, risk of injury recurrence, and limited long-term function recovery (Klepps et al., 2004; Krueger-Franke, Siebert, & Scherzer, 1995; Voleti, Buckley, & Soslowsky, 2012). Engineering strategies, which can apply a combination of cells, scaffolds, bioactive molecules, and *ex vivo* mechanical stimulation, may create functional replacements or increase the innate tendon healing capacities, improving healing quality to promote full restoration of tendon function.

CELLS

Cells widely used for tendon tissue engineering are tendon fibroblasts, dermal fibroblasts, and MSCs. Tendon fibroblasts (named tenocytes) synthesize constituents of ECM, such as collagen, proteoglycans, and glycoproteins, and have a great influence on collagen fiber formation (Canty & Kadler, 2005; Franchi, Trirè, Quaranta, Orsini, & Ottani, 2007) and they can upregulate the expression of Scx, Tnmd, and Dcn (Qiu et al., 2013) in response to growth factors. Tendon defects bridged by an autologous tenocyte-engineered tendon demonstrated improved mechanical strength and matrix deposition compared with cell free scaffold (Cao et al., 2002), and similar results were reported in another study with decellularized rabbit flexor tendons managed with autologous tenocytes, showing still a decreased ultimate stress (Chong et al., 2009) but reducing collagen degradation of the scaffold incubated in culture medium (Tilley, Chaudhury, Hakimi, Carr, & Czernuszka, 2012).

Despite the advances in tenocyte-based tendon tissue engineering, harvesting autologous tenocytes may cause secondary tendon defects at the donor site. For this reason, dermal fibroblasts have been considered as an alternative to address this limitation, given their availability with no donor site morbidity (Van Eijk et al., 2004).

In vivo dermal fibroblast- and tenocyte-engineered tendons were similar in terms of morphology, histology, and tensile stress. Dermal fibroblasts were used to produce human tendon-like neotissue under static strain, forming longitudinally aligned collagen fibers and spindle-shaped cells (C.F. Liu et al., 2011).

Adult MSCs are another promising cell source given their self-renewal and multilineage differentiation potential and can be found in several tissues including bone marrow, adipose tissue, and tendons (Tucker, Karamsadkar, Khan, & Pastides, 2010). Bone marrow MSCs (BMSCs) seeded in polylactide/glycolide suture material demonstrated higher collagen production compared with anterior cruciate ligament fibroblasts and skin fibroblasts (Van Eijk et al., 2004). Similar results were found in rabbit BMSCs seeded in a silk scaffold reporting significantly increased ECM expression, tenascin-C, and collagen types I and III, compared to anterior cruciate ligament fibroblast-seeded scaffolds (Liu, Fan, Toh, & Goh, 2008).

Less attention has been paid to adipose-derived mesenchymal stem cells (ASCs), which, when compared with BMSCs, results in less invasive harvest procedures, available in greater quantities, and similar potential to differentiate (Gimble, Katz, & Bunnell, 2007).

Park et al. showed that rat ASCs managed with GDF-5- expressed tendon-specifics markers such as Scx and Tnmd (Park et al., 2010), and when human ASCs are seeded onto a mesh, derived from hyaluronan (Hyalonect) and placed under mechanical stress, formed a vascularized tendon-like structure (Vindigni et al., 2013).

Resident adult stem cells seem to have a higher regenerative potential for the tissue where they reside (Lui & Chan, 2011), and therefore tendon-derived stem cells (TDSCs) have been identified and applied in tendon tissue engineering. Despite the potential advantages of TDSCs, the isolation of autologous cells would present the same donor site morbidities as tenocytes.

SCAFFOLDS

Scaffolds are another target in tendon tissue engineering, providing biomechanical support to healing tissue and preventing re-rupture by facilitating cell proliferation, promoting matrix production, and organizing matrix into functional tendon tissues (Liu et al., 2008). At present, there are three major categories of scaffolds: with native tendon matrices, made of synthetic polymers, and derived from naturally occurring proteins.

Before scaffolds can be used, they need to be free of native cells to prevent disease transmission and an immune response (Badylak, Freytes, & Gilbert, 2009; Deeken et al., 2011). Then, modified allografts or xenografts could be modified as functional delivery cells vehicles, gene therapy vectors, or for other biological agents. These grafts show highly preserved ECM proteins and growth factors (Ning et al., 2012; Pridgen et al., 2011), suggesting their potential biocompatibility and biofunctionality. Furthermore, extrinsic fibroblasts were cultured on a tri(n-butyl)phosphate-treated patellar tendon in vitro, developing viable tissue-engineered grafts (Cartmell & Dunn, 2004).

A great number of biodegradable and biocompatible polymers, in particular a-hydroxypolyesters, have been used for tendon tissue engineering, including polyglycolic acid (PGA), poly-L-lactic acid (PLLA), and their copolymer poly lactic-coglycolic acid (PLGA).

Recent studies demonstrated that PLGA with BMSCs show greater tensile strength and types I and III collagen deposition in a rabbit Achilles tendon injury model, compared to cell-free scaffolds (Ouyang, Goh, Thambyah, Teoh, & Lee, 2003). PGA, when seeded with mouse muscle-derived cells, develops tendon structure with mature collagen fibrils (Chen et al., 2012). In a similar study, aligned PLLA scaffolds supported cellular proliferation and tenogenic differentiation (Yin et al., 2010).

However, scaffolds, produced with synthetic polymers, exhibit several limitations, such as the absence of biochemical plot for cellular attachment and inability to fully regulate cell activity (Wan, Chen, Yang, Bei, & Wang, 2003). Probably, scaffolds made by natural proteins and their derivatives can address these issues; indeed, if based on collagen derivatives, in particular collagen type I, which is the major component in human tendons, scaffolds may result in more biocompatibility. Collagen derivatives also exhibit better biofunctionality supporting cell adhesion and cell proliferation better than polyester materials.

Scaffolds produced with MSCs in a collagen gel seem to improve histological and mechanical properties in a rabbit model (Juncosa-Melvin et al., 2006). Moreover, they are also further modifiable by cross-linking or cofabricating with other materials to improve mechanical strength and water resistance (Awad et al., 2003; Fessel, Gerber, & Snedeker, 2012; Panzavolta et al., 2011).

When mechanical stimulation in vitro is applied to human embryonic MSCs within a collagen-derived scaffold, they develop a tenocyte-like morphology and positively increase the expression of tendon-related gene markers (Chen et al., 2012).

A recent study reports that TDSCs within the same type of scaffold improve rabbit rotator cuff healing, exhibiting increased collagen deposition and better structural and biomechanical properties, as compared to the control group (Shen et al., 2012).

Topographical characteristics are an important goal that must be carefully considered when engineering tendon are prepared; for example, as tendons are primarily composed of parallel collagen fibers, alignment needs to be developed as a topographical characteristic to mimic correctly tendon tissue. Indeed, Scx and Tnmd were significantly increased on electrochemically aligned collagen threads when compared with randomly oriented collagen fibers (Kishore, Bullock, Sun, Van Dyke, & Akkus, 2012).

Aligned polyester materials have also been produced for tendon tissue engineering and, in a rotator cuff injury model, cell alignment, distribution, and EMC deposition were found reasonable to the nanofiber organization of a PLGA scaffold (Moffat et al., 2009). Aligned PLLA nanofibrous scaffolds with TDSCs showed increased tenogenesis and collagen deposition, with suppression of osteogenesis (Yin et al., 2010).

GROWTH FACTORS

Growth factors (GFs) are involved in tenogenesis and control of progenitor cell biology, and are divided in different families including transforming growth factors beta (TGF-ß1, TGF- ß2, and TGF-ß3), bone morphogenetic proteins (BMPs: BMP-12, BMP-13, and BMP-14), fibroblast GFs (FGF-2), vascular endothelial GF (VEGF), connective tissue GF (CTGF), platelet-derived GF (PDGF), and insulin-like GF 1 (IGF-1) (Longo, Franceschi, Berton, Maffulli, & Droena, 2012; Schneider, Angele, Järvinen, & Docheva, 2018).

GFs may drive the regenerative and reparative process of a wide variety of cells, including inflammatory cells, epithelial cells, fibroblasts, platelets, and tendon progenitor cells, binding to the external receptors on the cell membrane and leading to intracellular pathways involved in DNA synthesis and transcriptional expression of proteins allowing several cellular processes including proliferation, chemotaxis, matrix synthesis, and cell differentiation, all able to influence the healing of a damaged tissue.

In case of injury, tissues release GFs, firstly, from activated platelets, followed by recruiting inflammatory cells to the site of injury that, in turn, secrete additional GFs and amplify the inflammatory cascade. Further, mechanical loading placed on the injured tendon can modulate GFs production and their paracrine release activating the stem/progenitor tendon cells (Sharma & Maffulli, 2005a). A physiological tendon-inductive microenvironment requires multiple GFs over a specific temporal pattern (Aicale et al., 2020; Bisaccia et al., 2019; Oliva et al., 2019).

An optimal protocol for in vitro tenogenic differentiation remains to be determined, and several difficulties emerge in drawing an efficient in vitro approach considering the range of different culture conditions, time points, and experimental setups proposed in absence of robust conclusions. In this context, the in vitro role of GFs is further complicated by evidence that tenogenesis requires a combination of GFs with controlled concentrations and time of exposure.

Comparative studies of GFs teno-inductive capacity seem to suggest a central role of the TGFβ family: in particular, the expression of many tendon-related genes (such as Col I, Col III, Dcn, Tnc, and Scx) showed that TGF-β1 and EGF were the most efficient GFs to induce an early (at day 7) upregulation of Scx and Tnc, respectively (Citeroni et al., 2020).

At later stage, the influence of GFs can change: EGF controlled Col III expression at day 14, while TGF-β1 and PDGF-BB upregulated Col I at day 21. A recent study showed that the expression tendon-related genes are clearly stem cell source-dependent, suggesting a different ability in human amniotic fluid stem cells and adipose-derived MSCs (ADSCs) in undertaking in vitro tenogenic lineage commitment (Flamia, Zhdan, Martino, Castle, & Tamburro, 2004). Furthermore, TGF-β1 exposure has been recently identified as a potent tenogenic phenotype convertor in rat BMSCs promoting a greater upregulation of tenogenic-related genes and proteins compared to BMP-12, CTGF, and their combinations (Yin et al., 2016).

In addition, TGF-β1 in combination with BMPs appeared essential to preserve Scx-GFP expression levels in primary tenocytes after several days of culture (Maeda et al., 2011).

Another confirmation that TGFβ superfamily (in particular, TGF-β3) can be tendon inducer (inducing Scx overexpression) was reported testing the influence of different combinations of GFs (BMP-12, b-FGF, TGF-β3, CTGF, IGF-1) on human ADSCs, BMSCs, and TC during a two-step differentiation protocol in the presence of ascorbic acid (Perucca Orfei et al., 2019). This study confirmed an important role of TGF-β3 to promote an early teno-inductive effect and a late inhibitory influence on collagen fiber maturation on all cell typologies. Furthermore, BMP-12 seems to be subordinated to TGF-β3 in the induction of tendon-specific transcription factors with a late role in modulating the production of ECM (Perucca Orfei et al., 2019). In a recent study, TGF-β3 and BMP12 demonstrated a tenogenic effect in equine ADSCs cultured in monolayer or in 3D on decellularized tendon matrix (Roth et al., 2018).

A greater tendon-inductive influence was achieved by activating TGF-β2/3 signaling (Havis et al., 2014) that, additionally, played a role in suppressing cartilage marker Sox9 expression. However, no differences between TGF-β2 and TGF-β3 were found.

However, TGF-β needs an inductive surrounding microenvironment and while TGF-β3 or TGF-β3/BMP-12 upregulated Col IIa1, Col IIIa1, Tnc, Scx, and Mohawk in equine MSC cultures, they overexpressed Dcn and osteopontin by downregulating Smad8. These results allow to suggest that TGF-β3 may play a role as mediator for tenogenic induction, with BMP-12 as a modulator (Roth et al., 2018).

Regarding tenogenic differentiation action of BMP-14, TGF-β3, and VEGF, a recent study on rabbit BMSCs cultured in both 2D and 3D fibrin-based constructs, at 7 and 14 days, demonstrated that this association was the most effective in enhancing BMSC expression of Col Ia1, Col IIIa1, Tnc, and Tnmd in both cultures (Bottagisio et al., 2017).

CONCLUSION

In vitro teno-differentiation represents a prior step to treat in vivo tendon disorders with cell therapy or tissue engineering approaches to induce tissue repair or regeneration. Therefore it involves the use of a combination of key factors, such as cells, scaffolds, and biochemical/mechanical inputs, to produce a functional tissue-like construct. A combination of two or more techniques seems to be the best way to induce tendon differentiation in stem cells (J.L. Chen et al., 2015).

Cells, such as undifferentiated, predifferentiated, or differentiated stem cells, represent the building blocks of the engineered tissue. Many studies showed the involvement of different types of stem cells (embryonic, fetal, and adult stem cells) with promising results, but multiple studies focused on a single cell type. Scaffolds supply mechanical stability and provide a 3D support for cell growth and differentiation and the electrospinning technique has been shown to be able to generate 3D scaffolds with highly organized nanofibers, similar to fibers alignment in native tendon, improving structural organization of the construct during cell differentiation.

Taken together, all these elements contribute to the formation of a tissue-engineered substitute to be used as an in vitro model or to be applied in tissue replacement techniques in vivo. Several studies focus on a combined approach as a novel method for tendon tissue engineering, demonstrating how the cooperative effect of different factors improves results. For example, Testa et al. (2017) developed a C3H10T1/2 fibroblast cell line on a PEGylated–fibrinogen biomimetic matrix, exposing cells to TGF-β and a mechanical input, which reported that the proposed combined approach led to a highly organized neo-ECM, with Col I fibers parallel to the direction of stretching, reflecting the enhanced elastic modulus and endurance of the matrix. Govoni et al. fabricated a multiphase 3D construct composed of a hyaluronate elastic band merged with a fibrin hydrogel supplemented with human BMSCs and PLGA microcarriers. The constructs were loaded with human GDF-5 and the synergy between biochemical and mechanical inputs led to an increased expression of tenogenic markers, such as Col I, Col III, Dcn, Scx, and Tnc (Govoni et al., 2017).

In conclusion, in vitro techniques are essential to study tendon development, healing, and regeneration but only with a validated and successful in vitro model we will have a defined view of tendon biology and pathology to treat tendon disorders in vivo.

REFERENCES

Aicale, R., Bisaccia, R. D., Oliviero, A., Oliva, F., & Maffulli, N. (2020). Current pharmacological approaches to the treatment of tendinopathy. *Expert Opinion on Pharmacotherapy*, *21*(12), 1467–1477. Available from https://doi.org/10.1080/14656566.2020.1763306.

Aicale, R., Tarantino, D., Maccauro, G., Peretti, G. M., & Maffulli, N. (2019). Genetics in orthopaedic practice. *Journal of Biological Regulators and Homeostatic Agents*, *33*(1), 103–117. Available from http://www.biolifesas.org/contentsJBRHA.htm.

Aicale, R., Tarantino, D., & Maffulli, N. (2017). *Basic science of tendons. Bio-orthopaedics: A new approach* (pp. 249–273). Berlin Heidelberg: Springer. Available from https://doi.org/10.1007/978-3-662-54181-4_21.

Aicale, R., Tarantino, D., & Maffulli, N. (2018). Overuse injuries in sport: A comprehensive overview. *Journal of Orthopaedic Surgery and Research*, *13*(1), 309. Available from https://doi.org/10.1186/s13018-018-1017-5.

Atala, A., Irvine, D. J., Moses, M., & Shaunak, S. (2010). Wound healing vs regeneration: Role of the tissue environment in regenerative medicine. *MRS Bulletin, 35*(8), 597−606. Available from https://doi.org/10.1557/mrs2010.528.

Awad, H. A., Boivin, G. P., Dressler, M. R., Smith, F. N. L., Young, R. G., & Butler, D. L. (2003). Repair of patellar tendon injuries using a cell-collagen composite. *Journal of Orthopaedic Research, 21*(3), 420−431. Available from https://doi.org/10.1016/S0736-0266(02)00163-8.

Badylak, S. F., Freytes, D. O., & Gilbert, T. W. (2009). Extracellular matrix as a biological scaffold material: Structure and function. *Acta Biomaterialia, 5*(1), 1−13. Available from https://doi.org/10.1016/j.actbio.2008.09.013.

Bely, A. E., & Nyberg, K. G. (2010). Evolution of animal regeneration: Re-emergence of a field. *Trends in Ecology and Evolution, 25*(3), 161−170. Available from https://doi.org/10.1016/j.tree.2009.08.005.

Benazzo, F., Stennardo, G., Mosconi, M., Zanon, G., & Maffulli, N. (2001). Muscle transplant in the rabbit's achilles tendon. *Medicine and Science in Sports and Exercise, 33*(5), 696−701. Available from https://doi.org/10.1097/00005768-200105000-00003.

Bi, Y., Ehirchiou, D., Kilts, T. M., Inkson, C. A., Embree, M. C., Sonoyama, W., . . . Young, M. F. (2007). Identification of tendon stem/progenitor cells and the role of the extracellular matrix in their niche. *Nature Medicine, 13*(10), 1219−1227. Available from https://doi.org/10.1038/nm1630.

Bisaccia, D. R., Aicale, R., Tarantino, D., Peretti, G. M., & Maffulli, N. (2019). Biological and chemical changes in fluoroquinolone-associated tendinopathies: A systematic review. *British Medical Bulletin, 130* (1), 39−49. Available from https://doi.org/10.1093/bmb/ldz006.

Bottagisio, M., Lopa, S., Granata, V., Talò, G., Bazzocchi, C., Moretti, M., . . . Lovati, A. B. (2017). Different combinations of growth factors for the tenogenic differentiation of bone marrow mesenchymal stem cells in monolayer culture and in fibrin-based three-dimensional constructs. *Differentiation; Research in Biological Diversity, 95*, 44−53. Available from https://doi.org/10.1016/j.diff.2017.03.001.

Butler, D. L., Juncosa, N., & Dressler, M. R. (2004). Functional efficacy of tendon repair processes. *Annual Review of Biomedical Engineering, 6*, 303−329. Available from https://doi.org/10.1146/annurev.bioeng.6.040803.140240.

Canty, E. G., & Kadler, K. E. (2005). Procollagen trafficking, processing and fibrillogenesis. *Journal of Cell Science, 118*(7), 1341−1353. Available from https://doi.org/10.1242/jcs.01731.

Cao, Y., Liu, Y., Liu, W., Shan, Q., Buonocore, S. D., & Cui, L. (2002). Bridging tendon defects using autologous tenocyte engineered tendon in a hen model. *Plastic and Reconstructive Surgery, 110*(5), 1280−1289. Available from https://doi.org/10.1097/00006534-200210000-00011.

Cartmell, J. S., & Dunn, M. G. (2004). Development of cell-seeded patellar tendon allografts for anterior cruciate ligament reconstruction. *Tissue Engineering, 10*(7−8), 1065−1075. Available from https://doi.org/10.1089/ten.2004.10.1065.

Chen, B., Wang, B., Zhang, W. J., Zhou, G., Cao, Y., & Liu, W. (2012). In vivo tendon engineering with skeletal muscle derived cells in a mouse model. *Biomaterials, 33*(26), 6086−6097. Available from https://doi.org/10.1016/j.biomaterials.2012.05.022.

Chen, J. L., Zhang, W., Liu, Z. Y., Heng, B. C., Ouyang, H. W., & Dai, X. S. (2015). Physical regulation of stem cells differentiation into teno-lineage: Current strategies and future direction. *Cell and Tissue Research, 360*(2), 195−207. Available from https://doi.org/10.1007/s00441-014-2077-4.

Chisari, E., Rehak, L., Khan, W. S., & Maffulli, N. (2019). Tendon healing in presence of chronic low-level inflammation: A systematic review. *British Medical Bulletin, 132*(1), 97−116. Available from https://doi.org/10.1093/bmb/ldz035.

Chong, A. K. S., Riboh, J., Smith, R. L., Lindsey, D. P., Pham, H. M., & Chang, J. (2009). Flexor tendon tissue engineering: Acellularized and reseeded tendon constructs. *Plastic and Reconstructive Surgery, 123*(6), 1759−1766. Available from https://doi.org/10.1097/PRS.0b013e3181a65ae7.

Chuen, F. S., Chuk, C. Y., Ping, W. Y., Nar, W. W., Kim, H. L., & Ming, C. K. (2004). Immunohistochemical characterization of cells in adult human patellar tendons. *Journal of Histochemistry and Cytochemistry*, *52* (9), 1151−1157. Available from https://doi.org/10.1369/jhc.3A6232.2004.

Cipollaro, L., Trucillo, P., Bragazzi, N. L., Porta, G. D., Reverchon, E., & Maffulli, N. (2020). Liposomes for intra-articular analgesic drug delivery in orthopedics: State-of-art and future perspectives. insights from a systematic mini-review of the literature. *Medicina (Lithuania)*, *56*(9), 1−17. Available from https://doi.org/10.3390/medicina56090423.

Citeroni, M. R., Ciardulli, M. C., Russo, V., Porta, G. D., Mauro, A., Khatib, M. E., ... Barboni, B. (2020). In vitro innovation of tendon tissue engineering strategies. *International Journal of Molecular Sciences*, *21*, 6726.

De Albornoz, P. M., Aicale, R., Forriol, F., & Maffulli, N. (2018). Cell therapies in tendon, ligament, and musculoskeletal system repair. *Sports Medicine and Arthroscopy Review*, *26*(2), 48−58. Available from https://doi.org/10.1097/JSA.0000000000000192.

de Oliveira, R. R., Lemos, A., de Castro Silveira, P. V., da Silva, R. J., & de Moraes, S. R. A. (2011). Alterations of tendons in patients with diabetes mellitus: A systematic review. *Diabetic Medicine*, *28*(8), 886−895. Available from https://doi.org/10.1111/j.1464-5491.2010.03197.x.

Deeken, C. R., White, A. K., Bachman, S. L., Ramshaw, B. J., Cleveland, D. S., Loy, T. S., ... Grant, S. A. (2011). Method of preparing a decellularized porcine tendon using tributyl phosphate. *Journal of Biomedical Materials Research - Part B Applied Biomaterials*, *96*(2), 199−206. Available from https://doi.org/10.1002/jbm.b.31753.

Favata, M., Beredjiklian, P. K., Zgonis, M. H., Beason, D. P., Crombleholme, T. M., Jawad, A. F., ... Soslowsky, L. J. (2006). Regenerative properties of fetal sheep tendon are not adversely affected by transplantation into an adult environment. *Journal of Orthopaedic Research*, *24*(11), 2124−2132. Available from https://doi.org/10.1002/jor.20271.

Fessel, G., Gerber, C., & Snedeker, J. G. (2012). Potential of collagen cross-linking therapies to mediate tendon mechanical properties. *Journal of Shoulder and Elbow Surgery*, *21*(2), 209−217. Available from https://doi.org/10.1016/j.jse.2011.10.002.

Flamia, R., Zhdan, P. A., Martino, M., Castle, J. E., & Tamburro, A. M. (2004). AFM study of the elastin-like biopolymer poly(ValGlyGlyValGly). *Biomacromolecules*, *5*(4), 1511−1518. Available from https://doi.org/10.1021/bm049930r.

Franceschi, F., Papalia, R., Paciotti, M., Franceschetti, E., Martino, A. D., Maffulli, N., ... Denaro, V. (2014). Obesity as a risk factor for tendinopathy: A systematic review. *International Journal of Endocrinology*, *2014, 670262*. Available from https://doi.org/10.1155/2014/670262.

Franchi, M., Trirè, A., Quaranta, M., Orsini, E., & Ottani, V. (2007). Collagen structure of tendon relates to function. *The Scientific World Journal*, *7*, 404−420. Available from https://doi.org/10.1100/tsw.2007.92.

Gimble, J. M., Katz, A. J., & Bunnell, B. A. (2007). Adipose-derived stem cells for regenerative medicine. *Circulation Research*, *100*(9), 1249−1260. Available from https://doi.org/10.1161/01.RES.0000265074.83288.09.

Gott, M., Ast, M., Lane, L. B., Schwartz, J. A., Catanzano, A., Razzano, P., & Grande, D. A. (2011). Tendon phenotype should dictate tissue engineering modality in tendon repair: A review. *Discovery Medicine*, *12*(62), 75−84.

Govoni, M., Berardi, A. C., Muscari, C., Campardelli, R., Bonafè, F., Guarnieri, C., ... Della Porta, G. (2017). An engineered multiphase three-dimensional microenvironment to ensure the controlled delivery of cyclic strain and human growth differentiation factor 5 for the tenogenic commitment of human bone marrow mesenchymal stem cells. *Tissue Engineering - Part A*, *23*(15−16), 811−822. Available from https://doi.org/10.1089/ten.tea.2016.0407.

Havis, E., Bonnin, M. A., Olivera-Martinez, I., Nazaret, N., Ruggiu, M., Weibel, J., ... Duprez, D. (2014). Transcriptomic analysis of mouse limb tendon cells during development. *Development (Cambridge)*, *141* (19), 3683−3696. Available from https://doi.org/10.1242/dev.108654.

Juncosa-Melvin, N., Boivin, G. P., Gooch, C., Galloway, M. T., West, J. R., Dunn, M. G., . . . Butler, D. L. (2006). The effect of autologous mesenchymal stem cells on the biomechanics and histology of gel-collagen sponge constructs used for rabbit patellar tendon repair. *Tissue Engineering, 12*(2), 369−379. Available from https://doi.org/10.1089/ten.2006.12.369.

Kader, D., Saxena, A., Movin, T., & Maffulli, N. (2002). Achilles tendinopathy: Some aspects of basic science and clinical management. *British Journal of Sports Medicine, 36*(4), 239−249. Available from https://doi.org/10.1136/bjsm.36.4.239.

Kannus, P. (2000). Structure of the tendon connective tissue. *Scandinavian Journal of Medicine and Science in Sports, 10*(6), 312−320. Available from https://doi.org/10.1034/j.1600-0838.2000.010006312.x.

Khan, K. M., Cook, J. L., Bonar, F., Harcourt, P., & Åstrom, M. (1999). Histopathology of common tendinopathies: Update and implications for clinical management. *Sports Medicine, 27*(6), 393−408. Available from https://doi.org/10.2165/00007256-199927060-00004.

Kishore, V., Bullock, W., Sun, X., Van Dyke, W. S., & Akkus, O. (2012). Tenogenic differentiation of human MSCs induced by the topography of electrochemically aligned collagen threads. *Biomaterials, 33*(7), 2137−2144. Available from https://doi.org/10.1016/j.biomaterials.2011.11.066.

Klepps, S., Bishop, J., Lin, J., Cahlon, O., Strauss, A., Hayes, P., . . . Flatow, E. L. (2004). Prospective evaluation of the effect of rotator cuff integrity on the outcome of open rotator cuff repairs. *American Journal of Sports Medicine, 32*(7), 1716−1722. Available from https://doi.org/10.1177/0363546504265262.

Krueger-Franke, M., Siebert, C. H., & Scherzer, S. (1995). Surgical treatment of ruptures of the Achilles tendon: A review of long-term results. *British Journal of Sports Medicine, 29*(2), 121−125. Available from https://doi.org/10.1136/bjsm.29.2.121.

Li, Y., Xiao, Y., & Liu, C. (2017). The horizon of materiobiology: A perspective on material-guided cell behaviors and tissue engineering. *Chemical Reviews, 117*(5), 4376−4421. Available from https://doi.org/10.1021/acs.chemrev.6b00654.

Liu, C. F., Aschbacher-Smith, L., Barthelery, N. J., Dyment, N., Butler, D., & Wylie, C. (2011). What we should know before using tissue engineering techniques to repair injured tendons: A developmental biology perspective. *Tissue Engineering - Part B: Reviews, 17*(3), 165−176. Available from https://doi.org/10.1089/ten.teb.2010.0662.

Liu, H., Fan, H., Toh, S. L., & Goh, J. C. H. (2008). A comparison of rabbit mesenchymal stem cells and anterior cruciate ligament fibroblasts responses on combined silk scaffolds. *Biomaterials, 29*(10), 1443−1453. Available from https://doi.org/10.1016/j.biomaterials.2007.11.023.

Loebel, C., & Burdick, J. A. (2018). Engineering stem and stromal cell therapies for musculoskeletal tissue repair. *Cell Stem Cell, 22*(3), 325−339. Available from https://doi.org/10.1016/j.stem.2018.01.014.

Longo, U. G., Franceschi, F., Berton, A., Maffulli, N., & Droena, V. (2012). Conservative treatment and rotator cuff tear progression. *Medicine and Sport Science, 57*, 90−99. Available from https://doi.org/10.1159/000328910.

Longo, U. G., Ronga, M., & Maffulli, N. (2009). Achilles tendinopathy. *Sports Medicine and Arthroscopy Review, 17*(2), 112−126. Available from https://doi.org/10.1097/jsa.0b013e3181a3d625.

Lui, P. P. Y. (2013). Identity of tendon stem cells - how much do we know? *Journal of Cellular and Molecular Medicine, 17*(1), 55−64. Available from https://doi.org/10.1111/jcmm.12007.

Lui, P. P. Y., & Chan, K. M. (2011). Tendon-Derived Stem Cells (TDSCs): From basic science to potential roles in tendon pathology and tissue engineering applications. *Stem Cell Reviews and Reports, 7*(4), 883−897. Available from https://doi.org/10.1007/s12015-011-9276-0.

Maeda, T., Sakabe, T., Sunaga, A., Sakai, K., Rivera, A. L., Keene, D. R., . . . Sakai, T. (2011). Conversion of mechanical force into TGF-β-mediated biochemical signals. *Current Biology, 21*(11), 933−941. Available from https://doi.org/10.1016/j.cub.2011.04.007.

Maffulli, N., Barrass, V., & Ewen, S. W. B. (2000). Light microscopic histology of achilles tendon ruptures: A comparison with unruptured tendons. *American Journal of Sports Medicine, 28*(6), 857−863. Available from https://doi.org/10.1177/03635465000280061401.

Maffulli, N., Khan, K. M., & Puddu, G. (1998). Overuse tendon conditions: Time to change a confusing terminology. *Arthroscopy: the Journal of Arthroscopic & Related Surgery: Official Publication of the Arthroscopy Association of North America and the International Arthroscopy Association, 14*(8), 840−843. Available from https://doi.org/10.1016/S0749-8063(98)70021-0.

Maffulli, N., Oliva, F., Loppini, M., Aicale, R., Spiezia, F., & King, J. B. (2017). The Royal London Hospital test for the clinical diagnosis of patellar tendinopathy. *Muscles, Ligaments and Tendons Journal, 7*(2), 315−322. Available from https://doi.org/10.11138/mltj/2017.7.2.315.

Maffulli, N., & Peretti, G. M. (2020). Treatment decisions for acute Achilles tendon ruptures. *The Lancet, 395* (10222), 397−398. Available from https://doi.org/10.1016/S0140-6736(19)33133-2.

Maffulli, N., Sharma, P., & Luscombe, K. L. (2004). Achilles tendinopathy: Aetiology and management. *Journal of the Royal Society of Medicine, 97*(10), 472−476. Available from https://doi.org/10.1258/jrsm.97.10.472.

Maffulli, N., Via, A. G., & Oliva, F. (2015). Chronic achilles tendon disorders: Tendinopathy and chronic rupture. *Clinics in Sports Medicine, 34*(4), 607−624. Available from https://doi.org/10.1016/j.csm.2015.06.010.

Magra, M., Hughes, S., El Haj, A. J., & Maffulli, N. (2007). VOCCs and TREK-1 ion channel expression in human tenocytes. *American Journal of Physiology - Cell Physiology, 292*(3), C1053−C1060. Available from https://doi.org/10.1152/ajpcell.00053.2006.

Magra, M., & Maffulli, N. (2005a). Matrix metalloproteases: A role in overuse tendinopathies. *British Journal of Sports Medicine, 39*(11), 789−791. Available from https://doi.org/10.1136/bjsm.2005.017855.

Magra, M., & Maffulli, N. (2005b). Molecular events in tendinopathy: A role for metalloproteases. *Foot and Ankle Clinics, 10*(2), 267−277. Available from https://doi.org/10.1016/j.fcl.2005.01.012.

Magra, M., & Maffulli, N. (2006). Nonsteroidal antiinflammatory drugs in tendinopathy: Friend or foe. *Clinical Journal of Sport Medicine, 16*(1), 1−3. Available from https://doi.org/10.1097/01.jsm.0000194764.27819.5d.

McNeilly, C. M., Banes, A. J., Benjamin, M., & Ralphs, J. R. (1996). Tendon cells in vivo form a three dimensional network of cell processes linked by gap junctions. *Journal of Anatomy, 189*(3), 593−600.

Mendes, B. B., Gómez-Florit, M., Babo, P. S., Domingues, R. M., Reis, R. L., & Gomes, M. E. (2018). Blood derivatives awaken in regenerative medicine strategies to modulate wound healing. *Advanced Drug Delivery Reviews, 129*, 376−393. Available from https://doi.org/10.1016/j.addr.2017.12.018.

Mienaltowski, M. J., Adams, S. M., & Birk, D. E. (2013). Regional differences in stem cell/progenitor cell populations from the mouse achilles tendon. *Tissue Engineering - Part A, 19*(1−2), 199−210. Available from https://doi.org/10.1089/ten.tea.2012.0182.

Moffat, K. L., Kwei, A. S. P., Spalazzi, J. P., Doty, S. B., Levine, W. N., & Lu, H. H. (2009). Novel nanofiber-based scaffold for rotator cuff repair and augmentation. *Tissue Engineering - Part A, 15*(1), 115−126. Available from https://doi.org/10.1089/ten.tea.2008.0014.

Nichols, A. E. C., Best, K. T., & Loiselle, A. E. (2019). The cellular basis of fibrotic tendon healing: Challenges and opportunities. *Translational Research, 209*, 156−168. Available from https://doi.org/10.1016/j.trsl.2019.02.002.

Ning, L. J., Zhang, Y., Chen, X. H., Luo, J. C., Li, X. Q., Yang, Z. M., . . . Qin, T. W. (2012). Preparation and characterization of decellularized tendon slices for tendon tissue engineering. *Journal of Biomedical Materials Research - Part A, 100*(6), 1448−1456. Available from https://doi.org/10.1002/jbm.a.34083.

Oliva, F., Berardi, A. C., Misiti, S., & Maffulli, N. (2013). Thyroid hormones and tendon: Current views and future perspectives. Muscles, Ligaments and Tendons. *Journal, 3*(3), 201−203. Available from https://doi.org/10.11138/mltj/2013.3.3.201.

Oliva, F., Maffulli, N., Gissi, C., Veronesi, F., Calciano, L., Fini, M., . . . Berardi, A. C. (2019). Combined ascorbic acid and T3 produce better healing compared to bone marrow mesenchymal stem cells in an

Achilles tendon injury rat model: A proof of concept study. *Journal of Orthopaedic Surgery and Research*, *14*(1), 54. Available from https://doi.org/10.1186/s13018-019-1098-9.

Oliva, F., Misiti, S., & Maffulli, N. (2014). Metabolic diseases and tendinopathies: The missing link. *Muscles, Ligaments and Tendons Journal*, *4*(3), 273−274. Available from https://doi.org/10.11138/mltj/2014.4.3.273.

Oliva, F., Piccirilli, E., Berardi, A. C., Frizziero, A., Tarantino, U., & Maffulli, N. (2016). Hormones and tendinopathies: The current evidence. *British Medical Bulletin*, *117*(1), 39−58. Available from https://doi.org/10.1093/bmb/ldv054.

Oliva, F., Via, A. G., & Maffulli, N. (2012). Physiopathology of intratendinous calcific deposition. *BMC Medicine*, 10, 95. Available from https://doi.org/10.1186/1741-7015-10-95.

Ouyang, H. W., Goh, J. C. H., Thambyah, A., Teoh, S. H., & Lee, E. H. (2003). Knitted poly-lactide-coglycolide scaffold loaded with bone marrow stromal cells in repair and regeneration of rabbit achilles tendon. *Tissue Engineering*, *9*(3), 431−439. Available from https://doi.org/10.1089/107632703322066615.

O'Brien, M. (1997). Structure and metabolism of tendons. *Scandinavian Journal of Medicine and Science in Sports*, *7*(2), 55−61. Available from https://doi.org/10.1111/j.1600-0838.1997.tb00119.x.

Paavola, M., Paakkala, T., Kannus, P., & Järvinen, M. (1998). Ultrasonography in the differential diagnosis of achilles tendon injuries and related disorders: A comparison between pre-operative ultrasonography and surgical findings. *Acta Radiologica*, *39*(6), 612−619. Available from https://doi.org/10.3109/02841859809175485.

Panzavolta, S., Gioffrè, M., Focarete, M. L., Gualandi, C., Foroni, L., & Bigi, A. (2011). Electrospun gelatin nanofibers: Optimization of genipin cross-linking to preserve fiber morphology after exposure to water. *Acta Biomaterialia*, *7*(4), 1702−1709. Available from https://doi.org/10.1016/j.actbio.2010.11.021.

Park, A., Hogan, M. V., Kesturu, G. S., James, R., Balian, G., & Chhabra, A. B. (2010). Adipose-derived mesenchymal stem cells treated with growth differentiation factor-5 express tendon-specific markers. *Tissue Engineering - Part A*, *16*(9), 2941−2951. Available from https://doi.org/10.1089/ten.tea.2009.0710.

Parmar, C., & Hennessy, M. (2007). Achilles tendon rupture associated with combination therapy of levofloxacin and steroid in four patients and a review of the literature. *Foot and Ankle International*, *28*(12), 1287−1289. Available from https://doi.org/10.3113/FAI.2007.1287.

Perucca Orfei, C., Viganò, M., Pearson, J. R., Colombini, A., De Luca, P., Ragni, E., . . . de Girolamo, L. (2019). In vitro induction of tendon-specific markers in tendon cells, adipose- and bone marrow-derived stem cells is dependent on TGFβ3, BMP-12 and ascorbic acid stimulation. *International Journal of Molecular Sciences*, *20*(1), 149. Available from https://doi.org/10.3390/ijms20010149.

Pridgen, B. C., Woon, C. Y. L., Kim, M., Thorfinn, J., Lindsey, D., Pham, H., . . . Chang, J. (2011). Flexor tendon tissue engineering: Acellularization of human flexor tendons with preservation of biomechanical properties and biocompatibility. *Tissue Engineering - Part C: Methods*, *17*(8), 819−828. Available from https://doi.org/10.1089/ten.tec.2010.0457.

Qiu, Y., Wang, X., Zhang, Y., Carr, A. J., Zhu, L., Xia, Z., . . . Sabokbar, A. (2013). Development of a refined tenocyte differentiation culture technique for tendon tissue engineering. *Cells, Tissues, Organs*, *197*(1), 27−36. Available from https://doi.org/10.1159/000341426.

Riley, G. P., Curry, V., DeGroot, J., Van El, B., Verzijl, N., Hazleman, B. L., . . . Bank, R. A. (2002). Matrix metalloproteinase activities and their relationship with collagen remodelling in tendon pathology. *Matrix Biology*, *21*(2), 185−195. Available from https://doi.org/10.1016/S0945-053X(01)00196-2.

Roth, S. P., Schubert, S., Scheibe, P., Groß, C., Brehm, W., & Burk, J. (2018). Growth factor-mediated tenogenic induction of multipotent mesenchymal stromal cells is altered by the microenvironment of tendon matrix. *Cell Transplantation*, *27*(10), 1434−1450. Available from https://doi.org/10.1177/0963689718792203.

Ruzzini, L., Abbruzzese, F., Rainer, A., Longo, U. G., Trombetta, M., Maffulli, N., . . . Denaro, V. (2014). Characterization of age-related changes of tendon stem cells from adult human tendons. *Knee Surgery,*

Sports Traumatology, Arthroscopy, 22(11), 2856−2866. Available from https://doi.org/10.1007/s00167-013-2457-4.

Sayana, M. K., & Maffulli, N. (2007). Eccentric calf muscle training in non-athletic patients with Achilles tendinopathy. *Journal of Science and Medicine in Sport*, *10*(1), 52−58. Available from https://doi.org/10.1016/j.jsams.2006.05.008.

Schneider, M., Angele, P., Järvinen, T. A. H., & Docheva, D. (2018). Rescue plan for Achilles: Therapeutics steering the fate and functions of stem cells in tendon wound healing. *Advanced Drug Delivery Reviews*, *129*, 352−375. Available from https://doi.org/10.1016/j.addr.2017.12.016.

Schwellnus, M. P., Jordaan, G., & Noakes, T. D. (1990). Prevention of common overuse injuries by the use of shock absorbing insoles: A prospective study. *American Journal of Sports Medicine*, *18*(6), 636−641. Available from https://doi.org/10.1177/036354659001800614.

Sharma, P., & Maffulli, N. (2005a). Basic biology of tendon injury and healing. *The Surgeon: Journal of the Royal Colleges of Surgeons of Edinburgh and Ireland*, *3*(5), 309−316. Available from https://doi.org/10.1016/S1479-666X(05)80109-X.

Sharma, P., & Maffulli, N. (2005b). Tendon injury and tendinopathy: Healing and repair. *Journal of Bone and Joint Surgery - Series A*, *87*(1), 187−202. Available from https://doi.org/10.2106/JBJS.D.01850.

Shen, W., Chen, J., Yin, Z., Chen, X., Liu, H., Heng, B. C., . . . Ouyang, H. W. (2012). Allogenous tendon stem/progenitor cells in silk scaffold for functional shoulder repair. *Cell Transplantation*, *21*(5), 943−958. Available from https://doi.org/10.3727/096368911X627453.

Spanoudes, K., Gaspar, D., Pandit, A., & Zeugolis, D. I. (2014). The biophysical, biochemical, and biological toolbox for tenogenic phenotype maintenance in vitro. *Trends in Biotechnology*, *32*(9), 474−482. Available from https://doi.org/10.1016/j.tibtech.2014.06.009.

Testa, S., Costantini, M., Fornetti, E., Bernardini, S., Trombetta, M., Seliktar, D., . . . Gargioli, C. (2017). Combination of biochemical and mechanical cues for tendon tissue engineering. *Journal of Cellular and Molecular Medicine*, *21*(11), 2711−2719. Available from https://doi.org/10.1111/jcmm.13186.

Tilley, J. M. R., Chaudhury, S., Hakimi, O., Carr, A. J., & Czernuszka, J. T. (2012). Tenocyte proliferation on collagen scaffolds protects against degradation and improves scaffold properties. *Journal of Materials Science: Materials in Medicine*, *23*(3), 823−833. Available from https://doi.org/10.1007/s10856-011-4537-7.

Tucker, B. A., Karamsadkar, S. S., Khan, W. S., & Pastides, P. (2010). The role of bone marrow derived mesenchymal stem cells in sports injuries. *Journal of Stem Cells*, *5*, 155−166.

Van der Linden, P. D., Sturkenboom, M. C. J. M., Herings, R. M. C., Leufkens, H. M. G., Rowlands, S., & Stricker, B. H. C. (2003). Increased risk of Achilles tendon rupture with quinolone antibacterial use, especially in elderly patients taking oral corticosteroids. *Archives of Internal Medicine*, *163*(15), 1801−1807. Available from https://doi.org/10.1001/archinte.163.15.1801.

Van Eijk, F., Saris, D. B. F., Riesle, J., Willems, W. J., Van Blitterswijk, C. A., Verbout, A. J., . . . Dhert, W. J. A. (2004). Tissue engineering of ligaments: A comparison of bone marrow stromal cells, anterior cruciate ligament, and skin fibroblasts as cell source. *Tissue Engineering*, *10*(5−6), 893−903. Available from https://doi.org/10.1089/1076327041348428.

Vindigni, V., Tonello, C., Lancerotto, L., Abatangelo, G., Cortivo, R., Zavan, B., . . . Bassetto, F. (2013). Preliminary report of in vitro reconstruction of a vascularized tendonlike structure: A novel application for adipose-derived stem cells. *Annals of Plastic Surgery*, *71*(6), 664−670. Available from https://doi.org/10.1097/SAP.0b013e3182583e99.

Voleti, P. B., Buckley, M. R., & Soslowsky, L. J. (2012). Tendon healing: Repair and regeneration. *Annual Review of Biomedical Engineering*, *14*, 47−71. Available from https://doi.org/10.1146/annurev-bioeng-071811-150122.

Waldecker, U., Hofmann, G., & Drewitz, S. (2012). Epidemiologic investigation of 1394 feet: Coincidence of hindfoot malalignment and Achilles tendon disorders. *Foot and Ankle Surgery*, *18*(2), 119−123. Available from https://doi.org/10.1016/j.fas.2011.04.007.

Wan, Y., Chen, W., Yang, J., Bei, J., & Wang, S. (2003). Biodegradable poly(L-lactide)-poly(ethylene glycol) multiblock copolymer: Synthesis and evaluation of cell affinity. *Biomaterials*, *24*(13), 2195−2203. Available from https://doi.org/10.1016/S0142-9612(03)00107-8.

Yasuda, K., & Hayashi, K. (1999). Changes in biomechanical properties of tendons and ligaments from joint disuse. *Osteoarthritis and Cartilage*, *7*(1), 122−129. Available from https://doi.org/10.1053/joca.1998.0167.

Yin, Z., Chen, X., Chen, J. L., Shen, W. L., Hieu Nguyen, T. M., Gao, L., . . . Ouyang, H. W. (2010). The regulation of tendon stem cell differentiation by the alignment of nanofibers. *Biomaterials*, *31*(8), 2163−2175. Available from https://doi.org/10.1016/j.biomaterials.2009.11.083.

Yin, Z., Guo, J., Wu, T. Y., Chen, X., Xu, L. L., Lin, S. E., . . . Li, G. (2016). Stepwise differentiation of mesenchymal stem cells augments tendon-like tissue formation and defect repair in vivo. *Stem Cells Translational Medicine*, *5*(8), 1106−1116. Available from https://doi.org/10.5966/sctm.2015-0215.

Zafar, M. S., Mahmood, A., & Maffulli, N. (2009). Basic science and clinical aspects of Achilles tendinopathy. *Sports Medicine and Arthroscopy Review*, *17*(3), 190−197. Available from https://doi.org/10.1097/JSA.0b013e3181b37eb7.

Zhang, J., Pan, T., Liu, Y., & Wang, J. H. C. (2010). Mouse treadmill running enhances tendons by expanding the pool of Tendon Stem Cells (TSCs) and TSC-related cellular production of collagen. *Journal of Orthopaedic Research*, *28*(9), 1178−1183. Available from https://doi.org/10.1002/jor.21123.

Zhang, J., & Wang, J. H. C. (2010). Mechanobiological response of tendon stem cells: Implications of tendon homeostasis and pathogenesis of tendinopathy. *Journal of Orthopaedic Research*, *28*(5), 639−643. Available from https://doi.org/10.1002/jor.21046.

FURTHER READING

Chen, J. L., Yin, Z., Shen, W. L., Chen, X., Heng, B. C., Zou, X. H., . . . Ouyang, H. W. (2010). Efficacy of hESC-MSCs in knitted silk-collagen scaffold for tendon tissue engineering and their roles. *Biomaterials*, *31*(36), 9438−9451. Available from https://doi.org/10.1016/j.biomaterials.2010.08.011.

Liu, W., Chen, B., Deng, D., Xu, F., Cui, L., & Cao, Y. (2006). Repair of tendon defect with dermal fibroblast engineered tendon in a porcine model. *Tissue Engineering*, *12*(4), 775−788. Available from https://doi.org/10.1089/ten.2006.12.775.

Sharma, P., & Maffulli, N. (2006). Biology of tendon injury: Healing, modeling and remodeling. *Journal of Musculoskeletal Neuronal Interactions*, *6*(2), 181−190.

BIOMECHANICAL REQUIREMENTS FOR CERTIFICATION AND QUALITY IN MEDICAL DEVICES

25

Silvia Pianigiani[1] and Tomaso Villa[2,3]

[1]*Adler Ortho, Milano, Italy* [2]*IRCCS Istituto Ortopedico Galeazzi, Milan, Italy* [3]*Laboratory of Biological Structure Mechanics, Department of Chemistry, Materials and Chemical Engineering "Giulio Natta", Politecnico di Milano, Milan, Italy*

CERTIFICATION AND QUALITY OF AN ORTHOPEDIC MEDICAL DEVICE: REQUIREMENTS, REGULATIONS, LAWS, AND PROCEDURES

When we think of a medical device, we must consider, first of all, that this is an industrial product and that the industrial production is based on the ability to manufacture high numbers of identical devices. "Identical" means that all the devices must have the same design requirements and differ only for some aspects that do not affect their functionality: this particular industrial approach is based on the introduction and usage of technical standards and quality aspects. Technical standards set the performances that a device or process must have while quality refers to the fulfillment of the expected requirements: in this light, the manufacturer must clearly indicate the performances that the product owns and the quality system must guarantee that all the devices own those performances.

Those aspects that can be generally applied to any manufacturing process have some further peculiarities when the application in the medical device field is considered. As a matter of fact, the medical device market is a very strictly controlled one, due to the fact that the introduction into the market of a new device is subjected to a high number of regulatory activities whose main aim is to demonstrate that the device is safe for the patient, relatively to its prescribed use. Also, the specification related to the prescribed use is very peculiar: in the case of orthopedic devices, the use of the device is firstly related to the implantation performed by a medical surgeon. Hence, generally speaking, a prosthesis may be correctly designed to have all the requirements needed for substituting a diseased joint and quality may guarantee that all the manufactured prostheses have the same characteristics, but the final result on the patient is strongly mediated by the action of the surgeon: firstly, the surgeon must be informed by the manufacturer on the prescribed use of that device; secondly, he has to evaluate if that prescribed use is coherent to the clinical situation of the patient; and finally, he must use the device according its prescribed use. Only by fulfilling all the three above requirements, the prosthesis will be able to guarantee its functionality and the clinical outcome will be satisfactory.

In Europe, the framework that has helped defined the regulatory pathway for introducing a new medical device in the market is the (EU) 745/2017 Regulation, which starting from May 2021 has

Human Orthopaedic Biomechanics. DOI: https://doi.org/10.1016/B978-0-12-824481-4.00005-6

completely replaced the old 93/42/CE Directive. In such a framework, the manufacturer (which is sole responsible for the introduction into the market of a new device) must submit a proper technical documentation to a notified body that, on behalf of the competent authority of the country the notified body belongs and after assessing the compliance of the documents to the Regulation, allows the manufacturer to put the Conformitè Europëenne (CE) mark on the device: after getting the CE mark, the manufacturer is authorized to sell the device in all the countries belonging to European Community and in all the other countries that recognize the CE mark as necessary for medical device marketing in its territory (see Fig. 25.1).

The content of the technical file to be submitted to the notified body is also detailed in the Annex II of the Regulation and, at minimum, must include the following documents:

- product or trade name and a general description of the device including its intended purpose and intended users;
- a unique identification number, used to trace the device and eventually recall it from the market, if needed;
- the intended patient population and clinical application;
- the principles of operation of the device;
- the risk class of the device;
- an explanation of any novel features;
- a description of the accessories for the device, and of the various configurations/variants of the device;
- a description of the raw materials that have been used;
- technical specifications, such as features, dimensions, and performance attributes of the device.

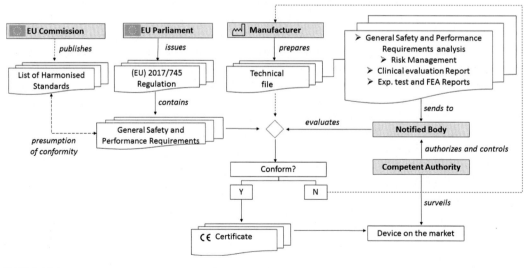

FIGURE 25.1

CE certification pathway.

All the documentation contained in the technical file is addressed to demonstrate that the new device is compliant to the General Safety and Performance Requirements described in the Annex I of the Regulation: these are general requirements that the manufacturer must take into account with specific references to the particular typology of device. All this path basically follows a risk assessment approach where risk must be considered as a combination of the probability that a certain harm may happen with the severity of such a harm.

In this light, the manufacturer has to submit a Risk Management Document in which he takes into account all the risks related to the usage of the device and demonstrates that, through the implementation of proper risk, control actions taken during the design of the device (e.g., experimental tests execution, finite element analysis simulations): (1) all risks have been lowered to an acceptable level by minimizing the probability of occurrence of all the possible harms and that (2) the benefit-residual risk analysis gives favorable results.

In the particular case of a device for orthopedic applications, risks are not only the ones that can arise after the surgery (failure of the implant for any reason with necessity of substituting it), but they must cover the whole life of the device, from the designing phase, to the manufacturing phase, to the surgical phase, even to the disposal phase.

A particular emphasis is put in the new 2017/745 Regulation to the demonstration of the clinical efficacy of the device and to the minimization of clinical risks: in the technical file a Clinical Evaluation Report must be presented to the Notified Body containing a deep insight on clinical data already available in literature on similar devices and, in case these data are judged insufficient, results of premarket clinical investigations (*clinical trials*) together with a plan for the postmarket clinical follow-up of the new device.

THE ROLE OF THE INTERNATIONAL STANDARDS IN THE CERTIFICATION PROCESS OF AN ORTHOPEDIC MEDICAL DEVICE

Standards are documents that are defined according to Regulation (EU) 1025/2012 as "technical specification," adopted by a recognized standardization body, for repeated or continuous application, with which compliance is not compulsory, and which is one of the following: (a) "international standard" means a standard adopted by an international standardization body (e.g., International Organization for Standardization − ISO); (b) "European standard" means a standard adopted by a European standardization organization; (c) "harmonized standard" means a European standard adopted on the basis of a request made by the Commission for the application of Union harmonization legislation; (d) "national standard" means a standard adopted by a national standardization body.

Hence, standards are documents that define the characteristics (dimensional, performance, environmental, quality, safety, etc.) of a product, process, or service, according to the state of the art.

The peculiar characteristics of the technical standards are that (1) they must be approved with the consent of those who participate in defining it, (2) all the concerned economic/social parties can participate the work and make observations in the process that precedes the final approval, (3) the standardization body reports the key stages in the approval process of a draft standard, keeping the project available to interested parties, (4) standards are a reference that the interested parties impose themselves spontaneously.

As long as medical device certification is concerned, the demonstration that a new orthopedic device fulfills the General Safety and Performance Requirements can be performed by the manufacturer using a specific way: the European Commission periodically publishes a list of harmonized standards edited by recognized standards organizations (case "c" of the above reported definition of "standard") that imply the so-called "presumption of conformity." This means that if a manufacturer follows the prescriptions of a harmonized standard to demonstrate a particular characteristic of the device, the fulfillment of the specific General Safety and Performance Requirement to which the standard addresses is automatically reached.

However, it must be underlined that the application of the technical standards (even the harmonized ones) is voluntary: they become mandatory if these are cited in the legislative acts. In fact, there are numerous laws that refer to technical standards, sometimes mandatorily, as the preferential way to comply with the law.

In the European Community, the law (EU Regulation or national laws) defines the essential health and safety requirements for medical devices and manufacturers can freely choose how to comply with these mandatory requirements: if they do it using the harmonized standards, as previously stated, the products automatically benefit from the presumption of conformity and can therefore freely circulate on the European market.

In the list of harmonized standards, those related to the design and demonstration of safety of orthopedic devices are unfortunately few. Most of them are not specifically designed for orthopedic devices and provide specification for general processes (e.g., standards for conducting a Risk Management process or a Clinical Investigation activity) or for side processes and aspects: processes generally applied in the manufacturing phase of any implantable medical device (cleaning, packaging, sterilization, transport, etc.) or related to how to provide information on the device to the final user (e.g., labeling, instruction for use editing, etc.) or related to the procedures for the demonstration of biocompatibility of materials (EN ISO 10993 series).

The only three harmonized standards related to orthopedic device manufacturing are listed below and are the only device for joint replacement, no other orthopedic device is considered:

1. EN ISO 21534:2009 Non-active surgical implants—Joint replacement implants—Particular requirements (EN ISO 21534:2009 *Non-active surgical implants—Joint replacement implants—Particular requirements*, n.d.);
2. EN ISO 21535:2009 Non-active surgical implants—Joint replacement implants—Specific requirements for hip-joint replacement implants (EN ISO 21535:2009 *Non-active surgical implants—Joint replacement implants—Specific requirements for hip-joint replacement implants*, n.d.);
3. EN ISO 21536:2009 Non-active surgical implants—Joint replacement implants—Specific requirements for knee-joint replacement implants (ISO 21536:2009 *Non-active surgical implants—joint replacement implants—specific requirements for knee-joint replacement implants*, n.d.)

The EN ISO 21534:2009 gives general requirements applicable to any joint replacement device. In particular, requirements related to the biomechanics of the devices include:

- the finishing requirements of articulating surfaces and their geometry;
- the allowable couplings related to the materials of different parts that come in contact;
- the preclinical evaluation requirements related to material biocompatibility, mechanical loads and related movements, fatigue and wear issues, the suitability of the dimensions and shape of the implant for the intended population, adhesion, and durability of coatings.

The EN ISO 21535:2009 is designed for hip prostheses and includes further and more specific requirements such as follows:

- the definition of the range of angular movements between head and cup;
- allowed tolerances and dimensions of taper connections and articulating surfaces;
- allowed thickness of ultra-high molecular weight polyethylene (UHMWPE) in acetabular components and bipolar heads;
- definition of endurance testing of femoral components, of head and neck region, and of pull-out performances of heads: in this case, further references to nonharmonized international standards (ISO 7206 series) that define details of these tests are reported (see also "Fatigue performances of a hip prosthesis stem: analysis of the available standards for an experimental and computational approach" for a deeper analysis);
- definition of wear testing protocols: in this case, surprisingly, no indication of a further standard is provided although a complete series of standards for this purpose exist since a long time (ISO 14242 series).

Similarly, the EN ISO 21536:2009 is designed for knee prostheses and includes further and more specific requirements such as:

- definition of the range of articular movements between femoral and tibial components;
- allowed thickness of UHMWPE tibial inserts;
- finishing of nonarticulating regions of metallic knee joint components;
- definition of endurance testing of tibial trays and of wear test protocols: in this case, both topics are supported by specific nonharmonized standards (ISO 14879 and ISO 14243, respectively) and details about wear testing can be found in "Wear of the tibial insert of a knee prosthesis: analysis of the available standards for an experimental and computational approach".

In this chapter, we have always referred to harmonized standards or standards edited by ISO, since those are the more indicated to be followed as far as certification of devices in the European Community is concerned. However, for the sake of completeness, another family of standards for the preclinical testing of orthopedic devices must be cited, namely those edited by the American Society for Testing and Materials (ASTM).

This family of standards actually covers a higher range of testing typologies: if we search the databases of ISO and ASTM using the term "hip prosthesis test," we can find 36 results in the ISO website and 491 results in the ASTM website, and most of the testing procedures considered by ASTM are not contemplated by any ISO standard.

The reason for this could be the inadequate details contained in different methods of certification: on one hand, ASTM tends to describe, even in great detail, the experimental setups and procedures that must be followed to investigate a particular issue, but, on the other hand, the acceptance criteria are seldom defined.

This is due to the fact that certification of medical devices in the United States is based on a demonstration of equivalence of the performances of a new device with a device already approved in US: this certification scenario is called 510k and is one of the certification routes that the Federal Drug Administration (FDA) allows for marketing in the United States In this light, manufacturers are asked to compare the performances of the two devices using the same methodology and performance criteria definition is not needed.

On the other side, the certification approach followed in the European Community based on the use of harmonized standards combined with further technical standards (as above described) needs a precise definition of the performance criteria because being compliant to these criteria automatically gives presumption of conformity to the Regulation and becomes a very powerful tool for manufacturer to get the CE mark.

Actually, in the everyday life of a biomechanical researcher, the two approaches are often used in combination, exploiting the wide range of experimental setups and procedures offered by ASTM standards and completing it trying to find reasonable acceptance criteria taking into consideration the prescribed use of the device.

EXAMPLES ON THE ROLE OF STANDARDS FOR THE DEMONSTRATION OF FULFILLMENT OF BIOMECHANICAL REQUIREMENTS FOR AN ORTHOPEDIC MEDICAL DEVICE

As described in previous chapters, several orthopedic devices are nowadays available for restoring musculoskeletal diseases and disorders in different anatomical sites. The devices are categorized based on the purpose they are designed for, i.e., they can be useful either for reestablishing a joint or for fixing an anatomical region after a trauma.

An orthopedic implant needs to be certified before it is being allowed to be implanted. The certification is in accordance with ISO and/or ASTM standards that are dictating the mechanical testing procedures with critical requirements to determine the lifetime of an implant and the load-bearing capacities. ISO and ASTM standards attempt to embody loads and boundary conditions once the device is implanted as explained in the previous paragraph.

Both static and dynamic loading conditions can be used to perform the mechanical tests. While static conditions are useful to test load-bearing capacities, dynamic loading procedures are suitable to predict the lifetime of a device.

As illustrated in the previous chapters, the orthopedic devices include soft-tissue restoration, joint replacement, spine stabilization, fixation for fractures, and reconstruction for tumors. Usually, only "on-shelf" available devices need to be certified, while customized implants can be implanted without being tested. Experimental tests coupled with numerical tests are used to test the devices based upon the requirements of the standards or to cover necessities of tests when the standards are missing.

Some examples are described below for the hip joint, the knee joint, and the shoulder.

FATIGUE PERFORMANCES OF A HIP PROSTHESIS STEM: ANALYSIS OF THE AVAILABLE STANDARDS FOR AN EXPERIMENTAL AND COMPUTATIONAL APPROACH

A hip prosthesis is considered as the sum of the stem plus the head for the femur. The head is going to articulate in the acetabular cup. Focusing on the stem, it can be generally made of one piece or two components, the stem plus the modular neck. For both cases, the mechanical resistance of the stem and the neck should be proved. The ISO standards (ISO 7206−4 2010) *(ISO 7206-4: 2010 Implants for surgery—Partial and total hip joint prostheses—Part 4: Determination of endurance properties and performance of stemmed femoral components, n.d.)* and ISO 7206−6:2013 *(ISO 7206−6:2013 Implants for surgery—Partial and total hip joint prostheses—Part 6: Endurance properties testing and performance requirements of neck region of stemmed femoral components, n.d.)* provide the description on how to test all potential designs of a stem for hip prosthesis.

In the ISO standards, the stem is mainly categorized depending on its length (the measurement between the center of the head and the distal stem) and its shape, standard, or anatomical. When the stem needs to be tested (ISO 7206−4) depending by the length, the stem is blocked starting from the distal part up to a determined level. Additionally, the stem is blocked with two described rotations in the frontal and sagittal planes. Also, the applied force on the femoral head is determined by the stem length.

The ISO describes the frequency with which the cycling load (between a minimum and maximum load) must be applied. Six specimens need to be tested for the same design configuration and all of them must not fail or deform to prove the mechanical resistance of that design of prosthesis. The stem is tested for 5 million cycles under a maximum load of 2300 N (1200 N for short stem). An example of testing configuration is described in Fig. 25.2.

Once the stem successfully passed the test under ISO 7206−4, the neck must also be checked. For this case, the block in which the stem is positioned, inclined as previously described, is covering the stem up to the level of bone resection. The stem is tested for 10 million cycles under a load of 5340 N.

Fig. 25.3 shows an example of testing configuration for ISO 7206−6.

Following the requirements of the standards, experimental tests are performed using servo-hydraulic machines that are applying the cycling loadings as described. A company that manufactures medical devices can own a testing machine or can ask to test the hip design to external certificated laboratories that will provide the certification of the results once the experimental tests are completed.

Usually, before performing the experimental tests that are cost and time consuming, theoretical analyses are made to select which is the worst configuration for a specific family of implant. In fact, different sizes are usually available for a stem design and the stem can be coupled with different head prostheses. The worst configuration is the combination of components that brings the maximum mechanical stresses or deformation. By proving the means of theoretical analyses, which is the worst combination, usually determined by the dimension of the components, the experimental test is only performed on that combination. The theoretical analysis usually schematizes the stem following the angle between the stem and the neck, with the dimensions of the stem and the neck up to the center of the head. The bidimensional scheme is oriented in the space following the rotations in the frontal and sagittal plane in order to axially load the center of the head. Changing the head size changes the lever arm affecting the section of the stem under observation (corresponding

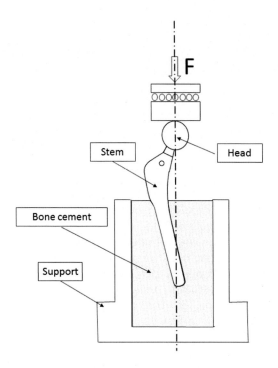

FIGURE 25.2

Loading hip stem. An example of testing configuration from ISO 7206−4 (2010).

From ISO 7206−4 (2010) (Implants for surgery – partial and total hip joint prostheses—part 4: determination of endurance properties and performance of stemmed femoral components. (n.d.). Retrieved February 27, 21 C.E., from https://www.iso.org/ standard/42769.html).

to the hypothetical level of the block). By performing the calculation for different sizes, the worst configuration can be determined.

The theoretical analysis can be supported by a numerical approach, for example, by means of finite element modeling. The numerical approach estimates not only the effect of the dimensions of the components, but also the contribution of the materials in use for the stem. Additionally, with the numerical approach, a three-dimensional analysis is performed instead of a bidimensional one allowing a more comprehensive evaluation.

Fig. 25.4 shows an example of model replicating ISO 72026−4 requirements (Pianigiani & Alemani, 2020).

Depending by its length, a prosthesis can be classified as "short" or "long"; thus, maximum 1200 N must be applied following the ISO 7206-4, and the numerical approach can help select the most appropriate one between two consecutive sizes, one that is classified short and the other that is classified normal (maximum applied load 2600 N). The model of the analysis will consider the three-dimensional original components, oriented in the space as described in the ISO: the block will be also considered. For the head, the neck, and the stem, the material in use for that prosthesis will be modeled together

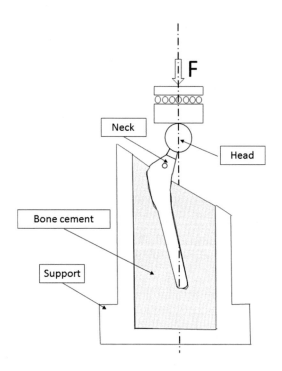

FIGURE 25.3

Load neck stem. An example of testing configuration from ISO 7206−6:2013.

From ISO 7206−6:2013 Implants for surgery—Partial and total hip joint prostheses—Part 6: Endurance properties testing and performance requirements of neck region of stemmed femoral components. (n.d.). https://www.iso.org/standard/51186.html.

with the cement usually used for the block (with values in a range according to the ISO standards). A static approach can be used during simulation to evaluate the effects due to the maximum load according to the ISO. Maximum stresses and the axial displacement of the head can be observed among the different simulated configuration to select the worst-case configuration.

The more theoretical analyses are annexed to the final certification together with the experimental test, the more the notify body can have a supported documentation for the evaluation of the design.

WEAR OF THE TIBIAL INSERT OF A KNEE PROSTHESIS: ANALYSIS OF THE AVAILABLE STANDARDS FOR AN EXPERIMENTAL AND COMPUTATIONAL APPROACH

The tibial insert of a total knee prosthesis is considered as the component that substitutes the menisci. It is a critical component in the way it is a plastic component between two metal components that articulate on it.

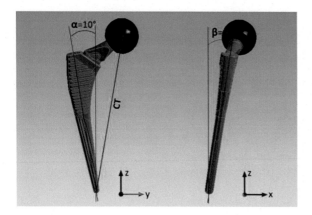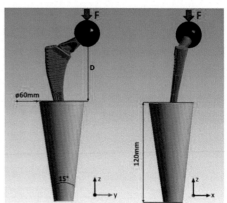

FIGURE 25.4

FEM hip. Model positioning and block dimension following the requirements of ISO 7206−4 for a finite element analysis.

From https://doi.org/10.1142/S0219519420500062

Different categories of inserts are available and, for each, different designs more or less congruent, with or without the posterior stabilization.

As it is a critical component and some literature is still reporting the wear of the polyethylene insert as one of the major motivation of a total knee arthroplasty failure, the insert wear must be observed and proved to let the knee design considered mechanically safe.

The two main standards available for testing the wear of a tibial insert are as follows:

- ISO 14243−1 (2009) *Implants for surgery—Wear of total knee-joint prostheses—Part 1: Loading and displacement parameters for wear-testing machines with load control and corresponding environmental conditions for test*, (n.d.);
- ISO 14243−3 (2014) *Implants for surgery—Wear of total knee-joint prostheses—Part 3: Loading and displacement parameters for wear-testing machines with displacement control and corresponding environmental conditions for test*, (n.d.).

The two standards are based upon literature studies that analyze and estimate the forces and displacement during gait task. In fact, both in load and in displacement control, the aim of the standards is to mimic gait task and to test the wear resistance for 5 million cycles without any breaking up or delamination of the articulating surfaces that disrupt normal function of the implant.

For the different degrees of freedom, the ISO dictates the relative movement and position of the femoral compartment and the tibial compartment following an initial reference system (Fig. 25.5).

As described for the hip, and also for the knee, the theoretical analyses support the selection of the size of the component to be experimentally tested among the available ones for a specific design.

Since the shapes of the knee joint are more complex than the one of the hip joint, the numerical approach based upon finite element analysis can help in supporting the analysis of potential

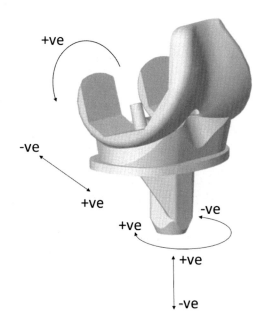

FIGURE 25.5

Knee relative positions. Sign convention for the forces and motions, shown for a left total knee replacement system in ISO 14243−3 (2014).

From ISO 14243−1 (2009) Implants for surgery—wear of total knee-joint prostheses—part 1: loading and displacement parameters for wear-testing machines with load control and corresponding environmental conditions for test. (n.d.). https://www.iso.org/standard/44262.html

malfunctionality. By means of static analysis, for example, the stress contours and the contact force on the polyethylene insert can be observed for different instant of the ones collected in the ISO, i.e., instants of the cycle for which, while the femoral component is rotated, the highest forces are applied. The effect of different femoral component designs for the same insert can be also observed Fig. 25.6 (Innocenti, Yagüe, Bernabé, & Pianigiani, 2015).

The observation of the numerical outputs, for example, for the smallest and biggest available sizes for a certain combination of femoral and insert components can help in determining the most wearing condition. The use of the numerical output as inputs of wear models, such as the Archard model, can also help in estimating the grade of polyethylene wear.

More sophisticated finite element models, with dynamic approach, better estimate the insert wear by also implementing the wear model. Usually, the model needs first to be verified against experimental outputs that were obtained for similar contact shapes (roll on plane, sphere on sphere, and dedicated experimental test on knee joint prosthesis) and materials in use. Once the model is able to numerically replicate the experimental test, the estimated wear factor can be also used for similar configurations, hence providing a more precise result during the estimation of the worst case.

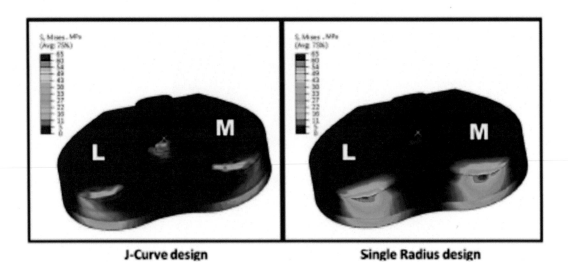

FIGURE 25.6

Stress on poly. Effect of different femoral components for the same insert in terms of stress.

From https://doi.org/10.1142/S0219519415400345

SHOULDER PROSTHESIS: IS THE CURRENT STANDARDIZATION ENOUGH?

By searching the official site of the ISO, there are two ISOs that are currently under development for the shoulder prosthesis.

The title of the two ISOs are already available: ISO/CD, 24085−1 *(Implants for surgery— Partial and total shoulder joint prosthesis—Part 1: Determination of resistance to static load of ceramic humeral heads and glenospheres*, n.d.) and ISO/CD, 24085−2 *(Implants for surgery— Partial and total shoulder joint prosthesis—Part 2: Determination of resistance to torque off head fixation of modular humeral prostheses*, n.d.).

However, several designs of shoulder prosthesis are already available on the market. Additionally, two different approaches for shoulder prostheses are used during the development of the design of the replacement.

Since standards are missing, literature studies bring a scientific contribution to help in understanding which aspects should be tested by a manufacturer.

For example, Anglin, Wyss, & Pichora (2000) show a biaxial test method to perform a laboratory testing of glenoid prostheses. The aim of the test is to provide improved designs, subsequently leading to a reduction in the incidence of clinical loosening. The concept behind the rocking-horse test was to load the superior and inferior rims of the glenoid alternately.

All components for the testing machine, loads, speed, and other boundary conditions are clearly described to replicate the test. Based upon the obtained results, the authors conclude that many design factors affect the performance of a prosthesis: the shape of the articulation surface (i.e., conformity, constraint, and size), the backing shape and surface, and the fixation design. However, the authors suggest two prosthesis changes of the ones analyzed: roughening the back surface and pegs and offering only a curved-backed design as opposed to a flat-backed design. The purpose of the study was to reduce the risk of glenoid loosening.

The research study of Smith et al. (2015) was aimed at reversing shoulder prostheses. It described a shoulder simulator that could allow weaknesses in current designs and let them improve. By comparing different designs with the same simulator, comparisons are then possible.

Since there were not ISO or ASTM to be followed, the manuscript describes as the first attempt for the simulator was based on the ISO for testing the hip joint. Then, based upon literature studies, all the parameters were set according to the performances of the shoulder joint.

Research studies as the ones reported above are fundamental to support the development and the marketing of prosthesis for which the regulatory is late.

Numerical simulations are also in this case of support both to test the prosthesis and to help in developing the experimental devices.

CONCLUSION AND FUTURE PERSPECTIVES

Certification of medical devices is an aspect that in the past was felt by manufacturers as a burden that did not give any additional value to their business: only since few years, manufacturers have realized that certification, even if more and more demanding especially with the introduction of the new EU Regulation, can become an opportunity to better focalize the device verification and validation activities toward a more efficient process.

In particular, the need of demonstrating the reliability of devices in terms of mechanical characteristics is mandatory to obtain the CE mark: in this light, also international standards have been updated and integrated in order to provide well-recognized schemes for this purpose, but some reliability aspects still remain uncovered by any standard, leaving the demonstration of the device reliability completely up to the manufacturer that has, in this case, to decide procedures, setups, and acceptability levels for the results.

This is quite normal in a market that presents a continuous and fast updating of new devices able to respond to new or different clinical needs brought by surgeons to the attention of manufacturer: the most striking example is the growing use of custom-made device that completely perturbs the certification pathways as planned and performed since now.

For these devices, which are only in part industrial products, the new Regulation has necessarily introduced new specifications that were almost absent in the 93/42 Directive but, nevertheless, the validation of their reliability performances still remains other than trivial: the most suitable technique for the reliability demonstration of a custom-made device is finite element analysis whose usage, conversely, is not taken into account by any international standard.

This is a big issue, considering that in other medical device applications, even more critical than the orthopedic ones (e.g., endovascular stents), the usage of finite element techniques is accepted in conjunction with more traditional experimental tests for the demonstration of the mechanical reliability of the device: time is certainly mature to standardize these procedures also for orthopedic applications, leaving experimental tests only for those aspects where numerical simulations are judged not sufficiently predictive of the real device performances.

REFERENCES

Anglin, C., Wyss, U. P., & Pichora, D. R. (2000). Mechanical testing of shoulder prostheses and recommendations for glenoid design. *Journal of Shoulder and Elbow Surger*, *9*(4), 323–331. Available from https://doi.org/10.1067/mse.2000.105451.

EN ISO 21534:2009 Non-active surgical implants – Joint replacement implants – Particular requirements. (n. d.). Available at: https://www.iso.org/standard/39024.html.

EN ISO 21535 2009 Non-active surgical implants – Joint replacement implants – Specific requirements for hip – joint replacement implants. (n.d.). Available at: https://www.iso.org/standard/40372.html.

Innocenti, B., Yagüe, H. R., Bernabé, R. A., & Pianigiani, S. (2015). Investigation on the effects induced by TKA features on tibio-femoral mechanics Part I: Femoral component designs. *Journal of Mechanics in Medicine and Biology*, *15*(02), 1540034. Available from https://doi.org/10.1142/S0219519415400345.

ISO 14243-1 (2009) Implants for surgery – wear of total knee-joint prostheses – part 1: loading and displacement parameters for wear-testing machines with load control and corresponding environmental conditions for test. (n.d.). Available at: https://www.iso.org/standard/44262.html.

ISO 14243-3 (2014) Implants for surgery – wear of total knee-joint prostheses – part 3: loading and displacement parameters for wear-testing machines with displacement control and corresponding environmental conditions for test. (n.d.). Available at: https://www.iso.org/standard/56649.html.

ISO 21536:2009 Non-active surgical implants – Joint replacement implants – Specific requirements for knee – joint replacement implants. (n.d.). Available at: https://www.iso.org/standard/40373.html.

ISO 7206-4 (2010) Implants for surgery – partial and total hip joint prostheses – part 4: determination of endurance properties and performance of stemmed femoral components. (n.d.). Retrieved February 27, 21 C.E., from https://www.iso.org/standard/42769.html.

ISO 7206–6:2013 Implants for surgery—Partial and total hip joint prostheses—Part 6: Endurance properties testing and performance requirements of neck region of stemmed femoral components. (n.d.). Available at: https://www.iso.org/standard/51186.html.

ISO/CD 24085-1 Implants for Surgery—Partial and total shoulder joint prosthesis—Part 1: Determination of resistance to static load of ceramic humeral heads and glenospheres. (n.d.). Available at: https://www.iso.org/standard/77774.html.

ISO/CD 24085-2 Implants for Surgery—Partial and total shoulder joint prosthesis—Part 2: Determination of resistance to torque off head fixation of modular humeral prostheses. (n.d.). Available at: https://www.iso.org/standard/77775.html.

Pianigiani, S., & Alemani, F. (2020). Evaluating the effects of experimental settings during ISO 7206-4 endurance and performance tests: A finite element analysis. *Journal of Mechanics in Medicine and Biology*, *20*(03), 2050006. Available from https://doi.org/10.1142/S0219519420500062.

Smith, S., Li, B., Buniya, A., Lin, S., Scholes, S., Johnson, G., ... Joyce, T. (2015). In vitro wear testing of a contemporary design of reverse shoulder prosthesis. *Journal of Biomechanics*, *48*(12), 3072–3079. Available from https://doi.org/10.1016/j.jbiomech.2015.07.022.

CLINICAL EVALUATION OF ORTHOPEDIC IMPLANTS

Sara Zacchetti
IRCCS Orthopedic Institute Galeazzi, Milano, Italy

OVERVIEW OF CLINICAL TRIALS OF MEDICAL DEVICES AND DEFINITIONS

After the preclinical studies, a new medical device is tested on humans before being marketed; such studies involving patients are called clinical trials. When a manufacturer decides to plan a clinical trial, the first crucial step is choosing the study design that fits the objective of the study and also serves the purpose of the manufacturer. Indeed, different kinds of clinical trials exist, and the choice of how to set up the trial is fundamental.

Clinical trials are classified based on the study object. The most common study objects in the field of orthopedics are as follows:

- pharmacological products;
- medical devices (only one covered in this chapter);
- procedures (surgical, medical, etc.).

Before starting with the descriptions of the clinical trials, it is important to present and define the entities involved:

- clinical center;
- promoter/sponsor;
- principal investigator (PI);
- patients/subjects.

CLINICAL CENTER

The clinical center is the place where patients to be studied are enrolled. Usually, centers are selected based on their expertise in the treatment of the pathology that could be treated with the medical device to be tested. Another factor that must be considered while making the selection of a clinical center is its ability to conduct the study; the selected institutes should have a good enrollment capacity and an efficient administrative setup to support the project.

The manufacturer should also consider the fact that some centers could not be authorized to enroll patients for clinical trials, and special permission should be taken depending on the national regulatory laws.

Human Orthopaedic Biomechanics. DOI: https://doi.org/10.1016/B978-0-12-824481-4.00002-0

PROMOTER/SPONSOR

The promoter is the entity that holds the data ownership. The promoter has the responsibility of:

* financing and managing the study;
* taking care of all the regulatory issues;
* ensuring the study follow the good clinical practices (based on the international regulation of clinical trials) under the supervision of certified external monitors, if necessary.

PRINCIPAL INVESTIGATOR

The PI is the person responsible for the conduct of a clinical trial in a specific trial site and is the leader of the team of individuals working on the trial (Directive 2001/20/EC, n.d.).

PATIENT/SUBJECT SELECTION

Clinical trials involve patients or healthy subjects, depending on the study design. In both cases, a careful selection based on the inclusion/exclusion criteria of the study protocol is mandatory. During the study, it could be necessary to exclude enrolled patients/subjects for different reasons and at different time-points of the study course.

Usually a promoter creates a "consort graph," which is a flow chart that usually includes selection, randomization, visits, and exams for each study group, and reasons and time-points for patient exclusion (Fig. 26.1).

The different types of clinical trials conducted for the validation of medical devices are as follows:

* pre- or postmarketing;
* sponsored or spontaneous studies;
* interventional and observational;
* retrospective, prospective, or cross-sectional;
* monocentric or multicenter.

PRE- AND POSTMARKETING STUDIES

To be allowed to sell a device for a clinical purpose, manufactures need to register their product as a medical device for one or more indications of use. This requires specific clinical trials, both pre- and postmarketing studies.

Premarketing trials are conducted for new devices either not already authorized for marketing as a medical device or already authorized but for a different indication of use. If the device is used for a different pathology or in a different way from what mentioned in the device dossier, its use is considered as premarket but not authorized for marketing.

Postmarketing trials involve devices that are already approved by the national authority as a medical device and are ready to be applied for the indication of use for which the device has been registered.

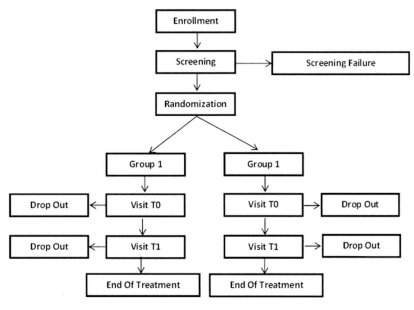

FIGURE 26.1

Example of a consort graph for RCT. *RCT*, Randomized clinical trial.

SPONSORED AND SPONTANEOUS STUDIES

The management of a trial may differ based on the kind of promoter (or sponsor).

A clinical trial is sponsored if the promoter of the study is other than the clinical center that recruited patients. The promoter can be the manufacturer, and in that case, it is referred to as a profit promoter. Besides, the study can also be promoted by another hospital or not-for-profit promoter.

In contrast, if the study is designed in the same clinical center that enrolls patients, it is called a spontaneous clinical trial. Usually, spontaneous studies are also not-for-profit.

A third possibility is a group study: In this case, several not-for-profit centers create an association (or group) with a specific statute, and the group becomes the promoter of the study.

If the manufacturer decides only to finance a study without taking the responsibility of being the promoter demanding them to oversee the work of the clinical center, the data ownership remains with the center and the manufacturer cannot access the data collected in the study before their publication.

INTERVENTIONAL AND OBSERVATIONAL STUDIES

Clinical trials are defined as interventional if they change the standard of care of enrolled patients. All premarketing clinical trials are interventional, but postmarketing trials also could be interventional if they change the clinical routine. Changing only part of the routine makes a clinical trial interventional, for example, the requirement of extra imaging examination or a clinical visit performed at different timings. If the medical device is not already used in the clinical center, the study must be defined interventional because it introduces a new device in the hospital.

A study can be defined as observational only if enrolled and not-enrolled patients are subjected to exactly the same procedure; the only difference is that enrolled patients can allow clinicians to collect and analyze their clinical data for the study, whereas the data from not-enrolled patients are not considered.

RETROSPECTIVE, PROSPECTIVE, AND CROSS-SECTIONAL STUDIES

One of the features used to classify a study is the timing of data collection.

A retrospective study is a simple collection of available data about procedures that have been performed before study approval.

Prospective trials include visits and procedures that will be done after the beginning of the study.

If data about a patient are collected in a single visit (the same visit in which patients consent to participate in the study), the study is called cross-sectional.

MONOCENTRIC OR MULTICENTER STUDIES

Another classification criterion is the number of centers participating in a study: if the study involves only one recruiting center, it is called monocentric; if there are many centers involved, the study is multicenter.

ASSIGNMENT OF THE PROCEDURE

If the study involves more than one group of patients that undergo different procedures to be compared, the trial is defined as a controlled study; the assignation of each patient to a group is done by randomization.

In a randomized clinical trial (RCT), subjects are allocated using random assignation done by a computer system. In the majority of cases, the randomization is made by an automatic system that exploits a randomization list. Sometimes the promoter prepares the assignations for consecutive patients in closed envelopes (one for each consecutive patient) that are given to the investigator: After enrollment, the investigator opens the envelope related to the patient and allocate him/her to the relative group. However, this procedure is quite rare because mistakes are relatively more frequent (due to, for example, losing or mixing up the envelopes).

The randomization is defined as blind if the clinician, the patient, or both are not allowed to know the procedure; this is very rare in clinical trials for medical devices in comparison with clinical trials on pharmaceutical products but it is performed in some cases, for example, by not communicating to a radiologist or a physiotherapist the medical device that is being tested.

LEVELS OF EVIDENCE

An additional classification criterion is "levels of evidence" (from I to V). The levels of evidence classify studies based on the methodological quality of design, the validity, and the applicability to

Table 26.1 Different classifications of the level of evidence.

Level of evidence	Type of study
1	Randomized control trails (RCTs)/ *meta*-analysis/systematic review of (homogeneous) RCT
2	Prospective comparative studies/cohort studies/outcome research/*meta*-analysis of low-quality RCTs/systematic review of (homogeneous) cohort studies
3	Case-control studies/retrospective studies/*meta*-analysis/systematic review of (homogeneous) case-control and retrospective studies
4	Cross-sectional studies/cart review/case series/low-quality cohort or case-control studies
5	Case report/expert opinion/clinical observation/animal and bench research

patient care. It could also be very important after the completion of the study while publishing the results.

In literature, there are different classification criteria available to determine the level of evidence of a clinical research work. Table 26.1 summarizes some of the most used level-of-evidence classification criteria.

CLASSIFICATION OF MEDICAL DEVICES

The Council Directive 90/385/EEC declared that a medical device "means any instrument, apparatus, appliance, material or other article, whether used alone or in combination, together with any accessories or software for its proper functioning, intended by the manufacturer to be used for human beings in:

1. the diagnosis, prevention, monitoring, treatment or alleviation of disease or injury,
2. investigation, replacement or modification of the anatomy or of a physiological process,
3. control of conception,
4. and which does not achieve its principal intended action by pharmacological, chemical, immunological or metabolic means, but which may be assisted in its function by such means" (Council Directive 90/385/EEC of 20 June 1990, 1990).

One important aspect to be taken into consideration by the manufacturers is they have to register each medical device under a specific risk classification. The classification is based on the complexity of the device, the indication of use, and the risk for patients who should be treated with it. The manufacturer is responsible for the declaration of the classification, taking into account specific issues:

1. how much the device is invasive for the patient;
2. how much the device is dependent of sources of energy;
3. how long it remains in contact with patient's body.

Manufacturers may also register a "kit" accompanying medical devices, and used, for example, as instrumentation that must be used together with devices during surgery.

INVASIVE AND NONINVASIVE DEVICES

Manufactures have to determine whether the device is:

1. noninvasive, that is, not penetrating the patient's body either through orifices or through the body surface;
2. invasive, that is, penetrated, even partially, into the body in different ways and remains in the patient's body for at least some days after surgery:
   a. passing the body surface through orifices;
   b. penetrating the body surface during surgical or other procedures;

TEMPORARY, SHORT-TERM, AND LONG-TERM DEVICES

Another classification criterion of medical devices is the duration of the use of the devices in each patient.

The duration of use can be temporary (usually less than 60 minutes), short term (usually no more than 30 days), or long term (more than 30 consecutive days) (Italian Health Ministry_Medical Device Overview, n.d.).

ACTIVE IMPLANTABLE DEVICES

If a device relies for its functioning on a source of electrical energy or any source of power other than that directly generated by the human body or gravity, it is called an active device (Council Directive 90/385/EEC of 20 June 1990, 1990).

Active implantable medical devices (AIMDs) are defined by the Council Directive 90/385/EEC as any active medical device that is intended to be totally or partially introduced, surgically or medically, into the human body or by medical intervention through a natural orifice, and is intended to remain there after the procedure. An example could be an internal, long-term blood pressure sensor. AIMD helps administering a medicinal product or incorporating, as an integral part, a substance that, if used separately, may be considered a medicinal product (Council Directive 90/385/EEC of 20 June 1990, 1990).

CLASSES OF RISK

Four main classes for medical devices that exist in Europe are as follows:

- Class I: medical devices with very low risk (e.g., a simple patch with no substances or actigraphies);
- Class IIa: medical devices with medium risk (e.g., a magnetic resonance device);
- Class IIb: medical devices with medium/high risk (e.g., an implantable micro bead);
- Class III: high-risk medical devices (e.g., a prosthesis) (Italian Health Ministry_Medical Device Overview, n.d.).

  In addition to the aforementioned classes, other classifications may exist in European countries, but if the manufacturer wants to register a product as a medical device with the CE mark, which authorizes products for marketing in the European market, it is necessary to classify the device in one of the aforementioned risk classes.

A similar classification model for medical devices also exists in the United States and is managed by the US Food and Drug Administration (FDA). The classification consists of three classes based on the level of control necessary to assure the safety and effectiveness of the device:

- Class I: general controls: with and without exemptions;
- Class II: general controls and special controls: with and without exemptions;
- Class III: general controls and premarket approval (FDA_Overview Medical Device Regulation, n.d.).

IN VITRO MEDICAL DEVICES

With the advancement of technology over the past few years, different types of medical devices have arrived in the market. An interesting example is the in vitro medical devices, defined by the Directive 98/79/EC (n.d.) as:

"any medical device which is a reagent, reagent product, calibrator, control material, kit, instrument, apparatus, equipment, or system, whether used alone or in combination, intended by the manufacturer to be used in vitro for the examination of specimens, including blood and tissue donations, derived from the human body, solely or principally for the purpose of providing information:

- — concerning a physiological or pathological state, or
- — concerning a congenital abnormality, or
- — to determine the safety and compatibility with potential recipients, or
- — to monitor therapeutic measures [. . .]" (Directive 98/79/EC, n.d.).

COMPETENT AUTHORITIES AND ETHICS COMMITTEE

Before the submission of a clinical trial, it is fundamental to understand the two authorities involved in the evaluation of clinical trials: ethics committee (EC) and national authority. The latter evaluates the trial only in select cases, described in detail in the following paragraphs.

ETHICS COMMITTEE

An EC is defined as "an independent body in a Member State, consisting of healthcare professionals and non-medical members, whose responsibility it is to protect the rights, safety and wellbeing of human subjects involved in a trial and to provide public assurance of that protection, by, among other things, expressing an opinion on the trial protocol, the suitability of the investigators and the adequacy of facilities, and on the methods and documents to be used to inform trial subjects and obtain their informed consent" (Directive 2001/20/EC, n.d.).

The main role of the EC is to control the ethical aspects of the trial, monitor the safety of patients, and verify the competence of the clinical center. In many cases, ECs evaluation different hospitals and institutions, but if the EC has been established by a private research institute, it could evaluate only its own center.

The approval of the competent ECs is required for each clinical center to enroll patients for the study; the competent EC is assigned according to the local rules. Usually, public and research hospitals have their own competent EC assigned since the beginning of their activities, while private clinics may require special procedures for the assignment of an EC. Moreover, private institutes may not be authorized to perform all types of clinical trials; for example, they may be allowed to conduct observational postmarket studies only. Thus, if the promoter decides to involve private clinical centers, then it is important for the promoter to ensure that an EC has already been assigned and a list of trials that would be performed in the clinic is in place.

NATIONAL COMPETENT AUTHORITY

The National Competent Authority (NCA) is defined as "a regulatory agency in an EU Member State or for medical devices, a *Competent Authority* is the organization with the authority to act on behalf of the government of a Member State to ensure that all medical devices meet the essential requirements laid down in the Directives prior to marketing authorisation" (EU Clinical Trials Register, n.d.).

The approval of the NCA is needed for clinical trials on premarketing medical devices; for postmarketing studies, a simple notification sent to the authority could be sufficient.

Each country has one or more competent authorities, so it is very important to identify them before submission. In some countries, more than one NCA that could take care of different kinds of studies exist, for example, one for pharmacological products and one for medical devices. The same authority may also be responsible for all kinds of studies, for example, the FDA.

INSTITUTIONAL REVIEW BOARD

In some countries such as the United States, another evaluation organ called the Institutional review board (IRB) exists. This is an administrative body internal to the clinical center, and is established to verify the rights and welfare of patients recruited in internal clinical studies.

FDA defined the IRB as "[...] an appropriately constituted group that has been formally designated to review and monitor biomedical research involving human subjects. In accordance with FDA regulations, an IRB has the authority to approve, require modifications in (to secure approval), or disapprove research. This group review serves an important role in the protection of the rights and welfare of human research subjects. The purpose of IRB review is to assure, both in advance and by periodic review, that appropriate steps are taken to protect the rights and welfare of humans participating as subjects in the research. To accomplish this purpose, IRBs use a group process to review research protocols and related materials (e.g., informed consent documents and investigator brochures) to ensure protection of the rights and welfare of human subjects of research" (FDA.gov, n.d.).

OTHER APPROVALS AND SUMMARY

In conclusion, for a single study, the promoter needs:

- the approval of the EC/IRB of each enrolled center and the approval of the competent authority of each country (premarket trials);
- the approval of the ECs/IRB of each enrolled center and a notification to be sent to the competent authority of each country (postmarket trials).

Another unofficial (but important) approval needed by the sponsor is the consent of the PI. Since the PI is the one responsible for the study in his/her Institute, the promoter needs to present the project to the PI before planning the official submission; the PI must then confirm the feasibility of the study in his institution. This is unofficial but is very important to ensure a study could be smoothly performed in a specific center.

PREMARKET STUDIES ON MEDICAL DEVICES

To obtain the CE mark on a medical device in the European Union, manufacturers have to perform clinical trials to test the product. Before, simply declaring that the device was similar to another that had already been marketed was sufficient to obtain the CE mark; however, with the new Medical Device Regulation, this is no longer possible.

STUDY DESIGN

Premarket studies are performed to demonstrate the performance of the device under normal conditions of use and to verify that the foreseeable risks and frequency of any adverse events are reduced to an acceptable minimum (also related to the benefits provided) (Italian Health Ministry_Medical device, n.d.). The "first time in humans" trial can be performed in a single patient group, with the aim of verifying the safety of the device. Then, the manufacturer can set up trials that compare the new device with others and the standard of care.

The most common study design is a randomized controlled study that compares the new device with the gold standard. Randomization is very important to compare two or more devices or to compare the new device with sham surgery, for example, when an active device is compared with the test device turned off; in such cases, randomization is crucial because it avoids bias and strengthens the study design. The randomization list can be prepared using dedicated software, but sometimes the investigator prefers simpler techniques such as a box of white/black papers. However, randomization software is preferred as it reduces the chance of human mistakes.

Usually, sponsors prefer a noninferiority study for the pilot trial, that is, the device is supposed to be at least as effective and safe as the gold standard, but a superiority design is also possible. A monocentric study could be enough in the first phase, but later multicenter studies will be needed.

Premarket studies are frequently sponsored by the manufacturer but sometimes are promoted by research units. This second case poses some difficulties because of licencing issues; besides, to present a premarket study, the Promoter needs specific documents about the device that only the manufacturer can produce. Therefore, the manufacturer is highly involved in premarket studies, at least to some extent.

Premarket trials can be performed only in research hospitals or in certified institutes: country-specific regulations determine which institutes are appropriate, so it is very important to verify local laws before starting the submission procedure. Furthermore, not all institutes have capabilities, expertise, and resources to perform this kind of studies; this could be verified by the manufacturer by conducting of a feasibility visit.

Monitoring patients is critical, and the promoter must notify to authorities every serious adverse event related to a medical device. The EC should also be notified as soon as the event happens. Not-serious events should be reported once a year.

Sample size

Special attention has to be paid on the calculation of the sample size. Although it could seem to be a secondary step, it is crucial to form a robust study design.

The number of patients to be recruited must be calculated (if possible by a biostatistician) based on the primary study objective. For example, if the study has been designed to compare between two groups, the most common primary aim could be evaluating the difference of a specific parameter at a specific time-point. In that case, the promoter needs to determine the difference before declaring that there exists a relevant clinical difference between the two groups, or the expected effect size. The promoter also needs to search in literature the mean value and the standard deviation found for the same parameter in previous studies. Using this information, the biostatistician can calculate the appropriate sample size.

Different statistical approaches used for different study designs exist, but all of them should have the following crucial information built in:

- data from literature of previous similar studies;
- expected effect/clinical significant effect to be examined with the new study.

If the above criteria are not fulfilled, then the new study could be defined as a "pilot" trial. In this case, different statistical approaches are used to determine sample size: For example, it is possible to estrapolate the appropriate sample size based on the prevision of the future clinical trials that could be designed based on the pilot study.

BUDGET FOR PREMARKET STUDIES

As we have already highlighted, a monocentric study could be sufficient for the first study. This could be a conservative strategy to reduce the initial budget, in order to allow the manufacturer to design a multicenter trial after the results of the pilot study are known.

If the sponsor decides to invest in a multicenter trial, the assistance of a contract research organization (CRO) should be considered. A CRO is a company that takes care of all the activities related to the preparation, submission, and conduction of a clinical trial (regulatory issues, monitoring, electronic data collection, device vigilance, etc.). The sponsor could delegate the CRO one or more activities related to the clinical trials. The services of such companies could be expensive, but also very valuable for big and articulated studies.

One of the most important roles of premarket studies is ensuring the safety of the patients, because it is the first time that the device will be used in humans. For that reason, the promoter has to manage proper insurance and monitoring activities.

If a comparative design has been chosen, the promoter has to provide for free not only the new devices but also the reference for comparison, that is, the gold standard. Besides, every procedure/

evaluation should be financially supported by the sponsor to pay the fees of ECs and competent authorities; those submissions are free of charge only for not-for-profit promoters.

In conclusion, premarket studies prove very expensive for manufactures because a specific insurance, monitoring activity, and coverage for all clinical performances (visit, intervention, exams, etc.) and devices are required; nothing is has to be paid by the patients, healthcare systems, or patients' insurances. Preparing and allocating a proper budget before starting a trial is therefore crucial.

APPROVAL AND MARK

Premarketing trials need to be approved by the NCA of each country and by the EC of each center that enrolls patients. The procedure for submission to national authorities are different in each country, so the manufacturers have to adapt their processes depending on the country in which the trial is performed.

In any case, the submission package includes not only the clinical protocol and documentation for patients (informed consent, letter to physician, etc.) but also the specific documentation of the device, such as instruction for use, technical documentation, risk assessment, and all results of preclinical studies. When the submission package is ready, it should be evaluated by the EC of each clinical center and by the NCA of each involved country. Approvals have to be obtained before starting with the trial; only after approval, the clinical centers can authorize the start of the enrollment.

Once the manufacturer collects all the needed data, the CE certification can be applied for. The certification reports the risk classification and the specific device model; thus, manufacturers mark together different sizes and models of the same device. The CE mark is required in the European Union but may not be sufficient for the commercialization of devices; in fact, some countries have specific mandatory databases, and each marked device needs to be registered in such databases to be marketable.

If the manufacturer wants to sell the device in extra-European countries, specific certifications of those countries need to be obtained. For example, in the United States, the approval for device commercialization has to be obtained from the FDA; besides, after the initial certification, the manufacturers are required to renew the registration of their device annually with the FDA (FDA Medical Device Registration, n.d.).

All procedures reported above should be followed also for medical devices already certified but for a different use, or in the case of any modifications carried out in the existing devices.

POSTMARKETING STUDIES ON MEDICAL DEVICES: INTERVENTIONAL STUDIES

Studies that involve medical devices that are already marked are called postmarketing clinical trials. They are very important to control the long-term outcomes and monitor the performance of commercially available devices. Moreover, those studies are crucial to gather additional information about safety, efficacy, and optimal use.

Studies on marked medical devices, designed according to the destination of use, should be approved by competent authorities, but the procedures are simplified with respect to premarket studies. If the study includes procedures that are out of the normal standard of care, or if patients have to be randomized, the study will be classified as interventional.

STUDY DESIGN AND TRIAL DOCUMENTATION

In the case of interventional studies, the most common design is the RCT. However, sometimes, the promoter chooses a single-group trial, but adds procedures to the standard of care (e.g., additional imaging examinations, etc.) or introduces the device in a hospital that does not use it.

First of all, the promoter should prepare the clinical protocol, considering the main objective, the sample size (with the same precautions as described above), the inclusion and exclusion criteria for enrollment, etc. During the trial, the promoter has to notify to NCAs and ECs of any serious adverse events related to the medical device; this aspect should be declared in the protocol or in other related documentation.

COST TO BE COVERED

The promoter must foster the safety of patients by subscribing to trial-specific insurance before the submission, the same as for premarketing trials. Both industrial and not-for-profit promoters have to guarantee the insurance coverage, the monitoring activities, and the preparation of the entire study package. The promoter also has to guarantee the financial coverage of all the procedures performed during the trial. With regard to those aspects, postmarketing interventional trials are very similar to premarketing ones.

To ensure that clinical data are collected according to good clinical practice (GCP), the promoter has to guarantee the control of the trial by delegating a clinical trial monitor. If the medical device is not used in the hospital where the clinical trial is performed, the promoter has to provide it for free for all enrolled subjects; control devices should also be provided, if relevant. Moreover, the promoter should also provide the documentation for patients (informed consent, letter to the physician/general practitioner, questionnaires, diary, etc.), the case report form (CRF), documents regarding the PI (curriculum vitae, declaration of conflict of interest), and any other documentation that may be necessary for the trial.

Profit promoters have to pay for the submission fee to the EC and NCA; submissions are free for not-for-profit promoters.

SUBMISSION

When the submission package is ready, it should be evaluated by the EC of each clinical center; the positive opinion has to be obtained before starting with enrollments. In Europe, the NCA should also be notified about the study. After approval, the promoter and the clinical center can sign the agreement and then the clinical center can authorize the start of the enrollments.

In the United States, postmarketing clinical trials have to be notified to FDA. The postmarketing study information includes the basic information that the FDA is committed to making available to the public (FDA Postmarketing Clinical Trials, n.d.).

Since postmarketing trials are conducted with the aim of monitoring all the possible issues related to the device even in the long term, it is fundamental that investigators do not lose patients at follow-up.

POSTMARKETING STUDIES ON MEDICAL DEVICES: OBSERVATIONAL STUDIES

To conduct long-term monitoring of medical devices, observational trials are fundamental. Those studies have usually a very long follow-up: in orthopedics, it is typically 10−20 years.
Observational studies could be:

- Prospective: data will be collected during the standard patient care and clinical follow-up.
- Retrospective: data will be collected only from available information related to procedures performed in the past.

DEFINITION OF OBSERVATIONAL STUDY

An observational study is defined as a study that *does not affect the clinical standard of care for patients*. This means that patients who agree to participate and patients who refuse will receive the same standard of care: The only difference is the data collection performed by the investigators.

Some small waivers are sometimes accepted: for example, the compilation of an extra questionnaire can be tolerated to maintain the definition of "observational"; small procedures (such as a blood sampling made at the same time as a standard sample) can be considered as "additional procedures" in an observational study.

SUBMISSION PACKAGE AND PRIVACY

Both prospective and retrospective observational studies need a very simple submission package that includes study protocol and synopsis, CRF, informed consent and other documentation related to the patients (such as questionaries and scores), and documentation of the PI (curriculum vitae, declaration of conflict of interest). Sometimes the EC also requires the documentation of the device.

Informed consent may not be necessary for retrospective studies; this depends on the privacy law in the specific country, so it is important to know the position of the EC about it before starting. The collection of additional consent for a retrospective study, which may be required by the EC, could be practically very difficult since it involves recontacting and obtaining a signed document from patients who may be living far from the clinical center or may wish not to be recontacted. In some cases, updated contact information also may not be available to the investigator.

Depending on the specific national regulations, these issues could be overcome with the complete anonymization of data, that is, making it impossible to discover the identity of specific patients by destroying all links between patients name and collected data.

COSTS FOR OBSERVATIONAL STUDIES

The study can be considered observational only if the medical device is already the standard of care in the hospital. In this case, the promoter does not have to provide the device for free. In the past, clinical trials had been used as a gimmick to introduce a new device in the standard of care; this is no longer tolerated by authorities today.

Since patients follow the standard of care, the insurance policy is not necessary and the promoter does not have to cover any clinical cost; monitoring activity is not mandatory, but is always recommended. Profit promoters have to pay the submission fee to the EC. Very often not-for-profit studies are designed as observational, because no budget is needed and surgeons can easily monitor the clinical situation of their patients.

Observational studies are less complex and less expensive than interventional trials, but the latter are still very important, especially for medical devices. In fact, many complications related to medical devices could arise many years after the implantation.

SUBMISSION

The submission of observational studies to the EC is very similar to that of interventional studies: sometimes the EC requires less documentation but the procedures are almost the same as for interventional postmarketing trials. If the study is retrospective, the EC could accept a simple notification. Usually, the notification of observational studies to the NCA is sufficient, without any kind of evaluation.

CLINICAL TRIALS ON MEDICAL DEVICES IN EUROPE: EU REGULATION (745/17)

In 2017, the new EU regulations about medical devices and about in vitro diagnostic medical devices were declared:

- Regulation (EU) 2017/745 of the European Parliament and of the Council of 5 April 2017 on medical devices, amending Directive 2001/83/EC, Regulation (EC) No 178/2002, and Regulation (EC) No 1223/2009 and repealing Council Directives 90/385/EEC and 93/42/EEC.
- Regulation (EU) 2017/746 of the European Parliament and of the Council of 5 April 2017 on in vitro diagnostic medical devices and repealing Directive 98/79/EC (n.d.) and Commission Decision 2010/227/EU (Regulation 2017: EC Europa, n.d.).

These regulations have been operational since May 26, 2021. The full application of the new regulation on medical devices is expected to take place after 3 years, whereas it is expected after 5 years for in vitro diagnostic medical devices (Regulation 2017, n.d.).

The main aims of the Medical Devices Regulation are as follows:

1. establishing a solid, regulatory framework for medical devices that ensures a high level of safety and health, supporting innovation;
2. ensuring proper management of the internal market for medical devices, protecting the health of patients and users;

3. establishing high-quality and safety parameters for medical devices, ensuring that the data obtained from clinical investigations are reliable and solid and that the safety of the subjects participating in such investigations is protected;
4. integrating many obligations of manufacturers defined in the previous directive only in the annexes into the provisions of the regulation to facilitate their implementation (FORMAFUTURA Presentation, n.d.).

WHAT IS NEW IN EUROPE WITH THE NEW REGULATION?

The most important difference from the past is medical devices are regulated by a "regulation" instead of a "directive." A regulation is more effective than a directive because it is directly valid at the national level without the need for transposition through specific national legislation. This allows greater legal stability and prevents variability in the adoption of rules relating to medical devices by the member states of the European Union (Regulation 2017, n.d.).

The regulation is very complex and will change the world of medical devices, likely impacting also extra-European countries by providing a reference. A crucial difference is that while within the actual Directive manufactures can obtain a *market extension* for a new device similar to another one that is already marked, with the new regulation this will not be possible anymore.

According to Regulation 745/17, clinical trials are classified as follows:

1. premarketing studies: clinical evaluation for conformity assessment purposes;
2. postmarketing studies: clinical investigations for devices already marked.

The risk classification of medical devices will change and manufactures will have to reassess their devices by modifying appropriately the risk class of already marked devices.

Other issues that are addressed in the new regulation are as follows:

- defining of the role of sponsors and CROs;
- introducing the "Unique Device Identifier" (UDI) to improve identification and traceability;
- strengthening supervision, conformity assessment, monitoring, and surprise inspections;
- introducing the unique database EUDAMED for the postmarket monitoring;
- open access to information for the public and healthcare professionals;
- limited autonomy of each member state with reference to civil liability;
- controlling the suitability of the investigators and clinical centers, and obtaining informed consent from subjects.

ETHICAL ISSUES RELATED TO CLINICAL TRIALS IN ORTHOPEDICS

When investigators plan a clinical trial, they have to consider several factors. If the purpose is limited to obtaining the CE marking, or verifying the performance, or selling the product, the promoter could, in principle, neglect less "scientific" aspects of the study, which are, however, equally important for its success.

First of all, it is fundamental to take into consideration the needs and comfort of patients. In addition to the collection of information important from a scientific point of view, it is equally important to also consider the wellness of the subjects who will be recruited. For example, asking patients to undergo a large number of imaging examinations or invasive procedures could limit their compliance and consequentially increase the number of dropouts, that is, patients who refuse to continue to support the study. Moreover, too many limitations could discourage people from giving consent to participate: for example, a long period of abstinence from physical activities could deter young subjects. It should also be noted that the EC may not approve the study if this is not ethically precautionary toward patients.

Even in studies that compare multiple devices, it must be ensured that all groups are equally protected: one product cannot be considered more advantageous than the other, and all subjects must receive the same level of assistance.

From an economic point of view, a patient cannot have a disadvantage because he/she consents to participate in a study. For this reason, the promoter has to cover all the costs related to procedures that are not the "standard of care." In some countries such as Italy, paying subjects who participate in a study is not allowed; however, it is possible to provide visits or clinical procedures for free and it is also possible to reimburse them for travel expenses or compensation for not working. This is a way of thanking subjects who have to follow many rules and indications and accept the risk of using an experimental device.

Another relevant aspect is to make subjects understand well what the clinician/promoter wants to investigate: the objective of the study must be precise and well-articulated, and the sample size must be calculated in a scientific manner. To avoid conflict between the desires of the promoter and clinicians in medical routine procedures, the best approach is to promote the interaction between clinicians and promoters. Sometimes promoters wish to include imaging investigations because they are more engaging and incisive; however, in some cases, the collection of clinical scores is preferable to radiological investigations. Without adequate discussion with medical experts, promoters risk to include exams that are not really useful, increasing the costs of the trial and discomfort for patients. It is also advisable to involve clinicians in the definition of the control group, especially if the promoter wants to compare two medical devices.

It should be noted that the clinicians that support the promoter in the design of the study should not be the same who will participate as investigators, in order to avoid a possible conflict of interest.

Furthermore, the promoter has to check the market potential of the device before planning a new clinical trial; if several similar devices are already commercially available, it is advisable to focus on a feature that distinguishes the new device from the others.

Finally, the promoter has to consider the requirements of the authorities, including the EC; contesting an authority that has asked for extra documentation is generally neither recommended nor productive. The promoter should clearly understand what is needed for a specific study and prepare everything in advance.

REFERENCES

Council Directive 90/385/EEC of 20 June 1990. (1990). https://eur-lex.europa.eu/LexUriServ/LexUriServ.do?uri = CELEX:31990L0385:en:HTML.

Directive 2001/20/EC. (n.d.). https://ec.europa.eu/health/sites/default/files/files/eudralex/vol-1/dir_2001_20/dir_2001_20_en.pdf.

Directive 98/79/EC. (n.d.). https://eur-lex.europa.eu/legal-content/EN/TXT/PDF/?uri = CELEX:31998L0079& from = EN.

EU Clinical Trials Register. (n.d.). https://www.clinicaltrialsregister.eu/doc/EU_Clinical_Trials_Register_Glossary.pdf.

FDA Medical Device Registration. (n.d.). https://www.fda.gov/industry/regulated-products/medical-device-overview#registration.

FDA Postmarketing Clinical Trials. (n.d.). https://www.fda.gov/vaccines-blood-biologics/biologics-post-market-activities/postmarketing-clinical-trials.

FDA.gov. (n.d.). https://www.fda.gov.

FDA_Overview Medical Device Regulation. (n.d.). https://www.fda.gov/medical-devices/overview-device-regulation/classify-your-medical-device.

FORMAFUTURA Presentation. (n.d.). Ricerca clinica con Device dopo il Regolamento UE 745/2017. https://www.health.formafutura.it/regolamento-ue-sperimentazione-dispositivi.html.

Italian Health Ministry_Medical Device Overview. (n.d.). https://www.salute.gov.it/portale/temi/p2_6.jsp?lingua=italiano&id=3&area=dispositivi-medici&menu=caratteristichegenerali.

Italian Health Ministry_Medical device. (n.d.). https://www.salute.gov.it/portale/temi/p2_6.jsp?lingua=italiano&id=4309&area=dispositivi-medici&menu=sperimentazione.

Regulation 2017. (n.d.). https://www.salute.gov.it/portale/temi/p2_6.jsp?lingua=italiano&id=4971&area=dispositivi-medici&menu=caratteristichegenerali.

Regulation 2017: EC Europa. (n.d.). https://ec.europa.eu/health/md_sector/overview_en.

COMPUTER-ASSISTED ORTHOPEDIC SURGERY

27

Nicola Francesco Lopomo

Department of Information Engineering, University of Brescia, Brescia, Italy

BACKGROUND

There is a necessity of utilizing and exploiting all the information available during the execution of surgical procedures, so as to increase the accuracy of the surgical gestures, improve control and flexibility over the operative workflow, and reduce the invasiveness and the variability of the surgery itself (Zheng & Nolte, 2015). Addressing the main goal of supporting and optimizing minimally invasive approaches (i.e., less number of incisions and increased amount of tissue spared compared to standard surgery), computer-assisted surgery (CAS) systems allow us to both minimize the risks of errors and adverse effects during surgery (i.e., decreased surgical morbidity) and reduce postoperative complications (e.g., shortened healing time, decreased possibility of infection, less pain, etc.), which translate into improved clinical outcomes (Kubicek et al., 2019; Zheng & Nolte, 2015). The concept of CAS is based on what is called "predictive, preventive, and personalized medicine" (Joskowicz, 2017).

In general, CAS exploits the integration of a broad range of methodologies and technologies for supporting/enhancing three main interconnected phases within the surgical workflow (Kubicek et al., 2019; Zheng & Nolte, 2015):

1. *planning* the surgical procedure (i.e., determining what to do and how to do it before entering the operating room and/or making the first incision, by exploiting digital imaging, computational simulations, and development of dedicated models, tools and instruments);
2. *guiding/assisting* the surgeon during the execution of each surgical step (by using dedicated and "smart" tools and instruments);
3. *assessing* that the obtained results meet the initially assumed requirements in terms of surgical outcomes (by also including both functional and biomechanical results).

A general representation of the main workflow is shown in Fig. 27.1.

CAS systems are used in a wide variety of surgical specialties, including neurosurgery, ENT (i.e., ear, nose and throat) surgery, visceral surgery, maxillofacial surgery, cardiac applications, radiosurgery, and orthopedics.

Computer-assisted orthopedic surgery (CAOS) represents CAS approaches in the context of traumatology and orthopedic surgery, aiming at the minimally invasive treatment of several pathologies affecting the musculoskeletal system. For instance, CAOS technologies are currently in use in

Human Orthopaedic Biomechanics. DOI: https://doi.org/10.1016/B978-0-12-824481-4.00010-X

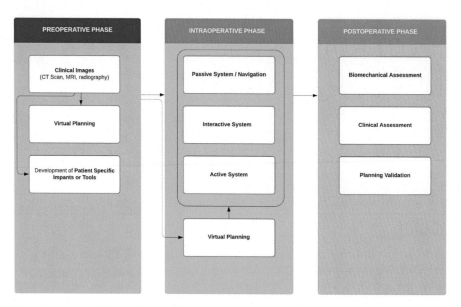

FIGURE 27.1

General workflow in computer-assisted surgery.

the accurate placement of a prosthetic component within the ankle (Berlet, Penner, Lancianese, Stemniski, & Obert, 2014), the precise resurfacing of a hip joint affected by osteoarthrosis (Sato et al., 2019), the exact drilling and placement of pedicle screws in a spine with complex diseases (Elmi-Terander et al., 2020), the personalized reconstruction of the anterior cruciate ligament (ACL) within the knee of a soccer player (Lopomo et al., 2008; Martelli et al., 2007), the ready fixation of a tibial fracture (Ma et al., 2018), or the optimized resection of a bone tumor (Siegel, Balach, Sweeney, Nystrom, & Colman, 2020).

Orthopedics represent one of the most suitable fields for the application of CAS methodologies. Bones and periarticular soft tissues can be easily identified and accurately evaluated by using specific imaging technologies such as radiography, fluoroscopy, computed tomography (CT), and magnetic resonance imaging (MRI); hence, bony and soft tissue structures can be reliably reconstructed to define three-dimensional representations and models that can be used in the preoperative planning or even in predicting the surgical outcomes (Hernandez, Garimella, Eltorai, & Daniels, 2017). Furthermore, bones present the inherent advantage of being considered as rigid structures (i.e., they do not deform significantly during the surgery, at least with respect to brain or abdominal organs), thus allowing the accurate tracking of their position and movement.

Over the past few decades, optimized technologies for localizing bones, components and tools, improved sets of dedicated instruments, and innovative software with enhanced interaction designs have led to achieving highly reliable clinical results, which—in the orthopedic domain—translate, for instance, into the reduction of aseptic loosening and the shortening of the rehabilitation period after arthroplasty, but also into optimized functional outcomes in ACL reconstruction. Indeed, CAOS systems are able to achieve improved outcomes with respect to conventional surgeries, including, but definitely not limited to, the measurement of the alignment of the prosthetic components with respect to tibia and femur bones in the knee (Pailhé, 2021; Siston, Giori, Goodman, & Delp, 2007) and hip arthroplasties and arthroscopies (Nakano, Audenaert, Ranawat, & Khanduja, 2018; Sato et al., 2019), the measurement of the mechanical axis of the lower limb in osteotomies (Murphy et al., 2015), or the measurement of the correction of the spine in complex deformities (Vadalà et al., 2020).

Furthermore, the advances made by CAOS technology (e.g., navigation, robotics, extended reality, 3D printing, 3D modeling, "smart" instrumentations, etc.) have led to the birth of novel approaches toward the overall management of the patients. These advancements definitely include the optimum utilization of all available information by introducing reliable biomechanical modeling, which can be applied throughout the whole surgical workflow, including the planning, the surgical steps, and the assessment of the main outcomes. The actual CAOS systems—commonly identified as "navigation" systems—are able to provide not only reliable guidance during the surgical intervention but also (near) real-time feedback concerning joint overall kinematics (Grassi et al., 2020; Safran et al., 2013; Zaffagnini, Urrizola, et al., 2016), joint statics and dynamic laxity (Bignozzi et al., 2010; Lopomo, Sun, Zaffagnini, Giordano, & Safran, 2010), impingement assessment (Tetsunaga et al., 2020), bone-to-bone distance mapping (Siegler et al., 2018; Signorelli et al., 2013), isometry and strain assessment of the ligaments (Grassi et al., 2020; Safran et al., 2013; Zaffagnini, Signorelli, et al., 2016), estimation of contact pressure distribution (Armand et al., 2004; Murphy et al., 2015), and so on, thus allowing clinicians to study the biomechanics associated with both the pathological condition and the treated one.

In this scenario, the capability of CAS to reliably perform a "biomechanically enhanced" surgery—on the basis of the available new technologies and methodologies—represents the potential to both advance the research in musculoskeletal biomechanics and introduce new optimized surgical approaches for the treatment of several kinds of musculoskeletal diseases, thus guiding the evolution of the surgery from the traditional models toward novel paradigms founded on CAOS.

In order to understand the importance of CAOS and its impact on orthopedics, it is fundamental to fully understand each and every step involved in the overall computer-assisted workflow, which involves different technologies and methodologies and how the principles of musculoskeletal biomechanics can be applied to this specific context.

MAIN FUNCTIONAL COMPONENTS

As previously highlighted, CAOS systems enable the surgeons to develop novel, more accurate, and less invasive surgical approaches by combining preoperative modeling and planning with intraoperative guidance and assessment tools. The broadest concept of CAOS includes many different instrumentations, devices, tools, sensors, robotic components, modeling, etc.; all these solutions are used to support the surgeons by enhancing the available information with quantitative data, measurements, and estimations.

Although clinical applications may be different, all the CAOS systems exploit the same basic design and general functional components, which are focused on the main hypothesis to create a sort of "digital twin" of the orthopedic patient within the specific surgical procedure (Picard, Deakin, Riches, Deep, & Baines, 2019; Zheng & Nolte, 2015).

For instance, we can virtually interact with a 3D image of a patient's knee coming from CT scan while identifying the correct size and position of the prosthetic components, simulate several achievable outcomes in term of alignments and balance, and then transfer this information to the operating room. Within this workflow, formally, we can identify two main classes of objects:

- *surgery objects*, which represent the targets of the surgery (i.e., the therapeutic object, such as the knee joint that should be replaced) and any tools required to realize the procedure;
- *virtual objects*, which are virtual representations of both the therapeutic objects and any further component used within the surgical workflow, including—for instance—an implant or an instrument.

Furthermore, we can also recognize a third fundamental element represented by any technology and methodology that is able to act as a *link* between the surgery and the virtual objects. For example, during the intraoperative execution, this element is represented by the navigation system or by the robotic solution.

GENERAL WORKFLOW

In order to correctly perform a surgery using COAS systems, COAS systems require the execution of several steps (Picard et al., 2019), whose order mainly depends on the design of the system (i.e., image-based or imageless approach, as described in the next section); in general, we have:

- *System setup*—during this phase, the system is initialized, the overall operating space is verified, and the tracking solutions (i.e., all the technologies that can be used to acquire the position, orientation, and the general movement of the surgery objects) are placed and tested for reliability in the operating theater.
- *Registration*—this step is based on the identification of all the anatomical landmarks (i.e., the basis for the definition of patient-specific reference frames), required either correctly associating preoperative and/or intraoperative clinical images, corresponding planning, and derived models for the therapeutic objects or for directly defining the correct anatomical features that can be used during the surgery itself (i.e., without the use of any prior acquired information). Thanks to the registration and due to the presence of the tracking technologies, it is possible to identify in real-time the position and orientation of the surgery objects and hence move the corresponding virtual objects. In image-based systems, several methods (e.g., identification of anatomical features, intorduction of fiducial markers, etc.) and algorithms (e.g., rigid, affine, shape-based or model-based transform) are commonly used for registration; on the other hand, in imageless systems, the identification of a predetermined set of anatomical landmarks is enough to define the anatomical reference frames. The registration step, together with the technology used for tracking and the overall quality of the clinical images (i.e., pixel size, slice thickness, geometric distortion, etc), significantly affects the overall accuracy of the CAOS system, which—for orthopedic applications—can be in the range of about $1-4$ mm (Picard et al., 2019). An example of the anatomical landmarks used for the definition of the anatomical landmarks for femur and tibia in ACL surgery is given in Fig. 27.2.
- *Planning*—this phase can be performed both before the procedure or in the operating room addressing the "ideal" surgical strategy (including, for instance, size/position/alignment of the

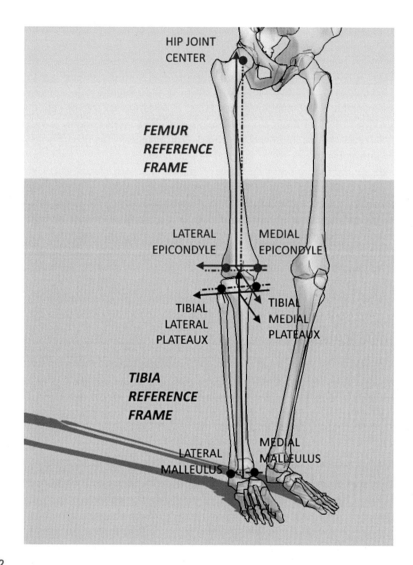

FIGURE 27.2

Example of the identification of anatomical landmarks and corresponding reference frames for femur and tibia, commonly used in CAOS for ACL surgery. *ACL*, Anterior cruciate ligament; *CAOS*, computer-assisted orthopedic surgery.

From Lopomo N.F., 2008, "Quantitative Assessment of Knee Stability during Surgery and Evaluation of Joint Restored Functionalities after the Reconstruction", PhD Thesis, PhD in Bioengineering, Politecnico di Milano, Milano, Italy.

prosthetic components, limb alignment, soft-tissue balance, etc.), exploiting the link created between the virtual and the surgery objects.

- *Execution*—once the planning of the surgery and verification of the correctness of the registration phase is performed, it is appropriate to proceed with the surgery itself. In this step, the activities can follow the planning and be performed at different levels, considering the system designs available in the operating room (i.e., passive, interactive or active systems, as hereinafter described). One of the main advantages of using CAOS systems is its potential to reduce the invasiveness of the approach, avoiding—for instance—the use of both intramedullary guides in knee arthroplasty.

- *Assessment*—during the execution of the surgery, the surgeon can always check for the achieved accuracy with respect to the planning, obtaining valuable real-time feedback, and having the possibility to accurately adjust the performance and thus to potentially eliminate the stacking of errors that can occur in any complex multistep surgical procedure. All the information acquired intraoperatively (e.g., limb alignment, implant position, etc.) can be quantitatively analyzed, recorded, and then compared for postoperative assessment (including radiological, clinical, functional outcomes, etc.).

SYSTEM PERFORMANCE

As previously underlined, the main objective of any CAOS system is to make the most accurate and least invasive surgery possible. For instance, in a passive navigation system, the required accuracy is 1 mm in positioning and 1 degree in orientation (for lines and planes) (Pei, 2019). In general, in order to define the performance of a CAOS system, first, it is necessary to identify the possible sources of error that can be classified into the following (Leardini, Belvedere, Ensini, Dedda, & Giannini, 2013):

- *Instrumentations and technologies*—examples include fixing of the technical frames for tracking, relative motion between these frames and the bony structures, occlusion, geometry/size of the technical frame, and algorithms used to identify the position/orientation of the technical frames.

- *Registration phase*—it is worth underlining that for the identification of specific landmarks associated with the patient's anatomy, over the past few decades, shape matching, bone-model morphing, anatomy-based methods, and functional approaches have been developed and proposed, each one with its own advantages and disadvantages (Kubicek et al., 2019; Zheng & Nolte, 2015).

- *Human component*—in all CAOS systems (even the active ones), there are several steps that must be performed manually, including the cuts made by using sawblades, the drilling, the bone final preparation, the cement application, and the positioning of the implants. Furthermore, the tests of the implants are usually performed with the trial components, which can introduce further variability with respect to the final outcomes.

For these reasons, it is mandatory to verify the final outcomes within the definitive configurations, and the complete CAOS procedure should be assessed in terms of repeatability (i.e., considering the same context) and reproducibility (i.e., including further users and different use cases) of the results (Bignozzi et al., 2008; Zheng & Nolte, 2015), and also in terms of biomechanical performance (Cartiaux, 2016).

SYSTEM DESIGNS

Several classifications of the different system designs have been proposed based on the functional characteristics of these systems (Hernandez et al., 2017; Pailhé, 2021; Schneider & Troccaz, 2001). In general, we can classify these system designs into three main classes: *passive*, *interactive*, and *active* systems.

PASSIVE SYSTEMS

These systems are not designed to actively perform any surgical action on the patient; however, they are a key component in providing reliable information to the surgeon during the intraoperative planning phase and real-time guidance. In general, the functioning of these systems is based on the use of a specific tracking technology (e.g., optoelectronics, electromechanics, electromagnetics, etc.) able to capture the position and orientation of the surgical tools and other instruments, with respect to the anatomy of the patient. Completely passive robotic articulated arms can be also used, but, the most common example of such an approach is the surgical navigation systems, which can operate both in imageless and in image-based modalities (as described in the following paragraphs). A typical application context concerning the use of image-free navigation system is shown in Fig. 27.3.

INTERACTIVE SYSTEMS

These systems are usually based on robotic platforms, and interface with the performing surgeon through specific levels of interaction; in general, two different strategies can be identified, which lead to the development of solutions called *synergistic* and *semiactive* systems. Synergistic systems are designed to express a programmable mechanical constraint while providing haptic feedback to the surgeon, whereas semiactive systems are programmed to actively perform very specific actions—such as moving a cutting guide or constraining the movement of a tool within a defined region/volume, without a direct direct advice to the operator. Usually, for these systems, the use of clinical images is recommended—although the last synergistic solutions allow for imageless approaches, and defined templates can be introduced to identify every necessary step and what the system can do within the surgical workflow (Pailhé, 2021; Siston et al., 2007). Like the passive ones, this kind of solution is not allowed to directly accomplish any surgical task on the patient. An example of semiactive systems is shown in Fig. 27.4.

Recently, *teleoperated* systems have been introduced in the interactive systems, which fundamentally are robotic solutions remotely controlled by a surgeon even from outside the operating room or the hospital (Pailhé, 2021).

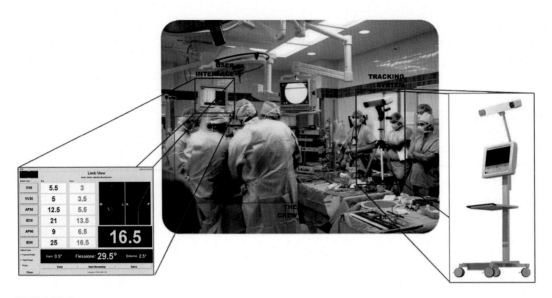

FIGURE 27.3

Example of a surgical navigation system used for anterior cruciate ligament (ACL) reconstruction, highlighting the optoelectronic tracking device and the user interface.

Unpublished material; courteously provided by the author.

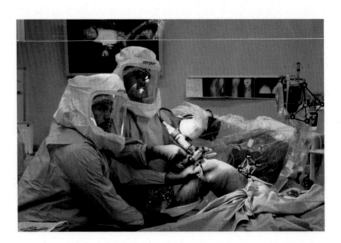

FIGURE 27.4

Example of the actual setup for the use of an interactive system in total knee arthroplasty (Mako by Stryker Orthopedics, Kalamazoo, MI, USA).

From (Grau, 2019). Grau, L., Lingamfelter, M., Ponzio, D., Post, Z., Ong, A., Le, D., & Orozco, F. (2019). Robotic arm assisted total knee arthroplasty workflow optimization, operative times and learning curve. Arthroplasty Today, 5(4), 465–470. https://doi.org/10.1016/j.artd.2019.04.007.

ACTIVE SYSTEMS

These technological solutions are real active robots that can perform several very specific pre- or intraoperatively planned surgical actions, such as drilling, without any direct intervention performed by the surgeon. For such systems, it is strictly required to define the surgical strategy that must be "programmed" within the system; in this case, the use of preoperative clinical images is highly recommended (Kubicek et al., 2019; Zheng & Nolte, 2015). Although the overall accuracy in specific surgeries was indeed improved by this solution (Pailhé, 2021), this kind of systems have not been widely used mainly due to their initial costs and quite high complexity. A paradigmatic representative of this kind of approaches is depicted in Fig. 27.5.

Most of the above-described systems can be further classified based on the type of information they use for the planning and execution of the surgery itself; in particular, the introduction of patient-specific clinical images—such as preoperative CT scans or intraoperative fluoroscopy or even the use of virtual anatomical models morphed on the current patient's features, and how all this information can be associated with the subject's actual status during surgery, leads us to define two specific working modalities (Pei, 2019), namely the image-based and the imageless (or image-free) systems.

FIGURE 27.5

Example of the actual setup for an active system (RoboDoc/TSolution-One by THINK Surgical Inc, Fremont, CA, USA).

From (Koenig, 2015).

Image-based system

In this kind of systems, the anatomical information is acquired by means of the processing and elaboration of specific clinical images, which can include CT scans, MRI, and 2D/3D fluoroscopy; thus, the available data can be used to model the anatomical structures, plan the surgical steps, and guide the surgeon during the procedure. In general, the systems that use preoperative images rely on anatomical models derived from CT images, or by morphing a generic model to match the bony geometry of a particular patient; the intraoperative adoption of image information commonly uses fluoroscopy. Image-based solutions usually require several processing steps of the clinical images to obtain the anatomical features related to both the bony and, eventually, soft-tissue components. The most common steps required are as follows (Kubicek et al., 2019; Zheng & Nolte, 2015):

- *Data acquisition*—as previously reported, the data can be acquired from several image sources, including CT scans, MRI, and fluoroscopy. It is mandatory to standardize the overall quality of all the available images.
- *Enhancement*—this step is required to improve the overall signal-to-noise ratio and optimize the identification of the structures of interest (e.g., bones). This step is usually accomplished by means of digital filters (e.g., Gaussian, average, median filter, etc.), interpolating strategies (e.g., wavelet transform, linear/cubic interpolation, vector quantization, etc.), and contrast enhancement techniques (e.g., contrast stretching, histogram equalization, etc.).
- *Segmentation*—in order to build the correct models of the structures of interest, it is necessary to partition the clinical image into 2D/3D segments. The segmentations can be based on a binary approach or a multiregional labeling (i.e., several labels are associated with the different structures). At present, there are several strategies that can be used for segmentation, including automatic, semiautomatic, or interactive methods, which exploit many different algorithms addressing thresholding, seed-based labeling, model-based labeling, meshing, etc.
- *Registration*—as previously underlined, this task is mainly necessary to align preoperative images—even those coming from different modalities (e.g., CT scan and MRI), intraoperatively with the therapeutic objects (i.e., the bony structure of the patients identified, for instance, with a "tracked" probe—as hereinafter described). Several approaches have been proposed to obtain an optimal transformation that ensures the best match between the objects of interest, including also 2D/2D, 2D/3D, or 3D/3D approaches. These approaches include rigid, affine, image-domain, model-based transformations, which can be performed automatically, semiautomatically, or interactively. Advance solutions for registration also allow us to automatically obtain specific landmark and anatomical feature (e.g., axes and planes) identification.

Imageless system

The design of this kind of systems implies that the whole surgical procedure can be realized without the use of any preoperative or intraoperative clinical images. In fact, the approach adopted by imageless solutions relies on both anatomical and functional information, which can be intraoperatively acquired and used for planning and supporting the surgeon during the procedure.

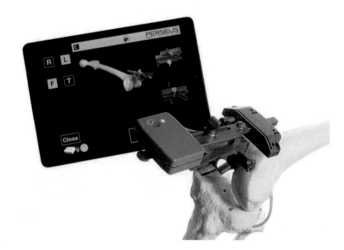

FIGURE 27.6

Example of a "smart" cutting guide for total knee arthroplasty (Perseus by Orthokey Italia S.r.l., Firenze, Italy).

Courteously provided by Orthokey Italia Srl.

In particular, imageless systems are able to retrieve the information about the required anatomical features without the need for a proper clinical image registration through:

- the use of a "tracked" probe for the direct identification of specific bony landmarks, such as knee epicondyles, tibial tuberosity, and ACL insertions;
- the exploitation of passive prototypical movements (e.g., pivoting rotation of the hip joint, flexion−extension of the knee joint, etc.) and dedicated optimization algorithms to obtain a functional identification of the joint centers (Bignozzi et al., 2010; Lopomo, Zaffagnini, Bignozzi, Visani, & Marcacci, 2010) and axes (Colle et al., 2012, 2014; Colle, Bruni, et al., 2016; Colle, Lopomo, Visani, Zaffagnini, & Marcacci, 2016; Grassi et al., 2020). An extremely interesting use of this approach is shown in Fig. 27.6, which represents a minimal navigation system able to guide the surgeon during total knee arthroplasty by using only inertial measurement units as tracking technologies;
- the integrated use of general anatomical models, a "tracked" probe for the identification of bony surfaces, and morphing algorithms able to adapt the general model to the current morphological feature of the patient; the estimated morphed model could then be checked for coherence with respect to the actual anatomy (Kubicek et al., 2019; Zheng & Nolte, 2015). An example of the process of bone morphing is shown in Fig. 27.7.

Further advances in this field are also represented by the use of statistical shape models, which allow us to reliably reconstruct a 3D anatomy of bony segments, even using 2D image data (Cerveri, Sacco, Olgiati, Manzotti, & Baroni, 2017; Zheng et al., 2009). An example of this approach is shown in Fig. 27.8.

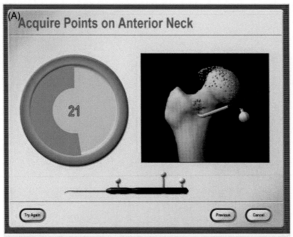

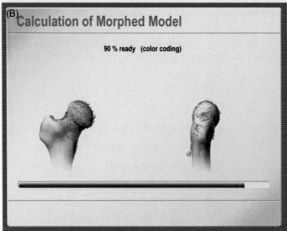

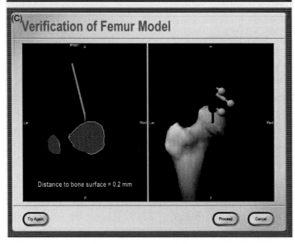

FIGURE 27.7

Process of bone morphing by (A) acquisition of the landmarks, (B) morphing of the model, and (C) verification.

From (Zheng, 2007).

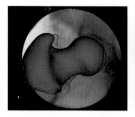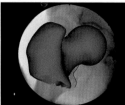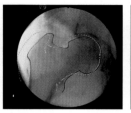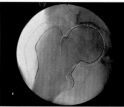

FIGURE 27.8

Example of 2D contour extraction from a 3D model for reconstructing a patient-specific surface model by using calibrated X-ray images.

From (Zheng et al., 2009).

HARDWARE ARCHITECTURES

In general, CAOS systems could present complex architectures, which integrate several devices and technologies, devoted to tracking the instruments, identifying the anatomical features of the patients and guiding/supporting the surgeon during the procedure, or actively performing a specific phase of the surgery by itself.

The core of each system is clearly the central unit, which comprises dedicated software tools, addressing the need of processing the available information (i.e., clinical images), defining the surgical parameters within the procedure (i.e., the planning), and guiding/supporting/accomplishing each step in the most reliable way, also providing correct (visual/acoustic) real-time feedback to the clinicians. From the perspective of the both imageless and image-bases systems, one of the most important components is the solution chosen to track the movements of the tools and the position and orientation of the anatomical segments. Additional devices and tools could be further integrated to ensure, for example, active intraoperative surgical execution by means of robotic components, clinical image registration, or calibration.

Minimal hardware configuration required for an imageless navigation system (Siston et al., 2007) is shown in Fig. 27.9.

TRACKING TECHNOLOGIES

The identification of the spatial position and orientation of each surgery object can be accomplished by exploiting the principle of dynamic referencing (i.e., tracking), so as to allow their "virtualization" within the CAOS procedure. Real-time measurements acomplished by the tracking technologies usually return information on rigid transformation (i.e., roto-translations). In interactive and active systems, the tracking issues can also be solved by integrating the use of electromechanical solutions and approaches based on inertial measurement units (IMUs). In passive systems, ultrasound-based tracking represents an early alternative. However—when focusing on navigation systems—two main kinds of tracking approaches are available, namely the optoelectronic and

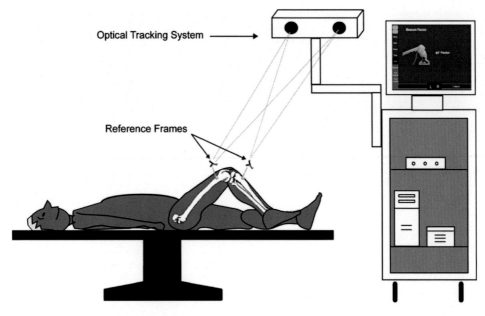

Optical Tracking System

Reference Frames

FIGURE 27.9

Main components of an imageless navigation system.

From (Siston et al., 2007).

electromagnetic technologies, with the first one by far the most commonly used in operating theaters (Kubicek et al., 2019; Zheng & Nolte, 2015).

- *Optoelectronic tracking*—this kind of solutions usually identify the position and orientation of inflexible reference frames/arrays that are rigidly fixed to the bony structures by means of mono- or bicortical pins/screws and/or attached to the surgical tools/instruments (Siston et al., 2007). Each frame can be embedded with either active or passive markers, which can emit or reflect infrared light, respectively; this allows their reliable 3D identification thanks to the use of, at least, a couple of optoelectronic cameras, which are usually mounted on a rigid structure, thus not requiring intraoperative calibration of the system itself. With this technology, an overall accuracy in positioning of up to 0.1 mm can be achieved (Pei, 2019). The main advantages of using active frames (mounting infrared LED) are related to their overall robustness and less sensitivity to the environmental conditions; the main drawback is the need for constant power supply. Most of the actual navigation systems are using passive solutions; in this case, the infrared light is emitted by the tracking system (usually circular LED crowns around each camera) and reflected by small spheres or 3D structures covered with reflector materials. The passive frames are indeed easy to use and can be easily replaced (i.e., they are disposable). The main downside of this solution is the sensitivity of the markers to contamination with fluids, and so their visibility can be reduced during surgery. The most common problem with an optoelectronic tracking system is related to the mandatory need of a direct line of sight between the cameras and the frames. This can be a critical issue in several operating room settings; for this reason, the use of different tracking systems has been proposed over the years. A typical example of an optoelectronic solution is shown in Fig. 27.10 (Swiatek-Najwer et al., 2008).

- *Electromagnetic tracking*—these systems are based on the use of a homogeneous electromagnetic (EM) field generator in order to detect the position and orientation of defined sensors (i.e., coils), which are fixed to both bones and surgical tools, as previously described. The use of small coils in the sensors enables the placement of dynamic reference frames via small incisions even under soft tissue or directly into bones without any occlusion problems. The achievable accuracy can be up to 3 mm in positioning (higher compared to the optoelectronic solutions) (Pei, 2019). Nevertheless, electromagnetic sources or the presence of ferromagnetic materials can interfere with the tracking process, so dedicated operating room setups and instrumentation are necessary for this kind of tracking technology. However, the progress made in this technology has led to the development of high-end solutions able to both statically and dynamically compensate for external disturbances and interferences. As previously highlighted, this kind of solutions are appropriate for a small operating volume, reduced segment dimensions, and with a compact line of sight.
- *Ultrasound tracking*—this solution is based on the use of at least three ultrasound emitters and three receivers placed on each frame, to retrieve information about its position and orientation (complete 6 degrees of freedom). The achievable overall accuracy is mainly based on the time-of-flight principle, which allows to reach up to 2−4 mm in CAOS applications (Pei, 2019). Unfortunately, this technology is highly sensitive to sources of environmental noise, including temperature and air.

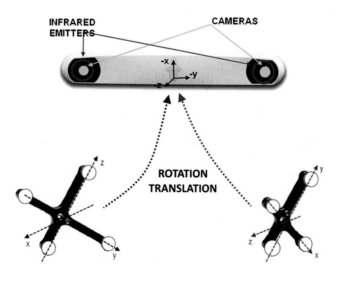

FIGURE 27.10

Example of an optoelectronic tracking system with passive frames.

- *Electromechanical tracking*—this is one of the first approaches used in CAOS systems, and its actual functionality is accomplished by several components present in the robotic solutions, i.e., the integration of rigid segments with instrumented joints. By knowing the geometry of the structure, measuring the angular displacement of each joint, and using a mathematical model able to describe the kinematics of the device, it is possible to identify the position and orientation of the end-effector or manipulator. This approach is quite stable but bulky and ensures an overall accuracy in positioning of about 2−3 mm (Pei, 2019).

As previously underlined, depending on the surgical technique involved, required implants, and given operating room settings, the optoelectronic tracking systems may be the first choice for most interventions; however, considering, instead, special indications, the last-generation EM tracking systems could be a better alternative.

In recent years, navigation systems based on IMUs have gained more and more attention, addressing above all the possibility of introducing reliable but simple solutions in the surgical workflow that are able to overcome the main issues related to the most conventional alignment methods, reducing the morbidity of the surgery itself, but avoiding the use of bulky and complex systems. This approach has been successfully applied to different clinical perspectives, including the placement of pedicle screws in spine surgery, osteotomy, and total knee arthroplasty (Zheng & Nolte, 2015). Unlike the optoelectronic devices, the line of sight is not an issue with this approach; however, IMU-based solutions deliver lower values of accuracy in terms of orientation (above 1 degree), and they are not reliable in terms of estimating the position of a tool or a bony segment. A recent example of an IMU-based approach in total knee arthroplasty is shown in Fig. 27.11.

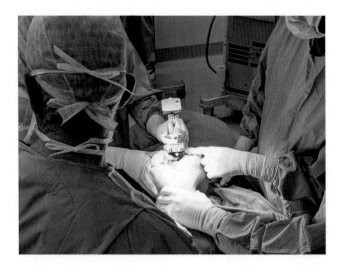

FIGURE 27.11

Example of an IMU-based navigation system for total knee arthroplasty. *IMU*, Inertial measurement unit.

Courteously provided by Orthokey Italia Srl.

CLINICAL APPLICATIONS

CAOS systems have been successfully used since the mid-1990s, addressing a variety of clinical applications, including the following:

- *Trauma surgery*—applications in trauma surgery are based on the need for the accuracy of the placement of implants and devices—and even for the management of small bony segments—which can be achieved by using image-based navigation systems (i.e., CT-based or fluoroscopy-based approaches). In this context, CAOS systems can be used in positioning percutaneous or cannulated screws and internal or external devices for fracture fixation (Hernandez et al., 2017).
- *Spinal surgery*—the complexity of the human spine and its associated pathologies have led to the development of image-based designs, so as to reduce morbidity and avoid neurovascular structures (Hernandez et al., 2017). CAOS systems have been successfully applied in different spinal surgeries, including pedicle screw placement in the lumbar and thoracic spine.
- *Joint surgery*—the application of CAOS systems in joint surgery covers a really wide range of clinical necessities, with the main indication of maintaining a minimally invasive procedure (Hernandez et al., 2017); in particular, a CAOS approach can be used in total hip replacement (with the focus on optimizing the cup placement, component alignment, limb alignment and offset, restoration of joint center) and hip resurfacing, and in total and unicompartmental knee replacement (to enhance limb and component alignment and positioning, reduce outliers, improve functional outcomes, optimize soft-tissue balancing, decrease morbidity due to the use of intramedullary guiding devices).
- *Soft tissues surgery*—navigation systems have been widely used in ACL reconstruction both for optimizing the surgical procedures (i.e., increasing the accuracy in tunnel placement, checking for graft isometry, etc.) (Hernandez et al., 2017) and for assessing the functional outcomes in terms of knee joint statics and dynamic laxity quantification (Bignozzi et al., 2010; Lopomo et al., 2008, 2009; Lopomo, Zaffagnini, et al., 2010; Zaffagnini et al., 2006).
- *Oncologic surgeries*—CAOS approaches (exploiting above all multimodal image-based systems) have proven to be extremely useful for oncologic surgery, addressing very different anatomical districts, including the treatment of complex pelvic and spinal tumors and long bone tumors (Fhkcos, Fhkcos, & Ortho, 2010; Cho, Kang, Kim, & Han, 2008); in these surgeries, the integration of preoperative clinical images, 3D modeling, accurate planning of resection, optimization of the implants and devices, and assessment of the achieved clearance—all in real-time—represents an added value by computer-assisted surgery.

BIOMECHANICALLY ENHANCED SURGERIES

From the biomechanical perspective, CAOS systems are indeed an extremely mature technology able to integrate different research and application domains such as robotics, image processing and analysis, computer science, physics, and mathematics.

Several of the concepts at the basis of the CAOS approach, such as the possibility to perform intraoperative measurements to improve the surgical accuracy, have the potential to enhance biomechanics research and introduce novel solutions for the treatment of several musculoskeletal diseases, with the further potential of quantifying the impact on patient-specific out-comes in terms of biomechanical performance (Cartiaux, 2016).

The first applications of a CAOS system were in fact related to the treatment of the bones, for their inherent nature (i.e., they can be approximated as rigid bodies, easy to track). Very recently, the application of CAOS approaches has expanded and also shifted toward studying and analyzing more complex biomechanical problems related to the whole musculoskeletal system. When consid-ering, for instance, the risk analysis of developing fractures, bones can no more be considered as rigid but should be modeled as a deformable continuum with a nonhomogeneous distribution of internal stresses (Payan, 2012). In this perspective, a patient-specific finite element model could be introduced to estimate internal stress at the interface between a prosthetic component and the bone, to understand the risk of fracture; even in this case, the bone should be modeled as a linear elastic material subjected to small deformations (Payan, 2012).

Moving toward soft-tissue surgeries, CAOS has to deal with very complex, time-dependent, nonlinear, inhomogeneous, and anisotropic behaviors, typical of ligaments, tendons, and muscles. In this context, it is therefore mandatory to introduce in the CAOS workflow the possibility to include into the model even large deformation and visco-hyperelastic constitutive laws (Payan, 2012).

The patient-specific approach represents the core of the new biomechanical approach built into CAOS systems. However, this perspective opens up novel challenges, including the need for simple and real-time exploitable models embedding tissues with inherent patient-specific properties and constitutive equations. In this frame, several methods have been proposed to design patient-specific models, including advanced modeling algorithms (Bucki, Lobos, & Payan, 2010; Wang & Ma, 2018), in-vivo tissue characterization (Schiavone, Boudou, Ohayon, & Payan, 2012), quasi-real-time simulations (Stavness, Lloyd, Payan, & Fels, 2011), and analysis of the impact of discretiza-tion errors (Duprez et al., 2018).

This kind of approach has led to positive outcomes concerning, for example, the biomechanical guidance in hip joint surgery, in which it is possible to monitor in real-time the fragment position, the joint range of motion, the peaks in contact pressure (Armand et al., 2004; Murphy et al., 2015), and distance maps reported in Fig. 27.12.

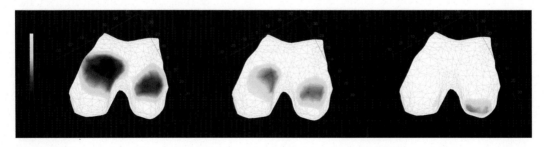

FIGURE 27.12

Femoral perspective of intrarticular distance maps estimated during knee flexion–extension movement, acquired during surgery.

Unpublished material; courteously provided by the author.

REFERENCES

Armand, M., Lepistö, J. V. S., Merkle, A. C., Tallroth, K., Liu, X., Taylor, R. H., & Wenz, J. (2004). Computer-aided orthopedic surgery with near-real-time biomechanical feedback. *Johns Hopkins APL Technical Digest (Applied Physics Laboratory), 25*(3), 242−252.

Berlet, G. C., Penner, M. J., Lancianese, S., Stemniski, P. M., & Obert, R. M. (2014). Total ankle arthroplasty accuracy and reproducibility using preoperative CT scan-derived, patient-specific guides. *Foot & Ankle International, 35*(7), 665−676. Available from https://doi.org/10.1177/1071100714531232.

Bignozzi, S., Lopomo, N., Zaffagnini, S., Martelli, S., Bruni, D., & Marcacci, M. (2008). Accuracy, reliability, and repeatability of navigation systems in clinical practice. *Operative Techniques in Orthopaedics, 18*(3), 154−157. Available from https://doi.org/10.1053/j.oto.2008.11.001.

Bignozzi, S., Zaffagnini, S., Lopomo, N., Fu, F. H., Irrgang, J. J., & Marcacci, M. (2010). Clinical relevance of static and dynamic tests after anatomical double-bundle ACL reconstruction. *Knee Surgery, Sports Traumatology, Arthroscopy, 18*(1), 37−42. Available from https://doi.org/10.1007/s00167-009-0853-6.

Bucki, M., Lobos, C., & Payan, Y. (2010). A fast and robust patient specific finite element mesh registration technique: Application to 60 clinical cases. *Medical Image Analysis, 14*(3), 303−317. Available from https://doi.org/10.1016/j.media.2010.02.003.

Cartiaux, O. (2016). The impact of accuracy of computer-aided orthopaedic surgery on patient-specific biomechanical outcomes. In *22nd Congress of the European Society of Biomechanics (ESB)* (pp. 2−3). Lyon, France. http://hdl.handle.net/2078.1/174700.

Cerveri, P., Sacco, C., Olgiati, G., Manzotti, A., & Baroni, G. (2017). 2D/3D reconstruction of the distal femur using statistical shape models addressing personalized surgical instruments in knee arthroplasty: A feasibility analysis. *International Journal of Medical Robotics and Computer Assisted Surgery, 13*(4). Available from https://doi.org/10.1002/rcs.1823.

Cho, H. S., Kang, H. G., Kim, H., & Han, I. (2008). Computer-assisted sacral tumor resection. A case report. *The Journal of Bone and Joint Surgery. American Volume, 90*, 1561−1566. Available from https://doi.org/10.2106/JBJS.G.00928.

Colle, F., Bignozzi, S., Lopomo, N., Zaffagnini, S., Sun, L., & Marcacci, M. (2012). Knee functional flexion axis in osteoarthritic patients: Comparison in vivo with transepicondylar axis using a navigation system. *Knee Surgery, Sports Traumatology, Arthroscopy, 20*(3), 552−558. Available from https://doi.org/10.1007/s00167-011-1604-z.

Colle, F., Bruni, D., Iacono, F., Visani, A., Zaffagnini, S., Marcacci, M., & Lopomo, N. (2016). Changes in the orientation of knee functional flexion axis during passive flexion and extension movements in navigated total knee arthroplasty. *Knee Surgery, Sports Traumatology, Arthroscopy, 24*(8), 2461−2469. Available from https://doi.org/10.1007/s00167-015-3816-0.

Colle, F., Lopomo, N., Bruni, D., Visani, A., Iacono, F., Zaffagnini, S., & Marcacci, M. (2014). Analysis of knee functional flexion axis in navigated TKA: Identification and repeatability before and after implant positioning. *Knee Surgery, Sports Traumatology, Arthroscopy, 22*(3), 694−702. Available from https://doi.org/10.1007/s00167-013-2780-9.

Colle, F., Lopomo, N., Visani, A., Zaffagnini, S., & Marcacci, M. (2016). Comparison of three formal methods used to estimate the functional axis of rotation: An extensive in-vivo analysis performed on the knee joint. *Computer Methods in Biomechanics and Biomedical Engineering, 19*(5), 484−492. Available from https://doi.org/10.1080/10255842.2015.1042464.

Duprez, M., Bordas, S. P. A., Bucki, M., Bui, H. P., Chouly, F., Lleras, V., . . . Tomar, S. (2018). Quantifying discretization errors for soft-tissue simulation in computer assisted surgery: A preliminary study. *Applied Mathematical Modelling, 77*, 709−723.

Elmi-Terander, A., Burström, G., Nachabé, R., Fagerlund, M., Ståhl, F., Charalampidis, A., . . . Gerdhem, P. (2020). Augmented reality navigation with intraoperative 3D imaging vs fluoroscopy-assisted free-hand surgery for spine fixation surgery: A matched-control study comparing accuracy. *Scientific Reports*, *10*(1), 707. Available from https://doi.org/10.1038/s41598-020-57693-5.

Fhkcos, Y. L., Fhkcos, K. M., & Ortho, F. (2010). Computer-assisted navigation in bone tumor surgery: Seamless workflow model and evolution of technique. *Clinical Orthopaedics and Related Research*, *468*, 2985−2991. Available from https://doi.org/10.1007/s11999-010-1465-7.

Grassi, A., Pizza, N., Lopomo, N. F., Marcacci, M., Capozzi, M., Maria, G., ... Zaffagnini, S. (2020). No differences in knee kinematics between active and passive flexion-extension movement: An intra-operative kinematic analysis performed during total knee arthroplasty. *Journal of Experimental Orthopaedics*, *7*, 12.

Grau, L, et al. (2019). Robotic arm assisted total knee arthroplasty workflow optimization, operative times and learning curve. *Arthroplasty Today*, *5*(4), 465−470. Available from https://doi.org/10.1016/j.artd.2019.04.007.

Hernandez, D., Garimella, R., Eltorai, A. E. M., & Daniels, A. H. (2017). Computer-assisted orthopaedic surgery. *Orthopaedic Surgery*, *9*(2), 152−158. Available from https://doi.org/10.1111/os.12323.

Joskowicz, L. (2017). Computer-aided surgery meets predictive, preventive, and personalized medicine. *EPMA Journal*, *8*(1), 1−4. Available from https://doi.org/10.1007/s13167-017-0084-8.

Koenig, J. H., et al. (2015). Available Robotic Platforms in Partial and Total Knee Arthroplasty. *Operative Techniques in Orthopaedics*, *25*(2), 85−94. Available from https://doi.org/10.1053/j.oto.2015.03.002.

Kubicek, J., Tomanec, F., Cerny, M., Vilimek, D., Kalova, M., & Oczka. (2019). Recent trends, technical concepts and components of computer-assisted orthopedic surgery systems: A comprehensive review. *Sensors (Basel)*, *19*, 1−33.

Leardini, A., Belvedere, C., Ensini, A., Dedda, V., & Giannini, S. (2013). *Accuracy of computer-assisted surgery. Knee surgery using computer assisted surgery and robotics* (pp. 3−20). Berlin Heidelberg: Springer. Available from https://doi.org/10.1007/978-3-642-31430-8_2.

Lopomo, N., Bignozzi, S., Martelli, S., Zaffagnini, S., Iacono, F., Visani, A., & Marcacci, M. (2009). Reliability of a navigation system for intra-operative evaluation of antero-posterior knee joint laxity. *Computers in Biology and Medicine*, *39*(3), 280−285. Available from https://doi.org/10.1016/j.compbiomed.2009.01.001.

Lopomo, N., Bignozzi, S., Zaffagnini, S., Giordano, G., Irrgang, J. J., Fu, F. H., & Marcacci, M. (2008). Quantitative correlation between IKDC score, static laxity, and pivot-shift test: A kinematic analysis of knee stability in anatomic double-bundle anterior cruciate ligament reconstruction. *Operative Techniques in Orthopaedics*, *18*(3), 185−189. Available from https://doi.org/10.1053/j.oto.2008.12.006.

Lopomo, N., Sun, L., Zaffagnini, S., Giordano, G., & Safran, M. R. (2010). Evaluation of formal methods in hip joint center assessment: An in vitro analysis. *Clinical Biomechanics*, *25*(3), 206−212. Available from https://doi.org/10.1016/j.clinbiomech.2009.11.008.

Lopomo, N., Zaffagnini, S., Bignozzi, S., Visani, A., & Marcacci, M. (2010). Pivot-shift test: Analysis and quantification of knee laxity parameters using a navigation system. *Journal of Orthopaedic Research*, *28*(2), 164−169. Available from https://doi.org/10.1002/jor.20966.

Ma, L., Zhao, Z., Zhang, B., Jiang, W., Fu, L., Zhang, X., & Liao, H. (2018). Three-dimensional augmented reality surgical navigation with hybrid optical and electromagnetic tracking for distal intramedullary nail interlocking. *The International Journal of Medical Robotics and Computer Assisted Surgery*, *14*(4), e1909. Available from https://doi.org/10.1002/rcs.1909.

Martelli, S., Zaffagnini, S., Bignozzi, S., Lopomo, N. F., Iacono, F., & Marcacci, M. (2007). KIN-Nav navigation system for kinematic assessment in anterior cruciate ligament reconstruction: Features, use, and perspectives. *Proceedings of the Institution of Mechanical Engineers, Part H: Journal of Engineering in Medicine*, *221*(7), 725−737. Available from https://doi.org/10.1243/09544119JEIM262.

Murphy, R. J., Armiger, R. S., Lepistö, J., Mears, S. C., Taylor, R. H., & Armand, M. (2015). Development of a biomechanical guidance system for periacetabular osteotomy. *International Journal of Computer Assisted Radiology and Surgery*, *10*(4), 497−508. Available from https://doi.org/10.1007/s11548-014-1116-7.

Nakano, N., Audenaert, E., Ranawat, A., & Khanduja, V. (2018). Review: Current concepts in computer-assisted hip arthroscopy. *International Journal of Medical Robotics and Computer Assisted Surgery*, *14*(6), 1−8. Available from https://doi.org/10.1002/rcs.1929.

Pailhé, R. (2021). Total knee arthroplasty: Latest robotics implantation techniques. *Orthopaedics and Traumatology: Surgery and Research*, *107*(1), 102780. Available from https://doi.org/10.1016/j.otsr.2020.102780.

Payan, Y. (2012). Biomechanics for computer-assisted surgery. *Computer Methods in Biomechanics and Biomedical Engineering*, *15*(Suppl. 1), 8−9. Available from https://doi.org/10.1080/10255842.2012.713595.

Pei, G. (2019). *Digital orthopedics*. *Digital orthopedics* (pp. 1−447). Springer. Available from https://doi.org/10.1007/978-94-024-1076-1.

Picard, F., Deakin, A. H., Riches, P. E., Deep, K., & Baines, J. (2019). Computer assisted orthopaedic surgery: Past, present and future. *Medical Engineering and Physics*, *72*, 55−65. Available from https://doi.org/10.1016/j.medengphy.2019.08.005.

Safran, M. R., Lopomo, N., Zaffagnini, S., Signorelli, C., Vaughn, Z. D., Lindsey, D. P., ... Marcacci, M. (2013). In vitro analysis of peri-articular soft tissues passive constraining effect on hip kinematics and joint stability. *Knee Surgery, Sports Traumatology, Arthroscopy*, *21*(7), 1655−1663. Available from https://doi.org/10.1007/s00167-012-2091-6.

Sato, R., Takao, M., Hamada, H., Sakai, T., Marumo, K., & Sugano, N. (2019). Clinical accuracy and precision of hip resurfacing arthroplasty using computed tomography-based navigation. *International Orthopaedics*, *43*(8), 1807−1814. Available from https://doi.org/10.1007/s00264-018-4113-6.

Schiavone, P., Boudou, T., Ohayon, J., & Payan, Y. (2012). In vivo measurement of the human soft tissues constitutive laws. Applications to computer aided surgery. *Computer Methods in Biomechanics and Biomedical Engineering*, *10*(Suppl. 1), 185−186. Available from https://doi.org/10.1080/10255840701479982.

Schneider, O., & Troccaz, J. (2001). A six-degree-of-freedom Passive Arm with Dynamic Constraints (PADyC) for cardiac surgery application: Preliminary experiments. *Computer Aided Surgery*, *6*(6), 340−351. Available from https://doi.org/10.1002/igs.10020.

Siegel, M. A., Balach, T., Sweeney, K. R., Nystrom, L. M., & Colman, M. W. (2020). Sacroiliac joint cut accuracy: Comparing new technologies in an idealized sawbones model. *Journal of Surgical Oncology*, *122*(6), 1218−1225. Available from https://doi.org/10.1002/jso.26124.

Siegler, S., Konow, T., Belvedere, C., Ensini, A., Kulkarni, R., & Leardini, A. (2018). Analysis of surface-to-surface distance mapping during three-dimensional motion at the ankle and subtalar joints. *Journal of Biomechanics*, *76*, 204−211. Available from https://doi.org/10.1016/j.jbiomech.2018.05.026.

Signorelli, C., Lopomo, N., Bonanzinga, T., Marcheggiani Muccioli, G. M., Safran, M. R., Marcacci, M., & Zaffagnini, S. (2013). Relationship between femoroacetabular contact areas and hip position in the normal joint: An in vitro evaluation. *Knee Surgery, Sports Traumatology, Arthroscopy*, *21*(2), 408−414. Available from https://doi.org/10.1007/s00167-012-2151-y.

Siston, R. A., Giori, N. J., Goodman, S. B., & Delp, S. L. (2007). Surgical navigation for total knee arthroplasty: A perspective. *Journal of Biomechanics*, *40*(4), 728−735. Available from https://doi.org/10.1016/j.jbiomech.2007.01.006.

Stavness, I., Lloyd, J. E., Payan, Y., & Fels, S. (2011). Coupled hard-soft tissue simulation with contact and constraints applied to jaw-tongue-hyoid dynamics. *International Journal for Numerical Methods in Biomedical Engineering*, *27*(3), 367−390. Available from https://doi.org/10.1002/cnm.1423.

Swiatek-Najwer, E., Bedzinski, R., Krowicki, P., Krysztoforski, K., Keppler, P., & Kozak, J. (2008). Improving surgical precision − Application of navigation system in orthopedic surgery. *Acta of Bioengineering and Biomechanics*, *10*(4), 55−62.

Tetsunaga, T., Yamada, K., Tetsunaga, T., Sanki, T., Kawamura, Y., & Ozaki, T. (2020). An accelerometer-based navigation system provides acetabular cup orientation accuracy comparable to that of computed tomography-based navigation during total hip arthroplasty in the supine position. *Journal of Orthopaedic Surgery and Research*, *15*(1), 147. Available from https://doi.org/10.1186/s13018-020-01673-y.

Vadalà, G., De Salvatore, S., Ambrosio, L., Russo, F., Papalia, R., & Denaro, V. (2020). Robotic spine surgery and augmented reality systems: A state of the art. *Neurospine*, *17*(1), 88–100. Available from https://doi.org/10.14245/ns.2040060.030.

Wang, M., & Ma, Y. (2018). A review of virtual cutting methods and technology in deformable objects. *International Journal of Medical Robotics and Computer Assisted Surgery*, *14*(5), 1–15. Available from https://doi.org/10.1002/rcs.1923.

Zaffagnini, S., Bignozzi, S., Martelli, S., Imakiire, N., Lopomo, N., & Marcacci, M. (2006). New intraoperative protocol for kinematic evaluation of ACL reconstruction: Preliminary results. *Knee Surgery, Sports Traumatology, Arthroscopy*, *14*(9), 811–816. Available from https://doi.org/10.1007/s00167-006-0057-2.

Zaffagnini, S., Signorelli, C., Bonanzinga, T., Lopomo, N., Raggi, F., Di Sarsina, T. R., ... Marcacci, M. (2016). Soft tissues contribution to hip joint kinematics and biomechanics. *HIP International*, *26*, 23–27. Available from https://doi.org/10.5301/hipint.5000407.

Zaffagnini, S., Urrizola, F., Signorelli, C., Grassi, A., Di Sarsina, T. R., Lucidi, G. A., ... Marcacci, M. (2016). Current use of navigation system in ACL surgery: A historical review. *Knee Surgery, Sports Traumatology, Arthroscopy*, *24*(11), 3396–3409. Available from https://doi.org/10.1007/s00167-016-4356-y.

Zheng, G., et al. (2007). Registration techniques for computer navigation. *Current Orthopaedics*, *21*(3), 170–179. Available from https://doi.org/10.1016/j.cuor.2007.03.002.

Zheng, G., Gollmer, S., Schumann, S., Dong, X., Feilkas, T., & González Ballester, M. A. (2009). A 2D/3D correspondence building method for reconstruction of a patient-specific 3D bone surface model using point distribution models and calibrated X-ray images. *Includes Special Section on Computational Biomechanics for Medicine*, *13*(6), 883–899. Available from https://doi.org/10.1016/j.media.2008.12.003.

Zheng, G., & Nolte, L. P. (2015). Computer-assisted orthopedic surgery: Current state and future perspective. *Frontiers in Surgery*, *2*, 6. Available from https://doi.org/10.3389/fsurg.2015.00066.

APPLICATIONS IN ORTHOPAEDIC BIOMECHANICS

EXPERIMENTAL ORTHOPEDIC BIOMECHANICS

Luigi La Barbera[1], Tomaso Villa[1,2], Bernardo Innocenti[3] and Fabio Galbusera[2]

[1]*Laboratory of Biological Structure Mechanics, Department of Chemistry, Materials and Chemical Engineering "Giulio Natta", Politecnico di Milano, Milan, Italy* [2]*IRCCS Istituto Ortopedico Galeazzi, Milan, Italy* [3]*Department of Bio Electro and Mechanical Systems (BEAMS), École Polytechnique de Bruxelles, Université Libre de Bruxelles, Brussels, Belgium*

EXPERIMENTAL TESTS AT THE ORGAN AND TISSUE LEVELS

This section presents an overview of the experimental tests suitable for characterizing the mechanical properties of hard, soft, biphasic, and poroelastic tissues at different scales.

HARD TISSUES

Mineralized tissues, like bone (Chapter 7), have a deeply organized hierarchical structure. Depending on the length scale, a variety of test methods can be used to characterize the mechanical and structural properties of the bones from macro- to nano-scale (Bailey & Vashishth, 2018; Hunt & Donnelly, 2016; Zysset, 2009).

Quasi-static macroscopic tests on whole bones

Long bones are typically tested using simple bending and torsional loads applied using established *testing machines*. Most machines can apply a *uniaxial* load/displacement and measure the resulting displacement/load (depending on the transducers and sensors built into the machine), where the load can be the force of a bending moment and displacement can be either a translation or a rotation. More complex testing machines can apply *multiaxial* loads/displacements, combining two or more loading modes (i.e., axial and torsional).

Numerous studies investigated the mechanical behavior of human long bones (i.e., femurs, tibias) to assess their macroscopic mechanical properties (Leppänen, Sievänen, & Järvinen, 2008; Martens, van Audekercke, de Meester, & Mulier, 1980, 1986; Papini, Zdero, Schemitsch, & Zalzal, 2007). Some author characterized the static strength of healthy and osteoporotic cadaveric human bones through quasi-static and impact ultimate failure strain tests on established testing machines while describing a variety of loading conditions, including simple flexion and torsion, but also more complex loadings (i.e., sideways fall of the proximal femur) (Bahaloo et al., 2018; Dragomir-Daescu et al., 2018; Jazinizadeh, Mohammadi, & Quenneville, 2020; Zani, Erani, Grassi, Taddei, & Cristofolini, 2015). Typical extrinsic mechanical properties can be analyzed by recording the force versus displacement curve: the structural stiffness (i.e., the slope of the linear portion of the curve); the peak load; the propagation to fracture (i.e., the area below the force—displacement curve), a measure of energy absorption; and the fracture load (i.e., the load associated with the onset of macroscopic cracks, leading to bone fracture).

Standardized synthetic bones, widely used in experimental orthopedic biomechanics, have also been developed to mimic the geometrical features and the mechanical response of the human bones (Cristofolini, Viceconti, Cappello, & Toni, 1996; Girardi et al., 2016; Papini et al., 2007), to avoid the problems related to reproducibility, availability, and conservation. Whole bone tests can provide important data useful in validating specimen-specific computational techniques capable of predicting fracture load and location under several loading conditions, and simulating sideways fall under impact and quasi-static conditions (Franceschini et al., 2020; Peleg et al., 2010; Schileo, Taddei, Malandrino, Cristofolini, & Viceconti, 2007).

Surface strain analysis on whole bones

Strain gauges (SGs) are a common strain transducer, based on a resistive grid, which is glued to a component. The gauge factor allows us to transduce a variation in length to a variation in electric resistance, which can be measured as a differential in the voltage across the grid. Uniaxial SGs have only one grid and are typically affected by misalignment issues; *SG rosettes* have three resistive grids (Fig. 28.1), allowing for the calculation of principal strain components and the orientation of the principal loading directions. SGs are used to assess the strain on cortical bony structures,

(A) Standard test on uniaxial testing machine

(B) In vitro test on spine simulator

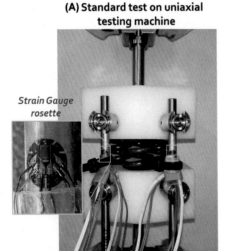

Strain Gauge rosette

FIGURE 28.1

Examples of two experimental set-ups to test the mechanical behavior of a posterior spine fixator according to standards on a uniaxial testing machine (A) and after implantation in a cadaveric lumbar spine segment tested on a 6 degrees-of-freedom spine simulator (B). Strain gauge rosettes, which consist of three resistive grids with a 45 degree angle in between, allowed direct measurement of the local principal strains on the spinal implant. Similar methods and measurement techniques are widely used in the field of orthopedic biomechanics.

such as long bones, during simulated stance and quasi-static sideways fall (Cristofolini et al., 2010; Cristofolini, Juszczyk, Taddei, & Field, et al., 2009; Zani et al., 2015) and the pelvic strain pattern under pseudo-physiological conditions (Ries, Pugh, Choy Au, Gurtowski, & Dee, 1989). SGs have also been applied on pelvic bones (Ries et al., 1989), on the cortical shell of human vertebrae, to study their capacity to bear compressive, tensile, and combined shear loadings (Cristofolini et al., 2013; Hongo et al., 1999; Shah, Hampson, & Jayson, 1978), and on the posterior lamina to study pars interarticularis fracture under asymmetric loadings (Bright, Tiernan, McEvoy, & Kiely, 2017, 2018). Such technology can only be used in the linear elastic strain range (small deformation) and on bone samples with adequate stiffness and thickness, but not to cause reinforcement effects related to the presence of the SG on the specimen. SGs provide a point-wise average measurement of the local strains (i.e., below the resistive grids); therefore, the size of the grid controls the resolution of the strain measurement. Local SG measurements can be very useful in calibrating and/or validating dedicated finite-element models by correctly describing the mechanical behavior of the tissue of interest (Ottardi, La Barbera, Pietrogrande, & Villa, 2016).

Digital image correlation (DIC) allows for a full-field noncontact evaluation of superficial displacements/rotations and strains of a component (Palanca, Tozzi, & Cristofolini, 2016; Sutton, Orteu, & Schreier, 2009). A random speckle pattern is typically obtained on the specimen to be tested (Lionello & Cristofolini, 2014); one or more cameras follow the pattern over time, and a cross-correlation algorithm is used to postprocess the images, to recognize specific areas, to track their displacements, and to calculate the strains over time (Palanca et al., 2016). The speckle pattern typically controls the resolution and the precision of the measurement, which also rely on postprocessing (Lionello & Cristofolini, 2014). Compared to point-wise SG analysis, DIC techniques allow for the full-field analysis over a wider and continuous surface, covering either small or large deformations. Since standard tests of whole-bone samples on uni- and multiaxial testing machines can only appreciate a macroscopic behavior, integration with DIC can provide useful insights into the initiation site and into the propagation of the fracture while continuously monitoring the strain field on the bone. DIC analysis has been used to integrate mechanical testing on a variety of human and animal tissues and organs (Palanca et al., 2016), in particular to assess the mechanical behavior of vertebrae (Palanca, Brugo, & Cristofolini, 2015; Palanca, Marco, Ruspi, & Cristofolini, 2018; Ruspi, Palanca, Faldini, & Cristofolini, 2017), pelvis (Karanika et al., 2016; Kourkoulis, Darmanis, Papadogoulas, & Pasiou, 2017), and long bones (Grassi et al., 2020).

Compression and tensile tests on bone samples

Compression and *tensile* tests are typically done on standard testing machines and allow us to characterize the apparent elastic modulus, strength of trabecular and cortical bone samples at the macro-scale, as well as time-dependent viscoelastic properties and fracture toughness (Burgers, Mason, Niebur, & Ploeg, 2008; Gong, Wang, Fan, Zhang, & Qin, 2016). Trabecular specimens are rather porous, heterogeneous, and anisotropic, with a preferential orientation of the trabeculae along the principal direction of loadings (i.e., Wolf's law) (Goldstein, 1987; Zhai, Nauman, Moryl, Lycke, & Chen, 2020); therefore, the dimension of the specimens, their porosities, and their bone mineral density significantly affect the resultant mechanical response (Carter, Schwab, & Spengler, 1980; Linde, Hvid, & Madsen, 1992; Nazarian, Muller, Zurakowski, Müller, & Snyder, 2007). The compressive mechanical response presents a plateau, typical of cellular solids, undergoing a progressive failure and compaction of the trabeculae. The tensile response is, instead, quite brittle with almost coincident yielding and ultimate properties (Bayraktar et al., 2004; Nazarian, Araiza Arroyo, Rosso, Aran, & Snyder, 2011). Both trabecular and cortical samples demonstrate an asymmetric mechanical response in tension and compression (Garcia, Zysset, Charlebois, & Curnier, 2009; Li, Demirci, & Silberschmidt, 2013). The mechanical properties of bone samples are known to be dependent on the strain rate (Linde, 1994; Linde, Nørgaard, Hvid, Odgaard, & Søballe, 1991) and vary significantly if the samples are tested in a wet or dry environment (Frank et al., 2018).

Indentation tests on bone samples

Indentation tests at the macro-, micro- and nano-scales allow for the mechanical characterization of elastic and inelastic properties of bone samples from the tissue scale down to the scale of the osteon, lamellae, and mineralized collagen fibril (Hoffler, Guo, Zysset, & Goldstein, 2005; Zysset, 2009). The tip shape (i.e., sphere [Brinell, Rockwell]; four-sided [Vickers] or three-sided [Berkovich] pyramidal; asymmetric pyramidal [Knooand]) and the indentation depth define the volume of material participating in the mechanical response, therefore affecting the scale length of the structure being tested (Guidoni, Swain, & Jäger, 2010; Paietta, Campbell, & Ferguson, 2011; Zysset, 2009).

The load–penetration depth curve provides important information (Oliver & Pharr, 2004): hardness (related to the penetration depth and correlated with density and tissue mineralization), indentation modulus (calculated as the slope of the unloading curve), and inelastic phenomena often related with damage (i.e., crack initiation and propagation). As the analytical formulae used to calculate specific parameters are based on the nominal area of the indenter on the specimen, it is often essential to correct some terms related to the effective areas to provide a more accurate estimate. Fracture toughness can also be quantified by measuring the extension of the crack propagation during the indentation test.

Indentation has been used to determine the toughness of bone samples (Gallant, Brown, Organ, Allen, & Burr, 2013) and to determine cortical bone elastic and inelastic properties. Viscoelastic time-dependent properties (Bandini et al., 2013; Carnelli, Vena, Dao, Ortiz, & Contro, 2013; Casanova et al., 2017) can also be quantified by repeating multiple indentation tests at different strain-rates or by keeping a long holding time (i.e., creep) at maximum load on the same spot (Fan & Rho, 2003; Isaksson et al., 2010). Nanoindentation tests have also been coupled with numerical models to characterize the role of damage mechanics at the lamellar level (Lucchini et al., 2011). Moreover, nanoindentation tests can provide relevant information on the microstructural composition of the bone down to its components (i.e., collagen fibers), which can be of interest to better understand growth (Lefèvre et al., 2019), osteoporosis (Ibrahim et al., 2020), and aging (Abraham, Agarwalla, Yadavalli, Liu, & Tang, 2016).

SOFT TISSUES

The vertebrate body includes several types of nonmineralized tissues, commonly known as soft tissues, with very diverse functions: connective tissues, which include, among others, ligaments and tendons (Chapter 9), cartilage (Chapter 10), meniscus (Chapter 11), and intervertebral discs (Chapter 12); epithelial tissue; nervous tissue; and muscle tissue (Chapter 8). All these nonmineralized soft tissues have been characterized from a mechanical point of view by means of experimental methods, which have been designed and optimized to allow for an accurate assessment of their peculiar material properties.

Quasi-static uniaxial and biaxial tensile tests

Soft tissues with a predominantly uniaxial nature such as tendons and several ligaments are commonly characterized by means of quasi-static tensile testing, which is typically performed with standard material testing systems. The output of such tests is the stress—strain curve, from which relevant biomechanical properties such as the elastic modulus, the yield strain, and the ultimate failure strain can be determined. Several challenges are associated with these tests, including achieving a proper gripping of the specimen, maintaining a physiological hydration, and measuring the cross-sectional area and the initial length; these issues and the relative solutions are described in detail in Chapter 9.

Tissues with a membrane-like rather than a string-like shape can be tested biaxially, that is, with devices able to stretch the specimen in two directions or to stretch in one direction while preventing retraction in the orthogonal direction (Lanir & Fung, 1974). Such devices have been used to test several connective tissues, including skin (Schneider, Davidson, & Nahum, 1984), pericardium (Chew, Yin, & Zeger, 1986), and ligaments (Claeson & Barocas, 2017).

Viscoelastic characterization in uniaxial testing

Biological soft tissues show a time-dependent mechanical response, which depends on their intrinsic viscoelastic properties as well as on their water content, which can vary during tissue deformation due to exudation or imbibition. Specimens subjected to uniaxial testing therefore show properties dependent on the loading rate, as well as energy dissipation, which can be observed as a hysteresis of the stress—strain curve under cyclic loading (Chapter 9). Studies on tendons and ligaments showed that high strain rates are associated with higher chances of tissue failure, whereas high strains applied at low rates more frequently determine the failure of the tissue insertion in bone (Crowninshield & Pope, 1976; Woo, Debski, Withrow, & Janaushek, 1999).

The time-dependent behavior can be characterized by performing two types of uniaxial tests, that is, creep and relaxation. In creep tests, a constant tensile load is applied to the specimen, and the elongation over time is measured; conversely, in relaxation testing, keeping the specimen length constant, the force necessary to maintain the length over time is measured. Experimental results obtained with creep and relaxation tests can be used to fit mathematical models of the viscoelastic response of the tissue, such as the quasilinear viscoelastic theory (Woo, 1982).

Contraction force, velocity, and length of skeletal muscle

In addition to the passive properties, which can be measured by means of standard uniaxial and biaxial tests, as done with ligaments and tendons, muscle tissue can also be subjected to experimental testing to determine its contraction capabilities. In general, such properties are investigated by suspending the specimen in a physiological solution, keeping constant the specimen length (isometric conditions) or the force applied to its ends by means of an actuator (isotonic conditions) while ensuring that the sample is subjected to a physiological pretension, and, finally, inducing contraction by using an electrode (Zdero, Borkowski, & Coirault, 2017). This experimental set-up allows performing two distinct types of tests: the "twitch" test in which a single, brief electrical excitation is applied and the "tetanus" test, which uses multiple pulses to determine a sustained contraction of the muscle. While the "twitch" test is normally performed only under isometric conditions, thus resulting in a force versus time curve, the "tetanus" test can be conducted with both isometric and isotonic constraints also allowing obtaining the length versus time as well as the velocity versus time curves (Chapter 8).

The experimental testing of muscle specimens shares many technical complexities and solutions with that of ligaments and tendons, such as achieving a sufficient gripping, measuring the specimen length and cross-sectional area, and ensuring a proper hydration and biochemical environment. The latter is made even more challenging by the fact that the specimen needs to be in equilibrium biochemically and that sufficient oxygen supply needs to be provided; the contractile muscle properties and, in turn, the measurements would be negatively affected if the test was performed under hypoxic conditions (Howlett & Hogan, 2007; Zdero et al., 2017).

Digital image correlation

DIC has been recently used to investigate the loadings on the surface of soft tissues, such as ligaments and tendons. In particular, the strain on the anterior longitudinal ligament, as well as the intervertebral disc in multiple vertebrae specimens, has been investigated applying quasi-static multidirectional loadings (Fig. 28.2) (Palanca et al., 2020; Ruspi et al., 2020). Due to the intrinsic viscoelastic behavior of the soft tissues involved, preconditioning is crucial to achieve a stable and repeatable mechanical response.

BIPHASIC CHARACTERIZATION OF SOFT TISSUES

In orthopedics, several soft tissues such as articular cartilage (Chapter 10), knee meniscus (Chapter 11), and intervertebral disc (Chapter 12) exhibit a biphasic nature characterized by a

FIGURE 28.2

Example of a complex test set-up for in vitro testing of a cadaveric lumbar spine segment based on a state-of-the-art spine tester integrated with DIC cameras and light sources to analyze the mechanical behavior of the anterior longitudinal ligament, vertebrae, and intervertebral discs. Details of the *white-on-blue* pattern obtained on the ventral spine are also shown, with the correlated area obtained after postprocessing.

high-water content and an evident time-dependent mechanical response. Since such tissues do not have shapes suitable for standard uniaxial testing, testing set-ups based on compression and indentation have been developed for their characterization.

Confined compression

One of the most common approaches to characterizing samples of cartilage, intervertebral disc, and meniscus is to subject them to creep testing in compression in a confined chamber. A cylindrical plug is typically extracted from the specimen, placed in a rigid chamber having the same diameter of the sample, and subjected to a constant compression force applied through a porous plate (Belkoff & Haut, 2008; Fung, 2013). During the test, part of the water content of the sample is exudated through the porous plate, determining a reduction of the height of the plug. The exudation continues throughout a transitory phase, the length of which depends on the permeability of the tissue. After a steady state is achieved, the aggregate modulus, which describes the stiffness of the material after the fluid outflow has terminated, can be calculated from the equilibrium stress–strain response. In order to calculate other biomechanical properties such as elastic modulus, Poisson ratio, and permeability, the measured creep response, or a part of it, can be fitted to a theoretical model describing the response of the tissue by means of standard numerical methods (Mow, Kuei, Lai, & Armstrong, 1980).

Unconfined compression

Compression tests conducted under unconfined conditions, that is, without the use of a loading chamber, which prevents lateral displacements, can also be used to characterize samples of biphasic soft tissues. Unconfined compression tests are performed by putting the specimen between two highly polished plates, which should theoretically allow for unconstrained radial deformation, and subjecting it to a constant strain (therefore realizing a relaxation test), or to a constant load (creep test). The time-dependent response is then fitted to the appropriate theoretical model for the calculation of the biomechanical properties of the tissue (Armstrong, Lai, & Mow, 1984; Hayes, 1997). Achieving a successful fit in unconfined compression appears to be more challenging than in confined compression, due to limitations in the most common analytical models regarding the simplified behavior of the collagen fibrils (Brown & Singerman, 1986). Although sophisticated models overcoming these limitations have been presented (Li, Soulhat, Buschmann, & Shirazi-Adl, 1999; Soltz & Ateshian, 2000), confined compression is commonly preferred to unconfined testing since it allows achieving satisfying results without the need for any complex numerical computation.

Dynamic mechanical analysis

An alternative technique to characterize the time-dependent behavior of biphasic soft tissues is to subject them to a sinusoidal load in a material testing system. By measuring the phase lag between the load and the displacement response, the viscoelastic properties can be assessed: purely elastic solids show a perfect in-phase behavior, ideal for viscous fluids with a 90-degree phase lag, while viscoelastic materials have an intermediate response. The measured phase lag angle is then used to calculate two parameters commonly used to describe viscoelastic materials, i.e., the storage modulus and the loss modulus, which represent the stored energy (elastic contribution) and the energy dissipated in form of heat (viscous contribution), respectively. Dynamic mechanical testing has been used to investigate cartilage (Cooke, Lawless, Jones, & Grover, 2018; Temple, Cederlund, Lawless, Aspden, & Espino, 2016) as well as bone-cartilage specimens (Lawless et al., 2017).

Indentation

Indentation has also been used to characterize biphasic soft tissues, within the same theoretical framework and by means of similar techniques to those used for characterizing bone samples; theoretical models purposely developed for soft tissues have also been reported (Hayes, Keer, Herrmann, & Mockros, 1972; Mak, Lai, & Mow, 1987). Depending on the chosen model, the calculation of the elastic properties may require the knowledge of the Poisson ratio, which is relatively variable and time-dependent in hydrated tissues; its value can be either assumed based on literature values or measured by means of special techniques (Jin & Lewis, 2004). Another required parameter is the thickness of the tissue, which can be measured experimentally using needle probes (Mow, Gibbs, Lai, Zhu, & Athanasiou, 1989). Indentation has been used to characterize the elastic properties of articular cartilage (Mow et al., 1989), knee meniscus (Fischenich, Lewis, Kindsfater, Bailey, & Haut Donahue, 2015), and intervertebral disc (Umehara et al., 1996). Published results show some inconsistencies with data from confined and unconfined compression studies, which have been attributed to the finite indenter size (Korhonen et al., 2002).

EXPERIMENTAL TESTS ON IMPLANTS AND PROSTHESES

This section is focused on the methodologies for characterizing the functional response of implants once inserted in an environment representative of the in vivo usage. In vitro tests could be used on cadaveric specimens, while several dedicated transducers and sensors could provide valuable information.

IN VITRO TESTING

In vitro cadaveric testing on human and animal tissue specimens is often used in experimental orthopedic biomechanics to analyze the behavior of implants and prostheses when interacting with tissues and organs under simplified controlled conditions. This is particularly useful when a specific loading mode, either static or dynamic, has been identified as crucial.

Resistive films: contact pressure

Pressure-sensitive films, such as TekScan sensors (TekScan, Inc., South Boston, MA, USA) or Fujifilms, can be used to record joint stresses under static and dynamic loading conditions. While Fujifilm Prescale Films are single-use, pressure-sensitive films can measure pressure (at different scales) through peak pressure snapshots, capturing a pressure profile via a color scale and revealing a pressure distribution. TekScan sensors are reusable sensors that allow direct measurement of interfacial contact area, contact pressure, and also contact point trajectory. Their application is quite broad in the industry, involving robotic, automotive, ergonomic, and biomedical applications (Wang, Chen, & Innocenti, 1992). In regard to biomechanical orthopedic applications, TekScan sensors were quite commonly used to measure changes in joint contact mechanics in healthy, pathologic, and artificial joints under quasi-static and dynamic loading conditions (Catani et al., 1986; Didden et al., 2010; Luyckx et al., 2009; van Jonbergen, Innocenti, Gervasi, Labey, & Verdonschot, 2012). As a result of the nonlinear behavior of the sensor, the conditioning, normalization, and calibration of the sensor are crucial to achieving correct measurements (Herregodts, Baets, Victor, & Verstraete, 2015). To guarantee the repeatability of the measurements, a proper sensor fixation in the joint is required. Moreover, special attention should be paid to the methodology (i.e., calibration) as it has a strong effect on the overall accuracy of the device. In the literature, a reduction of the accuracy from 3% up to 50% is reported as a result of the aforementioned effects (Herregodts et al., 2015).

Pressure sensors

Pressure sensors are particularly effective in measuring the hydrostatic pressure of hydrogel-like and liquid tissues like the nucleus pulposus of the intervertebral disc. They consist of a subtle needle with a small sensitive area on their tip, transducing the deflection of a thin layer, which is instrumented with miniaturized SGs, once exposed to an inward pressure (McNally, Adams, & Goodship, 1992). Such technology is rather established in spine biomechanics, and it has been successfully used to monitor the spine loadings on healthy subjects in vivo in a variety of functional tasks either static (i.e., relaxed standing, sitting, etc.) or dynamic (i.e., walking, climbing stairs, etc.) (Nachemson & Morris, 1963; Wilke, Neef, Caimi, Hoogland, & Claes, 1999). The same approach has been used during in vitro tests to investigate muscle contributions on spinal loads (Wilke, Wolf, Claes, Arand, & Wiesend, 1996) and on healthy and degenerated spine specimens (Wilke, Rohlmann, Bergmann, Graichen, & Claes, 2001), and how load sharing across the intervertebral structures is affected by degenerative changes (Adams, McNally, & Dolan, 1996). Pressure sensors can also provide valuable information on the stress-shielding phenomena occurring after posterior fixation with rigid and dynamic implants (Schilling et al., 2011; Schmoelz, Huber, Nydegger, Claes, & Wilke, 2006).

Linear variable differential transformer: micromotions at the bone—implant interface and implant migration

A linear variable differential transformer (LVDT) can translate a variation in length into an electric signal (i.e., voltage, current). They are based on magnets, where the magnetic coupling between two solenoids changes depending on their relative position, inducing a variable current in the secondary circuit. LVDTs are typically mounted on common testing machines, where they are used to monitor the position of the actuator applying the load, thus allowing to quantify the deflection of the specimen being tested and quantify its structural stiffness. LVDTs are often used in experimental orthopedic biomechanics to also monitor the relative position of one component compared to another (Newell, Carpanen, Grigoriadis, Little, & Masouros, 2019). The same principle can be applied to measure the relative positions of two parts, such as the displacement of an implant compared to the bone where it is implanted (Bieger, Ignatius, Reichel, & Dürselen, 2013; Fottner et al., 2017; Howie et al., 2007; Solitro, Whitlock, Amirouche, & Santis, 2016): this allows the investigation of the relative micromotions (i.e., migration) of the implants inside the bone, essential to quantify the primary stability of a prosthesis. Since an LVDT only provides a pointwise measurement, surface strain analyses are nowadays more commonly used to gather more detailed information over a wider area.

Surface strain analysis

SG technique is used to evaluate the local strain distributions on the bony structures before and after the implantation of the prostheses. Such an approach allows investigating stress-shielding phenomena following hip prosthesis implantation (Cristofolini et al., 2010; Cristofolini, Juszczyk, Taddei, & Viceconti, 2009; Zani et al., 2015; Finlay, Bourne, Landsberg, & Andreae, 1986; Miles & McNamee, 1989), as well as the deformation of the pelvis during insertion of press-fit acetabular cups (Kroeber et al., 2002), and the load sharing on the tibia following unicompartmental and total knee prostheses (Completo, Fonseca, & Simões, 2008; Ivarsson & Gillquist, 1992). SG is also adopted in orthopedic biomechanics to measure the strains locally supported by full-arch dental prostheses supported by multiple implants (Francetti et al., 2015), as well as in spine biomechanics to measure strains on implants such as the intervertebral cage (Kettler et al., 2006), interspinous devices (Kettler et al., 2008), and spinal fixators (Fig. 28.1) (La Barbera & Villa, 2016, 2017; La Barbera et al., 2018, 2020, 2021; La Barbera, Ottardi, & Villa, 2015; Villa, La Barbera, & Galbusera, 2014a, 2014b; La Barbera, 2018). This experimental information can be very useful to build, calibrate, and/or validate specific predictions of finite element models (La Barbera & Villa, 2016, 2017; La Barbera et al., 2015; Villa et al., 2014a, 2014b).

DIC is also used in orthopedic biomechanics for several applications such as the quantification of the relative micromotions of bone fragments following proximal humeral fixation with plates and intramedullary nails (Mathison et al., 2010), and the micromotions at the tibial plateau in different designs of primary and revised total knee arthroplasties (Small et al., 2016). The same approach allows us to study the effect of implant position on bone strain after knee replacement (Ali, Newman, Hooper, Davies, & Cobb, 2017), and the periacetabular bone strain following the insertion of press-fit acetabular cup (Ghosh, Gupta, Dickinson, & Browne, 2012). The main advantage of the DIC approach is providing a full-field continuous displacement and strain distribution over a much wider area compared to pointwise SG measurements. More recent studies coupled DIC analysis on the anterior spine and SG analysis on the posterior spinal fixator to quantify the load sharing on the ventral spine following posterior spinal fixation of highly unstable spinal osteotomies (La Barbera & Villa, 2016, 2017; La Barbera et al., 2015, 2018, 2020, 2021; Villa et al., 2014a, 2014b).

IN VIVO LOADS ON IMPLANTS

Measurement of loads acting in orthopedic joints and bony structures can be performed by means of instrumented orthopedic prostheses. In such devices, strain sensors are integrated into the body of the device. Patients carrying the device are asked to perform different daily activities, and the corresponding signals from the sensors are transmitted to an external data recorder unit by means of a telemetry circuit. The circuit, which comprises an alimentation source and an antenna, is generally located inside a housing designed to withstand external magnetic interactions, which can affect the accuracy of the final measurement, as well as to protect the electronic hardware from body fluids.

Hip prosthesis

Hip prostheses (Chapter 20) were the first devices to be instrumented with systems able to measure physiological loads. Apart from pioneering studies that made use of percutaneous cables to transmit measurements (Rydell, 2014), the first example of an instrumented hip prosthesis based on a telemetry system was proposed by Carlson, Mann, and Harris (1974) in order to measure cartilage surface pressure on the femoral head. In the late 80s, two groups (Bergmann, Graichen, Siraky, Jendrzynski, & Rohlmann, 1988; Davy et al., 1988) proposed a second-generation instrumented hip prosthesis able to measure the three spatial components of the loads acting on the device: additional sensors were also used to measure the temperature rise in the device. However, it was on the third generation of instrumented prostheses able to measure both loads and moments (Bergmann et al., 2001) that the foundations for building a wide database of measurements of various loads were laid: such a database is available today (http://www.orthoload.com) and includes a very large number of daily activities (standing, walking, stair climbing, weight lifting, chair sitting and standing up, etc.) for which the plots and data of the loads and moments acting in the three directions at the joint level are available.

Knee prosthesis

The models of instrumented knee prostheses (Chapter 21) were first proposed by D'Lima (D'Lima, Patil, Steklov, Slamin, & Colwell, 2006; D'Lima, Townsend, Arms, Morris, & Colwell, 2005) and Bergmann's group (Heinlein, Graichen, Bender, Rohlmann, & Bergmann, 2007; Kutzner et al., 2010): also, a standard prosthesis (tibial tray) was modified adding a measuring system based on SG technique, an alimentation source, and a transmitter. The differences between the two proposed devices lie in the positioning of the SGs: D'Lima initially put the sensors between two layers of the tibial tray to measure the distribution of the compressive loads on the four quadrants of the tibial tray; Heinlein and Kutzner chose the inner part of the tibial stem as the housing for the measuring tool; later Kirking proposed a revised version of the D'Lima device (Kirking, Krevolin, Townsend, Colwell, & D'Lima, 2006) where the strain sensors were also positioned, like Heinlein's solution, inside the tibial stem: the advantage of this configuration is the possibility of measuring the three components of both loads and moments, which was not possible in the earlier configuration by D'Lima. Like hip prosthesis, data related to knee prosthesis are also freely available (http://www.orthoload.com).

Spinal fixator and vertebral body replacement

For instrumented spinal implants (Chapter 23), two applications have been investigated, namely instrumented spinal fixators and instrumented vertebral body replacement (VBR) devices. The former was designed in 1994 (Rohlmann, Bergmann, & Graichen, 1994), and results of loads and moments acting on the device for different daily activities were collected and presented in the following years (Rohlmann, Arntz, Graichen, & Bergmann, 2001; Rohlmann, Bergmann, & Graichen, 1997; Rohlmann, Bergmann, & Graichen, 1999; Rohlmann, Graichen, & Bergmann, 2000): the device basically consists of a modified version of posterior bars where a measuring cartridge was integrated into the longitudinal rod containing six load sensor SGs and the alimentation/transmitting system. The instrumented VBR is a more recent innovation and consists of a commercially available vertebral body replacement device modified with a SG system able to measure the three force components and three moments acting on the implant (Rohlmann, Gabel, Graichen, Bender, & Bergmann, 2007): also, many different activities have been investigated (Rohlmann, Dreischarf, Zander, Graichen, & Bergmann, 2014; Rohlmann, Zander, Graichen, & Bergmann, 2013; Rohlmann, Zander, Graichen, Dreischarf, & Bergmann, 2011) and results are freely available (http://www.orthoload.com). Quantitative information about implant loading during different every day life activities can be very useful to build, calibrate, and/or validate computational models (La Barbera, Galbusera, Wilke, & Villa, 2016; La Barbera, Galbusera, Wilke, & Villa, 2017).

JOINT SIMULATORS

Compared to standard testing machines, which offer only limited degrees of freedom, joint simulators are machines specifically designed to reproduce the peculiar kinematics and dynamics of a specific joint. Specific research groups made significant efforts in designing and developing complex simulators for the major joints such as the hip (Chapter 13), the knee (Chapter 14), and the spine (Chapter 19).

HIP

Hip joint simulators have a long history of success, and their design and usage have been particularly focused on the evaluation of the wear of hip prostheses (Ali et al., 2016). In such machines, hip prosthesis articular components (namely the femoral head and acetabular liner/cup) are simultaneously subjected in long-duration tests (up to 5 million cyles) to the movements acting in the hip joint (abduction/adduction, internal/external rotation, and flexion/extension) and to a vertical load mimicking the bodyweight of the patient. In order to obtain proper lubricating conditions affecting wear debris production, the specimen is kept during the test in an environmental chamber at 37°C where it is immersed in a specific fluid with lubricating properties similar to that of the pseudo-synovial fluid acting in vivo after the implant. After the test, proper experimental methodologies are performed to measure the debris due to wear produced during the test.

The difference between several typologies of hip simulators developed over the years include the following:

- *Mechanical principle of operation*: simulators can be moved by means of hydraulic or electromechanical actuators.
- *Number of available stations*: older simulators were permitted to test only one specimen at a time (Dowson, Walker, Longfield, & Wright, 1970; Saikko, 1992), but since wear tests are very time-consuming (5 million cycles correspond approximately to a 60-day-long test), it is very useful to have the possibility of performing tests simultaneously on several specimens in a multistation machine available with 3 (Saikko, Ahlroos, Calonius, & Keränen, 2001), 5 (Smith & Unsworth, 2001), 6 (Kaddick & Wimmer, 2001), 10 (Barbour, Stone, & Fisher, 2000), or 12 (Affatato, Leardini, Jedenmalm, Ruggeri, & Toni, 2007; Essner, Sutton, & Wang, 2005) stations.
- *Number of degrees-of-freedom*: single-axis machines where only flexion—extension is simulated are the simplest and oldest (Dumbleton, Miller, & Miller, 1972), two-axis machines can simulate either flexion/extension and abduction/adduction (Goldsmith & Dowson, 1999) or flexion/extension and internal/external rotation (Saikko & Calonius, 2002), three-axis simulators are able to mimic all the three components of movement present in a natural hip joint (Saikko, 1996), thus representing most of the natural kinematics of the joint (Saikko & Calonius, 2002).

- *Load profile*: the standard load profile is specified in ISO14242 (ISO 14242:2014, 2014), which is the guideline researchers follow to perform standardized wear tests on hip simulators. As per the standard, a load profile with a double-peak load of 3000 N is proposed, while the flexion/extension angular movement ranges between 25 and −18 degrees, the abduction/adduction between 3 and −4 degrees, and the internal/external rotation between 2 and −10 degrees. (All the three movements have a sinusoidal pattern.) Despite this standardization, several research works have been conducted to assess the appropriateness of such a profile to reproduce a wear behavior representative of the in vivo one (Schwenke, Schneider, & Wimmer, 2006) or to evaluate different profiles derived from gait analysis measurements on patients carrying hip prostheses (Saikko & Calonius, 2002). Another discussion, not yet concluded, is whether wear tests must be performed replicating only low-intensity loading conditions but repeated for a high number of cycles (e.g., walking activity) or also including high-intensity activities (e.g., stairs climbing, squatting or stumbling) for a smaller number of cycles.

Although the hip simulators have been extensively used for the evaluation of wear, they have also been used to investigate different biomechanical aspects of artificial hip joint: evaluation of other mechanical phenomena (Ali et al., 2017) or device performances (Herrmann et al., 2015), insight studies on mechanical parameters related to wear debris production (Leslie et al., 2008; Nečas, Vrbka, Urban, Křupka, & Hartl, 2016; Wang, Essner, & Klein, 2001), and investigation of specific wear mechanisms (Affatato et al., 2002).

KNEE

The study of the biomechanics of the knee and, in particular, how the femur moves on the tibia started in 1836 with Weber and Weber (1992). They were the first researchers to describe the 2D movement on the medial side to be "like a cradle." Since then, several methods have been used to examine the kinematics of the human knee exploiting the advances in technology available to clinical research. As a matter of fact, recent studies have shown that the native human knee does not behave like a four-bar linkage, which had been considered an adequate description of its kinematics for a long time, but has a more complex 3D motion during both loaded and passive flexion–extension cycles (Iwaki, Pinskerova, & Freeman, 2000; Victor, Labey, Wong, Innocenti, & Bellemans, 2010).

Among the different approaches used in this field, knee joint simulators were one of the most common systems used by different researchers to investigate knee joint biomechanics during different activities. These simulators could be used to properly investigate not only 3D knee joint kinematics, in terms of both rotations and translations, but also tibio-femoral kinematics (Heesterbeek, Labey, Wong, Innocenti, & Wyemnga, 2012; Heyse et al., 2014; Labey et al., 2011, 2012; Victor et al., 2010; Victor, Van Doninck, et al., 2009), as well as in the investigation of the patellofemoral joint (Didden et al., 2010; Luyckx et al., 2009). Additionally, joint force, collateral ligament strain, and knee alignment and features could be also investigated using these methodologies (Arnout et al., 2015; Delport et al., 2015). In general, there are different devices that could be used as a knee simulator, but all of them could be subdivided in one of the two main categories: *robotic arms* (as Kuka robot) or *knee rig* (as Kansas knee simulator or Oxford Rig). The former are common electromechanical systems (as any industrial robot) that are programmed to simulate and record the motions and loads in a knee joint during knee motor tasks (i.e., walking, stairs, and squat) under boundary conditions, as reported in the Orthoload database (http://www.orthoload. com) or based on the results of previous numerical studies. Special attention should be paid to the patellofemoral joint (quadriceps motor), which should be added as an independent arm with respect to the femoral arm. The second category of devices are servo-hydraulic dynamic testing rigs, which can apply specific loads to a knee (usually quadriceps load) during a certain flexion.

The simulators could have different axes of control, and they could differ from the others in terms of which joint is loaded and which one is constrained. In some cases, the hip motor could provide the flexion–extension rotation and superior–inferior translation, thus having a constrained displacement at the ankle joint; in other cases, it is the hip joint that is constrained in the displacement and the ankle moves. Generally, the flexion–extension position of the knee is controlled by the quadriceps actuator, following a predetermined profile, while the remaining axes are left in load control. The quadriceps force could be controlled by a constant hip force (Kansas Rig) or by a constant ankle force (Oxford Rig). By applying loads at the hip (ankle), and quadriceps tendon, the simulator can perform many dynamic activities, including walking, stair climbing, and cutting maneuver profiles.

A knee joint simulator analyzes the full joint kinematics and kinetics in the native joint to highlight the eventual differences induced by a specific treatment (such as the removal of certain ligaments or the insertion of a certain knee implant). To improve the accuracy of kinematic analysis, optoelectronic markers are also integrated in the methodology. They are drilled directly into the specimen bone prior to the test, to avoid any eventual skin artifact, and identified with a preop CT (Heesterbeek et al., 2012; Heyse et al., 2014; Labey et al., 2011, 2012; Victor et al., 2010; Victor, Van Glabbeek, et al., 2009).

SPINE

Several testing systems designed to perform the mechanical characterization of cadaveric spine specimens have been described in the literature. Similar to other joint simulators, spine testers include actuators, which apply the loads, fixtures to block one end of the specimen while allowing for some degrees of freedom to the rest of the specimen, and load cells or displacement/rotation sensors. In the case of multilevel specimens, devices such as ultrasonic or optoelectronic systems able to capture the motion of the single vertebrae may be employed.

Since spine specimens typically show motion coupling, i.e., the motion plane may not coincide with the plane of the applied load, spine testers should permit unconstrained motion of the specimen in the three motion planes. Constrained loading devices such as standard uniaxial material testing systems are therefore not appropriate for testing spinal specimens, but, anyway, have been employed in numerous studies, in some cases in combination with add-ons allowing more refined boundary conditions (Holsgrove, Gheduzzi, Gill, & Miles, 2014).

A simple method to apply loads while allowing for unconstrained motion is by using dead weights connected to cables and then to actuators. The research group at ENSAM (Paris) developed a loading jig that uses loading bars pushing on the pedicles of the upper end of the specimen, fixed on its caudal side, to apply loads while leaving the motion free in all planes; linear and angular displacement sensors are then used to measure the vertebral motion. This device has been used in several studies aimed at characterizing the spine flexibility under intact conditions as well as with implants (Templier, Skalli, Lemaire, Diop, & Lavaste, 1999).

A spine tester able to apply pure moments to cadaveric specimens while leaving its motion unconstrained has been designed at Ulm University (Wilke, Claes, Schmitt, & Wolf, 1994), and serves as reference for several other spine testers based on the same principles (Schmoelz, Onder, Martin, & Von Strempel, 2009). This piece of equipment employs stepper motors to apply the pure moments by means of a gimbal attached to the cranial end of the specimen, while the caudal end is fixed to a 6-degree-of-freedom load cell and then to a rigid frame. The gimbal is connected to the frame through linear rails and is free to move in space, and its weight is compensated by means of a counterweight. Vertebral motions are captured by means of passive reflective markers and an optoelectronic motion analysis system. The stepper motors are controlled by a computer, which receives information from the motion analysis system and the load cell as feedback signals. Such spine simulators have been used in a number of applications in the field of spine to quantify the range of motion of spine segments following instrumentation with a variety of implants and techniques (Kettler et al., 2006, 2008), being easily integrated with SG (Fig. 28.1) and DIC analyses (Fig. 28.2), thus providing detailed insights into the loading distribution in implants and tissues (La Barbera et al., 2018, 2020, 2021; Palanca et al., 2020; Ruspi et al., 2020). This quantitative information can prove to be very useful in building, calibrating, and/or validating computational models (La Barbera et al., 2016).

Industrial robots equipped with load cells offer a practical alternative to test spine specimens; in this case, unconstrained conditions can be achieved by purposely programming the control system to minimize the reaction forces and moments in the directions in which the motion should be free (Hurschler, Pott, Gossé, & Wirth, 2005). This approach was further developed by a research group at Flinders University who designed a robot based on the "hexapod" concept (Lawless, Ding, Cazzolato, & Costi, 2014). The robot uses six linear actuators to accurately drive the position and orientation of a plate connected to the cranial end of the specimen, while the caudal end is connected to a load cell, which provides information about the reaction loads to the control system, and then to a rigid frame. This design has proven to be able to simulate high loading rates, up to values that were not achievable with the previous designs.

REFERENCES

Abraham, A. C., Agarwalla, A., Yadavalli, A., Liu, J. Y., & Tang, S. Y. (2016). Microstructural and compositional contributions towards the mechanical behavior of aging human bone measured by cyclic and impact reference point indentation. *Bone, 87,* 37−43. Available from https://doi.org/10.1016/j.bone.2016.03.013.

Adams, M. A., McNally, D. S., & Dolan, P. (1996). Stress' distributions inside intervertebral discs. The effects of age and degeneration. *The Journal of Bone and Joint Surgery. British Volume, 78*(6), 90052−90053. Available from https://doi.org/10.1016/0268-0033(94).

Affatato, S., Bersaglia, G., Foltran, I., Taddei, P., Fini, G., & Toni, A. (2002). The performance of gamma- and EtO-sterilised UHWMPE acetabular cups tested under severe simulator conditions. Part 1: Role of the third-body wear process. *Biomaterials, 23*(24), 4839−4846. Available from https://doi.org/10.1016/S0142-9612(02)00238-7.

Affatato, S., Leardini, W., Jedenmalm, A., Ruggeri, O., & Toni, A. (2007). Larger diameter bearings reduce wear in metal-on-metal hip implants. *Clinical Orthopaedics and Related Research, 456,* 153−158. Available from https://doi.org/10.1097/01.blo.0000246561.73338.68.

Ali, M., Al-Hajjar, M., Partridge, S., Williams, S., Fisher, J., & Jennings, L. M. (2016). Influence of hip joint simulator design and mechanics on the wear and creep of metal-on-polyethylene bearings. *Proceedings of the Institution of Mechanical Engineers, Part H: Journal of Engineering in Medicine*, 230(5), 389−397. Available from https://doi.org/10.1177/0954411915620454.

Ali, A. M., Newman, S. D. S., Hooper, P. A., Davies, C. M., & Cobb, J. P. (2017). The effect of implant position on bone strain following lateral unicompartmental knee arthroplasty: A biomechanical model using Digital Image Correlation. *Bone and Joint Research*, 6(8), 522−529. Available from https://doi.org/10.1302/2046-3758.68.BJR-2017-0067.R1.

Armstrong, C. G., Lai, W. M., & Mow, V. C. (1984). An analysis of the unconfined compression of articular cartilage. *Journal of Biomechanical Engineering*, 106(2), 165−173. Available from https://doi.org/10.1115/1.3138475.

Arnout, N., Vanlommel, L., Vanlommel, J., Luyckx, J. P., Labey, L., Innocenti, B., ... Bellemans, J. (2015). Post-cam mechanics and tibiofemoral kinematics: A dynamic in vitro analysis of eight posterior-stabilized total knee designs. *Knee Surgery, Sports Traumatology, Arthroscopy*, 23(11), 3343−3353. Available from https://doi.org/10.1007/s00167-014-3167-2.

Bahaloo, H., Enns-Bray, W. S., Fleps, I., Ariza, O., Gilchrist, S., Soyka, R. W., ... Helgason, B. (2018). On the failure initiation in the proximal human femur under simulated sideways fall. *Annals of Biomedical Engineering*, 46(2), 270−283. Available from https://doi.org/10.1007/s10439-017-1952-z.

Bailey, S., & Vashishth, D. (2018). Mechanical characterization of bone: State of the art in experimental approaches—What types of experiments do people do and how does one interpret the results? *Current Osteoporosis Reports*, 16(4), 423−433. Available from https://doi.org/10.1007/s11914-018-0454-8.

Bandini, A., Chicot, D., Berry, P., Decoopman, X., Pertuz, A., & Ojeda, D. (2013). Indentation size effect of cortical bones submitted to different soft tissue removals. *Journal of the Mechanical Behavior of Biomedical Materials*, 20, 338−346. Available from https://doi.org/10.1016/j.jmbbm.2013.02.011.

Barbour, P. S. M., Stone, M. H., & Fisher, J. (2000). A hip joint simulator study using new and physiologically scratched femoral heads with ultra-high molecular weight polyethylene acetabular cups. *Proceedings of the Institution of Mechanical Engineers, Part H: Journal of Engineering in Medicine*, 214(6), 569−576. Available from https://doi.org/10.1243/0954411001535598.

Bayraktar, H. H., Morgan, E. F., Niebur, G. L., Morris, G. E., Wong, E. K., & Keaveny, T. M. (2004). Comparison of the elastic and yield properties of human femoral trabecular and cortical bone tissue. *Journal of Biomechanics*, 37(1), 27−35. Available from https://doi.org/10.1016/S0021-9290(03)00257-4.

Belkoff, S. M., & Haut, R. C. (2008). *Springer handbook of experimental solid mechanics. Experimental methods in biological tissue testing* (pp. 871−890). Springer.

Bergmann, G., Deuretzbacher, G., Heller, M., Graichen, F., Rohlmann, A., Strauss, J., & Duda, G. N. (2001). Hip contact forces and gait patterns from routine activities. *Journal of Biomechanics*, 34((7)), 859−871. Available from https://doi.org/10.1016/S0021-9290(01)00040-9.

Bergmann, G., Graichen, F., Siraky, J., Jendrzynski, H., & Rohlmann, A. (1988). Multichannel strain gauge telemetry for orthopaedic implants. *Journal of Biomechanics*, 21(2), 169−176. Available from https://doi.org/10.1016/0021-9290(88)90009-7.

Bieger, R., Ignatius, A., Reichel, H., & Dürselen, L. (2013). Biomechanics of a short stem: In vitro primary stability and stress shielding of a conservative cementless hip stem. *Journal of Orthopaedic Research*, 31(8), 1180−1186. Available from https://doi.org/10.1002/jor.22349.

Bright, C., Tiernan, S., McEvoy, F., & Kiely, P. (2017). Strain distribution in the porcine lumbar laminae under asymmetric loading. *Proceedings of the Institution of Mechanical Engineers, Part H: Journal of Engineering in Medicine*, 231(10), 945−951. Available from https://doi.org/10.1177/0954411917719744.

Bright, C., Tiernan, S., McEvoy, F., & Kiely, P. (2018). Fatigue and damage of porcine pars interarticularis during asymmetric loading. *Journal of the Mechanical Behavior of Biomedical Materials*, 78, 505−514. Available from https://doi.org/10.1016/j.jmbbm.2017.12.008.

Brown, T. D., & Singerman, R. J. (1986). Experimental determination of the linear biphasic constitutive coefficients of human fetal proximal femoral chondroepiphysis. *Journal of Biomechanics*, *19*(8), 597−605. Available from https://doi.org/10.1016/0021-9290(86)90165-X.

Burgers, T. A., Mason, J., Niebur, G., & Ploeg, H. L. (2008). Compressive properties of trabecular bone in the distal femur. *Journal of Biomechanics*, *41*(5), 1077−1085. Available from https://doi.org/10.1016/j.jbiomech.2007.11.018.

Carlson, C. E., Mann, R. W., & Harris, W. H. (1974). A radio telemetry device for monitoring cartilage surface pressures in the human hip. *IEEE Transactions on Biomedical Engineering*, *21*(4), 257−264. Available from https://doi.org/10.1109/TBME.1974.324311.

Carnelli, D., Vena, P., Dao, M., Ortiz, C., & Contro, R. (2013). Orientation and size-dependent mechanical modulation within individual secondary osteons in cortical bone tissue. *Journal of the Royal Society Interface*, *10*(81), 20120953. Available from https://doi.org/10.1098/rsif.2012.0953.

Carter, D. R., Schwab, G. H., & Spengler, D. M. (1980). Tensile fracture of cancellous bone. *Acta Orthopaedica*, *51*(1−6), 733−741. Available from https://doi.org/10.3109/17453678008990868.

Casanova, M., Balmelli, A., Carnelli, D., Courty, D., Schneider, P., & Müller, R. (2017). Nanoindentation analysis of the micromechanical anisotropy in mouse cortical bone. *Royal Society Open Science*, *4*(2), 160971. Available from https://doi.org/10.1098/rsos.160971.

Catani, F., Innocenti, B., Belvedere, C., Labey., Ensini, A., & Leardini, A. (1986). The marc coventry award: Articular contact estimation in TKA using in vivo kinematics and finite element analysis. *Journal of Molecular and Cellular Cardiology*, *468*(1), 567−578. Available from https://doi.org/10.1007/s11999-009-0941-4.

Chew, P. H., Yin, F. C. P., & Zeger, S. L. (1986). Biaxial stress-strain properties of canine pericardium. *Journal of Molecular and Cellular Cardiology*, *18*(6), 567−578. Available from https://doi.org/10.1016/S0022-2828(86)80965-8.

Claeson, A. A., & Barocas, V. H. (2017). Planar biaxial extension of the lumbar facet capsular ligament reveals significant in-plane shear forces. *Journal of the Mechanical Behavior of Biomedical Materials*, *65*, 127−136. Available from https://doi.org/10.1016/j.jmbbm.2016.08.019.

Completo, A., Fonseca, F., & Simões, J. A. (2008). Strain shielding in proximal tibia of stemmed knee prosthesis: Experimental study. *Journal of Biomechanics*, *41*(3), 560−566. Available from https://doi.org/10.1016/j.jbiomech.2007.10.006.

Cooke, M. E., Lawless, B. M., Jones, S. W., & Grover, L. M. (2018). Loss of proteoglycan content primes articular cartilage for mechanically induced damage. *Osteoarthritis and Cartilage*, *26*(S371). Available from https://doi.org/10.1016/j.joca.2018.02.731.

Cristofolini, L., Brandolini, N., Danesi, V., Juszczyk, M. M., Erani, P., & Viceconti, M. (2013). Strain distribution in the lumbar vertebrae under different loading configurations. *Spine Journal*, *13*(10), 1281−1292. Available from https://doi.org/10.1016/j.spinee.2013.06.014.

Cristofolini, L., Conti, G., Juszczyk, M., Cremonini, S., Sint Jan, S. V., & Viceconti, M. (2010). Structural behaviour and strain distribution of the long bones of the human lower limbs. *Journal of Biomechanics*, *43*(5), 826−835. Available from https://doi.org/10.1016/j.jbiomech.2009.11.022.

Cristofolini, L., Juszczyk, M., Taddei, F., & Viceconti, M. (2009). Strain distribution in the proximal human femoral metaphysis. *Proceedings of the Institution of Mechanical Engineers, Part H: Journal of Engineering in Medicine*, *223*(3), 273−288. Available from https://doi.org/10.1243/09544119JEIM497.

Cristofolini, L., Juszczyk, M., Taddei, F., Field, R. E., Rushton, N., & Viceconti, M. (2009). Stress shielding and stress concentration of contemporary epiphyseal hip prostheses. *Proceedings of the Institution of Mechanical Engineers, Part H: Journal of Engineering in Medicine*, *223*(1), 27−44. Available from https://doi.org/10.1243/09544119JEIM470.

Cristofolini, L., Viceconti, M., Cappello, A., & Toni, A. (1996). Mechanical validation of whole bone composite femur models. *Journal of Biomechanics*, *29*(4), 525−535. Available from https://doi.org/10.1016/0021-9290(95)00084-4.

Crowninshield, R. D., & Pope, M. H. (1976). The strength and failure characteristics of rat medial collateral ligaments. *Journal of Trauma - Injury, Infection and Critical Care, 16*(2), 99–105. Available from https://doi.org/10.1097/00005373-197602000-00004.

D'Lima, D. D., Patil, S., Steklov, N., Slamin, J. E., & Colwell, C. W. (2006). Tibial forces measured in vivo after total knee arthroplasty. *Journal of Arthroplasty, 21*(2), 255–262. Available from https://doi.org/10.1016/j.arth.2005.07.011.

D'Lima, D. D., Townsend, C. P., Arms, S. W., Morris, B. A., & Colwell, C. W. (2005). An implantable telemetry device to measure intra-articular tibial forces. *Journal of Biomechanics, 38*(2), 299–304. Available from https://doi.org/10.1016/j.jbiomech.2004.02.011.

Davy, D. T., Kotzar, G. M., Brown, R. H., Heiple, K. G., Goldberg, V. M., Heiple, K. G., … Burstein, A. H. (1988). Telemetric force measurements across the hip after total arthroplasty. *Journal of Bone and Joint Surgery - Series A, 70*(1), 45–50. Available from https://doi.org/10.2106/00004623-198870010-00008.

Delport, H., Labey, L., Innocenti, B., De Corte, R., Vander Sloten, J., & Bellemans, J. (2015). Restoration of constitutional alignment in TKA leads to more physiological strains in the collateral ligaments. *Knee Surgery, Sports Traumatology, Arthroscopy, 23*(8), 2159–2169. Available from https://doi.org/10.1007/s00167-014-2971-z.

Didden, K., Luyckx, T., Bellemans, J., Labey, L., Innocenti, B., & Vandenneucker, H. (2010). Anteroposterior positioning of the tibial component and its effect on the mechanics of patellofemoral contact. *Journal of Bone and Joint Surgery - Series B, 92*(10), 1466–1470. Available from https://doi.org/10.1302/0301-620X.92B10.24221.

Dowson, D., Walker, P. S., Longfield, M. D., & Wright, V. (1970). A joint simulating machine for load-bearing joints. *Medical & Biological Engineering, 8*(1), 37–43. Available from https://doi.org/10.1007/BF02551747.

Dragomir-Daescu, D., Rossman, T. L., Rezaei, A., Carlson, K. D., Kallmes, D. F., Skinner, J. A., … Amin, S. (2018). Factors associated with proximal femur fracture determined in a large cadaveric cohort. *Bone, 116*, 196–202. Available from https://doi.org/10.1016/j.bone.2018.08.005.

Dumbleton, J. H., Miller, D. A., & Miller, E. H. (1972). A simulator for load bearing joints. *Wear, 20*(2), 165–174. Available from https://doi.org/10.1016/0043-1648(72)90379-1.

Essner, A., Sutton, K., & Wang, A. (2005). Hip simulator wear comparison of metal-on-metal, ceramic-on-ceramic and crosslinked UHMWPE bearings. *Wear, 259*(7–12), 992–995. Available from https://doi.org/10.1016/j.wear.2005.02.104.

Fan, Z., & Rho, J. Y. (2003). Effects of viscoelasticity and time-dependent plasticity on nanoindentation measurements of human cortical bone. *Journal of Biomedical Materials Research - Part A, 67*(1), 208–214. Available from https://doi.org/10.1002/jbm.a.10027.

Fischenich, K. M., Lewis, J., Kindsfater, K. A., Bailey, T. S., & Haut Donahue, T. L. (2015). Effects of degeneration on the compressive and tensile properties of human meniscus. *Journal of Biomechanics, 48*(8), 1407–1411. Available from https://doi.org/10.1016/j.jbiomech.2015.02.042.

Finlay, J. B., Bourne, R. B., Landsberg, R. P., & Andreae, P. (1986). Pelvic stresses in vitro – II. A study of the efficacy of metal-backed acetabular prostheses. *Journal of Biomechanics, 19*(9), 90195–90198. Available from https://doi.org/10.1016/0021-9290.

Fottner, A., Woiczinski, M., Kistler, M., Schröder, C., Schmidutz, T. F., Jansson, V., & Schmidutz, F. (2017). Influence of undersized cementless hip stems on primary stability and strain distribution. *Archives of Orthopaedic and Trauma Surgery, 137*(10), 1435–1441. Available from https://doi.org/10.1007/s00402-017-2784-x.

Franceschini, M., La Barbera, L., Anticonome, A., Ottardi, C., Tanaka, A., & Villa, T. (2020). Periprosthetic femoral fractures in sideways fall configuration: Comparative numerical analysis of the influence of femoral stem design. *HIP International, 30*(2), 86–93. Available from https://doi.org/10.1177/1120700020971312.

Francetti, L., Cavalli, N., Villa, T., La Barbera, L., Taschieri, S., Corbella, S., & Del Fabbro, M. (2015). Biomechanical in vitro evaluation of two full-arch rehabilitations supported by four or five implants. *International Journal of Oral and Maxillofacial Implants*, *30*(2), 419−426. Available from https://doi.org/10.11607/jomi.3767.

Frank, M., Marx, D., Nedelkovski, V., Fischer, J. T., Pahr, D. H., & Thurner, P. J. (2018). Dehydration of individual bovine trabeculae causes transition from ductile to quasi-brittle failure mode. *Journal of the Mechanical Behavior of Biomedical Materials*, *87*, 296−305. Available from https://doi.org/10.1016/j.jmbbm.2018.07.039.

Fung, Y. C. (2013). *Biomechanics: Mechanical properties of living tissues*. Springer-Verlag. Available from https://www.springer.com/gp/book/9780387979472.

Gallant, M. A., Brown, D. M., Organ, J. M., Allen, M. R., & Burr, D. B. (2013). Reference-point indentation correlates with bone toughness assessed using whole-bone traditional mechanical testing. *Bone*, *53*(1), 301−305. Available from https://doi.org/10.1016/j.bone.2012.12.015.

Garcia, D., Zysset, P. K., Charlebois, M., & Curnier, A. (2009). A three-dimensional elastic plastic damage constitutive law for bone tissue. *Biomechanics and Modeling in Mechanobiology*, *8*(2), 149−165. Available from https://doi.org/10.1007/s10237-008-0125-2.

Ghosh, R., Gupta, S., Dickinson, A., & Browne, M. (2012). Experimental validation of finite element models of intact and implanted composite hemipelvises using digital image correlation. *Journal of Biomechanical Engineering*, *134*(8), 081003. Available from https://doi.org/10.1115/1.4007173.

Girardi, B. L., Attia, T., Backstein, D., Safir, O., Willett, T. L., & Kuzyk, P. R. T. (2016). Biomechanical comparison of the human cadaveric pelvis with a fourth generation composite model. *Journal of Biomechanics*, *49*(4), 537−542. Available from https://doi.org/10.1016/j.jbiomech.2015.12.050.

Goldsmith, A. A. J., & Dowson, D. (1999). Development of a ten-station, multi-axis hip joint simulator. *Proceedings of the Institution of Mechanical Engineers, Part H: Journal of Engineering in Medicine*, *213*(4), 311−316. Available from https://doi.org/10.1243/0954411991535149.

Goldstein, S. A. (1987). The mechanical properties of trabecular bone: Dependence on anatomic location and function. *Journal of Biomechanics*, *20*, 90023−90026. Available from https://doi.org/10.1016/0021-9290.

Gong, H., Wang, L., Fan, Y., Zhang, M., & Qin, L. (2016). Apparent- and tissue-level yield behaviors of L4 vertebral trabecular bone and their associations with microarchitectures. *Annals of Biomedical Engineering*, *44*(4), 1204−1223. Available from https://doi.org/10.1007/s10439-015-1368-6.

Grassi, L., Kok, J., Gustafsson, A., Zheng, Y., Väänänen, S. P., Jurvelin, J. S., & Isaksson, H. (2020). Elucidating failure mechanisms in human femurs during a fall to the side using bilateral digital image correlation. *Journal of Biomechanics*, *106*, 109826. Available from https://doi.org/10.1016/j.jbiomech.2020.109826.

Guidoni, G., Swain, M., & Jäger, I. (2010). Nanoindentation of wet and dry compact bone: Influence of environment and indenter tip geometry on the indentation modulus. *Philosophical Magazine*, *90*(5), 553−565. Available from https://doi.org/10.1080/14786430903201853.

Hayes, W. C., Keer, L. M., Herrmann, G., & Mockros, L. F. (1972). A mathematical analysis for indentation tests of articular cartilage. *Journal of Biomechanics*, *5*(5), 541−551. Available from https://doi.org/10.1016/0021-9290(72)90010-3.

Hayes, W. V. (1997). *Basic orthopedic biomechanics*. Philadelphia: Lippincott-Raven.

Heesterbeek, P., Labey, L., Wong, P., Innocenti, B., & Wyemnga, A. (2012). Kinematics of an anatomically designed cruciate-retaining total knee arthroplasty implanted using a spacer-guided PCL balancing technique. *Orthopaedic Proceedings*, *94-B*, 100.

Heinlein, B., Graichen, F., Bender, A., Rohlmann, A., & Bergmann, G. (2007). Design, calibration and preclinical testing of an instrumented tibial tray. *Journal of Biomechanics*, *40*(1), S4. Available from https://doi.org/10.1016/j.jbiomech.2007.02.014.

Herregodts, S., Baets, P., Victor, J., & Verstraete, M. (2015). Use of Tekscan pressure sensors for measuring contact pressures in the human knee joint. *Sustainable Construction and Design*, 6. Available from https://doi.org/10.21825/SCAD.V6I2.1123.

Herrmann, S., Kähler, M., Grawe, R., Kluess, D., Woernle, C., & Bader, R. (2015). *Physiological-like testing of the dislocation stability of artificial hip joints, . Mechanisms and machine science* (Vol. 24, pp. 659−667). Kluwer Academic Publishers. Available from https://doi.org/10.1007/978-3-319-09411-3_70.

Heyse, T. J., El-Zayat, B. F., De Corte, R., Chevalier, Y., Scheys, L., Innocenti, B., ... Labey, L. (2014). UKA closely preserves natural knee kinematics in vitro. *Knee Surgery, Sports Traumatology, Arthroscopy*, *22*(8), 1902−1910. Available from https://doi.org/10.1007/s00167-013-2752-0.

Hoffler, C. E., Guo, X. E., Zysset, P. K., & Goldstein, S. A. (2005). An application of nanoindentation technique to measure bone tissue lamellae properties. *Journal of Biomechanical Engineering*, *127*(7), 1046−1053. Available from https://doi.org/10.1115/1.2073671.

Holsgrove, T. P., Gheduzzi, S., Gill, H. S., & Miles, A. W. (2014). The development of a dynamic, six-axis spine simulator. *Spine Journal*, *14*(7), 1308−1317. Available from https://doi.org/10.1016/j.spinee.2013.11.045.

Hongo, M., Abe, E., Shimada, Y., Murai, H., Ishikawa, N., & Sato, K. (1999). Surface strain distribution on thoracic and lumbar vertebrae under axial compression: The role in burst fractures. *Spine (Philadelphia, PA: 1986)*, *24*(12), 1197−1202. Available from https://doi.org/10.1097/00007632-199906150-00005.

Howie, D. W., Neale, S. D., Stamenkov, R., McGee, M. A., Taylor, D. J., & Findlay, D. M. (2007). Progression of acetabular periprosthetic osteolytic lesions measured with computed tomography. *Journal of Bone and Joint Surgery - Series A*, *89*(8), 1818−1825. Available from https://doi.org/10.2106/JBJS.E.01305.

Howlett, R. A., & Hogan, M. C. (2007). Effect of hypoxia on fatigue development in rat muscle composed of different fibre types. *Experimental Physiology*, *92*(5), 887−894. Available from https://doi.org/10.1113/expphysiol.2007.037291.

Hunt, H. B., & Donnelly, E. (2016). Bone quality assessment techniques: Geometric, compositional, and mechanical characterization from macroscale to nanoscale. *Clinical Reviews in Bone and Mineral Metabolism*, *14*(3), 133−149. Available from https://doi.org/10.1007/s12018-016-9222-4.

Hurschler, C., Pott, L., Gossé, F., & Wirth, C. J. (2005). Sensor-guided robotic spine motion-segment biomechanical testing: Validation against the pure moment apparatus. In *Transactions of the 51st Annual Meeting of the Orthopaedic Research Society* (Vol. 30), Washington, DC.

Ibrahim, A., Magliulo, N., Groben, J., Padilla, A., Akbik, F., & Abdel Hamid, Z. (2020). Hardness, an important indicator of bone quality, and the role of collagen in bone hardness. *Journal of Functional Biomaterials*, *11*(4), 85. Available from https://doi.org/10.3390/jfb11040085.

Isaksson, H., Nagao, S., Małkiewicz, M., Julkunen, P., Nowak, R., & Jurvelin, J. S. (2010). Precision of nanoindentation protocols for measurement of viscoelasticity in cortical and trabecular bone. *Journal of Biomechanics*, *43*(12), 2410−2417. Available from https://doi.org/10.1016/j.jbiomech.2010.04.017.

ISO 14242:2014. (2014). *Implants for surgery−wear of total hip-joint prostheses − Loading and displacement parameters for wear-testing machines and corresponding environmental conditions for test* (p. 14242) Switzerland: International Standard Organization. ISO.

Ivarsson, I., & Gillquist, J. (1992). The strain distribution in the upper tibia after insertion of two different unicompartmental prostheses. *Clinical Orthopaedics and Related Research*, *279*, 194−200. Available from https://doi.org/10.1097/00003086-199206000-00025.

Iwaki, H., Pinskerova, V., & Freeman, M. A. R. (2000). Tibiofemoral movement 1: The shape and relative movements of the femur and tibia in the unloaded cadaver knee. *Journal of Bone and Joint Surgery - Series B*, *82*(8), 1189−1195. Available from https://doi.org/10.1302/0301-620X.82B8.10717.

Jazinizadeh, F., Mohammadi, H., & Quenneville, C. E. (2020). Comparing the fracture limits of the proximal femur under impact and quasi-static conditions in simulation of a sideways fall. *Journal of the Mechanical Behavior of Biomedical Materials*, *103*, 103593. Available from https://doi.org/10.1016/j.jmbbm.2019.103593.

Jin, H., & Lewis, J. L. (2004). Determination of poisson's ratio of articular cartilage by indentation using different-sized indenters. *Journal of Biomechanical Engineering*, *126*(2), 138−145. Available from https://doi.org/10.1115/1.1688772.

Kaddick, C., & Wimmer, M. A. (2001). Hip simulator wear testing according to the newly introduced standard ISO 14242. *Proceedings of the Institution of Mechanical Engineers, Part H: Journal of Engineering in Medicine*, *215*((5)), 429−442. Available from https://doi.org/10.1243/0954411011536019.

Karanika, M., Georgiou, D., Darmanis, S., Papadogoulas, A., Pasiou, E. D., & Kourkoulis, S. K. (2016). *Assessing osteosynthesis techniques for pelvic fractures using Digital Image Correlation*, . Procedia structural integrity (Vol. 2, pp. 1252−1259). Elsevier B.V. Available from https://doi.org/10.1016/j.prostr.2016.06.160.

Kettler, A., Drumm, J., Heuer, F., Haeussler, K., Mack, C., Claes, L., & Wilke, H. J. (2008). Can a modified interspinous spacer prevent instability in axial rotation and lateral bending? A biomechanical in vitro study resulting in a new idea. *Clinical Biomechanics*, *23*(2), 242−247. Available from https://doi.org/10.1016/j.clinbiomech.2007.09.004.

Kettler, A., Niemeyer, T., Issler, L., Merk, U., Mahalingam, M., Werner, K., . . . Wilke, H. J. (2006). In vitro fixator rod loading after transforaminal compared to anterior lumbar interbody fusion. *Clinical Biomechanics*, *21*(5), 435−442. Available from https://doi.org/10.1016/j.clinbiomech.2005.12.005.

Kirking, B., Krevolin, J., Townsend, C., Colwell, C. W., & D'Lima, D. D. (2006). A multiaxial force-sensing implantable tibial prosthesis. *Journal of Biomechanics*, *39*(9), 1744−1751. Available from https://doi.org/10.1016/j.jbiomech.2005.05.023.

Korhonen, R. K., Laasanen, M. S., Töyräs, J., Rieppo, J., Hirvonen, J., Helminen, H. J., & Jurvelin, J. S. (2002). Comparison of the equilibrium response of articular cartilage in unconfined compression, confined compression and indentation. *Journal of Biomechanics*, *35*(7), 903−909. Available from https://doi.org/10.1016/S0021-9290(02)00052-0.

Kourkoulis, S. K., Darmanis, S., Papadogoulas., & Pasiou, E. D. (2017). 3D-DIC in the service of orthopaedic surgery: Comparative assessment of fixation techniques for acetabular fractures. *Engineering Fracture Mechanics*, *183*, 125−146. Available from https://doi.org/10.1016/j.engfracmech.2017.05.014.

Kroeber, M., Ries, M. D., Suzuki, Y., Renowitzky, G., Ashford, F., & Lotz, J. (2002). Impact biomechanics and pelvic deformation during insertion of press-fit acetabular cups. *Journal of Arthroplasty*, *17*(3), 349−354. Available from https://doi.org/10.1054/arth.2002.30412.

Kutzner, I., Heinlein, B., Graichen, F., Bender, A., Rohlmann, A., Halder, A., . . . Bergmann, G. (2010). Loading of the knee joint during activities of daily living measured in vivo in five subjects. *Journal of Biomechanics*, *43*(11), 2164−2173. Available from https://doi.org/10.1016/j.jbiomech.2010.03.046.

La Barbera, L., & Villa, T. (2016). ISO 12189 standard for the preclinical evaluation of posterior spinal stabilization devices − I: Assembly procedure and validation. *Proceedings of the Institution of Mechanical Engineers, Part H: Journal of Engineering in Medicine*, *230*(2), 122−133. Available from https://doi.org/10.1177/0954411915621587.

La Barbera, L., & Villa, T. (2017). Toward the definition of a new worst-case paradigm for the preclinical evaluation of posterior spine stabilization devices. *Proceedings of the Institution of Mechanical Engineers, Part H: Journal of Engineering in Medicine*, *231*(2), 176−185. Available from https://doi.org/10.1177/0954411916684365.

La Barbera, L., Brayda-Bruno, M., Liebsch, C., Villa, T., Luca, A., Galbusera, F., & Wilke, H. J. (2018). Biomechanical advantages of supplemental accessory and satellite rods with and without interbody cages implantation for the stabilization of pedicle subtraction osteotomy. *European Spine Journal*, *27*(9), 2357−2366. Available from https://doi.org/10.1007/s00586-018-5623-z.

La Barbera, L., Galbusera, F., Wilke, H. J., & Villa, T. (2016). Preclinical evaluation of posterior spine stabilization devices: Can the current standards represent basic everyday life activities? *European Spine Journal*, *25*(9), 2909−2918. Available from https://doi.org/10.1007/s00586-016-4622-1.

La Barbera, L., Galbusera, F., Wilke, H. J., & Villa, T. (2017). Preclinical evaluation of posterior spine stabilization devices: Can we compare in vitro and in vivo loads on the instrumentation? *European Spine Journal*, *26*(1), 200−209. Available from https://doi.org/10.1007/s00586-016-4766-z.

La Barbera, L., Ottardi, C., & Villa, T. (2015). Comparative analysis of international standards for the fatigue testing of posterior spinal fixation systems: The importance of preload in ISO 12189. *Spine Journal*, *15*(10), 2290−2296. Available from https://doi.org/10.1016/j.spinee.2015.07.461.

La Barbera, L. (2018). Chapter 17 − Fixation and fusion. In F. Galbusera, & H.-J. Wilke (Eds.), *Biomechanics of the spine* (pp. 301−327). Academic Press. Available from https://doi.org/10.1016/b978-0-12-812851-0.00017-3.

Lawless, I. M., Ding, B., Cazzolato, B. S., & Costi, J. J. (2014). Adaptive velocity-based six degree of freedom load control for real-time unconstrained biomechanical testing. *Journal of Biomechanics*, *47*(12), 3241−3247. Available from https://doi.org/10.1016/j.jbiomech.2014.06.023.

Leslie, I., Williams, S., Brown, C., Isaac, G., Jin, Z., Ingham, E., & Fisher, J. (2008). Effect of bearing size on the long-term wear, wear debris, and ion levels of large diameter metal-on-metal hip replacements − An in vitro study. *Journal of Biomedical Materials Research - Part B Applied Biomaterials*, *87*(1), 163−172. Available from https://doi.org/10.1002/jbm.b.31087.

La Barbera, L., Wilke, H. J., Liebsch, C., Villa, T., Luca, A., Galbusera, F., & Brayda-Bruno, M. (2020). Biomechanical in vitro comparison between anterior column realignment and pedicle subtraction osteotomy for severe sagittal imbalance correction. *European Spine Journal*, *29*(1), 36−44. Available from https://doi.org/10.1007/s00586-019-06087-x.

La Barbera, L., Wilke, H. J., Ruspi, M. L., Palanca, M., Liebsch, C., Luca, A., ... Cristofolini, L. (2021). Load-sharing biomechanics of lumbar fixation and fusion with pedicle subtraction osteotomy. *Scientific Reports*, *11*(1), 3595. Available from https://doi.org/10.1038/s41598-021-83251-8.

Labey, L., Bellemans, J., Chevalier, Y., El-Zayat, B., Fuchs-Winkelmann, S., Heesterbeek, P., ... Innocenti, B. (2012). In vitro kinematics of human native knees: A database of 60 specimens. *Journal of Biomechanics*, *45*, S394. Available from https://doi.org/10.1016/S0021-9290(12)70395-0.

Labey, L., Innocenti, B., Wong, P. D., Parizel, P. M., Victor, J., & Bellemans, J. (2011). Sensitivity of knee kinematics and soft tissues to quadriceps load near extension. *Journal of Orthopedics Translational Research and Clinical Application*, *3*(1), 27−37.

Lanir, Y., & Fung, Y. C. (1974). Two-dimensional mechanical properties of rabbit skin—II. Experimental results. *Journal of Biomechanics*, *7*(2), 171−182. Available from https://doi.org/10.1016/0021-9290(74)90058-x.

Lawless, B. M., Sadeghi, H., Temple, D. K., Dhaliwal, H., Espino, D. M., & Hukins, D. W. L. (2017). Viscoelasticity of articular cartilage: Analysing the effect of induced stress and the restraint of bone in a dynamic environment. *Journal of the Mechanical Behavior of Biomedical Materials*, *75*, 293−301. Available from https://doi.org/10.1016/j.jmbbm.2017.07.040.

Lefèvre, E., Farlay, D., Bala, Y., Subtil, F., Wolfram, U., Rizzo, S., ... Follet, H. (2019). Compositional and mechanical properties of growing cortical bone tissue: A study of the human fibula. *Scientific Reports*, *9*(1). Available from https://doi.org/10.1038/s41598-019-54016-1, Article number: 17629.

Leppänen, O. V., Sievänen, H., & Järvinen, T. L. N. (2008). Biomechanical testing in experimental bone interventions-May the power be with you. *Journal of Biomechanics*, *41*(8), 1623−1631. Available from https://doi.org/10.1016/j.jbiomech.2008.03.017.

Li, L. P., Soulhat, J., Buschmann, M. D., & Shirazi-Adl, A. (1999). Nonlinear analysis of cartilage in unconfined ramp compression using a fibril reinforced poroelastic model. *Clinical Biomechanics*, *14*(9), 673−682. Available from https://doi.org/10.1016/S0268-0033(99)00013-3.

Li, S., Demirci, E., & Silberschmidt, V. V. (2013). Variability and anisotropy of mechanical behavior of cortical bone in tension and compression. *Journal of the Mechanical Behavior of Biomedical Materials*, *21*, 109−120. Available from https://doi.org/10.1016/j.jmbbm.2013.02.021.

Linde, F. (1994). Elastic and viscoelastic properties of trabecular bone by a compression testing approach. *Danish Medical Bulletin*, *41*(2), 119−138.

Linde, F., Hvid, I., & Madsen, F. (1992). The effect of specimen geometry on the mechanical behaviour of trabecular bone specimens. *Journal of Biomechanics*, *25*(4). Available from https://doi.org/10.1016/0021-9290(92).

Linde, F., Nørgaard, P., Hvid, I., Odgaard, A., & Søballe, K. (1991). Mechanical properties of trabecular bone. Dependency on strain rate. *Journal of Biomechanics*, *24*(9), 803−809. Available from https://doi.org/10.1016/0021-9290(91)90305-7.

Lionello, G., & Cristofolini, L. (2014). A practical approach to optimizing the preparation of speckle patterns for digital-image correlation. *Measurement Science and Technology*, *25*(10), 107001. Available from https://doi.org/10.1088/0957-0233/25/10/107001.

Lucchini, R., Carnelli, D., Ponzoni, M., Bertarelli, E., Gastaldi, D., & Vena, P. (2011). Role of damage mechanics in nanoindentation of lamellar bone at multiple sizes: Experiments and numerical modeling. *Journal of the Mechanical Behavior of Biomedical Materials*, *4*(8), 1852−1863. Available from https://doi.org/10.1016/j.jmbbm.2011.06.002.

Luyckx, T., Didden, K., Vandenneucker, H., Labey, L., Innocenti, B., & Bellemans, J. (2009). Is there a biomechanical explanation for anterior knee pain in patients with patella alta? *The Journal of Bone and Joint Surgery. British Volume*, *91*(3), 344−350. Available from https://doi.org/10.1302/0301-620x.91b3.21592.

Martens, M., van Audekercke, R., de Meester, P., & Mulier, J. C. (1980). The mechanical characteristics of the long bones of the lower extremity in torsional loading. *Journal of Biomechanics*, *13*(8), 667−676. Available from https://doi.org/10.1016/0021-9290(80)90353-X.

Martens, M., van Audekercke, R., de Meester, P., & Mulier, J. C. (1986). Mechanical behaviour of femoral bones in bending loading. *Journal of Biomechanics*, *19*(6), 443−454. Available from https://doi.org/10.1016/0021-9290(86)90021-7.

Mak, A. F., Lai, W. M., & Mow, V. C. (1987). Biphasic indentation of articular cartilage-I. Theoretical analysis. *Journal of Biomechanics*, *20*(7), 703−714. Available from https://doi.org/10.1016/0021-9290(87)90036-4.

Mathison, C., Chaudhary, R., Beaupre, L., Reynolds, M., Adeeb, S., & Bouliane, M. (2010). Biomechanical analysis of proximal humeral fixation using locking plate fixation with an intramedullary fibular allograft. *Clinical Biomechanics*, *25*(7), 642−646. Available from https://doi.org/10.1016/j.clinbiomech.2010.04.006.

McNally, D. S., Adams, M. A., & Goodship, A. E. (1992). Development and validation of a new transducer for intradiscal pressure measurement. *Journal of Biomedical Engineering*, *14*(6), 495−498. Available from https://doi.org/10.1016/0141-5425(92)90102-Q.

Miles, A. W., & McNamee, P. B. (1989). Strain gauge and photoelastic evaluation of the load transfer in the pelvis in total hip replacement: The effect of the position of the axis of rotation. *Proceedings of the Institution of Mechanical Engineers, Part H: Journal of Engineering in Medicine*, *203*(2), 103−107. Available from https://doi.org/10.1243/PIME_PROC_1989_203_018_01.

Mow, V. C., Gibbs, M. C., Lai, W. M., Zhu, W. B., & Athanasiou, K. A. (1989). Biphasic indentation of articular cartilage-II. A numerical algorithm and an experimental study. *Journal of Biomechanics*, *22*(8−9), 853−861. Available from https://doi.org/10.1016/0021-9290(89)90069-9.

Mow, V. C., Kuei, S. C., Lai, W. M., & Armstrong, C. G. (1980). Biphasic creep and stress relaxation of articular cartilage in compression: Theory and experiments. *Journal of Biomechanical Engineering*, *102*(1), 73−84. Available from https://doi.org/10.1115/1.3138202.

Nachemson, A., & Morris, J. (1963). Lumbar discometry. Lumbar intradiscal pressure measurements in vivo. *Lancet*, *1*(7291). Available from https://doi.org/10.1016/s0140-6736, 91806−3.

Nazarian, A., Araiza Arroyo, F. J., Rosso, C., Aran, S., & Snyder, B. D. (2011). Tensile properties of rat femoral bone as functions of bone volume fraction, apparent density and volumetric bone mineral density. *Journal of Biomechanics*, *44*(13), 2482−2488. Available from https://doi.org/10.1016/j.jbiomech.2011.06.016.

Nazarian, A., Muller, J., Zurakowski, D., Müller, R., & Snyder, B. D. (2007). Densitometric, morphometric and mechanical distributions in the human proximal femur. *Journal of Biomechanics*, *40*(11), 2573−2579. Available from https://doi.org/10.1016/j.jbiomech.2006.11.022.

Nečas, D., Vrbka, M., Urban, F., Křupka, I., & Hartl, M. (2016). The effect of lubricant constituents on lubrication mechanisms in hip joint replacements. *Journal of the Mechanical Behavior of Biomedical Materials*, *55*, 295−307. Available from https://doi.org/10.1016/j.jmbbm.2015.11.006.

Newell, N., Carpanen, D., Grigoriadis, G., Little, J. P., & Masouros, S. D. (2019). Material properties of human lumbar intervertebral discs across strain rates. *Spine Journal*, *19*(12), 2013−2024. Available from https://doi.org/10.1016/j.spinee.2019.07.012.

Oliver, W. C., & Pharr, G. M. (2004). Measurement of hardness and elastic modulus by instrumented indentation: Advances in understanding and refinements to methodology. *Journal of Materials Research*, *19*(1), 3−20. Available from https://doi.org/10.1557/jmr.2004.19.1.3.

Ottardi, C., La Barbera, L., Pietrogrande, L., & Villa, T. (2016). Vertebroplasty and kyphoplasty for the treatment of thoracic fractures in osteoporotic patients: A finite element comparative analysis. *Journal of Applied Biomaterials and Functional Materials*, *14*(2), e197−e204. Available from https://doi.org/10.5301/jabfm.5000287.

Paietta, R. C., Campbell, S. E., & Ferguson, V. L. (2011). Influences of spherical tip radius, contact depth, and contact area on nanoindentation properties of bone. *Journal of Biomechanics*, *44*(2), 285−290. Available from https://doi.org/10.1016/j.jbiomech.2010.10.008.

Palanca, M., Brugo, T. M., & Cristofolini, L. (2015). Use of digital image correlation to investigate the biomechanics of the vertebra. *Journal of Mechanics in Medicine and Biology*, *15*((2), 1540004. Available from https://doi.org/10.1142/S0219519415400047, World Scientific Publishing Co. Pte Ltd.

Palanca, M., Marco, M., Ruspi, M. L., & Cristofolini, L. (2018). Full-field strain distribution in multi-vertebra spine segments: An in vitro application of digital image correlation. *Medical Engineering and Physics*, *52*, 76−83. Available from https://doi.org/10.1016/j.medengphy.2017.11.003.

Palanca, M., Ruspi, M. L., Cristofolini, L., Liebsch, C., Villa, T., Brayda-Bruno, M., . . . La Barbera, L. (2020). The strain distribution in the lumbar anterior longitudinal ligament is affected by the loading condition and bony features: An in vitro full-field analysis. *PLoS One*, *15*(1), e0227210. Available from https://doi.org/10.1371/journal.pone.0227210.

Palanca, M., Tozzi, G., & Cristofolini, L. (2016). The use of digital image correlation in the biomechanical area: A review. *International Biomechanics*, *3*(1), 1−21. Available from https://doi.org/10.1080/23335432.2015.1117395.

Papini, M., Zdero, R., Schemitsch, E. H., & Zalzal, P. (2007). The biomechanics of human femurs in axial and torsional loading: Comparison of finite element analysis, human cadaveric femurs, and synthetic femurs. *Journal of Biomechanical Engineering*, *129*(1), 12−19. Available from https://doi.org/10.1115/1.2401178.

Peleg, E., Beek, M., Joskowicz, L., Liebergall, M., Mosheiff, R., & Whyne, C. (2010). Patient specific quantitative analysis of fracture fixation in the proximal femur implementing principal strain ratios. Method and experimental validation. *Journal of Biomechanics*, *43*(14), 2684−2688. Available from https://doi.org/10.1016/j.jbiomech.2010.06.033.

Ries, M., Pugh, J., Choy Au, J., Gurtowski, J., & Dee, R. (1989). Normal pelvic strain pattern in vitro. *Journal of Biomedical Engineering*, *11*(5), 398−402. Available from https://doi.org/10.1016/0141-5425 (89)90103-9.

Rohlmann, A., Arntz, U., Graichen, F., & Bergmann, G. (2001). Loads on an internal spinal fixation device during sitting. *Journal of Biomechanics*, *34*(8), 989−993. Available from https://doi.org/10.1016/S0021-9290(01)00073-2.

Rohlmann, A., Bergmann, G., & Graichen, F. (1994). A spinal fixation device for in vivo load measurement. *Journal of Biomechanics*, *27*(7). Available from https://doi.org/10.1016/0021-9290(94), 90268−2.

Rohlmann, A., Bergmann, G., & Graichen, F. (1997). Loads on an internal spinal fixation device during walking. *Journal of Biomechanics*, *30*(1), 41−47. Available from https://doi.org/10.1016/S0021-9290(96)00103-0.

Rohlmann, A., Bergmann, G., & Graichen, F. (1999). Loads on internal spinal fixators measured in different body positions. *European Spine Journal*, *8*(5), 354−359. Available from https://doi.org/10.1007/s005860050187.

Rohlmann, A., Dreischarf, M., Zander, T., Graichen, F., & Bergmann, G. (2014). Loads on a vertebral body replacement during locomotion measured in vivo. *Gait and Posture*, *39*(2), 750−755. Available from https://doi.org/10.1016/j.gaitpost.2013.10.010.

Rohlmann, A., Gabel, U., Graichen, F., Bender, A., & Bergmann, G. (2007). An instrumented implant for vertebral body replacement that measures loads in the anterior spinal column. *Medical Engineering and Physics*, *29*(5), 580−585. Available from https://doi.org/10.1016/j.medengphy.2006.06.012.

Rohlmann, A., Graichen, F., & Bergmann, G. (2000). Influence of load carrying on loads in internal spinal fixators. *Journal of Biomechanics*, *33*(9), 1099−1104. Available from https://doi.org/10.1016/S0021-9290(00)00075-0.

Rohlmann, A., Zander, T., Graichen, F., & Bergmann, G. (2013). Lifting up and laying down a weight causes high spinal loads. *Journal of Biomechanics*, *46*(3), 511−514. Available from https://doi.org/10.1016/j.jbiomech.2012.10.022.

Rohlmann, A., Zander, T., Graichen, F., Dreischarf, M., & Bergmann, G. (2011). Measured loads on a vertebral body replacement during sitting. *Spine Journal*, *11*(9), 870−875. Available from https://doi.org/10.1016/j.spinee.2011.06.017.

Ruspi, M. L., Palanca, M., Cristofolini, L., Liebsch, C., Villa, T., Brayda-Bruno, M., . . . La Barbera, L. (2020). Digital image correlation (DIC) assessment of the non-linear response of the anterior longitudinal ligament of the spine during flexion and extension. *Materials*, *13*(2), 384. Available from https://doi.org/10.3390/ma13020384.

Ruspi, M. L., Palanca, M., Faldini, C., & Cristofolini, L. (2017). Full-field in vitro investigation of hard and soft tissue strain in the spine by means of digital image correlation. *Muscles, Ligaments and Tendons Journal*, *7*(4), 538−545. Available from https://doi.org/10.32098/mltj.04.2017.08.

Rydell, N. W. (2014). Forces acting on the femoral head-prosthesis: A study on strain gauge supplied prostheses in living persons. *Acta Orthopaedica Scandinavica*, *37*(Suppl. 88), 1−132. Available from https://doi.org/10.3109/ort.1966.37.suppl-88.01.

Saikko, V. O. (1996). A three-axis hip joint simulator for wear and friction studies on total hip prostheses. *Proceedings of the Institution of Mechanical Engineers, Part H: Journal of Engineering in Medicine*, *210*(3), 175−185. Available from https://doi.org/10.1243/PIME_PROC_1996_210_410_02.

Saikko, V. (1992). A simulator study of friction in total replacement hip joints. *Proceedings of the Institution of Mechanical Engineers, Part H: Journal of Engineering in Medicine*, *206*(4), 201−211. Available from https://doi.org/10.1243/PIME_PROC_1992_206_292_02.

Saikko, V., & Calonius, O. (2002). Slide track analysis of the relative motion between femoral head and acetabular cup in walking and in hip simulators. *Journal of Biomechanics*, *35*(4), 455−464. Available from https://doi.org/10.1016/S0021-9290(01)00224-X.

Saikko, V., Ahlroos, T., Calonius, O., & Keränen, J. (2001). Wear simulation of total hip prostheses with poly-ethylene against CoCr, alumina and diamond-like carbon. *Biomaterials*, *22*(12), 1507−1514. Available from https://doi.org/10.1016/S0142-9612(00)00306-9.

Schileo, E., Taddei, F., Malandrino, A., Cristofolini, L., & Viceconti, M. (2007). Subject-specific finite element models can accurately predict strain levels in long bones. *Journal of Biomechanics*, *40*(13), 2982−2989. Available from https://doi.org/10.1016/j.jbiomech.2007.02.010.

Schilling, C., Krüger, S., Grupp, T. M., Duda, G. N., Blömer, W., & Rohlmann, A. (2011). The effect of design parameters of dynamic pedicle screw systems on kinematics and load bearing: An in vitro study. *European Spine Journal*, *20*(2), 297−307. Available from https://doi.org/10.1007/s00586-010-1620-6.

Schmoelz, W., Huber, J. F., Nydegger, T., Claes, L., & Wilke, H. J. (2006). Influence of a dynamic stabilisation system on load bearing of a bridged disc: An in vitro study of intradiscal pressure. *European Spine Journal*, *15*(8), 1276−1285. Available from https://doi.org/10.1007/s00586-005-0032-5.

Schmoelz, W., Onder, U., Martin, A., & Von Strempel, A. (2009). Non-fusion instrumentation of the lumbar spine with a hinged pedicle screw rod system: An in vitro experiment. *European Spine Journal*, *18*(10), 1478−1485. Available from https://doi.org/10.1007/s00586-009-1052-3.

Schneider, D. C., Davidson, T. M., & Nahum, A. M. (1984). In vitro biaxial stress-strain response of human skin. *Archives of Otolaryngology*, *110*(5), 329−333. Available from https://doi.org/10.1001/archotol.1984.00800310053012.

Schwenke, T., Schneider, E., & Wimmer, M. A. (2006). Load profile and fluid composition influence the soak behavior of UHMWPE implants. *Journal of ASTM International*, *3*(8), 5. Available from https://doi.org/10.1520/JAI100255.

Shah, J. S., Hampson, W. G. J., & Jayson, M. I. V. (1978). The distribution of surface strain in the cadaveric lumbar spine. *Journal of Bone and Joint Surgery - Series B*, *60*(2), 246−251. Available from https://doi.org/10.1302/0301-620x.60b2.659474.

Small, S. R., Rogge, R. D., Malinzak, R. A., Reyes, E. M., Cook, P. L., Farley, K. A., & Ritter, M. A. (2016). Micromotion at the tibial plateau in primary and revision total knee arthroplasty: Fixed vs rotating platform designs. *Bone and Joint Research*, *5*(4), 122−129. Available from https://doi.org/10.1302/2046-3758.54.2000481.

Smith, S. L., & Unsworth, A. (2001). A five-station hip joint simulator. *Proceedings of the Institution of Mechanical Engineers, Part H: Journal of Engineering in Medicine*, *215*(1), 61−64. Available from https://doi.org/10.1243/0954411011533535.

Soltz, M. A., & Ateshian, G. A. (2000). A conewise linear elasticity mixture model for the analysis of tension-compression nonlinearity in articular cartilage. *Journal of Biomechanical Engineering*, *122*(6), 576−586. Available from https://doi.org/10.1115/1.1324669.

Solitro, G. F., Whitlock, K., Amirouche, F., & Santis, C. (2016). Measures of micromotion in cementless femoral stems-review of current methodologies. *Biomaterials and Biomechanics in Bioengineering*, *3*(2), 85−104. Available from https://doi.org/10.12989/bme.2016.3.2.085.

Sutton, M. A., Orteu, J. J., & Schreier, H. W. (2009). *Image correlation for shape, motion and deformation measurements* (1st ed.). Springer US. Available from https://doi.org/10.1007/978-0-387-78747-3.

Temple, D. K., Cederlund, A. A., Lawless, B. M., Aspden, R. M., & Espino, D. M. (2016). Viscoelastic properties of human and bovine articular cartilage: A comparison of frequency-dependent trends. *BMC Musculoskeletal Disorders*, *17*(1), 1−8. Available from https://doi.org/10.1186/s12891-016-1279-1.

Templier, A., Skalli, W., Lemaire, J. P., Diop, A., & Lavaste, F. (1999). Three-dimensional finite-element modelling and improvement of a bispherical intervertebral disc prosthesis. *European Journal of Orthopaedic Surgery and Traumatology*, *9*(1), 51−58. Available from https://doi.org/10.1007/bf01695723.

Umehara, S., Tadano, S., Abumi, K., Katagiri, K., Kaneda, K., & Ukai, T. (1996). Effects of degeneration on the elastic modulus distribution in the lumbar intervertebral disc. *Spine (Philadelphia, PA: 1986)*, *21*(7), 811−820. Available from https://doi.org/10.1097/00007632-199604010-00007.

van Jonbergen, H. P. W., Innocenti, B., Gervasi, G. L., Labey, L., & Verdonschot, N. (2012). Differences in the stress distribution in the distal femur between patellofemoral joint replacement and total knee replacement: A finite element study. *Journal of Orthopaedic Surgery and Research*, *7*(1), 28. Available from https://doi.org/10.1186/1749-799X-7-28.

Victor, J., Labey, L., Wong, P., Innocenti, B., & Bellemans, J. (2010). The influence of muscle load on tibiofemoral knee kinematics. *Journal of Orthopaedic Research*, *28*(4), 419−428. Available from https://doi.org/10.1002/jor.21019.

Victor, J., Van Doninck, D., Labey, L., Innocenti, B., Parizel, P. M., & Bellemans, J. (2009). How precise can bony landmarks be determined on a CT scan of the knee? *The Knee*, *16*(5), 358−365. Available from https://doi.org/10.1016/j.knee.2009.01.001.

Victor, J., Van Glabbeek, F., Sloten, J. V., Parizel, P. M., Somville, J., & Bellemans, J. (2009). An experimental model for kinematic analysis of the knee. *Journal of Bone and Joint Surgery - Series A*, *91*(6), 150−163. Available from https://doi.org/10.2106/JBJS.I.00498.

Villa, T., La Barbera, L., & Galbusera, F. (2014a). Comparative analysis of international standards for the fatigue testing of posterior spinal fixation systems. *Spine Journal*, *14*(4), 695−704. Available from https://doi.org/10.1016/j.spinee.2013.08.032.

Villa, T., La Barbera, L., & Galbusera, F. (2014b). Reply to the letter to the editor entitled: Response to "Comparative analysis of international standards for the fatigue testing of posterior spinal fixation systems.". *Spine Journal*, *14*(12), 3068. Available from https://doi.org/10.1016/j.spinee.2014.08.008.

Wang, H., Chen, T., & Innocenti, B. (1992). Tekscan measurements of interfacial contact area and stress in articulating joints. In *Experimental methods for orthopaedic biomechanics: A step-by-step practical manual*.

Wang, A., Essner, A., & Klein, R. (2001). Effect of contact stress on friction and wear of ultra-high molecular weight polyethylene in total hip replacement. *In Proceedings of the Institution of Mechanical Engineers, Part H: Journal of Engineering in Medicine*, *215*(2), 133−139. Available from https://doi.org/10.1243/0954411011533698.

Weber, W., & Weber, E. (1992). *Mechanics of the human walking apparatus*. Springer-Verlag.

Wilke, H. J., Claes, L., Schmitt, H., & Wolf, S. (1994). A universal spine tester for in vitro experiments with muscle force simulation. *European Spine Journal*, *3*(2), 91−97. Available from https://doi.org/10.1007/BF02221446.

Wilke, H. J., Neef, P., Caimi, M., Hoogland, T., & Claes, L. E. (1999). New in vivo measurements of pressures in the intervertebral disc in daily life. *Spine (Philadelphia, PA: 1986)*, *24*(8), 755−762. Available from https://doi.org/10.1097/00007632-199904150-00005.

Wilke, H. J., Rohlmann, A., Bergmann, G., Graichen, F., & Claes, L. E. (2001). Comparison of intradiscal pressures and spinal fixator loads for different body positions and exercises. *Ergonomics*, *44*(8), 781−794. Available from https://doi.org/10.1080/00140130120943.

Wilke, H. J., Wolf, S., Claes, L. E., Arand, M., & Wiesend, A. (1996). Influence of varying muscle forces on lumbar intradiscal pressure: An in vitro study. *Journal of Biomechanics*, *29*(4), 549−555. Available from https://doi.org/10.1016/0021-9290(95)00037-2.

Woo, S. L. Y. (1982). Mechanical properties of tendons and ligaments. I. Quasi-static and nonlinear viscoelastic properties. *Biorheology*, *19*(3), 385−396. Available from https://doi.org/10.3233/BIR-1982-19301.

Woo, S. L. Y., Debski, R. E., Withrow, J. D., & Janaushek, M. A. (1999). Biomechanics of knee ligaments. *American Journal of Sports Medicine*, *27*(4), 533−543. Available from https://doi.org/10.1177/03635465990270042301.

Zani, L., Erani, P., Grassi, L., Taddei, F., & Cristofolini, L. (2015). Strain distribution in the proximal Human femur during in vitro simulated sideways fall. *Journal of Biomechanics*, *48*(10), 2130−2143. Available from https://doi.org/10.1016/j.jbiomech.2015.02.022.

Zdero, R., Borkowski, M. M., & Coirault, C. (2017). *Measuring the contraction force, velocity, and length of skeletal muscle. Experimental methods in orthopaedic biomechanics* (pp. 363−378). Elsevier Inc. Available from https://doi.org/10.1016/B978-0-12-803802-4.00023-8.

Zhai, X., Nauman, E. A., Moryl, D., Lycke, R., & Chen, W. W. (2020). The effects of loading-direction and strain-rate on the mechanical behaviors of human frontal skull bone. *Journal of the Mechanical Behavior of Biomedical Materials*, *103*, 103597. Available from https://doi.org/10.1016/j.jmbbm.2019.103597.

Zysset, P. K. (2009). Indentation of bone tissue: A short review. *Osteoporosis International*, *20*(6), 1049−1055. Available from https://doi.org/10.1007/s00198-009-0854-9.

CHALLENGES IN THE SYSTEM MODELING OF THE MUSCULOSKELETAL APPARATUS

Serge Van Sint Jan[1] and Victor Sholukha[1,2]

[1]*Laboratory of Anatomy, Biomechanics and Organogenesis, Faculty of Medicine, Université Libre de Bruxelles, Brussels, Belgium* [2]*Department of Applied Mathematics, Peter the Great St. Petersburg Polytechnic University, St. Petersburg, Russia*

STATE-OF-THE-ART

The musculoskeletal system (or MSS) is essential for the performance of day-to-day functional tasks and thus leading a quality life. MSS pathologies can therefore have a significant impact on our daily life. The MSS is also involved in central nervous conditions (e.g., brain stroke, cerebral palsy, spinal cord damages, etc.) and is sensitive to cardiovascular disorders (e.g., decrease of blood supply towards skeletal muscles during heart stroke or vessel disease). Such conditions can lead to severe impairments of the MSS normal physiology, leading to severe disabilities. It is therefore not surprising to find MSS-related health problems at the top of global statistics on professional absenteeism or societal health costs (see *The Burden of Musculoskeletal Diseases in the United States*, AAOS). Recently, a couple of papers (Lozano, Naghavi, & Foreman, 2020; Murray, Vos, & Lozano, 2012) published in *The Lancet* reported results from a large comprehensive study about the worldwide impact of all major diseases and risk factors on disability and mortality; this study found "that musculoskeletal conditions affect more than 1.7 billion people worldwide and have the fourth greatest impact on the overall health of the world population, considering both death and disability (DALYs)." The same study also stated that musculoskeletal-related disorders has increased by 45% over the past 20 years and will continue to increase unless appropriate preventive action is taken. Despite the widely recognized important role of MSS in our daily functions and in maintaining quality of life, the current fundamental knowledge related to the individual components of the MSS architecture remains relatively limited. This limited knowledge with a lack of standardized data representation often proves insufficient in systems engineering and modeling applications. This limitation is a major obstacle in the progress of fundamental and applied research today. Efforts in bioengineering and biomechanics are needed to improve current simulation methods through a better integration of functional anatomy knowledge with improved and well-described system modeling. In fundamental research on MSS architecture, more effort should be devoted to gathering reliable statistical data on the architecture of each single muscle and quantify anatomical variations in the human species.

Human Orthopaedic Biomechanics. DOI: https://doi.org/10.1016/B978-0-12-824481-4.00001-9

This chapter focuses on compiling all the knowledge available on MSS architecture, muscle functions, and physiology in literature; there are some excellent textbooks available shedding light on MSS components from a fundamental perspective. The reader is expected to be well-informed about MSS anatomy and physiology in order to better understand the content of this chapter. This chapter presents the details with the aim of facilitating further research and bridging the gap between literature and current knowledge, and proposes research tracks be followed. It also presents a wider picture of several challenges lying ahead and tasks to be performed to achieve the goals of gaining objective fundamental knowledge on all MSS components in order to develop clinically useful MSS models.

Current MSS research, including modeling development, is based on fundamental knowledge available from a relatively obsolete literature as opposed to current research requirements. For example, current system modeling assumes that only minor differences exist between the 320 pairs of muscles found in the human body, while it is largely recognized that each muscle shows its own architecture and functional properties (Zatsiorsky & Prilutsky, 2012). Innovative modeling efforts (such as the Digital Patient) require quantitative data collected following validated protocols, rather than qualitative descriptions obtained from, for example, fundamental anatomy textbooks only. In summary, more advanced knowledge of MSS must be gathered to allow new frontiers to be reached in MSS-related bioengineering research and applied in medical research.

WHY IS DETAILED FUNCTIONAL MUSCULOSKELETAL SYSTEM KNOWLEDGE REQUIRED TODAY?

Current MSS biomechanics and simulation literature include complex modeling environments based on mathematical and engineering developments to enable predictions of various anatomical or physiological MSS parameters during clinical activities (Steele & van der Krogt, 2012). Despite the maturity of the underlying mathematical tools, the physiological relevance of the simulation results, through validation, has not been determined until today because of the simplistic system representation of the MSS architecture used to perform the predictions. This is mainly due to the limited availability of validated system models to describe each single MSS component processed during the attempted simulation. This lack of MSS knowledge and system integration is recognized as one of the main reasons for the limited applied results available today based on research carried out to date on MSS modeling (Van Sint Jan, 2005). Detailed system descriptions of muscle architecture and individual bone differences are yet to be obtained. Some efforts were made in the past, but results were fragmentary and lacked all required information (Sholukha, Snoeck, Moiseev, Rooze, & Van Sint Jan, 2008; Zatsiorsky & Prilutsky, 2012). It is also widely acknowledged that musculoskeletal geometry is the most critical element in MSS modeling, that is, calculated values of muscle force depend more heavily on estimates of muscle moment arms than on the muscle properties themselves (Pandy, 2001). This is why generic models are usually criticized (Epstein, 2006; Pandy, 2001) and why new efforts must be organized to model MSS architecture as accurately as possible, including anatomical variations. Recently, today's available knowledge has been openly criticized (Guimberteau & Delage, 2012): "The current scientific data cannot be satisfied with the simplistic mechanical explanations available from today's physiology books." This indicates that many mechanical theories and models related to the MSS physiology are obsolete because of a lack of new insights and novel knowledge related to MSS system architecture.

In summary, the only solution to overcome the above-mentioned shortcomings is to collect all required anatomical information related to specific subjects, to build statistical knowledge on anatomical variations, and to store the newly collected knowledge in detailed MSS system models. The variation descriptions in an entire population could then be used for carrying out objective measures and predictions using simulation tools. Novel methodological modeling pipelines should be developed, which will further enable researchers to create detailed MSS multiscale, multiorgan system models, including as many anatomical variations as possible. This chapter presents such integrative pipelines enabling detailed data collection and MSS multiscale modeling. It will hopefully give readers ideas for new lines of research.

One could wonder why so few fundamental MSS data are available. This is due to a combination of at least two factors. First, no need was felt for detailed system models related to MSS until recently. New integrative efforts and simulation environments for predictive research have come up only recently due to the mathematical and progresses in the information and computer technology field (ICT) made over the past decade (Fenner, Brook, & Clapworthy, 2008; Van Sint Jan, 2005; Van Sint Jan, Viceconti, & Clapworthy, 2007; Viceconti et al., 2006). The second factor is the fact that the MSS is often seen by funding agencies as a system built from well-known simple mechanical and architectural laws under the control of the central nervous system: This traditional "mind--body influence" theory (Astin, Shapiro, Eisenberg, & Forys, 2003) has guided the research in anatomy and physiology for centuries. It is therefore not surprising that the current literature does not include more detailed descriptions of each MSS component (i.e., independent muscle architecture, individual ligament properties, etc.). It is only recently that the MSS system has been recognized as having an influence on the brain cognition (Svare, 2006; Taylor, Goehler, Galper, Innes, & Bourguignon, 2010). In order to make progress in our understanding of how the MSS interacts with brain cognition, accurate descriptions of the MSS are needed and novel system descriptions are required to quantify the organizational architecture of the main MSS components.

The availability of advanced MSS modeling standards, such as the one this chapter is intended to develop, will open up new horizons for biomechanical and medical research. Current MSS-related prediction systems fail to provide extensive real-life results because of the unavailability of detailed system models that could be embedded into biomechanical simulations (Epstein, 2006; Pandy, 2001; Van Sint Jan, 2005). Novel MSS system models will enhance the utility of tools in a variety of fields aiming at better understanding human MSS architecture, muscle force production, or joint constraints. This will boost future clinical research and understanding related to neurological and orthopedic disorders (Steele & van der Krogt, 2012), surgery (Guimberteau & Delage, 2012), sport injuries (Thelen, Chumanov, & Hoerth, 2005), and ergonomics (Sironi, Standoli, Perego, & Andreoni, 2019). Advanced anatomical models will also be of high value as reference data for research in robotics, for example, to better mimic human anatomy in research on pneumatic muscles used for exoskeleton, or for humanoid robotics development and digital human modeling (Howard & Yang, 2012; Mooney, Rouse, & Herr, 2014).

GENERAL OBJECTIVES OF THE RESEARCH TO BE ORGANIZED

The proposed research lines should perform an extensive and accurate system modeling of individual bone variations and muscle architectural components. The presented methodology poses at least four challenges (Fig. 29.1). A large model repository must be created to shed new light on the

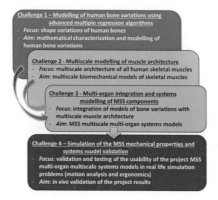

FIGURE 29.1

Challenges ahead (see text for explanations).

numerous human bone variations (Challenge 1, see below for description) and detailed multiscale human muscle architecture must be developed (Challenge 2). Further, from the results of the afore-mentioned two challenges, MSS multiscale, multiorgan system models should be built using novel mathematical representations and data fusion tools (Challenge 3). Usability of the newly created MSS system models in real-life applications must be validated through extensive in vivo simula-tions related to MSS mechanical behavior (Challenge 4). All newly acquired results must be avail-able to the Scientific Community by using sustainable digital data-sharing services and existing modeling tools.

METHODOLOGY

The proposed methodology, based on the four aforementioned Challenges, includes a total of 15 actions (Fig. 29.2).

This following research vision is based on the past experience of the authors and their team in fundamental anatomy, modeling, and applied biomechanical research related to the MSS. Such research work calls for involving experts from different fields, including anatomists, biomechanists, clinicians, mathematicians, and software engineers, and also involves medical image processing and multiscale system modeling. PhD students must be involved as well to work on various aspects of the workplan, encouraging them to pursue that kind of research in their career.

CHALLENGE 1: MODELING OF HUMAN BONE VARIATIONS USING ADVANCED MULTIPLE REGRESSION ALGORITHMS

Bone shape characterizes bone function(s) (e.g., the flat-shaped design of the scapula facilitates its gliding along the thorax, while the robustness of the iliac bone enables a stable but smooth gait). Bones also show many morphological variations. Although *qualitative* bone variations are found in

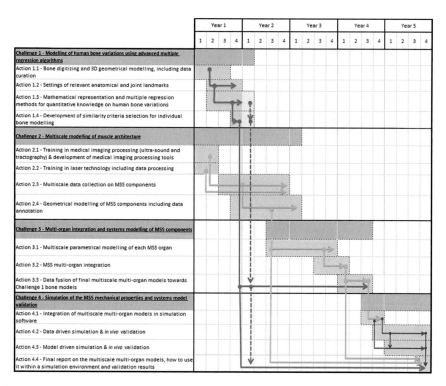

FIGURE 29.2

Challenges and tasks. All 4 Challenges and 15 Actions are indicated. *Arrows* indicate the methodology with *full arrows* indicating the data flow between the Actions. Specific details are given in the text.

the literature, few *quantitative* data are available. When available, data are often limited by the sample size, and models are usually generalized without taking into account variations. Such generalization is unsuitable for subject-specific simulation activities. More relevant results are expected from statistics obtained on a much larger sample, as envisaged by Challenge 1. Special modeling tools available from the literature should be improved (Sholukha et al., 2009, 2011). Such tools can be based on interactive similarity criteria selection during multiple regression procedures to access the available statistical population. This would prove to be highly innovative in the modeling literature where mostly relatively simple regression equations are used (Peng, Panda, Van Sint Jan, & Wang, 2015).

The skeleton is recognized as the foundation of the MSS because it allows other MSS components, such as muscles, ligaments, and aponeurosis, to act on joints to produce or to constrain skeletal displacements. Skeletal spatial displacement is an important physiological function in the human species, specific bone shape characterizes particular joint functions, and each bone type shows many morphological variations. These variations between individuals are directly linked to MSS performance (bone shape characterizes muscle moment arms; Pandy, 2001), MSS pathology (some variations of bone shapes are associated with limb deformities; Bobroff, Chambers, Sartoris, Wyatt,

& Sutherland, 1999), and MSS clinics (joint prosthesis customization is still a growing field; Van Sint Jan, 2005). Although qualitative descriptions are found in the literature, extensive quantitative data to measure the magnitude and details of the human bone morphological variations are not available. When available, data are unfortunately limited by the sample size (Sholukha et al., 2011), and models are usually generalized without taking into account the high variability of human anatomy. More statistically relevant results to be processed using multiple regression procedures are expected on a much larger sample in order to tackle Challenge 1. Novel procedures based on interactive similarity criteria selection must be developed to improve results from regressional equations. Because of the nature of similarity criteria (Sholukha et al., 2011), statistical subsamples are processed from a main sample to obtain specific sets of regression coefficients characterizing each processed bone subsample (i.e., bone variation). This approach is unconventional in the modeling literature where mostly relatively simple general regression equations are published. This reductionist approach is acceptable to obtain generic models, but less interesting to quantify human variations as required for bioengineering applications. The use of similarity criteria is therefore justified and highly recommended.

Challenge 1: implementation

Challenge 1 includes four actions aiming at quantifying the variations of most human bones and thus leading to detailed quantitative descriptions of the human osteology (Fig. 29.2). *Action 1.1*: Digitizing several hundreds of human specimens should be carried out to facilitate accurate three-dimensional (3D) geometric bone modeling. Models must be stored in standard formats (such as STL). *Action 1.2*: On each model obtained during Action 1.1, anatomical bone features such as key bony landmarks used in real-life applications and joint surface features should be located. The spatial coordinates of these features can be collected using so-called virtual palpation procedures (Van Sint Jan, 2007). *Action 1.3*: The anatomical features located during Action 1.2 on each bone model must be further processed to mathematically characterize each modeled specimen. Novel methods to mathematically define the numerous variations demonstrated by the various bones found in the human skeleton will have to be developed. It has been previously successfully demonstrated that the morphological features of individual bones can be predicted from anatomical landmarks by using advanced multiple regression methods based on mathematical modeling representations and similarity criteria (Sholukha et al., 2011) (Fig. 29.3). *Action 1.4*: A special access tool must be developed to allow further modeling and validation actions (in Challenges 3 and 4) through accessing models using similarity criteria and multiple regressions.

Innovative quantitative osteology knowledge, geometric and parametric bone models, multiple regression procedures, and access tools will be available after overcoming Challenge 1. These results can then be used in tackling Challenge 3 in order to be fused with system models related to specific muscle architecture (emerging from Challenge 2, see next).

Similar work has been carried out previously on a relatively limited amount of specimens (Sholukha et al., 2011). A much larger database including several hundreds, probably thousands, of specimens and more advanced regression algorithms is required to obtain more reliable statistical correlations. Furthermore, this effort must be extended to other human bones. The only way to fulfill this aim is through performing measurements in multiple internationally recognized centers that are well-known for their large and well-documented osteology collections in order to digitize the available collection using state-of-the-art 3D digitizing hardware. Detailed surface models should

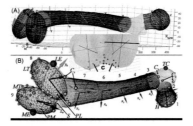

FIGURE 29.3

Mathematical modeling of the femoral bone morphology obtained from multiple regression algorithms. (A) Final result. (B) Final results and landmarks used to build the model (coloured spheres); the local frame of each bone component is visible and enable regression algorithms to scale and orientate the components according to input data. The method used here has proven to be able to reconstruct individual bone morphological characteristics from a few anatomical landmarks and a reference database.

be obtained at these locations and will have to go through a quality curation process. Once data related to a particular specimen have been positively curated, they must be further processed by trained anatomists to figure out key bony landmarks, bone morphology, and joint surfaces using well-validated methods (Action 1.2) (Van Sint Jan, 2007). All digitized morphological data would then be further processed to perform mathematical fitting (e.g., quadric and spherical representations) (Action 1.3) (Sholukha et al., 2009, 2011). Eventually, all available data and models will be annotated and stored in a database together with other individual data. The development of robust multiple regression equations using similarity criteria will enable the prediction of the spatial and morphological relationships between different elements based on the data available from the database (e.g., between particular bony anatomical landmarks and bone shape and joint surface orientation) (Salvia, Van Sint Jan, & Rooze, 2009; Sholukha et al., 2009, 2011).

The last action of Challenge 1 (Action 1.4) is ambitious, but results may be promising. It has been previously demonstrated that the best results from regression equations are obtained when similarity criteria are determined for each particular specimen (Sholukha et al., 2011). This leads to the following practical problem: Each specimen has its own set of regression equations obtained from the selection done following the similarity criteria procedure. The advantage of this representation is the high accuracy of the estimation, but the main problem is related to the fact that similarity criteria must be determined for each specimen; this makes it difficult to use the newly collected knowledge by third-party applications. Therefore, one should develop a special access pipeline that will enable generating similarity criteria on-the-fly. This pipeline will request the spatial location of anatomical landmarks (i.e., *3D coordinates of bony landmarks obtained locally from assessment activities similar to Challenge 4*) to perform the similarity criteria selection and to produce the regression equations for the current dataset. Finally, the access tool will generate a mathematical representation of the approximated bone data (*and of all muscle systems data attached to it during Challenge 3, see below*), and copies of all database specimens similar to the predicted values. This effort will call for an entirely novel regression processing pipeline to be developed to query the database and to perform regression analysis. To the author's knowledge, such a pipeline has never been created in the field of bioengineering, and it will be really challenging. However, once realized, it will make available new statistical resources for numerous applications (computational anatomy, biomechanics, and orthopedic research) aimed at anatomical feature estimation for simulation purposes. Challenge 1 will also allow a robust foundation to be built for the description of the muscular MSS components obtained in Challenge 2 and further integrated in Challenge 3.

CHALLENGE 2: MULTISCALE MODELING ON MUSCLE ARCHITECTURE

Architecture of each single muscle is a highly multiscale system made of various anatomical levels (Lieber, 2009; Zatsiorsky & Prilutsky, 2012) each of which is related to specific physiological parameters (Fig. 29.4): protein level (sarcomere length); cellular level (l^{CE} = muscle fiber length = optimal fiber length); tissue level (fascicule length, α_F = pennation angle, path of wrapping points, PCSA = physiological cross-sectional area); and organ level (origin[s] and insertion[s], muscle weight and volume, muscle length, l_0^{SE} = rest tendon length, and l^{APO} and β_A = aponeurosis length and angle). While system description is relatively well-advanced for other anatomical systems, accurate MSS multiscale modeling for each single muscle is still missing. This lack of knowledge is a limitation for a better understanding of, for example, the forces generated from the mechanical work produced by each single muscle to act on the skeleton through tendinous and aponeurotic structures to engender joint displacements. This explains the need for obtaining more accurate and quantified MSS models. The MSS shows a supplementary aspect compared to other anatomical systems: the individual architectural variations showed by each of the 320 pairs of human muscles (including their tendons and surrounding aponeurosis) (Lieber, 2009; Zatsiorsky & Prilutsky, 2012). As such, the multiscale aspect of each single muscle contributes to the complexity of the overall MSS, qualifying it as a *multiorgan* anatomical system. Better biomechanical modeling of each

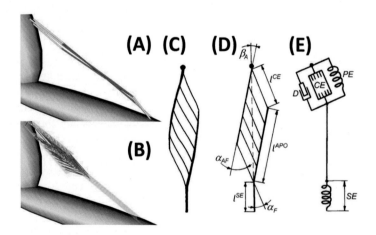

FIGURE 29.4

Geometrical and system muscle models. (A) Simplified muscle model used in most simulation methods available today. (B) Muscle model including collected multiscale data leading to more realistic muscle architecture. (C) Schematic structure of a penniforme muscle (other muscle configurations should be obtained as well). (D) System abstraction with denotation of multiscale muscle architecture properties. (E) Muscle model composition.

single muscle is necessary to improve our understanding on specific muscle action within this multiorgan MSS and improve current simulation tools.

Such complex multiscale, multiorgan organization is truly unique among all anatomical systems and makes MSS system modeling a real challenge. Better knowledge on muscle architectural variations and proper system modeling are fundamental in improving our current understanding on individual muscle functions within this multiorgan MSS model. Such progress is absolutely required to improve simulation tools that are currently lacking detailed anatomical realism. As mentioned above, the multiscale architecture of each individual MSS muscle is still poorly described in the contemporary literature; several application fields require more accurate knowledge on muscle architecture for their own developments (Van Sint Jan, 2005; Zatsiorsky & Prilutsky, 2012). There have been attempts to gain more detailed MSS knowledge, but, unfortunately, these efforts were limited in scope (i.e., they were not dealing with all aspects of the muscle multiscale architecture simultaneously) and did not include integration of the results into biomechanical system models that are easily usable by third-party engineering applications. For example, some physiological parameters have been collected for a few muscles using gross dissections, combined or not with laser technology (Delp, Suryanarayanan, Murray, Uhlir, & Triolo, 2001; Klein Horsman, Koopman, van der Helm, Prosé, & Veeger, 2006). Despite recognizing the value of these data, no significant work has been carried out on understanding the muscle and aponeurosis 3D architecture. Other authors concentrated only on the 3D morphology of a few muscles using either gross dissection (Kellis, Galanis, Natsis, & Kapetanos, 2010), ultrasound (Fry, Cough, & Shortland, 2004), or diffusion tensor imaging (Froeling, Nederveen, & Strijkers, 2012). Insights collected from such studies were promising, but were not combined with other relevant physiological parameters. A more extensive study was organized several years ago to collect MSS knowledge using a newly developed protocol that included data fusion from medical imaging (CT and MRI) and a novel stereophotogrammetry approach performed during gross anatomy dissections (Sholukha et al., 2008). A multiscale, multiorgan model was eventually obtained (Fig. 29.5) after data fusion of the collected heterogeneous data such as 3D geometries of bones, muscle origins and insertions, muscle fascicle length, tendon length, pennation angle, wrapping path, weight and volume, and joint kinematics. Although the collected data are probably one of the most detailed MSS-knowledge resources that are currently publicly available, they are still lacking several key aspects to explain the real 3D muscle architecture and mechanical behavior because of the limited data collection, which was mainly due to the lack of availability of proper resources and time.

In order to answer the above-mentioned needs, overcoming Challenge 2 should lead to the development of innovative modeling procedures to obtain models of the multiscale architecture of most human skeletal muscles. It must be organized around multimodal MSS data collection, including mathematical modeling representation and visualization. This requires dissection facilities, 3D digitizer, CT scan, laser and ultrasound equipment, MRI, and diffusion tensor imaging. Results from Challenge 2 will be further integrated into full MSS system representation (see Challenge 3).

Challenge 2: Implementation

The exhaustive multiscale muscle modeling work performed during Challenge 2 will request for a large-scale data collection effort.

Action 2.1 and Action 2.2 should focus on developing methodologies in medical imaging processing (tractography and ultrasound imaging) and laser technology, respectively. These actions

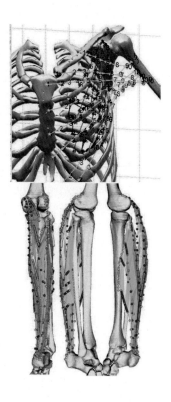

FIGURE 29.5

Multiscale, multiorgan data. These models (top: shoulder; bottom: shank) are among the few models in the literature combining physiological parameters with 3D architectural information (see text for more details).

will also include the development of data processing and modeling tools required for this particular challenge. Both actions will allow to have all expertise and tools required to start Action 2.3.

Action 2.3 protocol must include the following incremental steps applied ex vivo on cadaveric specimens. (1) Weight, length, gender, and medical conditions before death must be recorded. (2) The specimen must be prepared by inserting fiducial landmarks required for further data registration (Sholukha et al., 2008; Van Sint Jan, 2005) during the following Actions. (3) Average-gradient (minimum 3 T) MRI and diffusion tensor imaging must be done to collect *in-situ* spatial data (*muscle fiber orientation, muscle fiber length, tendon length, muscle/aponeurosis architecture, PCSA, pennation angle, and muscle wrapping path*). (4) CT imaging must then be performed to enable 3D reconstruction of the entire skeleton and anatomical landmarks to allow further data fusion (in Challenge 3). (5) Careful dissections must take place with stereophotogrammetry (*to collect spatial data on pennation angle, fascicle length, tendon length, wrapping path, origins, and insertions of all skeletal muscles*) and ultrasonography (*muscle fiber and aponeurosis 3D architecture, and pennation angle*). (6) After in-situ dissection (see step 5), each single skeletal muscle must be detached to obtain supplementary measurements required to complete system modeling (*muscle volume and weight of the muscle, high-gradient MRI [7 T] to obtain novel insights within the muscle intimate architecture and relationships with tendons and inner aponeurosis and sarcomere length, to determine the optimal muscle fiber length, using laser diffraction technique* (Klein Horsman et al., 2006) *on samples of muscle tissue*). (7) Kinematics of all major joints must be eventually analyzed using state-of-the-art motion analysis methods.

Data produced during Action 2.3 can be stored during Action 2.4 using modeling representation methods that are already available. This storage must take into account the requirements for further data fusion and multiscale system representations performed in Challenge 3 by adopting standard formats and proper data annotation. Challenge 2 models will be integrated during Challenge 3 with the regressional osteology knowledge obtained during Challenge 1. Such extensive multiscale MSS modeling is highly innovative and will allow the next project Challenges to produce novel MSS system models (Challenge 3) that will improve MSS simulation results (Challenge 4). Challenge 2 must process several human cadavers, both male and female. Ideally, several hundreds of specimens should be processed; due to the time-consuming nature of this kind of data collection (based on the authors' past experience, several weeks for one single specimen is required), a more realistic goal would be the processus of 5 specimens for each gender. One must keep in mind that the lower the amount of specimens, the lower the accuracy of the statistical sample compared to the entire human population and related variations.

CHALLENGE 3: MULTIORGAN INTEGRATION AND SYSTEM MODELING OF MUSCULOSKELETAL SYSTEM COMPONENTS

The previous Challenges 1 and 2 will allow obtaining one single specific model on each independent MSS component required to further build system models of the MSS multiscale (Action 3.1) and multiorgan (Action 3.2) architecture. Data emerging from the previous Challenges will be combined in Challenge 3 using novel data fusion methodologies that should be further developed (Fig. 29.6). The amount of data to combine, and their heterogeneous nature, makes this particular research challenging. However, once fully accomplished, detailed multiscale system models about the inner architecture of most human skeletal muscle, including the MSS multiorgan organization, will be made available for the first time. Further, the inner architecture of each muscle must be described according to bony features found in the osteology database obtained from Challenge 1. This important step is based on state-of-the-art data fusion (Action 3.3). It enables building anatomical relationships between muscle and bone MSS components to better define the muscle actions compared to the joint(s) being crossed. It means that each of the muscle system parameters collected during Challenge 2 should include information allowing its fusion with the Challenge 1 osteology information. This has never been done before at such a scale of anatomical details because of the complexity of such a heterogeneous data fusion.

Challenge 3: Implementation

The work to be performed in Action 3.1 should focus on the development of new modeling methods to describe each single MSS component using system modeling representations (see Fig. 29.4). Information to be processed is huge and highly heterogeneous. It will also include multiscale measurements related to each single muscle in the human body (origin, insertion, line-of-action, inner muscle architecture [tendon and muscle fiber organization], sarcomere and fascicle length, PCSA,

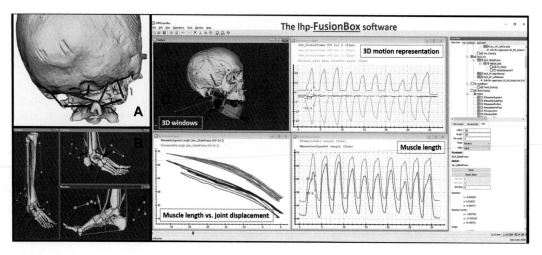

FIGURE 29.6

Heterogeneous data integration using data fusion and visualization methods. (A) Fusion of suboccipital muscles and joint kinematics data. (B) Multisegment foot model including fusion between CT data, gait analysis, and foot pressure plate. (C) Fusion combining medical imaging data (MRI) with motion capture data. Output graphs are obtained to quantify the behavior of anatomical structures included into the models, such as muscle and ligament behavior, and joint kinematics representation (including Grood & Suntay, OVP, Euler, helical axis). These modeling tools (included in the in-house lhpFusionBox software) are already available in the authors' group and should be further improved for more data generation. *MRI*, Magnetic resonance imaging.

weight, volume, pennation angle, wrapping path, aponeurotic relationships, etc.), and joint kinematics (Fig. 29.6).

The magnitude of the data fusion to be performed, the heterogeneous nature of the data, and the constraint to produce fully integrated biomechanical system representation make tackling Challenge 3 difficult. All must be combined (i.e., fused) together within a common reference system respecting the original morphology of the collected data. For some data fusion (e.g., fusion of motion data with muscle line-of-action or pennation angle), no major problems are expected thanks to methods available in the litterature (see Fig. 29.6). Other data combinations, however, request novel developments: for example, knowledge on the detailed muscle inner architecture has never been extensively combined with aponeurosis or motion data. The most advanced works currently available in the literature are limited to the tissue level (Wu, Ng-Thow-Hing, Singh, Agur, & McKee, 2007) without any further integration into multiscale system models. More work on data-driven integration is required to improve today's systems models. Action 3.1 will eventually enable building multiscale system models of each single skeletal muscle (Fig. 29.7, top).

Action 3.2 will integrate each single multiscale organ model obtained from Action 3.1 into one full MSS multiscale, multiorgan system representation (Fig. 29.7, bottom) using similar data fusion techniques to those used in Action 3.1.

Action 3.3 must combine the available multiscale, multiorgan system model (from Action 3.2) to the content of the osteology database collected during Challenge 1. This work must be performed

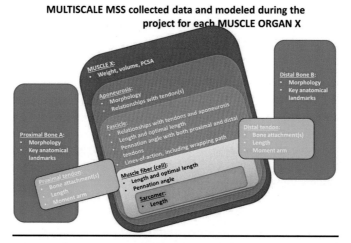

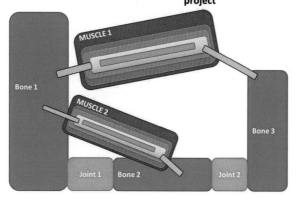

FIGURE 29.7

Multiscale modeling. Top: multiscale modeling of single muscle. Bottom: multiorgan MSS model obtained from the integration of single multiscale muscle models. The final system model obtained includes several hundred pairs of muscles and bones. The availability of such a complex integrated model will be highly innovative for third-party applications within the MSS simulation field. *MSS*, Musculoskeletal system.

based on the same data fusion protocol as above. The complexity of this particular stage will strongly rely on the preparation and proper annotation of the underlying multiorgan models during previous Actions. The main advantage of the multiscale system representation proposed in this chapter (see Fig. 29.7) is the flexibility it allows; if some anatomical levels are of poor quality in the model, it can either be eliminated or clearly identified (and tagged as less reliable in the annotation system) before fusion with the final Challenge 1 database. The final result will include a large collection of highly detailed system models related to a large spectrum of osteological variations and muscle configurations found in the human species. Such an achievement is currently not available in today's literature. The final results will be of unique value for the multibody simulations performed in Challenge 4 and for third-party researchers in need of system models related to human MSS architecture and related methods.

About muscle architecture modeling (in Actions 3.1 and 3.2). The MSS architectural relationships between contractile and connective tissue elements are important determinants of muscle function (Gans & Gaunt, 1991; Lieber & Ward, 2011). The arrangement of fiber bundles (FBs) and their attachments to tendons, bone, and other aponeurotic structures collected during Challenges 1 and 2 actions will enable determining the force-generating capability of the muscle, as well as the distance and velocity of excursion, and thus the range of force development (Gans & Gaunt, 1991; Zajac, 1989). The system modeling describing the FB organization will be characterized by muscle parameters (Lee et al., 2015) obtained during Actions 2.3 and 2.4 (such as optimized muscle fiber length, pennation angle variations, muscle volume, etc.). The available single muscle data must be further integrated into a multiscale system model using cubic uniform B-Spline, with clamped boundary conditions, and then stored as a nonuniform rational linear basis spline (NURBS) curve (Ravichandiran et al., 2009). An iterative closest point (ICP) algorithm could be used to position the reconstructed FB data for each single muscle, according to the 3D coordinates of the digitized data.

About muscle system modeling for prediction of passive and active muscle forces (in Actions 3.1 and 3.2). Hill-type muscle models are commonly used in biomechanical simulations to predict passive and active muscle forces during various movements (Epstein, 2006; Wu et al., 2007; Zatsiorsky & Prilutsky, 2012); such models enable muscle forces at the organ level and are therefore considered macroscopic muscle models using mathematical formalisms related to contractile and viscoelastic structures found within and around muscles. In conventional mechanics, Hill-type muscle models are classified as *0D* elements due to the lack of mass and inertia. Output of such models is a 1D force, which is applied to skeletal models between origin and insertion points with less attention to the real muscle—tendon pathway, which frequently shows a so-called wrapping around crossed anatomical structures (such as bone and joint surfaces or other muscles). Typically, Hill-type muscle models consist of at least four elements that can be configured in various ways according to the adopted system model: a contractile element incorporating force—length and force—velocity dependencies, a serial elastic element, a parallel elastic element, and a damping element (see Fig. 29.4) (Günther, Schmitt, & Wank, 2007; van Soest & Bobbert, 1993; Zajac, 1989). For complex biomechanical simulations of human movement (as the ones given in Challenge 4 below), both series and parallel elastic-damping and the characteristic concentric—eccentric force—velocity relations are considered. The determination of the muscle parameters used in Hill-type models highly depends on the data available for the model design. Indeed, Hill-type models are usually applied to optimize neurological stimulation patterns for human movement or to compute limb movements based on given muscle stimulation patterns (van Soest & Bobbert, 1993). The large number of muscle parameters to be integrated into the underlying system model usually makes the model designing difficult (Pandy, 2001; Van Sint Jan, 2005; Zatsiorsky & Prilutsky, 2012). Moreover, the specific architecture of each single muscle requests the development of a specific system model for each skeletal muscle found in the human anatomy. In practice, in most models, authors do not have access to detailed MSS data similar to the ones from Challenges 1 and 2; for this reason, authors obtain parameters from various literature sources, which are not linked to each other. Consequently, the physiological realism of the final models is questionable (Epstein, 2006). The protocol presented in this chapter should lead to detailed architectural data from well-controlled data collection (Challenges 1 and 2); this should enable researchers to develop and test various Hill-type configurations, and design improved Hill-type system models.

In summary, Action 3.1 focuses on developing novel and advanced multiscale Hill-based system models of each single muscle based on well-controlled and integrated data collection and modeling methodologies. Action 3.2 will fuse these multiscale models with each other to obtain a full MSS multiorgan system model that will enable the analysis of the intricacies of the spatial relationships between muscles, as well as the 3D arrangement of the FBs in relation to their tendons and aponeuroses, and individual bone variations (Action 3.3).

One of the motivations of such work is to allow biomechanical applications to access quality system models to further today's simulation research. An important requirement of these application fields is to obtain proper validation indicators together with the models. This is especially important for in vivo applications related to clinical prediction systems. An example of such a validation method is proposed during Challenge 4 (see below) to predict the MSS mechanical behavior within in vivo conditions. This behavior will be compared to noninvasive in vivo mechanical measurements to assess the quality of the prediction and therefore of the underlying MSS system models.

CHALLENGE 4: SIMULATION OF THE MUSCULOSKELETAL SYSTEM MECHANICAL PROPERTIES AND SYSTEM MODEL VALIDATION

The customizable multiscale, multiorgan MSS system models at the end of Challenge 3 should be ready to be implemented in dynamic simulation systems (Action 4.1). Challenge 4 produces a methodological pipeline focusing on system model customization for simulation purposes. The available models can be used in indirect (data-driven) and direct (physically based) simulations (Actions 4.2 and 4.3, respectively) of several real-life problems in order to determine the robustness and the usability of the developed system models. For example, two domains that are addressed for simulation activities are clinical motion analysis and ergonomy.

As such, data-driven simulations (Action 4.2) can be applied to the Challenge 3 system models using highly heterogenous in vivo measurements (*including manual palpation for bone dimensions and orientation, motion capture data for joint kinematics, external forces using dynamometry, electromyography for muscle recruitment, ultrasound for muscle fiber pennation angle variations during motion, MRI for aponeurosis architecture*) to drive the simulation. In this approach, joint force and torque are usually evaluated based on bush-structured joint kinematic measurements; this approach is, however, sensitive to the accuracy of the captured motion data and requests to constrain joint translations (i.e., reducing joints to ball-and-socket mechanisms) (Zatsiorsky & Prilutsky, 2012; Zinkovsky, Sholuha, & Ivanov, 1996). An alternative approach should be developed, for example, based on a mixed dynamics implementation, which includes constrained forward dynamics modeling of joint segment kinematics. Such an original approach should be possible thanks to the above Challenges leading to physiologically accurate MSS system models thanks to an extensive integrated data collection and system modeling workplan. Special attention should be paid to specific 6 degree-of-freedom joint mechanisms thanks to available methods (Besier, Fredericson, Gold, Beaupré, & Delp, 2009; Sholukha, Leardini, Salvia, Rooze, & Van Sint Jan, 2006). The available mixed dynamics must be then be used to assess in vivo joint kinematics. The reconstructed joint kinematics together with in vivo external forces (e.g., ground reaction force or any other measured force) and in vivo muscle information (e.g., pennation angle) can then be used to customize the available MSS system models to assess joint reaction forces and torques at various joint levels. Comparison of evaluated total torques from the simulation with torques obtained during in vivo assessment of the maximum joint torque will allow a preliminary evaluation of the muscle forces (which can be compared to the theoretical forces obtained from the muscle cross-sectional area). Once customized with in vivo input data, the MSS system model will correctly reproduce joint kinematics and satisfy external motions and force constraints, and the evaluated joint torques will enable predicting muscle forces.

In Action 4.3, results from the data-driven simulation (Action 4.2) will be used to identify key parameters to customize with in vivo data the innovative Hill-type system models that will be produced from Challenge 3 and subsequently used in model-based simulation activities. It has been previously mentioned that such models are based on a mathematical description of muscle contraction mechanism and include at least four elements: contractile element (CE), serial elastic element (SE), parallel elastic element (PE), and damping (D) element (Günther et al., 2007). The aim of Action 4.3 should be to determine muscle-specific parameter values based on in vivo experimental data from muscle contractions (from electromyography), muscle fiber orientation and pennation (from ultrasonography and MRI), bone orientation (from manual palpation), and muscle forces and joint torques (from Action 4.2). Usually, simulations based on Hill-type models employ muscle recruitment obtained from electromyography for a fixed set of muscle parameters. Action 4.3 will optimize a larger amount of in vivo input data and should therefore find most optimal solutions in terms of muscle recruitment, muscle force production, and joint constraints.

Challenge 4: Implementation

Action 4.1 focuses on the implementation and the customization of the multiscale, multiorgan MSS models available from Challenge 3 in existing dynamic simulation solutions. This will require adaptation of the models embedded into the available simulation software.

Action 4.2 uses data-driven simulations to produce measured external motion and force constraints to predict muscle forces during the following two daily activities: prehension activities and car pedal clutching. (*Note that this section describes the protocol for the pedal clutching only; simulation of prehension activities would be similar.*) The following in vivo measurements must take place on a volunteer who will be fully informed of the experimental protocol. Manual palpation will help determine the spatial locations of well-defined anatomical landmarks in order to define bone segment anatomical frames (Van Sint Jan, 2007) and to perform regression operations to obtain a customized MSS multiscale, multiorgan system model (using results from Challenges 1 to 3). The subject must be equipped with reflective markers required for motion analysis and electromyography electrodes for muscle recruitment measurements (Salvia et al., 2009). Joint maximum torque data for the main joints (e.g., hip, knee, and ankle joints) can be collected based on dynamometry. Supplementary subject-specific morphological data will be collected using medical imaging (ultrasonography and MRI). The following in vivo data will be finally available: joint range of motion; maximum joint torque; relationship between joint force level and muscle activity; maximum static external force; relationship between static external force level and muscle activity; relationship between quasi-static task posture and EMG; pennation angle variations; muscle volume and morphology; and bone and joint models. The evaluation of the joint torques during the analyzed tasks will request for the development and implementation of a mechanical model based on multibody dynamics approach. The model will

FIGURE 29.8

Data collection for validation procedures. Subject equipped with reflective markers and EMG sensors during measurements of maximum torques (left) and pedal clutching (right). The 3D MSS model (right) obtained from regression analysis of the subject's bone orientation (from Challenge 1) fused to multiscale, multiorgan system models (from Challenges 2 and 3).

be developed as a closed loop with 6 degree-of-freedom (6-DoFs, i.e., three rotations and three translations) joints connecting a tree-like structure (Zinkovsky et al., 1996) with five rigid bodies (pelvis, thigh, patella, shank, and foot in the case of pedal clutching). This model is controllable by applying external forces, torques, or motions on the pelvis and foot. Each joint will be modeled as a 6-DoF bush joint (XYZ translations and Euler angle rotation convention). Passive spring-dumper drivers will be implemented for each DoF to control the overall motion. Fig. 29.8 shows an example of the experimental setup and developed 3D multibody dynamic model that could be realized. Geometric and mass-inertia properties of the model will be determined from anthropometric measurements and regression estimation (de Leva, 1996).

The model motion simulation includes several steps. The initial pose of the model is defined from the motion-capture data. Then, by applying spring−dumper soft constraints at the joint level, each motion frame can be simulated and stored. Finally, by applying the stored joint kinematics, joint reaction forces and torques could be derived. Accurate results related to maximum force prediction and muscle forces produced by muscle activity are expected. At first, joint torques will be estimated from the modeled kinematics. Then, the relationship between joint torques and muscle activity will be compared with predicted values from maximum torque measurements. Finally, muscle forces estimated from joint torque and muscle moment arm behavior will be derived and compared to muscle activity. This protocol will be applied on the analysis of the upper limbs during prehension activities as well.

Action 4.3 concentrates on customizing the Hill models produced in Challenge 3 with in vivo experimental data related to muscle recruitment and output from Action 4.2 (muscle forces and joint torques). This study mainly focuses on muscle parameter determination for the shank muscles during different tasks involving isometric and concentric foot plantar flexions. For this purpose, a test device must be used to analyze joint kinematics, ground reaction force, and muscle recruitment (Fig. 29.9A). The subject is asked to lift with one leg a load slider against gravity with maximal

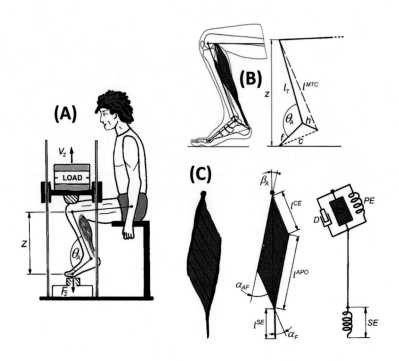

FIGURE 29.9

Experimental setup for validation. (A) Experimental setup allowing measurements related to kinematics and dynamics analysis. (B) Geometrical model and parameters of lower leg and foot MSS to calculate muscle contraction (the muscle here is the gastrocnemius; the demonstration can be extended to any body muscle). (C) (from left to right): Structure and components of the muscle model (including muscle fibers, aponeurosis, and tendon information), muscle system model abstraction with denotation of architecture properties, and muscle model composition. See text for details and acronyms.

voluntary efforts. For analysis of isometric contractions, the load slider will be rigidified. Subject-specific measurements must also be collected during the experiment (Fig. 29.9A and B): ankle joint displacement; shank length (l_T); distance between each tendon insertion on the heel and ankle joint center (h); distance between each tendon insertion on the heel and knee joint center (l^{MTC}); distance between joint center of the big toe and each tendon insertion (c); ground reaction force $F_Z(t)$; vertical position $z(t)$ and velocity $v_Z(t)$ of the slider; and muscle recruitment with surface electromyography (for the following muscles: gastrocnemius medialis muscle [GM], gastrocnemius lateralis muscle [GL], soleus muscle [SOL], tibialis anterior muscle [TA], and extensor digitorum longus muscle [EDL]).

On the basis of these data, contraction velocity (v^{MTC}) and force (F^{MTC}) of the analyzed muscles are calculated using the MSS system model available from Challenge 3 and customized to the subject undergoing the analysis. For this purpose, medium optimal fiber length $\left(l_0^{CE}\right)$ and pennation angle of the fibers (α_F) of the analyzed muscles will be estimated during the analyzed task. Results will be compared with in vivo measurements obtained using ultrasonic imaging (muscle fiber pennation angle [α_F], fiber angle in relation to the aponeurosis plane [α_{AF}], and angle of aponeurosis plane [β_A]). The muscle linear damping force will be defined by damping coefficient k_D, and elastic components will be described by quadratic functions. Pennation angle, length (l^{CE}), contraction velocity (v^{CE}), and force (F^{CE}) of the contractile element will be derived from the Challenge 3 system model (Fig. 29.9C).

Starting from a neural input $q(t)$ (i.e., normalized muscle stimulation level between 0 and maximum), sarcoplasmatic Ca^{2+} concentration (t) and muscle activation $Q(t)$ can be calculated using Eqs. (1)−(3) (Figs 29.10 and 29.11), whereas a time constant characterizes Ca^{2+} increase after stimulation.

The function considers the length dependence of attachment dynamics of actin and myosin. Due to the partial overlap of contractile filaments, the reachable force of the muscle fibers depends on particular sarcomere length (l^{CE}) defined in Eq. (4) (van Soest & Bobbert, 1993), where w is the relative length range around the optimal CE length, when muscle fibers are able to produce a maximal force. In the study, isometric, concentric, and eccentric muscle contractions must be investigated. The contractile element force (F^{CE}) is normalized in Eq. (5) to the maximal isometric force with respect to the degree of muscle activation Q and specific contractile element length ($L(l^{CE})$). Hill parameters a and b are normalized to F_0^{CE} and l_0^{CE} ($\hat{a}$ and $\hat{b}$), respectively in Eq. (5).

Equations	Eq. #
$\dot{\gamma} = \eta(\gamma_{MAX} \, q(t) - \gamma)$	1
$\rho(l^{CE}) = c_L \dfrac{l_{MAX} - 1}{\left(\dfrac{l_{MAX} \, l_0^{CE}}{l^{CE}}\right) - 1}$	2
$Q(q, l^{CE}, t) = \dfrac{Q_0 + \left(\rho(l^{CE})\gamma(q,t)\right)^2}{1 + \left(\rho(l^{CE})\gamma(q,t)\right)^2}$	3
$L(l^{CE}) = -\left(\dfrac{l^{CE}}{w}\right)^2 + \dfrac{2 l_0^{CE}}{w^2} l^{CE} - \left(\dfrac{l_0^{CE}}{w}\right)^2 + 1$	4
$v^{CE}(l^{CE}, F^{CE}, Q) = k_V(Q) \cdot l_0^{CE} \cdot \left[\dfrac{(L(l^{CE}) + \hat{a}(l^{CE}))\hat{b}}{\dfrac{F^{CE}}{Q \cdot F_0^{CE}} + \hat{a}(l^{CE})} - \hat{b}\right]$	5
$\hat{a}(l^{CE}) = \begin{cases} \dfrac{a}{F_0^{CE}} & , l^{CE} < l_0^{CE} \\ \dfrac{a L(l^{CE})}{F_0^{CE}} & , l^{CE} \geq l_0^{CE} \end{cases}$	6
$\hat{b} = \dfrac{b}{l_0^{CE}}$	7
$k_V(Q) = \begin{cases} 1 & , Q > 0.3 \\ 3.33Q & , Q \leq 0.3 \end{cases}$	8

FIGURE 29.10

Equations for contractile element modeling. See text for details and Fig. 2.11 for initial and border values.

Parameter	symbol	initial value	border values
available length range of CE related to l_0^{CE}	w	$0.40 \, l_0^{CE}$	$0.35 \, l_0^{CE} \dots 0.50 \, l_0^{CE}$
F_0^{CE}-normalised Hill-parameter a	$\hat{a}$	0.2	0.1 ... 0.6
l_0^{CE}-normalised Hill-parameter b	$\hat{b}$	2.5	1.0 ... 5.0
time constant of sarcoplasmatic Ca^{2+}-increase	η	$15.0s^{-1}$	$10.0s^{-1} \dots 25.0s^{-1}$

FIGURE 29.11

Supplementary optimization parameters. Initial and border values used in Fig. 2.10 are indicated.

Contraction velocity v^{CE} depends on muscle activation; this will be taken into account by the factor $k_V(Q)$ (Eq. 8).

Further optimization is required to predict CE contraction to a given time $\left(l^{CE}(t), v^{CE}(t), F^{CE}(t)\right)$. To achieve this, in vivo input data used are the muscle recruitment data obtained from electromyography (i.e., on GM, GL, SOL, etc.) and subject-specific F_0^{CE} determined from isometric measurements performed at an optimal muscle length. Fig. 29.11 shows supplementary muscle model parameters that will be used for this supplementary optimization. These values are to a high degree muscle specific and are difficult to determine. For this reason, an optimization routine must be applied to quantify these values.

The aim of optimization is to search for a muscle model parameter vector so that the model exactly describes a given time course of CE contraction under the condition that the difference between a stimulation input $q(t)$ into the muscle model and normalized EMG of the plantar flexors could be minimized (least-squares method). For this purpose, based on literature review, initial and border values for all parameters could be defined before running the optimization routine.

Cost function for the minimization of the weighted sum of three relatively independent components will be performed as follows: (1) maximum of the sum per unit time of muscle activation Q; (2) minimum of the sum per unit time of absolute dQ/dt, to avoid reaching the maximum sum of activation by exaggerated spikes in $Q(t)$; and (3) minimum difference between normalized measured electromyography signal $A_{EMG}(t)$ and estimated neural input $q(t)$. For each particular instant, the relationship between F^{CE}, v^{CE}, and $Q(t)$ can be found by substituting $k_V(Q)$ in Eq. (5) with Eq. (8), leading to a quadratic equation. From Eq. (3), $\gamma(t)$ can be extracted, and the neural input $q(t)$ can be computed after numerical differentiation of Eq. (1). Besides muscle model parameters, spline smoothing coefficients will be included as variables into the optimization procedure due to the necessity of numerical differentiation of $Q(t)$ and (t).

CONCLUSIONS

This chapter describes, in addition to the musculoskeletal architecture, a methodological pipeline based on real data collection performed in vivo and in vitro. The proposed method integrates all collected data into one single data fusion channel, leading to fully integrated models ready for third-party applications. Due to the magnitude of this integrative data collection, the available models will be of high quality and valuable for simulation testing.

A key issue is to raise funding for such fundamental research, which seems difficult. Today's funding priorities mainly go to ICT development, and not to real data generation, which is highly time consuming and requires the implementation of ICT tools. The effectiveness of ICT tools in real-life applications remains unclear because of the lack of extensive validations. Funding agencies seem more sensitive to visual results than fundamental research, dissection work, and digitization. This is counterproductive, because without full validation based on real-life experiments and detailed high-quality anatomical models, ICT solutions cannot be accepted in daily clinical activities (i.e., today clinical acceptance of MSS system modelling results is extremely low despite several decades of ICT developments). It is worth noting that funding agencies are now more reluctant to keep funding so-called *"innovative musculo-skeletal modeling projects,"* which are actually very similar from a technological point of view to projects funded 10−15 years ago. Until these projects really show their validity through extensive and true validation experiments based on extensive data collection and validation work, using a method similar to the method presented in this chapter, the authors truly believe that the clinical acceptance of MSS modeling work will be limited.

REFERENCES

Astin, J., Shapiro, S., Eisenberg, D., & Forys, K. (2003). Mind-body medicine: State of the science, implications for practice. *The Journal of the American Board of Family Medicine*, *16*(2), 131−147. Available from https://doi.org/10.3122/jabfm.16.2.131.

Besier, T., Fredericson, M., Gold, G., Beaupré, G., & Delp, S. (2009). Knee muscle forces during walking and running in patellofemoral pain patients and pain-free controls. *Journal of Biomechanics*, *42*(7), 898−905. Available from https://doi.org/10.1016/j.jbiomech.2009.01.032.

Bobroff, E., Chambers, H., Sartoris, D., Wyatt, M., & Sutherland, D. (1999). Femoral anteversion and neck-angle in children with cerebral palsy. *Clinical Orthopaedics and Related Research*, *364*, 194−204. Available from https://doi.org/10.1097/00003086-199907000-00025.

de Leva, P. (1996). Adjustments to Zatsiorsky-Seluyanov's segment inertia parameters. *Journal of Biomechanics*, *29*(9), 1223−1230. Available from https://doi.org/10.1016/0021-9290(95)00178-6.

Delp, S., Suryanarayanan, S., Murray, W., Uhlir, J., & Triolo, R. (2001). Architecture of the rectus abdominis, quadratus lumborum, and erector spinae. *Journal of Biomechanics*, *34*(3), 371−375. Available from https://doi.org/10.1016/S0021-9290(00)00202-5.

Epstein, M. (2006). Should tendon and aponeurosis be considered in series? *Journal of Biomechanics*, *39*(11), 2020−2025. Available from https://doi.org/10.1016/j.jbiomech.2005.06.011.

Fenner, J., Brook, B., & Clapworthy, G. (2008). The EuroPhysiome, STEP and a roadmap for the virtual physiological human. *Philosophical Transactions of the Royal Society*, *366*, 2979−2999. Available from https://doi.org/10.1098/rsta.2008.0089.

Froeling, M., Nederveen, A., & Strijkers, G. (2012). Diffusion-tensor MRI reveals the complex muscle architecture of the human forearm. *Journal of Magnetic Resonance Imaging*, *36*(1), 237−248. Available from https://doi.org/10.1002/jmri.23608.

Fry, N., Cough, M., & Shortland, A. (2004). Three-dimensional realisation of muscle morphology and architecture using ultrasound. *Gait & Posture*, *20*(2), 177−182. Available from https://doi.org/10.1016/j.gaitpost.2003.08.010.

Gans, C., & Gaunt, A. (1991). Muscle architecture in relation to function. *Journal of Biomechanics*, *24*, 53−65. Available from https://doi.org/10.1016/0021-9290(91)90377-y.

Guimberteau, J.-C., & Delage, J.-P. (2012). The multifibrillar network of the tendon sliding system. *Annales de Chirurgie Plastique Esthétique*, *57*(5), 467−481. Available from https://doi.org/10.1016/j.anplas.2012.07.002.

Günther, M., Schmitt, S., & Wank, V. (2007). High-frequency oscillations as a consequence of neglected serial damping in Hill-type muscle models. *Biological Cybernetics*, *97*(1), 63−79. Available from https://doi.org/10.1007/s00422-007-0160-6.

Howard, B., & Yang, J. (2012). A new stability criterion for human seated tasks with given postures. *International Journal of Humanoid Robotics*, *9*(3), 1250015. Available from https://doi.org/10.1142/S0219843612500156.

Kellis, E., Galanis, N., Natsis, K., & Kapetanos, G. (2010). Muscle architecture variations along the semitendinosus and biceps femoris (long head) length. *Journal of Electromyography and Kinesiology: Official*

Journal of the International Society of Electrophysiological Kinesiology, 20(6), 1237–1243. Available from https://doi.org/10.1016/j.jelekin.2010.07.012.

Klein Horsman, M., Koopman, H., van der Helm, F., Prosé, L., & Veeger, H. (2006). Morphological muscle and joint parameters for musculoskeletal modelling of the lower extremity. *Clinical Biomechanics, 22*(2), 239–247. Available from https://doi.org/10.1016/j.clinbiomech.2006.10.003.

Lee, D., Li, Z., Sohail, Q., Jackson, K., Fiume, E., & Agur, A. (2015). A three-dimensional approach to penna-tion angle estimation for human skeletal muscle. *Computer Methods in Biomechanics and Biomedical Engineering, 18*(13), 1474–1484. Available from https://doi.org/10.1080/10255842.2014.917294.

Lieber, R. (2009). *Skeletal muscle structure, function, and plasticity: The physiological basis of rehabilitation* (3rd ed.). Lippincott Williams and Wilkins.

Lieber, R., & Ward, S. (2011). Skeletal muscle design to meet functional demands. *Philosophical Transactions of the Royal Society, 366*, 1466–1476. Available from https://doi.org/10.1098/rstb.2010.0316.

Lozano, R., Naghavi, M., & Foreman, K. (2020). Global and regional mortality from 235 causes of death for 20 age groups in 1990 and 2010: A systematic analysis for the Global Burden of Disease Study 2010. *Lancet, 380*, 2095–2128. Available from https://doi.org/10.1016/S0140-6736(12)61728-0.

Mooney, L., Rouse, E., & Herr, H. (2014). Autonomous exoskeleton reduces metabolic cost of human walking during load carriage. *Journal of Neuroengineering and Rehabilitation, 11*, 80. Available from https://doi.org/10.1186/1743-0003-11-80.

Murray, C., Vos, T., & Lozano, R. (2012). Disability-adjusted life years (DALYs) for 291 diseases and injuries in 21 regions, 1990–2010: A systematic analysis for the Global Burden of Disease Study 2010. *Lancet, 380*, 2197–2223. Available from https://doi.org/10.1016/S0140-6736(12)61689-4.

Pandy, M. (2001). Computer modeling and simulation of human movement. *Annual Review of Biomedical Engineering, 3*, 245–273. Available from https://doi.org/10.1146/annurev.bioeng.3.1.245.

Peng, J., Panda, J., Van Sint Jan, S., & Wang, X. (2015). Methods for determining hip and lumbosacral joint centers. *Journal of Biomechanics, 48*(2), 396–400. Available from https://doi.org/10.1016/j.jbiomech.2014.11.040.

Ravichandiran, K., Ravichandiran, M., Oliver, M., Singh, K., McKee, N., & Ragur, A. (2009). Determining physiological cross-sectional area of extensor carpi radialis longus and brevis as a whole and by regions using 3D computer muscle models created from digitized fiber bundle data. *Computer Methods in Biomechanics and Biomedical Engineering, 95*(3), 203–212. Available from https://doi.org/10.1016/j.cmpb.2009.03.002.

Salvia, P., Van Sint Jan, S., & Rooze, M. (2009). Precision of shoulder anatomical landmark calibration by two approaches. *Gait & Posture, 29*(4), 587–591. Available from https://doi.org/10.1016/j.gaitpost.2008.12.013.

Sholukha, V., Chapman, T., Salvia, P., Moiseev, F., Euran, F., Rooze, M., & Van Sint Jan, S. (2011). Femur shape prediction by multiple regression based on quadric surface fitting. *Journal of Biomechanics, 44*, 712–718. Available from https://doi.org/10.1016/j.jbiomech.2010.10.039.

Sholukha, V., Leardini, A., Salvia, P., Rooze, M., & Van Sint Jan, S. (2006). Double-step registration of in vivo stereophotogrammetry with both in vitro 6-DOFs electrogoniometry and CT medical imaging. *Journal of Biomechanics, 39*(11), 2087–2095. Available from https://doi.org/10.1016/j.jbiomech.2005.06.014.

Sholukha, V., Snoeck, O., Moiseev, F., Rooze, M., & Van Sint Jan, S. (2008). Stereophotogrammetry for soft tissue 3D reconstruction of dissection and medical imaging. *Journal of Biomechanics, 41*, 223. Available from https://doi.org/10.1016/S0021-9290(08)70223-9.

Sholukha, V., Van Sint Jan, S., Snoeck, O., Salvia, P., Moiseev, F., & Rooze, M. (2009). Prediction of joint center location by customizable multiple regressions: Application to clavicle, scapula and humerus. *Journal of Biomechanics, 42*(3), 319–324. Available from https://doi.org/10.1016/j.jbiomech.2008.11.004.

Sironi, R., Standoli, C., Perego, P., & Andreoni, G. (2019). Digital human modelling and ergonomic design of sleeping systems. In S. Scataglini, & G. Paul (Eds.), *DHM and posturography* (pp. 385−396). Academic Press. Available from https://doi.org/10.1016/B978-0-12-816713-7.00028-3.

Steele, K., & van der Krogt, M. (2012). How much muscle strength is required to walk in a crouch gait? *Journal of Biomechanics*, *45*(15), 2564−2569. Available from https://doi.org/10.1016/j.jbiomech.2012.07.028.

Svare, H. (2006). *Body and practice in kant*. Springer. Available from https://doi.org/10.1007/1-4020-4119-5.

Taylor, A., Goehler, L., Galper, D., Innes, K., & Bourguignon, C. (2010). Top-down and bottom-up mechanisms in mind-body medicine: Development of an integrative framework for psychophysiological research. *Explore*, *6*(1), 29−41. Available from https://doi.org/10.1016/j.explore.2009.10.004.

Thelen, D., Chumanov, E., & Hoerth, D. (2005). Hamstring muscle kinematics during treadmill sprinting. *Medicine & Science in Sports & Exercise*, *37*(1), 108−114. Available from https://doi.org/10.1249/01.MSS.0000150078.79120.C8.

Van Sint Jan, S. (2005). Introducing anatomical and physiological accuracy in computerized anthropometry for increasing the clinical usefulness of modeling systems. *Critical Reviews in Physical and Rehabilitation Medicine*, *17*(4), 249−274. Available from https://doi.org/10.1615/CritRevPhysRehabilMed.v17.i4.10.

Van Sint Jan, S. (2007). *Color atlas of skeletal landmark definitions*. Elsevier.

Van Sint Jan, S., Viceconti, M., & Clapworthy, G. (2007). The EuroPhysiome: Towards an infrastructure for a more integrated research. *Journal of Biomechanics*, *40*, 282. Available from https://doi.org/10.1016/S0021-9290(07)70278-6.

van Soest, A., & Bobbert, M. (1993). The contribution of muscle properties in the control of explosive movements. *Biological Cybernetics*, *69*, 195−204.

Viceconti, M., Testi, D., Taddei, F., Martelli, S., Clapworthy, G., & Van Sint Jan, S. (2006). Biomechanics modeling of the musculoskeletal apparatus: Status and key issues. *Proceedings of the IEEE*, *94*(4), 725−739. Available from https://doi.org/10.1109/JPROC.2006.871769.

Wu, F., Ng-Thow-Hing, V., Singh, K., Agur, A., & McKee, N. (2007). Computational representation of the aponeuroses as NURBS surfaces in 3D musculoskeletal models. *Computer Methods and Programs in Biomedicine*, *88*(2), 112−122. Available from https://doi.org/10.1016/j.cmpb.2007.07.012.

Zajac, F. (1989). Muscle and tendon: Properties, models, scaling, and application to biomechanics and motor control. *Critical Reviews in Biomedical Engineering*, *17*(4), 359−411.

Zatsiorsky, V., & Prilutsky, B. (2012). *Biomechanics of skeletal muscles*. Human Kinetics.

Zinkovsky, A., Sholuha, V., & Ivanov, A. (1996). *Mathematical modelling and computer simulation of biomechanical systems*. World Scientific Publishing. Available from https://doi.org/10.1142/2874.

MEASURING JOINT KINEMATICS THROUGH INSTRUMENTED MOTION ANALYSIS

30

Lennart Scheys

KU Leuven, Department of Development and Regeneration, Institute for Orthopaedic Research and Training (IORT), Leuven, Belgium

INTRODUCTION

Orthopedics involves the treatment of injuries or diseases of the human musculoskeletal system. One of the primary roles of this musculoskeletal system—next to its more static function of supporting the body and protecting vital internal organs—involves allowing movement or locomotion. Therefore, the objective evaluation of human locomotion in both healthy and pathologic conditions provides critical insights for the development, implementation, and evaluation of orthopedic treatment strategies. Nevertheless, as previous chapters already mentioned, human locomotion—and as a result also its evaluation—has a high complexity as it results from a synergistic and well-synchronized cooperation between the different components of the musculoskeletal system, including the central nervous system, the skeletal system, the muscular system, and other connective tissues such as ligaments and tendons. Furthermore, orthopedic disorders and their treatment may affect one or more components of this musculoskeletal system with varying degrees of severity. Their effect on locomotion can thus range from very subtle to very notable and disabling. This objective evaluation of human locomotion in both normal and pathologic conditions encompasses the field of instrumented motion analysis. Although sometimes used interchangeably, *gait analysis* specifically focuses on the evaluation of movement patterns performed as part of human walking or gait (Silva, Stergiou, & Stergiou, 2020). Such evaluation typically integrates the analysis of (1) the relative movement of body segments, termed kinematics; (2) the forces and moments underlying the observed movements, termed kinetics; and (3) the muscle activations associated with that movement through electromyography (EMG) analysis.

In this chapter, we will specifically focus on the methodological aspects of kinematics through marker-based human motion analysis and not on its clinical application or interpretation. We will thereto cover the key technological and biomechanical components of contemporary human motion analysis systems while discussing possible error sources and their consequential effect on the accuracy of the resulting kinematic data.

Human Orthopaedic Biomechanics. DOI: https://doi.org/10.1016/B978-0-12-824481-4.00018-4

SOME FIRST BASIC DEFINITIONS, PRINCIPLES, AND ASSUMPTIONS

As the key component in the objective evaluation of human locomotion, kinematics targets the recording, analysis, and interpretation of the relative motion that occurs between the different bones of the skeleton as a function of time. Instrumented motion analysis typically considers the different bones that are positionally tracked as rigid body segments. The relative motion between different body segments is typically expressed as a combination of angular, Cardan-based measurements at the joints interconnecting these bones to quantify the temporal properties of movements of multiple body parts simultaneously (Singer et al., 2016). As such, translations that might occur at the level of the joints are typically not taken into account given the current limitations in terms of accuracy of state-of-the-art marker-based motion capture systems.

However, before being able to define relative positions between segments that come together at joints, we need to know the position of the individual segments. From basic physics, the position of at least three noncollinear landmarks of a rigid body fully defines its spatial position. This principle forms the basis of kinematic assessments in instrumented motion analysis: these landmarks are individually tracked through space and time by instrumenting them with positional markers that are tracked by an optical motion analysis system. So let's start by explaining the working principles of such a marker-based optical motion analysis system.

THE OPTICAL MOTION ANALYSIS SYSTEM

Fig. 30.1 illustrates the main components of common optical motion analysis systems.

Two or more digital video cameras make up a first essential component of such a system. As we will discuss in further detail hereunder, a minimum of two cameras should indeed simultaneously capture each marker at a given point in time to allow reconstructing the three-dimensional (3D) position for that time frame according to the well-established principles of stereo vision (Chen, Armstrong, & Raftopoulos, 1994).

For the same reason, these cameras should not be arbitrarily positioned around the measurement volume of interest, but rather in such a way that markers consistently remain within the common field of view of at least two cameras throughout the motion of the body segments on which they are positioned. Furthermore, occlusion by other static or moving objects within the measurement volume should also be prevented. As a result, more complex or multidirectional motions, as well as movements with more extreme ranges of motion, can become gradually more challenging to capture successfully with a limited number of available cameras.

Once these cameras are positioned in the motion analysis laboratory, a calibration procedure should be performed to define the relative position O_{camera_number} and orientation X_{camera_number}, Y_{camera_number}, Z_{camera_number} of each connected camera with respect to each other camera, and with respect to the laboratories reference frame O_{lab}, X_{lab}, Y_{lab}, Z_{lab}. Hereto, a calibration object is used that is equipped with at least three markers at known, predefined distances from each other. By recording the movement of the markers on this object throughout a predefined number of captured frames (*dynamic calibration phase*, Di Marco et al., 2017), an optimization procedure can be performed to retrieve those camera parameters that minimize the 3D reconstruction errors in terms of

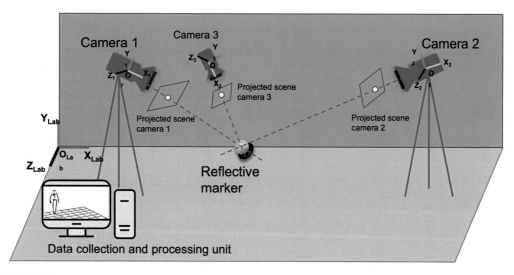

FIGURE 30.1

Basic components and stereo vision-based 3D reconstruction principle of optical motion analysis systems.
A typical 3D optical motion analysis system contains the following components: (i) two or more optically and
spatially calibrated camera's, (ii) spherical marker(s) with a coating that reflects the infrared light emitted by
LEDs mounted on the cameras, and (iii) a data collection and processing unit where the 2D image data from the
optical cameras is integrated to allow reconstructing the 3D position of markers detected by two or more cameras
through the typical mathematical concepts of stereo vision.

intermarker distances with respect to the known ground-truth distances over all collected time
frames. Next to this spatial calibration, this procedure typically also includes an optical calibration
step in which the cameras' optical parameters such as the lens distortion are being defined.
However, this falls outside the scope of this chapter. For more technical details on the underlying
calibration procedure known as *direct linear transformation*, the reader is referred to Abdel-Aziz
and Karara (2015) and Hatze (1988). In a second, static calibration phase, the calibration object is
then positioned within the laboratory to define its origin and reference axes, Olab, Xlab, Y_{lab}, Zlab.
Please note that some manufacturers encase multiple cameras within a rigid frame or body. In that
case, the relative position of these cameras is fixed and a priori known and a dynamic calibration
phase for positional calibration is therefore not required.

   A second component consists of spherical markers. Depending on the manufacturer, these markers can either be passive markers, as illustrated in Fig. 30.1, or active markers. As their name indicates, a reflective coating applied on such passive markers simply reflects the infrared light used to
detect their spatial position, directly along the line of incidence (Whittle, 1996). As illustrated in
the figure, the infrared light itself is typically being emitted by a ring of light-emitting diodes
mounted around the lens of each camera. Active markers instead emit the light themselves and thus
require powering. Although the need for a power source, cables might make them more "invasive"
for the movement being analyzed, they come with the advantage that strobing of the markers can
be exploited to uniquely identify individual markers in each camera's view which in turn facilitates

the definition of correspondences between cameras. On the other hand, passive markers provide more flexibility in terms of size, positioning, and clustering at a lower cost per marker.

A third and final component consists of the data collection unit where the 2D images from all cameras are being collected in a time synchronized manner. On the same unit, dedicated software performs the necessary calculations to allow 3D reconstruction of the spatial position of each marker that is being detected by two or more cameras (Chen et al., 1994) as well as their further integration and postprocessing towards kinematics.

FROM TRACKING MARKERS TO TRACKING BODY SEGMENTS

As discussed above, the assumption of rigidity in kinematic assessments allows the position and orientation of each body segment to be defined through three or more noncollinear points on that segment. Simultaneously, these three points allow defining a body segment reference axis system. Although in theory the positioning of these points could be random as long as they are noncollinear, their selection is a crucial step towards accurate, reliable, and clinically useful kinematics. Because of the important role of motion analysis for the biomechanics community and, more specifically, the above importance of the selection of anatomical points, the International Society of Biomechanics (ISB) defined a standard set of local coordinate systems for the articulating body segments of all major joints of the human body (Wu et al., 2002, 2005). As a first example, let's take a closer look at the ISB recommendation for the pelvis (Fig. 30.2).

For the pelvis, the mediolateral or Z-axis is defined by the line connecting the right and left anterior superior iliac spines (ASISs), with direction pointing to the right. The anterior-posterior or X-axis is defined orthogonal to the Z-axis and lies within the plane defined by both ASISs and the midpoint between both posterior superior iliac spines (PSISs). By definition, it points anteriorly. Finally, the superior-inferior axis is defined mutually perpendicular to both the Z- and X-axis and points superiorly. When translating this definition of the pelvic body axis system to instrumented motion analysis, the ISB recommendation would thus require positioning four reflective markers, that is, on both ASIS and both PSIS anatomical landmarks of the pelvis, to uniquely track its position in the lab space. Markers that are directly positioned over anatomical landmarks that define the respective body axis system are called *anatomical markers* (Stief, 2018).

As a second example, let's look at the femur (Fig. 30.2). The superior-inferior axis of the femur is defined by the line connecting the center of the femoral head and the midpoint between both femoral epicondyles (FE). The former, functionally coinciding with the hip joint's center of rotation, also serves as the recommended mutual origin of the femoral and pelvic reference systems. Importantly, this anatomical landmark cannot be reliably palpated and therefore also not reliably instrumented with a skin-mounted reflective marker due to its nonexternal position, deeper within the subject. It should thus be defined in another way. Two common approaches exist for such non-palpable anatomical landmarks that have in common that they derive the position of the specific landmarks based on the positions of neighboring landmarks that can be instrumented with markers. A first approach, the so-called *regression-based approach* estimates the position of the landmark based on its average anatomical location within another reference system of which the anatomical landmarks are known. Applied to the femoral head/hip joint center (HJC), this approach defines its

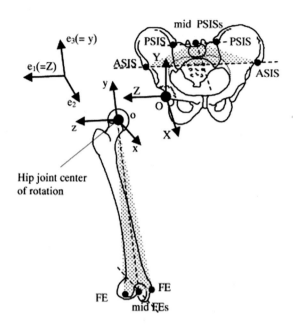

FIGURE 30.2

ISB-recommended joint axis system for the hip joint. ISB recommended joint axis system for the hip joint and the associated local body segment axis systems for pelvis and femur. PSIS and ASIS stand for posterior and anterior superior iliac spine, respectively. Together, they constitute the anatomical landmarks that uniquely define the pelvic body segment axis system. FE stands for femoral epicondyle. Both the lateral and medial femoral epicondyles, along with the femoral head center constitute the anatomical landmarks that uniquely define the femoral body segment axis system.

From Wu, G., Siegler, S., Allard, P., Kirtley, C., Leardini, A., Rosenbaum, D., Whittle, M., D'Lima, D. D., Cristofolini, L., Witte, H., Schmid, O., & Stokes, I. (2002). ISB recommendation on definitions of joint coordinate system of various joints for the reporting of human joint motion − Part I: Ankle, hip, and spine. Journal of Biomechanics, 35(4), 543−548. https://doi.org/10.1016/S0021-9290(01)00222-6.

position within the pelvic reference frame as a linear function of easily measurable anatomical dimensions of the subjects, such as pelvic width (Bell, Pedersen, & Brand, 1990) or pelvic width and leg length (Davis, Õunpuu, Tyburski, & Gage, 1991). For example, the method by Bell et al. defines the HJC as a percentage of the distance between both ASIS markers, at 30% distal, 14% medial, and 22% posterior to the ASIS based on the average position in a cohort of adult subjects (Bell et al., 1990). By capturing and defining this position during a static trial with the subject standing upright and still, the position of this landmark is simultaneously defined with respect to the other femoral landmarks, that is, both FEs, allowing the further definition of the femoral axis system. A second approach, *the functional approach*, instead defines the location of such a landmark based on functional knowledge of the joint to which the body segment belongs. Applied to the femur and its HJC, this method bases itself on the assumption that the hip functions as a ball-in-socket joint. Consequently, hip joint motion would imply that any anatomical landmark on the femur would move on a sphere with the HJC at its center. By capturing the spherical trajectories of anatomical markers on the femur during a dynamic trial wherein the subject is performing hip joint motion, the center of these spheres can be calculated with respect to the other femoral and pelvic landmarks, as described in further detail in Ehrig, Taylor, Duda, and Heller (2006), again allowing further definition of the femoral axis system. Herein, the mediolateral (Z-axis, Fig. 30.2) axis is defined by the plane containing this HJC and both FEs, with direction running to the right perpendicular to the above-defined superior-inferior axis, whereas the final anterior-posterior (X-axis, Fig. 30.2) axis is again the mutually perpendicular, anterior axis.

Now, let's look again at translating this femoral reference axis system to instrumented motion analysis. Whereas the position of the HJC is defined at a fixed position within the pelvic reference system, two reflective anatomical markers, one on each FE, are required to uniquely track the position of the femurs in the lab space. Although this marker setup is practically feasible during a static standing trial, the marker on the medial FE could be easily knocked off, impede the movement of the subject, or get lost in terms of visibility by the motion capture system. To prevent such practical problems, most marker models additionally make use of so-called *technical markers* that allow removing such practically problematic markers during dynamic trials. In contrast with anatomical markers, these markers do not typically coincide with a specific anatomical location but merely aim at ensuring the requirement that at least three noncollinear points are tracked in a practically feasible, well-trackable location on the body segmentduring dynamic trials (Stief, 2018). In some marker models, these technical markers themselves define a technical axis system for the body segment. However, by simultaneously capturing the technical and anatomical markers of a given marker model during a static trial, an assumed-to-be constant transformation can be defined between any given technical axis system and the above ISB-recommended anatomical axis system, which allows expressing relative body segment position throughout subsequent dynamic motions according to the relevant anatomical definitions as outlined in the next section.

FROM TRACKING BODY SEGMENTS TO CALCULATING JOINT KINEMATICS

The above definition of the objective evaluation of human locomotion states that it does not aim at documenting the position of the different body segments or bones of the skeleton as a function of time as described in the previous section, but rather the relative position of adjacent body segments at joints, that is, joint kinematics. As was already stated above, this relative motion between adjacent body segments is typically expressed as a combination of angular, Cardan-based measurements at the joints, disregarding possible relative translations (Singer et al., 2016).

Let's illustrate this further by building further on our example of the hip, although the same Cardan decomposition is similarly applied at other joints (Baker, Leboeuf, Reay, & Sangeux, 2017). As schematically illustrated in Fig. 30.3, a given change in relative position between femur and pelvis from the reference position, that is, the position in which the reference axis systems of both body segments coincide and all joint angles are thus defined to be zero, will be decomposed by:

1. flexion/extension of the femur as a rotation about the mediolateral axis of the proximal segment, that is, the pelvic Z-axis;

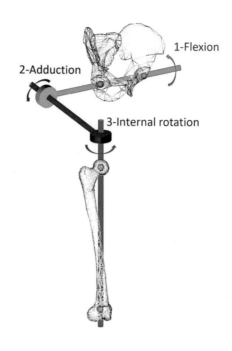

FIGURE 30.3

Conventional Cardan decomposition for hip joint motion. Schematic representation of the Cardan decomposition of a given change in hip joint position from the reference position. This change can be defined as subsequent rotations about three different and mutually perpendicular axes, namely as a sequence of hip flexion (1), hip adduction (2), and hip internal rotation (3).

From Sangeux, M. (2019). Biomechanics of the hip during gait. In S. Alshryda, J. Howard, J. Huntley, & J. Schoenecker (Eds), The pediatric and adolescent hip. Springer. https://doi.org/10.1007/978-3-030-12003-0_3.

2. abduction/adduction of the femur as a rotation about a *floating* axis defined mutually orthogonal to the mediolateral axis of the pelvis and the now repositioned superior-inferior axis of the distal segment, that is, the femur's Y-axis after the application of the flexion/extension; hence, the term floating;

3. internal/external rotation of the femur about the repositioned superior-inferior Y-axis of the femur.

Translations are indeed not integrated in the conventional expression of joint kinematics, implying that the origin of the pelvic and femoral reference axis systems will always remain coinciding throughout the movement or, in other words, joints are typically pinned. Furthermore, kinematics typically expresses the position of the more distal body segment with respect to the more proximal body segment, the latter thus serving as the "parent frame." Herein, the expression of pelvic movement typically forms an exception as it is conventionally expressed in relation to the laboratory reference axis system (Baker et al., 2017).

As already mentioned above, the same Cardan decomposition is very similarly applied at other joints. For example at the knee joint, this results in the relative position of the tibia with respect to the femur being expressed in close accordance with the Grood and Suntay convention (Grood & Suntay, 1983), decomposing motion at the knee joint as a sequence of flexion/extension about a mediolateral axis of the proximal, femoral segment, followed by varus/valgus about a floating axis which is mutually perpendicular to the superior-inferior reference axis of the tibia and the just-mentioned mediolateral axis of the femur and ending with internal/external rotation of the tibia about its superior-inferior axis.

Finally, motion analysis aims at quantifying the temporal properties of movements at multiple joints simultaneously (Singer et al., 2016). When for example analyzing lower limb motion, kinematics of the pelvis, hip, knee, and ankle joints are simultaneously reported. Thereto, multijoint kinematics are conventionally based on calculating a hierarchical sequence of relative orientations that starts at the most proximal segment, for example at the lower limbs expressing the pelvic orientation with respect to the laboratory, and proceeds joint by joint with defining the orientation of the distal neighboring body segment with respect to its proximal body segment under the above assumption of pinned joints, that is, only rotations and no translations are occurring (Schmitz et al., 2015). For further methodological details, the reader is kindly referred to Davis et al. (1991) and Kadaba, Ramakrishnan, and Wootten (1990).

SOURCES OF ERROR AND VARIABILITY

Two methodological sources of error primarily limit the accuracy of joint kinematics in marker-based motion analysis and induce variability, defined by McGinley, Baker, Wolfe, and Morris (2009) in this context as "the extent to which *kinematic* measurements are consistent or free from variation." A first important source of errors pertains to inaccuracies in anatomical landmark identification. This source of error is often termed "palpation error" given the fact that this identification is mostly achieved through manual palpation of the associated bony segment. Nevertheless, as discussed above, certain anatomical landmarks, such as for example the HJC, cannot be defined through palpation, but similarly suffer from inaccuracies. As these anatomical landmarks form the foundation of the body segment reference axis systems, inaccuracies in their identification will inevitably result in erroneous locations and/or orientations of these axis systems, and consequently also in erroneous joint kinematics. The magnitude of these errors as well as their consequential effect on joint kinematics is obviously landmark-dependent and has been most extensively documented for the HJC (Stief, 2018), where it can exceed 2 cm (Assi et al., 2016). A first important general consideration for this specific type of errors is that the above-described hierarchical setup of most conventional kinematic models will result in the downstream propagation of these errors (Schmitz et al., 2015). Or, in other words, erroneous landmark identification at the pelvis will not only affect the calculated kinematics of the pelvis itself but will also result in erroneous kinematics at the hip, knee, and ankle joint. A second general consideration is that this type of errors also results in so-called kinematic crosstalk. Indeed, the resulting malalignment of the reference axes will lead to kinematic changes that anatomically occurred for a given degree of freedom of the joint (for example knee flexion/extension) being misinterpreted as kinematic changes in another degree of freedom at the joint (e.g., knee varus/valgus) (Piazza & Cavanagh, 2000; Stief, 2018).

Next to the evident proper training and informing of motion analysis assessors, methodological improvements in regression-based and functional landmark identification methods continue to provide possible pathways to reduce landmark identification errors. See, for example, Leboucher et al. (2021), Sandau et al. (2015), or Kainz, Carty, Modenese, Boyd, and Lloyd (2015). Finally, medical imaging is often used as ground-truth data to evaluate and improve landmark identification errors, as for example in Sandau et al. (2015) or Assi et al. (2016). Similarly, various medical imaging modalities have also been suggested in the literature to replace, correct, or complement other landmark identification approaches, including magnetic resonance imaging (Scheys, Desloovere, Spaepen, Suetens, & Jonkers, 2011), biplanar radiography (Severijns et al., 2020), or ultrasound (Telfer, Woodburn, & Turner, 2014).

A second source of error—and the most critical one—is the skin-motion or soft-tissue artifact (STA). It relates to the fact that the markers used to track the position of individual body segments are typically positioned on the subject's skin and that this skin can move with respect to the underlying bones during motion due to the occurence of muscle contractions, deformation of the skin, inertial motions of underlying adipose tissues, etc. (Fiorentino, Atkins, Kutschke, Bo Foreman, & Anderson, 2020). As a result, these markers move with respect to the anatomical bony landmark they are supposed to track and represent, as well as with respect to the other markers with whom they define a body segment reference axis system. These STA will thus again result in erroneous locations and/or orientations of these axis systems, and consequently also in erroneous joint kinematics.

In the literature, STA and its effect on joint kinematics have been studied with both invasive and noninvasive techniques. Invasive techniques make use of bone pin-mounted markers, external fixators, percutaneous trackers, or Roentgen photogrammetry, as each described in more detail in Leardini, Chiari, Croce, and Cappozzo (2005). However, given the practical and ethical limitations associated with these invasive techniques, the focus in the STA literature shifted from 2005 onwards towards noninvasive, primarily medical imaging-based techniques that specifically allow to simultaneously capture the 3D position of the skin-mounted markers and the underlying bones (Peters, Galna, Sangeux, Morris, & Baker, 2010). Videofluoroscopy has gained significant popularity—see, for example, Tsai, Lu, Kuo, and Hsu (2009) or Akbarshahi et al. (2010)—as it allows participants to move relatively freely in an upright standing posture compared to the supine position and constrained space available within the bore of a magnetic resonance imaging device as in, for example, Sangeux, Marin, Charleux, Dürselen, and Ho Ba Tho (2006). Recently, the mounting of fluoroscopes on moving platforms, as for example in List et al. (2017), has even further extended the possibilities of fluoroscopy in terms of movement types that can be analyzed. Nevertheless, alternative and rapidly developing imaging modalities such as dynamic MRI and ultrasound imaging (see for example Rouhandeh, Joslin, Qu, & Ono, 2014) do remain interesting options given the fact they do not rely on ionizing radiation.

Now, what have we learned from these studies? As a first general consideration, the magnitude of STA is location-dependent, but additionally depends on the motor task being performed and the subject being measured (Leardini et al., 2005). For the lower limbs, typical STA magnitudes on the pelvis reach up to about 25 mm (Fiorentino et al., 2017), while they exceed 30 mm on the femoral body segment and reach up to 15 mm on the tibial segment (Peters et al., 2010). Furthermore, the STA tends to increase as the marker is positioned closer to the knee joint (Akbarshahi et al., 2010). As such, they exceed by far the errors associated with today's optical motion capture systems achieving submillimeter accuracy even in large capture volumes (Aurand, Dufour, & Marras, 2017). In terms of their effect on calculated joint kinematics, the STA is reported to be especially detrimental for out-of-sagittal plane joint rotations; a fortiori since the amount of motion occurring in these directions is typically smaller than in the sagittal plane (Leardini et al., 2005). At the hip, Fiorentino et al. (2017) reported that STAs during dynamic activities lead to a kinematic error with respect to a dual fluoroscopy-based ground truth that reaches an average increased extension of 1.9 degrees, an average increased adduction 0.6 degrees, and an average increased internal rotation of 5.8 degrees. At the knee, Akbarshahi et al. (2010) reported that STAs induce maximal root mean square errors during open-chain knee flexion of 24.3, 17.8, and 14.5 degrees for flexion/extension, internal/external rotation, and abduction/adduction, respectively. At the foot and ankle, Schallig et al. (2021) very recently reported that STAs induce errors in multisegment foot kinematics up to 6.7 degrees on average. Although these studies give an indication on what absolute errors can be expected, it remains important to take into account that they are inherently dependent on the movement type, subject characteristics, and the applied marker protocol and kinematic model. Furthermore, these absolute errors are often best interpreted in the context of the intended use. For example, do these errors exceed the expected intervention effect or the minimal clinically important difference?

As a further general consideration, STAs are typically difficult to eliminate using traditional noise filtering techniques as it has frequencies similar to those of the joint motion being measured and shows nonlinear relationships with those motions (Tsai et al., 2009). Nevertheless, several optimization approaches have been introduced in the literature to minimize the influence of STA on joint kinematics. Because most marker models rely on more than three markers per body segment, they often exploit this redundancy to further increase the accuracy. Although a detailed overview and explanation on possible STA-compensation methods is beyond the scope of this chapter, example approaches include but are not limited to Kalman smoothing (De Groote, De Laet, Jonkers, & De Schutter, 2008), global optimization approaches (Lu & O'Connor, 1999) or singular value decomposition-based approaches (Cereatti, Della Croce, & Cappozzo, 2006). For a more detailed overview, the reader is kindly referred to Peters et al. (2010).

To conclude, both inaccuracies in anatomical landmark identification and STAs are *extrinsic* or procedural contributors to the variability of kinematic analysis according to the definition and interpretative framework proposed by Schwartz, Trost, and Wervey (2004). *Intrinsic sources* also typically contribute to variability in joint kinematics. This intrinsic variability reflects the fact that both healthy and pathologic subjects will not perform a given motor task in the exact same way (Schwartz et al., 2004). For a more detailed discussion on the reliability and repeatability of movement analysis, the reader is kindly referred to McGinley et al. (2009).

CONCLUSION

This chapter aimed at introducing the main methodological aspects of instrumented motion analysis and its specific use for the calculation of joint kinematics. In addition to a detailed discussion on the main sources of error, this chapter showed that, although important challenges and pitfalls remain to be associated with instrumented motion analysis, it is a very dynamic and even more multidisciplinary aspect of orthopedic research. Nevertheless, future users, whether for research purposes or clinical use, should be aware of its strengths and weaknesses.

REFERENCES

Abdel-Aziz, Y. I., & Karara, H. M. (2015). Direct linear transformation from comparator coordinates into object space coordinates in close-range photogrammetry. *Photogrammetric Engineering & Remote Sensing, 81*(2), 103−107. Available from https://doi.org/10.14358/PERS.81.2.103.

Akbarshahi, M., Schache, A. G., Fernandez, J. W., Baker, R., Banks, S., & Pandy, M. G. (2010). Non-invasive assessment of soft-tissue artifact and its effect on knee joint kinematics during functional activity. *Journal of Biomechanics, 43*(7), 1292−1301. Available from https://doi.org/10.1016/j.jbiomech.2010.01.002.

Assi, A., Sauret, C., Massaad, A., Bakouny, Z., Pillet, H., Skalli, W., & Ghanem, I. (2016). Validation of hip joint center localization methods during gait analysis using 3D EOS imaging in typically developing and cerebral palsy children. *Gait & Posture, 48*, 30−35. Available from https://doi.org/10.1016/j.gaitpost.2016.04.028.

Aurand, A. M., Dufour, J. S., & Marras, W. S. (2017). Accuracy map of an optical motion capture system with 42 or 21 cameras in a large measurement volume. *Journal of Biomechanics, 58*, 237−240. Available from https://doi.org/10.1016/j.jbiomech.2017.05.006.

Baker, R., Leboeuf, F., Reay, J., & Sangeux, M. (2017). *The conventional gait model − Success and limitations* (pp. 1−19). Springer International Publishing. Available from https://doi.org/10.1007/978-3-319-30808-1_25-2.

Bell, A. L., Pedersen, D. R., & Brand, R. A. (1990). A comparison of the accuracy of several hip center location prediction methods. *Journal of Biomechanics, 23*(6), 617−621. Available from https://doi.org/10.1016/0021-9290(90)90054-7.

Cereatti, A., Della Croce, U., & Cappozzo, A. (2006). Reconstruction of skeletal movement using skin markers: Comparative assessment of bone pose estimators. *Journal of Neuroengineering and Rehabilitation, 3*, 7. Available from https://doi.org/10.1186/1743-0003-3-7.

Chen, L., Armstrong, C. W., & Raftopoulos, D. D. (1994). An investigation on the accuracy of three-dimensional space reconstruction using the direct linear transformation technique. *Journal of Biomechanics, 27*(4), 493−500. Available from https://doi.org/10.1016/0021-9290(94)90024-8.

Davis, R. B., Õunpuu, S., Tyburski, D., & Gage, J. R. (1991). A gait analysis data collection and reduction technique. *Human Movement Science, 10*(5), 575−587. Available from https://doi.org/10.1016/0167-9457(91)90046-Z.

De Groote, F., De Laet, T., Jonkers, I., & De Schutter, J. (2008). Kalman smoothing improves the estimation of joint kinematics and kinetics in marker-based human gait analysis. *Journal of Biomechanics, 41*(16), 3390−3398. Available from https://doi.org/10.1016/j.jbiomech.2008.09.035.

Di Marco, R., Rossi, S., Castelli, E., Patanè, F., Mazzà, C., & Cappa, P. (2017). Effects of the calibration procedure on the metrological performances of stereophotogrammetric systems for human movement analysis. *Measurement, 101*, 265−271. Available from https://doi.org/10.1016/j.measurement.2016.01.008.

Ehrig, R. M., Taylor, W. R., Duda, G. N., & Heller, M. O. (2006). A survey of formal methods for determining the centre of rotation of ball joints. *Journal of Biomechanics, 39*(15), 2798−2809. Available from https://doi.org/10.1016/j.jbiomech.2005.10.002.

Fiorentino, N. M., Atkins, P. R., Kutschke, M. J., Bo Foreman, K., & Anderson, A. E. (2020). Soft tissue artifact causes underestimation of hip joint kinematics and kinetics in a rigid-body musculoskeletal model. *Journal of Biomechanics, 108*, 109890. Available from https://doi.org/10.1016/j.jbiomech.2020.109890.

Fiorentino, N. M., Atkins, P. R., Kutschke, M. J., Goebel, J. M., Foreman, K. B., & Anderson, A. E. (2017). Soft tissue artifact causes significant errors in the calculation of joint angles and range of motion at the hip. *Gait & Posture, 55*, 184−190. Available from https://doi.org/10.1016/j.gaitpost.2017.03.033.

Grood, E., & Suntay, W. J. (1983). A joint coordinate system for the clinical description of three-dimensional motions: Application to the knee. *Journal of Biomechanical Engineering, 105*, 136−144. Available from https://doi.org/10.1115/1.3138397.

Hatze, H. (1988). High-precision three-dimensional photogrammetric calibration and object space reconstruction using a modified DLT-approach. *Journal of Biomechanics*, *21*(7), 533−538. Available from https://doi.org/10.1016/0021-9290(88)90216-3.

Kadaba, M. P., Ramakrishnan, H. K., & Wootten, M. E. (1990). Measurement of lower extremity kinematics during level walking. *Journal of Orthopaedic Research*, *8*(3), 383−392. Available from https://doi.org/10.1002/jor.1100080310.

Kainz, H., Carty, C. P., Modenese, L., Boyd, R. N., & Lloyd, D. G. (2015). Estimation of the hip joint centre in human motion analysis: A systematic review. *Clinical Biomechanics*, *30*(4), 319−329. Available from https://doi.org/10.1016/j.clinbiomech.2015.02.005.

Leardini, A., Chiari, L., Croce, U. D., & Cappozzo, A. (2005). Human movement analysis using stereophotogrammetry: Part 3. Soft tissue artifact assessment and compensation. *Gait & Posture*, *21*(2), 212−225. Available from https://doi.org/10.1016/j.gaitpost.2004.05.002.

Leboucher, J., Salami, F., Öztürk, O., Heitzmann, D. W. W., Götze, M., Dreher, Th, & Wolf, S. I. (2021). Focusing on functional knee parameter determination to develop a better clinical gait analysis protocol. *Gait & Posture*, *84*, 127−136. Available from https://doi.org/10.1016/j.gaitpost.2020.10.033.

List, R., Postolka, B., Schütz, P., Hitz, M., Schwilch, P., Gerber, H., ... Taylor, W. R. (2017). A moving fluoroscope to capture tibiofemoral kinematics during complete cycles of free level and downhill walking as well as stair descent. *PLoS One*, *12*(10), e0185952. Available from https://doi.org/10.1371/journal.pone.0185952.

Lu, T.-W., & O'Connor, J. J. (1999). Bone position estimation from skin marker co-ordinates using global optimisation with joint constraints. *Journal of Biomechanics*, *32*(2), 129−134. Available from https://doi.org/10.1016/S0021-9290(98)00158-4.

McGinley, J. L., Baker, R., Wolfe, R., & Morris, M. E. (2009). The reliability of three-dimensional kinematic gait measurements: A systematic review. *Gait & Posture*, *29*(3), 360−369. Available from https://doi.org/10.1016/j.gaitpost.2008.09.003.

Peters, A., Galna, B., Sangeux, M., Morris, M., & Baker, R. (2010). Quantification of soft tissue artifact in lower limb human motion analysis: A systematic review. *Gait & Posture*, *31*(1), 1−8. Available from https://doi.org/10.1016/j.gaitpost.2009.09.004.

Piazza, S. J., & Cavanagh, P. R. (2000). Measurement of the screw-home motion of the knee is sensitive to errors in axis alignment. *Journal of Biomechanics*, *33*(8), 1029−1034. Available from https://doi.org/10.1016/S0021-9290(00)00056-7.

Rouhandeh, A., Joslin, C., Qu, Z., & Ono, Y. (2014). *Non-invasive assessment of soft-tissue artefacts in hip joint kinematics using motion capture data and ultrasound depth measurements*. In 2014 36th Annual International Conference of the IEEE Engineering in Medicine and Biology Society (pp. 4342−4345). https://doi.org/10.1109/EMBC.2014.6944585

Sandau, M., Heimbürger, R. V., Villa, C., Jensen, K. E., Moeslund, T. B., Aanæs, H., ... Simonsen, E. B. (2015). New equations to calculate 3D joint centres in the lower extremities. *Medical Engineering & Physics*, *37*(10), 948−955. Available from https://doi.org/10.1016/j.medengphy.2015.07.001.

Sangeux, M., Marin, F., Charleux, F., Dürselen, L., & Ho Ba Tho, M. C. (2006). Quantification of the 3D relative movement of external marker sets vs. bones based on magnetic resonance imaging. *Clinical Biomechanics*, *21*(9), 984−991. Available from https://doi.org/10.1016/j.clinbiomech.2006.05.006.

Schallig, W., Streekstra, G. J., Hulshof, C. M., Kleipool, R. P., Dobbe, J. G. G., Maas, M., ... van den Noort, J. C. (2021). The influence of soft tissue artifacts on multi-segment foot kinematics. *Journal of Biomechanics*, *120*, 110359. Available from https://doi.org/10.1016/j.jbiomech.2021.110359.

Scheys, L., Desloovere, K., Spaepen, A., Suetens, P., & Jonkers, I. (2011). Calculating gait kinematics using MR-based kinematic models. *Gait & Posture*, *33*(2), 158−164. Available from https://doi.org/10.1016/j.gaitpost.2010.11.003.

Schmitz, A., Buczek, F., Bruening, D., Rainbow, M., Cooney, K., & Thelen, D. (2015). Comparison of hierarchical and six degrees-of-freedom marker sets in analyzing gait kinematics. *Computer Methods in Biomechanics and Biomedical Engineering*, *19*, 1−9. Available from https://doi.org/10.1080/10255842.2015.1006208.

Schwartz, M. H., Trost, J. P., & Wervey, R. A. (2004). Measurement and management of errors in quantitative gait data. *Gait & Posture*, *20*(2), 196−203. Available from https://doi.org/10.1016/j.gaitpost.2003.09.011.

Severijns, P., Overbergh, T., Thauvoye, A., Baudewijns, J., Monari, D., Moke, L., ... Scheys, L. (2020). A subject-specific method to measure dynamic spinal alignment in adult spinal deformity. *The Spine Journal*, *20*(6), 934−946. Available from https://doi.org/10.1016/j.spinee.2020.02.004.

Silva, L. M., Stergiou, N., & Stergiou, N. (2020). *Chapter 7 − The basics of gait analysis* (pp. 225−250). Academic Press. Available from https://doi.org/10.1016/B978-0-12-813372-9.00007-5.

Singer, H. S., Mink, J. W., Gilbert, D. L., Jankovic, J., Singer, H. S., Mink, J. W., ... Jankovic, J. (2016). *Chapter 5 − Motor assessments* (pp. 57−66). Academic Press. Available from https://doi.org/10.1016/B978-0-12-411573-6.00005-X.

Stief, F. (2018). *Variations of marker sets and models for standard gait analysis* (pp. 509−526). Springer International Publishing. Available from https://doi.org/10.1007/978-3-319-14418-4_26.

Telfer, S., Woodburn, J., & Turner, D. E. (2014). An ultrasound based non-invasive method for the measurement of intrinsic foot kinematics during gait. *Journal of Biomechanics*, *47*(5), 1225−1228. Available from https://doi.org/10.1016/j.jbiomech.2013.12.014.

Tsai, T.-Y., Lu, T.-W., Kuo, M.-Y., & Hsu, H.-C. (2009). Quantification of three-dimensional movement of skin markers relative to the underlying bones during functional activities. *Biomedical Engineering - Applications, Basis and Communications*, *21*(03), 223−232. Available from https://doi.org/10.4015/S1016237209001283.

Whittle, M. W. (1996). Clinical gait analysis: A review. *Human Movement Science*, *15*(3), 369−387. Available from https://doi.org/10.1016/0167-9457(96)00006-1.

Wu, G., Siegler, S., Allard, P., Kirtley, C., Leardini, A., Rosenbaum, D., ... Stokes, I. (2002). ISB recommendation on definitions of joint coordinate system of various joints for the reporting of human joint motion − Part I: Ankle, hip, and spine. *Journal of Biomechanics*, *35*(4), 543−548. Available from https://doi.org/10.1016/S0021-9290(01)00222-6.

Wu, G., Van Der Helm, F. C. T., Veeger, H. E. J., Makhsous, M., Van Roy, P., Anglin, C., ... Buchholz, B. (2005). ISB recommendation on definitions of joint coordinate systems of various joints for the reporting of human joint motion − Part II: Shoulder, elbow, wrist and hand. *Journal of Biomechanics*, *38*(5), 981−992. Available from https://doi.org/10.1016/j.jbiomech.2004.05.042.

MEASUREMENT OF JOINT KINEMATICS UTILISING VIDEO-FLUOROSCOPY

31

Alexander Cleveland Breen

Centre for Biomechanics Research, AECC University College, Bournemouth, United Kingdom

INTRODUCTION

HISTORY OF FLUOROSCOPY

Fluoroscopy has the ability to provide real-time imaging of anatomic structures during dynamic activities. It can trace its origins back to Wilhelm Conrad Röntgen, a German mechanical engineer who in 1895 first produced and detected a form of electromagnetic radiation now known as X-rays. Within months of the discovery of X-rays, the fluoroscope was produced. Originally consisting of an X-ray source and a screen coated in a thin layer of metal salt, the screen would emit a faint light (fluoresce) when it interacted with X-rays producing shadowy images of objects positioned between the two. However, the images obtained were very faint and not permanent. Although fluoroscopy has its own advantages, the higher image clarity and longevity of plain film radiograph images of plain film radiography makes it to supersede fluoroscopy as the more common medical imaging modality synonymous with X-ray.

In fluoroscopy's early development, the large doses of radiation required and the lack of ability to record the captured images constrained the application of fluoroscopy. With the advent of the image intensifier in the 1950s and 1960s image quality has been increased, while minimizing the radiation dose to the patient. Subsequent technical advancements have allowed the recording of fluoroscopy procedures using digital cameras (Schueler, 2000). Fluoroscopic imaging is widely used in many types of examinations and procedures, such as guiding surgical procedures, barium X-rays, placement of catheters, and arthrography (visualization of a joint or joints). Fluoroscopy may be used in biomechanics research, visualizing and measuring the movement of the skeletal bones. Despite many recent developments in magnetic resonance and computer tomography, fluoroscopy remains the principal imaging method for continuous analysis of the skeleton in motion.

Many names have been used over the years to describe the generation of moving images with X-rays: fluorography, cinefluorography, photofluorography, fluororadiography, kymography, electrokymography, roentgenkymography, cineradiography, videofluorography, videofluoroscopy, and digital motion X-ray. However, fluoroscopy has remained the most common.

Human Orthopaedic Biomechanics. DOI: https://doi.org/10.1016/B978-0-12-824481-4.00020-2

EQUIPMENT

Fluoroscopic equipment is available in many different configurations for use in a wide variety of clinical applications. There are many different designs of fluoroscope but broadly divided into fixed or mobile systems. Variants of both detect incoming X-rays and produce images either using an image intensifier or using a flat-panel detector.

The components included in a modern fluoroscopic imaging system are shown in Fig. 31.1. The components that produce and filter X-rays are similar to those included in plane film radiography, whereas other components related to the capturing of images are unique to fluoroscopy. Typically,

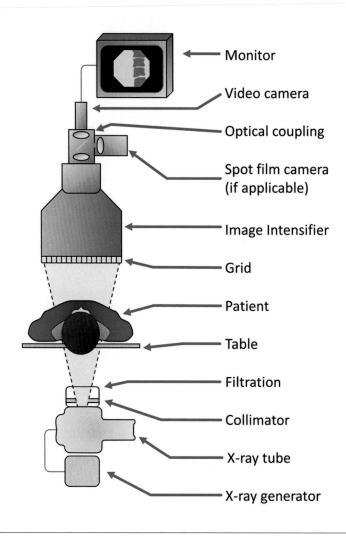

FIGURE 31.1

Fluoroscopic imaging chain and its components.

additional apparatus built in to the system to allow for image recording. Coupling optics connect the output from the image intensifier to a charge-coupled device (CCD) camera. This converts light into an electronic signal which can be shown on a monitor or recorded for later use.

X-RAY TUBE AND GENERATOR

A stable, high-voltage power supply is required for X-ray production. It is most often achieved through a multiphase generator built into the system to ensure a constant current to the X-ray tube and subsequently maintain image quality. While the X-ray tube design will vary between systems, focusing on mobility, cooling, and/or ability to focus the generated beams, in general X-ray tubes have the same basic design. An X-ray tube converts electrical energy into X-rays by accelerating electrons with a high voltage (50–150 kV), to strike a rotating (Tungsten) metal target electrode (the anode) that is maintained at a positive potential difference relative to the cathode (McCollough, 1997). The vast majority of energy released is in the form of heat with a small fraction being in the form of diagnostically useful X-rays. For optimum conversion of the energy into X-rays, minimizing heat generation, and ensuring minimal X-ray leakage, the tube housing, the cathode, and tungsten target (anode) are kept in a vacuum glass tube shielded from all sides. A cross-sectional view of a typical X-ray tube is shown in Fig. 31.2.

Fluoroscopy systems often have three modes: digital radiography that uses a single (relatively) long exposure of X-ray light to acquire a single static image. The images generated are very similar to that of a plane film radiograph, but due to the lack of sharpness and increased noise inherent to

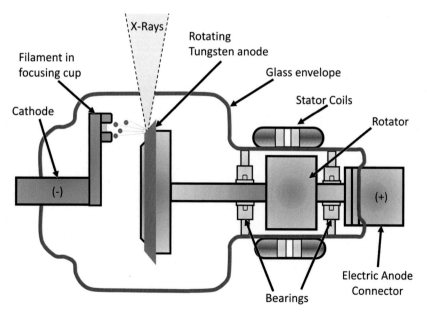

FIGURE 31.2

X-ray tube cross-section. X-ray tube assembly with major components.

fluoroscopic image capture, they are inferior to those generated with plane film radiographs. The second mode is a continuous exposure to X-rays and the original form of fluoroscopy. The X-ray tube runs at a constant but very low current, and images are quickly captured at 8–60 frames per second. The last mode is known as "pulsed fluoroscopy," where X-rays are released from the tube in pulses, on and off. This has the added benefit of reducing the overall dose to the patient and reducing the blurring between captured images due to the motion of the observed objects between frames.

IMAGE RECEPTORS

There are two main imaging receptor systems used in fluoroscopy: "image intensifiers" and "flat-panel detectors." Both allow for real-time image capture. Image intensifiers are still the most common detector found in fluoroscopic systems. Within these image intensifiers, several conversions and multipliers take place to increase the gain of detected X-rays. In brief, X-rays initially strike the input window of the image intensifier tube, pass into the vacuum envelope, and interact with the input phosphor. When the input phosphor is struck, the energy of the X-ray is converted into multiple lower energy photons of light (Balter, 1999). When illuminated by the phosphor, a photocathode that is bonded to the input phosphor emits electrons which in turn ejects electrons with every interaction, this is known as photoemission (Bushong, 2020). These electrons are accelerated and guided by electron optics towards the output phosphor. The focusing of the incoming electrons as well as their acceleration means that the output from the output phosphor is far brighter than the first (Fig. 31.3). This illumination increase, known as brightness gain, ranges from a 5000 to 20,000. The images generated by the output phosphor are either saved as a sequence of image data or as a digital video (Acker et al., 2011; Balter, 1999).

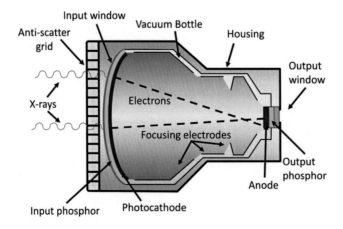

FIGURE 31.3

Cross-sectional schematic of an image intensifier shows its major components. X-rays are filtered at the antiscatter grid to reduce noise. The X-rays that enter the intensifier are converted into multiple electrons and accelerated towards a detector that converts this signal into an image that can be recorded.

Flat-panel detectors are more common in fixed systems due to their thinner build. These systems operate by having a grid of individual detector elements, which directly or indirectly convert the X-ray light that can be detected and processed as images. Indeed, many modern fluoroscopy systems include flat-panel detectors that have a large operational dynamic range, approximately 60 times greater than that of image intensifier systems (Nickoloff, 2011). However, for the application of human motion analysis these are limited in image capture rate compared to image intensifiers with most clinical systems only capable of speeds up to 30 FPS with limited spatial resolution. Comparing this with bespoke systems that use image intensifiers are capable of reaching speeds of 1000 FPS using continuous fluoroscopy and 120 FPS using pulsed X-ray (Anderst, Zauel, Bishop, Demps, & Tashman, 2009; Ivester et al., 2015).

Image quality

There are three resolutions that relate to the quality of the image recorded and the amount of radiation (dose) that the patient, participant, or object that is being examined will receive. When establishing the exposure profiles for biomechanical analysis one must inevitably trade-off between the three resolutions. The three resolutions are *spatial* (the minimum size of objects or distance between objects that can be visualized), *contrast* (the amount of difference in object relative density that can be visualized), and *temporal* (the amount of exposure time over which an individual image is captured).

Which factors are most important? The ability of detecting the shapes and borders of objects in an image (*spatial resolution*). The ability to distinguish between objects of similar opacity (*contrast resolution*), or the ability to capture an object in motion (*temporal resolution*). To ensure that good resolutions are obtained within captured images, most fluoroscopic systems utilize an automatic exposure device in conjunction with parameters gained from exposure tables to adjust the beam output from the X-ray tube and aspects of the image receptor. Exposure tables are predetermined for specific regions of the body or type of examination (orthopedics, procedures involving contrast agents, biomechanics, etc.), while automatic exposure devices adjust the exposure setting along these tables based off the image received at the receptor. Postprocessing of the output images can also be performed to enhance contrast and brightness of the displayed image.

SAFETY AND PROTECTION DURING A FLUOROSCOPIC ACQUISITION

Getting the right images from the right person to produce the most useful information and of course minimizing the radiation dose to the participant. This has led to the ALARP principle, to acquire useful images with a dose which is "As Low As Reasonably Practicable." As with all ionizing radiation imaging sources, consideration of radiation dose must be taken into account, not only to the participant, but also to all personnel that are required to be in the room during imaging.

This is done in three ways: time, distance, and shielding. The amount of time that radiation is being exposed is minimized. As the amount of X-ray received from a radiation source diminishes as a function of distance, everyone must be as far away from the radiation source as is practical. Wherever possible a form of shielding that can block or diminish X-rays should be used. This may be in the form of some sort of dense, radiopaque material in the form of a barium plaster wall or a piece of protective (lead lined) clothing which would block incoming radiation before it interacts with organic matter.

WHY USE FLUOROSCOPY FOR MECHANICAL MEASUREMENTS?

The change from analog fluoroscopic imaging of the past to digital and virtual modeling has revolutionized how we use medical imaging to measure interjoint motion, promising greater information about in-vivo biomechanics. An accurate knowledge of continuous joint motion has provided researchers for the first time the ability to evaluate both normative biomechanics, in real time, in real people as they perform functional tasks and may have important clinical implications in diagnosis of joint instability and evaluation of prosthetic implants. Such knowledge is critical for our understanding of musculoskeletal disorders induced by trauma or degenerative diseases and can also be useful for planning rehabilitation regimens for patients after conservative/surgical treatments.

SKIN SURFACE MEASUREMENTS

While fluoroscopy is an accurate way of visualizing and assessing continuous joint motion, is fluoroscopy the most *effective* way of acquiring the information we need? As mentioned before, persons working in medical imaging must always adhere to the ALARP principle. If the data needed could be acquired without the use of ionizing radiation, it is always preferential to do so. Surface-based motion tracking allows for the reconstruction of marker locations, where either retroreflective markers or inertial measurement units are adhered to the participant's skin (or on tight-fitting clothing) over the body segment to be analyzed. As well as being easy to apply, skin-based measures have proven highly useful in providing high-quality real-time measures of body motion (Burton & Tillotson 1988). However, while a versatile and inexpensive way of acquiring positional data, it is erroneous to assume that markers attached to the skin are rigidly connected to the underlying bones. Indeed, it is well established that markers on the surface of the body do not move with respect to the underlying bones because of the soft tissues which lie in the intervening space. This leads to two major sources of error in motion capture, the difficulty locating the positions of the bone and soft tissue artifact. The latter has been established to be the greater of the two sources of error in human motion analysis (Andriacchi & Alexander 2000).

In the measurement of the thigh rotational errors larger than 12 degrees have been detected (Benoit et al., 2006; Stagni, Fantozzi, Cappello, & Leardini, 2005; Tsai, Lu, Kuo, & Lin, 2011). In measures of spinal kinematics accurate clinical measurements are provided only for large spinal tracts; as a result, it is very difficult to recognize a specific interjoint disorder (Anderson & Sweetman 1975; Pearcy, 1986). To compound the issue, a second source of error is the difficulty in correctly identifying the underlying anatomy leading to landmark misplacement (Della Croce, Leardini, Chiari, & Cappozzo, 2005).

These errors mean that joint translations and the complex motions of multisegmental joint, such as the individual segments of the spine, are largely ignored in biomechanical analysis using marker-based motion capture. However, accurate measurement of in-vivo joint kinematics is imperative for the understanding of normal and pathological human motion and for evaluating the outcome of surgical procedures.

2D TRACKING

MANUAL VERSUS AUTOMATED TRACKING

Until the emergence of technologies that could accurately quantify dynamic joint motion using fluoroscopy, the standard approach to evaluating the mechanics of osseous linkages in vivo, such as the spine, has remained a pair of plain radiographs taken at the end of bending range. With the quantitative fluoroscopy techniques becoming more prevalent, this allowed for the establishment of quasistatic radiographic motion assessments of osseous structures uncovering new knowledge and understanding of the natural motion of joints not previously accessible from static radiographic images. Historically, because the image quality and computational power were not yet available, objects visualized by fluoroscopic imaging were manually identified in each image. These techniques have demonstrated good results in the shoulder, cervical, and lumbar spine as well as leading to the measurement of complex mechanics in total knee replacements. However, the manual tracking techniques are laborious and time consuming and resulted in these processes not being highly adopted. With the advent of higher quality image processing and object tracking techniques, various automated approaches have been proposed in order to reduce operator workload and limit the reliance on the operator for the manual registration of objects within radiographic images.

When measuring intersegmental spinal mechanics, recent methods have aimed towards automated vertebrae recognition in the attempt to reduce the reliance on the operator and to improve the accuracy and repeatability of measures. Common approaches to measure the motion of vertebra during a bend are confined to a single plane of motion (e.g., flexion-extension) and require the that assumption of no out-of-plane or coupled motion takes place (Anderst et al., 2009; Ivester et al., 2015). These two-dimensional (2D) (uniplanar) motion studies are based on template matching techniques (Bifulco, Cesarelli, Allen, Sansone, & Bracale, 2001; Cerciello, Romano, Bifulco, Cesarelli, & Allen, 2011; Muggleton & Allen, 1997), vertebral body outline descriptors (McCane, King, & Abbott, 2006; Zheng, Nixon, & Allen, 2004), or Bayesian estimators (Lam, McCane, & Allen, 2009). These are highly accurate systems with precision for tracking bone movements of less than 1 mm and 1 degree.

2D–3D REGISTRATION

There is a limitation when measuring osseous joint motion with single plane fluoroscopy to ignore motion that is not in the plane of imaging, that is, rotations and translations. To date, this has been overcome by performing tasks constrained to a single plane of motion. However, during functional tasks (walking, bending to pick up an object or climbing stairs) constrained motion does not accurately represent reality (Acker et al., 2011; Fregly, Rahman, & Banks, 2005; Hirokawa, Hossain, Kihara, & Ariyoshi, 2008). Indeed, particularly in the measurement of lower limb mechanics, translations play an important part of how the normal healthy joint functions. When implants or bone travel out of plane during single plane fluoroscopic examinations, they appear subtly larger or smaller in the image. By using fluoroscopy combined with model-based image registration (MBIR), an accurate estimate on joint poses can be generated by projecting a three-dimensional (3D) model of an implant or bone onto the 2D X-ray image and adjusting its position until the projection aligns with the image. The first key study to use this technique was carried out by Banks et al. (1992) to calculate and quantify the 6 degrees of freedom of total knee replacements. Reported accuracy of this method was 1 mm translation and 0.5-degree rotation with respect to in plane motions (Banks & Hodge 1996). While efforts have been made to improve upon this, deviations from ground truth measures remain larger in the Z translation direction (along the axis of the X-ray beam) as the registration methods are much less sensitive in that direction (Mahfouz et al., 2003).

Biplane fluoroscopy (or dual fluoroscopy) seeks to overcome this by using two synchronized fluoroscopes with overlapping imaging views. These systems have been constructed by modifying two C-arms (Brainerd et al., 2010; Liu et al., 2016; Torry et al., 2011) or from custom-designed components (Anderst, Baillargeon, Donaldson, Lee, & Kang, 2011; Tashman & Anderst 2003) in orthogonal positions. Originally, to estimate bone posed in 3D, a technique known as roentgen stereophotogrammetry was used. This was an invasive procedure that involves the insertion of metal markers rigidly attached to bones via surgery, which were imaged using biplanar radiographic measurements (Keller, Hurschler, & Schwarze, 2021; Olsson, Selvik, & Willner, 1977). These systems were highly accurate with the ability to measure subdegree rotation of joints (Kage et al., 2020), and being able to measure motion in all three planes by utilizing biplanar radiographs or biplanar fluoroscopy of the joints of interest in multiple positions. However, due to the need to surgically attach these metal markers to the bone, it is limited by its ethical implications to studies which involve surgical interventions.

With the advent of the MBIR techniques, previously mentioned, the need for invasive surgical procedures has become less necessary. Recent studies have used high-resolution computed tomography (CT) or magnetic resonance imaging (MRI) scans to generate 3D subject-specific models of each bone to be measured (Lin et al., 2020; Zhou et al., 2020). These bone models are then registered to the biplane radiographs using an automated or manual registration process to determine bone location and orientation within each pair of synchronized radiographs allowing for the 3D poses to be established in each frame using a manual or semiautomated registration process (Fig. 31.4). To ensure high temporal resolution, some research groups have developed bespoke biplane X-ray systems that have taken components from clinical fluoroscopes and mounted them within custom housings to maximize positioning capability. They have incorporated high speed video cameras to allow the recording of dynamic activities with higher frame rates (Brainerd et al., 2010; Campbell, Wilson, LaPrade, & Clanton, 2016; Torry et al., 2011). Other research groups have taken this one step further and built fully custom biplane X-ray systems. These are typically capable of high-speed video radiography which uses very fast pulsed X-rays of speeds of up to 150 FPS that are time synchronised with high speed cameras (Anderst et al., 2009; Ivester et al., 2015).

In addition, as a consequence of using two radiological sources, the radiation dose to participants is greatly increased. Compounding this, the doses required to generate bone models from CT scans are nearly four times greater than the fluoroscopic data acquisition itself. In a MBIR study using biplanar fluoroscopy and CT scans to record the motion of the lower spine during a lifting task, the effective radiation dose of the fluoroscopic sequences, across three repeated measures, equated to 3.6 mSv (milliSieverts). However, the maximum estimated effective radiation dose from

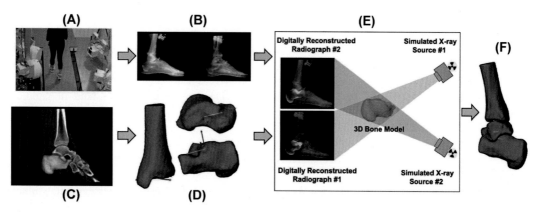

FIGURE 31.4

Model-based image registration data collection and processing workflow. (A) Participants perform dynamic tasks within the biplane radiographic imaging system while (B) synchronized biplane radiographs are collected at 100 Hz. (C) Bilateral CT scans were collected and used to create subject-specific 3D bone models. (D) Anatomical landmarks were placed onto the bone models to establish coordinate systems. (E) 3D bone models were matched to the biplane radiographs using a validated process, and (F) relative translations and rotations were calculated.

From Yang, S., Canton, S. P., Hogan, M. V., & Anderst, W. (2021). Healthy ankle and hindfoot kinematics during gait: Sex differences, asymmetry and coupled motion revealed through dynamic biplane radiography. Journal of Biomechanics, 116, *110220.*
https://doi.org/10.1016/j.jbiomech.2020.110220.

CT was approximately 12.33 mSv (giving a total of 15.93 mSv per participant) (Aiyangar, Zheng, Anderst, & Zhang, 2015). In comparison in a uniplanar study using template matching of the same part of the body, similar tasks were performed. The median effective dose was 0.66 mSv (0.78 mSv upper third quartile) (Breen, Hemming, Mellor, & Breen, 2019).

CLINICAL BIOMECHANICS UTILITY/JOINT MOTION

Accurate knowledge of osseous kinematics at the specific joint level is critical for the evaluation of mechanical disorders induced by trauma or joint degeneration at single or multiple segments. Moreover, the evaluation of joint mechanics in both asymptomatic and symptomatic individuals can also be useful for planning treatment regimens for patients before and after conservative/surgical interventions. While fluoroscopy can provide kinematic information of most joints across the body, the three main areas of focus have been the foot, knee, and spine kinematics. This is predominantly because few centers have the specialist knowledge and equipment to be able to investigate *in-vivo* bone mechanics at the joint level with high accuracy. However, there is a drive to increase the capacity in this area of research, primarily due to limitations with conventional surface-based motion analysis systems and the errors (previously mentioned) that are associated with their positioning and tracking. Ankle instability has been hypothesized to contribute to mature osteoarthritis due to altered joint kinematics and its effect on joint function. Fluoroscopy has been used to asses normal joint function and determine the kinematic response to treatments such as ligament repair or association with footwear type and ankle joint mechanics. The knee has similarly been investigated for a myriad of joint dysfunctions including instability; however, research of in-vivo knee mechanics has focused on the assessment of knee arthroplasty mechanics, and its association with patient-reported outcomes. Lastly, unlike previously mentioned joints, the spine has multiple joints both in parallel (each functional spinal unit consisting of multiple joint surfaces) and in series (24 successive functional spinal units), which are susceptible to a wide range of degenerative traumatic causes of aberrant mechanics. Research into this structure has been split into three main paths: firstly the establishment of what normal spine mechanics is during specific tasks, secondly the alterations in spine mechanics due to functional or pathological conditions, and lastly to assess the effects of spinal prosthesis such as spine fusions and disk replacements to restore or restrain intersegmental spinal motion. State-of-the-art biomechanical analysis for the planning of treatment interventions requires the generation of patient-specific models and accurate biomechanical representation of the joints. With the advances in medical imaging, models can be generated that utilize subject-specific geometry from CT or MRI and subject-specific joint mechanics with high accuracy from fluoroscopic imaging. Using finite element analysis and rigid body musculoskeletal models in vivo, evaluations of muscular forces and joint loads during static and dynamic activities can be estimated.

THE FUTURE

The acquisition and data processing of these sequences remain costly and require specialist knowledge and large amounts of operator interactions. There is a need for improvements in image enhancement, automated data extraction, and translation of the mechanical outputs to the patient and clinical population. In an effort to establish more functional measure of osseous kinematics, while still rare, a few labs have developed systems to investigate this using fluoroscopy. While not an exhaustive list, some of the notable labs are as follows. The AECC University College in the United Kingdom (Breen et al., 2019) uses constrained guided motion in an effort to remove behavioral variations during a task and remove the need for biplanar radiography. The University of Pittsburgh's Biodynamics Laboratory (bdl.pitt.edu/research/) has performed research studies that span most joints in the body using a dynamic stereo X-ray system (Aiyangar et al., 2015). Recently, moving fluoroscopy systems have been developed. ETH Zurich has adapted a single-plane fluoroscope to track a volunteer as they walk through a room (List et al., 2017). The system tracks the volunteer by using a wire sensor attached to the leg to ensure the participants knee stays within the imaging field of view. The University of Melbourne has built a versatile mobile biplane X-ray (MoBiX) system, which, similarly, can follow a participant as they walk along a walk way but uses a high-speed camera to track the participant's movement (Guan, Gray, Keynejad, & Pandy, 2016). While systems that have the ability to move with a participant provide the opportunity to further our understanding of in-vivo kinematics during complex tasks are far more complex compared to the stationary systems. This increased complexity can lead to errors in image capture, such as misalignment of the X-ray tubes and intensifiers due to ground vibration or variations in gearing.(Yang, Canton, Hogan, & Anderst, 2021).

Moving forward, the primary drive for the use of fluoroscopy in biomechanics research is to investigate the relationships between dynamic joint function and joint disease or injury, in an effort to improve diagnosis and treatment of orthopedic conditions. These efforts will require strong collaborations between clinicians, basic scientists and industry to develop innovative technologies and methods for assessing the dynamic function of joints and musculoskeletal tissues.

REFERENCES

Acker, S., Li, R., Murray, H., John, P. S., Banks, S., Mu, S., . . . Deluzio, K. (2011). Accuracy of single-plane fluoroscopy in determining relative position and orientation of total knee replacement components. *Journal of Biomechanics, 44*(4), 784−787.

Aiyangar, A., Zheng, L., Anderst, W., & Zhang, X. (2015). Apportionment of lumbar L2-S1 rotation across individual motion segments during a dynamic lifting task. *Journal of Biomechanics, 48*(13), 3709−3715. Available from https://doi.org/10.1016/j.jbiomech.2015.08.022.

Anderson, J., & Sweetman, B. (1975). A combined flexi-rule/hydrogoniometer for measurement of lumbar spine and its sagittal movement. *Rheumatology, 14*(3), 173−179.

Anderst, W., Zauel, R., Bishop, J., Demps, E., & Tashman, S. (2009). Validation of three-dimensional model-based tibio-femoral tracking during running. *Medical Engineering & Physics, 31*(1), 10−16.

Anderst, W. J., Baillargeon, E., Donaldson, W. F., III, Lee, J. Y., & Kang, J. D. (2011). Validation of a non-invasive technique to precisely measure in vivo three-dimensional cervical spine movement. *Spine (Philadelphia, PA: 1986), 36*(6), E393.

Andriacchi, T. P., & Alexander, E. J. (2000). Studies of human locomotion: Past, present and future. *Journal of Biomechanics, 33*(10), 1217−1224.

Balter, S. (1999). X-ray image intensifier. *Catheterization and Cardiovascular Interventions: Official Journal of the Society for Cardiac Angiography & Interventions, 46*(2), 238−244.

Banks, S. A. (1992). *Model based 3D kinematic estimation from 2D perspective silhouettes: Application with total knee prostheses*. Massachusetts Institute of Technology.

Banks, S. A., & Hodge, W. A. (1996). Accurate measurement of three-dimensional knee replacement kinematics using single-plane fluoroscopy. *IEEE Transactions on Biomedical Engineering, 43*(6), 638−649.

Benoit, D. L., Ramsey, D. K., Lamontagne, M., Xu, L., Wretenberg, P., & Renström, P. (2006). Effect of skin movement artifact on knee kinematics during gait and cutting motions measured in vivo. *Gait & Posture, 24*(2), 152−164.

Bifulco, P., Cesarelli, M., Allen, R., Sansone, M., & Bracale, M. (2001). Automatic recognition of vertebral landmarks in fluoroscopic sequences for analysis of intervertebral kinematics. *Medical and Biological Engineering and Computing, 39*(1), 65−75.

Brainerd, E. L., Baier, D. B., Gatesy, S. M., Hedrick, T. L., Metzger, K. A., Gilbert, S. L., & Crisco, J. J. (2010). X-ray reconstruction of moving morphology (XROMM): Precision, accuracy and applications in comparative biomechanics research. *Journal of Experimental Zoology Part A: Ecological Genetics and Physiology, 313*(5), 262−279.

Breen, A., Hemming, R., Mellor, F., & Breen, A. (2019). Intrasubject repeatability of in vivo intervertebral motion parameters using quantitative fluoroscopy. *European Spine Journal, 28*(2), 450−460. Available from https://doi.org/10.1007/s00586-018-5849-9.

Burton, A. K., & Tillotson, K. M. (1988). Reference values for 'normal' regional lumbar sagittal mobility. *Clinical Biomechanics, 3*(2), 106−113.

Bushong, S. (2020). *Radiologic science for technologists e-book: Physics, biology, and protection*. Mosby.

Campbell, K. J., Wilson, K. J., LaPrade, R. F., & Clanton, T. O. (2016). Normative rearfoot motion during barefoot and shod walking using biplane fluoroscopy. *Knee Surgery, Sports Traumatology, Arthroscopy*, *24*(4), 1402−1408.

Cerciello, T., Romano, M., Bifulco, P., Cesarelli, M., & Allen, R. (2011). Advanced template matching method for estimation of intervertebral kinematics of lumbar spine. *Medical Engineering & Physics*, *33*(10), 1293−1302.

Della Croce, U., Leardini, A., Chiari, L., & Cappozzo, A. (2005). Human movement analysis using stereophotogrammetry: Part 4: Assessment of anatomical landmark misplacement and its effects on joint kinematics. *Gait & Posture*, *21*(2), 226−237.

Fregly, B. J., Rahman, H. A., & Banks, S. A. (2005). Theoretical accuracy of model-based shape matching for measuring natural knee kinematics with single-plane fluoroscopy. *Journal of Biomechanical Engineering*, *127*(4), 692−699. Available from https://doi.org/10.1115/1.1933949.

Guan, S., Gray, H. A., Keynejad, F., & Pandy, M. G. (2016). Mobile biplane X-Ray imaging system for measuring 3D dynamic joint motion during overground gait. *IEEE Transactions on Medical Imaging*, *35*, 326−336. Available from https://doi.org/10.1109/TMI.2015.2473168.

Hirokawa, S., Hossain, M. A., Kihara, Y., & Ariyoshi, S. (2008). A 3D kinematic estimation of knee prosthesis using X-ray projection images: Clinical assessment of the improved algorithm for fluoroscopy images. *Medical & Biological Engineering & Computing*, *46*(12), 1253−1262.

Ivester, J. C., Cyr, A. J., Harris, M. D., Kulis, M. J., Rullkoetter, P. J., & Shelburne, K. B. (2015). A reconfigurable high-speed stereo-radiography system for sub-millimeter measurement of in vivo joint kinematics. *Journal of Medical Devices*, *9*(4), 041009.

Kage, C. C., Akbari-Shandiz, M., Foltz, M. H., Lawrence, R. L., Brandon, T. L., Helwig, N. E., & Ellingson, A. M. (2020). Validation of an automated shape-matching algorithm for biplane radiographic spine osteokinematics and radiostereometric analysis error quantification. *PLoS One*, *15*(2), e0228594.

Keller, M. C., Hurschler, C., & Schwarze, M. (2021). Experimental evaluation of precision and accuracy of RSA in the lumbar spine. *European Spine Journal*, *30*, 2060−2068. Available from https://doi.org/10.1007/s00586-020-06672-5.

Lam, S. C. B., McCane, B., & Allen, R. (2009). Automated tracking in digitized videofluoroscopy sequences for spine kinematic analysis. *Image and Vision Computing*, *27*(10), 1555−1571.

Lin, C.-C., Lu, T.-W., Li, J.-D., Kuo, M.-Y., Kuo, C.-C., & Hsu, H.-C. (2020). An Automated three-dimensional bone pose tracking method using clinical interleaved biplane fluoroscopy systems: Application to the knee. *Applied Sciences*, *10*(23), 8426.

List, R., Postolka, B., Schütz, P., Hitz, M., Schwilch, P., Gerber, H., ... Taylor, W. R. (2017). A moving fluoroscope to capture tibiofemoral kinematics during complete cycles of free level and downhill walking as well as stair descent. *PLoS One*, *12*(10), e0185952. Available from https://doi.org/10.1371/journal.pone.0185952.

Liu, Z., Tsai, T.-Y., Wang, S., Wu, M., Zhong, W., Li, J.-S., ... Li, G. (2016). Sagittal plane rotation center of lower lumbar spine during a dynamic weight-lifting activity. *Journal of Biomechanics*, *49*(3), 371−375.

Mahfouz, M. R., Hoff, W. A., Komistek, R. D., & Dennis, D. A. (2003). A robust method for registration of three-dimensional knee implant models to two-dimensional fluoroscopy images. *IEEE transactions on medical imaging*, *22*(12), 1561−1574. Available from https://doi.org/10.1109/tmi.2003.820027.

McCane, B., King, T. I., & Abbott, J. H. (2006). Calculating the 2D motion of lumbar vertebrae using splines. *Journal of Biomechanics*, *39*(14), 2703−2708.

McCollough, C. H. (1997). The AAPM/RSNA physics tutorial for residents. X-ray production. *Radiographics: A Review Publication of the Radiological Society of North America, Inc*, *17*(4), 967−984.

Muggleton, J., & Allen, R. (1997). Automatic location of vertebrae in digitized videofluoroscopic images of the lumbar spine. *Medical Engineering & Physics*, *19*(1), 77−89.

Nickoloff, E. L. (2011). AAPM/RSNA physics tutorial for residents: Physics of flat-panel fluoroscopy systems: Survey of modern fluoroscopy imaging: Flat-panel detectors vs image intensifiers and more. *Radiographics: A Review Publication of the Radiological Society of North America, Inc, 31*(2), 591−602.

Olsson, T. H., Selvik, G., & Willner, S. (1977). Mobility in the lumbosacral spine after fusion studied with the aid of roentgen stereophotogrammetry. *Clinical Orthopaedics and Related Research, 129,* 181−190.

Pearcy, M. (1986). Measurement of back and spinal mobility. *Clinical Biomechanics, 1*(1), 44−51.

Schueler, B. A. (2000). The AAPM/RSNA physics tutorial for residents general overview of fluoroscopic imaging. *Radiographics: A Review Publication of the Radiological Society of North America, Inc, 20*(4), 1115−1126.

Stagni, R., Fantozzi, S., Cappello, A., & Leardini, A. (2005). Quantification of soft tissue artefact in motion analysis by combining 3D fluoroscopy and stereophotogrammetry: A study on two subjects. *Clinical Biomechanics, 20*(3), 320−329.

Tashman, S., & Anderst, W. (2003). In-vivo measurement of dynamic joint motion using high speed biplane radiography and CT: Application to canine ACL deficiency. *Journal of Biomechanical Engineering, 125* (2), 238−245.

Torry, M. R., Shelburne, K. B., Peterson, D. S., Giphart, J. E., Krong, J. P., Myers, C., ... Woo, S. (2011). Knee kinematic profiles during drop landings: A biplane fluoroscopy study. *Medicine and Science in Sports and Exercise, 43*(3), 533−541.

Tsai, T.-Y., Lu, T.-W., Kuo, M.-Y., & Lin, C.-C. (2011). Effects of soft tissue artifacts on the calculated kinematics and kinetics of the knee during stair-ascent. *Journal of Biomechanics, 44*(6), 1182−1188.

Yang, S., Canton, S. P., Hogan, M. V., & Anderst, W. (2021). Healthy ankle and hindfoot kinematics during gait: Sex differences, asymmetry and coupled motion revealed through dynamic biplane radiography. *Journal of Biomechanics, 116,* 110220. Available from https://doi.org/10.1016/j.jbiomech.2020.110220.

Zheng, Y., Nixon, M. S., & Allen, R. (2004). Automated segmentation of lumbar vertebrae in digital videofluoroscopic images. *IEEE Transactions on Medical Imaging, 23*(1), 45−52.

Zhou, C., Wang, H., Wang, C., Tsai, T.-Y., Yu, Y., Ostergaard, P., ... Cha, T. (2020). Intervertebral range of motion characteristics of normal cervical spinal segments (C0-T1) during in vivo neck motions. *Journal of Biomechanics, 98,* 109418.

FURTHER READING

Anderst, W. J., Donaldson, W. F., III, Lee, J. Y., & Kang, J. D. (2015). Three-dimensional intervertebral kinematics in the healthy young adult cervical spine during dynamic functional loading. *Journal of Biomechanics, 48*(7), 1286−1293.

Breen, A., & Breen, A. (2018). Uneven intervertebral motion sharing is related to disc degeneration and is greater in patients with chronic, non-specific low back pain: An in vivo, cross-sectional cohort comparison of intervertebral dynamics using quantitative fluoroscopy. *European Spine Journal, 27*(1), 145−153.

FINITE ELEMENT ANALYSIS IN ORTHOPEDIC BIOMECHANICS

32

Markus O. Heller

Department of Mechanical Engineering, University of Southampton, Southampton, United Kingdom

FINITE ELEMENT ANALYSIS AS A METHOD

Finite element analysis (FEA) is a numerical analysis method used for solving a multitude of engineering problems related to structural analysis and fluid flow (Zienkiewicz et al., 2005). In a structural analysis related to orthopedics, FEA might be used to determine the strains in a bone (Duda et al., 1998, 2003; Rohlmann, Mössner, Bergmann, & Kölbel, 1982; Maldonado et al., 2003), the stresses in an implant (Duda et al., 2003; Heller et al., 2011), or the micromotion between bone and implant (Bah, Shi, Browne et al., 2015; Bah, Shi, Heller et al., 2015). At the core of its success lies the fact that the partial differential equations, which often describe these engineering problems, can principally be solved using the FE method and that indeed computers do so very efficiently in practice. Here the attractiveness of the FE method is also rooted in the underlying approach to split up larger, more complex problems into many smaller problems (using many small, but finite elements; Fig. 32.1) for which at least an approximate solution can be found.

When considering the human body and its organs and structures, it is clear that key assumptions underlying classical analytical mechanical analysis methods regarding, for example, the response of the material to load or even just their shapes may not hold. Such deviations may arise from the often rather complex mechanical characteristics of biological tissues (Cowin & Doty, 2007; Fung, 1993). Importantly, the beautiful shapes of musculoskeletal structures such as the bones tend to not only differ from the more basic shapes underlying traditional engineering designs and for which analytical models may have been developed but also exhibit substantial variation between individuals (Bah et al., 2015). As such, the widespread use of FEA in orthopedics (Taylor & Prendergast, 2014) as a method to estimate the mechanical behavior of a complex structure with an organic shape following a divide and conquer approach does not come as a surprise. Here we provide a concise introduction into essential considerations and some best practices in the use of FEA in applications to orthopedic biomechanics.

GENERAL CONSIDERATIONS FOR CONDUCTING FEA

The development of any model benefits from starting out from a more simple model with limited complexity to then later, as the understanding of the problem and how to best capture its nature in

Human Orthopaedic Biomechanics. DOI: https://doi.org/10.1016/B978-0-12-824481-4.00026-3

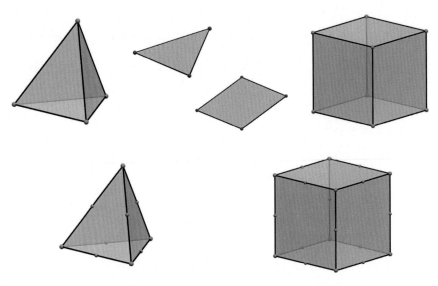

FIGURE 32.1

Finite elements. For many orthopedic applications three-dimensional solid finite elements such as tetrahedra (left) or hexahedra (right) are used, while shell elements (top, middle) might be considered to more efficiently model the behavior of thin structures such as a cartilage layer covering the joint surface. The examples at the top represent linear finite elements with nodes (*blue spheres*) located at each of the corners of the elements [element faces of the solid elements are rendered semitransparent (*gray*) to reveal visible and hidden edges (shown in *red* and *orange*, respectively)]. A linear tetrahedral element therefore has four nodes, while a linear hexahedral element has eight nodes. Quadratic elements (bottom) add further nodes which are typically located at the midpoints of the element edges (*pink spheres*) to better capture nonlinear deformations.

a computational model grows, add further complexity. Such a staged "crawl, walk, run" approach is certainly to be recommended in the development of any FEA.

After having successfully navigated the essential steps for setting up and running an FEA, the results need to be carefully analyzed and scrutinized. To that end, a basic understanding of the expected mechanical behavior of the structure is indispensable, and in addition to using first principles and estimates informed by analytical approaches to confirm whether the FEA results match reasonable expectations.

Steps to prepare a FEA will typically include:

- mesh generation,
- defining material properties,
- defining boundary conditions,
- defining loading conditions.

MESH CONVERGENCE ANALYSIS AND MODEL VALIDATION

As for any numerical analysis, it is important also for FEA to ensure that the results obtained do indeed provide a valid approximation of the sought solution. A critical step here is a mesh convergence analysis, which helps ensure that the extent of the subdivision of the structure studied and the choice of finite elements are appropriate to obtain a solution of the required accuracy.

Even if an FEA passes all these checks, the only way to establish how well a model can, for example, really predict the failure load of a femur is to directly compare and validate the predictions of such an outcome from FEA against experimentally determined data. Whilst studies without such validation may very well serve to provide valuable insights and can be instrumental to develop a general understanding of the load transfer of a structure, it is generally accepted that thorough, direct validation of FEA results is not just a nice to have feature but is an absolute must for any clinical application of the method (de Ruiter et al., 2021; Eberle, Gottlinger, & Augat, 2013; Falcinelli & Whyne, 2020; Kok, Grassi, Gustafsson, & Isaksson, 2021; Rohlmann, Mössner, Bergmann, & Kölbel, 1982; Schileo, Pitocchi, Falcinelli, & Taddei, 2020).

A CASE STUDY

In the following, we will elaborate on the essential steps described above, and to further illustrate key aspects of importance for the successful application of FEA in orthopedic biomechanics, we will follow a case study. In this case study, we seek to characterize the strain distribution in the femoral diaphysis under typical loads associated with activities of daily living (Bergmann et al., 2001; Heller et al., 2001, 2005). The geometry of the femur considered in the case study was derived from computed tomography (CT) scans of a 64-year-old man who underwent medical imaging for nonorthopedic related matters (Table 32.1) (Bah, Shi, Browne et al., 2015). No further information on the subject was available and for the purpose of the current study their body mass was assumed to be 85 kg. All FEAs related to the case study here were performed using FEBio v.3.2 (Maas, Ellis, Ateshian, & Weiss, 2012).

MESH GENERATION

The extraction of a surface model from medical image data is often the starting point for generating FE models (Fig. 32.2). Here a CT scan was used to semi-automatically extract the external (periosteal) surface of the femur (ITK-SNAP v3.8; Yushkevich et al., 2006). The periosteal bone surface generated here was further processed to obtain an estimation of the thickness distribution of the cortical bone (Stradview v6.12; Pearson & Treece, 2017; Poole et al., 2012; Treece, Gee, Mayhew, & Poole, 2010;

Table 32.1 Morphological parameters of the femur used in the finite element analyses. Key morphological parameters to characterize the size and shape of the femur studied here were determined using an automated process.

Characteristic	Value
Bicondylar length	448.3 mm
Femoral head radius	23.7 mm
Anteversion angle	11.7 degrees
Neck-shaft or caput-collum-diaphyseal (CCD) angle	122.9 degrees

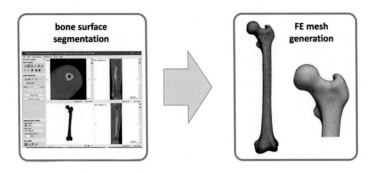

FIGURE 32.2

From medical image data to surfaces and volumetric meshes. The femoral bone surface was extracted (segmented) from a CT scan using a semiautomatic process (ITK-SNAP v3.8; left). The resulting bone surface is represented by triangles, which formed the input for a mesh generation tool (Tetgen v1.6) with which the femur was subdivided in many tetrahedral FE elements ("mesh"; right) for the subsequent analysis.

Treece & Gee, 2014), an important step to capture its contribution to the overall bone stiffness (Schileo et al., 2020; Zdero et al., 2010). The periosteal and endosteal bone surfaces obtained in that way served as inputs to a mesh generation tool. Mesh generation divides the structure of interest into many small but finitely sized elements, a crucial step in any FE model development, for which a range open source tools are available (Dapogny, Dobrzynski, & Frey, 2014; Geuzaine & Remacle, 2009; Schöberl, 1997). For the purpose of the current case study, the femur was meshed using linear (four nodes) tetrahedral elements using Tetgen v.1.6 (Si, 2015) (Fig. 32.2).

What exactly the size of the finite elements should be in order for the solution to be independent of the characteristics of the mesh requires a more detailed mesh convergence analysis as detailed below. Indeed before we can run any such analysis, we need to define material properties, boundary conditions, and loads.

MATERIAL PROPERTIES

To study the response of a structure to load, an appropriate material model and its properties need to be chosen. These properties may be as simple as an elastic module and a Poisson number for an isotropic linear elastic material but more complex material behavior including multiphasic conditions (solid−fluid interactions) might be considered, for example, when studying soft tissues such as cartilage (Halloran et al., 2012), ligaments (Ristaniemi, Tanska, Stenroth, Finnilä, & Korhonen, 2021), muscle (Chi et al., 2010), or the earlier stages of a healing bone fracture (Isaksson, Wilson, van Donkelaar, Huiskes, & Ito, 2006).

Bone, the organ and tissue of interest here is known to be a hierarchically structured material offering fantastic mechanical performance over a range of conditions (Fratzl, Gupta, Paschalis, & Roschger, 2004). Which aspects of its behavior need to be considered is primarily dependent upon the nature of the questions that the FE model attempts to address. While analyzing the response of bone to typical, physiological loads that result from activities of daily living, any load rate-dependent viscosity-related effects of bone tissue might be neglected, but such simplifications may

Table 32.2 Material properties of the femur finite element models.

Material	Elastic modulus (MPa)	Poisson number
Cortical bone	17,000	0.3
Trabecular bone	1500	0.3
Medullary cavity	15	0.3

Note: *For studying long bone mechanics under physiological loading associated to activities of daily living bone was considered to exhibit linear elastic material properties which were assumed to be homogeneously distributed.*

not be valid for studies in fall/impact load scenarios (Dall'Ara et al., 2013; Haider, Speirs, & Frei, 2013).

It is now widely accepted that studies on the interactions between bone and an implant (joint replacement devices, screws, etc.) should consider the local variation in bone material properties (Falcinelli & Whyne, 2020; Taddei et al., 2006). Methods to estimate local variations and inhomogeneity in terms of the elastic modulus of linear elastic material models from image data for use in FEA are well established and dedicated software to determine such data are readily available (Schileo et al., 2020; Taddei et al., 2014). The identification of direction-dependent material properties from routine medical image data remains challenging (Pahr & Zysset, 2016) and with some notable exceptions, only limited data on the spatial distribution of such properties are accessible (Rohrbach et al., 2015).

Material properties considered in the case study

Whilst our understanding of bone's detailed, multiscale behavior is still being actively developed, it appears that key insights into the overall deformation and strain pattern of the femur under physiological loading are likely more dependent upon an accurate description of the cortical thickness than considering its detailed anisotropy and inhomogeneous distribution (Duda et al., 1998; Peng, Bai, Zeng, & Zhou, 2005; Schileo et al., 2020; Zdero et al., 2010). For the purpose of the case study, linear elastic material properties and a homogeneous distribution were considered for both cortical and trabecular bone (Table 32.2, Fig. 32.3), consistent with related previous work (Duda et al., 2003; Heller et al., 2011; Kleemann et al., 2003; Speirs et al., 2006; Szwedowski et al., 2012).

BOUNDARY CONDITIONS

In static FEA, the model needs to be constrained such that at least rigid body motions are not possible and the model does indeed remain static when loads are applied, which means translations along each of the three coordinate axes as well as three rotations around these axes need to be constrained. Thus in static FEA, a minimum of 6 degrees of freedom (DoFs) need to be constrained to remove three-dimensional (3D) rigid body motion. Quite often models published in the literature are overconstrained, that is, more than the minimum number of 6 DoFs is fixed (Speirs et al., 2006). Although there might well exist conditions where the structure is truly very much constrained and capturing that in the FE model would indeed be desirable, the introduction of a very constrained boundary that is not consistent with reality is not to be recommended. Conceptually, the introduction of a rigidly fixed boundary such as a rigidly fixed, distally clamped femur (Fig. 32.4, right) is equivalent to considering the plane of fixation to be a symmetry plane for the

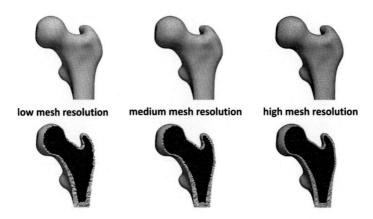

FIGURE 32.3

Continuous mesh refinement by reducing element size to study mesh convergence. To demonstrate that the results computed in the FEA are independent of the chosen mesh size, multiple meshes need to be generated and used for analysis. The proximal femur for three exemplary meshes used for the convergence analysis of our case study ranging from coarse (left, corresponding to M2) to detailed (right, M6) is shown. The bottom row of images exposes the internal structure of the mesh by peeling away data in from of a plane approximately parallel to the femoral neck. Colors indicate the type of material assigned to the elements for our case study, where elements shown in gray represent cortical bone, those in red represent trabecular bone, and green elements represent the medullary cavity.

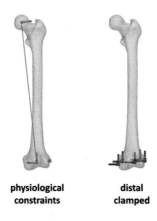

FIGURE 32.4

Boundary conditions. When using physiological constraints (left) only the minimum number of 6 degrees of freedom (DoFs) required to remove rigid body motion is constrained. Here all 3 DoFs are constrained at a node of an element located at the knee center (*green arrows*), while a further node at the lateral condyle is restricted from moving anterior-posteriorly (1 DoF). A node at the femoral head center is constrained to move along a line (brown) towards the knee center only (2 DoFs). Alternative methods to constrain the femur include the distal clamped condition (right) where typically all DoFs of multiple nodes are constrained (*gray arrows*).

structure. Whether the anatomy and structures below the knee are consistent with a mirror image of the femur is at least questionable. Critically, except from very specific conditions (Duda et al., 1998) the exact choice of constraint conditions influences the deformation patterns. Variations in that choice of constraint conditions may result in vastly different deformations of the bone and in consequence also strain magnitudes and patterns.

Boundary conditions considered in the case study

To illustrate the potentially dramatic effect of the choice of boundary conditions, our case study will compare the overall bone deformations and strains in the diaphysis of the femur under walking loads in FE models where either physiological constraints (Spears et al., 2007) or distally clamped conditions are applied (Fig. 32.4). Here the physiological constraint conditions represent not only a minimum constrained condition (only 6 DoFs are constrained) but also try to consider that in vivo, medial movement of the femoral head is limited by the acetabulum that acts as a barrier and forces the femoral head to move primarily downwards.

LOADING CONDITIONS

Typically, a FEA seeks to examine some aspect of the load transfer through or the load sharing between structures. Therefore the loading conditions applied to the structures are of central importance for FEA. In orthopedic applications, there is often a need to consider either typical or critical loads. Determining which conditions constitute the latter is dependent on the specific question at hand but may also be difficult to establish a priori, though an understanding of the former (i.e., typical) conditions might prove useful for developing a strategy to identify the latter.

Typical joint loading conditions

The internal musculoskeletal loading conditions of the major joints and of the long bones during activities of daily living are governed by the forces produced by the muscles, which contract to either allow movement to occur or to actively stabilize a joint. The muscles crossing a joint are the key contributors to the contact forces exerted at the joint. These joint contact forces are known to take on values of several multiples of body weight (BW) during activities of daily living (Bergmann et al., 2001; Heller, Bergmann, Deuretzbacher, Claes et al., 2001; Heller, Bergmann, Deuretzbacher, Dürselen et al., 2001; Heller, Bergmann et al., 2005). While the contact forces at the knee and hip during slow walking might be as low as 2 BW, these forces can easily reach magnitudes of over 5 BW during slow jogging. These data on the joint forces already provide strong evidence that the internal forces acting on the skeletal structures are very much activity-dependent and can vary by at least a factor between activities. Yet evidence from stumbling events suggests that in vivo joint forces at the hip can reach even higher force levels with data measured at over 9 BW during such conditions (Bergmann et al., 2001, 2014; Bergmann, Bender, Dymke, Duda, & Damm, 2016).

Which conditions constitute critical loading conditions?

Considerations regarding what constitutes a critical load for use in FEA routinely take the load magnitude as a key determinant of criticality and peak loads are often thought to be most critical. However, conditions with smaller forces that act at more unusual and oblique angles might prove rather more critical: in the absence of sufficient axial compression, the detrimental effect of a shear force on the stability of a device may well be more pronounced. There is thus growing consensus for the need to study a wider range of loading conditions to actively assess which loading scenario may prove most critical (Berahmani et al., 2016; Quevedo González et al., 2019).

The orientation of the joint forces may be a key factor to be considered while seeking to determine critical loading conditions for use in FEAs for the hip. Here the loading conditions during stair climbing, though an activity less frequent than walking (Morlock et al., 2001), exhibit a substantially increased anterior-posterior (AP) joint force component (Bergmann, Graichen, & Rohlmann, 2004) because of the more flexed hip and the more oblique orientation of the muscles to the femur (Bergmann et al., 2001; Heller, Bergmann, Deuretzbacher, Dürselen et al., 2001; Heller, Bergmann et al., 2005). These increased AP joint forces in turn result in substantially increased axial torque around the femoral shaft and might challenge the fixation of a joint replacement (Heller, Kassi, Perka, & Duda, 2005; Kassi, Heller, Stoeckle, Perka, & Duda, 2004; Kassi, Heller, Stoeckle, Perka, & Duda, 2006; Bergmann et al., 1995).

The fact that the impact forces during sideways fall present a rather high risk of bone fracture in the elderly underlines that unusual forces acting at unusual locations have a high potential to be critical to the integrity of the bones (Choi, Cripton, & Robinovitch, 2015). While internal forces may vary significantly in magnitude, the force directions vary relatively less, because the way the muscles cover the skeleton and skeletal structures they are well adapted to take such loads. The deformation of the femoral neck induced by an impact force on the other hand creates a strain distribution that is rather different from typical conditions and may thus result in bone failure (Haider et al., 2013; Iori et al., 2020; Kok et al., 2021; Schileo, Balistreri, Grassi, Cristofolini, & Taddei, 2014). Under such impact conditions, the loading rate, as discussed earlier in relation to material properties, may be a factor to consider when attempting to tease out which loading conditions are most critical and should be considered in an FEA.

Application of physiological loads to FE models

There is evidence to suggest that the application of the relevant joint contact together with key muscle forces is critical to maintaining the intrinsic link between muscle and joint contact forces to re-create tissue-level mechanics in long bones resembling physiological conditions (Duda et al., 1998; Rohlmann, Mössner, Bergmann, & Kölbel, 1982). Such muscle and joint contact forces can be derived on a subject-specific basis using appropriate computer models, factors influencing variation in physiological loading conditions between individuals. Patients with musculoskeletal conditions may exhibit variations in muscle coordination patterns such as increased muscle cocontraction levels that result in more substantial variation in musculoskeletal loading conditions between subjects and between patients and healthy individuals (Trepczynski et al., 2012, 2018). In addition to the specific way in which an activity is performed such as the speed with which a person walks, intersubject variations in morphology such as the femoral offset, and femoral anteversion all have the potential to modulate muscle and joint forces to some extent (Heller et al., 2001; 2007). Therefore the subject characteristics such as specifics of neuromuscular control and detailed anatomical configuration would ideally be considered in validated, subject-specific musculoskeletal models that provide a personalized estimate of the musculoskeletal loading conditions (joint and muscle forces) for use in an FEA (Duda et al., 2003; Heller et al., 2011; Kleemann, Heller, Stoeckle, Taylor, & Duda, 2003; Speirs, Heller, Taylor, Duda, & Perka, 2006; Szwedowski et al.,

2012). In a fully integrated workflow, muscle attachments in the musculoskeletal model and the FE model match so there is minimal if any error when mapping the forces onto the FE model. When such detailed analyses are not possible, a viable fallback solution would refer to validated predictions of typical joint and muscle forces (Heller, Kassi, et al., 2005). More advanced approaches for such situation consider the specific bone morphology and model the likely spectrum of musculoskeletal loads that can arise from variations in task execution (Taddei et al., 2014).

Loading conditions considered in the case study

For the purpose of the case study, we considered two load cases comprising muscle and joint contact forces for the instances of peak loading during a walking and stair climbing cycle as previously published (Heller, Kassi, et al., 2005). Hip contact force magnitudes for these two load cases differed only slightly (stair climbing 6% larger than walking), while the AP force component of that joint contact force in stair climbing was 1.8 times larger than during walking. The stair climbing load case was also characterized by a substantially increased quadriceps force, which amounted to more than four times (4.3) the force value during walking. All muscle and hip joint contact forces were applied as concentrated point loads.

MESH CONVERGENCE ANALYSIS

To ensure that the results generated by the FE method are sound, it is critical to demonstrate that the numerical solution obtained is independent of the specifics of the mesh. To that end a mesh convergence analysis is performed, where the results for key parameters of interest for multiple analyses with different meshes are compared to identify mesh characteristics for which the required error level is obtained while minimizing computational effort and solution time. The purpose of a mesh convergence analysis is therefore to determine the characteristics of a mesh (element size, order of the shape functions) which provide for a solution of the desired accuracy. To that end an approach sometimes referred to by h-refinement is often considered where the characteristic edge length "h" of the tetrahedral elements is continuously refined (reduced) and the mesh density is increased. Alternatively, one may consider "p-refinement" where the degree of the shape functions of the elements is increased to obtain a better approximation of the solution. A p-refinement step could mean a change from using linear to using quadratic elements. A combination of the two methods (h-p-refinement) is also possible.

The mesh convergence analysis needs to support the chosen mesh resolution for the specific question and depending on the nature of the problem different quantities may be considered as readout parameters to assess whether and at which refinement level convergence has occurred. Maximum deformation is often a quantity of interest; however, stresses and strains are derived from a gradient of the deformations and thus convergence with respect to deformation is not sufficient to verify convergence in relation to a stress or strain related readout. Studies related to implant performance might therefore be interested in a stress-based convergence criterion such as the maximum von Mises stress in the device, while analysis into the behavior of bone could consider equivalent (von Mises) strain as a related comprehensive measure for strain. To verify whether convergence has occurred, it is often useful to plot the readout parameters of interest over the logarithm of the DoFs of the different meshes considered. Here, an asymptotic behavior of the readouts would be indicative of convergence to a stable solution. In order to be able to observe such asymptotic behavior, it is often suggested to consider at least four levels of refinement. By using predefined, problem-specific error conditions, and then checking whether these conditions are exactly fulfilled, it is possible to determine whether an increase in mesh density will change the readout parameter by less than the error threshold and whether any convergence has occurred. Furthermore, if the problem is of such a nature that a continuous stress and strain distribution is to be expected, a visual check of the distribution can help observe whether the computed stress/strain field is continuous and whether any convergence has occurred, or if it has not occurred, whether the distribution shows clear signs of discontinuity and step changes and therefore further refinement of the mesh is required.

In the discussion of mesh convergence and the consideration of peak values of the readouts, one also needs to consider effects that may occur at so-called singularities. Such singularities can exist, for example, at a sharp corner or at points of load application. Here, mesh refinement will result in divergence where stress and strain readouts will theoretically approach infinity. Mesh refinement is thus not a suitable approach to better estimate the actual mechanics for such conditions which would typically require some rounding-off of sharp corners or the distribution of nodal loads to a larger surface area. It is important to be aware of such conditions to be able to avoid that any numerical artifacts arising from these singularities impact the mesh convergence analysis.

Details of the mesh convergence analysis considered in the case study

To illustrate aspects of mesh convergence, a total of seven meshes composed entirely of linear tetrahedral elements with approximately 65,000 and 1,100,000 DoFs were created and considered in the case study (Table 32.3, Fig. 32.3) following an h-refinement approach where the characteristic (mean) element edge length was reduced from a starting value of 3.6 mm to a final value of 1.21 mm. As we are interested in the deformation pattern of the bone and the distribution of the strains in the diaphysis of the femur, the readout quantities to assess mesh convergence are the maximum deformation of the femur and the equivalent strains on the surface of the bone. As the femoral diaphysis is reasonably expected to be loaded in bending and compression (Duda et al., 1998; Heller et al., 2001; Rohlmann, Mössner, Bergmann, & Kölbel, 1982), we are focusing on the surface strains where both compressive and tensile strains will peak. A difference of 5% change in equivalent strains is sufficient to demonstrate convergence for the purpose of our case study and will plot both maximum deformation and equivalent strain as a function of the mesh DoFs to visualize convergence behavior. A surface plot of the diaphyseal strain distribution will further help to visually assess convergence or lack thereof. Since applying all forces to the model as point loads

Table 32.3 Characteristics of meshes used in convergence analysis. A range of meshes with ever decreasing element edge length and thus increasing number of degrees of freedom was created to study mesh.

Characteristic/ reference	M1	M2	M3	M4	M5	M6	M7
DoFs	65,727	112,260	139,521	238,614	539,598	774,216	1,108,233
Mean edge length (SD) (mm)	3.60 (0.85)	2.98 (0.61)	2.75 (0.66)	2.23 (0.66)	1.60 (0.62)	1.41 (0.52)	1.21 (0.51)

concentrated on selected nodes (see loading conditions below), we will illustrate the effect of divergence at such singularities when mesh density is increased.

CASE STUDY DATA ANALYSIS

The following sections present key results of the case study to illustrate the analysis of mesh convergence, and demonstrate the effect of boundary conditions on the deformation and strain distribution in the femur before examining the differences in the diaphyseal surface strain distribution when applying walking versus stair climbing loads.

Mesh convergence analysis

The plot of the normalized maximum deformation and the maximum equivalent strain in the femoral diaphysis (Fig. 32.5) shows that the error of the strains (27%) in the coarsest mesh is substantially larger than the error of the maximum deformation (4%). The first step of mesh refinement brings the results for both measures considered here to within 2.5% of those of the most refined mesh, thus satisfying our predefined error threshold of 5%. When using linear elements to approximate curved shapes such as a bone reducing the element size will result in a better approximation of the actual shape. Such variation in the discretization error will influence the smoothness of such plots and is one reason why the use of quadratic elements might be preferred.

In line with the formal evaluation of convergence (Fig. 32.5), visual inspection of the surface strain fields for different mesh sizes (Fig. 32.6) confirms that the strain distribution for the coarsest mesh (M1), though largely comparable in the overall pattern to the conditions for the somewhat refined mesh M2 and the very fine mesh M7, has very clear discontinuities. These discontinuities present in mesh M1 provide strong indications that the expected smooth strains cannot be correctly

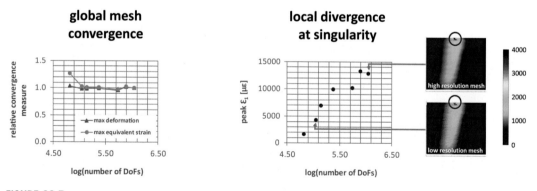

FIGURE 32.5

Analysis of mesh convergence. Left: Plotting key readout measures (maximum deformation and maximum equivalent strain) helps to ascertain when and if convergence of the results to the correct solution occurs while continuously refining the mesh. Here readouts are normalized to the conditions for the most refined mesh analyzed to compare the convergence characteristics of deformations and strains. Right: Maximum principal (tensile) strains around the load application area of the quadriceps muscle force as a function of element size provide an illustration of divergence around a singularity.

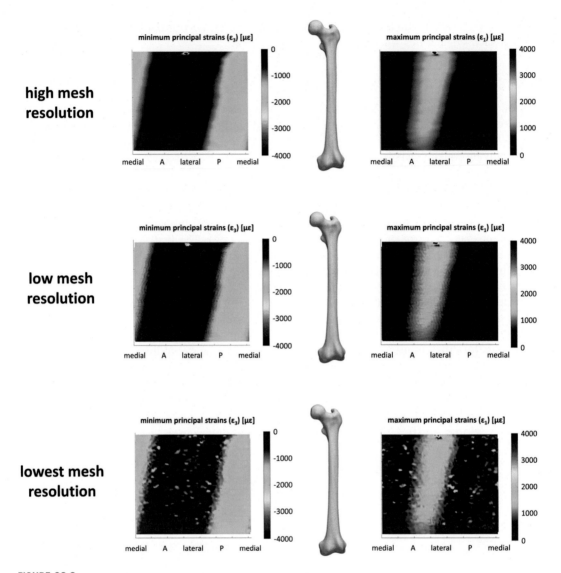

FIGURE 32.6

Influence of mesh density on the surface strains in the femoral diaphysis. Maximum principal strains (ε1, tensile) and minimum principal strains (ε3, compressive) across the surface of the femoral diaphysis comparing the conditions for a very refined mesh (top row: M7) to those for the two coarsest meshes (middle row: M2, bottom row: M1). Strain data for elements at the surface of the bone are interpolated and evaluated at a regularly spaced grid in a cylindrical coordinate system. The surface is "cut open" from distal to proximal along the medial edge and then "unrolled" allowing a detailed comparison of the results (A: anterior, P: posterior). All strains are reported in units of microstrain ($\mu\varepsilon$).

captured by that mesh. In contrast to that, the distribution for the mesh of the next refinement level (M2) appears already smooth. While some variation in strain magnitudes exists for the various mesh resolutions, the overall pattern and distribution are characterized by higher peak compressive than tensile strains, and a distribution where peak strains in the proximal diaphysis appear lateral (ε_1, tensile) and medial (ε_3, compressive) while locations of peak strains in the distal diaphysis are found more anterior (ε_1, tensile) and more posterior (ε_3, compressive), respectively.

This finding of the visual inspection of the strain distribution is consistent with the formal evaluation that indicates that mesh M2 can provide an approximation of the diaphyseal strains within the defined error threshold of 5%.

Visual inspection of the surface strain plots further reveals a singularity at the center, top of each of the plots being most prominent in the visualization of the maximum principle (tensile) strains ε1 (Fig. 32.6). The location where this singularity can be observed in the area where a point load associated with the quadriceps muscle forces is applied. A plot of the maximum principle (tensile) strains ε1 at the location of the singularity (Fig. 32.5, right) shows that the strains here continue to increase and do not converge to a stable solution. Such conditions of divergence will always occur at singularities such as nodes where point loads are applied or at sharp corners. Data at the location of such singularities may mask the overall mesh convergence behavior if not appropriately accounted for.

Influence of boundary conditions on femoral deformation and femoral strain distribution

Deformation patterns are distinctly different for the two sets of boundary conditions studied (Fig. 32.7): for the physiological boundary conditions overall only small and primarily distally directed head deflections can be observed, whilst the dominant deformation of the femoral head

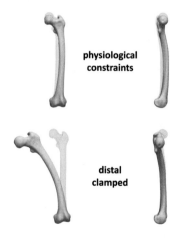

physiological constraints

distal clamped

FIGURE 32.7

Influence of boundary conditions on the deformation of the femur. Deformation patterns of the femur (10 times magnified) for physiological (top row) and distal clamped (bottom row) boundary conditions of femur under walking loads. The femur in its reference, the undeformed condition, is shown in blue (semitransparent), while the deformed configuration is shown in gray (opaque).

under distal clamped conditions is medial and much more substantial. For the physiological boundary conditions, the mid diaphysis of the bone is moving somewhat anterior and more substantially laterally, whilst the diaphysis under distal clamped conditions moves somewhat posterior and rather substantially medially.

The patterns of the strain distribution (Fig. 32.8) are substantially different with peak strains for the distal clamped conditions more than double those when using physiological constraints. The strain distribution for the distal clamped conditions exhibits a pattern where peak strains over the entire diaphysis consistently occur lateral ($\varepsilon 1$, tensile) and medial ($\varepsilon 3$, compressive), conditions indicative of a situation with more dominant bending in the frontal plane than is the case for the conditions when using physiological constraints for which some bending in the sagittal plane is also evident, especially distally. The analysis of the deformation and strains clearly demonstrate the significant effect that the specific choice of boundary conditions has. Although loads, mesh, and

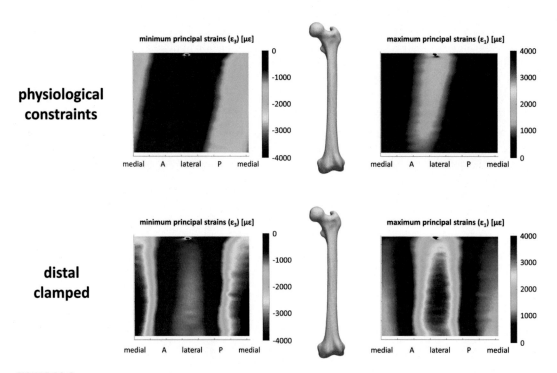

FIGURE 32.8

Influence of boundary conditions on the surface strains in the femoral diaphysis. Maximum principal strains ($\varepsilon 1$, tensile) and minimum principal strains ($\varepsilon 3$, compressive) across the surface of the femoral diaphysis comparing the conditions when using physiological constraints (top row) with those using distal clamping (bottom row). Strain data for elements at the surface of the bone are interpolated and evaluated at a regularly spaced grid in a cylindrical coordinate system. The surface is "cut open" from distal to proximal along the medial edge and then "unrolled" allowing a detailed comparison of the results (A: anterior, P: posterior). All strains are reported in units of microstrain ($\mu\varepsilon$).

material properties are identical between the two conditions, the deformations and strains (magnitude and patterns) are not. Careful consideration of the choice of the boundary conditions is thus key to enable a valid interpretation of the physiological straining of a long bone.

The stark differences to the conditions with physiological constraints, which are particularly prominent distally, show that the often repeated argument that the principle of St Venant (decreasing influence of artifacts the further away from the source of the artefact one gets) may allow valid interpretation of the strains at some distance from the clamped nodes may not at all hold. Clamped conditions as used here affect the deformation and strain pattern throughout the entire bone and though any artifacts associated with singularities at the constraint nodes might have indeed faded at a short distance away from them, the substantial overestimation of frontal plane bending does not. Any FEA interested in the distal mechanics of a long bone should thus carefully consider which boundary conditions and loads to apply (Duda et al., 2003; Märdian, Schaser, Duda, & Heyland, 2015; Taylor et al., 1996).

Influence of activity-dependent loading conditions on the femoral strain distribution

When considering physiological boundary conditions for the FEA, both walking and stair climbing loads result in a strain distribution that is indicative of bending and compression though the magnitude of the peak strains is substantially increased under stair climbing loads where minimum principal strains of -4546 $\mu\varepsilon$ and maximum principal strains of 2994 $\mu\varepsilon$ were recorded (Fig. 32.9). Peak minimum principal strains for walking loads occur in a posteromedial location in the distal diaphysis (-2005 $\mu\varepsilon$), while maximum tensile strains occur in the lateral proximal diaphysis (1639 $\mu\varepsilon$). From proximal to distal, the location of the peak strains rotates by about 60 degrees such that peak tensile strains in the distal diaphysis occur antero-lateral. Consistent with the more oblique muscle force direction for the more flexed hip during stair climbing and the associated, increased AP component of the hip joint contact force and the linked increased sagittal plane bending moment, the peak tensile strains in the proximal diaphysis occur more anterior and the peak compressive strains appear more posterior for the stair climbing than for the walking loads (Fig. 32.9). Although hip contact force magnitudes are similar between activities, the additional quadriceps forces during stair climbing result in higher strains in the femoral diaphysis during stair climbing.

MODEL VALIDATION

Although the focus of the chapter is on the practical aspects directly related to the computational aspects of FEA, we need to stress how experimental validation of FE predictions, in particular those that are to be used in the context of clinical decision making such as the prediction a bone fracture load, require a careful validation against experimental data. Although strain gauges to measure surface strains on bones have been used extensively for validation of computational models since the pioneering work on the role of muscle forces for long bone mechanics by Rohlmann et al. (1982) and are still widely used today (Cristofolini, Juszczyk, Taddei, & Viceconti, 2009; Eberle et al., 2013), more recently developed methods such as digital image correlation (DIC) offer even further opportunities for model validation as they can capture the strains across a more substantial and continuous surface area thereby overcoming the limitation of strain gauges which are restricted to

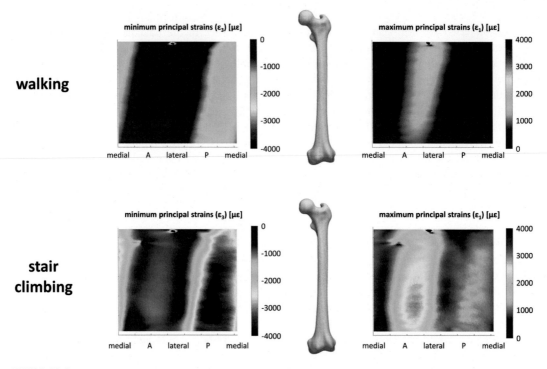

FIGURE 32.9

Influence of loading conditions on the surface strains in the femoral diaphysis. Maximum principal strains ($\varepsilon 1$, tensile) and minimum principal strains ($\varepsilon 3$, compressive) across the surface of the femoral diaphysis comparing the conditions under muscle and hip contact force loading consistent with the instances of peak hip contact force during walking (top row) and stair climbing (bottom row) (Heller, Bergmann et al., 2005; Heller, Kassi, et al., 2005). Strain data for elements at the surface of the bone are interpolated and evaluated at a regularly spaced grid in a cylindrical coordinate system. The surface is "cut open" from distal to proximal along the medial edge and then "unrolled" allowing a detailed comparison of the results (A: anterior, P: posterior). All strains are reported in units of microstrain ($\mu\varepsilon$).

measurements at only a few select surface locations (Collins, Yang, Crenshaw, & Ploeg, 2021; de Ruiter et al., 2021; Woods, Heller, & Browne, 2017; Kok et al., 2021). Similarly, if not, even more exciting opportunities for detailed model validation arise from techniques such as digital volume correlation, which offers a way to assess conditions throughout the analysis volume thus removing the restriction to surface measurements inherent to strain gauges and DIC (Comini, Palanca, Cristofolini, & Dall'Ara, 2019; Roberts, Perilli, & Reynolds, 2014; Ryan, Oliviero, Costa, Wilkinson, & Dall'Ara, 2020; Tavana, Clark, Newell, Calder, & Hansen, 2020; Turunen et al., 2020). Given the crucial computational aspect in the working principle of the latter method, anyone with a truly keen interest in FEA should also be keenly interested in making good use of such advanced techniques for FE model validation.

SUMMARY AND CONCLUSION

Although there is growing consensus that for robustly assessing orthopedic problems with FEA rather more comprehensive analyses are required that take into account the multiple sources of variability typically encountered here (bone anatomy, material properties, loading conditions, surgeon-related factors, etc.) and thus require hundreds if not thousands of individual FEAs (Bah, Shi, Browne, et al., 2015; Bah, Shi, Heller, et al., 2015; Bryan et al., 2012), the journey begins with setting up an individual analysis and doing so well.

This chapter introduced key steps required for setting up a FE model and provided essential background based on which the reader can develop a good understanding of essential considerations regarding mesh generation, the choice of material properties, boundary and loading conditions, and avoid common pitfalls. A good grasp of essential principles of how to perform and assess a mesh convergence study together with a rigorous approach to scrutinize results using first principles and tried and tested analytical approaches on one hand and an awareness of the rich spectrum of experimental methods available to validate an FE models' prediction on the other hand form an excellent basis for developing studies that move the field forward. Paired with a mindset and the realization that such verification and validations steps are not just nice to have but constitute an essential prerequisite for the successful application of FEAs that can influence practice in orthopedics, the reader is well placed to breeze through the crawl and walking stages of FEA.

REFERENCES

Bah, M. T., Shi, J., Browne, M., Suchier, Y., Lefebvre, F., Young, P., King, L., Dunlop, D. G., & Heller, M. O. (2015). Exploring inter-subject anatomic variability using a population of patient-specific femurs and a statistical shape and intensity model. *Medical Engineering & Physics*, 37(10), 995−1007. Available from https://doi.org/10.1016/j.medengphy.2015.08.004.

Bah, M. T., Shi, J., Heller, M. O., Suchier, Y., Lefebvre, F., Young, P., King, L., Dunlop, D. G., Boettcher, M., Draper, E., & Browne, M. (2015). Inter-subject variability effects on the primary stability of a short cementless femoral stem. *Journal of Biomechanics*, 48(6), 1032−1042. Available from https://doi.org/10.1016/j.jbiomech.2015.01.037.

Berahmani, S., Janssen, D., Wolfson, D., de Waal Malefijt, M., Fitzpatrick, C. K., Rullkoetter, P. J., & Verdonschot, N. (2016). FE analysis of the effects of simplifications in experimental testing on micromotions of uncemented femoral knee implants. *Journal of Orthopaedic Research: Official Publication of the Orthopaedic Research Society*, 34(5), 812−819. Available from https://doi.org/10.1002/jor.23074.

Bergmann, G., Bender, A., Dymke, J., Duda, G., & Damm, P. (2016). Standardized loads acting in hip implants. *PLoS One*, 11(5), e0155612. Available from https://doi.org/10.1371/journal.pone.0155612.

Bergmann, G., Bender, A., Graichen, F., Dymke, J., Rohlmann, A., Trepczynski, A., & Kutzner, I. (2014). Standardized loads acting in knee implants. *PLoS One*, 9(1), e86035. Available from https://doi.org/10.1371/journal.pone.0086035.

Bergmann, G., Deuretzbacher, G., Heller, M., Graichen, F., Rohlmann, A., Strauss, J., & Duda, G. N. (2001). Hip contact forces and gait patterns from routine activities. *Journal of Biomechanics*, 34(7), 859−871. Available from http://www.ncbi.nlm.nih.gov/pubmed/11410170.

Bergmann, G., Graichen, F., & Rohlmann, A. (1995). Is staircase walking a risk for the fixation of hip implants? *Journal of Biomechanics*, 28(5), 535−553. Available from https://doi.org/10.1016/0021-9290(94)00105-D.

Bergmann, G., Graichen, F., & Rohlmann, A. (2004). Hip joint contact forces during stumbling. *Langenbeck's Archives of Surgery/Deutsche Gesellschaft fur Chirurgie*, *389*(1), 53–59. Available from https://doi.org/10.1007/s00423-003-0434-y.

Bryan, R., Nair, P. B., & Taylor, M. (2012). Influence of femur size and morphology on load transfer in the resurfaced femoral head: A large scale, multi-subject finite element study. *Journal of Biomechanics*, *45*(11), 1952–1958. Available from https://doi.org/10.1016/j.jbiomech.2012.05.015.

Chi, S. W., Hodgson, J., Chen, J. S., Reggie Edgerton, V., Shin, D. D., Roiz, R. A., & Sinha, S. (2010). Finite element modeling reveals complex strain mechanics in the aponeuroses of contracting skeletal muscle. *Journal of Biomechanics*, *43*(7), 1243–1250. Available from https://doi.org/10.1016/j.jbiomech.2010.01.005.

Choi, W. J., Cripton, P. A., & Robinovitch, S. N. (2015). Effects of hip abductor muscle forces and knee boundary conditions on femoral neck stresses during simulated falls. *Osteoporosis International: A Journal Established as Result of Cooperation Between the European Foundation for Osteoporosis and the National Osteoporosis Foundation of the USA*, *26*(1), 291–301. Available from https://doi.org/10.1007/s00198-014-2812-4.

Collins, C. J., Yang, B., Crenshaw, T. D., & Ploeg, H.-L. (2021). Evaluation of experimental, analytical, and computational methods to determine long-bone bending stiffness. *Journal of the Mechanical Behavior of Biomedical Materials*, *115*, 104253. Available from https://doi.org/10.1016/j.jmbbm.2020.104253.

Comini, F., Palanca, M., Cristofolini, L., & Dall'Ara, E. (2019). Uncertainties of synchrotron microCT-based digital volume correlation bone strain measurements under simulated deformation. *Journal of Biomechanics*, *86*, 232–237. Available from https://doi.org/10.1016/j.jbiomech.2019.01.041.

Cowin, S. C., & Doty, S. B. (2007). *Tissue mechanics*. Springer. Available from http://site.ebrary.com/id/10230555.

Cristofolini, L., Juszczyk, M., Taddei, F., & Viceconti, M. (2009). Strain distribution in the proximal human femoral metaphysis. *Proceedings of the Institution of Mechanical Engineers. Part H, Journal of Engineering in Medicine*, *223*(3), 273–288. Available from https://doi.org/10.1243/09544119jeim497.

Dall'Ara, E., Luisier, B., Schmidt, R., Kainberger, F., Zysset, P., & Pahr, D. (2013). A nonlinear QCT-based finite element model validation study for the human femur tested in two configurations in vitro. *Bone*, *52*(1), 27–38. Available from https://doi.org/10.1016/j.bone.2012.09.006.

Dapogny, C., Dobrzynski, C., & Frey, P. (2014). Three-dimensional adaptive domain remeshing, implicit domain meshing, and applications to free and moving boundary problems. *Journal of Computational Physics*, *262*, 358–378. Available from https://doi.org/10.1016/j.jcp.2014.01.005.

de Ruiter, L., Rankin, K., Browne, M., Briscoe, A., Janssen, D., & Verdonschot, N. (2021). Decreased stress shielding with a PEEK femoral total knee prosthesis measured in validated computational models. *Journal of Biomechanics*, *118*, 110270. Available from https://doi.org/10.1016/j.jbiomech.2021.110270.

Duda, G. N., Heller, M., Albinger, J., Schulz, O., Schneider, E., & Claes, L. (1998). Influence of muscle forces on femoral strain distribution. *Journal of Biomechanics*, *31*(9), 841–846. Available from http://www.ncbi.nlm.nih.gov/pubmed/9802785.

Duda, G. N., Mandruzzato, F., Heller, M., Schutz, M., Claes, L., & Haas, N. P. (2003). Mechanical borderline indications in the treatment of unreamed tibial nailing. *Der Unfallchirurg*, *106*(8), 683–689. Available from https://doi.org/10.1007/s00113-003-0633-6.

Eberle, S., Gottlinger, M., & Augat, P. (2013). Individual density-elasticity relationships improve accuracy of subject-specific finite element models of human femurs. *Journal of Biomechanics*, *46*(13), 2152–2157. Available from https://doi.org/10.1016/j.jbiomech.2013.06.035.

Falcinelli, C., & Whyne, C. (2020). Image-based finite-element modeling of the human femur. *Computer Methods in Biomechanics and Biomedical Engineering*, *23*(14), 1138–1161. Available from https://doi.org/10.1080/10255842.2020.1789863.

Fratzl, P., Gupta, H. S., Paschalis, E. P., & Roschger, P. (2004). Structure and mechanical quality of the collagen−mineral nano-composite in bone. *Journal of Materials Chemistry*, *14*(14), 2115−2123. Available from https://doi.org/10.1039/B402005G.

Fung, Y. C. (1993). *Biomechanics: Mechanical properties of living tissues.* New York: Springer. Available from https://books.google.co.uk/books?id = zo2McgAACAAJ.

Geuzaine, C., & Remacle, J.-F. (2009). Gmsh: A 3-D finite element mesh generator with built-in pre- and post-processing facilities. *International Journal for Numerical Methods in Engineering*, *79*(11), 1309−1331. Available from https://doi.org/10.1002/nme.2579.

Haider, I. T., Speirs, A. D., & Frei, H. (2013). Effect of boundary conditions, impact loading and hydraulic stiffening on femoral fracture strength. *Journal of Biomechanics*, *46*(13), 2115−2121. Available from https://doi.org/10.1016/j.jbiomech.2013.07.004.

Halloran, J. P., Sibole, S., van Donkelaar, C. C., van Turnhout, M. C., Oomens, C. W. J., Weiss, J. A., Guilak, F., & Erdemir, A. (2012). Multiscale mechanics of articular cartilage: Potentials and challenges of coupling musculoskeletal, joint, and microscale computational models. *Annals of Biomedical Engineering*, *40*(11), 2456−2474. Available from https://doi.org/10.1007/s10439-012-0598-0.

Heller, M. O., Bergmann, G., Deuretzbacher, G., Claes, L., Haas, N. P., & Duda, G. N. (2001). Influence of femoral anteversion on proximal femoral loading: Measurement and simulation in four patients. *Clinical Biomechanics (Bristol, Avon)*, *16*(8), 644−649. Available from http://www.ncbi.nlm.nih.gov/pubmed/11535345.

Heller, M. O., Bergmann, G., Deuretzbacher, G., Dürselen, L., Pohl, M., Claes, L., Haas, N. P., & Duda, G. N. (2001). Musculo-skeletal loading conditions at the hip during walking and stair climbing. *Journal of Biomechanics*, *34*(7), 883−893. Available from http://www.ncbi.nlm.nih.gov/pubmed/11410172.

Heller, M. O., Bergmann, G., Kassi, J. P., Claes, L., Haas, N. P., & Duda, G. N. (2005). Determination of muscle loading at the hip joint for use in pre-clinical testing. *Journal of Biomechanics*, *38*(5), 1155−1163. Available from https://doi.org/10.1016/j.jbiomech.2004.05.022.

Heller, M. O., Kassi, J. P., Perka, C., & Duda, G. N. (2005). Cementless stem fixation and primary stability under physiological-like loads in vitro. *Biomedizinische Technik (Berl)*, *50*(12), 394−399. Available from https://doi.org/10.1515/BMT.2005.054.

Heller, M. O., Mehta, M., Taylor, W. R., Kim, D. Y., Speirs, A., Duda, G. N., & Perka, C. (2011). Influence of prosthesis design and implantation technique on implant stresses after cementless revision THR. *Journal of Orthopaedic Surgery and Research*, *6*(1), 20. Available from https://doi.org/10.1186/1749-799X-6-20.

Heller, M. O., Schröder, J. H., Matziolis, G., Sharenkov, A., Taylor, W. R., Perka, C., & Duda, G. N. (2007). Muskuloskeletale Belastungsanalysen. Biomechanische Erklärung klinischer Resultate − und mehr? *Orthopäde*, *36*(3), 188−194. Available from https://doi.org/10.1007/s00132-007-1054-y.

Iori, G., Peralta, L., Reisinger, A., Heyer, F., Wyers, C., van den Bergh, J., Pahr, D., & Raum, K. (2020). Femur strength predictions by nonlinear homogenized voxel finite element models reflect the microarchitecture of the femoral neck. *Medical Engineering & Physics*, *79*, 60−66. Available from https://doi.org/10.1016/j.medengphy.2020.03.005.

Isaksson, H., Wilson, W., van Donkelaar, C. C., Huiskes, R., & Ito, K. (2006). Comparison of biophysical stimuli for mechano-regulation of tissue differentiation during fracture healing. *Journal of Biomechanics*, *39*(8), 1507−1516. Available from https://doi.org/10.1016/j.jbiomech.2005.01.037.

Kassi, J.-P., Heller, M. O., Stoeckle, U., Perka, C., & Duda, G. N. (2004). Stair climbing is more critical than walking in pre-clinical assessment of primary stability in cementless THA in vitro. *Journal of Biomechanics*, *38*(5), 1143−1154. Available from https://doi.org/10.1016/j.jbiomech.2004.05.023.

Kassi, J. P., Heller, M. O., Stoeckle, U., Perka, C., & Duda, G. N. (2006). Response to: "Stair climbing is more critical than walking in pre-clinical assessment of primary stability in cementless THA in vitro" [J. Biomech. 2005;38: 1143−1154]. *Journal of Biomechanics*, *39*(16), 3087−3090. Available from https://doi.org/10.1016/j.jbiomech.2006.09.012.

Kleemann, R. U., Heller, M. O., Stoeckle, U., Taylor, W. R., & Duda, G. N. (2003). THA loading arising from increased femoral anteversion and offset may lead to critical cement stresses. *Journal of Orthopaedic Research, 21*(5), 767−774. Available from https://doi.org/10.1016/S0736-0266(03)00040-8.

Kok, J., Grassi, L., Gustafsson, A., & Isaksson, H. (2021). Femoral strength and strains in sideways fall: Validation of finite element models against bilateral strain measurements. *Journal of Biomechanics, 122,* 110445. Available from https://doi.org/10.1016/j.jbiomech.2021.110445.

Maas, S. A., Ellis, B. J., Ateshian, G. A., & Weiss, J. A. (2012). FEBio: Finite elements for biomechanics. *Journal of Biomechanical Engineering, 134*(1), 011005. Available from https://doi.org/10.1115/1.4005694.

Maldonado, Z. M., Seebeck, J., Heller, M. O., Brandt, D., Hepp, P., Lill, H., & Duda, G. N. (2003). Straining of the intact and fractured proximal humerus under physiological-like loading. *Journal of Biomechanics, 36*(12), 1865−1873. Available from https://www.ncbi.nlm.nih.gov/pubmed/14614940.

Märdian, S., Schaser, K.-D., Duda, G. N., & Heyland, M. (2015). Working length of locking plates determines interfragmentary movement in distal femur fractures under physiological loading. *Clinical Biomechanics, 30*(4), 391−396. Available from https://doi.org/10.1016/j.clinbiomech.2015.02.006.

Morlock, M., Schneider, E., Bluhm, A., Vollmer, M., Bergmann, G., Müller, V., & Honl, M. (2001). Duration and frequency of every day activities in total hip patients. *Journal of Biomechanics, 34*(7), 873−881.

Pahr, D. H., & Zysset, P. K. (2016). Finite element-based mechanical assessment of bone quality on the basis of in vivo images. *Current Osteoporosis Reports, 14*(6), 374−385. Available from https://doi.org/10.1007/s11914-016-0335-y.

Pearson, R. A., & Treece, G. M. (2017). Measurement of the bone endocortical region using clinical CT. *Medical Image Analysis, 44*(Suppl. C), 28−40. Available from https://doi.org/10.1016/j.media.2017.11.006.

Peng, L., Bai, J., Zeng, X., & Zhou, Y. (2005). Comparison of isotropic and orthotropic material property assignments on femoral finite element models under two loading conditions. *Medical Engineering & Physics, 28*(3), 227−233. Available from https://doi.org/10.1016/j.medengphy.2005.06.003.

Poole, K. E. S., Treece, G. M., Mayhew, P. M., Vaculík, J., Dungl, P., Horák, M., Štěpán, J. J., & Gee, A. H. (2012). Cortical thickness mapping to identify focal osteoporosis in patients with hip fracture. *PLoS One, 7*(6), e38466. Available from https://doi.org/10.1371/journal.pone.0038466.

Quevedo González, F. J., Lipman, J. D., Lo, D., De Martino, I., Sculco, P. K., Sculco, T. P., Catani, F., & Wright, T. M. (2019). Mechanical performance of cementless total knee replacements: It is not all about the maximum loads. *Journal of Orthopaedic Research: Official Publication of the Orthopaedic Research Society, 37*(2), 350−357. Available from https://doi.org/10.1002/jor.24194.

Ristaniemi, A., Tanska, P., Stenroth, L., Finnilä, M. A. J., & Korhonen, R. K. (2021). Comparison of material models for anterior cruciate ligament in tension: From poroelastic to a novel fibril-reinforced nonlinear composite model. *Journal of Biomechanics, 114,* 110141. Available from https://doi.org/10.1016/j.jbiomech.2020.110141.

Roberts, B. C., Perilli, E., & Reynolds, K. J. (2014). Application of the digital volume correlation technique for the measurement of displacement and strain fields in bone: A literature review. *Journal of Biomechanics, 47*(5), 923−934. Available from https://doi.org/10.1016/j.jbiomech.2014.01.001.

Rohlmann, A., Mössner, U., Bergmann, G., & Kölbel, R. (1982). Finite-element-analysis and experimental investigation of stresses in a femur. *Journal of Biomedical Engineering, 4*(3), 241−246. Available from https://doi.org/10.1016/0141-5425(82)90009-7.

Rohrbach, D., Grimal, Q., Varga, P., Peyrin, F., Langer, M., Laugier, P., & Raum, K. (2015). Distribution of mesoscale elastic properties and mass density in the human femoral shaft. *Connective Tissue Research, 56*(2), 120−132. Available from https://doi.org/10.3109/03008207.2015.1013627.

Ryan, M. K., Oliviero, S., Costa, M. C., Wilkinson, J. M., & Dall'Ara, E. (2020). Heterogeneous strain distribution in the subchondral bone of human osteoarthritic femoral heads, measured with digital volume correlation. *Materials, 13*(20), 4619. Available from https://doi.org/10.3390/ma13204619.

Schileo, E., Balistreri, L., Grassi, L., Cristofolini, L., & Taddei, F. (2014). To what extent can linear finite element models of human femora predict failure under stance and fall loading configurations? *Journal of Biomechanics*, *47*(14), 3531−3538. Available from https://doi.org/10.1016/j.jbiomech.2014.08.024.

Schileo, E., Pitocchi, J., Falcinelli, C., & Taddei, F. (2020). Cortical bone mapping improves finite element strain prediction accuracy at the proximal femur. *Bone*, *136*, 115348. Available from https://doi.org/10.1016/j.bone.2020.115348.

Schöberl, J. (1997). NETGEN An advancing front 2D/3D-mesh generator based on abstract rules. *Computing and Visualization in Science*, *1*(1), 41−52. Available from https://doi.org/10.1007/s007910050004.

Si, H. (2015). TetGen, a Delaunay-based quality tetrahedral mesh generator. *ACM Transactions on Mathematical Software*, *41*(2). Available from https://doi.org/10.1145/2629697, Article 11.

Speirs, A. D., Heller, M. O., Duda, G. N., & Taylor, W. R. (2006). Physiologically based boundary conditions in finite element modelling. *Journal of Biomechanics*, *40*(10), 2318−2323. Available from https://doi.org/10.1016/j.jbiomech.2006.10.038.

Speirs, A. D., Heller, M. O., Taylor, W. R., Duda, G. N., & Perka, C. (2006). Influence of changes in stem positioning on femoral loading after THR using a short-stemmed hip implant. *Clinical Biomechanics*, *22*(4), 431−439. Available from https://doi.org/10.1016/j.clinbiomech.2006.12.003.

Szwedowski, T. D., Taylor, W. R., Heller, M. O., Perka, C., Muller, M., & Duda, G. N. (2012). Generic rules of mechano-regulation combined with subject specific loading conditions can explain bone adaptation after THA. *PLoS One*, *7*(5), e36231. Available from https://doi.org/10.1371/journal.pone.0036231.

Taddei, F., Palmadori, I., Taylor, W. R., Heller, M. O., Bordini, B., Toni, A., & Schileo, E. (2014). European Society of Biomechanics S.M. Perren Award 2014: Safety factor of the proximal femur during gait: A population-based finite element study. *Journal of Biomechanics*, *47*(14), 3433−3440. Available from https://doi.org/10.1016/j.jbiomech.2014.08.030.

Taddei, F., Schileo, E., Helgason, B., Cristofolini, L., & Viceconti, M. (2006). The material mapping strategy influences the accuracy of CT-based finite element models of bones: An evaluation against experimental measurements. *Medical Engineering & Physics*, *29*(9), 973−979. Available from https://doi.org/10.1016/j.medengphy.2006.10.014.

Tavana, S., Clark, J. N., Newell, N., Calder, J. D., & Hansen, U. (2020). In vivo deformation and strain measurements in human bone using digital volume correlation (DVC) and 3T clinical MRI. *Materials*, *13*(23), 5354. Available from https://doi.org/10.3390/ma13235354.

Taylor, M., & Prendergast, P. J. (2014). Four decades of finite element analysis of orthopaedic devices: Where are we now and what are the opportunities? *Journal of Biomechanics*, *48*(5), 767−778. Available from https://doi.org/10.1016/j.jbiomech.2014.12.019.

Taylor, M. E., Tanner, K. E., Freeman, M. A., & Yettram, A. L. (1996). Stress and strain distribution within the intact femur: Compression or bending? *Medical Engineering & Physics*, *18*(2), 122−131. Available from https://doi.org/10.1016/1350-4533(95)00031-3.

Treece, G. M., & Gee, A. H. (2014). Independent measurement of femoral cortical thickness and cortical bone density using clinical CT. *Medical Image Analysis*, *20*(1), 249−264. Available from https://doi.org/10.1016/j.media.2014.11.012.

Treece, G. M., Gee, A. H., Mayhew, P. M., & Poole, K. E. S. (2010). High resolution cortical bone thickness measurement from clinical CT data. *Medical Image Analysis*, *14*(3), 276−290. Available from https://doi.org/10.1016/j.media.2010.01.003.

Trepczynski, A., Kutzner, I., Kornaropoulos, E., Taylor, W. R., Duda, G. N., Bergmann, G., & Heller, M. O. (2012). Patellofemoral joint contact forces during activities with high knee flexion. *Journal of Orthopaedic Research: Official Publication of the Orthopaedic Research Society*, *30*(3), 408−415. Available from https://doi.org/10.1002/jor.21540.

Trepczynski, A., Kutzner, I., Schwachmeyer, V., Heller, M. O., Pfitzner, T., & Duda, G. N. (2018). Impact of antagonistic muscle co-contraction on in vivo knee contact forces. *Journal of Neuroengineering and Rehabilitation*, *15*(1), 101. Available from https://doi.org/10.1186/s12984-018-0434-3.

Turunen, M. J., Le Cann, S., Tudisco, E., Lovric, G., Patera, A., Hall, S. A., & Isaksson, H. (2020). Sub-trabecular strain evolution in human trabecular bone. *Scientific Reports*, *10*(1), 13788. Available from https://doi.org/10.1038/s41598-020-69850-x.

Woods, C. J., Heller, M. O., & Browne, M. (2017). Assessing the effect of unicondylar knee arthroplasty on proximal tibia bone strains using digital image correlation. *International Journal of Biomedical Engineering and Science*, *4*(3), 01–06. Available from https://doi.org/10.5121/ijbes.2017.4301.

Yushkevich, P. A., Piven, J., Hazlett, H. C., Smith, R. G., Ho, S., Gee, J. C., & Gerig, G. (2006). User-guided 3D active contour segmentation of anatomical structures: Significantly improved efficiency and reliability. *Neuroimage*, *31*(3), 1116–1128. Available from https://doi.org/10.1016/j.neuroimage.2006.01.015.

Zdero, R., Bougherara, H., Dubov, A., Shah, S., Zalzal, P., Mahfud, A., & Schemitsch, E. H. (2010). The effect of cortex thickness on intact femur biomechanics: A comparison of finite element analysis with synthetic femurs. *Proceedings of the Institution of Mechanical Engineers. Part H, Journal of Engineering in Medicine*, *224*(7), 831–840. Available from https://doi.org/10.1243/09544119JEIM702.

Zienkiewicz, O. C., Taylor, R. L., Zhu, J. Z., Zienkiewicz, O. C., Taylor, R. L., & Zhu, J. Z. (2005). *The finite element method: Its basis and fundamentals* (7th ed.). Butterworth-Heinemann p. iii. Available from https://doi.org/10.1016/B978-1-85617-633-0.00020-4.

RIGID-BODY AND MUSCULOSKELETAL MODELS 33

Michael Skipper Andersen

Department of Materials and Production, Aalborg University, Aalborg, Denmark

INTRODUCTION

Forces may accelerate and/or deform bodies. Therefore, if we wish to understand why some (bio) mechanical system moves in a specific way, or why some materials, for example, tissues, are deforming in a certain manner, we must gain insights into the underlying forces that cause these. Within mechanical systems, we may be able to measure these forces directly by incorporating force sensors. However, within the human body, direct measurements of forces of the muscles, ligaments, cartilage, etc. are highly invasive and not something that can be accomplished on a regular basis. Therefore, the only practically viable approach to obtain force information is to estimate them, and this is where rigid-body and musculoskeletal modeling comes into play. Such models are designed to estimate the internal forces based on a mathematical model description of the (bio)mechanical system, and some provided inputs, for example, measured movements, external forces, and, in case of musculoskeletal models, muscle activation patterns via so-called electromyography measurements.

Regardless of whether we are investigating a purely mechanical system or a musculoskeletal system, developing mechanical models of the system has an array of benefits. By having a mathematical model of the relationship between the inputs and the outputs of the system, we can do the following:

1. systematically investigate the causal relationships between inputs and outputs;
2. gain insights into quantities that we may not be able to measure experimentally; and/or
3. alter the system to obtain a desired performance.

These are just a few potential applications of such models and, later in the chapter, we will look into these in more detail. This we will do by looking into how changes to a bicycle affect the internal loads in the rider and we will use this model to alter the mechanics of the bicycle with the aim of reducing the musculoskeletal loads. But, before we get to that example, we will first review some of the fundamental physics that we will build our models on.

FUNDAMENTS OF RIGID-BODY AND MUSCULOSKELETAL MODELING

Rigid-body and musculoskeletal modeling is founded on Newton's laws of motion. In their classic form, these laws express how forces affect the movement of a single particle and how, if a force is applied to a body, that the body will apply an equal and opposite force to back on the body that applied the force. But what is a force? This may seem like a simple question, but it is not straightforward to accurately define. Qualitatively, we can say that a force:

1. is a quantity that can cause a body to accelerate (e.g., a free fall);
2. is an action of one system that affects another system;
3. is a quantity that can deform a body (e.g., a spring).

Formally, we will define a force as a 2D or 3D vector depending on whether we are doing analysis in 2D or 3D. This means that a force has both a magnitude and a direction.

The two first observations above have been more accurately expressed in Newton's laws, whereas the third is described by Hook's law in case of linear elastic materials. I am sure that you have heard these before, but let's review them so that we are all on the same page going forward.

Newton's first law of motion states that, unless acted upon by a net external force, a particle at rest will remain at rest and a particle in constant motion will remain at constant motion. This law actually follows as a natural consequence of Newton's second law of motion.

Newton's second law of motion relates the resultant force acting on the particle, the mass of the particle, and the acceleration of the particle:

$$\sum_{i=1}^{N} \mathbf{F}_i = m\mathbf{a} \tag{33.1}$$

where m is the particle's mass and $\mathbf{a}$ is the linear acceleration vector of the particle. This equation holds for both 2D and 3D and F_i is the ith force applied to the particle.

Finally, *Newton's third law* states that all forces between two particles are equal in magnitude but opposite in direction and, hereby, explains how the forces are transferred between adjacent particles. This is also sometimes referred to as the *law of action and reaction*.

The classical form of Newton's second law of motion describes the relationship between the net external forces acting on a particle and its linear acceleration. However, for most applications, we are interested in analyzing systems that are not only composed of independent and infinitely small particles. Instead, we are interested in bodies, which we will describe as a collection of infinitely small particles. Therefore, we need to extend the above laws to bodies instead. This can be done for both rigid and deformable bodies, but in this chapter, we will focus only on rigid bodies and what we mean by this is that the body will remain the same shape independent of how large forces the body is exposed to. From a practical point-of-view, no real body is rigid but there are many applications, where the deformations are so small that they can be neglected.

Whereas a particle can only translate, a body can both translate and rotate. In 2D, the body can rotate about one axis and in 3D, it can rotate about three axis. The force vector can cause linear acceleration as per Newton's second law, but it can also cause rotation of the body. To describe this, we need to introduce the moment vector, which is defined as:

$$\mathbf{M}^O = \mathbf{r} \times \mathbf{F} \tag{33.2}$$

where $\mathbf{M}^O$ is the moment vector around the point O and $\mathbf{r}$ is a vector from the point O to the line of action of the force vector, $\mathbf{F}$. This is only defined for 3D vectors but can of course also be applied to 2D problems by setting the coordinate that points out of the plane to zero. If we assume that this is the z-axis, then we will find that the moment vector will end up having only a z-component and the x- and y-components are always zero. Therefore, the moment in 2D is typically expressed as a scalar rather than a vector as the moment is always around the axis pointing out of the 2D plane.

In terms of the relationship between the resultant moment vector and the rotation of the body, it depends on whether the system is modeled in 2D or 3D. In 2D, the equation is:

$$\sum_{i=1}^{N_M} M_i^O = J\alpha \tag{33.3}$$

where M_i^O is the ith moment around the center of mass, J is the mass moment of inertia referring to the center of mass, and α is the angular acceleration of the body around the axis pointing out of the 2D plane. N^M is the number of moments. In 3D, the equation, referred to as Euler's equation, is:

$$\sum_{i=1}^{N_M} \mathbf{M}_i^O = \mathbf{J}'\dot{\omega}' + \omega' \times \mathbf{J}'\omega' \tag{33.4}$$

where $\mathbf{M}_i^O$ is the ith moment vector on the body, $\mathbf{J}'$ is the inertial tensor in body-fixed coordinates, denoted by the apostrophe, and given with respect to the body's center of mass, and ω'_i is the angular velocity vector of the body measured relative to its body-fixed coordinate system. $\dot{\omega}'$ is the angular acceleration vector of the body computed as the time derivative of the angular velocity vector. The dot denotes the derivative with respect to time.

Therefore, for each body, we can set up three equations for 2D problems and six equations for 3D problems.

These equations are called the Newton–Euler equations. If the system consists of more than one body, then the interaction between the bodies is captured by introducing reaction forces and moments between the bodies in accordance with Newton's third law.

For the remaining part of the chapter, we will develop the equations for 3D analysis. To allow modeling of a system of bodies, we need to decide on a terminology to refer to them. To this end, we will apply the so-called full Cartesian formulation, where the position and orientation of the ith body are described by $\mathbf{q}_i = \begin{bmatrix} \mathbf{r}_i^T & \mathbf{p}_i^T \end{bmatrix}^T$, where $\mathbf{r}_i$ is the position vector of the body's center of mass relative to the global coordinate system and $\mathbf{p}_i = \begin{bmatrix} e_{i0} & e_{i1} & e_{i2} & e_{i3} \end{bmatrix}^T$ are four Euler parameters used to describe the orientation of the body relative to the global coordinate system. Euler parameters are selected to avoid issues with singularities but other rotational coordinates might equally well have been used, such as Cardan angles, direction cosines, etc.

A rigid body in 3D has only three rotational degrees of freedom, so with four Euler parameters introduced, a constraint equation must be introduced to ensure that the vector $\mathbf{p}_i$ has length unity, that is, $\mathbf{p}_i^T\mathbf{p}_i - 1 = 0$ for each body. A thorough description of Euler parameters can be found in other studies, for instance Nikravesh (1987).

The velocity of the ith body is described by $\mathbf{v}_i = \begin{bmatrix} \dot{\mathbf{r}}_i^T & \omega_i'^T \end{bmatrix}^T$. To keep the equations concise, the time-dependency of these variables is not explicitly written, for example, when we write $\mathbf{q}$, it is to be understood as $\mathbf{q}(t)$ and similar for $\mathbf{v}$, etc. Finally, we assemble the position and velocity vectors of all bodies in vectors $\mathbf{q} = \begin{bmatrix} \mathbf{q}_1^T & \mathbf{q}_2^T & \cdots & \mathbf{q}_n^T \end{bmatrix}^T$ and $\mathbf{v} = \begin{bmatrix} \mathbf{v}_1^T & \mathbf{v}_2^T & \cdots & \mathbf{v}_n^T \end{bmatrix}^T$, respectively.

The bodies of a mechanical system are often connected with mechanical joints, for example, revolute, spherical, and universal joints. To model these, kinematic constraint equations are frequently introduced on the form:

$$\Phi(\mathbf{q}) = 0 \tag{33.5}$$

How these constraint equations are formulated depend on the type of joint connecting the bodies, and the details can be found in other studies (Andersen, 2021; Nikravesh, 1987).

This set of nonlinear constraint equations must be solved together with the Newton–Euler equations, but exactly how this is accomplished depend on whether a forward or inverse dynamic approach is taken. In the section "Inverse dynamic analysis" later in this chapter, we will expand on how it is handled in the case of inverse dynamic analysis. In general, however, the reaction forces and moments induced on the whole system by the constraints can be computed as $\boldsymbol{\Phi}_q^T \boldsymbol{\lambda}$, where the subscript denotes the partial derivative with respect to $\mathbf{q}$ and $\boldsymbol{\lambda}$ are the generalized reaction forces (Damsgaard, Rasmussen, Christensen, Surma, & de Zee, 2006; Nikravesh, 1987).

To have the equations in a condensed form, the Newton–Euler equations for the ith body can be written as follows, which comes directly from Eqs. (33.1) and (33.4):

$$\begin{bmatrix} m_i \mathbf{I} & 0 \\ 0 & \mathbf{J}_i' \end{bmatrix} \dot{\mathbf{v}}_i + \begin{bmatrix} 0 \\ \sim \boldsymbol{\omega}_i' \mathbf{J}_i' \boldsymbol{\omega}_i' \end{bmatrix} = \mathbf{g}_i^{(\text{ext})} \tag{33.6}$$

where m_i is the mass of the ith body, $\mathbf{I}$ is the identity matrix, $\mathbf{J}_i'$ is the inertial tensor for the ith body, and $\mathbf{g}_i^{(\text{ext})}$ is the vector of all forces (top three components) and moments (bottom three components) acting to the body. The tilde denotes the skew-symmetric matrix.

Denoting the mass matrix by $\mathbf{M}_i$ and the second term, containing the gyroscopic terms, by $\mathbf{b}_i$, we get:

$$\mathbf{M}_i \dot{\mathbf{v}}_i + \mathbf{b}_i = \mathbf{g}_i^{(\text{ext})} \tag{33.7}$$

This can be written up for all n bodies in the mechanical system as:

$$\begin{aligned} \mathbf{M}_1 \dot{\mathbf{v}}_1 + \mathbf{b}_1 &= \mathbf{g}_1^{(\text{ext})} \\ \mathbf{M}_2 \dot{\mathbf{v}}_2 + \mathbf{b}_2 &= \mathbf{g}_2^{(\text{ext})} \\ &\vdots \\ \mathbf{M}_n \dot{\mathbf{v}}_n + \mathbf{b}_n &= \mathbf{g}_n^{(\text{ext})} \end{aligned} \tag{33.8}$$

which on condensed form can be written as:

$$\mathbf{M}\dot{\mathbf{v}} + \mathbf{b} = \mathbf{g}^{(\text{ext})} \tag{33.9}$$

where

$$\mathbf{M} = \begin{bmatrix} \mathbf{M}_1 & & & 0 \\ & \mathbf{M}_2 & & \\ & & \ddots & \\ 0 & & & \mathbf{M}_n \end{bmatrix}, \mathbf{b} = \begin{bmatrix} \mathbf{b}_1 \\ \mathbf{b}_2 \\ \vdots \\ \mathbf{b}_n \end{bmatrix}, \mathbf{g}^{(\text{ext})} = \begin{bmatrix} \mathbf{g}_1^{(\text{ext})} \\ \mathbf{g}_2^{(\text{ext})} \\ \vdots \\ \mathbf{g}_n^{(\text{ext})} \end{bmatrix} \tag{33.10}$$

It follows from the above equations that if we know the resultant force and moment, we can compute the movements of the bodies. This approach is called forward dynamics. On the other hand, if we know the acceleration and angular velocity vector, we can compute the resultant forces and moments that cause the movements, which is called inverse dynamics. If neither are known, we have to apply so-called predictive methods to compute both the movement and the forces, which is typically done by introducing a performance criterion. This type of simulation is beyond the scope of this chapter and we refer you to relevant literature (Anderson & Pandy, 1999; Davoudabadi Farahani, Andersen, de Zee, & Rasmussen, 2016; Falisse et al., 2019; Farahani, Bertucci, Andersen, Zee, & Rasmussen, 2015).

Forward dynamics in its pure form, where only forces and moments are provided as inputs to the simulations, is never used in biomechanical models but is the common approach when studying pure mechanical systems. For musculoskeletal models, on the other hand, measured kinematics, measured external forces and, in some models, Electromyography (EMG) are typically used as inputs and the internal forces estimated as output. So-called forward dynamics-based tracking methods (Seth & Pandy, 2007; Thelen, Anderson, & Delp, 2003) or inverse dynamics-based methods (Crowninshield & Brand, 1981; Damsgaard et al., 2006; Rasmussen, Damsgaard, & Voigt, 2001) can be applied to accomplish this. In the remaining of this chapter, we will look into an inverse dynamics-based method.

MUSCULOSKELETAL MODELING

A musculoskeletal model is a mechanical description of bones, muscles, and joints, and it can be applied to any living organism consisting of bones, joints, and muscles. Rigid bodies (that we for the rest of the chapter will refer to as segments to avoid confusion with the human body) are used to model either the segments of the organism or rigid structural elements of devices that are interacting with the organism. In cases where the deformations of the segments are of interests, finite-element models of the segments are typically developed and the forces estimated by the musculoskeletal model applied as boundary conditions (Cronskär, Rasmussen, & Tinnsten, 2015). To model the mechanical properties of the muscles, the phenomenological model of Hill (Zajac, 1989) is typically applied. This model describes both the elastic properties of the tissues and models the contractile element to account for the force-generating capacity's dependency on muscle length and contraction velocity. The joints are either modeled through kinematic constraint equations to capture the gross joint movements (Begon, Andersen, & Dumas, 2018) or through more advanced joint models with elastic descriptions of both ligaments and cartilage (Andersen, de Zee, Damsgaard, Nolte, & Rasmussen, 2017; Dejtiar et al., 2020).

HUMAN—BICYCLE MODEL

To make the ideas and concepts more clear to you, we will base the whole description of the musculoskeletal modeling approaches on an example. The example we will look into is a model of a human on a bicycle and to simplify matters, we will keep the model more or less in 2D. Biking is a very common form of transportation and exercise. For instance, when knee osteoarthritis patients get to the stage in the decease progression, where the pain prevents them from running or doing other activities with high loads in the knees, biking is a good alternative due to the relatively lower loads on the knees even in cases of high cardiovascular loads.

When we have built this model of a human on a bicycle, we will investigate how we can optimize the bicycle to reduce the musculoskeletal loads. We have all experienced how the pedals are harder to push when they are in the up—down position as compared to other positions. This is because the human body in this position has lower potential to generate forces in the tangential direction of the crank, leading to relatively higher loads on the body to generate the same crank moment. These positions are referred to as the top and bottom dead centers. At the end of the chapter, we will investigate the possibility of reducing the musculoskeletal loads by introducing springs to help overcome the top bottom dead centers. The model is shown in Fig. 33.1.

FIGURE 33.1

Human−bicycle model. Illustration of the human−bicycle model.

2D LOWER EXTREMITY MODEL

The first step to develop the model is to create a model of the anatomy that we are interested in studying. How much of the anatomy to include, and to what level of detail to model it, depends on the question of interest. Creating a full-body, full 3D model of a human is a very extensive task and for this reason, most of the software systems available today for musculoskeletal modeling comes with more or less completed and/or detailed models of human body. As the purpose of this chapter is to introduce the principles of how musculoskeletal modeling can be performed, we will use a much simplified version of the lower extremity. However, the presented principles also apply to much more detailed models.

The model we will look into is the so-called *2D leg model* from the AnyBody Managed Model Repository (AMMR) version 2.3.0 (Lund et al., 2021) and the simulations will be performed with the AnyBody Modeling Systems v. 7.3 (AnyBody Technology A/S, Aalborg, Denmark). The model is illustrated in Fig. 33.2.

The model consists of the pelvis and the left and right thigh, shank and foot segments. The total body mass of the subject being modeled is 75 kg, which is distributed to each segment based on regression equations (Winter, 2009), that is, the masses are 14.2%, 10.0%, 4.6%, and 1.4% for pelvis, thigh, shank, and foot, respectively of the total body mass. The mass moments of inertia relative to each segment's center of mass are estimated by assuming that the segments all are cylindrical and have a density of 1000 kg/m^3.

The hips, knees, and ankles are all modeled as revolute joints that are all perpendicular to each other. Hence, even though AMS allows 3D modeling, each leg of this particular model is only allowed to move in a single plane, making it effectively a 2D model.

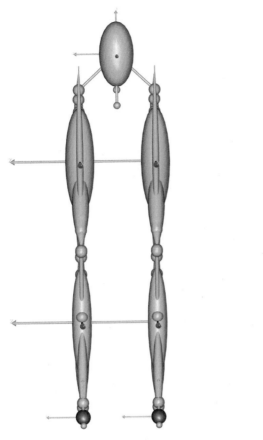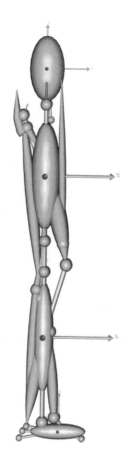

FIGURE 33.2

2D leg model. Illustration of the 2D leg model.

Eight muscles are included for each leg, namely the hamstrings, biceps femoris short head, gluteus maximus, rectus femoris, vastii, gastrocnemius, soleus, and tibialis anterior. For simplicity, the muscles are modeled with constant strength independent of muscle length and contraction velocity. In more advanced models, the muscle's strength dependency on the muscle length and contraction velocity is typically included and how this can be accomplished will be discussed later in the chapter.

BICYCLE MODEL

The bicycle model is illustrated in Fig. 33.1 and is simplified compared to a complete bicycle. The purpose of the model is to be able to investigate how the bicycle parameters, such as saddle height, handlebar height, gear ratio, pedal arm length and width affect the human biomechanics when cycling on a bicycle trainer. Therefore, the frame, saddle, handlebar and front wheel are modeled as one combined rigid segment. Additionally, the crank, including the pedal arms, and the rear wheel are modeled as one rigid segment each to allow different rotation velocity of the rear wheel and the crank. The pedals are not explicitly modeled but will be captured through the connection between the feet and the pedal arms modeled in the next section.

The bicycle frame is rigidly connected to ground, that is, six kinematic constraints, capturing that the bicycle cannot translate or rotate relative to the global coordinate system. Additionally, three reaction forces and three reaction moments are introduced between the bicycle frame and ground—one per kinematic constraint. The crank and the rear wheel are each connected to the bicycle frame with a revolute joint, hereby allowing rotation around one axis each. The angular velocity of the rear wheel is set equal to the angular velocity of the crank times a gearing ratio. Hereby, the bicycle has one remaining degrees of freedom (DOF), namely the rotation of the crank and this DOF we will control as part of creating the connection between the human and the bicycle.

HUMAN–BICYCLE INTERACTION MODEL

The final modeling step before we can start performing analysis is to connect the human model to the bicycle model. To accomplish this, it is important that we keep track of the number of DOFs that our human model has after we have connected all its segments with our joint models for the anatomical joints and how many DOFs are left in the external mechanical system, in our case the bicycle, that we wish to connect to the human model.

To this end, it is important that we learn how to count DOFs and this we will look into before we make the human–bicycle connection model.

So, DOF of a mechanical system is the independent way of motion that the system has. The best way to determine the number of DOFs of a 3D mechanical system is through the following equation:

$$n^{\text{DOF}} = 6n^{\text{segments}} - n^{\text{constraints}}$$

(33.11)

where n^{DOF} is the number of DOFs for the mechanical system, n^{segments} is the number of segments included in the model, and $n^{\text{constraints}}$ is the total number of constraints included.

With this equation, let us count up how many DOFs our mechanical system has before we start connecting the human and bicycle (Table 33.1).

Hereby, we are currently in the situation that we have to provide 13 more kinematic constraints and/or driver equations before all movements in the model are specified. To this end, we will make a rigid connection between the pelvis and the saddle, taking away six more DOFs. In reality, there is some movement between the saddle and pelvis, but, for simplicity, we neglect this in the model. This model could be improved by including experimentally observed relative movements between pelvis and the saddle.

The next connection, we will model, is the connection between the feet and the pedals. Here, we introduce two constraints between each foot and pedal, specifying that the anterior/posterior and superior/inferior coordinate of the contact point on the foot and the contact point on the pedal must be the same, taking away a total of four DOFs. Hereby, we have three DOFs left to specify. Since we, so far, only have constrained each leg in two DOFs, we still have one DOF left in each leg. This DOF in each leg, we will specify by making a model of how the ankle angle varies during a revolution of the crank. The model is illustrated in Fig. 33.3. This model is based on experimentally observed ankle angles and approximated with a Fourier series (Farahani et al., 2015).

Table 33.1 Overview of the human model and bicycle DOFs (degrees of freedom) before the two models are connected.

	Model	Description
	Human model	
6×7 segments	42	Seven segments (pelvis and left and right thigh, shank and foot) each introducing six DOFs
-6 revolute joints $\times$ 5 constraints	-30	Six revolute joints (hips, knees, and ankles) each constraining five relative DOFs between the segments they connect.
	Bicycle model	
6×3 segments	18	Three segments (frame, crank, and rear wheel) each introducing six DOFs.
$-1 \times$ grounding	-6	Grounding of the bicycle frame, locking all six relative DOFs between the frame and ground.
$-2 \times$ revolute joints	-10	Two revolute joints, connecting the crank to the frame and the rear wheel to the frame.
$-1 \times$ coupling of the crank and rear wheel rotations	-1	Coupling of the crank and rear wheel rotations to capture the effect of the two sprockets and the chain.
Number of remaining DOFs	13	

Hereby, we only have one DOF left, which is the crank rotation. For simplicity, we assume that the crank rotation velocity is constant over the crank rotation and consider it to be 80 rounds per minute (RPMs). Additionally, we will assume that a constant crank moment is provided by the rider independent of crank angle. This also implies symmetry between the two legs of the rider. However, the crank velocity and moment vary cyclically and have previously been modeled using Fourier series and could be improved if a more accurate model is desired. For more accurate modeling of the crank rotation and moment, refer to Farahani et al. (2015).

KINEMATIC ANALYSIS

Now that we have set up the kinematic constraint equations, we can continue to the analysis itself. Since we have set up exactly as many equations as we have unknowns in our problem, we can compute the positions and orientations of all the segments in our model from the kinematic constraint equations alone. This we call a *kinematically determinate system* and is the foundation for the basic form of inverse dynamic analysis. If we had fewer kinematic constraint equations, they would have infinitely many solutions and we call this a *kinematically underdeterminate system*. In this case, we cannot solve kinematics without consideration of the forces that cause the movement and this is the foundation for the basic form of forward dynamic analysis. The last situation, we may encounter, is the situation, where we have more kinematic constraints than we strictly need. This happens frequently in biomechanics when the movement input to the model is obtained from motion capture experiments. General solutions to this situation have been developed; referr to Andersen et al. (2017), Dejtiar et al. (2020), and Andersen, Jin, Li, & Chen (2021) for details of that case, which is called *kinematically overdeterminate system*.

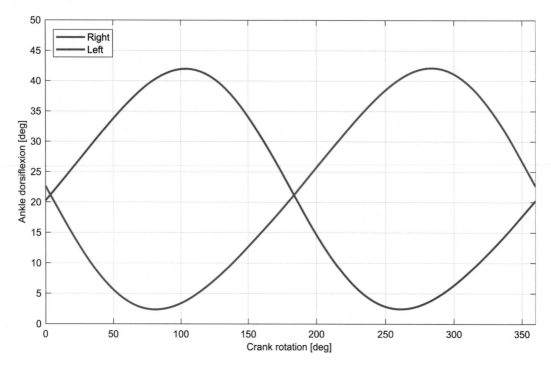

FIGURE 33.3

Ankle angles, Modeled ankle angles for the left and right leg over the duration of the crank rotation cycle.

For the remaining part of the chapter, we will focus solely on *kinematically determinate systems*. In this case, which matches what we have set up for the human–bicycle model above, we have as many unknowns in our problem as we have set up kinematic constraint equations, leading to a nonlinear system of equations as:

$$\mathbf{\Phi}(\mathbf{q}, t) = 0 \tag{33.12}$$

To solve the above system of equations, we need to compute $\mathbf{q}$, for all desired time steps, t, which can be accomplished using any nonlinear equation solver, for example, a Newton–Raphson algorithm (see, e.g., Nikravesh 1987 for details).

However, as seen earlier, the Newton–Euler equations depend not only on the position and orientations of the bodies, but also depend on the angular velocity (in 3D) and linear and angular accelerations of the segments. To compute the velocities and accelerations, we differentiate the kinematic constraint equations in Eq. (33.12) with respect to time once or twice to get:

$$\dot{\mathbf{\Phi}} = \mathbf{\Phi}_{\mathbf{q}}\dot{\mathbf{q}} + \mathbf{\Phi}_t = 0 \tag{33.13}$$

and

$$\ddot{\mathbf{\Phi}} = \mathbf{\Phi}_{\mathbf{q}}\ddot{\mathbf{q}} + \left(\mathbf{\Phi}_{\mathbf{q}}\dot{\mathbf{q}}\right)_{\mathbf{q}} + 2\ \mathbf{\Phi}_{\mathbf{q}t}\dot{\mathbf{q}} + \mathbf{\Phi}_{tt} = 0 \tag{33.14}$$

$\mathbf{\Phi_q}$ is the so-called Jacobian matrix, consisting of the partial derivatives of the constraint equations with respect to the coordinates, and $\mathbf{q}$. $\mathbf{\Phi}_t$ denotes the partial derivative of the constraint equations with respect to t. Similarly in the acceleration equation (Eq. 33.14), where the equations have been differentiated one more time, for example, $\mathbf{\Phi_{qt}}$ denotes the partial double derivatives by $\mathbf{q}$ and t and similarly for the other terms. These are linear equations in $\dot{\mathbf{q}}$ and $\ddot{\mathbf{q}}$, respectively, and can be solved with standard linear algebra.

Therefore, from Eq. (33.13), we can compute the segment velocities, $\dot{\mathbf{q}}$, when $\mathbf{q}$ is known from the solution of Eq. (33.12). Subsequently, we can compute the segment accelerations, $\ddot{\mathbf{q}}$, from Eq. (33.14) after substituting the known positions and velocities into the equation. Hereby, the kinematic analysis is complete and we can continue to determine the forces and moments that are necessary to produce this movements.

In 3D, however, simpler and shorter expressions are typically obtained if the velocity equations are expressed in terms of $\mathbf{v}$ instead of $\dot{\mathbf{q}}$ and can be accomplished through a coordinate transformation. This is an advanced topic and is beyond the scope of this chapter; refer to Damsgaard et al. (2006) and Michael Skipper Andersen et al. (2017) for more details.

INVERSE DYNAMIC ANALYSIS

To compute the forces and moments, we have to return to the Newton–Euler equations in Eq. (33.9). In its current form, all external forces and moments are contained in $\mathbf{g}^{(ext)}$ including the joint reaction and muscles forces that we wish to compute. Therefore, if we expand the right-hand side, we get:

$$\mathbf{M\dot{v}} + \mathbf{b} = \mathbf{C}^{(M)}\mathbf{f}^{(M)} + \mathbf{C}^{(R)}\mathbf{f}^{(R)} + \mathbf{g}^{(app)} \tag{33.15}$$

where $\mathbf{C}^{(M)}$ is the coefficient matrix for the muscle forces, $\mathbf{f}^{(M)}$ is the vector of muscle forces, $\mathbf{C}^{(R)}$ is the coefficient matrix for the joint reaction forces, $\mathbf{f}^{(R)}$ is the vector of joint reaction forces, and $\mathbf{g}^{(app)}$ includes all applied forces and moments. To compute the muscle and joint reaction forces, we isolate their terms on one side of the equation system:

$$\mathbf{C}^{(M)}\mathbf{f}^{(M)} + \mathbf{C}^{(R)}\mathbf{f}^{(R)} = \mathbf{M\dot{v}} + \mathbf{b} - \mathbf{g}^{(app)} \tag{33.16}$$

All the terms on the right-hand side are known after the kinematic analysis has been completed, and we denote them by $\mathbf{d} = \mathbf{M\dot{v}} + \mathbf{b} - \mathbf{g}^{(app)}$. The coefficient matrices for the muscles and joint reaction forces are grouped in one matrix:

$$\begin{bmatrix} \mathbf{C}^{(M)} & \mathbf{C}^{(R)} \end{bmatrix} \begin{bmatrix} \mathbf{f}^{(M)} \\ \mathbf{f}^{(R)} \end{bmatrix} = \mathbf{d} \tag{33.17}$$

and by defining $\mathbf{C} = \begin{bmatrix} \mathbf{C}^{(M)} & \mathbf{C}^{(R)} \end{bmatrix}$ and $\mathbf{f} = \begin{bmatrix} \mathbf{f}^{(M)} & \mathbf{f}^{(R)} \end{bmatrix}^{T}$, we have the dynamic equilibrium equations in the following condensed form:

$$\mathbf{Cf} = \mathbf{d} \tag{33.18}$$

As mentioned earlier, the joint reaction forces coming from kinematic constraints are given by the Jacobian matrix multiplied by the generalized reaction forces, which is also expressed in the equations above. In general, the coefficient matrix for the reaction forces, $\mathbf{C}^{(R)}$, is given as the partial derivative of the constraint equations with respect to a set of coordinates that corresponds to $\mathbf{v}$. However, if all constraint equations from Eq. (33.12) are included when setting up the reaction forces, we would introduce movement providers that would take up the actions of the muscles. Therefore, we have to exclude those kinematic constraint equations, that is, driver equations, that specify movements that the muscles must perform. In our case for the human–bicycle model, these are the crank rotation and the ankle angle drivers. $\mathbf{C}^{(M)}$ can be computed based on the partial derivative of the origin to insertion length of the muscle with respect to a set of coordinates that corresponds to $\mathbf{v}$ (Damsgaard et al., 2006).

For typical musculoskeletal models, including our simple 2D leg model on the bicycle, we have more unknown muscle and joint reaction forces than we have dynamic equilibrium equations. In the bicycle model, we have three DOFs that must be controlled by the muscles (one at the hip, knee, and ankle) but we have introduced eight muscles. So even in our highly simplified model, we have more unknowns than we can compute from the Newton–Euler equations alone. Therefore, we have to introduce a model of how the load is distributed, and, to this end, we will assume that the muscles are recruited optimally and found through solving an optimization problem:

$$
\begin{aligned}
\min \quad & H\big(\mathbf{f}^{(M)}\big) \\
\text{s.t.} \quad & \mathbf{Cf} = \mathbf{d} \\
& 0 \le \mathbf{f}^{(M)} \le \mathbf{s}^{(M)}
\end{aligned}
\tag{33.19}
$$

where H is a scalar objective function formulated as a function of the muscle forces that is minimized while ensuring that the dynamic equilibrium equations are fulfilled and that the muscles can only pull (i.e., are positive) and that the maximum muscle force remains below the instantaneous strength of the muscle, $\mathbf{s}^{(M)}$. It is an open research question whether the central nervous system actually does apply an optimal criterion and, if it does, which one it is. An array of different suggestions for this criterion can be found in the literature (Rasmussen et al., 2001), but the most common is a polynomial criterion (Rasmussen et al., 2001):

$$
H\big(\mathbf{f}^{(M)}\big) = \sum_{i=1}^{n^{(M)}} \left(\frac{\mathbf{f}_i^{(M)}}{\mathbf{s}_i^{(M)}} \right)^p
\tag{33.20}
$$

where p is the polynomial power and $\mathbf{s}_i^{(M)}$ is the instantaneous muscle strength for the ith muscle. $n^{(M)}$ is the number of muscles in the model. In our bicycle model, $\mathbf{s}_i^{(M)}$ is the maximum isometric force for the ith muscle as we in that model, for simplicity, assume that the muscle strength does not depend on the working conditions. However, the so-called contraction dynamics for the muscles can be included in this approach by computing the muscle strength as a function of the muscle length and contraction velocity that are known after the kinematic analysis. Details of how to accomplish this can be found in Andersen et al. (2021). The muscle force divided by its strength is called the muscle activity.

A polynomial power of either two or three is most frequently used. Recent studies have applied variations of the polynomial criterion by introducing a weight factor to account for the muscle volume (Andersen, 2018; Happee & Van der Helm, 1995; Marra et al., 2015; Thelen, Won Choi, & Schmitz, 2014), muscle physiological cross sectional area (Andersen, 2018), or a factor to account for subdivided muscles (Holmberg & Klarbring, 2012).

ANALYSIS OF HUMAN–BICYCLE DYNAMICS

Now, finally, we have both the human–bicycle model developed and all the theory required to perform inverse dynamic simulations including muscle recruitment, so let us look into the outputs of the model. To this end, we will investigate how we can reduce the musculoskeletal loads by storing and releasing elastic energy in springs to assist around the top and bottom dead centers in the crank cycle.

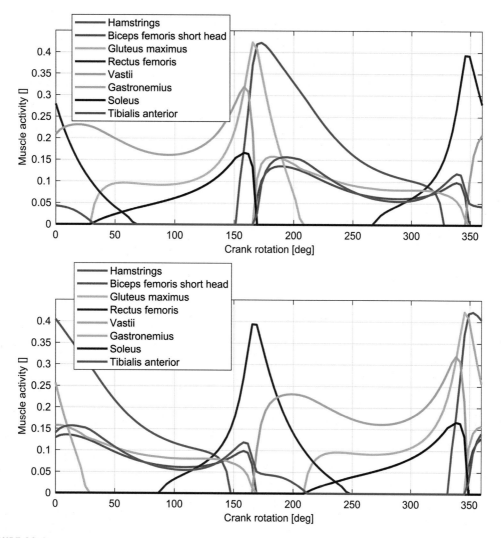

FIGURE 33.4

Baseline muscle activities. Muscle activities for the right (top) and left (bottom) over the crank rotation cycle with the baseline bicycle parameters.

To start with, we perform a baseline simulation with the model before we introduce the springs. The simulated muscle activities are shown in Fig. 33.4, and the resultant joint reaction forces at the left and right knee are shown in Fig. 33.5.

In these figures, we observe (1) phase-shifted results between the left and right leg but otherwise identical results. (2) For both the muscle forces and the joint reaction forces, we see distinct increase in variables around the top and bottom dead centers in the crank rotation cycle, which occur at 0 degree (and 360 degrees that is the same as 0 degree) and 180 degrees of crank rotation. Note that 0 degree corresponds to the right leg being in the top position and the left leg being at the bottom position. To get a better overview of the muscular load, at each simulated frame, we compute the maximum muscle activity over all muscles and plot it against the crank rotation angle (Fig. 33.6). The plot of the maximum muscle activity also clearly shows how the muscular load varies cyclically with variations from 17% to 43%.

To improve on this uneven load, we will explore the idea of storing elastic energy in the parts of the cycle, where we have relatively small muscular loads, that is subsequently released again in the parts of the cycle, where we currently have relatively large muscular loads. This approach will likely lead to a decrease in the peak muscular loads but also come at the expense that the minimum loads increase.

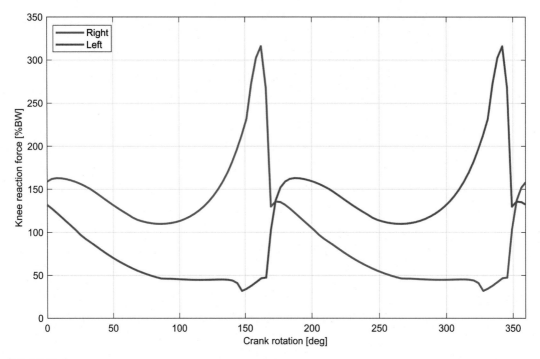

FIGURE 33.5

Baseline knee joint reaction forces. Resultant knee joint reaction forces of the right and left leg over the duration of the crank rotation cycle with the baseline bicycle parameters.

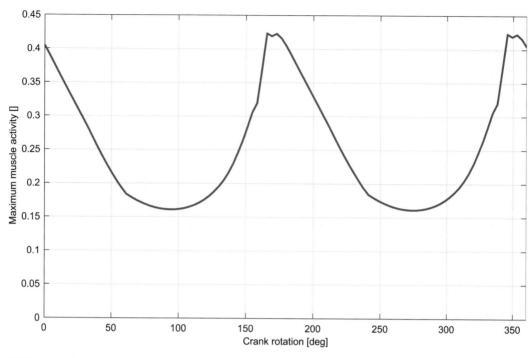

FIGURE 33.6

Baseline maximum muscle activity. Maximum muscle activity plotted as a function of the crank rotation angle under the baseline load.

To this end, we will introduce two springs as illustrated as the green thick lines in Fig. 33.7.

To determine the exact attachments as well as the spring properties (stiffness and slack length), an optimization problem was solved. The goal of this optimization problem was to minimize the peak of the maximum muscle activity while varying the coordinates of the attachment on the frame, the coordinates of the attachment on the crank, the spring slack length and the spring stiffness. The resultant maximum muscle activity is shown in Fig. 33.8. These results show a large impact of the springs on the muscular loads. In particular, we note how the load varies remarkably less than before and that the peak load is now reduced to 24%, while the minimum load has only increased to 17%.

All muscle forces and the joint reaction forces are shown in Figs. 33.9 and 33.10, respectively.

The muscle activity results show that all the muscle activities are affected but with the largest reductions occurring in gluteus maximum, the hamstrings, and rectus femoris. Additionally, increases in the muscle activities are also observed, when we are not near the top and bottom dead centers.

Similar to the muscle activities, the knee joint reaction forces also show a large reduction in the peak loads (Fig. 33.10). Also here, however, we see a small increase in the forces, when we are not near the top and bottom dead centers.

This concludes our analysis of the human–bicycle model. The model itself allows many more interesting investigations, such as the saddle height, pedal arm dimensions, and the effect of load

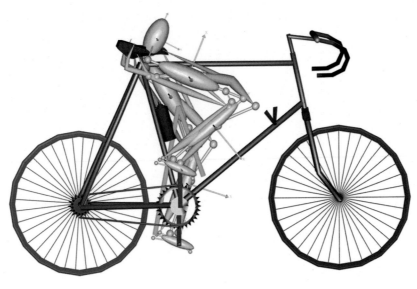

FIGURE 33.7

Human–bicycle model with assistive springs. Illustration of the bicycle model with two linear springs (the green lines) introduced. The springs attach at the same point on the bicycle frame and to different points on the crank.

increase, which can readily be analyzed. Additionally, we can extend the model to a more complete model of the lower extremity, for example, based on the Twente Lower Extermity 2.0 data set (Carbone et al., 2015) that is already available in the AMMR. Additionally, we could extend the bicycle model to 3D and include steering effects by modeling the handle bar and front wheel as separate segments. Exactly what to include depends on the research question, but all the theory and all the methods you have been introduced to in this chapter can directly be applied.

OTHER APPLICATIONS

While we looked specifically at the human–bicycle model in this chapter, the possibilities with rigid and musculoskeletal models are endless. In the following, we look at a few other application examples.

One potential use of the model is, as previously mentioned, to gain insight into quantities that we cannot measured directly. This was, for instance, explored in the study by Mellon et al. (2013). They investigated the direction of the hip joint reaction force for metal-on-metal hip resurfacing patients and used the distance to the edge of the cup of the hip joint reaction force to explain why some patients with malaligned cups show high levels of metal ions while others do not.

Another example in this regard is the study of knee osteoarthritis patients by Dell'Isola, Smith, Andersen, and Steultjens (2017), which was conducted on varus malaligned knees and investigated differences between patients with the disease exclusively in the medial compartment and those with the disease both in the medial and lateral departments. This study suggests the existence of an increased medial contact force phenotype as the results showed larger medial forces for the patient group with exclusively medial compartment disease.

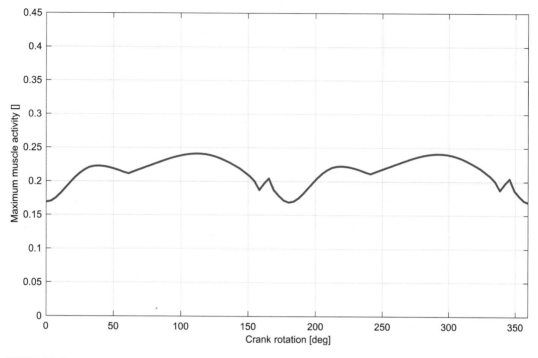

FIGURE 33.8

Maximum muscle activity with assistive springs. The maximum muscle activity with the optimized springs as a function of the crank rotation angle.

The model can also be applied to investigate relationships that may not be experimentally feasible. An example of this is the study by Smith, Brandon, and Thelen (2019), who investigated how neuromuscular coordination changes affect knee biomechanics in anterior cruciate ligament deficient knees. More specifically, they investigated whether or not there exist a change in neuromuscular coordination that can restore normal knee biomechanics by simulating 10,000 perturbations in the neuromuscular coordination and found that none of the coordinations could restore normal knee biomechanics. Due to the large amount of required retraining of the neuromuscular coordination, such a study would not be possible to conduct experimentally.

In a similar manner, Simonsen et al. (2019) investigated how tibialis posterior muscle pain, simulated as reduction in the tibialis posterior muscle strength, affects the muscle recruitment and found that flexor digitorum longus and flexor hallucis longus compensate for the reduced force in tibialis posterior and that this compensation strategy increases the ankle joint forces. Due to the attachment sites of the flexor digitorum longus and flexor hallucis longus, this compensation strategy could also be a contributing factor to the forefoot deformity observed in rheumatoid arthritis patients.

Another possible application of musculoskeletal models is to create realistic boundary conditions for more detailed modeling of subsystems, for instance, modeling using the finite-element method. An example of this is the study conducted by Halonen et al. (2017), where a subject-specific

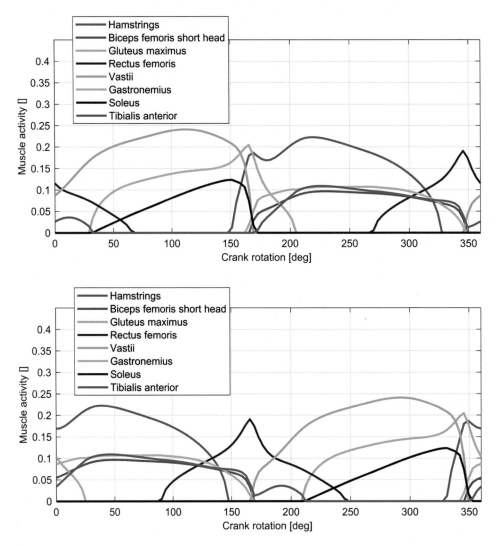

FIGURE 33.9

Muscle activities with assistive springs. Muscle activities for the right (top) and left (bottom) over the crank rotation cycle with the assistive springs included.

musculoskeletal model was applied to estimate the muscle and joint reaction forces, which were subsequently applied to a finite-element knee model. This combination enabled investigation of how gait modifications and lateral insoles, which are common early interventions for medial knee osteoarthritis patients, affect the stress, fibril strain, and fluid pressures of the articular cartilage of the knee.

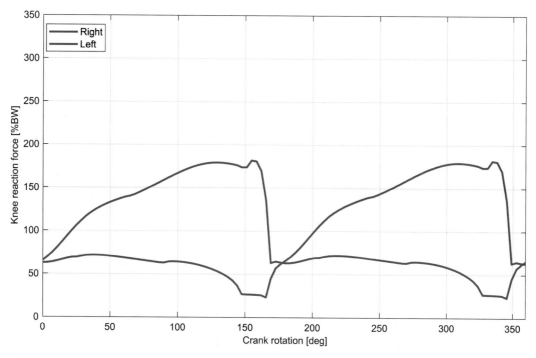

FIGURE 33.10

Knee joint reaction forces with assistive springs. Resultant knee joint reaction forces of the right and left leg over the duration of the crank rotation cycle with the assistive springs included.

There are many other application types, but a type worth mentioning here is the one our human—bicycle model falls into, namely the design of equipment that interacts with the human body. Besides the human—bicycle model, other applications include exoskeletons, joint replacements, artificial limps, etc. Here, the equipment and the human model is cosimulated to evaluate how changes to the equipment affects the human body and, hereby, enable optimization of the equipment to induce a desired change to the human biomechanics. An example of this can be found in Zhou, Bai, Andersen, and Rasmussen (2015), who optimized springs in a passive shoulder and arm orthosis to compensate simulated lesions to the brachial plexus nerves.

With enabling design optimization, co-simulation of the human body and interacting equipment has attracted a lot of attention, and we expect to see significant progress in this area in the coming years.

CONCLUDING REMARKS

To conclude, in this chapter, you got introduced to the fundaments for rigid-body and musculoskel-etal modeling. The overall concepts for inverse dynamic analysis, including muscle recruitment, were introduced. We then looked at a simplified human model on a bicycle and investigated how the muscular loads can be reduced by introducing assistive springs to smooth out the otherwise uneven muscular load caused by the top and bottom dead centers. Finally, you got introduced to a couple of other application examples to demonstrate the vast amount of possibilities rigid body and musculoskeletal models offer.

REFERENCES

Andersen, M. S. (2018). How sensitive are predicted muscle and knee contact forces to normalization factors and polynomial order in the muscle recruitment criterion formulation? *International Biomechanics*, *5*(1), 88−103. Available from https://doi.org/10.1080/23335432.2018.1514278.

Andersen, M. S. (2020). Introduction to musculoskeletal modelling Computational Modelling of Biomechanics and Biotribology in the Musculoskeletal System : Biomaterils and Tissues. Jin, Z., Li, J. & Chen, Z. (red.). 2 udg. Woodhead Publishing, s. 41−80.

Andersen, M. S., Jin, Z., Li, J., & Chen, Z. (2021). *4 − Introduction to musculoskeletal modelling. Woodhead publishing series in biomaterials* (pp. 41−80). Woodhead Publishing. Available from https://doi.org/10.1016/B978-0-12-819531-4.00004-3.

Andersen, M. S., de Zee, M., Damsgaard, M., Nolte, D., & Rasmussen, J. (2017). Introduction to force-dependent kinematics: Theory and application to mandible modeling. *Journal of Biomechanical Engineering*, *139*(9). Available from https://doi.org/10.1115/1.4037100.

Anderson, F. C., & Pandy, M. G. (1999). A dynamic optimization solution for vertical jumping in three dimensions. *Computer Methods in Biomechanics and Biomedical Engineering*, *2*(3), 201−231.

Begon, M., Andersen, M. S., & Dumas, R. (2018). Multibody kinematics optimization for the estimation of upper and lower limb human joint kinematics: A systematized methodological review. *Journal of Biomechanical Engineering*, *140*(3). Available from https://doi.org/10.1115/1.4038741.

Carbone, V., Fluit, R., Pellikaan, P., van der Krogt, M. M., Janssen, D., Damsgaard, M., ... Verdonschot, N. (2015). TLEM 2.0 − A comprehensive musculoskeletal geometry dataset for subject-specific modeling of lower extremity. *Journal of Biomechanics*, *48*(5), 734−741. Available from https://doi.org/10.1016/j.jbiomech.2014.12.034.

Cronskär, M., Rasmussen, J., & Tinnsten, M. (2015). Combined finite element and multibody musculoskeletal investigation of a fractured clavicle with reconstruction plate. *Computer Methods in Biomechanics and Biomedical Engineering*, *18*(7), 740−748. Available from https://doi.org/10.1080/10255842.2013.845175.

Crowninshield, R. D., & Brand, R. A. (1981). A physiologically based criterion of muscle force prediction in locomotion. *Journal of Biomechanics*, *14*(11), 793−801.

Damsgaard, M., Rasmussen, J., Christensen, S. T., Surma, E., & de Zee, M. (2006). Analysis of musculoskeletal systems in the AnyBody Modeling System. *Simulation Modelling Practice and Theory*, *14*(8), 1100−1111.

Davoudabadi Farahani, S., Andersen, M. S., de Zee, M., & Rasmussen, J. (2016). Optimization-based dynamic prediction of kinematic and kinetic patterns for a human vertical jump from a squatting position. *Multibody System Dynamics*, *36*(1), 37−65. Available from https://doi.org/10.1007/s11044-015-9468-5.

Dejtiar, D. L., Dzialo, C. M., Pedersen, P. H., Jensen, K. K., Fleron, M., & Andersen, M. S. (2020). Development and evaluation of a subject-specific lower limb model with an 11 DOF natural knee model using MRI and EOS during a quasi-static lunge. *Journal of Biomechanical Engineering*, *142*(6), 061001. Available from https://doi.org/10.1115/1.4044245.

Dell'Isola, A., Smith, S. L., Andersen, M. S., & Steultjens, M. (2017). Knee internal contact force in a varus malaligned phenotype in knee osteoarthritis (KOA). *Osteoarthritis and Cartilage, 25*(12), 2007−2013. Available from https://doi.org/10.1016/j.joca.2017.08.010.

Falisse, A., Serrancolí, G., Dembia, C. L., Gillis, J., Jonkers, I., & De Groote, F. (2019). Rapid predictive simulations with complex musculoskeletal models suggest that diverse healthy and pathological human gaits can emerge from similar control strategies. *Journal of the Royal Society, Interface, 16*(157), 20190402. Available from https://doi.org/10.1098/rsif.2019.0402.

Farahani, S. D., Bertucci, W., Andersen, M. S., de Zee, M., & Rasmussen, J. (2015). Prediction of crank torque and pedal angle profiles during pedaling movements by biomechanical optimization. *Structural and Multidisciplinary Optimization, 51*(1), 251−266. Available from https://doi.org/10.1007/s00158-014-1135-6.

Halonen, K. S., Dzialo, C. M., Mannisi, M., Venäläinen, M. S., de Zee, M., & Andersen, M. S. (2017). Workflow assessing the effect of gait alterations on stresses in the medial tibial cartilage − Combined musculoskeletal modelling and finite element analysis. *Scientific Reports, 7*(1), 17396. Available from https://doi.org/10.1038/s41598-017-17228-x.

Happee, R., & Van der Helm, F. C. (1995). The control of shoulder muscles during goal directed movements, an inverse dynamic analysis. *Journal of Biomechanics, 28*(10), 1179−1191.

Holmberg, L. J., & Klarbring, A. (2012). Muscle decomposition and recruitment criteria influence muscle force estimates. *Multibody System Dynamics, 28*(3), 283−289. Available from https://doi.org/10.1007/s11044-011-9277-4.

Lund, M. E., Tørholm, S., De Pieri, E., Simonsen, S. T., Iversen, K., & Engelund, B. K. (2021). AnyBody Managed Model Repository (version 2.3.2). *Zenodo*. Available from https://doi.org/10.5281/zenodo.4305559.

Marra, M. A., Vanheule, V., Fluit, R., Koopman, B. H. F. J. M., Rasmussen, J., Verdonschot, N., & Andersen, M. S. (2015). A subject-specific musculoskeletal modeling framework to predict in vivo mechanics of total knee arthroplasty. *Journal of Biomechanical Engineering, 137*(2), 020904. Available from https://doi.org/10.1115/1.4029258.

Mellon, S., Grammatopoulos, G., Andersen, M. S., Pegg, E., Pandit, H., Murray, D. & Gill, H. (2013). Individual motion patterns during gait and sit-to-stand contribute to edge-loading risk in metal-on-metal hip resurfacing. *Proceedings of the Institution of Mechanical Engineers, Part H: Journal of Engineering in Medicine. 227*(7), 799−810.

Nikravesh, P. E. (1987). *Computer-aided analysis of mechanical systems*. Englewood Cliffs, NJ: Prentice Hall.

Rasmussen, J., Damsgaard, M., & Voigt, M. (2001). Muscle recruitment by the min/max criterion − A comparative numerical study. *Journal of Biomechanics, 34*(3), 409−415.

Seth, A., & Pandy, M. G. (2007). A neuromusculoskeletal tracking method for estimating individual muscle forces in human movement. *Journal of Biomechanics, 40*(2), 356−366.

Simonsen, M. B., Yurtsever, A., Næsborg-Andersen, K., Leutscher, P. D. C., Hørslev-Petersen, K., Hirata, R. P., & Andersen, M. S. (2019). A parametric study of effect of experimental tibialis posterior muscle pain on joint loading and muscle forces-Implications for patients with rheumatoid arthritis? *Gait & Posture, 72*, 102−108. Available from https://doi.org/10.1016/j.gaitpost.2019.06.001.

Smith, C. R., Brandon, S. C. E., & Thelen, D. G. (2019). Can altered neuromuscular coordination restore soft tissue loading patterns in anterior cruciate ligament and menisci deficient knees during walking? *Journal of Biomechanics, 82*, 124−133. Available from https://doi.org/10.1016/j.jbiomech.2018.10.008.

Thelen, D. G., Anderson, F. C., & Delp, S. L. (2003). Generating dynamic simulations of movement using computed muscle control. *Journal of Biomechanics, 36*(3), 321−328.

Thelen, D. G., Won Choi, K., & Schmitz, A. M. (2014). Co-simulation of neuromuscular dynamics and knee mechanics during human walking. *Journal of Biomechanical Engineering, 136*(2), 021033. Available from https://doi.org/10.1115/1.4026358.

Winter, D. A. (2009). *Biomechanics and motor control of human movement* (4th ed.). John Wiley & Sons, Inc.

Zajac, F. E. (1989). Muscle and tendon: Properties, models, scaling, and application to biomechanics and motor control. *Critical Reviews in Biomedical Engineering, 17*(4), 359−411.

Zhou, L., Bai, S., Andersen, M. S., & Rasmussen, J. (2015). Modeling and design of a spring-loaded, cable-driven, wearable exoskeleton for the upper extremity. *Modeling, Identification and Control (Online Edition), 36*(3), 167−177.

THE USE OF COMPUTATIONAL MODELS IN ORTHOPEDIC BIOMECHANICAL RESEARCH

Bernardo Innocenti[1], Edoardo Bori[1], Federica Armaroli[1], Benedikt Schlager[2], René Jonas[2], Hans-Joachim Wilke[2] and Fabio Galbusera[3]

[1]Department of Bio Electro and Mechanical Systems (BEAMS), École Polytechnique de Bruxelles, Université Libre de Bruxelles, Brussels, Belgium [2]Institute of Orthopaedic Research and Biomechanics, Ulm University, Ulm, Germany [3]IRCCS Istituto Ortopedico Galeazzi, Milan, Italy

INTRODUCTION

In the last few decades, the importance of computational modeling in the field of orthopedics has been continuously increasing. Reported applications covered all the major joints and anatomical regions, such as the lower limb, including hip, knee and ankle joint, the upper limb as well as the spine. For each of these anatomical regions, a large number of computational studies investigating it under native conditions, either healthy or showing degenerative features, as well as after the most common surgical procedures and related devices aimed at treating degenerative disorders, have been published.

Most computational models used in orthopedic biomechanics can be classified into (1) multi-body models, based on dynamics of rigid bodies and (2) finite element (FE) models, based on the equations of continuum mechanics.

Multibody models are typically used to calculate muscular and joint contact forces during daily activities such as walking, running, and carrying loads based on inverse dynamics simulations, commonly driven by motion analysis data, and provide a valuable tool to study the biomechanical function of the musculoskeletal system (Stops, Wilcox, & Jin, 2012). Within this approach, the kinematics of the body segments of interest is first acquired by means of systems employing visible markers or inertial measurement units applied on subjects performing such tasks and then used to calculate the muscle forces which, in combination with external loads, would determine the measured kinematics. This approach requires a biomechanical model of the body part, which typically consists of rigid segments with mass and moment of inertia corresponding to those of the body segments, articulated through joints such as cylindrical and spherical hinges, sleeves, and universal joints (Delp et al., 2007). Considering muscles as the actuators, the inverse dynamics problem is then solved by first calculating forces and moments at the joints, and then by using the inverse dynamics equations of motion to determine the actuator loads which would result in the measured kinematics. The outputs of multibody models, therefore, consist of a set of muscle forces, as well as the forces and moments in each joint.

Human Orthopaedic Biomechanics. DOI: https://doi.org/10.1016/B978-0-12-824481-4.00003-2

Multibody modeling of the human musculoskeletal system is directly available on specialized software platforms such as the proprietary AnyBody Technology (Aalborg, Denmark) and the free OpenSim (https://opensim.stanford.edu/). Both platforms offer ready-to-use biomechanical models of the human body which could be scaled to fit different body sizes or customized to introduce implants or more realistic models of specific anatomical regions. Besides, general-purpose software for rigid body analysis used in other engineering fields such as MSC Adams (MSC Software, Newport Beach, CA, USA) can also be employed for biomechanical simulations, but provide neither models of the human body nor dedicated biomechanical functions such as those to handle muscle redundancy and wrapping.

The FE method is based on the discretization of complex mechanical structures into finite numbers of separate components with simple geometry, called elements, allowing for the conversion of complex nonlinear problems into problems that can be solved numerically. FE models can be employed to investigate issues such as local stresses and strains, contact between tissues and implants, as well as material nonlinearity, damage, and failure, commonly under the action of simplified loads mimicking a physiological loading scenario (Huiskes & Chao, 1983).

FE software platforms used in biomechanical research, such as ABAQUS (Dassault Sistèmes, Vélizy-Villacoublay, France) and Ansys (Canonsburg, PA, USA), include a large library of material models, ranging from linear elasticity to complex hyperelastic anisotropic materials; methods to simulate fracture, failure, fatigue, crack propagation, and damage accumulation are also offered. COMSOL Multiphysics (COMSOL Inc., Stockholm, Sweden) permits the solution of user-provided systems of partial differential equations besides standard applications such as structural mechanics and computational fluid dynamics. Regarding free software, FEBio (https://febio.org/) offers extensive capabilities particularly indicated to simulate soft structures such as cartilage, ligaments, tendons, and intervertebral discs (IVDs) undergoing large deformations, under the open-source MIT license.

THE HIP JOINT

As the primary link between the trunk and the lower limb, the hip joint plays an important role in force transmission and generation during daily and athletic activities alike (Polkowski & Clohisy, 2010). Life quality may critically decrease in the eventuality of hip pathologies or trauma, which commonly affect this joint (Nogier et al., 2010); osteoarthritis (OA) and fracture of the femoral neck have indeed a high prevalence. Although the etiology of hip OA is not clear, its mechanics have been implicated as one of the primary factors concerning risk of fracture due to the high loads this joint supports.

Besides, total hip replacement has high clinical relevance and is one of the most frequent and successful orthopedic interventions (Callaghan, Albright, Goetz, Olejniczak, & Johnston, 2000).

The possibility of investigating the biomechanical behavior of the hip joint in silico supported the development of treatments for hip disorders by allowing the prediction of the surgical outcome, of bone remodeling after surgery, as well as possible causes of implant failure. The literature body about numerical models of the hip joint and hip replacement is very large. Topics covered in scientific publications include the design of hip prostheses, the use of bone cement to ensure proper primary stability, the materials used in the implants, as well as the biomechanics of femoral neck fractures and their treatments, which are, however, not described in the next paragraphs.

THE HEALTHY AND DEGENERATED HIP JOINT

The quantification of the mechanical parameters of the native and surgically treated hip joint such as loads and stresses in ligaments, muscles, bones, and implants is a central topic for understanding pathological conditions and improving their treatment. To reproduce in a realistic way, the joint mechanics in a computational model require deep knowledge about the mechanical properties and geometry of the biological tissues, which have been investigated in several studies.

Muscle loads were mostly investigated by means of musculoskeletal models based on optimization techniques and frequently focused on the simulation of the gait cycle. Correa, Crossley, Kim, & Pandy (2010) studied the contribution of the different muscles to the hip contact force, relying on an optimization-based numerical model. The model revealed that the gluteus medius and maximus, along with the hamstrings and vasti muscles, significantly contribute (approximately 92%) to the predicted peak contact force (4.3BW for normal walking speeds). Interestingly, the vasti muscles do not span the hip joint but are instead involved in the knee motion; the model, therefore, demonstrated the contribution of nonhip joint muscles to the contact force at the hip. Jonkers, Spaepen, Papaioannou, & Stewart (2002) employed musculoskeletal modeling, muscle-driven forward simulation, and gradient-based optimization techniques to analyze the kinematics of the lower limb in the stance phase of gait. The normalized activation patterns of 22 muscles were used as inputs for the simulations, which were carried out on two distinct musculoskeletal models with different degrees of freedom. One year later, the research group expanded the study by analyzing the swing phase of gait and estimating the contribution of each muscle to the gait pattern, by artificially excluding each muscle activation and predicting the resulting joint kinematics (Jonkers, Stewart, & Spaepen, 2003).

Bone geometry plays a fundamental role in numerical models of the hip since it directly affects the mechanical loads acting in the joint. Besides, since the geometry of the femoral head determines geometrically the center of the hip joint, it therefore also influences the peak and distribution of the contact pressure. Using a simple multibody model, Lengsfeld investigated a possible correlation between the peak of the joint forces and the position of the center of the hip, finding that if the latter was moved laterally, superiorly, or posteriorly from its neutral position, and then the forces were increased; however, the direction of those forces was mostly unaffected. By performing discrete element simulations on a population of both males and females, Genda et al. (2001) showed that the femoral head radii are inversely proportional to contact pressures and this leads to higher pressures in women, even if the magnitude of the joint does not significantly differ among sexes. Another important feature of the geometry of the hip joint is the femoral neck: its length and angle may vary significantly from patient to patient, with implications on the hip loads. Lenaerts performed a sensitivity analysis in order to evaluate the effects of these parameters on muscle activation and hip contact force during the stance phase of gait. Results showed relatively low importance of the neck−shaft angle, but the hip contact force markedly increased while increasing the neck length (Lenaerts et al., 2008).

Abnormal bone morphologies and their influence on the strain in the biological tissues were also investigated with numerical models. Cooper et al. developed and analyzed 20 patient-specific models of the hip to investigate cam geometry of femoral-acetabular impingement, demonstrating the capability of such numerical models to quantify the effect of different abnormal deformities on bone, cartilage, and potential tissue damage. The authors suggested that a detailed 3D knowledge of hip anatomy could be useful to stratify patients when considering treatment options (Cooper, Williams, Mengoni, & Jones, 2018).

THE IMPLANTED HIP JOINT

Numerical analysis gave a great contribution to the field of total hip replacement, allowing foreseeing stresses, strains, and wear issues associated with hip loads and the choice of materials and designs. Lengsfeld, Bassaly, Boudriot, Pressel, & Griss (2000) developed FE models aimed at comparing the strain energy between an implanted and a native contralateral hip on 11 patients who underwent a total hip replacement, while Martelli, Taddei, Cristofolini, Gill, & Viceconti (2011) performed a sensitivity test of a new resurfacing design by implementing it in multiple models, each one different from the other for bone size and properties, surgical parameters, and positioning. Ong, Lehman, Notz, Santner, & Bartel (2006) found that the design of the acetabular cup and the accuracy of the surgical implantation deeply influenced the outcome of the implant, while interpatient variability had only a minor role in the success; they were moreover able to identify an ideal design that reduced the periprosthetic joint space by using a validated optimization scheme. Marco Viceconti, Brusi, Pancanti, & Cristofolini (2006) focused on the primary stability of a cementless hip stem and its correlation with anatomy, material properties, gaps at the bone—implant surface, and patient weight.

Focusing on the prosthesis itself, Ruggiero, Merola, & Affatato (2018) simulated a gait cycle and analyzed the deformations and stress intensity on the polyethylene insert, further comparing two different frictional conditions (dry and wet). Fatigue effects and crack growth were then analyzed in a study by Colic et al. (2016), exploiting the extended finite element method (XFEM). As wear of the acetabular cup and the relative adverse tissue reaction to wear debris are responsible for the loosening and implant failure, several numerical models investigated lubrication and wear modeling in the hip prosthesis, considering different lubrication models, boundary conditions, materials, and implant designs (Wu, Hung, Shu, & Chen, 2003).

Another application of numerical simulation in orthopedics is the possibility of performing a preliminary check of an implant design prior to preclinical and clinical testing; an example could be identified in Colic et al., who investigated static and fatigue crack growth in a hip replacement. Such numerical investigations do not aim at completely replacing preclinical experimental testing, but have the potential to facilitate the detection of designs with poor performance, thus reducing development costs and marketing times (Colic et al., 2016).

THE KNEE JOINT

Among the human joints, the knee is one of the most complex joints, and, as it is characterized by longer level arms (the femur and the tibia) that link the knee respectively to the hip and the ankle, it is extremely vulnerable to trauma and lesions due to the higher forces and torques that could be generated. Both tibiofemoral and patellofemoral joints enable daily activities, such as walking, running, sitting, and kneeling, which produce complex mechanical loads. Therefore, to be able to investigate and predict the 3D behavior of a knee, the analysis of stress and strain distribution, contact kinematics, and joint kinetics by means of numerical modeling is of paramount importance. Numerical models of the knee have been useful in the clinical practice (Catani et al., 2010; Peña, Calvo, Martínez, & Doblaré, 2006), industrial implant design, in basic research (Adouni, Shirazi-Adl, & Shirazi, 2012; Donahue, Hull, Rashid, & Jacobs, 2002).

The available literature about numerical simulations for the study of healthy knee and knee replacement is vast, but it was not until recently that many papers from this category had been cited as compared to papers related to hip joint. Published papers cover several topics such as the biomechanics of soft tissues (cartilage, meniscus, ligaments), knee osteoarthritis, ligament reconstruction, patient-specific models, as well as design, characteristics, and materials of total knee replacements (TKR).

THE HEALTHY AND DEGENERATIVE KNEE JOINT

The development of a FE knee model requires dealing with a high number of assumptions and hypotheses related to geometry, material properties, loading and boundary conditions, together with long computational times due to the nonlinearities and the complexity of the contact problems involved. For instance, articular cartilage or the menisci are rarely considered poroelastic materials (Halonen et al., 2014), and generally modeled as 2D shell models. While in the past knee biomechanics was mainly limited to a simplistic approach, such as linear bi-dimensional models investigating only single structures or tissue types (Huiskes & Chao, 1983), nowadays 3D, nonlinear, multi-structures and tissues, kinetic and kinematic driven, and even multi-joint systems can be also simulated (Kiapour et al., 2014; Peña et al., 2006).

Similar to the other joints, the geometry of a knee model could be based on a "generic" or "patient-specific" shape. The first approach is used when general knee outputs are required and standard bone shapes as the "standardize tibia" or the "standardized femur" (Viceconti, Ansaloni, Baleani, & Toni, 2003) are used. However, when the analysis of a patient-specific knee is required, the generation of the 3D geometry is usually based on the digitization of an array of 2D images of the patient or of a cadaveric specimen; this approach is in general labor-intensive and time-consuming (Pianigiani, Croce, D'Aiuto, Pascale, & Innocenti, 2018), especially if cartilage layers and menisci need to be integrated (Donahue et al., 2002). Usually, CT scans are used to reconstruct the bone geometries and to find the reference landmarks (Victor et al., 2009) in order to describe joint axes, coordinate systems, and bones position (Innocenti, Bori, & Piccolo, 2020; Victor et al., 2009). MRI scans are also a good alternative to obtain patient-specific data not involving ionizing radiation, especially when soft tissues, such as cartilage and menisci, should also be implemented; in these cases, MRI scans become fundamental in order to evaluate their geometries, insertion points, and additional conditions (Pianigiani et al., 2018). However, most FE models do not include the tendons and muscles and the boundary conditions are usually simplified in concentrated action applied on specific points and fixing specific regions (as in Donahue et al., 2002; Innocenti et al., 2014; Peña et al., 2006) (Fig. 34.1).

Once the full geometrical model is defined, each part must be characterized by a specific material model with proper mechanical properties. In general, bone can be considered a rigid structure

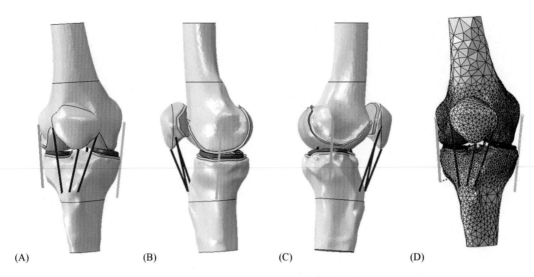

FIGURE 34.1

Full knee finite element model: (A) frontal view, (B) medial view, (C) lateral view, and (D) mesh of the frontal view. The colors represent the different parts of the model: white (cortical bone), pink (articular cartilage), dark pink (menisci), red (patellar tendon), light blue (*MCL*, medial collateral ligament), green (*LCL*, lateral collateral ligament), yellow (*ACL*, anterior cruciate ligament), and blue (*PCL*, posterior cruciate ligament).

(Peña et al., 2006; Peña, Calvo, Martínez, & Doblaré, 2008; Peña, Calvo, Martínez, & Doblaré, 2007) as a homogeneous material or mainly subdivided its cortical and trabecular structures, using an isotropic constitutive law or, more recently, also transverse isotropic laws (Donahue et al., 2002; Innocenti et al., 2014; Innocenti, Bellemans, & Catani, 2016; Innocenti, Pianigiani, Ramundo, & Thienpont, 2016). The properties of bone may also vary to simulate healthy and pathologic conditions, such as osteoporosis or osteopenia.

In terms of soft tissues, several material models have been used to represent knee ligaments, ranging from elastic unidimensional elements to complex hyperelastic 3D structures with anatomically realistic shapes. An extensive discussion of this issue could be found in Galbusera et al. (2014), and only a summary is here reported. Due to the complex anatomy and structural diversity of the knee joint, some FE studies represented the knee ligaments with a one-dimensional (1D) truss or beam (Adouni et al., 2012; Innocenti et al., 2014; Innocenti, 2020; Innocenti, Pianigiani, et al., 2016; Penrose, Holt, Beaugonin, & Hose, 2002) or spring elements (Beillas, Papaioannou, Tashman, & Yang, 2004; Donahue et al., 2002; Moglo & Shirazi-Adl, 2003) with simplified material properties. Reference strains, also called pre-strains, are also sometimes implemented in the model (Adouni et al., 2012; Innocenti et al., 2014; Innocenti, 2020) but they are difficult to estimate experimentally, and assumptions were often made in order to circumvent this problem and to correctly simulate the initial state of knee extension (Wismans, Veldpaus, Janssen, Huson, & Struben, 1980). As reported by Galbusera et al. (2014) and Pianigiani et al. (2018) the correct selection of the proper material property is one of the most critical aspects in knee modeling as a small change in the properties could induce a variation up to 50% on the contact condylar pressure.

In terms of constraints and boundary conditions, the performed knee motor task can be statically or dynamically replicated and, generally, can be force- or displacement-controlled. Boundary conditions, forces, and kinematics inputs can have different origins, such as the one replicating different static or dynamic experimental protocol (Kiapour et al., 2014; Mootanah et al., 2014; Pianigiani et al., 2018), by rigid body simulations or in-vivo fluoroscopic or gait analysis results (Catani et al., 2010) or from ISO standard or similar normative, such as the ISO 14243−1 used to simulate gait (Innocenti, 2020).

THE IMPLANTED KNEE JOINT

If an implanted knee FE model needs to be developed, additional information needs to be specified and integrated into the process. Indeed, the implant TKR geometry needs to be provided and it should come from the original CAD file as an STL 3D reconstruction may have quality issues similar to those of CT or MRI reconstructions. Accurate TKR geometry and position is fundamental as the bone stress and, in general, the knee performance is strongly dependent on these aspects (Baldwin et al., 2012; Innocenti et al., 2014; Innocenti, Bellemans, et al., 2016; Perillo-Marcone, Barrett, & Taylor, 2000).

Metallic knee components are modeled as linear elastic (Innocenti et al., 2014) or rigid components (Baldwin et al., 2012), while polymeric components can be assumed to be linear elastic or nonlinear elastoplastic (Halloran, Petrella, & Rullkoetter, 2005). If a cemented implant needs to be analyzed, bone cement can be modeled as linear elastic material with properties obtained from the manufacture or the literature, and the bone−cement interactions could be also implemented (Waanders, Janssen, Miller, Mann, & Verdonschot, 2009). Once materials are modeled, the proper friction coefficient simulating the presence of the synovial joint liquid should also be considered between each material couple.

Numerical modeling could also be a valid tool to predict the clinical outcome of TKR, such as the risk of stress shielding (Barink, Verdonschot, & de Waal Malefijt, 2003; Perillo-Marcone et al., 2000), the risk of fracture (Zalzal, Backstein, Gross, & Papini, 2006), and the wear of the polyethylene component.

THE SPINE

The application of numerical methods aimed at answering research questions in the spine field began in the 1980s. One of the first published studies investigated the stress distribution on a vertebra after laminectomy and Harrington rod procedures. Soon after, studies started to analyze the possible causes for low back pain by modeling functional spinal units. With the development of more sophisticated models and an increase in computer capacity, the applications of numerical models of the spine continued to expand. Currently, numerical models of the spine are used in various fields such as the calculation of load and stress distributions in the tissues, the analysis and optimization of surgical treatments, the investigation of etiology of diseases such as idiopathic scoliosis and disc degeneration, the prediction of the risk of vertebral fractures, the healing and remodeling of bone after fixation, the development and analysis of spinal implants, and the effect of conservative treatments such as bracing and plaster casting.

The relevant body of literature covers bone mechanics, spinal loads, interbody fusion, posterior fixation, trauma and fractures, disc stresses and intradiscal pressure, disc degeneration, and deformities. The literature includes a large number of most cited papers, demonstrating that this research field is very active and clinically relevant.

THE HEALTHY AND DEGENERATED SPINE

To date, a large number of numerical models of the healthy and the degenerated spine have been described in publications. Each model was developed to investigate distinct research questions at specific scales, ranging from mechanobiology to trabecular bone structure and IVD to the spinal segment, and finally to modeling the overall complex musculoskeletal system.

The IVD is one of the most investigated structures in the whole orthopedic research, as it is involved in various pathologies such as disc herniation and degeneration, deformities, as well as back pain. Numerous studies tried to model this structure in detail by spending effort in representing the composite fiber network of the annulus and simulate the stress distribution (Malandrino, Noailly, & Lacroix, 2013), as well as its time-dependent response (Galbusera, Schmidt, Neidlinger-Wilke, Gottschalk, & Wilke, 2011) (Fig. 34.2). Indeed, the IVD is a fluid-filled porous structure, the mechanical response of which shows creep and relaxation phenomena related to the motion of fluid inside the IVD and through its endplates. A milestone in this field was the FE model published by Argoubi & Shirazi-Adl (1996), who employed a poroelastic constitutive law to investigate the time-dependent response of the IVD under compressive loads. Other poroelastic FE models aimed at investigating the effect of degenerative morphological changes of the IVD taking into account the effect of alterations in fluid flow and the swelling capability. For example, Galbusera, Schmidt, Neidlinger-Wilke, & Wilke (2011) and Galbusera, Schmidt, Neidlinger-Wilke, Gottschalk, et al. (2011) simulated the effect of several degenerative parameters, such as disc height, endplate sclerosis, and water content. Poroelastic FE models also investigated IVD nutrition in association with fluid flow, such as the metabolism of oxygen and lactate (Malandrino et al., 2014).

Bone tissue is a heterogeneous material with a complex structure, which is commonly the subject of numerical investigations of the fracture risk in physiological and pathological conditions, for example, in the case of osteoporosis. Among the numerous papers addressing the mechanics of vertebral bone, Silva & Gibson (1997) investigated the effects of microstructural changes in the trabecular bone architecture associated with aging, demonstrating the mechanical importance of the number of trabeculae. Dall'Ara, Pahr, Varga, Kainberger, & Zysset (2012) used quantitative CT images to build models incorporating the local material properties of bone aimed at evaluating the fracture risk, which proved superior to the assessment based on dual-energy X-ray absorptiometry. The influence of size and location of metastatic lesions within the vertebral body was investigated to identify the risk of spinal fractures (Tschirhart, Nagpurkar, & Whyne, 2004).

Regarding the study of the biomechanics of the whole motion segment, that is, two adjacent vertebrae, IVD, and spinal ligaments, one of the first nonlinear 3D FE models of a complete L2-L3 segment was published by Shirazi-Adl, Ahmed, & Shrivastava (1986). The IVD was modeled considering a composite-like behavior of the annulus fibrosus and a hydrostatic nucleus pulposus. In this study, flexion—extension moments were simulated, investigating the range of motion, the

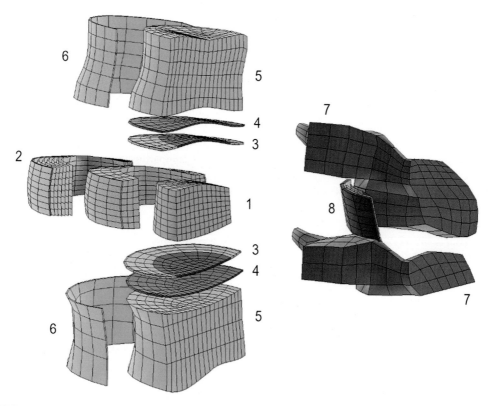

FIGURE 34.2

Exploded view of a finite element model of the L4-L5 motion segment, depicting all anatomical structures: (1) nucleus pulposus; (2) annulus fibrosus; (3) cartilaginous endplates; (4) bony endplates; (5) trabecular bone in the vertebral bodies; (6) cortical shell around the vertebral bodies; (7) posterior elements; and (8) facet joints.

From Galbusera, F., Schmidt, H., Neidlinger-Wilke, C., Gottschalk, A., & Wilke, H. (2011). The mechanical response of the lumbar spine to different combinations of disc degenerative changes investigated using randomized poroelastic finite element models. European Spine Journal: Official Publication of the European Spine Society, the European Spinal Deformity Society, and the European Section of the Cervical Spine Research Society, 20(4), 563–71. https://doi.org/10.1007/s00586-010-1586-4.

intradiscal pressure, and the strain behavior of the soft tissues. Similarly, the model of L4-L5 by Schmidt et al. (2007) was used to investigate different load combinations and their effect on the deformation of the lumbar IVD and was calibrated against a large set of experimental measurements of specimens subjected to incremental defects. Regarding the cervical region, Clausen et al. introduced a FE model of a C5-C6 motion segment in 1997 (Clausen, Goel, Traynelis, & Scifert, 1997), which included a fiber-reinforced annulus fibrosus as well as Luschka joints, which have been neglected in most successive studies although this model highlighted their biomechanical importance. The effect of morphological features on the biomechanics of the lumbar spine was investigated by Niemeyer, Wilke, & Schmidt (2012), who identified that the range of motion in the lumbar spine is mainly influenced by the disc height, thickness in articular cartilage, and the pedicle length.

Models covering not only a single motion segment but also a whole spinal region, or even the complete spine, have also been published. Also, there are numerous FE models covering the lumbar region completely, some of which have been quantitatively compared in 2014 (Dreischarf et al., 2014). Although different research groups employed different data to generate each model, all models represented well the range of motion in comparison with in-vitro data. The most pronounced differences regarded forces in the facet joints and intradiscal pressures.

One of the first 3D FE models of the thoracolumbar spine, as well as the rib cage, was published by Andriacchi, Schultz, Belytschko, & Galante (1974) and was based on simplified anatomy consisting of rigid and flexible beam elements. The authors aimed at investigating the mechanical response of their model to a variety of different loads, and at comparing the stability of the healthy spine with that of a scoliotic model. Recent numerical models generated directly from computed tomography data represented in detail the anatomy of the overall spine, pelvis, and rib cage. These models have been for instance used to simulate pathomechanism of spinal deformities or conservative treatments (Cobetto, Parent, & Aubin, 2018).

Concerning the cervical spine, one of the first 3D FE models, which represented the whole region, was presented by Kleinberger in 1993, and aimed at investigating the mechanics of cervical injuries during automotive crashes (Kleinberger, 1993). Successive models have been used to simulate injury mechanisms, investigate parametric influences, or test and compare different implants and surgical techniques (Wheeldon, Khouphongsy, Kumaresan, Yoganandan, & Pintar, 2000).

FE models of spinal segments and regions most commonly try to replicate the mechanical response of the passive structures of the spine but exclude the muscles. However, several numerical models including muscles have been presented, based on the FE method (Shirazi-Adl, Sadouk, Parnianpour, Pop, & El-Rich, 2002) and multibody dynamics. As such multibody models continue to improve and be validated more and more extensively, they have become powerful tools to evaluate for instance the intradiscal pressure during different lifting tasks and daily activities (Bassani, Stucovitz, Qian, Briguglio, & Galbusera, 2017).

THE IMPLANTED SPINE

A driving motivation for FE models in orthopedics is to evaluate surgical strategies and the stresses acting in the implants and the surrounding anatomical regions, in order to identify the risk of mechanical failure of the implant itself or its interface with the adjacent anatomy. A major goal of computational modeling is indeed to simulate different treatment options prior to orthopedic intervention in order to identify the "optimal" surgical strategy, potentially leading to an improvement of the outcome and a reduction of the complication rates. To date, the limited predictive capability of numerical models has restricted their use in a "bench-to-bedside" setting. However, by combining validated models with experimental testing and by performing extensive parametric and sensitivity studies, clinically relevant results about surgical treatments can be successfully extracted from numerical models.

The body of literature about computational modeling of spine surgery and spinal implants is very large, includes papers about various implants for the treatment of degenerative disorders, deformities, trauma, and tumors, and covers various scales from bone and IVD microstructure to the whole spine from head to pelvis. This paragraph briefly describes computational models that have been used to investigate the different types of spinal implants, such as pedicle screws, rods, hooks, and wires, for the treatment of degenerative disorders and the correction of deformities, bone cement for the fixation of vertebral fractures, and motion-preserving implants such as artificial discs and dynamic stabilization systems used in degenerative surgery.

Instrumented fusion is the treatment of choice for symptomatic spinal degeneration, and is commonly achieved by the implantation of pedicle screws and rods. The pedicle screw-rod construct reduces the flexibility of the instrumented segment and sustains 20%−40% of the loads acting in the spine (Galbusera, Bellini, et al., 2011). Pedicle screws and rods have been modeled either in a simplified way by means of beam elements (Schmidt, Heuer, & Wilke, 2009) or with high geometrical and mechanical details explicitly representing threads and connected parts employing contact definitions (Chen, Lin, & Chang, 2003). Modeling threads and contacts in the bone−implant interface allows for an accurate estimation of the interaction loads and stresses, thus permitting the investigation of clinically relevant phenomena such as screw loosening; on the contrary, simplified beam models do not allow for such evaluations but are deemed acceptable when the focus is on the global behavior of the construct, for example, the residual flexibility of the implanted spine. The same two approaches, that is, simplified versus geometrically accurate models, have been also used for the study of pedicle-based dynamic stabilization systems, which do not aim at a bony fusion but rather at reducing the spinal flexibility, as well as for interspinous devices used for the treatment of spinal stenosis (Galbusera, Schmidt, Neidlinger-Wilke, Gottschalk, et al., 2011; Jahng, Kim, & Moon, 2013).

In addition to degenerative disorders, posterior fixation is used for the correction of spinal deformities such as adolescent idiopathic scoliosis, sagittal imbalance, and adult deformities. The treatment of deformities commonly requires long instrumentations, deciding on the instrumentation length and selecting vertebrae that need to be stabilized are challenging depend on the experience of the surgeon. The issue was investigated in several studies by means of multibody models (Aubin et al., 2008) as well as FE models (Gardner-Morse & Stokes, 1994). Alternative anchoring techniques such as sublaminar hooks and wiring were also investigated with numerical models (Clin, Aubin, Parent, Sangole, & Labelle, 2010).

Posterior instrumentation is often supplemented by devices implanted in the intervertebral space aimed at providing anterior support and stability. These implants, named cages, can be inserted through different surgical approaches (posterior, anterior, transforaminal, lateral), which have been investigated in several numerical studies (Fantigrossi, Galbusera, Raimondi, Sassi, & Fornari, 2007; Galbusera, Schmidt, & Wilke, 2012; Kim & Vanderby, 2000). These models replicated the removal of the IVDs and cartilage endplates which is performed during the surgical implantation and modeled the interaction between implants and bony endplate by means either of tied contacts, which simulate a successful bony fusion, or contact definitions including friction, sliding and separation, which replicate the immediate postoperative condition and investigate phenomena such as the instability in extension.

Anterior fixation is commonly performed in the cervical spine due to the easy and safe anterior access to vertebral bodies, which facilitates the implantation of grafts, cages, and plates. Anterior cervical discectomy and fusion (ACDF) is used for the treatment of cervical disc herniation and involves the removal of the IVD, its replacement with a graft or cage, and, in many cases, fixation with a plate. FE models of ACDF are relatively rare and employed the same techniques used for posterior fixation, including tied or frictional contact pairs to model screw threads and the bone−implant interface (Natarajan, Chen, An, & Andersson, 2000).

Total disc replacements are used for the treatment of degenerative disc disease in both the cervical and the lumbar spine and aim at preserving the physiological mobility of the motion segment. Most of these implants consist of two or three contacting surfaces, which govern the resulting motion; their simulation with a numerical model is rather challenging since their behavior is dominated by contact pairs. Numerous studies addressing total disc replacement have been published, targeting their kinematics (Goel, Faizan, Palepu, & Bhattacharya, 2012), their effect on load-sharing patterns on the adjacent segments (Denozière & Ku, 2006), failure and wear (Bhattacharya, Goel, Liu, Kiapour, & Serhan, 2011), as well as the biomechanics of multi-level implantation (Schmidt, Galbusera, Rohlmann, Zander, & Wilke, 2012).

THE SHOULDER JOINT

The human shoulder is a perfect compromise between mobility and stability (Veeger & van der Helm, 2007). Stability is mainly provided by active muscle actions with a minor contribution from the passive stabilizers, such as glenohumeral capsule, labrum, and ligaments (Zheng, Zou, Bartolo, Peach, & Ren, 2017). The glenohumeral joint allows for the widest movement range of the shoulder: the articular surface of the glenoid is considerably smaller than the humerus one, which facilitates the large range of movement of the joint (Haering, Raison, & Begon, 2014). However, the poor congruency of the glenohumeral articular surface may cause joint dislocations.

The shoulder has been widely investigated since the 1990s by means of numerical models, although the number of available papers tends to be lower with respect to the hip and the knee joints. Models were developed to investigate several issues in the field of shoulder biomechanics, including trauma and fractures, especially in the proximal humerus, the motion and loads in the joints, as well as the design of total shoulder replacements (TSRs).

THE HEALTHY AND DEGENERATED SHOULDER JOINT

Although biological materials usually have a complex mechanical behavior exhibiting anisotropic, nonhomogeneous, nonlinear, and viscoelastic properties, in numerical models of the shoulder these properties have been frequently simplified or even neglected, with either a variable degree of impact on the accuracy of the results based on the specific research question, reviewed in Zheng et al. (2017). Similarly to hip and knee models, in many FE models of the shoulder bones were assumed rigid or isotropic linear elastic material. In early models, muscles and tendons were modeled as isotropic linear elastic materials, while later nonlinear models became more common (Büchler, Ramaniraka, Rakotomanana, Iannotti, & Farron, 2002; Favre et al., 2012). Joint capsules were frequently modeled as isotropic hyperelastic (Debski, Weiss, Newman, Moore, & McMahon, 2005; Ellis, Debski, Moore, McMahon, & Weiss, 2007) and the labrum was assumed to be an isotropic material (Yeh, Lintner, & Luo, 2005). Detailed transverse isotropic material properties were employed in recent studies, which compared well to the experiment measurements (Gatti, Maratt, Palmer, Hughes, & Carpenter, 2010). The articular cartilages were considered rigid in some FE models (Adams, Baldwin, Laz, Rullkoetter, & Langenderfer, 2007; Gatti et al., 2010), whereas in other studies cartilage was modeled as homogeneous linear elastic (Inoue, Chosa, Goto,

& Tajima, 2013; Quental, Fernandes, Monteiro, & Folgado, 2014; Sano, Wakabayashi, & Itoi, 2006; Seki et al., 2008; Wakabayashi et al., 2003). Boundary and loading conditions show a large variability among the various studies. Some models only considered part of the complex and simulated the neglected parts using boundary or loading conditions (Gatti et al., 2010). Some recent models include realistic active muscles instead of simplified, schematic loading conditions (Favre et al., 2012).

Numerical models were frequently used to investigate the biomechanical effect of shoulder disorders. Büchler et al. (2002) developed a numerical model to quantify the influence of the humeral head shape on the stress distribution in the scapula, in order to compare the biomechanics of a native shoulder and an osteoarthritic shoulder with a primary degenerative disease. They found out that one possible cause of the glenoid loosening is the eccentric loading of the glenoid component due to the translation of the humeral head. Terrier, Reist, Vogel, & Farron (2007) investigated the supraspinatus deficiency, which could reduce shoulder stability and lead to osteoarthritis. Two models were considered: a normal shoulder and a shoulder without supraspinatus. The results of the simulations revealed that the deficiency increased the upward migration of the humerus, the eccentric loading, and the joint and muscle forces, which could cause a limitation of active abduction, osteoarthritis and rotator cuff tear. Walia, Miniaci, Jones, & Fening (2013) investigated the effects on shoulder stability of both glenoid and humeral defects in the glenohumeral joint, such as a Hill-Sachs lesion or a Bankart defect; they found that glenohumeral joint stability was largely affected by the lesions.

Over the years, FE models of the glenoid labrum have also been developed. Gatti et al. (2010) developed and validated a FE model of the glenoid labrum for humeral head translation to understand the mechanical environment of the labrum and its connection to pathology. Klemt et al. (2017) investigated passive glenohumeral stability by means of subject-specific FE models validated against in-vitro measurements, which demonstrated the importance of the glenoid labrum in ensuring stability under physiological loading.

SHOULDER JOINT REPLACEMENT

A TSR involves the replacement of the humeral head and of the glenoid with prostheses that perform as a new joint. Despite a generally good success, glenoid component loosening commonly affects this type of surgical procedures and, indeed, most FE models of TSR aimed at investigating this issue. Through FE simulation analyses, the biomechanical effects on the risk of loosening of different key design parameters, such as implant shape, positioning, fixation ("keeled" or "pegged"), material, orientation, use of bone cement, and articular conformity, were examined.

Besides, Couteau et al. (2001) analyzed the mechanical effect of some of the surgical variables of TSR, investigating the effect of one eccentric load case, cement thickness, and conformity. The importance of the humeral head centering in the horizontal plane was underlined. Low conformity could cause an increase of stresses and a thick cement mantle did not reduce the stresses at the cement−bone interface. Büchler & Farron (2004) studied the influence of the shape of the prosthetic humeral head to evaluate the benefits of an anatomical reconstruction of the humeral head. Two prosthetic designs were examined with respect to an intact shoulder: a state-of-the-art universal prosthesis and a patient-specific anatomical implant. It was found that a patient-specific reconstruction of the prosthetic humeral head restricts eccentric loading of the glenoid, and thus could also prevent glenoid component loosening.

Two TSR designs are most commonly employed in clinical practice. The conservative TSR mimics the shape of the two extremities (humeral sphere and glenoid cavity), which constitute the shoulder joint, while the reverse TSR is characterized by a design opposite to the anatomical structure of the shoulder. While the humeral head is resected to accommodate a stem with a concave end, the scapular glenoid is reconstructed by providing a glenosphere to be articulated with the humerus. Both solutions were investigated with FE models; for example, Hopkins et al. (2006) developed a 3D model of a conservative TSR and analyzed the effects of several design parameters upon glenohumeral interaction. It was found that conformity was the principal factor responsible for the magnitude of humeral head translation, but not for the loads required to destabilize the joint. Regarding the reverse TSR, Virani et al. (2008) developed physical models of a reverse shoulder design, using high-strength polyurethane foam blocks, to represent the glenoid bone/baseplate connection, as well as a FE model. Seven glenospheres were tested and the results from FE analysis were in strong agreement with the mechanical testing, which revealed no significant difference between the configurations. Later, Nigro, Gutiérrez, & Frankle (2013) employed a FE model to study the additional implant−bone contact achieved when the glenosphere undersurface is in contact with the glenoid. More recently, Chae et al. (2016) estimated the effect of the orientation of the glenoid component on primary stability in reverse TSR, namely its inferior tilt. The 10-degree inferior tilt fixation showed higher relative micromotions and bone stress than the neutral tilt fixation, demonstrating that inferior tilt fixation could negatively affect primary stability and longevity after a reverse TSR. The effect of backside bone support on the risk of failure of a glenoid component was recently investigated by Verhaegen et al. for two types of glenoid components: cemented all polyethylene and metal-backed. Decreasing backside bone support for an anatomical glenoid component leads to an increased risk of fixation and bone failure, smaller for all-poly components than in metal-backed components. Verhaegen, Campopiano, Debeer, Scheys, & Innocenti (2020) concluded that an anatomical glenoid component should always be implanted while maximizing backside bone support.

THE ANKLE JOINT

The ankle joint has been the subject of several FE investigations, either isolated or in combination with the rest of the foot complex. Most models have been created based on CT or MRI images of a single subject, typically healthy, or a combination of scans from different subjects (Gefen, Megido-Ravid, Itzchak, & Arcan, 2000); the resulting models have then been modified if the study addressed a specific pathology or surgical treatment. In comparison with the spine as well as hip and knee joints, the body of literature about the numerical models of the ankle joint is relatively limited. The topics covered in the computational papers include the biomechanics of the foot in general, of trauma and fractures of the ankle, the loads acting in the joint, the fixation of ankle fractures, and models of total ankle replacements.

The fundamental components of ankle FE models are the relevant bones (at least the distal tibia and the talus, whereas some models included also the fibula and the subtalar joint, that is, the calcaneus, and the navicular bone), which can be either modeled as rigid or deformable, possibly covered by a soft cartilage layer, and ligaments (the deltoid ligament on the medial side, comprising the anterior and posterior tibiotalar ligaments, as well as the tibiocalcaneal ligament and tibionavicular ligament in the models including the whole foot or hindfoot; the talofibular and calcaneofibular ligaments on the lateral side; the anterior and posterior tibiofibular ligaments).

THE HEALTHY AND DEGENERATED ANKLE JOINT

Since the kinematics of the ankle joint is guided by the combined action of the contact between the articulating surfaces and ligament tensioning, a proper modeling approach should in first place replicate well the physics of these two phenomena. In comparison with other anatomical regions, the mechanical response of bone and cartilage is less critical, especially if the focus of the investigation is the joint motion; as a matter of facts, these materials have been frequently simply modeled as linear isotropic elastic (Cheung, Zhang, Leung, & Fan, 2005; Gefen et al., 2000; Tao et al., 2009; Yu et al., 2008). Contact between the articular surfaces was most frequently modeled by means of surface-to-surface algorithms, either frictionless (Alonso-Rasgado, Jimenez-Cruz, & Karski, 2017; Park, Lee, Yoon, & Chae, 2019; Tao et al., 2009; Yu et al., 2008) or with a low friction coefficient (Mondal & Ghosh, 2017; Zhu et al., 2016). Ligaments were frequently modeled as tension-only linear elements such as trusses, links, or springs (Alonso-Rasgado et al., 2017; Cheung & Zhang, 2005; Mondal & Ghosh, 2017; Tao et al., 2009; Yu et al., 2008), whereas unidimensional nonlinear elements (Gefen et al., 2000; Xu, Liu, Li, Wang, & Li, 2017) or solid elements (Wang, Wong, Tan, Li, & Zhang, 2019) were employed in other studies. Recent papers described a sophisticated calibration procedure based on in-vitro tests on cadaveric specimens, also including failure properties, to build realistic models of the ankle ligaments including the single fiber bundles (Nie, Forman, et al., 2017; Nie, Panzer, et al., 2017). Models covering the whole foot also usually included soft tissue that pads the bony surfaces, typically modeled as a hyperelastic solid (Cheung & Zhang, 2005; Cheung et al., 2005; Gefen et al., 2000; Yu et al., 2008).

As for boundary and loading conditions, most papers simulated the simple standing posture, which consisted in constraining the plantar side of the model, applying a compressive load to the tibia mimicking the body weight and a traction load replicating the action of the Achilles tendon (Cheung & Zhang, 2005; Morales-Orcajo, Souza, Bayod, & Barbosa de Las Casas, 2017; Yu et al., 2008). Other studies considered various phases of the gait cycle, each modeled with specific loads and positions (Gefen et al., 2000), or pure moments to simulate internal and external rotations (Zhu et al., 2016).

FE models of the ankle joint and the foot were used to investigate several clinically relevant research questions. Alonso-Rasgado et al. (2017) developed a model to study the biomechanical effect of malunion in posterior malleolar fractures. They conducted a sensitivity analysis of the size and position of the bone fragment, so that the widely acknowledged hypothesis stating that fracture fixation is necessary if the fragment is larger than 25%−33% of the tibial plafond can be verified by means of numerical simulations. Interestingly, the study showed that the fracture induced an increase in the contact area in the ankle joint as opposed to previous studies (van den Bekerom, Lamme, Hogervorst, & Bolhuis, 2007), and no joint instability was observed even with the largest fragment sizes; therefore, the previous hypothesis cannot be confirmed.

Zhu et al. investigated the stability of the ankle joint in case of injuries of the posterolateral ligaments, that is, the posterior talofibular ligament and the posterior inferior tibiofibular ligament, based on the clinical observation that posterolateral instability of the ankle is relatively frequent. Based on simulations of plantar flexion, dorsiflexion, inversion, and eversion, the authors demonstrated that both ligaments played a major role in determining the ankle stability, whereas the posterior inferior tibiofibular ligament was also critical for the subtalar stability (Zhu et al., 2016).

Yu et al. built a model of the female foot to investigate the effect of high heel shoes on loads and stresses in the ankle—foot complex. This study revealed significant increases in the stresses in the anterior part of the foot, namely in the first metatarsophalangeal joint which may lead to hallux valgus deformity, while the strain in the plantar fascia was reduced, confirming the effectiveness of heels for the reduction of the symptoms of plantar fasciitis. As a matter of facts, the ankle joint was only marginally affected by the use of high heels (Yu et al., 2008).

ANKLE JOINT REPLACEMENT

The use of total ankle replacement, especially for the treatment of patients suffering from severe osteoarthritis, is becoming more and more widespread (Barg et al., 2015). The first generation of prostheses was associated with high rates of complications, such as subsidence, osteolysis, and loosening, and was soon abandoned (Gougoulias, Khanna, & Maffulli, 2010). Technical innovation brought to the appearance of new designs featuring uncemented components requiring less bone resection, based on various technical solutions (two or three components, with mobile, semi-constrained or constrained axis of rotation). However, clinical studies soon showed that these designs still suffered from complications such as wear of the articulating surfaces (typically a polymer-on-metal coupling) as well as possible impingement and dislocation (Gougoulias et al., 2010; Guyer & Richardson, 2008). Recent models, commonly named third-generation replacements, are supposed to further improve the clinical outcome, but a comprehensive clinical validation still needs to be achieved; indeed, FE modeling provided fundamental support for the investigation of the recent designs (Martinelli et al., 2017; Reggiani, Leardini, Corazza, & Taylor, 2006; Sopher, Amis, Calder, & Jeffers, 2017), posing the basis for further improvements. Nevertheless, total ankle replacement is still considered to be at a lower level of clinical achievement with respect to hip and knee arthroplasties, with revision rates approximately double of that of hip and knee implants (Jastifer & Coughlin, 2015; Jay Elliot, Gundapaneni, & Goswami, 2014; Labek, Thaler, Janda, Agreiter, & Stöckl, 2011).

Several FE models aiming at the investigation of total ankle replacement have been published and were mainly used in the evaluation of the most critical issues such as the risk of wear. Some studies were based on models of the ankle joint or the foot, which were modified to simulate the implantation of the prosthesis (Falsig, Hvid, & Jensen, 1986; Sopher et al., 2017; Terrier, Larrea, Guerdat, & Crevoisier, 2014; Wang et al., 2019) (Fig. 34.3). Other models only included the device itself and the ankle ligaments, with no explicit modeling of the bony structures and thus not allowing for the calculation of the stresses and loads at the bone—implant interface (Espinosa, Walti, Favre, & Snedeker, 2010; Jay Elliot et al., 2014; Martinelli et al., 2017; Reggiani et al., 2006).

Reggiani et al. (2006) developed a FE model of a third-generation three-component prosthesis including the major ankle ligaments, each one modeled as an array of five fibers and having a nonlinear mechanical behavior. The authors used the model to predict the kinematics of the implanted

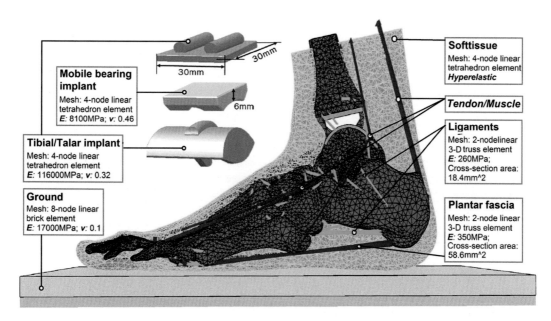

Mobile bearing implant
Mesh: 4-node linear tetrahedron element
E: 8100MPa; *v*: 0.46

Tibial/Talar implant
Mesh: 4-node linear tetrahedron element
E: 116000MPa; *v*: 0.32

Ground
Mesh: 8-node linear brick element
E: 17000MPa; *v*: 0.1

Softtissue
Mesh: 4-node linear tetrahedron element
Hyperelastic

Tendon/Muscle

Ligaments
Mesh: 2-nodelinear 3-D truss element
E: 260MPa;
Cross-section area: 18.4mm^2

Plantar fascia
Mesh: 2-node linear 3-D truss element
E: 350MPa;
Cross-section area: 58.6mm^2

30mm
30mm
6mm

FIGURE 34.3

Finite element model of the foot and ankle with total ankle arthroplasty, and parameters of material properties and mesh.

From Wang, Y., Wong, D. W., Tan, Q., Li, Z., & Zhang, M. (c.2019). Total ankle arthroplasty and ankle arthrodesis affect the biomechanics of the inner foot differently. Scientific Reports, *9(1), 13334. https://doi.org/10.1038/s41598-019-50091-6.*

ankle in passive sagittal dorsiflexion and during the stance phase of gait, based on literature data (Procter & Paul, 1982; Seireg & Arvikar, 1975). The FE model predicted a motion pattern compatible with that of the intact ankle joint and relatively low peak contact stresses, thus suggesting a favorable wear behavior. Jay Elliot et al. (2014) compared seven different models of ankle replacements spanning two generations, in order to investigate their wear pattern resulting from repetitive loading. The polymeric liner was characterized as a viscoelastic material based on experimental data, and wear was estimated based on Hertzian contact and Archard's law. Ligaments and bones were not considered in the models, which were loaded in order to replicate a gait cycle. Results showed generally higher stresses in the articulating surfaces, compatible with significant wear rates which could be responsible for a poor clinical outcome. Increasing thickness and contact area appeared to be feasible solutions to decrease wear of the polymeric component. Martinelli et al. (2017) conducted a similar study on a recent semi-constrained device, also including ankle ligaments in the FE models, and found slightly lower stresses, which could be further decreased by increasing the thickness of the liner. Espinosa et al. (2010) tested the sensitivity of predicted contact stresses to misalignment of the components, concluding that deviations from the recommendations of the manufacturer larger than 5 degrees may increase such stresses to values higher than the yield stress of the material, thus demonstrating the crucial importance of correct surgical positioning.

A number of studies investigated the interaction between the prosthesis and the surrounding tissues, aiming at predicting the risk of other failure modalities such as loosening and subsidence. Cui et al. (2018) developed a sophisticated model of the interaction between the talus and the talar component to test its risk of subsidence and found that medial tilting of the device may occur in case of poor bone quality. Sturnick et al. (2018) investigated the risk of loosening of the tibial component in a semi-constrained design, which is a common complication after total ankle replacement, by creating a FE model including bony components and loading it with simple loads (compression, anterior shear, pure moment). The model identified areas of low stresses, which can be a source of stress shielding and thus loosening, in the proximity of the fixation keel. Sopher et al. (2017) compared three different modern designs in terms of bone strain and micromotion at the bone—implant interface, finding conspicuous differences between the three models and a clear negative effect of prosthesis malpositioning. Finally, Wang et al. (2019) investigated the effect of a total ankle replacement on the foot, by means of a comprehensive FE model (Fig. 34.3). The study revealed a large increase of contact pressure at the medial cuneonavicular joint, which may contraindicate the use of ankle prostheses in patients suffering from disorders in that region.

VERIFICATION, VALIDATION, AND CALIBRATION OF ORTHOPEDIC COMPUTATIONAL MODELS

Verification, validation, and calibration are important steps in the numerical modeling process. Verification aims at ensuring that the solutions of the computational model accurately reflect the solution of the underlying mathematical model, that is, "solving the equations right" (Viceconti, Olsen, Nolte, & Burton, 2005). The term validation involves quantifying the accuracy of a model by comparing the numerical solutions to experimental data, either existing or purposely acquired (Henninger, Reese, Anderson, & Weiss, 2010). In other words, validation aims at assessing that the mathematical model of the phenomenon of interest is an accurate representation of its physics, that is, "solving the right equations" (Viceconti et al., 2005). Calibration consists of adjusting and fitting the numerical model (e.g., material properties) to experimental data in order to maximize its accuracy (Schmidt et al., 2007). Depending on the research question, multiple datasets can be used to either calibrate or validate the computational model; data used to calibrate a model must be different from the data used to validate it (Henninger et al., 2010).

When using commercial or free computational software which has been previously tested and benchmarked, the user can safely assume that the implementation of the numerical methods is free of programming bugs and verification of the software is not needed. On the contrary, the accuracy and robustness of software developed by the researchers themselves need to be extensively tested. Apart from code verification, the verification of a linear FE model generally consists of assessing that the solution is independent of the discretization, that is, a mesh sensitivity study needs to be performed. Nonlinear models, including, for example, contact pairs and nonlinear materials, may require additional verification steps.

Calibration and validation are among the most challenging tasks in creating numerical models in biomechanics since a limited amount of data exists and available data are often heterogeneous and naturally present a large standard deviation (Galbusera et al., 2014). Besides, the access to quantities that can be used for model validation can be challenging since due to the anatomical complexity of each joint, the implantation of sensors in cadaveric specimens may disrupt the anatomical continuity of the structure and thus influence the value of the quantity object of the measurements.

Among the different options to estimate joint in-vivo loads for the sake of model validation, measurements obtained from instrumented implants are currently used by several researchers as paramount information. Public datasets are accessible on the OrthoLoad website (https://orthoload.com) and include loading data acquired with telemetrized hip, knee, and shoulder replacements, as well as spinal internal fixators and vertebral body replacements (Bergmann et al., 2001; Rohlmann et al., 2013).

Regarding specific FE models of the spine, calibration and validation are often performed using range of motion data from in-vitro experiments. Besides, models have been validated using the intradiscal pressure measurement from in-vitro or in-vivo experiments or the reaction forces on facet joints and ligaments (Ayturk & Puttlitz, 2011). Several models that included soft tissue and disc calibrated the material properties against stepwise resection data (Heuer, Schmidt, Claes, & Wilke, 2007).

For the validation of hip models, as previously mentioned data acquired by means of instrumented implants are frequently employed. For example, Heller et al. exploited available data (Bergmann et al., 2001; Heller et al., 2001) to calculate the muscle forces needed to replicate the real joint kinetics by means of computational modeling. Koivumäki et al. (2012) used cadaveric specimens to validate models of failure. Bessho et al. (2004) focused on superficial strains and thus used strain gauges to validate their models, Pettersen, Wik, & Skallerud (2009) took into account implant micromotion.

Similar to the hip, the validation of a knee model is frequently based on experimental studies performed on knee specimens having the same geometries, boundary conditions, and, eventually, implant design (Baldwin et al., 2012; Catani et al., 2010; Halloran et al., 2005; Innocenti et al., 2014; Mootanah et al., 2014). In-vivo gait-lab analysis has also been used to compare the model outputs for similar boundary conditions (Shu et al., 2018).

Regarding the ankle joint, contact pressure profiles measured in in-vitro tests have been frequently used for validation purposes. For example, Anderson et al. (2007) comprehensively validated a numerical model by comparing the contact pressure in the tibiotalar joint measured with a pressure transducer applied to human specimens. Several studies included a comparison with purposely collected in-vitro measurements (Nie, Forman, et al., 2017; Yu et al., 2008; Z.-J. Zhu et al., 2016) or data from the literature. In-vivo measures, such as motion analysis and contact pressure measurements, were also employed for model validation with good success. For instance, Tao et al. (2009) built a FE model of the foot based on MRI scans and performed in vivo investigations on the same subject for the sake of validation, that is, motion analysis in combination with a pressure platform as well as anatomical measurements on radiographs of the foot.

Similar approaches have been used also for models of the shoulder joint. For example, Couteau et al. (2001) calibrated a FE model of a glenoid replacement by adjusting the mechanical properties of the bone tissue and then validated it through a comparison with experimental measurements of surface strains. Luo, Wang, Tai, Chen, & Shih (2015) developed a physical shoulder model manufactured by 3D printing of the bone geometry and the mold for tissue-mimicking silicone to do an experimental validation of the model.

As the application of computational models within the medical field continues to grow, regulation authorities (ASME, FDA) have started to incorporate guidelines for the verification and validation of numerical models for medical investigations.

LIMITATIONS

As previously mentioned, all computational models are associated with certain limitations, originating from the uncertainty on material models, boundary conditions, inter-subject variability, geometries, and others that could dramatically reduce the accuracy and validity of the results obtained. With the increase of complexity of modeling the human musculoskeletal system at different scales, the degree of uncertainty increases by adding more structures and the number of influential parameters to the model. We recommend clearly describing the limitations in all papers to better express the applicability and validity of each model, especially when the model is used to draw clinically relevant conclusions (Viceconti et al., 2005).

Many publications describing the development of a computational model lack information about the calibration and validation procedure. Although some quantities such as contact pressures of strains in tissues are indeed difficult to measure, many alternative data sources which could be employed for validation purposes exist, such as fluoroscopic analyses describing the joint kinematics (Cenni et al., 2012) or retrieval studies showing wear patterns (Currier et al., 2019). In summary, we recommend that any paper in which a computational model is used to extract clinically relevant information must incorporate detailed information about its validity against experimental and/or in-vivo measurements (Viceconti et al., 2005).

A large number of studies employ simplified constitutive models for the biological tissues, especially soft tissues (e.g., reviewed in Galbusera et al., 2014 about knee ligaments). Indeed, viscoelastic and poroelastic properties, which are fundamental in hydrated tissues such as articular cartilage, IVD, and knee meniscus, are usually neglected. With some notable exceptions (e.g., Helgason, Viceconti, Rúnarsson, & Brynjólfsson, 2008), spatially varying properties are also commonly not taken into account. The alterations in tissue mechanics due to degenerative processes are sometimes not taken into consideration or modeled with simplistic assumptions; the scarcity of available experimental data for such tissue play a determinant role in that respect.

Another critical aspect appears to be a general lack of standardization under the boundary and loading conditions used for the simulations, which are rather heterogeneous and commonly referring to a small number of old studies. Although papers aimed at defining simple standardized protocols exist (e.g., Wilke, Wenger, & Claes, 1998), several studies employ different loading and boundary conditions for various reasons, for example, in order to improve the similarity to the physiological condition or because they aim at replicating patient-specific motions or complex exercises (walking, stair climbing, etc.). This heterogeneity determines a lack of comparability of the various studies, which in turn involves difficulties in the interpretation of the results and the extrapolation of clinically relevant conclusions.

FUTURE DEVELOPMENTS

Personalized medicine is nowadays emerging as the next frontier of medical research. Patient-specific models implement such an approach in the field of computational modeling and are indeed gaining importance in biomechanics. As a matter of fact, a quick generation of a personalized model is already achievable in terms of anatomy and geometry, by exploiting medical imaging and the recent developments in automated image segmentation and 3D reconstruction made possible by artificial intelligence techniques. The personalized approach to medicine employing numerical models has been effectively summarized in the concept of the "digital twin," a high-resolution model of an individual subject which may be used, for example, to virtually test and select the most appropriate treatment for a disease. However, aiming at achieving a fully patient-specific computational model or "digital twin," some research questions have not been extensively investigated; for example, measuring the personalized material properties remains an open problem, although methods which could potentially be employed to this purpose, such as ultrasound (Sigrist, Liau, Kaffas, Chammas, & Willmann, 2017) and MRI elastography (Dong, White, & Kolipaka, 2018), are nowadays used in clinical practice. Although research toward the determination of patient-specific loading and boundary conditions is ongoing, for example, by coupling musculoskeletal and FE models (Meena et al., 2016; Scarton et al., 2018; Zhu et al., 2013), such a solution still needs to gain widespread acceptance, and the use of simplified standardized loading conditions remains the gold standard for most applications.

Artificial intelligence and predictive models, such as autonomous driving and robotics, have a major impact in most research, business, and industrial fields,. In the medical area, such technologies are bringing a revolution to radiology, since computer vision techniques extract information from images in an accurate, precise, and repeatable manner (Hosny, Parmar, Quackenbush, Schwartz, & Aerts, 2018). Furthermore, artificial intelligence is used to predict the outcome of treatments, detect patients with a higher risk of complications, and guide clinical decisions (Galbusera, Casaroli, & Bassani, 2019). In orthopedic biomechanics, artificial intelligence has been used for the design optimization of hip replacements (Cilla, Borgiani, Martínez, Duda, & Checa, 2017) and to determine subject-specific loading conditions (Garijo, Verdonschot, Engelborghs, García-Aznar, & Pérez, 2017). Although the impact of such methods in biomechanics is still relatively of lower importance, the appearance of hybrid computational—artificial intelligence models combining multiphysics simulations and data-driven predictions is easy to foresee in the next future. The potential impact of such groundbreaking methods is arguably enormous.

In recent years, the concept of in-silico clinical trials, that is, simulated trials conducted on computational models of various patients aiming at the evaluation of novel treatments and implants, has gained a wide recognition (Viceconti, Henney, & Morley-Fletcher, 2016). Computational models for the simulation of the biomechanical behavior of joints and implants play a cardinal role within this approach since they constitute one of the main tools on which such simulated trials are based, together with artificial intelligence models which may further improve their predictive power.

Finally, although some model repositories already exist (e.g., SimTK, https://simtk.org/, and International Society of Biomechanics, http://isbweb.org/data/), in the authors' opinion the creation of an open, platform-independent, community-driven repository for validated computational models accessible would be valuable in providing momentum to research in the field and would foster knowledge sharing among the scientific community.

CONCLUSIONS

From this literature analysis, several recurring topics that are becoming increasingly relevant for all the considered anatomical regions emerged. These topics included the trends toward patient-specific models and multi-scale approaches, which would allow for a stronger impact of numerical analysis in the orthopedic field by implementing the so-called personalized medicine paradigm, with direct consequences on the improvement of treatment options and clinical outcomes for specific patients. On the other hand, the limited access to patient-specific material properties, or even of tissues in the different stages of the degenerative process, together with the undeniable difficulties in calibrating and validating personalized models cast doubts on this vision, at least in the immediate future. Although patient-specific modeling and personalized medicine should remain as long-term goals, upcoming research should be rather focused on strengthening the basic knowledge about orthopedic biomechanics which is a prerequisite for such future developments.

REFERENCES

Adams, C. R., Baldwin, M. A., Laz, P. J., Rullkoetter, P. J., & Langenderfer, J. E. (2007). Effects of rotator cuff tears on muscle moment arms: A computational study. *Journal of Biomechanics*, *40*(15), 3373−3380.

Adouni, M., Shirazi-Adl, A., & Shirazi, R. (2012). Computational biodynamics of human knee joint in gait: From muscle forces to cartilage stresses. *Journal of Biomechanics*, *45*(12), 2149−2156. Available from https://doi.org/10.1016/j.jbiomech.2012.05.040.

Alonso-Rasgado, T., Jimenez-Cruz, D., & Karski, M. (2017). 3-D computer modelling of malunited posterior malleolar fractures: Effect of fragment size and offset on ankle stability, contact pressure and pattern. *Journal of Foot and Ankle Research*, *10*, 13. Available from https://doi.org/10.1186/s13047-017-0194-5.

Anderson, D. D., Goldsworthy, J. K., Li, W., James Rudert, M., Tochigi, Y., & Brown, T. D. (2007). Physical validation of a patient-specific contact finite element model of the ankle. *Journal of Biomechanics*, *40*(8), 1662−1669.

Andriacchi, T., Schultz, A., Belytschko, T., & Galante, J. (1974). A model for studies of mechanical interactions between the human spine and rib cage. *Journal of Biomechanics*, *7*(6), 497−507.

Argoubi, M., & Shirazi-Adl, A. (1996). Poroelastic creep response analysis of a lumbar motion segment in compression. *Journal of Biomechanics*, *29*(10), 1331−1339.

Aubin, C. E., Labelle, H., Chevrefils, C., Desroches, G., Clin, J., & Eng, A. B. M. (2008). Preoperative planning simulator for spinal deformity surgeries. *Spine (Philadelphia, PA: 1986)*, *33*(20), 2143−2152. Available from https://doi.org/10.1097/BRS.0b013e31817bd89f.

Ayturk, U. M., & Puttlitz, C. M. (2011). Parametric convergence sensitivity and validation of a finite element model of the human lumbar spine. *Computer Methods in Biomechanics and Biomedical Engineering*, *14*(8), 695−705. Available from https://doi.org/10.1080/10255842.2010.493517.

Baldwin, M. A., Clary, C. W., Fitzpatrick, C. K., Deacy, J. S., Maletsky, L. P., & Rullkoetter, P. J. (2012). Dynamic finite element knee simulation for evaluation of knee replacement mechanics. *Journal of Biomechanics*, *45*(3), 474−483. Available from https://doi.org/10.1016/j.jbiomech.2011.11.052.

Barg, A., Wimmer, M. D., Wiewiorski, M., Wirtz, D. C., Pagenstert, G. I., & Valderrabano, V. (2015). Total ankle replacement. *Deutsches Arzteblatt International*, *112*(11), 177−184. Available from https://doi.org/10.3238/arztebl.2015.0177.

Barink, M., Verdonschot, N., & de Waal Malefijt, M. (2003). A different fixation of the femoral component in total knee arthroplasty may lead to preservation of femoral bone stock. *Proceedings of the Institution of Mechanical Engineers. Part H, Journal of Engineering in Medicine*, *217*(5), 325−332.

Bassani, T., Stucovitz, E., Qian, Z., Briguglio, M., & Galbusera, F. (2017). Validation of the AnyBody full body musculoskeletal model in computing lumbar spine loads at L4L5 level. *Journal of Biomechanics*, *58*, 89−96. Available from https://doi.org/10.1016/j.jbiomech.2017.04.025.

Beillas, P., Papaioannou, G., Tashman, S., & Yang, K. H. (2004). A new method to investigate in vivo knee behavior using a finite element model of the lower limb. *Journal of Biomechanics, 37*(7), 1019–1030.

Bergmann, G., Deuretzbacher, G., Heller, M., Graichen, F., Rohlmann, A., Strauss, J., & Duda, G. N. (2001). Hip contact forces and gait patterns from routine activities. *Journal of Biomechanics, 34*(7), 859–871.

Bessho, M., Ohnishi, I., Okazaki, H., Sato, W., Kominami, H., Matsunaga, S., & Nakamura, K. (2004). Prediction of the strength and fracture location of the femoral neck by CT-based finite-element method: A preliminary study on patients with hip fracture. *Journal of Orthopaedic Science : Official Journal of the Japanese Orthopaedic Association, 9*(6), 545–550.

Bhattacharya, S., Goel, V. K., Liu, X., Kiapour, A., & Serhan, H. A. (2011). Models that incorporate spinal structures predict better wear performance of cervical artificial discs. *The Spine Journal : Official Journal of the North American Spine Society, 11*(8), 766–776. Available from https://doi.org/10.1016/j.spinee.2011.06.008.

Büchler, P., & Farron, A. (2004). Benefits of an anatomical reconstruction of the humeral head during shoulder arthroplasty: A finite element analysis. *Clinical Biomechanics (Bristol, Avon), 19*(1), 16–23.

Büchler, P., Ramaniraka, N. A., Rakotomanana, L. R., Iannotti, J. P., & Farron, A. (2002). A finite element model of the shoulder: Application to the comparison of normal and osteoarthritic joints. *Clinical Biomechanics (Bristol, Avon), 17*(9–10), 630–639.

Callaghan, J. J., Albright, J. C., Goetz, D. D., Olejniczak, J. P., & Johnston, R. C. (2000). Charnley total hip arthroplasty with cement. Minimum twenty-five-year follow-up. *The Journal of Bone and Joint Surgery. American Volume, 82*(4), 487–497.

Catani, F., Innocenti, B., Belvedere, C., Labey, L., Ensini, A., & Leardini, A. (2010). The Mark Coventry Award: Articular contact estimation in TKA using in vivo kinematics and finite element analysis. *Clinical Orthopaedics and Related Research, 468*(1), 19–28. Available from https://doi.org/10.1007/s11999-009-0941-4.

Cenni, F., Leardini, A., Belvedere, C., Bugané, F., Cremonini, K., Miscione, M. T., & Giannini, S. (2012). Kinematics of the three components of a total ankle replacement: In vivo fluoroscopic analysis. *Foot & Ankle International, 33*(4), 290–300. Available from https://doi.org/10.3113/FAI.2012.0290.

Chae, S.-W., Lee, H., Kim, S. M., Lee, J., Han, S.-H., & Kim, S.-Y. (2016). Primary stability of inferior tilt fixation of the glenoid component in reverse total shoulder arthroplasty: A finite element study. *Journal of Orthopaedic Research : Official Publication of the Orthopaedic Research Society, 34*(6), 1061–1068. Available from https://doi.org/10.1002/jor.23115.

Chen, S.-I., Lin, R.-M., & Chang, C.-H. (2003). Biomechanical investigation of pedicle screw-vertebrae complex: A finite element approach using bonded and contact interface conditions. *Medical Engineering & Physics, 25*(4), 275–282.

Cheung, J. T.-M., & Zhang, M. (2005). A 3-dimensional finite element model of the human foot and ankle for insole design. *Archives of Physical Medicine and Rehabilitation, 86*(2), 353–358.

Cheung, J. T.-M., Zhang, M., Leung, A. K.-L., & Fan, Y.-B. (2005). Three-dimensional finite element analysis of the foot during standing — A material sensitivity study. *Journal of Biomechanics, 38*(5), 1045–1054.

Cilla, M., Borgiani, E., Martínez, J., Duda, G. N., & Checa, S. (2017). Machine learning techniques for the optimization of joint replacements: Application to a short-stem hip implant. *PLoS One, 12*(9), e0183755. Available from https://doi.org/10.1371/journal.pone.0183755.

Clausen, J. D., Goel, V. K., Traynelis, V. C., & Scifert, J. (1997). Uncinate processes and Luschka joints influence the biomechanics of the cervical spine: Quantification using a finite element model of the C5-C6 segment. *Journal of Orthopaedic Research : Official Publication of the Orthopaedic Research Society, 15*(3), 342–347.

Clin, J., Aubin, C.-E., Parent, S., Sangole, A., & Labelle, H. (2010). Comparison of the biomechanical 3D efficiency of different brace designs for the treatment of scoliosis using a finite element model. *European Spine Journal : Official Publication of the European Spine Society, the European Spinal Deformity Society, and the European Section of the Cervical Spine Research Society, 19*(7), 1169–1178. Available from https://doi.org/10.1007/s00586-009-1268-2.

Cobetto, N., Parent, S., & Aubin, C.-E. (2018). 3D correction over 2years with anterior vertebral body growth modulation: A finite element analysis of screw positioning, cable tensioning and postoperative functional activities. *Clinical Biomechanics (Bristol, Avon), 51,* 26−33. Available from https://doi.org/10.1016/j.clinbiomech.2017.11.007.

Colic, K., Sedmak, A., Grbovic, A., Burzić, M., Hloch, S., & Sedmak, S. (2016). Numerical simulation of fatigue crack growth in hip implants. *International Conference on Manufacturing Engineering and Materials, ICMEM 2016, 6−10 June 2016, Nový Smokovec, Slovakia, 149,* 229−235. Available from https://doi.org/10.1016/j.proeng.2016.06.661.

Cooper, R. J., Williams, S., Mengoni, M., & Jones, A. C. (2018). Patient-specific parameterised cam geometry in finite element models of femoroacetabular impingement of the hip. *Clinical Biomechanics (Bristol, Avon), 54,* 62−70. Available from https://doi.org/10.1016/j.clinbiomech.2018.03.007.

Correa, T. A., Crossley, K. M., Kim, H. J., & Pandy, M. G. (2010). Contributions of individual muscles to hip joint contact force in normal walking. *Journal of Biomechanics, 43*(8), 1618−1622. Available from https://doi.org/10.1016/j.jbiomech.2010.02.008.

Couteau, B., Mansat, P., Estivalèzes, E., Darmana, R., Mansat, M., & Egan, J. (2001). Finite element analysis of the mechanical behavior of a scapula implanted with a glenoid prosthesis. *Clinical Biomechanics (Bristol, Avon), 16*(7), 566−575.

Cui, Y., Hu, P., Wei, N., Cheng, X., Chang, W., & Chen, W. (2018). Finite element study of implant subsidence and medial tilt in agility ankle replacement. *Medical Science Monitor : International Medical Journal of Experimental and Clinical Research, 24,* 1124−1131.

Currier, B. H., Hecht, P. J., Nunley, J. A., Mayor, M. B., Currier, J. H., & Van Citters, D. W. (2019). Analysis of failed ankle arthroplasty components. *Foot & Ankle International, 40*(2), 131−138. Available from https://doi.org/10.1177/1071100718802589.

Dall'Ara, E., Pahr, D., Varga, P., Kainberger, F., & Zysset, P. (2012). QCT-based finite element models predict human vertebral strength in vitro significantly better than simulated DEXA. *Osteoporosis International : A Journal Established as Result of Cooperation between the European Foundation for Osteoporosis and the National Osteoporosis Foundation of the USA, 23*(2), 563−572. Available from https://doi.org/10.1007/s00198-011-1568-3.

Debski, R. E., Weiss, J. A., Newman, W. J., Moore, S. M., & McMahon, P. J. (2005). Stress and strain in the anterior band of the inferior glenohumeral ligament during a simulated clinical examination. *Journal of Shoulder and Elbow Surgery, 14*(Suppl 1), 24S−31S.

Delp, S. L., Anderson, F. C., Arnold, A. S., Loan, P., Habib, A., John, C. T., ... Thelen, D. G. (2007). OpenSim: Open-source software to create and analyze dynamic simulations of movement. *IEEE Transactions on bio-medical Engineering, 54*(11), 1940−1950.

Denozière, G., & Ku, D. N. (2006). Biomechanical comparison between fusion of two vertebrae and implantation of an artificial intervertebral disc. *Journal of Biomechanics, 39*(4), 766−775.

Donahue, T. L. H., Hull, M. L., Rashid, M. M., & Jacobs, C. R. (2002). A finite element model of the human knee joint for the study of tibio-femoral contact. *Journal of Biomechanical Engineering, 124*(3), 273−280.

Dong, H., White, R. D., & Kolipaka, A. (2018). Advances and future direction of magnetic resonance elastography. *Topics in Magnetic Resonance Imaging : TMRI, 27*(5), 363−384. Available from https://doi.org/10.1097/RMR.0000000000000179.

Dreischarf, M., Zander, T., Shirazi-Adl, A., Puttlitz, C. M., Adam, C. J., Chen, C. S., ... Schmidt, H. (2014). Comparison of eight published static finite element models of the intact lumbar spine: Predictive power of models improves when combined together. *Journal of Biomechanics, 47*(8), 1757−1766. Available from https://doi.org/10.1016/j.jbiomech.2014.04.002.

Ellis, B. J., Debski, R. E., Moore, S. M., McMahon, P. J., & Weiss, J. A. (2007). Methodology and sensitivity studies for finite element modeling of the inferior glenohumeral ligament complex. *Journal of Biomechanics*, *40*(3), 603–612.

Espinosa, N., Walti, M., Favre, P., & Snedeker, J. G. (2010). Misalignment of total ankle components can induce high joint contact pressures. *The Journal of Bone and Joint Surgery. American Volume*, *92*(5), 1179–1187. Available from https://doi.org/10.2106/JBJS.I.00287.

Falsig, J., Hvid, I., & Jensen, N. C. (1986). Finite element stress analysis of some ankle joint prostheses. *Clinical Biomechanics (Bristol, Avon)*, *1*(2), 71–76. Available from https://doi.org/10.1016/0268-0033(86)90078-1.

Fantigrossi, A., Galbusera, F., Raimondi, M. T., Sassi, M., & Fornari, M. (2007). Biomechanical analysis of cages for posterior lumbar interbody fusion. *Medical Engineering & Physics*, *29*(1), 101–109.

Favre, P., Senteler, M., Hipp, J., Scherrer, S., Gerber, C., & Snedeker, J. G. (2012). An integrated model of active glenohumeral stability. *Journal of Biomechanics*, *45*(13), 2248–2255. Available from https://doi.org/10.1016/j.jbiomech.2012.06.010.

Galbusera, F., Bellini, C. M., Anasetti, F., Ciavarro, C., Lovi, A., & Brayda-Bruno, M. (2011). Rigid and flexible spinal stabilization devices: A biomechanical comparison. *Medical Engineering & Physics*, *33*(4), 490–496. Available from https://doi.org/10.1016/j.medengphy.2010.11.018.

Galbusera, F., Casaroli, G., & Bassani, T. (2019). Artificial intelligence and machine learning in spine research. *JOR Spine*, *2*(1), e1044. Available from https://doi.org/10.1002/jsp2.1044.

Galbusera, F., Freutel, M., Dürselen, L., D'Aiuto, M., Croce, D., Villa, T., . . . Innocenti, B. (2014). Material models and properties in the finite element analysis of knee ligaments: A literature review. *Frontiers in Bioengineering and Biotechnology*, *2*, 54. Available from https://doi.org/10.3389/fbioe.2014.00054.

Galbusera, F., Schmidt, H., & Wilke, H.-J. (2012). Lumbar interbody fusion: A parametric investigation of a novel cage design with and without posterior instrumentation. *European Spine Journal : Official Publication of the European Spine Society, the European Spinal Deformity Society, and the European Section of the Cervical Spine Research Society*, *21*(3), 455–462. Available from https://doi.org/10.1007/s00586-011-2014-0.

Galbusera, F., Schmidt, H., Neidlinger-Wilke, C., & Wilke, H.-J. (2011). The effect of degenerative morphological changes of the intervertebral disc on the lumbar spine biomechanics: A poroelastic finite element investigation. *Computer Methods in Biomechanics and Biomedical Engineering*, *14*(8), 729–739. Available from https://doi.org/10.1080/10255842.2010.493522.

Galbusera, F., Schmidt, H., Neidlinger-Wilke, C., Gottschalk, A., & Wilke, H.-J. (2011). The mechanical response of the lumbar spine to different combinations of disc degenerative changes investigated using randomized poroelastic finite element models. *European Spine Journal : Official Publication of the European Spine Society, the European Spinal Deformity Society, and the European Section of the Cervical Spine Research Society*, *20*(4), 563–571. Available from https://doi.org/10.1007/s00586-010-1586-4.

Gardner-Morse, M., & Stokes, I. A. (1994). Three-dimensional simulations of the scoliosis derotation maneuver with Cotrel-Dubousset instrumentation. *Journal of Biomechanics*, *27*(2), 177–181.

Garijo, N., Verdonschot, N., Engelborghs, K., García-Aznar, J. M., & Pérez, M. A. (2017). Subject-specific musculoskeletal loading of the tibia: Computational load estimation. *Journal of the Mechanical Behavior of Biomedical Materials*, *65*, 334–343. Available from https://doi.org/10.1016/j.jmbbm.2016.08.026.

Gatti, C. J., Maratt, J. D., Palmer, M. L., Hughes, R. E., & Carpenter, J. E. (2010). Development and validation of a finite element model of the superior glenoid labrum. *Annals of Biomedical Engineering*, *38*(12), 3766–3776. Available from https://doi.org/10.1007/s10439-010-0105-4.

Gefen, A., Megido-Ravid, M., Itzchak, Y., & Arcan, M. (2000). Biomechanical analysis of the three-dimensional foot structure during gait: A basic tool for clinical applications. *Journal of Biomechanical Engineering*, *122*(6), 630–639.

Genda, E., Iwasaki, N., Li, G., MacWilliams, B. A., Barrance, P. J., & Chao, E. Y. (2001). Normal hip joint contact pressure distribution in single-leg standing—effect of gender and anatomic parameters. *Journal of Biomechanics, 34*(7), 895—905.

Goel, V. K., Faizan, A., Palepu, V., & Bhattacharya, S. (2012). Parameters that effect spine biomechanics following cervical disc replacement. *European Spine Journal : Official Publication of the European Spine Society, the European Spinal Deformity Society, and the European Section of the Cervical Spine Research Society, 21*(Suppl 5), S688—S699. Available from https://doi.org/10.1007/s00586-011-1816-4.

Gougoulias, N., Khanna, A., & Maffulli, N. (2010). How successful are current ankle replacements?: A systematic review of the literature. *Clinical Orthopaedics and Related Research, 468*(1), 199—208. Available from https://doi.org/10.1007/s11999-009-0987-3.

Guyer, A. J., & Richardson, G. (2008). Current concepts review: Total ankle arthroplasty. *Foot & Ankle International, 29*(2), 256—264. Available from https://doi.org/10.3113/FAI.2008.0256.

Haering, D., Raison, M., & Begon, M. (2014). Measurement and description of three-dimensional shoulder range of motion with degrees of freedom interactions. *Journal of Biomechanical Engineering, 136*(8). Available from https://doi.org/10.1115/1.4027665.

Halloran, J. P., Petrella, A. J., & Rullkoetter, P. J. (2005). Explicit finite element modeling of total knee replacement mechanics. *Journal of Biomechanics, 38*(2), 323—331.

Halonen, K. S., Mononen, M. E., Jurvelin, J. S., Töyräs, J., Salo, J., & Korhonen, R. K. (2014). Deformation of articular cartilage during static loading of a knee joint—experimental and finite element analysis. *Journal of Biomechanics, 47*(10), 2467—2474. Available from https://doi.org/10.1016/j.jbiomech.2014.04.013.

Helgason, B., Viceconti, M., Rúnarsson, T. P., & Brynjólfsson, S. (2008). On the mechanical stability of porous coated press fit titanium implants: A finite element study of a pushout test. *Journal of Biomechanics, 41*(8), 1675—1681. Available from https://doi.org/10.1016/j.jbiomech.2008.03.007.

Heller, M. O., Bergmann, G., Deuretzbacher, G., Dürselen, L., Pohl, M., Claes, L., ... Duda, G. N. (2001). Musculo-skeletal loading conditions at the hip during walking and stair climbing. *Journal of Biomechanics, 34*(7), 883—893.

Henninger, H. B., Reese, S. P., Anderson, A. E., & Weiss, J. A. (2010). Validation of computational models in biomechanics. *Proceedings of the Institution of Mechanical Engineers. Part H, Journal of Engineering in Medicine, 224*(7), 801—812.

Heuer, F., Schmidt, H., Claes, L., & Wilke, H.-J. (2007). Stepwise reduction of functional spinal structures increase vertebral translation and intradiscal pressure. *Journal of Biomechanics, 40*(4), 795—803.

Hopkins, A. R., Hansen, U. N., Amis, A. A., Taylor, M., Gronau, N., & Anglin, C. (2006). Finite element modelling of glenohumeral kinematics following total shoulder arthroplasty. *Journal of Biomechanics, 39*(13), 2476—2483.

Hosny, A., Parmar, C., Quackenbush, J., Schwartz, L. H., & Aerts, H. J. W. L. (2018). Artificial intelligence in radiology. *Nature Reviews. Cancer, 18*(8), 500—510. Available from https://doi.org/10.1038/s41568-018-0016-5.

Huiskes, R., & Chao, E. Y. (1983). A survey of finite element analysis in orthopedic biomechanics: The first decade. *Journal of Biomechanics, 16*(6), 385—409.

Innocenti, B. (2020). High congruency MB insert design: Stabilizing knee joint even with PCL deficiency. *Knee Surgery, Sports Traumatology, Arthroscopy : Official Journal of the ESSKA, 28*(9), 3040—3047. Available from https://doi.org/10.1007/s00167-019-05764-0.

Innocenti, B., Bellemans, J., & Catani, F. (2016). Deviations from optimal alignment in TKA: Is there a biomechanical difference between femoral or tibial component alignment? *The Journal of Arthroplasty, 31*(1), 295—301. Available from https://doi.org/10.1016/j.arth.2015.07.038.

Innocenti, B., Bilgen, Ö. F., Labey, L., van Lenthe, G. H., Sloten, J. V., & Catani, F. (2014). Load sharing and ligament strains in balanced, overstuffed and understuffed UKA. A validated finite element analysis. *The Journal of Arthroplasty, 29*(7), 1491−1498. Available from https://doi.org/10.1016/j.arth.2014.01.020.

Innocenti, B., Bori, E., & Piccolo, S. (2020). Development and validation of a robust patellar reference coordinate system for biomechanical and clinical studies. *The Knee, 27*(1), 81−88. Available from https://doi.org/10.1016/j.knee.2019.09.007.

Innocenti, B., Pianigiani, S., Ramundo, G., & Thienpont, E. (2016). Biomechanical effects of different varus and valgus alignments in medial unicompartmental knee arthroplasty. *The Journal of Arthroplasty, 31*(12), 2685−2691. Available from https://doi.org/10.1016/j.arth.2016.07.006.

Inoue, A., Chosa, E., Goto, K., & Tajima, N. (2013). Nonlinear stress analysis of the supraspinatus tendon using three-dimensional finite element analysis. *Knee Surgery, Sports Traumatology, Arthroscopy : Official Journal of the ESSKA, 21*(5), 1151−1157. Available from https://doi.org/10.1007/s00167-012-2008-4.

Jahng, T.-A., Kim, Y. E., & Moon, K. Y. (2013). Comparison of the biomechanical effect of pedicle-based dynamic stabilization: A study using finite element analysis. *The Spine Journal : Official Journal of the North American Spine Society, 13*(1), 85−94. Available from https://doi.org/10.1016/j.spinee.2012.11.014.

Jastifer, J. R., & Coughlin, M. J. (2015). Long-term follow-up of mobile bearing total ankle arthroplasty in the United States. *Foot & Ankle International, 36*(2), 143−150. Available from https://doi.org/10.1177/1071100714550654.

Jay Elliot, B., Gundapaneni, D., & Goswami, T. (2014). Finite element analysis of stress and wear characterization in total ankle replacements. *Journal of the Mechanical Behavior of Biomedical Materials, 34*, 134−145. Available from https://doi.org/10.1016/j.jmbbm.2014.01.020.

Jonkers, I., Spaepen, A., Papaioannou, G., & Stewart, C. (2002). An EMG-based, muscle driven forward simulation of single support phase of gait. *Journal of Biomechanics, 35*(5), 609−619.

Jonkers, I., Stewart, C., & Spaepen, A. (2003). The study of muscle action during single support and swing phase of gait: Clinical relevance of forward simulation techniques. *Gait & Posture, 17*(2), 97−105.

Kiapour, A., Kiapour, A. M., Kaul, V., Quatman, C. E., Wordeman, S. C., Hewett, T. E., ... Goel, V. K. (2014). Finite element model of the knee for investigation of injury mechanisms: Development and validation. *Journal of Biomechanical Engineering, 136*(1), 011002. Available from https://doi.org/10.1115/1.4025692.

Kim, Y., & Vanderby, R. (2000). Finite element analysis of interbody cages in a human lumbar spine. *Computer Methods in Biomechanics and Biomedical Engineering, 3*(4), 257−272.

Kleinberger, M. (1993). Application of finite element techniques to the study of cervical spine mechanics. *SAE International*. Available from https://doi.org/10.4271/933131.

Klemt, C., Nolte, D., Grigoriadis, G., Di Federico, E., Reilly, P., & Bull, A. M. J. (2017). The contribution of the glenoid labrum to glenohumeral stability under physiological joint loading using finite element analysis. *Computer Methods in Biomechanics and Biomedical Engineering, 20*(15), 1613−1622. Available from https://doi.org/10.1080/10255842.2017.1399262.

Koivumäki, J. E. M., Thevenot, J., Pulkkinen, P., Kuhn, V., Link, T. M., Eckstein, F., & Jämsä, T. (2012). Ct-based finite element models can be used to estimate experimentally measured failure loads in the proximal femur. *Bone, 50*(4), 824−829. Available from https://doi.org/10.1016/j.bone.2012.01.012.

Labek, G., Thaler, M., Janda, W., Agreiter, M., & Stöckl, B. (2011). Revision rates after total joint replacement: Cumulative results from worldwide joint register datasets. *The Journal of Bone and Joint Surgery. British Volume, 93*(3), 293−297. Available from https://doi.org/10.1302/0301-620X.93B3.25467.

Lenaerts, G., De Groote, F., Demeulenaere, B., Mulier, M., Van der Perre, G., Spaepen, A., & Jonkers, I. (2008). Subject-specific hip geometry affects predicted hip joint contact forces during gait. *Journal of Biomechanics, 41*(6), 1243−1252. Available from https://doi.org/10.1016/j.jbiomech.2008.01.014.

Lengsfeld, M., Bassaly, A., Boudriot, U., Pressel, T., & Griss, P. (2000). Size and direction of hip joint forces associated with various positions of the acetabulum. *The Journal of Arthroplasty*, *15*(3), 314−320.

Luo, Y., Wang, Y., Tai, B. L., Chen, R. K., & Shih, A. J. (2015). Bone geometry on the contact stress in the shoulder for evaluation of pressure ulcers: Finite element modeling and experimental validation. *Medical Engineering & Physics*, *37*(2), 187−194. Available from https://doi.org/10.1016/j.medengphy.2014.11.006.

Malandrino, A., Lacroix, D., Hellmich, C., Ito, K., Ferguson, S. J., & Noailly, J. (2014). The role of endplate poromechanical properties on the nutrient availability in the intervertebral disc. *Osteoarthritis and Cartilage*, *22*(7), 1053−1060. Available from https://doi.org/10.1016/j.joca.2014.05.005.

Malandrino, A., Noailly, J., & Lacroix, D. (2013). Regional annulus fibre orientations used as a tool for the calibration of lumbar intervertebral disc finite element models. *Computer Methods in Biomechanics and Biomedical Engineering*, *16*(9), 923−928. Available from https://doi.org/10.1080/10255842.2011.644539.

Martelli, S., Taddei, F., Cristofolini, L., Gill, H. S., & Viceconti, M. (2011). Extensive risk analysis of mechanical failure for an epiphyseal hip prothesis: A combined numerical-experimental approach. *Proceedings of the Institution of Mechanical Engineers. Part H, Journal of Engineering in Medicine*, *225*(2), 126−140.

Martinelli, N., Baretta, S., Pagano, J., Bianchi, A., Villa, T., Casaroli, G., & Galbusera, F. (2017). Contact stresses, pressure and area in a fixed-bearing total ankle replacement: A finite element analysis. *BMC Musculoskeletal Disorders*, *18*(1), 493. Available from https://doi.org/10.1186/s12891-017-1848-y.

Meena, V. K., Kumar, M., Pundir, A., Singh, S., Goni, V., Kalra, P., & Sinha, R. K. (2016). Musculoskeletal-based finite element analysis of femur after total hip replacement. *Proceedings of the Institution of Mechanical Engineers. Part H, Journal of Engineering in Medicine*, *230*(6), 553−560. Available from https://doi.org/10.1177/0954411916638381.

Moglo, K. E., & Shirazi-Adl, A. (2003). Biomechanics of passive knee joint in drawer: Load transmission in intact and ACL-deficient joints. *The Knee*, *10*(3), 265−276.

Mondal, S., & Ghosh, R. (2017). A numerical study on stress distribution across the ankle joint: Effects of material distribution of bone, muscle force and ligaments. *Journal of Orthopaedics*, *14*(3), 329−335. Available from https://doi.org/10.1016/j.jor.2017.05.003.

Mootanah, R., Imhauser, C. W., Reisse, F., Carpanen, D., Walker, R. W., Koff, M. F., ... Hillstrom, H. J. (2014). Development and validation of a computational model of the knee joint for the evaluation of surgical treatments for osteoarthritis. *Computer Methods in Biomechanics and Biomedical Engineering*, *17*(13), 1502−1517. Available from https://doi.org/10.1080/10255842.2014.899588.

Morales-Orcajo, E., Souza, T. R., Bayod, J., & Barbosa de Las Casas, E. (2017). Non-linear finite element model to assess the effect of tendon forces on the foot-ankle complex. *Medical Engineering & Physics*, *49*, 71−78. Available from https://doi.org/10.1016/j.medengphy.2017.07.010.

Natarajan, R. N., Chen, B. H., An, H. S., & Andersson, G. B. (2000). Anterior cervical fusion: A finite element model study on motion segment stability including the effect of osteoporosis. *Spine (Philadelphia, PA: 1986)*, *25*(8), 955−961.

Nie, B., Forman, J. L., Panzer, M. B., Mait, A. R., Donlon, J.-P., & Kent, R. W. (2017). Fiber-based modeling of in situ ankle ligaments with consideration of progressive failure. *Journal of Biomechanics*, *61*, 102−110. Available from https://doi.org/10.1016/j.jbiomech.2017.07.005.

Nie, B., Panzer, M. B., Mane, A., Mait, A. R., Donlon, J.-P., Forman, J. L., & Kent, R. W. (2017). Determination of the in situ mechanical behavior of ankle ligaments. *Journal of the Mechanical Behavior of Biomedical Materials*, *65*, 502−512. Available from https://doi.org/10.1016/j.jmbbm.2016.09.010.

Niemeyer, F., Wilke, H.-J., & Schmidt, H. (2012). Geometry strongly influences the response of numerical models of the lumbar spine − A probabilistic finite element analysis. *Journal of Biomechanics*, *45*(8), 1414−1423. Available from https://doi.org/10.1016/j.jbiomech.2012.02.021.

Nigro, P. T., Gutiérrez, S., & Frankle, M. A. (2013). Improving glenoid-side load sharing in a virtual reverse shoulder arthroplasty model. *Journal of Shoulder and Elbow Surgery*, *22*(7), 954−962. Available from https://doi.org/10.1016/j.jse.2012.10.025.

Nogier, A., Bonin, N., May, O., Gedouin, J.-E., Bellaiche, L., Boyer, T., & Lequesne, M. (2010). Descriptive epidemiology of mechanical hip pathology in adults under 50 years of age. Prospective series of 292 cases: Clinical and radiological aspects and physiopathological review. *Orthopaedics & Traumatology, Surgery & Research : OTSR*, *96*(Suppl 8), S53−S58. Available from https://doi.org/10.1016/j.otsr.2010.09.005.

Ong, K. L., Lehman, J., Notz, W. I., Santner, T. J., & Bartel, D. L. (2006). Acetabular cup geometry and bone-implant interference have more influence on initial periprosthetic joint space than joint loading and surgical cup insertion. *Journal of Biomechanical Engineering*, *128*(2), 169−175.

Park, S., Lee, S., Yoon, J., & Chae, S.-W. (2019). Finite element analysis of knee and ankle joint during gait based on motion analysis. *Medical Engineering & Physics*, *63*, 33−41. Available from https://doi.org/10.1016/j.medengphy.2018.11.003.

Peña, E., Calvo, B., Martínez, M. A., & Doblaré, M. (2006). A three-dimensional finite element analysis of the combined behavior of ligaments and menisci in the healthy human knee joint. *Journal of Biomechanics*, *39*(9), 1686−1701.

Peña, E., Calvo, B., Martínez, M. A., & Doblaré, M. (2007). Effect of the size and location of osteochondral defects in degenerative arthritis. A finite element simulation. *Computers in Biology and Medicine*, *37*(3), 376−387.

Peña, E., Calvo, B., Martínez, M. A., & Doblaré, M. (2008). Computer simulation of damage on distal femoral articular cartilage after meniscectomies. *Computers in Biology and Medicine*, *38*(1), 69−81.

Penrose, J. M. T., Holt, G. M., Beaugonin, M., & Hose, D. R. (2002). Development of an accurate three-dimensional finite element knee model. *Computer Methods in Biomechanics and Biomedical Engineering*, *5*(4), 291−300.

Perillo-Marcone, A., Barrett, D. S., & Taylor, M. (2000). The importance of tibial alignment: Finite element analysis of tibial malalignment. *The Journal of Arthroplasty*, *15*(8), 1020−1027.

Pettersen, S. H., Wik, T. S., & Skallerud, B. (2009). Subject specific finite element analysis of implant stability for a cementless femoral stem. *Clinical Biomechanics (Bristol, Avon)*, *24*(6), 480−487. Available from https://doi.org/10.1016/j.clinbiomech.2009.03.009.

Pianigiani, S., Croce, D., D'Aiuto, M., Pascale, W., & Innocenti, B. (2018). Sensitivity analysis of the material properties of different soft-tissues: Implications for a subject-specific knee arthroplasty. *Muscles, Ligaments and Tendons Journal*, *7*(4), 546−557. Available from https://doi.org/10.11138/mltj/2017.7.4.546.

Polkowski, G. G., & Clohisy, J. C. (2010). Hip biomechanics. *Sports Medicine and Arthroscopy Review*, *18*(2), 56−62. Available from https://doi.org/10.1097/JSA.0b013e3181dc5774.

Procter, P., & Paul, J. P. (1982). Ankle joint biomechanics. *Journal of Biomechanics*, *15*(9), 627−634.

Quental, C., Fernandes, P. R., Monteiro, J., & Folgado, J. (2014). Bone remodelling of the scapula after a total shoulder arthroplasty. *Biomechanics and Modeling in Mechanobiology*, *13*(4), 827−838. Available from https://doi.org/10.1007/s10237-013-0537-5.

Reggiani, B., Leardini, A., Corazza, F., & Taylor, M. (2006). Finite element analysis of a total ankle replacement during the stance phase of gait. *Journal of Biomechanics*, *39*(8), 1435−1443.

Rohlmann, A., Dreischarf, M., Zander, T., Graichen, F., Strube, P., Schmidt, H., & Bergmann, G. (2013). Monitoring the load on a telemeterised vertebral body replacement for a period of up to 65 months. *European Spine Journal : Official Publication of the European Spine Society, the European Spinal Deformity Society, and the European Section of the Cervical Spine Research Society*, *22*(11), 2575−2581. Available from https://doi.org/10.1007/s00586-013-3057-1.

Ruggiero, A., Merola, M., & Affatato, S. (2018). Finite element simulations of hard-on-soft hip joint prosthesis accounting for dynamic loads calculated from a musculoskeletal model during walking. *Materials (Basel, Switzerland)*, *11*(4), 574. Available from https://doi.org/10.3390/ma11040574.

Sano, H., Wakabayashi, I., & Itoi, E. (2006). Stress distribution in the supraspinatus tendon with partial-thickness tears: An analysis using two-dimensional finite element model. *Journal of Shoulder and Elbow Surgery, 15*(1), 100−105.

Scarton, A., Guiotto, A., Malaquias, T., Spolaor, F., Sinigaglia, G., Cobelli, C., ... Sawacha, Z. (2018). A methodological framework for detecting ulcers' risk in diabetic foot subjects by combining gait analysis, a new musculoskeletal foot model and a foot finite element model. *Gait & Posture, 60*, 279−285. Available from https://doi.org/10.1016/j.gaitpost.2017.08.036.

Schmidt, H., Galbusera, F., Rohlmann, A., Zander, T., & Wilke, H.-J. (2012). Effect of multilevel lumbar disc arthroplasty on spine kinematics and facet joint loads in flexion and extension: A finite element analysis. *European Spine Journal : Official Publication of the European Spine Society, the European Spinal Deformity Society, and the European Section of the Cervical Spine Research Society, 21*(Suppl 5), S663−S674. Available from https://doi.org/10.1007/s00586-010-1382-1.

Schmidt, H., Heuer, F., & Wilke, H.-J. (2009). Which axial and bending stiffnesses of posterior implants are required to design a flexible lumbar stabilization system? *Journal of Biomechanics, 42*(1), 48−54. Available from https://doi.org/10.1016/j.jbiomech.2008.10.005.

Schmidt, H., Heuer, F., Drumm, J., Klezl, Z., Claes, L., & Wilke, H.-J. (2007). Application of a calibration method provides more realistic results for a finite element model of a lumbar spinal segment. *Clinical Biomechanics (Bristol, Avon), 22*(4), 377−384.

Seireg, A., & Arvikar. (1975). The prediction of muscular lad sharing and joint forces in the lower extremities during walking. *Journal of Biomechanics, 8*(2), 89−102.

Seki, N., Itoi, E., Shibuya, Y., Wakabayashi, I., Sano, H., Sashi, R., ... Shimada, Y. (2008). Mechanical environment of the supraspinatus tendon: Three-dimensional finite element model analysis. *Journal of Orthopaedic Science : Official Journal of the Japanese Orthopaedic Association, 13*(4), 348−353. Available from https://doi.org/10.1007/s00776-008-1240-8.

Shirazi-Adl, A., Ahmed, A. M., & Shrivastava, S. C. (1986). A finite element study of a lumbar motion segment subjected to pure sagittal plane moments. *Journal of Biomechanics, 19*(4), 331−350.

Shirazi-Adl, A., Sadouk, S., Parnianpour, M., Pop, D., & El-Rich, M. (2002). Muscle force evaluation and the role of posture in human lumbar spine under compression. *European Spine Journal : Official Publication of the European Spine Society, the European Spinal Deformity Society, and the European Section of the Cervical Spine Research Society, 11*(6), 519−526.

Shu, L., Yamamoto, K., Yao, J., Saraswat, P., Liu, Y., Mitsuishi, M., & Sugita, N. (2018). A subject-specific finite element musculoskeletal framework for mechanics analysis of a total knee replacement. *Journal of Biomechanics, 77*, 146−154. Available from https://doi.org/10.1016/j.jbiomech.2018.07.008.

Sigrist, R. M. S., Liau, J., Kaffas, A. E., Chammas, M. C., & Willmann, J. K. (2017). Ultrasound elastography: Review of techniques and clinical applications. *Theranostics, 7*(5), 1303−1329. Available from https://doi.org/10.7150/thno.18650.

Silva, M. J., & Gibson, L. J. (1997). Modeling the mechanical behavior of vertebral trabecular bone: Effects of age-related changes in microstructure. *Bone, 21*(2), 191−199.

Sopher, R. S., Amis, A. A., Calder, J. D., & Jeffers, J. R. T. (2017). Total ankle replacement design and positioning affect implant-bone micromotion and bone strains. *Medical Engineering & Physics, 42*, 80−90. Available from https://doi.org/10.1016/j.medengphy.2017.01.022.

Stops, A., Wilcox, R., & Jin, Z. (2012). Computational modelling of the natural hip: A review of finite element and multibody simulations. *Computer Methods in Biomechanics and Biomedical Engineering, 15*(9), 963−979. Available from https://doi.org/10.1080/10255842.2011.567983.

Sturnick, D., Saito, G., Deland, J., Demetracopoulos, C., Chen, X., & Rodriguez, S. (2018). Finite element analysis of tibial bone-implant load transfer and bone strains with a modern fixed-bearing total ankle replacement. *Foot & Ankle Orthopaedics, 3*(3). Available from https://doi.org/10.1177/2473011418S00118, 2473011418S00118.

Tao, K., Wang, D., Wang, C., Wang, X., Liu, A., Nester, C. J., & Howard, D. (2009). An in vivo experimental validation of a computational model of human foot. *Journal of Bionic Engineering, 6*(4), 387–397. Available from https://doi.org/10.1016/S1672-6529(08)60138-9.

Terrier, A., Larrea, X., Guerdat, J., & Crevoisier, X. (2014). Development and experimental validation of a finite element model of total ankle replacement. *Journal of Biomechanics, 47*(3), 742–745. Available from https://doi.org/10.1016/j.jbiomech.2013.12.022.

Terrier, A., Reist, A., Vogel, A., & Farron, A. (2007). Effect of supraspinatus deficiency on humerus translation and glenohumeral contact force during abduction. *Clinical Biomechanics (Bristol, Avon), 22*(6), 645–651.

Tschirhart, C. E. C. E., Nagpurkar, A., & Whyne, C. M. C. M. (2004). Effects of tumor location, shape and surface serration on burst fracture risk in the metastatic spine. *Journal of Biomechanics, 37*(5), 653–660.

van den Bekerom, M. P. J., Lamme, B., Hogervorst, M., & Bolhuis, H. W. (2007). Which ankle fractures require syndesmotic stabilization? *The Journal of Foot and Ankle Surgery : Official Publication of the American College of Foot and Ankle Surgeons, 46*(6), 456–463.

Veeger, H. E. J., & van der Helm, F. C. T. (2007). Shoulder function: The perfect compromise between mobility and stability. *Journal of Biomechanics, 40*(10), 2119–2129.

Verhaegen, F., Campopiano, E., Debeer, P., Scheys, L., & Innocenti, B. (2020). How much bone support does an anatomic glenoid component need? *Journal of Shoulder and Elbow Surgery, 29*(4), 743–754. Available from https://doi.org/10.1016/j.jse.2019.09.019.

Viceconti, M., Ansaloni, M., Baleani, M., & Toni, A. (2003). The muscle standardized femur: A step forward in the replication of numerical studies in biomechanics. *Proceedings of the Institution of Mechanical Engineers. Part H, Journal of Engineering in Medicine, 217*(2), 105–110.

Viceconti, M., Brusi, G., Pancanti, A., & Cristofolini, L. (2006). Primary stability of an anatomical cementless hip stem: A statistical analysis. *Journal of Biomechanics, 39*(7), 1169–1179.

Viceconti, M., Henney, A., & Morley-Fletcher, E. (2016). In silico clinical trials: How computer simulation will transform the biomedical industry. *International Journal of Clinical Trials, 3*(2), 37–46. Available from https://www.ijclinicaltrials.com/index.php/ijct/article/view/105.

Viceconti, M., Olsen, S., Nolte, L.-P., & Burton, K. (2005). Extracting clinically relevant data from finite element simulations. *Clinical Biomechanics (Bristol, Avon), 20*(5), 451–454.

Victor, J., Van Doninck, D., Labey, L., Innocenti, B., Parizel, P. M., & Bellemans, J. (2009). How precise can bony landmarks be determined on a CT scan of the knee? *The Knee, 16*(5), 358–365. Available from https://doi.org/10.1016/j.knee.2009.01.001.

Virani, N. A., Harman, M., Li, K., Levy, J., Pupello, D. R., & Frankle, M. A. (2008). In vitro and finite element analysis of glenoid bone/baseplate interaction in the reverse shoulder design. *Journal of Shoulder and Elbow Surgery, 17*(3), 509–521. Available from https://doi.org/10.1016/j.jse.2007.11.003.

Waanders, D., Janssen, D., Miller, M. A., Mann, K. A., & Verdonschot, N. (2009). Fatigue creep damage at the cement-bone interface: An experimental and a micro-mechanical finite element study. *Journal of Biomechanics, 42*(15), 2513–2519. Available from https://doi.org/10.1016/j.jbiomech.2009.07.014.

Wakabayashi, I., Itoi, E., Sano, H., Shibuya, Y., Sashi, R., Minagawa, H., & Kobayashi, M. (2003). Mechanical environment of the supraspinatus tendon: A two-dimensional finite element model analysis. *Journal of Shoulder and Elbow Surgery, 12*(6), 612–617.

Walia, P., Miniaci, A., Jones, M. H., & Fening, S. D. (2013). Theoretical model of the effect of combined glenohumeral bone defects on anterior shoulder instability: A finite element approach. *Journal of Orthopaedic Research : Official Publication of the Orthopaedic Research Society, 31*(4), 601–607. Available from https://doi.org/10.1002/jor.22267.

Wang, Y., Wong, D. W.-C., Tan, Q., Li, Z., & Zhang, M. (2019). Total ankle arthroplasty and ankle arthrodesis affect the biomechanics of the inner foot differently. *Scientific Reports, 9*(1), 13334. Available from https://doi.org/10.1038/s41598-019-50091-6.

Wheeldon, J., Khouphongsy, P., Kumaresan, S., Yoganandan, N., & Pintar, F. A. (2000). Finite element model of human cervical spinal column. *Biomedical Sciences Instrumentation, 36,* 337–342.

Wilke, H. J., Wenger, K., & Claes, L. (1998). Testing criteria for spinal implants: Recommendations for the standardization of in vitro stability testing of spinal implants. *European Spine Journal : Official Publication of the European Spine Society, the European Spinal Deformity Society, and the European Section of the Cervical Spine Research Society, 7*(2), 148–154.

Wismans, J., Veldpaus, F., Janssen, J., Huson, A., & Struben, P. (1980). A three-dimensional mathematical model of the knee-joint. *Journal of Biomechanics, 13*(8), 677–685.

Wu, J. S.-S., Hung, J.-P., Shu, C.-S., & Chen, J.-H. (2003). The computer simulation of wear behavior appearing in total hip prosthesis. *Computer Methods and Programs in Biomedicine, 70*(1), 81–91.

Xu, C., Liu, H., Li, M., Wang, C., & Li, K. (2017). A three-dimensional finite element analysis of displaced intra-articular calcaneal fractures. *The Journal of Foot and Ankle Surgery : Official Publication of the American College of Foot and Ankle Surgeons, 56*(2), 319–326. Available from https://doi.org/10.1053/j.jfas.2016.09.018.

Yeh, M.-L., Lintner, D., & Luo, Z.-P. (2005). Stress distribution in the superior labrum during throwing motion. *The American Journal of Sports Medicine, 33*(3), 395–401.

Yu, J., Cheung, J. T.-M., Fan, Y., Zhang, Y., Leung, A. K.-L., & Zhang, M. (2008). Development of a finite element model of female foot for high-heeled shoe design. *Clinical Biomechanics (Bristol, Avon), 23*(Suppl 1), S31–S38.

Zalzal, P., Backstein, D., Gross, A. E., & Papini, M. (2006). Notching of the anterior femoral cortex during total knee arthroplasty characteristics that increase local stresses. *The Journal of Arthroplasty, 21*(5), 737–743.

Zheng, M., Zou, Z., Bartolo, P. J. D. S., Peach, C., & Ren, L. (2017). Finite element models of the human shoulder complex: A review of their clinical implications and modelling techniques. *International Journal for Numerical Methods in Biomedical Engineering, 33*(2), e02777. Available from https://doi.org/10.1002/cnm.2777.

Zhu, R., Zander, T., Dreischarf, M., Duda, G. N., Rohlmann, A., & Schmidt, H. (2013). Considerations when loading spinal finite element models with predicted muscle forces from inverse static analyses. *Journal of Biomechanics, 46*(7), 1376–1378. Available from https://doi.org/10.1016/j.jbiomech.2013.03.003.

Zhu, Z.-J., Zhu, Y., Liu, J.-F., Wang, Y.-P., Chen, G., & Xu, X.-Y. (2016). Posterolateral ankle ligament injuries affect ankle stability: A finite element study. *BMC Musculoskeletal Disorders, 17,* 96. Available from https://doi.org/10.1186/s12891-016-0954-6.

Index

Note: Page numbers followed by "*f*" and "*t*" refer to figures and tables, respectively.

Printed in the United States
by Baker & Taylor Publisher Services